60 Years of the Casimir Effect

Journal of Physics: Conference Series Volume 161

Brasilia, Brazil
23-27 June 2008

Editors:

Gabriel Barton
Victor Dodonov
Giovanni Carugno
Margarita Man'ko

ISBN: 978-1-61738-373-1

Printed from e-media with permission by:

Curran Associates, Inc.
57 Morehouse Lane
Red Hook, NY 12571
www.proceedings.com

Some format issues inherent in the e-media version may also appear in this print version.

Printed by Curran Associates, Inc. (2010)

Institute of Physics
Dirac House, Temple Back
Bristol BS1 6BE UK

Phone: 44 1 17 929 7481
Fax: 44 1 17 920 0979

techtracking@iop.org

TABLE OF CONTENTS

CASIMIR FORCES: CURRENT STATUS OF THE THEORY AND EXPERIMENT

CASIMIR EFFECT, QUANTUM FIELD THEORY AND GRAVITATION

DYNAMICAL CASIMIR EFFECT AND MOTION INDUCED RADIATION

NEW MANIFESTATIONS AND APPLICATIONS OF THE CASIMIR EFFECT

PREFACE

In 1948 Hendrick Casimir published a short article predicting that (neutral) ideal metallic plates attract each other. This attraction is widely ascribed to the quantum vacuum fluctuations of the electromagnetic field (even though away from the limit of ideal metals it depends demonstrably on the physics of the charge carriers vanishing when they cease to carry). Casimir's remarkable discovery, nowadays called the Casimir effect, has charmed several generations of physicists. In the last decade alone, more than a thousand publications have addressed its many consequences, generalizations, and possible applications in different areas from particle physics to cosmology. Interest in the field is still growing driven by impressive progress in experimental skills and its importance for the recently opened-up area of micro- and nano-electromechanical systems: according to the Thompson ISI Web of Science database, in 2005 the number of papers related to the Casimir effect or to Casimir forces jumped to over 125, compared to approximately 60 in 2000 and 30 in 1995. The increase continues, with more than 170 papers in 2008.

The International Workshop '60 Years of the Casimir Effect' took place on 23–27June 2008, in Brasilia (Brazil) organized by the International Center for Condensed Matter Physics (ICCMP). The purpose was to celebrate this anniversary of Casimir's pioneering paper by inviting the leading specialists in the area, both theorists and experimentalists, together with young researchers and post-graduate students interested in hearing about the most recent achievements in the field. The Workshop was attended by 65 participants from 14 countries, who presented 41 talks and 12 posters.

These Proceedings contain extended versions of almost all the talks and some posters, plus several papers by authors who had planned to attend but for various reasons could not.

The contributions are divided (with some inevitable arbitrariness) into four groups. The largest one consists of work devoted to the current status of the theory and measurements related to Casimir forces. Readers must be warned that some topics in this field of research remain controversial (especially the dependence on temperature): they can and do generate debates that sometimes become quite heated. These controversies are reflected in the papers. We believe that at present it is not the business of conference organisers to adjudicate such issues, and hope that detailed expositions of different approaches and different points of view will help readers to formulate their own, and will eventually lead to a better understanding of the problems and of the solutions proposed.

The other three groups contain contributions bearing on (1) topics related to causes and consequences of Casimir effects in quantum field theory and gravitation; (2) the so-called dynamical (or nonstationary) Casimir effect and motion-induced radiation, (3) some new manifestations and applications of the Casimir effect.

We are grateful to the authors for making their papers so interesting; to the referees for their careful reading of the initial versions, and for their many helpful comments and suggestions; to the Institute of Physics for its kindness in offering to publish these Proceedings in *Journal of Physics: Conference Series*; and to the Institute of Physics office at the Lebedev Physical Institute in Moscow for essential help in the preparation of this volume. On behalf of the participants of the workshop, we thank the direction and staff of the ICCMP for their splendid organization of the event. Finally we acknowledge the support of the Brazilian scientific funding agencies FAP-DF and CNPQ, which covered the local and travel expenses of many participants.

The Editors
Gabriel Barton (University of Sussex, Brighton, UK)
Giovanni Carugno (INFN - Sezione di Padova, Italy)
Victor Dodonov (University of Brasilia, Brazil)
Margarita Man'ko (Lebedev Physical Institute, Moscow, Russia)

Workshop Organizers
Gabriel Barton and Victor Dodonov

International Advisory Committee of the Workshop

Workshop on 60 Years of Casimir Effect
ICCMP - UnB BRASILIA, June 23rd - 27th, 2008

Journal of Physics: Conference Series **161** (2009) 012001 doi:10.1088/1742-6596/161/1/012001

Recent Developments in the Casimir Effect

K A Milton[1]

Oklahoma Center for High Energy Physics, and H. L. Dodge Department of Physics and Astronomy, University of Oklahoma, Norman, OK 73019 USA

E-mail: `milton@nhn.ou.edu`

Abstract. In this paper I review various developments in the past year concerning quantum vacuum energy, the Casimir effect. In particular, there has been continuing controversy surrounding the temperature correction to the Lifshitz formula for the Casimir force between real materials, be they metals or semiconductors. Consensus has emerged as to how Casimir energy accelerates in a weak gravitational field; quantum vacuum energy, including the divergent parts which renormalize the masses of the Casimir plates, accelerates indeed according to the equivalence principle. Significant development has been forthcoming in applying the multiple scattering formalism to describe the interaction between nontrivial objects. In weak coupling, closed-form expressions for the Casimir force between the bodies, which for example reveal significant discrepancies from the naive proximity force approximation, can be achieved in many cases.

1. Introduction

There has been burgeoning interest in the Casimir effect [1], the quantum force between macroscopic bodies due to zero-point fluctuations, as witnessed by the multitude of international conferences and workshops devoted to the subject, including two in Brazil back-to-back. This interest reflects a great deal of progress in the subject both from the experimental and theoretical sides. This paper will describe three topics my collaborators and I have been involved with during the past several months, which mirror these developments: the controversy concerning the temperature dependence of real materials, such as metals and semiconductors; the question of how the Casimir energy associated with parallel Casimir plates is accelerated by gravity; and the achievement of exact closed-form results for the interaction of bodies subject to weak coupling.

2. Temperature Controversy

2.1. Introduction

There is a continuing controversy surrounding the question of how to incorporate thermal corrections into the Casimir force between real metal plates [2, 3]. The procedure to do this seems straightforward. The thermal Green's functions must be periodic in imaginary time, with period $\beta = 1/T$. This implies a Fourier series decomposition, rather than a Fourier transform, where we have a sum over Matsubara frequencies $\zeta_m = 2\pi m T$.

[1] The work reported here was done in collaboration with I Brevik, S Å Ellingsen and J S Høye (section 2); S Fulling and A Romeo (section 3); I Cavero-Peláez, P Parashar, K V Shajesh and J Wagner (sections 3 and 4).

Journal of Physics: Conference Series **161** (2009) 012001 doi:10.1088/1742-6596/161/1/012001

Thus, the Casimir (Lifshitz) [4] pressure between parallel semi-infinite nonmetallic plates separated by a distance a is given by

$$P^T = -\frac{T}{\pi} \sum_{i=\mathrm{TE},\mathrm{TM}} \sum_{m=0}^{\infty}{}' \int_{\zeta_m}^{\infty} \mathrm{d}\kappa\, \kappa^2 \frac{1}{(r^i)^{-2} \mathrm{e}^{2\kappa a} - 1}, \tag{1}$$

where the prime means that the $m = 0$ term is counted with half weight, in terms of the reflection coefficients

$$r^{\mathrm{TE}} = \frac{-\sqrt{1 + \frac{\zeta^2}{\kappa^2}(\varepsilon(\mathrm{i}\zeta) - 1)} + 1}{\sqrt{1 + \frac{\zeta^2}{\kappa^2}(\varepsilon(\mathrm{i}\zeta) - 1)} + 1}, \tag{2a}$$

$$r^{\mathrm{TM}} = \frac{-\sqrt{1 + \frac{\zeta^2}{\kappa^2}(\varepsilon(\mathrm{i}\zeta) - 1)} + \varepsilon(\mathrm{i}\zeta)}{\sqrt{1 + \frac{\zeta^2}{\kappa^2}(\varepsilon(\mathrm{i}\zeta) - 1)} + \varepsilon(\mathrm{i}\zeta)}. \tag{2b}$$

Let us rewrite the Lifshitz formula at finite temperature in the form

$$P^T = \sum_{m=0}^{\infty}{}' f_m = \int_0^{\infty} \mathrm{d}m\, f(m) - \sum_{k=0}^{\infty} \frac{B_{2k}}{(2k)!} f^{(2k-1)}(0), \tag{3}$$

where the second equality uses the Euler-Maclaurin sum formula, in terms of

$$f(m) = -\frac{1}{2\pi\beta} \sum_i \int_0^{\infty} \mathrm{d}k_\perp^2\, \kappa(\zeta_m)(d_m^i)^{-1}, \tag{4}$$

where

$$d_m^i = (r^i)^{-2} \mathrm{e}^{2\kappa a} - 1, \tag{5}$$

and $\kappa^2 = k_\perp^2 + \zeta^2$.

Evidently, for the Drude model, where the permittivity has the form

$$\varepsilon(\mathrm{i}\zeta) = 1 + \frac{\omega_p^2}{\zeta(\zeta + \gamma)}, \tag{6}$$

or more generally, whenever

$$\lim_{\zeta \to 0} \zeta^2 [\varepsilon(\mathrm{i}\zeta) - 1] = 0, \tag{7}$$

the TE zero mode vanishes,

$$f^{(\mathrm{TE})}(0) = 0. \tag{8}$$

This leads to a physical discontinuity in the TE mode, between the $m = 0$ and $m = 1$ Matsubara frequencies. Naively, this implies an additional linear term in the pressure at low temperatures:

$$P^T \sim P^{T=0} + \frac{\zeta(3)}{8\pi a^3} T, \qquad aT \ll 1. \tag{9}$$

Exclusion of the TE zero mode will also reduce the linear temperature dependence expected at high temperatures,

$$P^T \sim -\frac{\zeta(3)}{8\pi a^3} T, \qquad aT \gg 1, \tag{10}$$

one-half the usual ideal metal result. This anomaly was first pointed out by Boström and Sernelius [5]. See also [6].

Several other recent papers also lend support to our point of view. Jancovici and Šamaj [7,8] and Buenzli and Martin [9, 10] have examined the Casimir force between ideal-conductor walls with emphasis on the high-temperature limit. Not surprisingly, ideal inert boundary conditions are shown to be inadequate, and fluctuations within the walls, modeled by the classical Debye-Hückel theory, determine the high temperature behavior. The linear in temperature behavior of the Casimir force is found to be reduced by a factor of two from the behavior predicted by an ideal metal, just as seen above in (10). This is precisely the signal of the omission of the $m = 0$ TE mode. Further support for our conclusions can be found in the work of Sernelius [11], who calculated the van der Waals-Casimir force between gold plates using the Lindhard or random phase approximation dielectric function. The central theme of his work is to describe the thermal Casimir effect in terms of spatial dispersion. He also finds that for large separations the force is one-half that of the ideal metal. Sernelius shows that, for arbitrary separation between the plates, the spatial-dispersion results nearly exactly coincide with the local dissipation-based results.

Thus, it is very hard to see how the corresponding modification of the low-temperature behavior can be avoided. However, because a linear T dependence in the pressure implies a linear T dependence in the free energy, this seems to imply a violation of the Nernst heat theorem, the third law of thermodynamics, because then a naive calculation implies $S(T = 0) \neq 0$. However, we [12] and others have shown that for real metals, where γ has a residual value at zero temperature, the free energy per area has a vanishing slope at the origin. Indeed, in the Drude model, the free energy per area has the behavior [13, 14]

$$F = F_0 + T^2 \frac{\omega_p^2}{48\gamma}(2 \ln 2 - 1), \tag{11}$$

for sufficiently low temperatures. The $T^{5/2}$ and T^3 corrections to this were calculated last year in [12, 13] This is illustrated in figure 1, taken from [12], which shows the numerical evaluation of the free energy for gold.

2.2. Experimental constraints

The difficulty is that, experimentally, it is not easy to perform Casimir force measurements at other than room temperature, so current constraints on the theory all come from room temperature experiments. Then all one can do is compare the theory at room temperature with the experimental results, which must be corrected for a variety of effects, such as surface roughness, finite conductivity, and patch potentials. *There are no direct experiments which observe T dependence in metals!* (The temperature dependence of the Casimir-Polder force between a Bose-Einstein condensate and a dielectric substrate was measured in 2006 [15].)

Although the earlier experiment by Lamoreaux [16] was carried out at such large distances ($\sim 1\,\mu$m) where the temperature correction proposed here would be large ($\sim 15\%$) [17]. there are reasons to think that the uncertainties due to such things as surface roughness and patch potentials were so large as to afford only limited accuracy in his experiment. There has been a great number of high quality experiments during the last decade, revealing a number of facets of the quantum vacuum energy [18–27]. As for the temperature effect, the most stringent test is given by the precision experiments of Decca *et al* [28, 29], where the agreement of the data with the zero-temperature theory seems to rule out a large thermal correction such as Boström and Sernelius [5] and we have proposed. It seems to us that there may be a number of effects that have not been properly accounted for before one can conclude that no large temperature effect is present, such as the shielding effect recently proposed by Pitaevskii [30]. We should

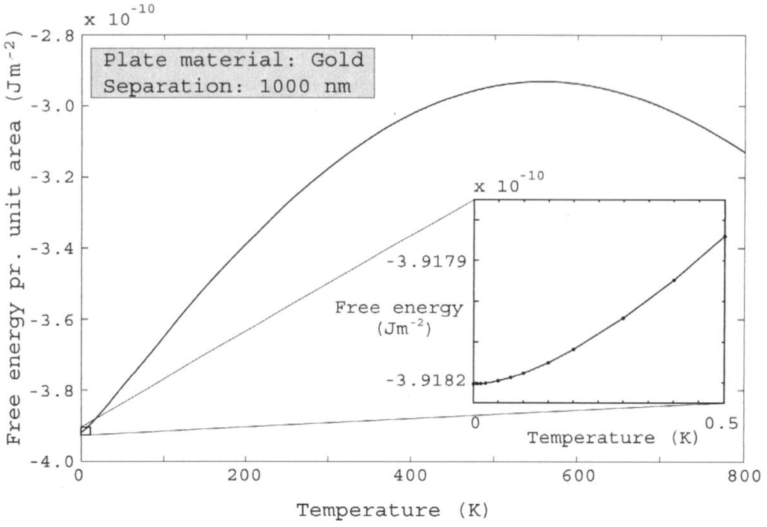

Figure 1. Numerical evaluation of the free energy between two gold halfspaces as a function of temperature. The inset gives details for low T.

also remind the reader that doubts have been repeatedly expressed [2] concerning the accuracy and precision of the Purdue experiment [28, 29], particularly in view of the claim that their measurement simultaneously limits both thermal effects and Yukawa forces.

2.3. Anomaly for semiconductors

Recently Klimchitskaya, Geyer and Mostepanenko [31–34] have observed there is a similar anomaly, now affecting the TM reflection coefficient, for dielectrics and semiconductors. Here is a simple way to understand that argument. Suppose we model a dielectric with some small conductivity by the permittivity function

$$\varepsilon(i\zeta) = 1 + \frac{\bar{\varepsilon}-1}{1+\zeta^2/\omega_0^2} + \frac{4\pi\sigma}{\zeta}. \tag{12}$$

(The only essential point is that as $\zeta \to 0$, $\varepsilon \to \bar{\varepsilon}$ if $\sigma = 0$, otherwise $\varepsilon \to \infty$.)

For a dielectric the TE reflection coefficient is continuous, but if there is vanishingly small (but not zero) conductivity the TM coefficient is discontinuous:

$$r^{\mathrm{TM}}(\zeta = 0) = 1, \quad \lim_{\zeta \to 0} r^{\mathrm{TM}} = \frac{\bar{\varepsilon}-1}{\bar{\varepsilon}+1}. \tag{13}$$

This implies a linear term in the pressure

$$\begin{aligned} P_T^{\mathrm{TM}} &= \frac{T}{2\pi}\int_0^\infty d\kappa\, \kappa^2 \left[\frac{1}{\left(\frac{\bar{\varepsilon}+1}{\bar{\varepsilon}-1}\right)^2 e^{2\kappa a}-1} - \frac{1}{e^{2\kappa a}-1}\right] \\ &= \frac{T}{2\pi}\sum_{n=1}^\infty \left[\left(\frac{\bar{\varepsilon}-1}{\bar{\varepsilon}+1}\right)^{2n} - 1\right]\int_0^\infty d\kappa\, \kappa^2 e^{-2n\kappa a} = \frac{T}{8\pi a^3}\left[\mathrm{Li}_3\left(\left[\frac{\bar{\varepsilon}-1}{\bar{\varepsilon}+1}\right]^2\right) - \zeta(3)\right], \end{aligned} \tag{14}$$

where the polylogarithmic function is

$$\mathrm{Li}_n(\xi) = \sum_{k=1}^{\infty} \frac{\xi^k}{k^n}. \tag{15}$$

Note that the linear term vanishes for $\bar{\varepsilon} \to \infty$. The free energy term F is obtained from this by multiplying by $a/2$. Thus at zero temperature, the entropy is nonzero,

$$S(0) = -\left(\frac{\partial F}{\partial T}\right)_V = -\frac{1}{16\pi a^2}\left[\mathrm{Li}_3\left(\left[\frac{\bar{\varepsilon}-1}{\bar{\varepsilon}+1}\right]^2\right) - \zeta(3)\right]. \tag{16}$$

Moreover, besides this thermodynamical contradiction, recent experiments by Mohideen's group [27, 35–37] show that apparently the conductivity effect must be excluded for semiconductors with very low carrier concentrations, but included for semiconductors with high carrier concentrations. This conundrum has not yet been resolved, although there has been some progress in understanding what happens when the conductivity remains small but nonzero at zero temperature [38] —see also Simen Ådnøy Ellingsen's report in this meeting [39]. We should also mention our joint work on temperature effects and anomalies [40] which, while not addressing the semiconductor anomaly, attempts to survey the underpinnings of thermal effects in the Casimir effect.

3. How Does Casimir Energy Fall?

In a series of papers last year [41–43] we showed that the total Casimir energy of an apparatus consisting of parallel plates, including the *divergent parts* of the energy, which renormalize the masses of the plates, possesses the gravitational mass demanded by the equivalence principle.

This result puts the lie to the naive presumption that zero-point energy is not observable. On the other hand, because of the severe divergence structure of the theory, controversy has surrounded it from the beginning. Sharp boundaries give rise to divergences in the local energy density near the surface, which may make it impossible to extract meaningful self-energies of single objects, such as the perfectly conducting sphere considered by Boyer [44]. These objections have recently been most forcefully presented by Graham, Jaffe, *et al* [45], but they date back to Deutsch and Candelas [46]. In fact, it now appears that these surface divergences can be dealt with successfully in a process of renormalization, and that finite self-energies in the sense of Boyer may be extracted [47, 48].

Gravity couples to the local energy-momentum tensor, and such surface divergences promise serious difficulties. We first ask how does the completely finite Casimir interaction energy of a pair of parallel conducting plates, as well as the divergent self-energies of non-ideal plates, couple to gravity? Disparate answers had been given in the past [49, 50]; thus, the question, and its answer, turn out to be surprisingly less straightforward than the reader might suspect!

We will not here address what gravitational field results from the (in general divergent) Casimir energy. For a beginning of the renormalization of Einstein's equations resulting from singular Casimir surface energy densities see [51].

Brown and Maclay [52] showed that, for parallel perfectly conducting plates separated by a distance a in the z-direction, the electromagnetic stress tensor acquires the vacuum expectation value between the plates

$$\langle T^{\mu\nu}\rangle = \frac{\mathcal{E}_c}{a}\mathrm{diag}(1,-1,-1,3), \qquad \mathcal{E}_c = -\frac{\pi^2}{720a^3}\hbar c. \tag{17}$$

Outside the plates the value of $\langle T^{\mu\nu}\rangle = 0$.

Because there are some subtleties here, let us review the derivation of (17) for the case of a conformally coupled scalar (the electromagnetic case differs by a factor of two). Actually, the result between the plates, $0 < z < a$, is given in great detail in [53] (ξ is the conformal parameter)

$$\langle T^{\mu\nu}\rangle = (u_0 + u)\mathrm{diag}(1,-1,-1,3) + (1-6\xi)s(z)\mathrm{diag}(1,-1,-1,0), \tag{18}$$

where

$$u_0 = -\frac{1}{12\pi^2}\int_0^\infty \mathrm{d}\kappa\,\kappa^3, \qquad u = -\frac{\pi^2}{1440a^4}, \tag{19}$$

Note that u_0 is a divergent constant, independent of a, and is present (as we shall see) both inside and outside the plates, so it does not contribute to any observable force or energy (the force on the plates is given by the discontinuity of $\langle T_{zz}\rangle$), and so may be simply disregarded (as long as we are not concerned with dark energy). Similarly, in terms of the Hurwitz zeta function,

$$s(z) = -\frac{1}{16\pi^2 a^4}[\zeta(4,z/a) + \zeta(4,1-z/a)], \tag{20}$$

which exhibits the universal surface divergence near the plates,

$$s(z) \sim -\frac{1}{16\pi^2 z^4}, \qquad z \to 0+, \tag{21}$$

is also unobservable (if we disregard gravity) because it does not contribute to the force on the plates, nor does it contribute to the total energy, since the integral over $s(z)$ between the plates is independent of the plate separation. Of course, the best way to eliminate that term is to choose the conformal value $\xi = 1/6$.

Since the exterior calculation does not appear to be referred to in [53], let us sketch the calculation here: Consider parallel Dirichlet plates at $z = 0$ and $z = a$. The reduced Green's function satisfies

$$\left(-\frac{\mathrm{d}^2}{\mathrm{d}z^2} + \kappa^2\right) g(z,z') = \delta(z-z'), \tag{22}$$

where $\kappa^2 = k^2 - \omega^2 = k^2 + \zeta^2$. The solution for $z, z' < 0$ is

$$g(z,z') = -\frac{1}{\kappa}e^{\kappa z_<}\sinh \kappa z_>. \tag{23}$$

where $z_{<,>}$ means the lesser or greater of z and z'. It is very straightforward to calculate the one-loop expectation value of the stress tensor from

$$\langle T^{\mu\nu}\rangle = \left(\partial^\mu\partial'^\nu - \frac{1}{2}g^{\mu\nu}\partial^\lambda\partial'_\lambda\right)\frac{1}{\mathrm{i}}G(x,x')\bigg|_{x'=x} - \xi(\partial^\mu\partial^\nu - g^{\mu\nu}\partial^2)\frac{1}{\mathrm{i}}G(x,x), \tag{24}$$

which follows from

$$\langle\phi(x)\phi(x')\rangle = \frac{1}{\mathrm{i}}G(x,x'). \tag{25}$$

After integrating over $\omega = \mathrm{i}\zeta$ and $\mathbf{k}$, we find the result ($z < 0$)

$$\langle T^{\mu\nu}\rangle = u_0\mathrm{diag}(1,-1,-1,3) - \frac{(1-6\xi)}{16\pi^2|z|^4}\mathrm{diag}(1,-1,-1,0). \tag{26}$$

This is exactly as expected. The u_0 term is the same as inside the box, so is just the vacuum value, and the second term is the universal surface divergence seen in (21) (independent of plate separation), which can be eliminated by setting $\xi = 1/6$.

Thus, we conclude that the physical stress tensor VEV is just that found by Brown and Maclay [52]:

$$\langle T^{\mu\nu}\rangle = u\,\mathrm{diag}(1,-1,-1,3)\theta(z)\theta(a-z). \tag{27}$$

in terms of the usual step function.

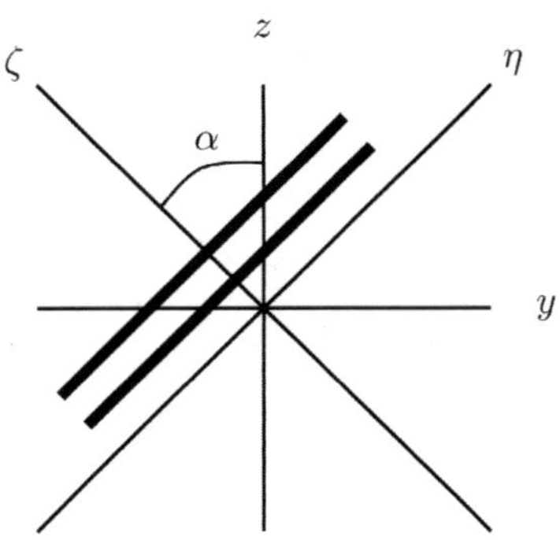

Figure 2. Relation between two Cartesian coordinate frames: One attached to the earth (x, y, z), where $-z$ is the direction of gravity, and one attached to the parallel-plate Casimir apparatus (ζ, η, χ), where ζ is in the direction normal to the plates. The parallel plates are indicated by the heavy lines parallel to the η axis. The $x = \chi$ axis is perpendicular to the page.

3.1. Variational principle

Now we turn to the question of the gravitational interaction of this Casimir apparatus. It seems this question can be most simply addressed through use of the gravitational definition of the energy-momentum tensor,

$$\delta W_m \equiv -\frac{1}{2}\int (\mathrm{d}x)\sqrt{-g}\,\delta g^{\mu\nu}T_{\mu\nu} = \frac{1}{2}\int (\mathrm{d}x)\sqrt{-g}\,\delta g_{\mu\nu}T^{\mu\nu}. \tag{28}$$

For a weak field,

$$g_{\mu\nu} = \eta_{\mu\nu} + 2h_{\mu\nu}, \tag{29}$$

so if we think of turning on the gravitational field as the perturbation, we can ignore $\sqrt{-g}$. The gravitational energy, for a static situation, is therefore given by $(\delta W = -\int \mathrm{d}t\,\delta E)$

$$E_g = -\int (\mathrm{d}\mathbf{x}) h_{\mu\nu}T^{\mu\nu}. \tag{30}$$

Here we use the gravity-free electromagnetic Casimir stress tensor (27), with u now replaced by $\mathcal{E}_c/a$ for the electromagnetic situation.

The Fermi metric describes an inertial coordinate system, as we will discuss further below in section 3.2:

$$h_{00} = -gz, \quad h_{0i} = h_{ij} = 0. \tag{31}$$

Let us consider a Casimir apparatus of parallel plates separated by a distance a, with transverse dimensions $L \gg a$. Let the apparatus be oriented at an angle α with respect to the direction of gravity, as shown in figure 2. The Cartesian coordinate system attached to the earth is (x, y, z), where z is the direction of $-\mathbf{g}$. The Cartesian coordinates associated with the Casimir apparatus are (ζ, η, χ), where ζ is normal to the plates, and η and χ are parallel to the plates.

The relation between the two sets of coordinates is

$$z = \zeta\cos\alpha + \eta\sin\alpha, \tag{32a}$$
$$y = \eta\cos\alpha - \zeta\sin\alpha, \tag{32b}$$
$$x = \chi. \tag{32c}$$

Let the center of the apparatus be located at

$$(\zeta = \zeta_0, \eta = 0, \chi = 0). \tag{33}$$

Now we calculate the gravitational energy from (30),

$$E_g = \int (\mathrm{d}\mathbf{x}) gz\langle T^{00}\rangle = \frac{\mathcal{E}_c}{a} gL \int_{-L/2}^{L/2} \mathrm{d}\eta \int_{\zeta_0-a/2}^{\zeta_0+a/2} \mathrm{d}\zeta(\zeta\cos\alpha + \eta\sin\alpha) = g\mathcal{E}_c L^2 \zeta_0 \cos\alpha + K, \quad (34)$$

where K is a constant, independent of the center of the apparatus ζ_0, $z_0 = \zeta_0 \cos\alpha$. Thus, the gravitational force per area $A = L^2$ on the apparatus is independent of orientation

$$\mathcal{F} = \frac{F}{A} = -\frac{\partial E_g}{A \partial z_0} = -g\mathcal{E}_c = \epsilon \mathcal{F}_c, \quad (35)$$

a small upward push. Here $\epsilon = \frac{1}{3} ga$ is a measure of the gravitational force relative to the Casimir force $\mathcal{F}_c = -\pi^2/(240a^4)$. Note that on the earth's surface, the dimensionless number ϵ is very small. For a plate separation of 1μm,

$$\epsilon = \frac{ga}{3c^2} \sim 10^{-22}, \quad (36)$$

so the considerations here would appear to be only of theoretical interest. The effect is far smaller than the Casimir forces between the plates.

It is a bit simpler to use the energy formula (30) to calculate the force by considering the variation in the gravitational energy directly, as we illustrate by considering a mass point at the origin:

$$T^{\mu\nu} = m\delta(\mathbf{r})\delta^{\mu 0}\delta^{\nu 0}. \quad (37)$$

If we displace the particle rigidly upward by an amount δz_0, the change in the metric is $\delta h_{00} = -g\delta z_0$. This implies a change in the energy, exactly as expected:

$$\delta E_g = -m\left(-g\delta z_0\right) = mg\delta z_0. \quad (38)$$

Now we repeat this calculation for the Casimir apparatus. The gravitational force per area on the rigid apparatus is

$$\mathcal{F} = -\frac{\delta \mathcal{E}_g}{\delta z_0} = -g\mathcal{E}_c, \quad (39)$$

again the same result found in (35), which is 1/4 that found by Bimonte *et al* [50]. It does, however, agree with one of the results found in the earlier paper by the same collaboration [49], who now completely agree with our calculation [54]. Our result is further completely consistent with the principle of equivalence, and with one result of Jaekel and Reynaud [55].

In electrodynamics, one defines the field (the 4-vector potential) by

$$\delta W = \int (\mathrm{d}x)\delta J^\mu A_\mu. \quad (40)$$

From this, one can compute the force on a charge by considering the displacement of a static charge,

$$J^\mu = e\delta^{\mu 0}\delta(\mathbf{r}), \quad (41)$$

that is,

$$\delta J^\mu = e\delta^{\mu 0}\left(-\delta\mathbf{r}\cdot\boldsymbol{\nabla}\right)\delta(\mathbf{r}). \quad (42)$$

Inserting this in the above and integrating by parts we get

$$\delta W = \int \mathrm{d}t \int (\mathrm{d}\mathbf{r}) e\left(-\boldsymbol{\nabla} A^0\right)\cdot\delta\mathbf{r}\,\delta(\mathbf{r}). \quad (43)$$

Because the electric field is $E^i = F^{0i} = \partial^0 A^i - \partial^i A^0$, and a total derivative on time is irrelevant, we deduce the force on the charge to have the expected form:

$$-\frac{\delta E}{\delta \mathbf{r}} = \mathbf{F} = e\mathbf{E}. \tag{44}$$

We therefore should be able to proceed similarly with vacuum energy, starting from the definition of the gravitational field

$$\delta W = \int (\mathrm{d}x) \delta T^{\mu\nu} h_{\mu\nu}. \tag{45}$$

Again, check this for the force on a mass point, described by (37), so

$$\delta E_g = -\int (\mathrm{d}\mathbf{r}) m \left(-\delta\mathbf{r}\cdot\boldsymbol{\nabla}\right)\delta(\mathbf{r}) h^{00} = -m\delta\mathbf{r}\cdot\boldsymbol{\nabla} h^{00}. \tag{46}$$

Since $h^{00} = -gz$, we conclude

$$-\frac{\delta E}{\delta \mathbf{r}} = \mathbf{F} = -mg\hat{z}. \tag{47}$$

For the same constant field the force on a Casimir apparatus is obtained from the change in the energy density

$$\langle T^{00}\rangle = \frac{\mathcal{E}_c}{a}\theta(a/2 - \zeta + \zeta_0)\theta(\zeta - \zeta_0 + a/2), \tag{48}$$

that is, recalling that $z_0 = \zeta_0 \cos\alpha$, we obtain the change of the energy density under a rigid displacement δz_0:

$$\delta\langle T^{00}\rangle = \frac{\mathcal{E}_c}{a}\delta z_0 \frac{1}{\cos\alpha}\left[\delta(\zeta - \zeta_0 - a/2) - \delta(\zeta - \zeta_0 + a/2)\right], \tag{49}$$

which yields the gravitational force, since $h^{00} = -g(\zeta\cos\alpha + \eta\sin\alpha)$,

$$\mathcal{F} = -\frac{\delta E_g}{A\delta z_0} = \frac{F}{A} = \frac{\mathcal{E}_c}{a}\frac{1}{\cos\alpha}h^{00}\Big|_{\zeta=\zeta_0-a/2}^{\zeta=\zeta_0+a/2} = -g\mathcal{E}_c. \tag{50}$$

3.2. Inertial coordinates

However, one might think that the above metric, while sufficing for massive Newtonian objects, might seem inappropriate for photons, near the surface of the earth. Rather, shouldn't we use the perturbation of the Schwarzchild metric, for weak fields ($GM/r \ll 1$) either in its original or isotropic form? Doing so would give rise to a different force on the Casimir apparatus. The reason we get different answers in different coordinate systems reflects the fact that our starting point is not gauge invariant. Under a coordinate redefinition, which for weak fields is a gauge transformation of $h_{\mu\nu}$ [56]

$$h_{\mu\nu} \to h_{\mu\nu} + \partial_\mu \xi_\nu + \partial_\nu \xi_\mu, \tag{51}$$

where ξ_μ is a vector field, the interaction W is invariant only if the stress tensor is conserved, $\partial_\mu T^{\mu\nu} = 0$. Otherwise, there is a change in the action,

$$\Delta W = -2\int (\mathrm{d}x)\xi_\nu \partial_\mu T^{\mu\nu}. \tag{52}$$

Now in our case (where we make the finite size of the plate explicit, but ignore edge effects because $L \gg a$)

$$T^{\mu\nu} = \frac{\mathcal{E}_c}{a}\text{diag}(1,-1,-1,3)\theta\left(\zeta-\zeta_0+\frac{a}{2}\right)\theta\left(\frac{a}{2}-\zeta+\zeta_0\right) \times\theta(\eta+L/2)\theta(L/2-\eta)\theta(\chi+L/2)\theta(L/2-\chi). \tag{53}$$

Thus the nonzero components of $\partial_\mu T^{\mu\nu}$ are

$$\partial_\mu T^{\mu\zeta} = \frac{3\mathcal{E}_c}{a}\left[\delta(\zeta-\zeta_0+a/2)-\delta(\zeta-\zeta_0-a/2)\right]\theta\ldots, \tag{54a}$$

$$\partial_\mu T^{\mu\eta} = -\frac{\mathcal{E}_c}{a}\left[\delta(\eta+L/2)-\delta(\eta-L/2)\right]\theta\ldots, \tag{54b}$$

$$\partial_\mu T^{\mu\chi} = -\frac{\mathcal{E}_c}{a}\left[\delta(\chi+L/2)-\delta(\chi-L/2)\right]\theta\ldots, \tag{54c}$$

where $\theta\ldots$ refer to the remaining step functions. Therefore, the change in the energy obtained from ΔW is

$$\begin{aligned}\Delta E_g = \frac{6\mathcal{E}_c}{a}&\int \mathrm{d}\eta\,\mathrm{d}\chi\,[\xi_\zeta(\zeta_0-a/2,\eta,\chi)-\xi_\zeta(\zeta_0+a/2,\eta,\chi)]\\ &-\frac{2\mathcal{E}_c}{a}\int \mathrm{d}\zeta\,\mathrm{d}\chi\,[\xi_\eta(\zeta,-L/2,\chi)-\xi_\eta(\zeta,L/2,\chi)]\\ &-\frac{2\mathcal{E}_c}{a}\int \mathrm{d}\zeta\,\mathrm{d}\eta\,[\xi_\chi(\zeta,\eta,-L/2)-\xi_\chi(\zeta,\eta,L/2)]\,.\end{aligned} \tag{55}$$

Since we have demonstrated that the gravitational force on a Casimir apparatus is not a gauge-invariant concept, we must ask if there is any way to extract a physically meaningful result. There seem to be two possible ways to proceed. Either we add another interaction, say a fluid exerting a pressure on the plates, resulting in a total stress tensor that is conserved, or we find a physical basis for believing that one coordinate system is more realistic than another. The former procedure is undoubtedly more physical, but will yield model-dependent results. The latter apparently has a natural solution.

A Fermi coordinate system is the general relativistic generalization of an inertial coordinate frame. Such a system has been given in [57] for a resting observer in the field of a static mass distribution. For our case, where the gravitational potential is $\phi = gz$ (apart from an irrelevant additive constant), we recover the Fermi coordinate metric (31) for a gravitating body,

$$\mathrm{d}s^2 = -(1+2gz)\mathrm{d}t^2+\mathrm{d}\mathbf{r}^2. \tag{56}$$

This is *a priori* obvious, because in this coordinate system, coordinate lengths don't depend on z. The Fermi coordinate system may be immediately derived from the Schwarzchild coordinates, far from the gravitating body (so that $GM/r \ll 1$), by a simple coordinate redefinition.

3.3. Alternative derivation

For an alternative derivation of the force on the Casimir apparatus, we start from the variational definition of the stress tensor (28), and consider a general coordinate transformation,

$$x'^\mu = x^\mu + \delta x^\mu, \tag{57}$$

so that

$$g'_{\mu\nu}(x') = g'_{\mu\nu}(x)+\partial_\lambda g_{\mu\nu}\delta x^\lambda = g_{\mu\nu}(x)+\delta g_{\mu\nu}(x), \tag{58}$$

where

$$\delta g_{\mu\nu} = \delta x^\lambda \partial_\lambda g_{\mu\nu} + g_{\alpha\nu}\frac{\partial \delta x^\alpha}{\partial x^\mu} + g_{\mu\beta}\frac{\partial \delta x^\beta}{\partial x^\nu}. \tag{59}$$

For a rigid translation, δx^λ is a constant, so only the first term here is present, which gives the result found above in section 3.1. However, if we don't make this restriction, we obtain

$$\delta W = -\int_\Omega (\mathrm{d}x)\left[\partial_\nu\left(\sqrt{-g}T^\nu{}_\lambda\right) - \frac{1}{2}\sqrt{-g}T^{\mu\nu}\partial_\lambda g_{\mu\nu}\right]\delta x^\lambda + \int_{\partial\Omega}\mathrm{d}\Sigma_\nu\sqrt{-g}T^\nu{}_\lambda\delta x^\lambda, \tag{60}$$

where the surface term results from integration by parts. Again, notice for δx^λ constant, the first and third terms identically cancel. If we take the region Ω to be all space the surface term vanishes (and therefore so does the first term for a rigid displacement). A simple calculation indeed shows that the surface term cancels.

3.4. Rindler coordinates

Relativistically, uniform acceleration is described by hyperbolic motion

$$t = \xi\sinh\tau, \quad z = \xi\cosh\tau, \tag{61}$$

which corresponds to the metric

$$\mathrm{d}t^2 - \mathrm{d}z^2 - \mathrm{d}x_\perp^2 = \xi^2\mathrm{d}\tau^2 - \mathrm{d}\xi^2 - \mathrm{d}x_\perp^2. \tag{62}$$

The d'Alembertian operator has the following form in Rindler coordinates:

$$-\left(\frac{\partial}{\partial t}\right)^2 + \left(\frac{\partial}{\partial z}\right)^2 + \nabla_\perp^2 = -\frac{1}{\xi^2}\left(\frac{\partial}{\partial\tau}\right)^2 + \frac{1}{\xi}\frac{\partial}{\partial\xi}\left(\xi\frac{\partial}{\partial\xi}\right) + \nabla_\perp^2. \tag{63}$$

In this subsection, we will consider scalar fields and so-called semitransparent plates, that is, ones described by a δ-function potential

$$V = \lambda\delta(\xi - \xi_a), \tag{64}$$

where ξ_a is the position of the plate. For a single semitransparent plate at ξ_1, the Green's function can be written as

$$G(x,x') = \int\frac{\mathrm{d}\omega}{2\pi}\frac{\mathrm{d}^2k}{(2\pi)^2}\mathrm{e}^{-\mathrm{i}\omega(\tau-\tau')}\mathrm{e}^{\mathrm{i}\mathbf{k}\cdot(\mathbf{r}-\mathbf{r}')_\perp}g(\xi,\xi'), \tag{65}$$

where the reduced Green's function satisfies

$$\left[-\frac{\omega^2}{\xi^2} - \frac{1}{\xi}\frac{\partial}{\partial\xi}\left(\xi\frac{\partial}{\partial\xi}\right) + k^2 + \lambda\delta(\xi-\xi_1)\right]g = \frac{1}{\xi}\delta(\xi-\xi'), \tag{66}$$

which we recognize as just the semitransparent cylinder problem with $m \to \zeta = -\mathrm{i}\omega$ and $\kappa \to k$ [48]. The reduced Green's function for single plate thus is

$$g(\xi,\xi') = I_\zeta(k\xi_<)K_\zeta(k\xi_>) - \frac{\lambda\xi_1 K_\zeta^2(k\xi_1)I_\zeta(k\xi)I_\zeta(k\xi')}{1+\lambda\xi_1 I_\zeta(k\xi_1)K_\zeta(k\xi_1)}, \qquad \xi,\xi' < \xi_1, \tag{67a}$$

$$= I_\zeta(k\xi_<)K_\zeta(k\xi_>) - \frac{\lambda\xi_1 I_\zeta^2(k\xi_1)K_\zeta(k\xi)K_\zeta(k\xi')}{1+\lambda\xi_1 I_\zeta(k\xi_1)K_\zeta(k\xi_1)}, \qquad \xi,\xi' > \xi_1. \tag{67b}$$

In the limit $\lambda \to \infty$ we recover the well-known Green's function subject to Dirichlet boundary conditions.

If we use the uniform asymptotic expansion (UAE), based on the limit

$$\xi \to \infty, \quad \xi_1 \to \infty, \quad \xi - \xi_1 \text{ finite}, \quad \zeta = \hat{\zeta}\xi_1 \to \infty, \quad \hat{\zeta} \text{ finite}, \tag{68}$$

we recover the Green's function for a single plate in Minkowski space,

$$\xi_1 g(\xi, \xi') \to \frac{\mathrm{e}^{-\kappa|\xi-\xi'|}}{2\kappa} - \frac{\lambda}{2\kappa} \frac{\mathrm{e}^{-\kappa(|\xi-\xi_1|+|\xi'-\xi_1|)}}{\lambda + 2\kappa}, \tag{69}$$

where $\kappa = \sqrt{k^2 + \zeta^2}$, $\omega = \mathrm{i}\zeta$.

The canonical energy-momentum for a scalar field is given by

$$T_{\mu\nu} = \partial_\mu \phi \partial_\nu \phi + g_{\mu\nu} \frac{1}{\sqrt{-g}} \mathcal{L}, \tag{70}$$

where the Lagrange density includes the δ-function potential. Using the equations of motion, we find the energy density to be

$$T_{00} = \frac{1}{2}\left(\frac{\partial \phi}{\partial \tau}\right)^2 - \frac{1}{2}\phi \frac{\partial^2}{\partial \tau^2}\phi + \frac{\xi}{2}\frac{\partial}{\partial \xi}\left(\phi \xi \frac{\partial}{\partial \xi}\phi\right) + \frac{\xi^2}{2}\boldsymbol{\nabla}_\perp \cdot (\phi \boldsymbol{\nabla}_\perp \phi). \tag{71}$$

From this and similar expressions for the spatial components of $T^{\mu\nu}$ we can obtain $\langle T_{\mu\nu} \rangle$ by applying a differential operator to the Green's function, in view of the one-loop connection (25).

The force density is given by

$$f_\lambda = -\frac{1}{\sqrt{-g}} \partial_\nu (\sqrt{-g} T^\nu{}_\lambda) + \frac{1}{2} T^{\mu\nu} \partial_\lambda g_{\mu\nu}, \tag{72}$$

which follows from (60), or

$$f_\xi = -\frac{1}{\xi} \partial_\xi (\xi T^{\xi\xi}) - \xi T^{00}. \tag{73}$$

When we integrate over all space to get the force,

$$F = \int (\mathrm{d}\mathbf{x}) \sqrt{-g} f_\xi, \tag{74}$$

the first term in (73) is a surface term which does not contribute, as noted in (60), so the gravitational force on the system is

$$\mathcal{F} = \int \mathrm{d}\xi \xi f_\xi = -\int \frac{\mathrm{d}\xi}{\xi^2} T_{00}, \tag{75}$$

which when multiplied by the gravitational acceleration g is the gravitational force/area on the Casimir energy. Using the expression (71) for the energy density, and rescaling we see that

$$\mathcal{F} = \int \mathrm{d}\xi \xi \int \frac{\mathrm{d}\hat{\zeta}\, \mathrm{d}^2 k}{(2\pi)^3} \hat{\zeta}^2 g(\xi, \xi) \quad (\zeta = \hat{\zeta}\xi). \tag{76}$$

This result is an immediate consequence of the general formula [58]

$$E_c = \frac{1}{2\mathrm{i}} \int (\mathrm{d}\mathbf{r}) \int \frac{\mathrm{d}\omega}{2\pi} 2\omega^2 \mathcal{G}(\mathbf{r}, \mathbf{r}), \tag{77}$$

in terms of the frequency transform of the Green's function,

$$G(x,x') = \int_{-\infty}^{\infty} \frac{\mathrm{d}\omega}{2\pi} \mathrm{e}^{-i\omega(t-t')} \mathcal{G}(\mathbf{r},\mathbf{r}'). \tag{78}$$

Alternatively, we can start from the following formula for the force density for a single semitransparent plate,

$$f_\xi = \frac{1}{2}\phi^2 \partial_\xi \lambda \delta(\xi - \xi_1), \tag{79}$$

or, in terms of the Green's function,

$$\mathcal{F} = -\lambda \frac{1}{2} \int \frac{\mathrm{d}\zeta\, \mathrm{d}^2 k}{(2\pi)^3} \partial_{\xi_1} [\xi_1 g(\xi_1, \xi_1)]. \tag{80}$$

For example, the force on a single plate is given by

$$\mathcal{F} = -\partial_{\xi_1} \frac{1}{2} \int \frac{\mathrm{d}\zeta\, \mathrm{d}^2 k}{(2\pi)^3} \ln[1 + \lambda I_\zeta(k\xi_1) K_\zeta(k\xi_1)], \tag{81}$$

Expanding this about some arbitrary point ξ_0, with $\zeta = \hat{\zeta}\xi_0$, and using the UAE, we get

$$\mathcal{F} = -\frac{1}{96\pi^2 a^3} \int_0^\infty \frac{\mathrm{d}y\, y^2}{1 + y/\lambda a}, \tag{82}$$

which is just the negative of the (divergent) quantum vacuum energy of a single plate.

For two plates at ξ_1, ξ_2, for $\xi, \xi' < \xi_1 < \xi_2$,

$$g(\xi,\xi') = I_< K_> - \frac{\lambda_1 \xi_1 K_1^2 + \lambda_2 \xi_2 K_1^2 - \lambda_1 \lambda_2 \xi_1 \xi_2 K_1 K_2 (K_2 I_1 - K_1 I_2)}{\Delta} I I_\prime, \tag{83a}$$

while for $\xi, \xi' > \xi_2 > \xi_1$,

$$g(\xi,\xi') = I_< K_> - \frac{\lambda_1 \xi_1 I_1^2 + \lambda_2 \xi_2 I_2^2 + \lambda_1 \lambda_2 \xi_1 \xi_2 I_1 I_2 (I_2 K_1 - I_1 K_2)}{\Delta} K K_\prime, \tag{83b}$$

and between the plates, $\xi_1 < \xi, \xi' < \xi_2$,

$$g(\xi,\xi') = I_< K_> - \frac{\lambda_2 \xi_2 K_2^2 (1 + \lambda_1 \xi_1 K_1 I_1)}{\Delta} I I_\prime - \frac{\lambda_1 \xi_1 I_1^2 (1 + \lambda_2 \xi_2 K_1 I_1)}{\Delta} K K_\prime + \frac{\lambda_1 \lambda_2 \xi_1 \xi_2 I_1^2 K_2^2}{\Delta} (I K_\prime + K I_\prime), \tag{83c}$$

where

$$\Delta = 1 + \lambda_1 \xi_1 K_1 I_1 + \lambda_2 \xi_2 I_2 K_2 - \lambda_1 \lambda_2 \xi_1 \xi_2 I_1 K_2 (I_1 K_2 - I_2 K_1), \tag{84}$$

and where we have used the abbreviations $I_a = I_\zeta(k\xi_a)$, $I = I_\zeta(k\xi)$, $I_\prime = I_\zeta(k\xi')$, etc.

In the weak acceleration limit, the Green's function reduces to exactly the expected result, for $\xi_1 < \xi, \xi' < \xi_2$ $(a = \xi_2 - \xi_1)$

$$\xi_0 g(\xi,\xi') \to \frac{1}{2\kappa} \mathrm{e}^{-\kappa|\xi-\xi'|} + \frac{1}{2\kappa\tilde{\Delta}} \left[\frac{\lambda_1 \lambda_2}{4\kappa^2} 2\cosh\kappa(\xi - \xi') - \frac{\lambda_1}{2\kappa}\left(1 + \frac{\lambda_2}{2\kappa}\right) \mathrm{e}^{-\kappa(\xi+\xi'-2\xi_2)} - \frac{\lambda_2}{2\kappa}\left(1 + \frac{\lambda_1}{2\kappa}\right) \mathrm{e}^{\kappa(\xi+\xi'-2\xi_1)} \right], \tag{85}$$

with

$$\tilde{\Delta} = \left(1 + \frac{\lambda_1}{2\kappa}\right)\left(1 + \frac{\lambda_2}{2\kappa}\right) e^{2\kappa a} - \frac{\lambda_1 \lambda_2}{4\kappa^2}, \tag{86}$$

The flat space limit also holds outside the plates.

In general, we have two alternative forms for the force on the two-plate system:

$$\mathcal{F} = -(\partial_{\xi_1} + \partial_{\xi_2})\frac{1}{2}\int \frac{d\zeta\, d^2k}{(2\pi)^3} \ln \Delta, \tag{87}$$

which is equivalent to (76). From either of these two formulæ, we find the gravitational force on the Casimir energy to be in the $\xi \to \infty$ limit

$$\mathcal{F} = -\frac{1}{4\pi^2}\int_0^\infty d\kappa\, \kappa^2 \ln \Delta_0 \tag{88}$$

where $\Delta_0 = e^{-2\kappa a}\tilde{\Delta}$, $\tilde{\Delta}$ given in (86). From this we get the explicit force

$$\begin{aligned}\mathcal{F} &= \frac{1}{96\pi^2 a^3}\int_0^\infty dy\, y^3 \frac{1 + \frac{1}{y+\lambda_1 a} + \frac{1}{y+\lambda_2 a}}{\left(\frac{y}{\lambda_1 a}+1\right)\left(\frac{y}{\lambda_2 a}+1\right)e^y - 1} \\ &\quad - \frac{1}{96\pi^2 a^3}\int_0^\infty dy\, y^2 \left[\frac{1}{\frac{y}{\lambda_1 a}+1} + \frac{1}{\frac{y}{\lambda_2 a}+1}\right] = -(\mathcal{E}_c + \mathcal{E}_{d1} + \mathcal{E}_{d2}),\end{aligned} \tag{89}$$

which is just the negative of the Casimir energy of the two semitransparent plates, *including divergent parts associated with each plate.* The divergent terms simply renormalize the masses m_i per area of each plate:

$$\mathcal{E}_{\text{total}} = m_1 + m_2 + \mathcal{E}_{d1} + \mathcal{E}_{d2} + \mathcal{E}_c = M_1 + M_2 + \mathcal{E}_c, \tag{90}$$

$M_i = m_i + \mathcal{E}_{di}$, and thus the gravitational force on the entire apparatus obeys the equivalence principle

$$g\mathcal{F} = -g(M_1 + M_2 + \mathcal{E}_c). \tag{91}$$

3.5. Centripetal Acceleration

Consider finally a Casimir apparatus undergoing centripetal acceleration with angular velocity ω ($\omega r \ll 1$), as illustrated in figure 3. A detailed calculation [59] shows that the force on the Casimir apparatus is just as expected, including the divergent pieces which renormalize the masses per area A of the plates:

$$\frac{\mathbf{F}}{A} = -\frac{\omega^2}{A}\int d^3x\, \mathbf{r}\, t^{00}(\mathbf{r}) = -\omega^2 \mathbf{r}_{\text{CM}}(m_1 + m_2 + \mathcal{E}_{d1} + \mathcal{E}_{d2} + \mathcal{E}_c), \tag{92}$$

where $\mathbf{r}_{\text{CM}}$ is the center of energy of the entire system. Nontrivial is the fact that this result is independent of the orientation of the plates.

4. Multiple Scattering Technique

In the past few years, there has been a tremendous resurgence of interest in the multiple-scattering technique. This is largely inspired by the desire to improve the comparison with experiment, such comparison heretofore having been restricted to the use of the "proximity force approximation," which describes the Casimir interaction between bodies with curved surfaces by pairwise interactions between small plane elements of the surfaces [60]. The multiple-scattering

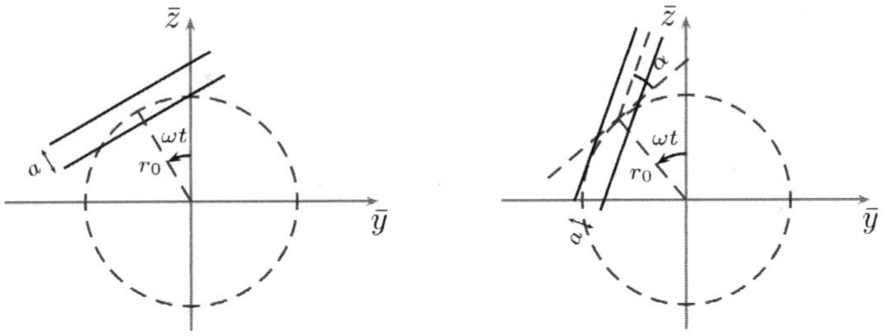

Figure 3. Casimir apparatus undergoing uniform circular motion with angular velocity ω. The figure on the left has the Casimir plates tangent to the circle of revolution, while on the right the plates make an angle α with respect to that tangent.

method certainly dates back to at least the Krein formula [61], and was first used in connection with the Casimir effect by Renne [62]. A most famous variation were the multiple reflection calculations of Balian and Duplantier [63]. In the last few years there has been an explosion of publications on variations of this technique, as physicists realized it could be practically used to obtain numerical results for forces between bodies of arbitrary shapes, for example [64–76]. We may remark that much of this work was inspired by the ingenious worldline approach [77].

We have also contributed to this redevelopment, giving the general formulation for scalar fields [78], and applying it to bodies described by δ-function potentials [79], and, after generalizing to electrodynamics, to dielectric bodies [80]. The methodology has also been applied to non-contact gears [81, 82].

The multiple scattering approach starts from the well-known formula for the vacuum energy or Casimir energy (for simplicity here we first restrict attention to a massless scalar field)(τ is the "infinite" time that the configuration exists)

$$E = \frac{\mathrm{i}}{2\tau}\mathrm{Tr}\ln G \to \frac{\mathrm{i}}{2\tau}\mathrm{Tr}\ln GG_0^{-1}, \tag{93}$$

where G (G_0) is the Green's function,

$$(-\partial^2 + V)G = 1, \quad -\partial^2 G_0 = 1, \tag{94}$$

subject to appropriate boundary conditions at infinity. We will use the Feynman or causal Green's functions.

Now we define the T-matrix [83]

$$T = S - 1 = V(1 + G_0 V)^{-1}. \tag{95}$$

If the potential has two disjoint parts, $V = V_1 + V_2$ it is easy to derive

$$T = (V_1 + V_2)(1 - G_0 T_1)(1 - G_0 T_1 G_0 T_2)^{-1}(1 - G_0 T_2), \tag{96}$$

where

$$T_i = V_i(1 + G_0 V_i)^{-1}, \quad i = 1, 2. \tag{97}$$

Thus, we can write the general expression for the interaction between the two bodies (potentials) in two equivalent forms:

$$E_{12} = -\frac{\mathrm{i}}{2\mathcal{T}}\mathrm{Tr}\ln(1 - G_0T_1G_0T_2) \tag{98a}$$

$$= -\frac{\mathrm{i}}{2\mathcal{T}}\mathrm{Tr}\ln(1 - V_1G_1V_2G_2), \tag{98b}$$

where $G_i = (1 + G_0V_i)^{-1}G_0, \quad i = 1, 2.$

4.1. Multipole expansion

To proceed to apply this method to general bodies, we use an even older technique, the multipole expansion. Let's illustrate this with a 2 + 1 dimensional version, which allows us to describe cylinders with parallel axes. We seek an expansion of the free Green's function

$$G_0(\mathbf{R} + \mathbf{r}' - \mathbf{r}) = \frac{\mathrm{e}^{\mathrm{i}|\omega||\mathbf{r}-\mathbf{R}-\mathbf{r}'|}}{4\pi|\mathbf{r} - \mathbf{R} - \mathbf{r}'|} = \int \frac{\mathrm{d}k_z}{2\pi}\mathrm{e}^{\mathrm{i}k_z(z-Z-z')}g_0(\mathbf{r}_\perp - \mathbf{R}_\perp - \mathbf{r}'_\perp). \tag{99}$$

Here, the reduced Green's function is ($\omega = \mathrm{i}\zeta$)

$$g_0(\mathbf{r}_\perp - \mathbf{R}_\perp - \mathbf{r}'_\perp) = \int \frac{(\mathrm{d}^2k_\perp)}{(2\pi)^2}\frac{\mathrm{e}^{-\mathrm{i}\mathbf{k}_\perp\cdot\mathbf{R}_\perp}\mathrm{e}^{\mathrm{i}\mathbf{k}_\perp\cdot(\mathbf{r}_\perp-\mathbf{r}'_\perp)}}{k_\perp^2 + k_z^2 + \zeta^2}. \tag{100}$$

As long as the two potentials do not overlap, so that we have $\mathbf{r}_\perp - \mathbf{R}_\perp - \mathbf{r}'_\perp \neq 0$, we can write an expansion in terms of modified Bessel functions:

$$g_0(\mathbf{r}_\perp - \mathbf{R}_\perp - \mathbf{r}'_\perp) = \sum_{m,m'} I_m(\kappa r)\mathrm{e}^{\mathrm{i}m\phi}I'_m(\kappa r')\mathrm{e}^{-\mathrm{i}m'\phi'}\tilde{g}^0_{m,m'}(\kappa R), \quad \kappa^2 = k_z^2 + \zeta^2. \tag{101}$$

By Fourier transforming, and using the definition of the Bessel function

$$\mathrm{i}^m J_m(kr) = \int_0^{2\pi}\frac{\mathrm{d}\phi}{2\pi}\,\mathrm{e}^{-\mathrm{i}m\phi}\mathrm{e}^{\mathrm{i}kr\cos\phi}, \tag{102}$$

we easily find

$$\tilde{g}^0_{m,m'}(\kappa R) = \frac{1}{2\pi}\int\frac{\mathrm{d}k\,k}{k^2+\kappa^2}J_{m-m'}(kR)\frac{J_m(kr)J_m(kr')}{I_m(\kappa r)I_m(\kappa r')} = \frac{(-1)^{m'}}{2\pi}K_{m-m'}(\kappa R). \tag{103}$$

Thus we can derive an expression for the interaction between two bodies, in terms of discrete matrices,

$$\mathfrak{E} \equiv \frac{E_{\mathrm{int}}}{L} = \frac{1}{8\pi^2}\int \mathrm{d}\zeta\,\mathrm{d}k_z\ln\det\left(1 - \tilde{g}^0\tilde{T}_1\tilde{g}^{0\top}\tilde{T}_2\right), \tag{104}$$

where $\top$ denotes transpose, and the $\tilde{T}$ matrix elements are given by

$$\tilde{T}_{mm'} = \int \mathrm{d}r\,r\,\mathrm{d}\phi\int\mathrm{d}r'\,r'\,\mathrm{d}\phi' I_m(\kappa r)\mathrm{e}^{-\mathrm{i}m\phi}I_{m'}(\kappa r')\mathrm{e}^{\mathrm{i}m'\phi'}T(r,\phi;r',\phi'). \tag{105}$$

We will apply this 2+1 dimensional formalism to the interaction between two cylinders, as illustrated in figure 4. Also illustrated there is the idea behind the proximity force approximation, which applies when the separation between the bodies is much smaller than the radii of curvature of the bodies.

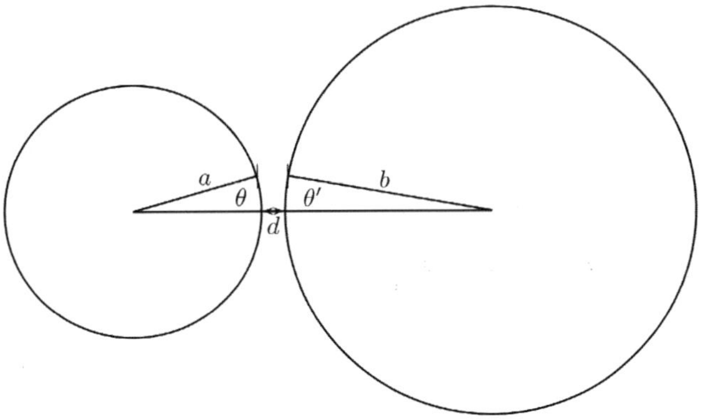

Figure 4. Geometry of two cylinders (or two spheres) with radii a and b, respectively, and distances between their centers of $R > a + b$. The proximity force approximation applies when the distance of closest approach $d = R - a - b \ll a, b$. The approximation consists in assuming that the interaction is dominated by the interaction of adjacent surface elements, as shown.

Consider two parallel semitransparent cylinders, of radii a and b, respectively, lying outside each other, described by the potentials

$$V_1 = \lambda_1 \delta(r - a), \quad V_2 = \lambda_2 \delta(r' - b), \tag{106}$$

with the separation between the centers R satisfying $R > a + b$. It is easy to work out the scattering matrix in this situation,

$$(t_1)_{mm'} = 2\pi\lambda_1 a \delta_{mm'} \frac{I_m^2(\kappa a)}{1 + \lambda_1 a I_m(\kappa a) K_m(\kappa a)}. \tag{107}$$

Thus the Casimir energy per unit length is

$$\mathfrak{E} = \frac{1}{4\pi}\int_0^\infty \mathrm{d}\kappa\, \kappa \,\mathrm{tr} \ln(1 - A), \tag{108}$$

where $A = B(a)B(b)$, in terms of the matrices

$$B_{mm'}(a) = K_{m+m'}(\kappa R) \frac{\lambda_1 a I_{m'}^2(\kappa a)}{1 + \lambda_1 a I_{m'}(\kappa a) K_{m'}(\kappa a)}. \tag{109}$$

The strong coupling limit, the case of Dirichlet cylinders, has been considered in [75], for example. In contrast, here we will consider the situation of weak coupling, $\lambda_1 a \ll 1$, $\lambda_2 b \ll 1$, where the formula for the interaction energy between two cylinders becomes

$$\mathfrak{E} = -\frac{\lambda_1 \lambda_2 a b}{4\pi R^2} \sum_{m,m'=-\infty}^{\infty} \int_0^\infty \mathrm{d}x\, x\, K_{m+m'}^2(x) I_m^2(xa/R) I_{m'}^2(xb/R). \tag{110}$$

One merely exploits the small argument expansion for the modified Bessel functions $I_m(xa/R)$ and $I_{m'}(xb/R)$:

$$I_m^2(x) = \left(\frac{x}{2}\right)^{2|m|} \sum_{n=0}^{\infty} Z_{|m|,n} \left(\frac{x}{2}\right)^{2n}, \tag{111}$$

where the coefficients $Z_{m,n}$ are

$$Z_{m,n} = \frac{2^{2(m+n)}\,\Gamma\left(m+n+\frac{1}{2}\right)}{\sqrt{\pi}\; n!\;(2m+n)!\;\Gamma(m+n+1)}. \tag{112}$$

In this case we get an amazingly simple result

$$\mathfrak{E} = -\frac{\lambda_1 a \lambda_2 b}{4\pi R^2}\frac{1}{2}\sum_{n=0}^{\infty}\left(\frac{a}{R}\right)^{2n} P_n(\mu), \tag{113}$$

where $\mu = b/a$, and where by inspection we identify the binomial coefficients

$$P_n(\mu) = \sum_{k=0}^{n}\binom{n}{k}^2 \mu^{2k}. \tag{114}$$

Remarkably, it is possible to perform the sums, so we obtain the following closed form for the interaction between two weakly-coupled cylinders:

$$\mathfrak{E} = -\frac{\lambda_1 a \lambda_2 b}{8\pi R^2}\left[\left(1-\left(\frac{a+b}{R}\right)^2\right)\left(1-\left(\frac{a-b}{R}\right)^2\right)\right]^{-1/2}. \tag{115}$$

We note that in the limit $R-a-b=d\to 0$, d being the distance between the closest points on the two cylinders, we recover the proximity force theorem in this case

$$U(d) = -\frac{\lambda_1\lambda_2}{32\pi}\sqrt{\frac{2ab}{R}}\frac{1}{d^{1/2}}, \qquad d \ll a, b. \tag{116}$$

The rate of approach is given by

$$\frac{\mathfrak{E}}{U} \approx 1 - \frac{1+\mu+\mu^2}{4\mu}\frac{d}{R} \approx 1 - \frac{R^2 - aR + a^2}{4a(R-a)}\frac{d}{R}. \tag{117}$$

The comparison between the exact result, and the proximity force approximation (116) is given in figure 5, for equal radii, and in figure 6 for unequal radii. Evidently, the proximity force approximation fails badly as the cylinders are pulled apart.

We can similarly consider the interaction between a cylinder and a plane. By the method of images, we can find the interaction between a semitransparent cylinder and a Dirichlet plane to be

$$\mathfrak{E} = \frac{1}{4\pi}\int_0^\infty \kappa\, d\kappa\, \mathrm{tr}\,\ln(1-B(a)), \tag{118}$$

where $B(a)$ is given in (109). In the strong-coupling limit this result agrees with that given by Bordag and Nikolaev [72], because

$$\mathrm{tr}\, B^s = \mathrm{tr}\, \tilde{B}^s, \quad \tilde{B}_{mm'} = \frac{1}{K_m(\kappa a)} K_{m+m'}(\kappa R) I_{m'}(\kappa a). \tag{119}$$

In exactly the way given above, we can obtain a closed-form result for the interaction energy between a Dirichlet plane and a weakly-coupled cylinder of radius a separated by a distance $R/2$. The result is again quite simple:

$$\mathfrak{E} = -\frac{\lambda a}{4\pi R^2}\left[1-\left(\frac{2a}{R}\right)^2\right]^{-3/2}. \tag{120}$$

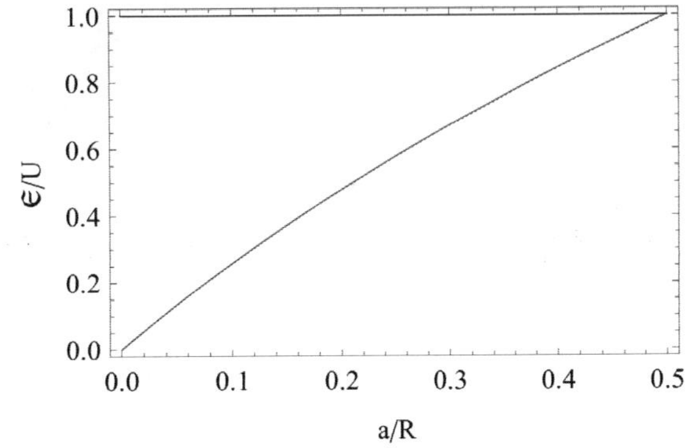

Figure 5. Plotted is the ratio of the exact interaction energy of two equal radii weakly-coupled cylinders to the proximity force approximation. Here, and in the following figures, the horizontal line represents the proximity force approximation, that is, if the proximity force theorem result were applied for all distances.

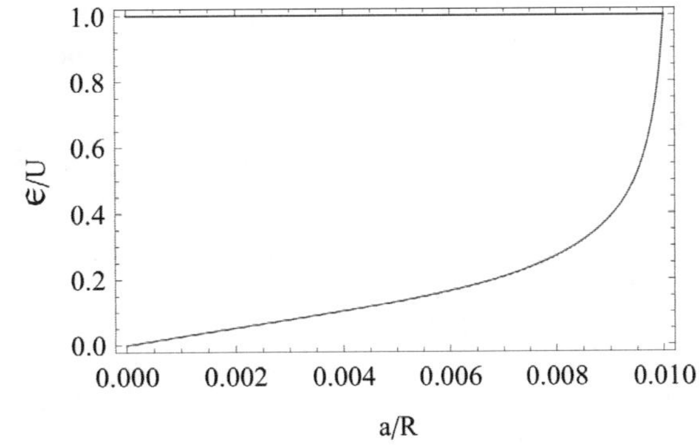

Figure 6. Plotted is the ratio of the exact interaction energy of two weakly-coupled cylinders with ratio of radii $b/a = 99$ to the proximity force approximation.

In the limit as $d \to 0$, this agrees with the PFA for this situation:

$$U(d) = -\frac{\lambda}{64\pi}\frac{\sqrt{2a}}{d^{3/2}}. \tag{121}$$

Note again that this form is ambiguous: the proximity force theorem is equally well satisfied if we replace a by $R/2$, for example, in $U(d)$. The exact result is compared with the PFA in figure 7.

4.2. 3-dimensional formalism

The three-dimensional formalism is very similar. In this case, the free Green's function has the representation

$$G_0(\mathbf{R}+\mathbf{r}'-\mathbf{r}) = \sum_{lm,l'm'} j_l(\mathrm{i}|\zeta|r)j_{l'}(\mathrm{i}|\zeta|r')Y^*_{lm}(\hat{\mathbf{r}})Y_{l'm'}(\hat{\mathbf{r}}')g_{lm,l'm'}(\mathbf{R}). \tag{122}$$

The reduced Green's function can be written in the form

$$g^0_{lm,l'm'}(\mathbf{R}) = (4\pi)^2\mathrm{i}^{l'-l}\int\frac{(\mathrm{d}\mathbf{k})}{(2\pi)^3}\frac{\mathrm{e}^{\mathrm{i}\mathbf{k}\cdot\mathbf{R}}}{k^2+\zeta^2}\frac{j_l(kr)j_{l'}(kr')}{j_l(\mathrm{i}|\zeta|r)j_{l'}(\mathrm{i}|\zeta|r')}Y_{lm}(\hat{\mathbf{k}})Y^*_{l'm'}(\hat{\mathbf{k}}). \tag{123}$$

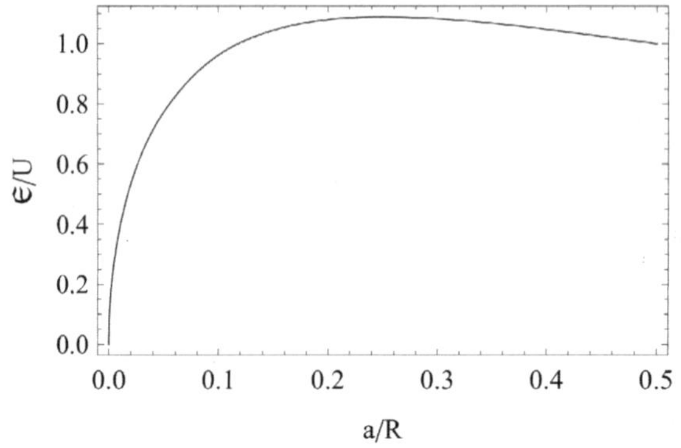

Figure 7. Interaction energy between a weakly coupled δ-function cylinder and a Dirichlet plane. Plotted is the ratio between the energy and the PFA approximation (121).

Now we use the plane-wave expansion

$$e^{i\mathbf{k}\cdot\mathbf{R}} = 4\pi \sum_{l''m''} i^{l''} j_{l''}(kR) Y_{l''m''}(\hat{\mathbf{R}}) Y^*_{l''m''}(\hat{\mathbf{k}}), \tag{124}$$

so now we encounter something new, an integral over three spherical harmonics,

$$\int d\hat{\mathbf{k}} Y_{lm}(\hat{\mathbf{k}}) Y^*_{l'm'}(\hat{\mathbf{k}}) Y^*_{l''m''}(\hat{\mathbf{k}}) = C_{lm,l'm',l''m''}, \tag{125}$$

where

$$C_{lm,l'm',l''m''} = (-1)^{m'+m''} \sqrt{\frac{(2l+1)(2l'+1)(2l''+1)}{4\pi}} \begin{pmatrix} l & l' & l'' \\ 0 & 0 & 0 \end{pmatrix} \begin{pmatrix} l & l' & l'' \\ m & m' & m'' \end{pmatrix}. \tag{126}$$

The three-j symbols (Wigner coefficients) here vanish unless $l + l' + l''$ is even. This fact is crucial, since because of it we can write $j_{l''}(kR)$ in terms of Hankel functions of the first and second kind, using the reflection property of the latter,

$$h^{(2)}_{l''}(kR) = (-1)^{l''} h^{(1)}_{l''}(-kR), \tag{127}$$

and then extending the k integral over the entire real axis to a contour integral closed in the upper half plane. This leads to

$$g^0_{lm,l'm'}(\mathbf{R}) = 4\pi i^{l'-l} \sqrt{\frac{2|\zeta|}{\pi R}} \sum_{l''m''} C_{lm,l'm',l''m''} K_{l''+1/2}(|\zeta|R) Y_{l''m''}(\hat{\mathbf{R}}). \tag{128}$$

For the case of two semitransparent spheres that are totally outside each other,

$$V_1(r) = \lambda_1 \delta(r-a), \qquad V_2(r') = \lambda_2 \delta(r'-b), \tag{129}$$

in terms of spherical coordinates centered on each sphere, it is again very easy to calculate the scattering matrices,

$$T_1(\mathbf{r},\mathbf{r}') = \frac{\lambda_1}{a^2} \delta(r-a)\delta(r'-a) \sum_{lm} \frac{Y_{lm}(\hat{\mathbf{r}}) Y^*_{lm}(\hat{\mathbf{r}}')}{1 + \lambda_1 a K_{l+1/2}(|\zeta|a) I_{l+1/2}(|\zeta|a)}, \tag{130}$$

and then the harmonic transform is very similar to that seen for the cylinder, ($k = \mathrm{i}|\zeta|$)

$$\begin{aligned}(t_1)_{lm,l'm'} &= \int (\mathrm{d}\mathbf{r})(\mathrm{d}\mathbf{r}') j_l(kr) Y^*_{lm}(\hat{\mathbf{r}}) j_{l'}(kr') Y_{l'm'}(\hat{\mathbf{r}}') T_1(\mathbf{r},\mathbf{r}') \\ &= \delta_{ll'}\delta_{mm'}(-1)^l \frac{\lambda_1 a\pi}{2|\zeta|} \frac{I^2_{l+1/2}(|\zeta|a)}{1+\lambda_1 a K_{l+1/2}(|\zeta|a) I_{l+1/2}(|\zeta|a)}. \end{aligned} \tag{131}$$

Let us suppose that the two spheres lie along the z-axis, that is, $\mathbf{R} = R\hat{\mathbf{z}}$. Then we can simplify the expression for the energy somewhat by using $Y_{lm}(\theta = 0) = \delta_{m0}\sqrt{(2l+1)/4\pi}$. The formula for the energy of interaction becomes

$$E = \frac{1}{2\pi}\int_0^\infty \mathrm{d}\zeta \,\mathrm{tr}\ln(1-A), \tag{132}$$

in terms of the matrix

$$A_{lm,l'm'} = \delta_{m,m'} \sum_{l''} B_{ll''m}(a) B_{l''l'm}(b) \tag{133}$$

with

$$\begin{aligned} B_{ll'm}(a) &= \frac{\sqrt{\pi}}{\sqrt{2\zeta R}} \mathrm{i}^{-l+l'} \sqrt{(2l+1)(2l'+1)} \sum_{l''} (2l''+1) \\ &\times \begin{pmatrix} l & l' & l'' \\ 0 & 0 & 0 \end{pmatrix} \begin{pmatrix} l & l' & l'' \\ m & -m & 0 \end{pmatrix} \frac{K_{l''+1/2}(\zeta R)\lambda_1 a I^2_{l'+1/2}(\zeta a)}{1+\lambda_1 a I_{l'+1/2}(\zeta a) K_{l'+1/2}(\zeta a)}. \end{aligned} \tag{134}$$

Note that the phase always cancels in the trace.

Again, the strong coupling limit was considered in [75]. For weak coupling, a major simplification results because of the orthogonality property,

$$\sum_{m=-l}^{l} \begin{pmatrix} l & l' & l'' \\ m & -m & 0 \end{pmatrix} \begin{pmatrix} l & l' & l''' \\ m & -m & 0 \end{pmatrix} = \delta_{l''l'''} \frac{1}{2l''+1}, \qquad l \le l', \tag{135}$$

and then we find for the interaction energy

$$\begin{aligned} E = -\frac{\lambda_1 a \lambda_2 b}{4R} \int_0^\infty \frac{dx}{x} \sum_{l l' l''} (2l+1)(2l'+1)(2l''+1) \begin{pmatrix} l & l' & l'' \\ 0 & 0 & 0 \end{pmatrix}^2 \\ \times K^2_{l''+1/2}(x) I^2_{l+1/2}(xa/R) I^2_{l'+1/2}(xb/R). \end{aligned} \tag{136}$$

As with the cylinders, we expand the modified Bessel functions of the first kind in power series in $a/R, b/R < 1$. This expansion yields the infinite series

$$E = -\frac{\lambda_1 a \lambda_2 b}{4\pi R} \frac{ab}{R^2} \sum_{n=0}^{\infty} \frac{1}{n+1} \sum_{m=0}^{n} D_{n,m} \left(\frac{a}{R}\right)^{2(n-m)} \left(\frac{b}{R}\right)^{2m}, \tag{137}$$

where by inspection of the first several $D_{n,m}$ coefficients we can identify them as

$$D_{n,m} = \frac{1}{2}\binom{2n+2}{2m+1}, \tag{138}$$

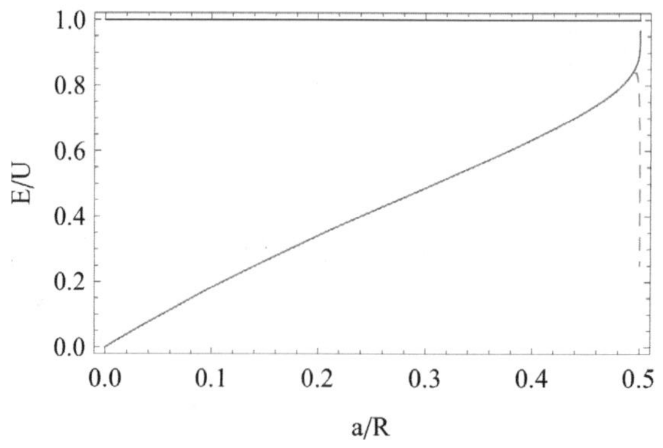

Figure 8. Plotted is the ratio of the exact interaction energy of two weakly-coupled spheres of equal radii to the proximity force approximation.

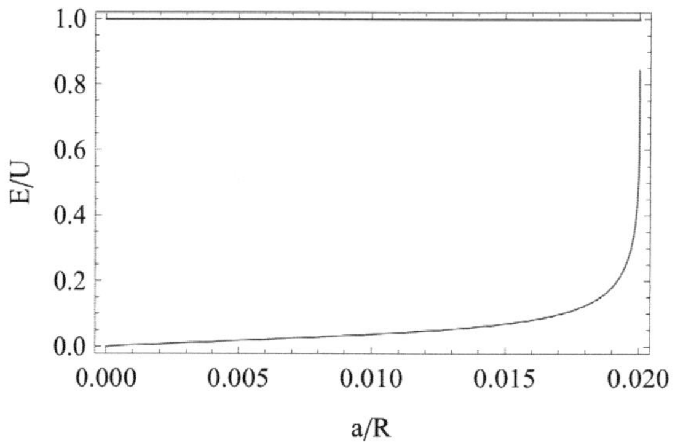

Figure 9. Plotted is the ratio of the exact interaction energy of two weakly-coupled spheres, with ratio of radii $b/a = 49$, to the proximity force approximation.

and now we can immediately sum the expression for the Casimir interaction energy to give the closed form

$$E = \frac{\lambda_1 a \lambda_2 b}{16\pi R} \ln \left(\frac{1 - \left(\frac{a+b}{R}\right)^2}{1 - \left(\frac{a-b}{R}\right)^2} \right). \tag{139}$$

Again, when $d = R - a - b \ll a, b$, the proximity force theorem is reproduced:

$$U(d) \sim \frac{\lambda_1 \lambda_2 ab}{16\pi R} \ln(d/R), \qquad d \ll a, b. \tag{140}$$

However, as figure 8 and figure 9 demonstrate, the approach is not very smooth, even for equal-sized spheres. The ratio of the energy to the PFA is ($b/a = \mu$)

$$\frac{E}{U} = 1 + \frac{\ln[(1+\mu)^2/2\mu]}{\ln d/R}, \qquad d \ll a, b. \tag{141}$$

In just the way indicated above, we can obtain a closed-form result for the interaction energy between a weakly-coupled sphere and a Dirichlet plane. Using the simplification that

$$\sum_{m=-l}^{l} (-1)^m \begin{pmatrix} l & l & l' \\ m & -m & 0 \end{pmatrix} \begin{pmatrix} l & l & l' \\ 0 & 0 & 0 \end{pmatrix} = \delta_{l'0}, \tag{142}$$

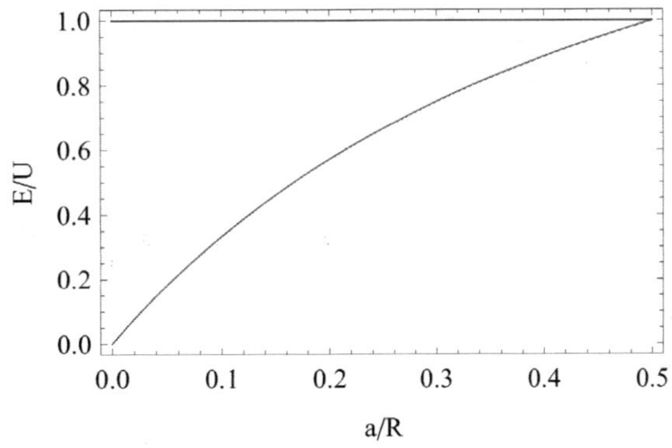

Figure 10. Plotted is the ratio of the exact interaction energy of a weakly-coupled sphere above a Dirichlet plane to the PFA.

we can write the interaction energy in the form

$$E = -\frac{\lambda a}{2\pi R}\int_0^\infty dx \sum_{l=0}^\infty \sqrt{\frac{\pi}{2x}}(2l+1)K_{1/2}(x)I^2_{l+1/2}\left(x\frac{a}{R}\right). \tag{143}$$

Then in terms of $R/2$ as the distance between the center of the sphere and the plane, the exact interaction energy is

$$E = -\frac{\lambda}{2\pi}\left(\frac{a}{R}\right)^2 \frac{1}{1-(2a/R)^2}, \tag{144}$$

which as $a \to R/2$ reproduces the proximity force limit, contained in the (ambiguously defined) PFA formula

$$U = -\frac{\lambda}{8\pi}\frac{a}{d}. \tag{145}$$

The ratio of the exact interaction to the PFA is plotted in figure 10.

4.3. Exact Results—Weak Coupling

In weak coupling it is possible to derive the exact (scalar) interaction between two potentials in general forms,

$$2D: \quad \mathfrak{E} = \frac{E}{L} = -\frac{1}{32\pi^3}\int (d\mathbf{r}_\perp)(d\mathbf{r}'_\perp)\frac{V_1(\mathbf{r}_\perp)V_2(\mathbf{r}'_\perp)}{|\mathbf{r}-\mathbf{r}'|^2}, \tag{146a}$$

$$3D: \quad E = -\frac{1}{64\pi^3}\int (d\mathbf{r})(d\mathbf{r}')\frac{V_1(\mathbf{r})V_2(\mathbf{r}')}{|\mathbf{r}-\mathbf{r}'|^3}. \tag{146b}$$

The results found above in sections 4.1 and 4.2 can be directly derived from these pairwise summation formulæ. For exact results for δ-function plates, see the contribution to this Proceedings by Jef Wagner.

It is straightforward to adapt the multiple scattering formalism to the far more relevant situation of electromagnetism. For weak coupling, this reduces to the (retarded dispersion) van der Waals (vdW) potential between polarizable molecule given by

$$V = -\frac{23}{4\pi}\frac{\alpha_1\alpha_2}{r^7}, \qquad \alpha = \frac{\varepsilon - 1}{4\pi N}. \tag{147}$$

This allows us to consider in the same vein (electromagnetic) interactions between distinct dilute dielectric bodies of arbitrary shape. (Of course, this potential represents the retarded limit of the van der Waals force. The non-retarded limit of the van der Waals force between bodies has long been studied, for example between dielectric cylinders [84,85], and between a sphere and a plane [86].)

This vdW potential may be directly derived from the multiple scattering formula

$$W = -\frac{\mathrm{i}}{2}\mathrm{Tr}\,\ln\frac{\mathbf{\Gamma}}{\mathbf{\Gamma}_0} \approx -\frac{\mathrm{i}}{2}\mathrm{Tr}\,V_1\mathbf{\Gamma}_0V_2\mathbf{\Gamma}_0, \tag{148}$$

where where have omitted self-action terms, and

$$\mathbf{\Gamma}_0 = \boldsymbol{\nabla}\times\boldsymbol{\nabla}\times\mathbf{1}\frac{e^{-|\zeta||\mathbf{r}-\mathbf{r}'|}}{4\pi|\mathbf{r}-\mathbf{r}'|} + \mathbf{1}. \tag{149}$$

The quantity $V = \varepsilon - 1$ is the potential representing the dielectric body.

We first apply this approach to compute the force between a dilute dielectric slab and an infinite dilute dielectric plate, as shown in figure 11. If the cross sectional area of the finite slab

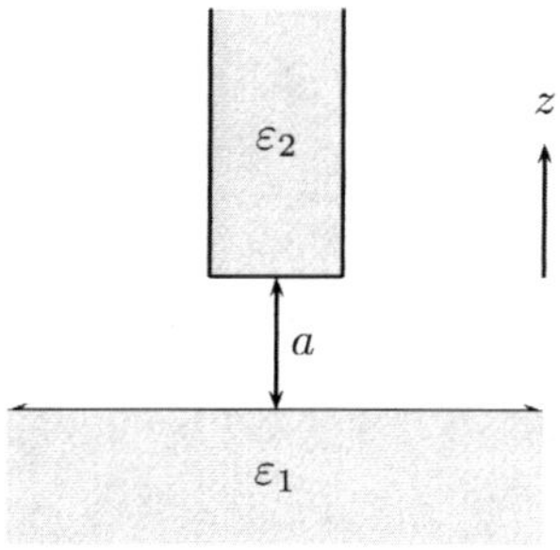

Figure 11. A plate of cross section A, a distance a above an infinite plate. Both plates are of semi-infinite extent in the direction perpendicular to both plates.

is A, the force between the slabs is

$$\frac{F}{A} = -\frac{23}{640\pi^2}\frac{1}{a^4}(\varepsilon_1 - 1)(\varepsilon_2 - 1), \tag{150}$$

which is the Lifshitz formula for infinite (dilute) slabs. Note that there is no correction due to the finite area of the upper slab.

In the same way we can immediately compute the force between a dilute dielectric sphere of radius a a distance Z away from an infinite slab, as illustrated in figure 12. The interaction energy of this system is

$$E = -\frac{23}{640\pi^2}(\varepsilon_1 - 1)(\varepsilon_2 - 1)\frac{4\pi a^3/3}{Z^4}\frac{1}{(1 - a^2/Z^2)^2}, \tag{151}$$

which agrees with the PFA in the short separation limit, $Z - a = \delta \ll a$, where the force is

$$F_{\mathrm{PFA}} = 2\pi a\mathcal{E}_{\parallel}(\delta) = -\frac{23}{640\pi^2}(\varepsilon_1 - 1)(\varepsilon_2 - 1)\frac{2\pi a}{3\delta^3}, \tag{152}$$

with an exact correction, intermediate between that for scalar averaged Dirichlet and Neumann [71,72] and electromagnetic perfectly-conducting boundaries [73].

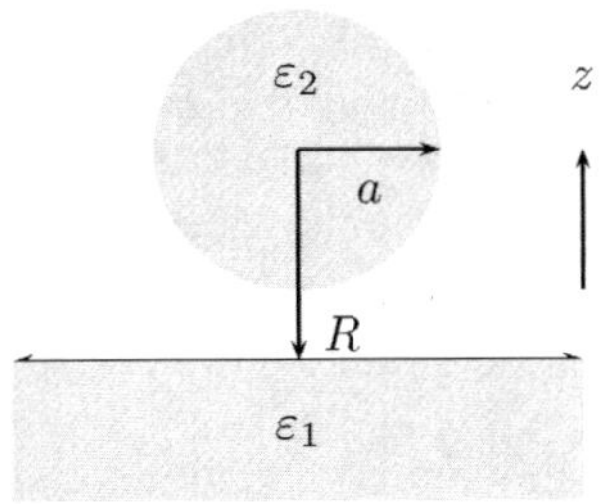

Figure 12. A dilute dielectric sphere, of radius a, the center of which is a distance Z above an infinite dilute dielectric plate.

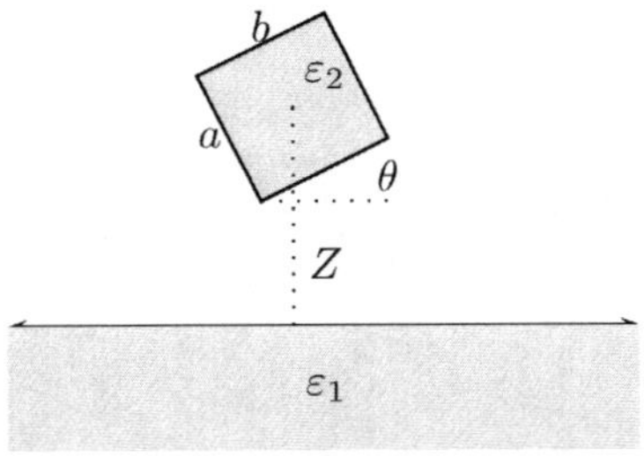

Figure 13. Rectangular solid (with dielectric constant ε_2) of side a, b, and c (perpendicular to the plane, not shown) a distance Z above an infinite plate (with dielectric constant ε_1) extending to $z = -\infty$. The side b makes an angle θ with respect to the plate.

We can also consider torques between bodies. For example, consider a rectangular dilute dielectric solid, with sides a, b, and c, the center of mass of which is a distance Z above a dilute dielectric plate, but which is tilted through an angle θ relative to the plate, as shown in figure 13. Generically, the shorter side wants to align with the plate, which is obvious geometrically, since that (for fixed center of mass position) minimizes the energy. However, if the slab has square cross section, the equilibrium position occurs when a corner is closest to the plate, also obvious geometrically. But if the two sides are close enough in length, a nontrivial equilibrium position between these extremes can occur, as well illustrated in figure 14. For large enough separation, the shorter side wants to face the plate, but for

$$Z < Z_0 = \frac{a}{2}\sqrt{\frac{2a^2 + 5b^2 + \sqrt{9a^4 + 20a^2b^2 + 20b^4}}{5\,(a^2 - b^2)}}, \qquad a > b, \tag{153}$$

the equilibrium angle increases, until finally at $Z = D = \sqrt{a^2 + b^2}/2$ the slab touches the plate at an angle $\theta = \arctan b/a$, that is, the center of mass is just above the point of contact, about which point there is no torque.

Finally, once again we can get exact results for the vdW interaction between parallel cylinders

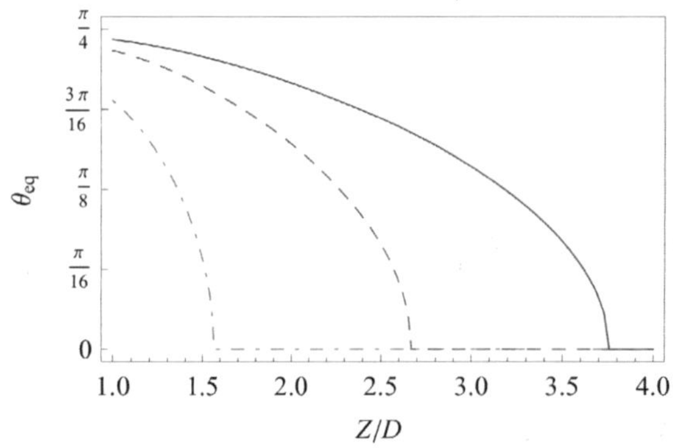

Figure 14. The stable equilibrium angle of a slab above an infinite plate for given b/a ratios 0.95, 0.9, and 0.7, respectively given by solid, dashed, and dot-dashed lines.

(radii a, b, center separation R):

$$\mathfrak{E} = \frac{E}{L} = -\frac{23}{60\pi}(\varepsilon_1 - 1)(\varepsilon_2 - 1)\frac{a^2b^2}{R^6}\frac{1 - \frac{1}{2}\left(\frac{a^2+b^2}{R^2}\right) - \frac{1}{2}\left(\frac{a^2-b^2}{R^2}\right)^2}{\left[\left(1-\left(\frac{a+b}{R}\right)^2\right)\left(1-\left(\frac{a-b}{R}\right)^2\right)\right]^{5/2}}, \tag{154}$$

and between spheres:

$$E = -\frac{23}{1920\pi}\frac{(\varepsilon_1 - 1)(\varepsilon_2 - 1)}{R}\left\{\ln\left(\frac{1-\left(\frac{a-b}{R}\right)^2}{1-\left(\frac{a+b}{R}\right)^2}\right)\right.$$
$$\left. + \frac{4ab}{R^2}\frac{\frac{a^6-a^4b^2-a^2b^4+b^6}{R^6} - \frac{3a^4-14a^2b^2+3b^4}{R^4} + 3\frac{a^2+b^2}{R^2} - 1}{\left[\left(1-\left(\frac{a-b}{R}\right)^2\right)\left(1-\left(\frac{a+b}{R}\right)^2\right)\right]^2}\right\}. \tag{155}$$

This last expression, which is rather ugly, may be verified to yield the proximity force theorem:

$$E \to U = -\frac{23}{640\pi}\frac{a(R-a)}{R\delta^2}, \quad \delta = R - a - b \ll a, b. \tag{156}$$

It also, in the limit $b \to \infty$, $R \to \infty$ with $R - b = Z$ held fixed, reduces to the result (151) for the interaction of a sphere with an infinite plate.

The same methodology may be used to derive the force between corrugated planes and corrugated concentric cylinders, which calculations may have relevance to nanomechanics. See the contribution to this Proceedings by Cavero-Peláez, and [81, 82].

5. Comments and Prognosis

In this paper I have described some recent exciting progress in the practical development of quantum vacuum energy. I have concentrated on the work of the Oklahoma group, without meaning to denigrate the work of many others of equal or greater importance. I will summarize the status of the topics I discussed briefly here:

- The temperature controversy can only be resolved by dedicated experiments, designed to measure Casimir forces at different temperatures between metal surfaces
- The semiconductor anomaly needs to be examined more closely both theoretically and experimentally.

- Casimir energies, including their divergent contributions, exhibit the inertial and gravitational masses expected: $m = E/c^2$. These gravitational interactions need to be examined for more complicated geometries, such as for the Casimir energy of a perfectly conducting spherical shell, with the hope of understanding the physical significance of the divergent contributions.
- The multiple-scattering methods recently developed are in fact not particularly novel, and illustrate the ability of physicists to continually rediscover old methods. What is new is the recognition that one can evaluate continuum determinants (or infinitely dimensional discrete ones) accurately numerically, and in some cases even exactly in closed form. This is making it possible to compute Casimir forces and torques for geometries previously inaccessible.
- It is indeed remarkable, if perhaps not surprising in retrospect, to see that closed form expressions can be obtained for the interaction between spheres, between parallel cylinders, and for other geometries in weak coupling. These results demonstrate most conclusively the unreliability of the proximity force approximation (of course, the proximity force theorem holds true).
- This methodology has been used to obtain new results for non-contact gears: See the report by Inés Cavero-Peláez.
- Jef Wagner has presented new results for edge effects for semitransparent potentials.
- Undoubtedly these developments will lead to improved conceptual understanding, and to better comparison with experiment.

Acknowledgments
We thank the US National Science Foundation, grant number PHY-0554826, and the US Department of Energy, grant number DE-FG-04ER41305, for partial support of the research reported here. We especially thank Victor Dodonov for organizing such a successful meeting on "60 Years of the Casimir Effect," and the International Center for Condensed Matter Physics at the University of Brasilia, and the College of Arts and Sciences and the office of the Vice President for Research at the University of Oklahoma for travel support. I thank my collaborators for allowing me to report on our joint work.

References

[1] Casimir H B G 1948 *Proc. Kon. Ned. Akad. Wetensch.* **51** 793
[2] Brevik I, Ellingsen S A and Milton K A 2006 *New J. Phys.* **8** 236 (*Preprint* quant-ph/0605005)
[3] Klimchitskaya G L and Mostepanenko V M 2006 *Comtemp. Phys.* **47** 131 (*Preprint* quant-ph/0609145)
[4] Lifshitz E M 1955 *Zh. Eksp. Teor. Fiz.* **29** 94
[5] Boström M and Sernelius B E 2000 *Phys. Rev. Lett.* **84** 4757
[6] Høye J S, Brevik I, Aarseth J B and Milton K A 2003 *Phys. Rev. E* **67** 056116 (*Preprint* quant-ph/0212125)
[7] Jancovici B and Šamaj L 2004 *J. Stat. Mech.* P08006 (*Preprint* cond-mat/0407392)
[8] Jancovici B and Šamaj L 2005 *Europhys. Lett.* **72** 35 (*Preprint* cond-mat/0506363)
[9] Buenzli P R and Martin Ph A 2005 *Europhys. Lett.* **72** 42 (*Preprint* cond-mat/0506303)
[10] Buenzli P R and Martin Ph A 2008 *Phys. Rev. E* **77** 011114 (*Preprint* 0709.4194 [quant-ph])
[11] Sernelius Bo E 2006 *J. Phys. A: Math. Gen.* **39** 6741
[12] Brevik I, Ellingsen S A, Høye J S and Milton K A 2008 *J. Phys. A: Math. Theor.* **41** 164017 (*Preprint* 0710.4882 [quant-ph])
[13] Høye J S, Brevik I, Ellingsen S A and Aarseth J B 2007 *Phys. Rev. E* **75** 051127 (*Preprint* quant-ph/0703174)
[14] Brevik I, Aarseth J B, Høye J S and Milton K A 2004 *Proc. 6th Workshop on Quantum Field Theory under the Influence of External Conditions* ed K A Milton (Paramus, NJ: Rinton Press) p 54 (*Preprint* quant-ph/0311094)
[15] Obrecht J M, Wild R J, Antezza M, Pitaevskii L P, Stringari S and Cornell E A 2007 *Phys. Rev. Lett.* **98** 063201 (*Preprint* physics/0608074)
[16] Lamoreaux S K 1997 *Phys. Rev. Lett.* **87** 5
[17] Brevik I and Aarseth J B 2006 *J. Phys. A: Math. Gen.* **39** 6187 (*Preprint* quant-ph/0511037)

[18] Mohideen U and A. Roy A 1998 *Phys. Rev. Lett.* **81** 4549 (*Preprint* physics/9805038)
[19] Roy A, Lin C-Y, and Mohideen U 1999 *Phys. Rev. D* **60** 111101(R) (*Preprint* quant-ph/9906062)
[20] Harris B W, Chen F, and Mohideen U 2000 *Phys. Rev. A* **62** 052109 (*Preprint* quant-ph/0005088)
[21] Ederth T 2000 *Phys. Rev. A* **62** 062104 (*Preprint* quant-ph/0008009)
[22] Chan H B *et al* 2001 *Phys. Rev. Lett.* **87** 211801 (*Preprint* quant-ph/0109046); Chan H B *et al* 2001 *Science* **291** 1941
[23] Chen F, Mohideen U, Klimchitskaya G L, and Mostepanenko V M 2002 *Phys. Rev. Lett.* **88** 101801 (*Preprint* quant-ph/0201087)
[24] Bressi G, Carugno G, Onofrio R, and Ruoso G 2002 *Phys. Rev. Lett.* **88** 041804 (*Preprint* quant-ph/0203002)
[25] Iannuzzi D, Lisanti M, and Capasso F 2004 *Proc. Natl. Acad. Sci. USA* **101** 4019 (*Preprint* quant-ph/0403142)
[26] Lisanti M, Iannuzzi D, and Capasso F 2005 *Proc. Natl. Acad. Sci. USA* **102** 11989 (*Preprint* quant-ph/0502123)
[27] Chen F, Klimchitskaya G L, Mostepanenko V M, and Mohideen U 2007 *Phys. Rev. B* **76** 035338 (*Preprint* 0707.4390)
[28] Decca R S *et al* 2005 *Ann. Phys. NY* **318** 37 (*Preprint* quant-ph/0503105)
[29] Bezerra V B *et al* 2006 *Phys. Rev. E* **73** 028101 (*Preprint* quant-ph/0503134)
[30] L. P. Pitaevskii 2008 *Preprint* 0801.0656v3
[31] Geyer B, Klimchitskaya G L and Mostepanenko V M 2005 *Phys. Rev. D* **72** 085009 (*Preprint* quant-ph/0510054)
[32] Klimchitskaya G L, Geyer B, and Mostepanenko V M 2006 *J. Phys. A: Math. Gen.* **39** 6495 (*Preprint* quant-ph/0602097)
[33] Geyer B, Klimchitskaya G L, and Mostepanenko VM 2006 *Int. J. Mod. Phys. A* **21** 5007 (*Preprint* 0704.1040)
[34] Klimchitskaya G L and Geyer B 2008 *J. Phys. A: Math. Gen.* **41** 164032 (*Preprint* 0802.3729)
[35] Klimchitskaya G L, Mohideen U and Mostepanenko V M 2007 *J. Phys. A: Math. Theor.* **40** F841 (*Preprint* 0707.2734)
[36] Castillo-Garza R, Chang C-C, Jimenez D, Klimchitskaya G L, Mostepanenko V M and Mohideen U 2007 *Phys. Rev. A* **75** 062114 (*Preprint* 0705.1358 [quant-ph])
[37] Chen F, Klimchitskaya G L, Mostepanenko V M and Mohideen U 2007 *Optics Express* **15** 4823 (*Preprint* quant-ph/0610094)
[38] Ellingsen S A, Brevik I, Høye J S and Milton K 2008 *Phys. Rev. E* **76** 051127 (*Preprint* 0805.3065)
[39] Ellingsen S Å, Brevik I, Høye J S and Milton K 2008 *Preprint* 0809:0763
[40] Brevik I and Milton K A 2008 *Phys. Rev. E* **78** 011124 (*Preprint* 0802.2542 [quant-ph])
[41] Fulling S A, Milton K A, Parashar P, Romeo A, Shajesh K V and Wagner J 2007 *Phys. Rev. D* **76** 025004 (*Preprint* hep-th/0702091)
[42] Milton K A, Parashar P, Shajesh K V and Wagner J 2007 *J. Phys. A: Math. Theor.* **40** 10935 (*Preprint* 0705.2611 [hep-th])
[43] Milton K A, Fulling S A, Parashar P, Romeo A, Shajesh K V and Wagner J A 2008 *J. Phys. A: Math. Theor.* **41** 164052 (*Preprint* 0710.3841 [hep-th])
[44] Boyer T H 1968 *Phys. Rev.* **174** 1764
[45] Graham N, Jaffe R L, Khemani V, Quandt M, Schroeder O and Weigel H 2004 *Nucl. Phys. B* **677** 379 (*Preprint* hep-th/0309130), and references therein
[46] Deutsch D and Candelas P 1979 *Phys. Rev. D* **20** 3063
[47] Cavero-Peláez I, Milton K A and Wagner J 2006 *Phys. Rev. D* **73** 085004 (*Preprint* hep-th/0508001)
[48] Cavero-Peláez I, Milton K A and Kirsten K 2007 *J. Phys. A: Math. Theor.* **40** 3607 (*Preprint* hep-th/0607154)
[49] Calloni E *et al* 2002 *Phys. Lett. A* **297** 328 (*Preprint* hep-th/0302082)
[50] Bimonte G *et al* 2006 *Phys. Rev. D* **74** 085011 (*Preprint* hep-th/0606042)
[51] Estrada R, Fulling S A, Liu Z, Kaplan L, Kirsten K and Milton K A 2008 *J. Phys. A: Math. Theor.* **41** 164055
[52] Brown L S and Maclay G J 1969 *Phys. Rev.* **184** 1272
[53] Milton K A 2001 *The Casimir Effect: Physical Manifestations of Zero-Point Energy* (Singapore: World Scientific), Sec. 11.1
[54] Bimonte G, Esposito G and Rosa L 2008 *Phys. Rev. D* **78** 024010 (*Preprint* 0804.2839 [hep-th])
[55] Jaekel M T and Reynaud S 1993 *Journal de Physique I* **3** 1093
[56] Schwinger J 1970 *Particles, Sources, and Fields* (Reading, MA: Addison-Wesley)
[57] Marzlin K P 1994 *Phys. Rev. D* **50** 888
[58] Milton K A 2004 *J. Phys. A: Math. Gen.* **37** R209 (*Preprint* hep-th/0406024)
[59] Shajesh K V, Milton K A, Parashar P and Wagner J A 2008 *J. Phys. A: Math. Theor.* **41** 164058 (*Preprint* 0711.1206 [hep-th])
[60] Blocki J, Randrup J, Świątecki W J and Tsang C F 1977 *Ann. Phys. NY* **105** 427

[61] Krein M G 1953 *Mat. Sb. (N.S.)* **33** 597; 1962 *Dokl. Akad. Nauk SSSR* **144** 268 [*Sov. Math.-Dokl.* **3** 707]; Birman M Sh and Krein M G 1962 *Dokl. Akad. Nauk SSSR* **144** 475 [*Sov. Math.-Dokl.* **3** 740]
[62] Renne M J 1971 *Physica* **56** 125
[63] Balian R and Duplantier B 1977 *Ann. Phys. NY* **104** 300; Balian R and Duplantier B 1978 **112** 165; Balian R and Duplantier B 2004 *Preprint* quant-ph/0408124
[64] Büscher R and Emig T 2005 *Phys. Rev. Lett.* **94** 133901; Emig T, Jaffe R L, Kardar M, and Scardicchio A 2006 *Phys. Rev. Lett.* **96** 080403 (*Preprint* cond-mat/0601055)
[65] Reynaud S, Maia Neto P A, and Lambrecht A 2008 *J. Phys. A: Math. Theor.* **41** 164004 (*Preprint* 0709.5452) (*Preprint* quant-ph/0410101); Maia Neto P A, Lambrecht A, and Reynaud S 2005 *Europhys. Lett.* **69** 924; Maia Neto P A, Lambrecht A, and Reynaud S 2005 *Phys. Rev. A* **72** 012115 (*Preprint* quant-ph/0505086); Lambrecht A, Maia Neto PA, and Reynaud S, 2006 *New J. Phys.* **8** 243 (*Preprint* quant-ph/0611103)
[66] Bulgac A, Marierski P, and Wirzba A 2006 *Phys. Rev. D* **73** 025007 (*Preprint* hep-th/0511056); Wirzba A, Bulgac A and Magierski P 2006 *J. Phys. A* **39** 6815 (*Preprint* quant-ph/0511057); Wirzba A 2008 *J. Phys. A: Math. Theor.* **41** 164003 (*Preprint* 0711.2395)
[67] Dalvit D A R, Lombardo F C, Mazzitelli F D, and Onofrio R 2006 *Phys. Rev. A* **74** 020101(R) (*Preprint* quant-ph/0608033); Dalvit D A R, Lombardo F C, Mazzitelli F D, and Onofrio R 2006 *New J. Phys.* **8** 240 (*Preprint* quant-ph/0610181)
[68] Bordag M 2006 *Phys. Rev. D* **73** 125018 (*Preprint* hep-th/0602295); Bordag M 2007 *Phys. Rev. D* **75** 065003 (*Preprint* quant-ph/0611243)
[69] Kenneth O and Klich I 2008 *Phys. Rev. B* **78** 014103 (*Preprint* 0707.4017 [quant-ph])
[70] Kenneth O and Klich I 2006 *Phys. Rev. Lett.* **97** 160401 (*Preprint* quant-ph/0601011)
[71] Wirzba A 2008 *J. Phys. A: Math. Theor.* **41** 164003 (*Preprint* 0711.2395 [quant-ph])
[72] Bordag M and Nikolaev V 2008 *J. Phys. A: Math. Theor.* **41** 164002 (*Preprint* 0802.3633 [hep-th])
[73] Maia Neto P A, Lambrecht A, and Reynaud S 2008 *Phys. Rev. A* **78** 012115 (*Preprint* 0803.2444)
[74] Emig T, Graham N, Jaffe R L and Kardar M 2007 *Phys. Rev. Lett.* **99** 170403 (*Preprint* 0707.1862 [cond-mat.stat-mech])
[75] Emig T, Graham N, Jaffe R L and Kardar M 2008 *Phys. Rev. D* **77** 025005 (*Preprint* 0710.3084 [cond-mat.stat-mech])
[76] Emig T and Jaffe R L 2008 *J. Phys. A: Math. Theor.* **41** 164001 (*Preprint* 0710.5104 [quant-ph])
[77] Gies H and Klingmüller K 2006 *Phys. Rev. Lett.* **96** 220401 (*Preprint* quant-ph/0601094); *Phys. Rev. Lett.* **97** 220405 (*Preprint* quant-ph/0606235); *Phys. Rev. D* **74** 045002 (*Preprint* quant-ph/0605141); Gies H, Langfeld K and Moyaerts L 2003 *JHEP* **0306** 018 (*Preprint* hep-th/0303264)
[78] Milton K A and Wagner J 2008 *J. Phys. A: Math. Theor.* **41** 155402 (*Preprint* 0712.3811 [hep-th])
[79] Milton K A and Wagner J 2008 *Phys. Rev. D* **77** 045005 (*Preprint* 0711.0774 [hep-th])
[80] Milton K A, Parashar P and Wagner J 2008 *Phys. Rev. Lett.* **101** 160402 (*Preprint* 0806.2880 [hep-th])
[81] Cavero-Peláez I, Milton K A, Parashar P and Shajesh K V 2008 *Phys. Rev. D* **78** 065018 (*Preprint* 0805.2776 [hep-th])
[82] Cavero-Peláez I, Milton K A, Parashar P and Shajesh K V 2008 *Phys. Rev. D* **78** 065019 (*Preprint* 0805.2777 [hep-th])
[83] Lippmann B and Schwinger J 1950 *Phys. Rev.* **79** 469
[84] Mahanty J and Ninham B W 1976 *Dispersion Forces* (London: Academic Press)
[85] Nyland G H and Brevik I 1994 *Physica A* **202** 81
[86] Noguez C, Román-Velázquez C E, Esquivel-Sirvent R and Villarreal C 2004 *Europhys. Lett.* **67** 191 (*Preprint* quant-ph/0310068); Román-Velázquez C E, Noguez C, Villarreal C, Esquivel-Sirvent R 2004 *Phys. Rev. A* **69** 042109 (*Preprint* quant-ph/0303172)

Journal of Physics: Conference Series **161** (2009) 012002 doi:10.1088/1742-6596/161/1/012002

Problems and paradoxes of the Lifshitz theory

G L Klimchitskaya[1]

Center of Theoretical Studies and Institute for Theoretical Physics, Leipzig University, Vor dem Hospitaltore 1, 100920, D-04009, Leipzig, Germany

E-mail: galina.klimchitskaya@itp.uni-leipzig.de

Abstract. The problems and paradoxes of the Lifshitz theory in application to real dielectric and semiconductor materials are reviewed. It is shown that the inclusion of drift current of conduction electrons into the model of dielectric response results in contradictions with both thermodynamics and experimental data of different experimental groups. Physical reasons why the problems and paradoxes arise are analyzed and found to be connected with the violation of basic applicability condition of the Lifshitz theory, the thermal equilibrium. A recent alternative approach to the resolution of the problems based on the inclusion of screening effects and diffusion current is considered and demonstrated to be thermodynamically and experimentally inconsistent. It is argued that the inclusion of the diffusion current leads to an even deeper violation of thermal equilibrium. Phenomenologically, the Lifshitz theory with role of drift and diffusion currents neglected is shown to be free of problems and in agreement with both thermodynamics and all available experimental data.

1. Introduction

It is common knowledge that the Lifshitz theory gives a description of the van der Waals and Casimir forces between two plane parallel plates (semispaces) in terms of the frequency-dependent dielectric permittivities of the plate materials $\varepsilon(\omega)$ [1–3]. Starting in 2000, problems and paradoxes with the application of the Lifshitz theory to real materials (at first to metals [4] and then to dielectrics [5]) were recognized. Specifically, for metals described by the Drude dielectric function a large thermal effect was predicted at short separations [4] in qualitative disagreement with the case of ideal metals. This prediction was excluded experimentally at a 99.9% confidence level [6,7]. At large separations, for the Drude metals one half of the Casimir force acting between ideal metals was obtained [4]. The same result was obtained in [8,9] by modelling a metal with a set of nonrelativistic hard-core or point-like particles confined by walls. In [10] it was shown that the substitution of the Drude dielectric function into the Lifshitz formulas results in a violation of the third law of thermodynamics for metals with perfect crystal lattices. In [11] it was demonstrated that for metals with impurities the Nernst heat theorem is satisfied. This, however, does not solve the problem because in accordance with quantum statistical physics the Nernst theorem must be satisfied for metals with perfect crystal lattices as well. The situation is even more dramatic for dielectrics, the materials for which the Lifshitz theory was originally formulated. It was shown that the inclusion of the dc conductivity arising in all dielectrics at nonzero temperature radically changes the theoretical predictions and leads to disagreement between the Lifshitz theory and thermodynamics [5,12,13].

[1] On leave from North-West Technical University, St.Petersburg, Russia

In this paper we review recent paradoxial applications of the Lifshitz theory to real materials (mostly to dielectrics) and analyze the physical reasons of why the paradoxes appear. We show that difficulties arise each time when the drift or diffusion currents are included in the model of the dielectric response. In this connection we discuss recent experiments with semiconductors [14,15] and dielectric [16] test bodies and show that the measurement results are in disagreement with theory taking the dc conductivity into account. Recent theoretical approaches attempting to solve problems of the Lifshitz theory by accounting for screening effects and diffusion currents [17,18] are demonstrated to be in contradiction with both thermodynamic and experimental data.

Our analysis leads to the conclusion that the roots of the problems are in the violation of the applicability conditions of the Lifshitz theory when the role of free charge carriers is included in the model of the dielectric response. We argue that in this case one violates the condition of thermal equilibrium by admitting a unidirectional flux of heat from the Casimir plates to the heat reservoir.

In Sec. 2 we discuss the consistency of the Lifshitz theory with thermodynamics. Sec. 3 is devoted to the experimental tests of the Lifshitz theory under different assumptions about the model of the dielectric response. Sec. 4 contains the formulation of a phenomenological approach of how to apply theory in agreement with both thermodynamics and the experimental data. In Sec. 5 we provide thermodynamic and experimental tests for the proposed modification of the reflection coefficients due to screening effects and diffusion current. Our discussion and conclusions are presented in Sec. 6.

2. Consistency of the Lifshitz theory with thermodynamics

The main result of the Lifshitz theory is the expression for the free energy of dispersion interaction per unit area of two semispaces at a separation distance a, at temperature T in thermal equilibrium with an environment [1–3]

$$\mathcal{F}(a,T) = \frac{k_BT}{2\pi}\sum_{l=0}^{\infty}{}' \int_0^\infty k_\perp dk_\perp \left\{\ln\left[1 - r_{\rm TM}^{(1)}(\mathrm{i}\xi_l,k_\perp) r_{\rm TM}^{(2)}(\mathrm{i}\xi_l,k_\perp)\,\mathrm{e}^{-2aq_l}\right] \right.$$
$$\left. + \ln\left[1 - r_{\rm TE}^{(1)}(\mathrm{i}\xi_l,k_\perp) r_{\rm TE}^{(2)}(\mathrm{i}\xi_l,k_\perp)\,\mathrm{e}^{-2aq_l}\right]\right\}. \tag{1}$$

Here, $k_\perp$ is the magnitude of a wave vector in the plane of boundary planes (perpendicular to the z-axis), prime adds a multiple 1/2 to the term $l = 0$, $\xi_l = 2\pi k_BTl/\hbar$ ($l = 0, 1, 2, \ldots$) are the Matsubara frequencies, k_B is the Boltzmann constant, and the Fresnel reflection coefficients along the imaginary frequency axis are given by ($n = 1, 2$)

$$r_{\rm TM}^{(n)}(\mathrm{i}\xi_l,k_\perp) = \frac{\varepsilon_l^{(n)} q_l - k_l^{(n)}}{\varepsilon_l^{(n)} q_l + k_l^{(n)}}, \qquad r_{\rm TE}^{(n)}(\mathrm{i}\xi_l,k_\perp) = \frac{q_l - k_l^{(n)}}{q_l + k_l^{(n)}}, \tag{2}$$

where

$$q_l^2 = k_\perp^2 + \frac{\xi_l^2}{c^2}, \qquad {k_l^{(n)}}^2 = k_\perp^2 + \varepsilon_l^{(n)}\frac{\xi_l^2}{c^2}, \qquad \varepsilon_l^{(n)} \equiv \varepsilon^{(n)}(\mathrm{i}\xi_l). \tag{3}$$

The free energy of atom-wall interaction, where $\varepsilon_l^{(1)}$ is the dielectric permittivity of wall material, can be expressed as

$$\mathcal{F}^A(a,T) = -k_BT\sum_{l=0}^{\infty}{}'\alpha(\mathrm{i}\xi_l)\int_0^\infty k_\perp dk_\perp q_l \mathrm{e}^{-2aq_l} \tag{4}$$
$$\times\left\{2r_{\rm TM}^{(1)}(\mathrm{i}\xi_l,k_\perp) - \frac{\xi_l^2}{q_l^2c^2}\left[r_{\rm TM}^{(1)}(\mathrm{i}\xi_l,k_\perp) + r_{\rm TE}^{(1)}(\mathrm{i}\xi_l,k_\perp)\right]\right\},$$

where $\alpha(\mathrm{i}\xi_l)$ is the dynamic atomic polarizability.

For dielectrics, the dielectric permittivity along the imaginary frequency axis is commonly presented in the form [19]

$$\varepsilon(\mathrm{i}\xi) = 1 + \sum_j \frac{g_j}{\omega_j^2 + \xi^2 + \gamma_j \xi} \approx 1 + \sum_j \frac{g_j}{\omega_j^2 + \xi^2}, \tag{5}$$

where $\omega_j \neq 0$ are the oscillator frequencies, g_j are the oscillator strength and γ_j are the relaxation parameters. This representation assumes some finite value of the static dielectric permittivity

$$\varepsilon_0 \equiv \varepsilon(0) = 1 + \sum_j \frac{g_j}{\omega_j^2} < \infty. \tag{6}$$

Using this model one can find the asymptotic behavior of the free energy of the Casimir interaction between two similar dielectric plates with $\varepsilon^{(1)} = \varepsilon^{(2)} = \varepsilon$ at low temperature $T \ll T_{\text{eff}} \equiv \hbar c/(2ak_B)$ [5]

$$\mathcal{F}(a,T) \approx E(a) - \frac{\hbar c}{32a^3}\zeta(3) r_0^2(\varepsilon_0 + 1)\left(\frac{T}{T_{\text{eff}}}\right)^3. \tag{7}$$

Here, $E(a)$ is the Casimir energy at zero temperature, $\zeta(z)$ is the Riemann zeta function, and $r_0 \equiv (\varepsilon_0 - 1)/(\varepsilon_0 + 1)$ is the value of the TM reflection coefficient at zero frequency. A similar asymptotic expression can be obtained from (4) for the free energy of the atom-wall interaction [20]

$$\mathcal{F}^A(a,T) \approx E^A(a) - \frac{\hbar c \pi^3}{240a^4}\alpha(0) C_D(\varepsilon_0)\left(\frac{T}{T_{\text{eff}}}\right)^4, \tag{8}$$

where $C_D(\varepsilon_0)$ is some constant found in [20].

Equations (7) and (8) allow one to find the asymptotic behavior of the Casimir entropy in the plate-plate and atom-plate configurations

$$S(a,T) = -\frac{\partial \mathcal{F}(a,T)}{\partial T}, \qquad S^A(a,T) = -\frac{\partial \mathcal{F}^A(a,T)}{\partial T} \tag{9}$$

at low temperature. Thus, from (7) and (8) one obtains

$$\begin{aligned} S(a,T) &\approx \frac{3k_B}{16\pi a^2}\zeta(3) r_0^2(\varepsilon_0 + 1)\left(\frac{T}{T_{\text{eff}}}\right)^2, \\ S^A(a,T) &\approx \frac{\pi^3 k_B}{30a^3}\alpha(0) C_D(\varepsilon_0)\left(\frac{T}{T_{\text{eff}}}\right)^3. \end{aligned} \tag{10}$$

As is seen from (10), in both cases the Casimir entropy goes to zero when T vanishes, i.e., the Lifshitz theory with the dielectric permittivity (5) satisfies the Nernst heat theorem.

The model of the dielectric response (5) is, however, not exact. It does not take into account that at $T \neq 0$ all dielectric materials possess small but physically real dc conductivity $\sigma(T)$. For all dielectric materials, the general behavior of $\sigma(T)$ with vanishing temperature is given by

$$\sigma(T) \sim \exp\left(-\frac{C}{k_B T}\right) \to 0 \quad \text{when } T \to 0, \tag{11}$$

where the physical meaning of the coefficient C is different for different classes of dielectrics [21,22]. The dc conductivity of dielectrics is usually taken into account by means of the Drude-like term (see, e.g., [23])

$$\tilde{\varepsilon}(\mathrm{i}\xi, T) = \varepsilon(\mathrm{i}\xi) + \frac{4\pi\sigma(T)}{\xi}. \tag{12}$$

It would be reasonable to expect that the substitution of the dielectric permittivity (12), instead of (5), into the Lifshitz formulas (1) and (4) leads to only minor corrections to the asymptotic results (7) and (8). In reality, however, this substitution radically changes the asymptotic behavior of the free energy at low temperature:

$$\tilde{\mathcal{F}}(a,T) = \mathcal{F}(a,T) - \frac{k_BT}{16\pi a^2}\left[\zeta(3) - \mathrm{Li}_3(r_0^2) + R(T)\right],$$
$$\tilde{\mathcal{F}}^A(a,T) = \mathcal{F}^A(a,T) - \frac{k_BT}{4a^3}\left[(1-r_0)\alpha(0) + \tilde{R}(T)\right]. \quad (13)$$

Here, $\mathrm{Li}_n(z)$ is the polylogarithm function, $\mathcal{F}(a,T)$ and $\mathcal{F}^A(a,T)$ are given by (7) and (8), respectively, and $R(T)$, $\tilde{R}(T)$ are exponentially vanishing functions when $T \to 0$. In this case the definition of the Casimir entropy (9) results in the following nonzero values at zero temperature [5,20]

$$\tilde{S}(a,0) = \frac{k_B}{16\pi a^2}\left[\zeta(3) - \mathrm{Li}_3(r_0^2)\right] > 0,$$
$$\tilde{S}^A(a,0) = \frac{k_B}{4a^3}(1-r_0)\alpha(0) > 0. \quad (14)$$

This is in violation of the third law of thermodynamics (the Nernst theorem) because not only are these values not equal to zero, but they also depend on the parameters of the system (separation distance, static dielectric permittivity and atomic polarizability).

The obtained result can be considered as a puzzle because instead of more accurate asymptotic expressions one arrives to the conclusion of an apparent contradiction of the Lifshitz theory with the fundamental principle of thermodynamics.

3. Experimental tests

Rapid progress in the measurements of the Casimir force is very helpful for the resolution of the theoretical controversies. As mentioned in the Introduction, the experiment [6,7] excluded the use of the Drude model to describe the thermal Casimir force between metal plates. It is important to note, however, that a much smaller thermal effect, as predicted by the generalized plasma model [24,25], was not registered in that experiment. Here we consider the results of two other experiments involving semiconductor and dielectric materials [14–16]. Both of them are accurate enough to resolve between the theoretical approaches including and neglecting the dielectric dc conductivity. What is more, one of them [16] provides the first experimental observation of the thermal effects in Casimir physics.

We start with the observation of modulation of the Casimir force between a gold coated sphere and a Si plate with laser light [14,15]. In this experiment, a B doped p-type Si plate was illuminated with laser pulses and the difference of the Casimir forces between a sphere and a plate in the presence and in the absence of light has been measured. In the absence of laser pulse the concentration of charge carriers in the Si plate was $\tilde{n} = 5 \times 10^{14}\,\mathrm{cm}^{-3}$ and in the presence of light $n_1 = 2.1 \times 10^{19}\,\mathrm{cm}^{-3}$ or $n_2 = 1.4 \times 10^{19}\,\mathrm{cm}^{-3}$ for the two different absorbed powers $P_{w1} = 9.3\,\mathrm{mW}$ and $P_{w2} = 4.7\,\mathrm{mW}$. The directly measured quantity was, thus, $\Delta F(a) = F^L(a) - F(a)$, where F^L and F are the Casimir forces in the presence and in the absence of laser light on the plate, respectively. The experimental data were compared with the Lifshitz theory either neglecting or including dc conductivity of the high-resistivity Si in the dark phase.

In Fig. 1(a,b) the mean experimental data $\langle\Delta F(a)\rangle$ are shown as dots for the absorbed powers P_{w1} and P_{w2}, respectively. The calculational results for $\Delta F(a)$, using the Lifshitz theory with neglect of the dc conductivity of the high-resistivity Si (dark phase), are shown as the solid

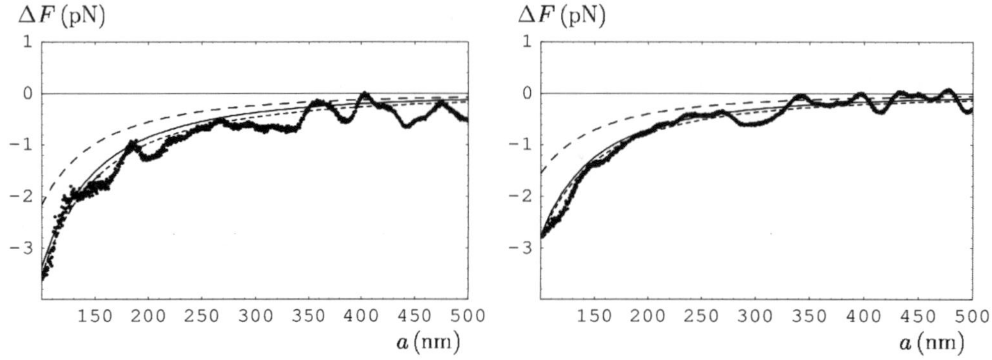

(a) (b)

Figure 1. Differences of the Casimir forces in the presence and in the absence of light versus separation for different absorbed powers: (a) 9.3 mW and (b) 4.7 mW. The mean measured differences are shown as dots. Differences calculated using the Lifshitz formula at $T = 300\,\mathrm{K}$ and $T = 0$ are shown as the solid and short-dashed lines, respectively, and those calculated including the dc conductivity of high-resistivity Si as the long-dashed lines.

lines. Similar calculational results using the Lifshitz theory with included dc conductivity (12) in the dark phase are shown as the long-dashed lines. As is seen in Fig. 1(a,b) the solid lines are consistent with the data whereas the long-dashed lines are excluded over the wide separation region. In the same figure, the short-dashed lines show the computational results obtained from the Lifshitz formula at zero temperature. As is seen in the figure, the computational results at $T = 0$ are consistent with the data. These results do not depend on whether the dc conductivity was included or neglected. The experiment under consideration is precise enough to exclude the large thermal effect, as obtained when the dc conductivity in the dark phase is included, but is not of sufficient precision to measure the small thermal effect obtained in the case when the dc conductivity in the dark phase is neglected.

For illustrative purpose, in Fig. 2 we demonstrate the comparison between experiment and theory over a more narrow separation interval from 100 to 150 nm where only each third experimental point is plotted together with its error bars $[\pm\Delta a, \pm\Delta^{\rm tot}(\Delta F^{\rm expt})]$ shown as crosses (the absorbed power is 8.5 mW). Note that these errors are entirely experimental. They are determined at a 95% confidence level as a combination of random and systematic errors and are not connected with any theory. The theoretical force differences $\Delta F^{\rm theor}$ computed at $T = 300\,\mathrm{K}$ with neglected and included dc conductivity in the dark phase are shown as the solid and dashed lines, respectively. From Fig. 2 one concludes that inclusion of the dc conductivity in the dark phase is excluded by the experimental data at a 95% confidence level within the separation region from 100 to 150 nm.

Note that computations of the Casimir force between an Au coated sphere and Si plate were done using the proximity force approximation [26]

$$F(a,T) = 2\pi R\mathcal{F}(a,T), \tag{15}$$

where $\mathcal{F}(a,T)$ is given by (1). The dielectric permittivities $\varepsilon^{(1)}({\rm i}\xi)$ and $\varepsilon^{(2)}({\rm i}\xi)$ for Si and Au, respectively, were obtained from the tabulated optical data [23] by means of the Kramers-Kronig relations. Practically the same results can be obtained using the analytical representations for $\varepsilon^{(n)}({\rm i}\xi)$. For Au high precision six-oscillator model is considered in [7]. For Si, the imaginary part of the dielectric permittivity can be modeled as a constant within some frequency region [27]

$$\mathrm{Im}\,\varepsilon^{(1)}(\omega) = \begin{cases} \bar{\varepsilon}, & \omega_0 \leq \omega \leq \omega_1, \\ 0, & \omega < \omega_0, \ \omega > \omega_1, \end{cases} \tag{16}$$

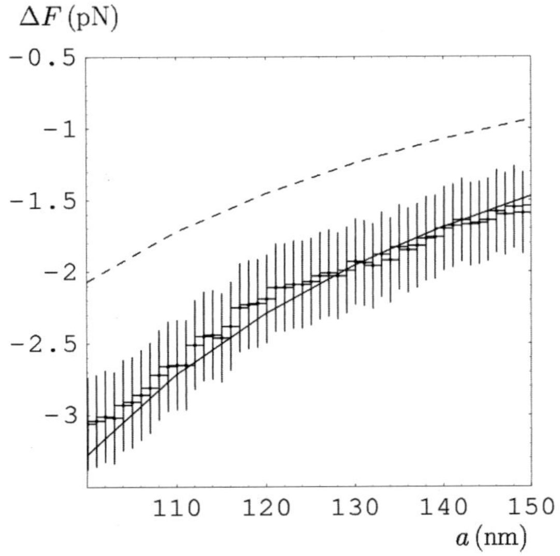

Figure 2. Mean experimental differences of the Casimir forces with their experimental errors are shown as crosses. Solid and dashed lines represent the theoretical differences computed at $T = 300\,\mathrm{K}$ using the model of dielectric permittivity of high-resistivity Si neglecting and including the dc conductivity, respectively.

where $\omega_0 = 3.22\,\mathrm{eV}$, $\omega_1 = 4.62\,\mathrm{eV}$, $\bar{\varepsilon} = 48$. Using the Kramers-Kronig relation, this results in

$$\varepsilon^{(1)}(\mathrm{i}\xi) = 1 + \frac{\bar{\varepsilon}}{\pi}\ln\left(\frac{\omega_1^2+\xi^2}{\omega_0^2+\xi^2}\right). \tag{17}$$

The use of the approximation (17) instead of the tabulated optical data [23] leads to less than 1% deviations in the Casimir force (15) between an Au sphere and Si plate. At separations $a = 50$, 100, 200, 400 and 600 nm this deviation is equal to −0.58%, 0.30%, 0.70%, 0.84% and 0.86%, respectively.

Another analytic formula representing the dielectric permittivity of Si along the imaginary frequency axis takes the form [28]

$$\varepsilon^{(1)}(\mathrm{i}\xi) = \varepsilon^{(1)}(\mathrm{i}\infty) + \frac{[\varepsilon^{(1)}(\mathrm{i}0) - \varepsilon^{(1)}(\mathrm{i}\infty)]\omega_0^2}{\omega_0^2+\xi^2}, \tag{18}$$

where $\omega_0 = 6.6 \times 10^{15}\,\mathrm{rad/s}$, $\varepsilon^{(1)}(\mathrm{i}0) = 11.87$ and $\varepsilon^{(1)}(\mathrm{i}\infty) = 1.035$. This representation also leads to less than 1% differences in the case of Au sphere above Si plate in comparison to computations using the tabulated optical data for Si. Thus, at separations $a = 50$, 100, 200, 400 and 600 nm the relative deviations between both computations are equal to 0.75%, 0.74%, 0.62%, 0.55% and 0.52%. Because of this, both the tabulated optical data for Si and different analytical representations for its dielectric permittivity can be used in the computations within 1% precision.

Another important experimental test is the measurement of the thermal Casimir-Polder force through center-of-mass oscillations of the trapped Bose-Einstein condensate [16]. In this experiment, not only were different models of the dielectric response compared with the data [16,29], but the thermal effect was measured for the first time in Casimir physics. The condensate of Rb atoms was at a distance from 7 to 11 μm from a fused SiO_2 dielectric substrate. In the original publication this substrate was considered as a dielectric material with the static dielectric permittivity $\varepsilon^{(1)}(0) = 3.81$. However, at room temperature (in an equilibrium situation the temperature of the substrate and environment were equal to $T_S = T_E = 310\,\mathrm{K}$) SiO_2 has nonzero conductivity which is ionic in nature and varies from $10^{-9}\,\mathrm{s}^{-1}$ to $10^2\,\mathrm{s}^{-1}$ depending on the concentration of alkali ions which are always present as trace constituents [30,31]. This conductivity was taken into account in [29] in accordance with (12).

The directly measured quantity in the experiment [16] was the relative shift of the frequency of center-of-mass oscillations of the Bose-Einstein condensate due to the presence of the Casimir-Polder force

$$\gamma_z = \frac{|\omega_0 - \omega_z|}{\omega_0} = \frac{1}{2m\omega_0^2}\,|\langle\partial_z F^A(z,T)\rangle|. \tag{19}$$

Here, $m = 1.443 \times 10^{-25}\,\mathrm{kg}$ is the mass of the Rb atom, ω_0 is the unperturbed trap frequency $\omega_0 = 2\pi \times 229\,\mathrm{rad/s}$, $F^A(z,T)$ in the equilibrium situation is given by the negative derivative of $\mathcal{F}^A(z,T)$ defined in (4) with respect to z, and the averaging is done over the period of condensate oscillations and over the volume of the condensate cloud.

The experimental data for the relative frequency shift γ_z are presented as crosses in Fig. 3. The arms of the crosses show the total experimental errors determined in [16] at a 67% confidence level individually for each data point. The results of theoretical computations using (19) with neglected dc conductivity of SiO_2 substrate are shown by the solid line [16,29]. As is seen in the figure, theory neglecting the dc conductivity of fused silica is in a very good agreement with the data. In the same figure, the results of theoretical computations by (19) with dc conductivity of SiO_2 included are shown by the dashed line [29]. As is seen in the figure, the dashed line is in evident disagreement with the first two experimental points.

The experiment was repeated for two more times in situations out of thermal equilibrium ($T_S = 479\,\mathrm{K}$, $T_E = 310\,\mathrm{K}$ and $T_S = 605\,\mathrm{K}$, $T_E = 310\,\mathrm{K}$), i.e., with the temperature of the plate higher than the environment temperature. In this case the Casimir-Polder force $F^A(z,T)$ in (19) should be replaced with

$$F^A(z,T_S,T_E) = F^A(z,T_E) + F^{\mathrm{neq}}(z,T_S) - F^{\mathrm{neq}}(z,T_E), \tag{20}$$

where the explicit form of the function $F^{\mathrm{neq}}(z,T)$ can be found in [16,29,32].

In Fig. 4 the experimental data for γ_z are shown for (a) $T_S = 479\,\mathrm{K}$ and (b) $T_S = 605\,\mathrm{K}$ as crosses with their arms indicating the experimental errors. As in Fig. 3, the theoretical results

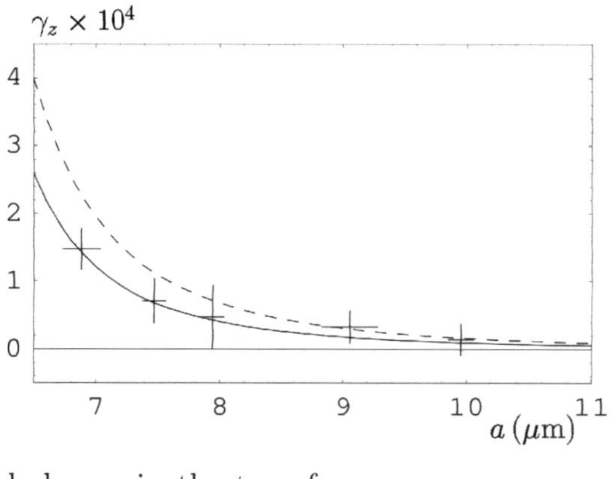

Figure 3. Fractional change in the trap frequency versus separation in thermal equilibrium with $T_S = T_E = 310\,\mathrm{K}$ computed by neglecting (solid line) and including (dashed line) the conductivity of the dielectric substrate. The experimental data are shown as crosses.

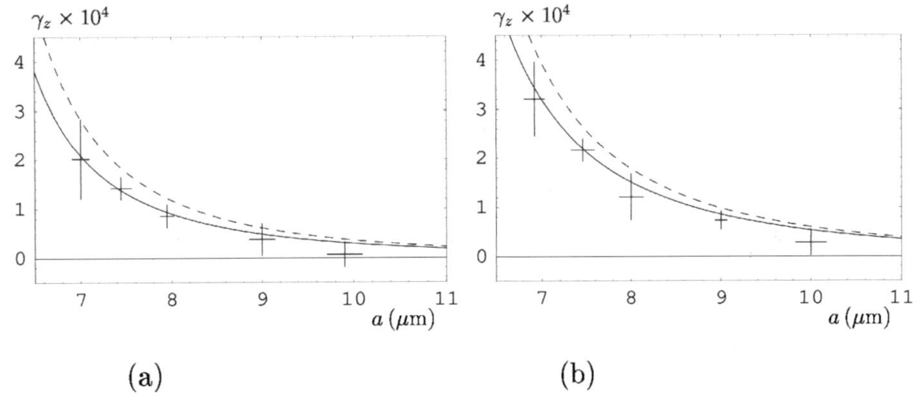

Figure 4. Fractional change in the trap frequency versus separation out of thermal equilibrium with (a) $T_S = 479\,\mathrm{K}$, $T_E = 310\,\mathrm{K}$ and (b) $T_S = 605\,\mathrm{K}$, $T_E = 310\,\mathrm{K}$. Computations are done by neglecting (solid line) and including (dashed line) the conductivity of the dielectric substrate. The experimental data are shown as crosses.

computed with the dc conductivity of SiO_2 plate neglected and included are shown by the solid and dashed lines, respectively. As is seen in Fig. 4(a), the dashed line is in disagreement with the three experimental points, and the other two only touch it. In Fig. 4(b) all data points exclude the theoretical approach represented by the dashed line. Thus, the inclusion of the dc conductivity of the dielectric substrate into the Lifshitz theory leads to disagreement with the experimental data. The calculational results using the Lifshitz theory with the dc conductivity neglected are in very good agreement with the experimental data. We emphasize that the theoretical results obtained with dc conductivity included do not depend on the value of conductivity, but only of the fact that it is nonzero. Importantly, the inclusion of conductivity does not influence the nonequilibrium contributions to γ_z. Thus, the puzzling effect comes only from the contribution to γ_z given by the Lifshitz theory.

4. Phenomenological approach and its justification

In the above, both theoretical and experimental problems are elucidated arising in the Lifshitz theory when applied to dielectric materials at nonzero temperature with dc conductivitincludedy. Recently, these problems were discussed also in [33] but no solution was found. In [33] only "weakly conducting" materials were considered which possess nonzero conductivity at zero temperature. These are in fact metallic-type semiconductors with a dopant concentration above critical. To satisfy the Nernst heat theorem, it was suggested that there is nonzero residual relaxation at $T = 0$, i.e., the crystal lattice is not perfect. Materials of such type have nothing to do with dielectrics whose conductivity vanishes when temperature goes to zero. The formalism of [33] is also in contradiction with precision experiments dealing with metallic test bodies [6,7].

To avoid contradictions of the Lifshitz theory with the principles of thermodynamics and experimental results, one can propose the following phenomenological prescription. In the applications of the Lifshitz theory to all materials having zero conductivity at zero temperature, the presence of free charge carriers at nonzero temperature should be disregarded. The dielectric permittivity of such materials should be described by equation (5). In the applications of the Lifshitz theory to all materials having nonzero conductivity at zero temperature, free charge carriers should be included in the framework of the generalized plasma-like model

$$\varepsilon_{\mathrm{gp}}(\mathrm{i}\xi) = 1 + \frac{\omega_p^2}{\xi^2} + \sum_j \frac{g_j}{\omega_j^2 + \xi^2 + \gamma_j \xi}, \tag{21}$$

where all $\omega_j \neq 0$. This prescription is consistent with thermodynamics and all available experimental data [22,24,25,34].

The physical reasons of why the Lifshitz theory does not allow the inclusion of real conductivity processes can be understood as follows. Lifshitz derived his famous formulas under the condition of thermal equilibrium. This means that not only T = const, but also all irreversible processes connected with the dissipation of energy into heat have already been terminated [35,36]. The Drude-like dielectric function (12) is derived from the Maxwell equations with a real drift current of conduction electrons $\boldsymbol{j} = \sigma_0\boldsymbol{E}$ initiated by the external electric field $\boldsymbol{E}$ [24]. The drift current is an irreversible process which takes a system out of thermal equilibrium. This current leads to Joule heating of the Casimir plates (Ohmic losses) [37]. To preserve the temperature constant, one should admit that there exists an unidirectional flux of heat from the medium to the heat reservoir [38]. Such interactions between a system and a heat reservoir are prohibited by the definition of thermal equilibrium [39]. One can conclude that the substitution of the Drude-like dielectric permittivities into the Lifshitz formulas violate the applicability conditions under which this theory was developed. Then it is not surprising that such substitution results in problems with thermodynamics and contradictions with experimental data. From this point of view, the Lifshitz theory is free of problems and paradoxes which arise only if it is applied incorrectly.

Another question to be answered is the following. One could agree with the fact that the conductivity processes in real materials connected with the drift current of conduction electrons violate the applicability conditions of the Lifshitz theory and, thus, are not described by this theory. These processes, however, exist in reality. Because of this should they be taken into account in some future general theory of dispersion forces? At the moment we have no answer to this question. All computations on the basis of the Lifshitz theory for dielectrics with dc conductivity neglected and for metals by using the plasma-like dielectric permittivity (21) are in agreement with both thermodynamics and experimental data. This suggests that the conductivity properties are not related to dispersion forces. More precise experiments will probably bring additional information to bear on this subject.

5. Approaches based on the modification of reflection coefficients due to screening effects

Recently it was proposed [17,18] that the problems of the Lifshitz theory described above may have an alternative solution if the screening effects due to free charge carriers are taken into account. This approach includes both the drift and diffusion currents of free charge carriers through use of the transport Boltzmann equation [18]. It uses the standard Lifshitz formulas (1) and (4) for the free energy of wall-wall and atom-wall interaction with the TE reflection coefficient, $\tilde{r}_{\rm TE}$, as defined in (2) within the Drude model approach, but with the modified TM reflection coefficient

$$\tilde{r}_{\rm TM}(\mathrm{i}\xi,k_\perp)=\frac{\tilde{\varepsilon}(\mathrm{i}\xi)q-k-\frac{k_\perp^2}{\eta(\mathrm{i}\xi)}\,\frac{\tilde{\varepsilon}(\mathrm{i}\xi)-\varepsilon(\mathrm{i}\xi)}{\varepsilon(\mathrm{i}\xi)}}{\tilde{\varepsilon}(\mathrm{i}\xi)q+k+\frac{k_\perp^2}{\eta(\mathrm{i}\xi)}\,\frac{\tilde{\varepsilon}(\mathrm{i}\xi)-\varepsilon(\mathrm{i}\xi)}{\varepsilon(\mathrm{i}\xi)}}. \tag{22}$$

Here,

$$\tilde{\varepsilon}(\mathrm{i}\xi)=\varepsilon(\mathrm{i}\xi)+\frac{\omega_p^2}{\xi(\xi+\gamma)}, \qquad k^2=k_\perp^2+\tilde{\varepsilon}(\mathrm{i}\xi)\frac{\xi^2}{c^2}, \tag{23}$$

where $\varepsilon(\mathrm{i}\xi)$ is the permittivity of the bound core electrons given in (5). The quantity $\eta(\mathrm{i}\xi)$ is defined as

$$\eta(\mathrm{i}\xi)=\left[k_\perp^2+\kappa^2\frac{\varepsilon(0)}{\varepsilon(\mathrm{i}\xi)}\,\frac{\tilde{\varepsilon}(\mathrm{i}\xi)}{\tilde{\varepsilon}(\mathrm{i}\xi)-\varepsilon(\mathrm{i}\xi)}\right]^{1/2}, \tag{24}$$

where $1/\kappa$ is the screening length. If the charge carriers of density n are described by the classical Maxwell-Boltzmann statistics, as assumed in [18], one gets the Debye-Hückel screening length

[40,41]

$$R_D = \frac{1}{\kappa} = \left(\frac{\varepsilon_0 k_B T}{4\pi e^2 n}\right)^{1/2}. \tag{25}$$

Assuming Fermi-Dirac statistics, one arrives at the Thomas-Fermi screening length [40,41]

$$R_{TF} = \frac{1}{\kappa} = \left(\frac{\varepsilon_0 E_F}{6\pi e^2 n}\right)^{1/2}. \tag{26}$$

Here, e is the electron charge, $\varepsilon_0 = \varepsilon(0)$ is the dielectric constant due to the core electrons, and $E_F = \hbar\omega_p$ is the Fermi energy.

Note that the paper [18] applies this approach to intrinsic semiconductor media. Bearing in mind, however, that the classical Drude dielectric permittivity is exploited, it is justified to apply it for metals and doped semiconductors (both dielectric and metallic type) as well, by substituting the respective screening length defined by (25) or (26).

Now we consider whether or not the approach taking the screening effects into account is consistent with thermodynamics. We concentrate our attention on the case of dielectric materials whose conductivity goes to zero when the temperature vanishes. We start from the case of two similar dielectric semispaces $[\varepsilon^{(1)} = \varepsilon^{(2)} = \tilde{\varepsilon}]$. The Casimir free energy per unit area is given by (1) with the reflection coefficients $r_{\rm TE}^{(1)} = r_{\rm TE}^{(2)} = \tilde{r}_{\rm TE}$ from (2), (23), and $r_{\rm TM}^{(1)} = r_{\rm TM}^{(2)} = \tilde{r}_{\rm TM}$ from (22).

The derivation of the asymptotic behavior of the Casimir free energy and entropy at low temperature is performed in perfect analogy with the derivation of (7) and (13) in [5]. First we introduce the small parameter ($\xi \geq \xi_1$)

$$\beta = \frac{4\pi\sigma(T)}{\xi}, \tag{27}$$

which goes to zero due to (11) when the temperature vanishes. Then we expand the reflection coefficients $\tilde{r}_{\rm TM,TE}$ at all nonzero Matsubara frequencies in powers of the parameter β

$$\begin{aligned}
\tilde{r}_{\rm TM}({\rm i}\zeta, y) &= r_{\rm TM}({\rm i}\zeta, y) + \beta\frac{\varepsilon y[2y^2 + (\varepsilon-2)\zeta^2]}{\sqrt{y^2+(\varepsilon-1)\zeta^2}[\varepsilon y + \sqrt{y^2+(\varepsilon-1)\zeta^2}]^2} + O(\beta^2),\\
\tilde{r}_{\rm TE}({\rm i}\zeta, y) &= r_{\rm TE}({\rm i}\zeta, y) + \beta\frac{y[y - \sqrt{y^2+(\varepsilon-1)\zeta^2}]}{\sqrt{y^2+(\varepsilon-1)\zeta^2}[y + \sqrt{y^2+(\varepsilon-1)\zeta^2}]} + O(\beta^2).
\end{aligned} \tag{28}$$

Here, the reflection coefficients $r_{\rm TM,TE}$ are defined by (2) with the dielectric permittivity (5). They lead to the asymptotic behavior of the Casimir free energy per unit area given in (7). The dimensionless variables ζ and y are connected with other dimensional ones by the equations

$$y = 2aq, \qquad \zeta = \frac{\xi}{\omega_c} = \frac{2a\xi}{c}, \tag{29}$$

where $q = q(\xi)$ is defined in (3).

Now we substitute (28) in (1) and arrive at the following expression for the Casimir free energy per unit area taking the screening effects into account

$$\tilde{\mathcal{F}}(a,T) = \mathcal{F}(a,T) + \frac{k_B T}{16\pi a^2}\left\{\int_0^\infty y\,dy \ln\left[1 - \tilde{r}_0^2(y){\rm e}^{-y}\right] + {\rm Li}_3(r_0^2) + Q(T)\right\}, \tag{30}$$

where

$$\tilde{r}_0(y) \equiv \frac{\varepsilon_0\sqrt{y^2+(2a\kappa)^2} - y}{\varepsilon_0\sqrt{y^2+(2a\kappa)^2} + y} \tag{31}$$

and $Q(T)$ vanishes exponentially when $T \to 0$. The free energy $\mathcal{F}(a,T)$ in (30) is given by (7), and κ in (31) is defined in (25). Calculating the negative derivative of both sides of (30) with respect to T, we obtain the asymptotic behavior of the Casimir entropy at low temperature

$$\tilde{S}(a,T) = S(a,T) - \frac{k_B}{16\pi a^2}\left\{\int_0^\infty y\,dy\ln\left[1-\tilde{r}_0^2(y)\mathrm{e}^{-y}\right] + \mathrm{Li}_3(r_0^2)\right. \tag{32}$$

$$\left. -8a^2\varepsilon_0 T\frac{\partial\kappa^2}{\partial T}\int_0^\infty dy\frac{y^2\tilde{r}_0(y)}{\mathrm{e}^y-\tilde{r}_0^2(y)}\frac{1}{\sqrt{y^2+(2a\kappa)^2}[\varepsilon_0\sqrt{y^2+(2a\kappa)^2}+y]^2} + Q(T) + TQ'(T)\right\},$$

where $S(a,T)$ is defined in (10). It can be easily seen that the last three terms in the curly brackets on the right-hand side of this equation go to zero when T goes to zero for any dielectric material.

The behavior of the first two terms in the curly brackets on the right-hand side of (32) when T goes to zero is more involved. If $n(T)$ exponentially decays to zero with vanishing temperature (as is true for pure insulators and intrinsic semiconductors), then according to (25) so does κ. As a result, $\tilde{r}_0(y) \to r_0 = (\varepsilon_0-1)/(\varepsilon_0+1)$ and the first two terms in the curly brackets cancel. Then the Casimir entropy $\tilde{S}(a,T)$ goes to zero when T vanishes following the same law as $S(a,T)$, i.e., as T^2. This means that for insulators and intrinsic semiconductors the formalism under consideration is in agreement with the Nernst heat theorem.

However, there is a wide class of dielectric materials (such as doped semiconductors with dopant concentration below critical and solids with ionic conductivity) for which n does not go to zero when T goes to zero. In fact, conductivity can be presented as [41]

$$\sigma(T) = n\,|e|\,\mu, \tag{33}$$

where μ is the mobility of charge carriers. Although $\sigma(T)$ goes to zero exponentially fast for all dielectrics when T goes to zero, for most of them this happens due to the vanishing mobility. For instance, as mentioned in Sec. 3, the conductivity of SiO_2 used as the plate material in the experiment [16] is ionic in nature and is determined by the concentration of impuritues. For all such materials, in accordance with (25), $\kappa \to \infty$ when $T \to 0$. As a result, $\tilde{r}_0(y) \to 1$ when T goes to zero in accordance with (31). In this case we obtain from (32)

$$\tilde{S}(a,0) = \frac{k_B}{16\pi a^2}\left[\zeta(3) - \mathrm{Li}_3(r_0^2)\right] > 0 \tag{34}$$

in violation of the Nernst heat theorem. This means that for a wide class of dielectric materials the proposed approach taking the screening effects into account is thermodynamically inconsistent in the same way as the standard Lifshitz theory with the dc conductivity included [compare with (14)].

We emphasize that the existence of dielectric materials for which n does not go to zero but μ does go to zero when T vanishes demonstrates that the reflection coefficient (22) at $\xi = 0$ is ambiguous. In reality, for such materials $\tilde{r}_{\mathrm{TM}}(0,k_\perp) \to 1$ when T and μ simultaneously vanish. This is because $\kappa \to \infty$ when $T \to 0$ in disagreement with physical intuition that there should be no screening at zero mobility. This ambiguity is connected with the break of continuity of the reflection coefficient $\tilde{r}_{\mathrm{TM}}(\mathrm{i}\xi,k_\perp)$ at the point $\xi = 0$, $T = 0$. If one takes the limit $T \to 0$ first, keeping $\xi = \mathrm{const} \neq 0$, the standard Fresnel reflection coefficients r_{TM} from (2) with no screening is reproduced. This property is preserved in the subsequent limiting transition $\xi \to 0$. As was noted in [42] on general grounds, the violation of the Nernst heat theorem is caused by the breaks in the continuity of the reflection coefficients at the origin of the (ξ,T)-plane.

Journal of Physics: Conference Series **161** (2009) 012002 doi:10.1088/1742-6596/161/1/012002

The theoretical approach taking the screening effects into account can also be applied to describe the Casimir-Polder atom-wall interaction. In this case it is also inconsistent with thermodynamics for dielectric materials with a nonvanishing density of charge carriers. To find the asymptotic behavior of the free energy at low temperature, one should substitute into (4) the reflection coefficients $r_{\rm TM,TE}^{(1)} = \tilde{r}_{\rm TM,TE}$, as defined in (28). The result is

$$\tilde{\mathcal{F}}^A(a,T) = \mathcal{F}^A(a,T) - \frac{k_B T\alpha(0)}{8a^3}\left[\int_0^\infty y^2\,dy\,\tilde{r}_0(y)\mathrm{e}^{-y} - 2r_0 + Q_1(T)\right], \tag{35}$$

where the asymptotic behavior of $\mathcal{F}^A(a,T)$ is presented in (8), and $Q_1(T)$ vanishes exponentially when $T \to 0$. Calculating the negative derivative of both sides of (35) with respect to T, one obtains

$$\tilde{S}^A(a,T) = S^A(a,T) + \frac{k_B\alpha(0)}{8a^3}\left[\int_0^\infty y^2\,dy\,\tilde{r}_0(y)\mathrm{e}^{-y} - 2r_0 \right. \tag{36}$$
$$\left. +4a^2\varepsilon_0 T\frac{\partial\kappa^2}{\partial T}\int_0^\infty dy\frac{y^3\mathrm{e}^{-y}}{\sqrt{y^2+(2a\kappa)^2}(\varepsilon_0\sqrt{y^2+(2a\kappa)^2}+y)^2} + Q_1(T) + TQ_1'(T)\right],$$

where $S^A(a,T)$ is defined in (10). For dielectric materials with exponentially decaying $n(T)$ in the limit of low T, $\tilde{r}_0(y) \to r_0$ and all terms on the right-hand side of (36) added to $S^A(a,T)$ go to zero when $T \to 0$. As a result, $\tilde{S}^A(a,T)$ goes to zero as T^3 and the Nernst theorem is satisfied. On the contrary, for all dielectric materials with nonvanishing n we get $\tilde{r}_0(y) \to 1$ when $T \to 0$ and in this case (36) results in

$$\tilde{S}^A(a,0) = \frac{k_B\alpha(0)}{4a^3}(1-r_0) > 0. \tag{37}$$

This means that the Nernst theorem is violated in the same way as in (14) for the standard Lifshitz theory with the dc conductivity included.

The theoretical predictions following from the approach proposed in [18] can be compared with the measurement data of the experiment on the modulation of the Casimir force with light (see Sec. 3 for details). In Fig. 5(a) and 5(b) we present the experimental data of this experiment obtained for the absorbed powers P_{w1} and P_{w2}, respectively, with the experimental errors of force measurements determined at 70% confidence level. The theoretical difference Casimir force, computed using the standard Lifshitz theory with the dc conductivity of high-resistivity Si in the dark phase neglected, is shown by the solid lines. The theoretical difference Casimir force computed by including the screening effects at all Matsubara frequencies, as described above, is shown by the dashed lines. It can be seen that the standard Lifshitz theory is consistent with data, whereas the theoretical approach taking the screening effects into account is excluded by the data at a 70% confidence level.

In Sec. 4 the contradictions with thermodynamics and the experimental data of the standard Lifshitz theory with the dc conductivity included are explained based on the violation of its applicability condition, i.e., of thermal equilibrium, in the presence of drift current. In a similar way, the approach taking the screening effects into account applies the Lifshitz theory to physical phenomena involving both the drift and diffusion currents. The latter is caused by the nonequilibrium distribution of charge carriers in an external field, i.e., by a physical situation out of thermal equilibrium. This is because the diffusion current is determined by a nonzero gradient of charge carrier density, whereas for homogeneous systems in thermal equilibrium the charge carrier density must be homogeneous. Thus, the problems and paradoxes arising in this case are also explained by the use of the Lifshitz theory outside of its defined application region.

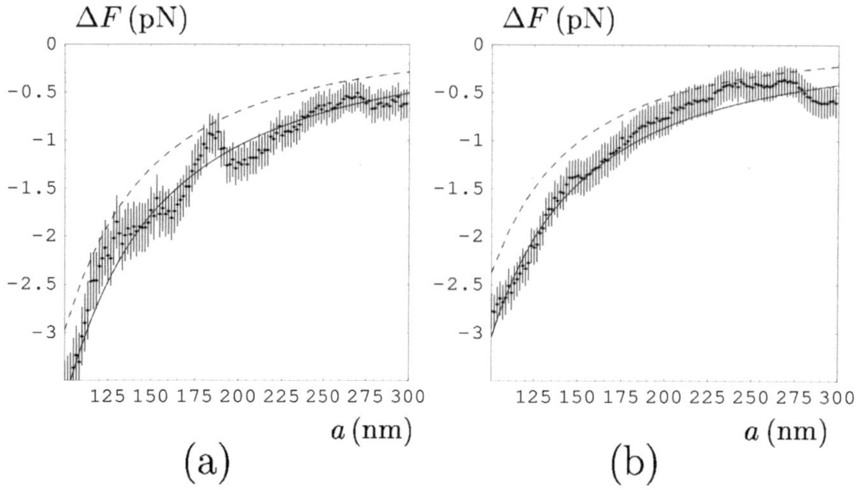

Figure 5. Differences of the Casimir forces between an Au-coated sphere and a Si plate, in the presence and in the absence of laser light on the plate, versus separation for the absorbed power of (a) 9.3 mW and (b) 4.7 mW. The experimental data are shown as crosses. Solid and dashed lines indicate the theoretical results calculated using the standard Lifshitz theory with the dc conductivity of Si in the dark phase neglected, and the generalization of this theory [17,18], respectively.

An earlier attempt to incorporate the screening effects in the theory of dispersion forces was undertaken in [17]. In this approach the effect of screening was taken into account only for the static field in the interaction of an atom with a dielectric wall. Bearing in mind that in accordance with (4) the TE reflection coefficient at zero frequency does not contribute to the result, the single modified coefficient was [17]

$$r_{\rm TM}^{\rm mod}(0,k_\perp)=\frac{\varepsilon_0\sqrt{k_\perp^2+\kappa^2}-k_\perp}{\varepsilon_0\sqrt{k_\perp^2+\kappa^2}+k_\perp}, \tag{38}$$

where κ is given in (25). It is easily seen that (38) coincides with the general TM reflection coefficient $\tilde{r}_{\rm TM}({\rm i}\xi,k_\perp)$ defined in (22) if one puts $\xi=0$.

The reflection coefficient (38) can be obtained by the formal replacement of the dielectric permittivities depending only on the frequency with the permittivities depending both on frequency and the wave vector in the Fresnel reflection coefficient of an uniaxial crystal. To confirm that this is the case, one should introduce nonisotropic permittivities $\varepsilon_x=\varepsilon_y\neq\varepsilon_z$ and replace the coefficients $r_{\rm TM}$ and $r_{\rm TE}$ in (2) with

$$r_{\rm TM}^{\rm mod}({\rm i}\xi,k_\perp)=\frac{\sqrt{\varepsilon_x({\rm i}\xi)\varepsilon_z({\rm i}\xi)}q-k_z}{\sqrt{\varepsilon_x({\rm i}\xi)\varepsilon_z({\rm i}\xi)}q+k_z}, \qquad r_{\rm TE}^{\rm mod}({\rm i}\xi,k_\perp)=\frac{q-k_x}{q+k_x}. \tag{39}$$

Here, k_z and k_x are defined by (3) with the replacement of $\varepsilon({\rm i}\xi)$ by $\varepsilon_z({\rm i}\xi)$ and $\varepsilon_x({\rm i}\xi)$, respectively. For ε_x and ε_z depending only on the frequency, (39) is commonly used for the description of uniaxial crystals (see, e.g., [43,44]). Then (38) follows from (39) if one puts

$$\varepsilon_x(0)=\varepsilon_0, \quad \varepsilon_z(0)=\varepsilon_z(0,k_\perp)=\varepsilon_0\left(1+\frac{\kappa^2}{k_\perp^2}\right), \tag{40}$$

where now $\varepsilon_z(0)$ depends on the wave vector, i.e., spatial dispersion is present. Thus, to obtain (38) one substitutes into the Fresnel reflection coefficient (39) derived from the standard continuity boundary conditions a dielectric permittivity depending on the wave vector (see [45,46] why this is inappropriate).

We emphasize that the general reflection coefficients taking the screening effects into account [the transverse electric from (2) with the dielectric permittivity (23) and the transverse magnetic from (22)] cannot be represented in the form of (39). To prove this statement, we note that from the equality $\tilde{r}_{\rm TE} = r_{\rm TE}^{\rm mod}$ it follows that $\tilde{\varepsilon}({\rm i}\xi) = \varepsilon_x({\rm i}\xi)$. Thus, in accordance with (23), $\varepsilon_x({\rm i}\xi) \to \infty$ when $\xi \to 0$. Keeping in mind that $\varepsilon_z(0) > 0$, from (39) one obtains $r_{\rm TM}^{\rm mod}(0, k_\perp) = 1$. This is in contradiction with the above result that at zero frequency the reflection coefficient (22) coincides with the reflection coefficient (38). This means that the general approach taking the screening effects into account [18] cannot be described in terms of any dielectric permittivity which includes spatial dispersion.

The approach based on the modification of only one reflection coefficient (38) meets the same thermodynamical difficulties as the general approach considered above. Thus, if one computes the Casimir entropy with the reflection coefficient (38) and all the other standard ones, as given by (2), it goes to zero when the temperature vanishes for dielectrics with exponentially vanishing charge carrier density n. However, for dielectrics with nonvanishing n the result (37) is reproduced, i.e., the Nernst heat theorem is violated [20].

The modification of the TM reflection coefficient in accordance with (38) can be verified experimentally by using the measurement data of the optical modulation experiment discussed in Sec. 3. In theoretical computations at zero frequency the expression (38) was used. At all nonzero Matsubara frequencies the standard reflection coefficients (2) were substituted into the Lifshitz formula. The computational results using the standard reflection coefficients at all nonzero Matsubara frequencies coincide with those obtained with the general reflection coefficients $\tilde{r}_{\rm TM}({\rm i}\xi, k_\perp)$ and $\tilde{r}_{\rm TM}({\rm i}\xi, k_\perp)$. Then we return to the dashed lines in Fig. 5(a,b) representing the difference Casimir force computed with the inclusion of screening effects. More generally, it can be easily seen that for materials with sufficiently small charge carrier density (for instance for Si with $n \leq 10^{17}\,{\rm cm}^{-3}$) the general reflection coefficient (22) including the screening effects leads to approximately the same values at room temperature, as the standard one at all nonzero Matsubara frequencies. Thus, the theoretical results computed by using the modified TM reflection coefficient (38) are excluded by data at a 70% confidence level.

6. Conclusions and discussion

In the above we have considered the application of the Lifshitz theory to real metals, dielectrics and semiconductors. When the conductivity processes, connected with the presence of free charge carriers, are included into the model of the dielectric response, problems and paradoxes in the application of the Lifshitz theory are immediately apparent. They are manifested in the contradictions of the Lifshitz theory with both thermodynamics and experiment. We have listed several measurements performed by different experimental groups with metals, dielectrics and semiconductors which are in contradiction with the theoretical results taking into account the drift current of conduction electrons.

In searching for reasons of why the problems and paradoxes arise, it was emphasized that the Lifshitz theory was formulated for systems in thermal equilibrium. As a result, this theory is not applicable to physical systems where processes go on which violate thermal equilibrium. Specifically, in thermal equilibrium all processes of heat transfer, diffusion and others connected with the dissipation of energy are terminated. A question then arises if the drift current of charge carriers which leads to a unidirectional flux of Joule heat from the Casimir plates to the heat reservoir is compatible with the state of thermal equilibrium. Our conclusion is that it is not compatible with thermal equilibrium and, thus, cannot be considered in the framework of the Lifshitz theory. The inclusion of conductivity properties in the Lifshitz theory is in violation of its basic applicability condition.

We have analyzed recent attempts to solve problems and paradoxes using screening effects and diffusion processes. We have shown that these attempts are in conflict with thermodynamics and in contradiction with experiment. This is because diffusion processes make the deviations from thermal equilibrium even more dramatic.

One may conclude that the Lifshitz theory, if used within its application region, is free of problems and paradoxes. Problems and paradoxes arise when the basic applicability condition of the Lifshitz theory, thermal equilibrium, is not taken into account by using the Drude model or introducing the diffusion current. What is more, phenomenologically the Lifshitz theory leads to both thermodynamically and experimentally consistent results if the processes of electric conductivity connected with both drift and diffusion currents are simply disregarded. Future progress should show if these processes have any relation to the phenomenon of dispersion forces.

Acknowledgments

This work was supported by Deutsche Forschungsgemeinschaft, Grant No 436 RUS 113/789/0–4. The author is grateful to the Center of Theoretical Studies and Institute of Theoretical Physics of Leipzig University, where this work was performed, for their kind hospitality.

[1] Lifshitz E M 1956 *Zh. Eksp. Teor. Fiz.* **29** 94 (*Sov. Phys. JETP* **2** 73)
[2] Dzyaloshinskii I E, Lifshitz E M and Pitaevskii L P 1961 *Sov. Phys. Usp.* **4** 153
[3] Lifshitz E M and Pitaevskii L P 1984 *Statistical Physics*, Pt II (Oxford: Pergamon Press)
[4] Boström M and Sernelius B E 2000 *Phys. Rev. Lett.* **84** 4757
[5] Geyer B, Klimchitskaya G L and Mostepanenko V M 2005 *Phys. Rev.* D **72** 085009
[6] Decca R S, López D, Fischbach E, Klimchitskaya G L, Krause D E and Mostepanenko V M 2007 *Phys. Rev.* D **75** 077101
[7] Decca R S, López D, Fischbach E, Klimchitskaya G L, Krause D E and Mostepanenko V M 2007 *Eur. Phys. J.* C **51** 963
[8] Jancovici B and Šamaj L 2005 *Europhys. Lett.* **72** 35
[9] Buenzli P R and Martin Ph A 2008 *Phys. Rev.* E **77** 011114
[10] Bezerra V B, Klimchitskaya G L, Mostepanenko V M and Romero C 2004 *Phys. Rev.* A **69** 022119
[11] Brevik I, Ellingsen S A, Høye J S and Milton K A 2008 *J. Phys. A: Math. Theor.* **41** 164017
[12] Geyer B, Klimchitskaya G L and Mostepanenko V M 2006 *Int. J. Mod. Phys.* A **21** 5007
[13] Geyer B, Klimchitskaya G L and Mostepanenko V M 2008 *Ann. Phys. NY* **323** 291
[14] Chen F, Klimchitskaya G L, Mostepanenko V M and Mohideen U 2007 *Optics Express* **15** 4823
[15] Chen F, Klimchitskaya G L, Mostepanenko V M and Mohideen U 2007 *Phys. Rev.* B **76** 035338
[16] Obrecht J M, Wild R J, Antezza M, Pitaevskii L P, Stringari S and Cornell E A 2007 *Phys. Rev. Lett.* **98** 063201
[17] Pitaevskii L P 2008 *Phys. Rev. Lett.* **101** 163202
[18] Dalvit D A R and Lamoreaux S K 2008 *Phys. Rev. Lett.* **101** 163203
[19] Parsegian V A 2005 *Van der Waals forces: A Handbook for Biologists, Chemists, Engineers, and Physicists* (Cambridge: Cambridge University Press)
[20] Klimchitskaya G L, Mohideen U and Mostepanenko V M 2008 *J. Phys. A: Math. Theor.* **41** 432001
[21] Shklovskii B I and Efros A L 1984 *Electronic Properties of Doped Semiconductors. Solid State Series*, v.45 (Berlin: Springer)
[22] Klimchitskaya G L and Geyer B 2008 *J. Phys. A.: Mat. Theor.* **41** 164032
[23] Palik E D (ed) 1985 *Handbook of Optical Constants of Solids* (New York: Academic)
[24] Geyer B, Klimchitskaya G L and Mostepanenko V M 2007 *J. Phys. A: Math. Theor.* **40** 13485
[25] Mostepanenko V M and Geyer B 2008 *J. Phys. A: Mat. Theor.* **41** 164014
[26] Bordag M, Mohideen U and Mostepanenko V M 2001 *Phys. Rep.* **353** 1
[27] Inui N 2006 *J. Phys. Soc. Jap.* **75** 24004
[28] Lambrecht A, Pirozhenko I, Duraffourg L and Andreucci Ph 2007 *Europhys. Lett.* **77** 44006
[29] Klimchitskaya G L and Mostepanenko V M 2008 *J. Phys. A: Math. Theor.* **41** 312002
[30] Shackelford J F and Alexander W (eds) 2001 *Material Science and Engineering Handbook* 3rd edn (Boca Raton: CRC Press)
[31] Bansal N P and Doremus R H 1986 *Handbook of Glass Properties* (New York: Academic)

[32] Antezza M, Pitaevskii L P and Stringari S 2005 *Phys. Rev. Lett.* **95** 113202
[33] Ellingsen S A, Brevik I, Høye J S and Milton K A 2008 *Phys. Rev.* E **78** 021117
[34] Klimchitskaya G L, Mohideen U and Mostepanenko V M 2007 *J. Phys. A.: Mat. Theor.* **40** F339
[35] Rumer Yu B and Ryvkin M S 1980 *Thermodynamics, Statistical Physics, and Kinetics* (Moscow: Mir).
[36] Kondepugi D and Prigogine I 1998 *Modern Thermodynamics* (New York: Wiley)
[37] Geyer B, Klimchitskaya G L and Mostepanenko V M 2003 *Phys. Rev. A* **67** 062102
[38] Bryksin V V and Petrov M P 2008 *Fiz. Tverdogo Tela* **50** 222 (*Phys. Solid State* **50** 229)
[39] Landau L D and Lifshitz E M 1980 *Statistical Physics* Pt I (Oxford: Pergamon Press)
[40] Chazalviel J-N 1999 *Coulomb Screening of Mobile Charges: Applications to Material Science, Chemistry and Biology* (Boston: Birkhauser)
[41] Ashcroft N W and Mermin N D 1976 *Solid State Physics* (Philadelphia: Saunders College)
[42] Intravaia F and Henkel C 2008 *J. Phys. A: Math. Theor.* **41** 164018
[43] Greenaway D L, Harbeke G, Bassani F and Tosatti E 1969 *Phys. Rev.* **178** 1340
[44] Blagov E V, Klimchitskaya G L and Mostepanenko V M 2005 *Phys. Rev.* B **71** 235401
[45] Barash Yu S and Ginzburg V L 1975, *Usp. Fiz. Nauk* **116** 5 (*Sov. Phys. Uspekhi* **18** 305)
[46] Klimchitskaya G L and Mostepanenko V M 2006 *Phys. Rev.* B **75** 036101

Journal of Physics: Conference Series **161** (2009) 012003 doi:10.1088/1742-6596/161/1/012003

Experiment, theory and the Casimir effect

V M Mostepanenko[1]

Center of Theoretical Studies and Institute for Theoretical Physics, Leipzig University, Vor dem Hospitaltore 1, 100920, D-04009, Leipzig, Germany

E-mail: vladimir.mostepanenko@itp.uni-leipzig.de

Abstract. Several problems at the interface between the field-theoretical description of the Casimir effect and experiments on measuring the Casimir force are discussed. One of these problems is connected with the definition of the Casimir free energy in ideal metal rectangular boxes satisfying the general physical requirements. It is shown that the consideration of rectangular boxes with a partition (piston) does not negate the previously known results obtained for boxes without a piston. Both sets of results are found to be in mutual agreement. Another problem is related to the use of the proximity force approximation for the interpretation of the experimental data and to the search of analytical results beyond the PFA based on the first principles of quantum field theory. Next, we discuss concepts of experimental precision and of the measure of agreement between experiment and theory. The fundamental difference between these two concepts is clarified. Finally, recent approach to the thermal Casimir force taking screening effects into account is applied to real metals. It is shown that this approach is thermodynamically and experimentally inconsistent. The physical reasons of this inconsistency are connected with the violation of thermal equilibrium which is the basic applicability condition of the Lifshitz theory.

1. Introduction

For almost 50 years since Casimir's discovery [1], theory existed independent of rare experiments [2,3]. A lot of theoretical work has been done during this period. However, the relationship to reality of some theoretical models, such as an ideal metal spherical shell, an ideal metal rectangular box, a dielectric ball etc., remains unclear up to now. In the last ten years scientific investigations in the field of the Casimir effect have experienced an interaction between experiment and theory. This has revealed that the application of some basic theories to real experimental situations is highly nontrivial and even leads to communication difficulties between theorists and experimentallists.

In this paper we summarize some experience of the interaction between "high theory" and real world experimental details in the last ten years. Different points of view are considered on such problems as agreement between experiment and theory and applicability of some ideal models and approximate methods in real experimental situations. It is shown that in some cases confusion arises from an inadequate use of terminology.

In Sec. 2 we discuss an old problem of the thermal Casimir force in ideal metal rectangular boxes and suggest a new solution which satisfies general physical criteria. It is shown that the case of an isolated box is independent of a box with a partition (piston). The results for the Casimir force obtained for each of these configurations are in mutual agreement.

[1] On leave from Noncommercial Partnership "Scientific Instruments", Moscow, Russia

Section 3 briefly reviews the proximity force approximation including its justification from the first principles of quantum field theory and experimental applications. In this respect a new representation for the Casimir energy in terms of the functional determinants and scattering matrices is considered. The application of this representation to real material bodies is still problematic.

In Sec. 4 we consider the problem of the reliability of experiments. It is underlined that the experimental error is an independent characteristic of the experimental precision which should not be confused with the measure of agreement between experiment and theory.

Section 5 is devoted to the comparison between experiment and theory in the measurements of the Casimir force. In this respect different approaches to the theoretical description of the Casimir force between real metals are compared with the most precise indirect measurement of the Casimir pressure between two parallel plates by means of micromechanical torsional oscillator [4,5]. Special attention is paid to uncertainties which might be introduced in the computations due to deviations of the tabulated optical data from the data particular to the metallic films actually used.

In Sec. 6 we consider a recent theoretical approach to the thermal Casimir force taking into account the screening effects and diffusion currents. We apply this approach to the case of real metals and analyze its consistency with the principles of thermodynamics and experimental data. Specifically we show that for metals with perfect crystal lattices the inclusion of screening effects results in violation of the Nernst heat theorem. The experimental data of the experiment [4,5] exclude this approach at a 99.9% confidence level.

Section 7 contains our conclusions and discussion.

2. Thermal Casimir force in ideal metal rectangular boxes

Ideal metal rectangular boxes were first considered by Lukosz [6], Mamayev and Trunov [7,8] and Ambjørm and Wolfram [9]. This configuration attracted much attention because it was found that the electromagnetic Casimir force in rectangular boxes can be both attractive and repulsive depending on the ratio of sides a_x, a_y and a_z along the x, y and z axes. The nonrenormalized Casimir energy of the box is equal to (for simplicity we consider the massless scalar field with Dirichlet boundary conditions)

$$E_0(a_x, a_y, a_z) = \frac{\hbar}{2} \sum_{n,l,p=1}^{\infty} \omega_{nlp}, \tag{1}$$

where

$$\omega_{nlp} = \pi c \left[\left(\frac{n}{a_x}\right)^2 + \left(\frac{l}{a_y}\right)^2 + \left(\frac{p}{a_z}\right)^2 \right]^{1/2}. \tag{2}$$

The regularization of (1) can be performed, e.g., using the Epstein zeta function or the cut-off method [10]. The latter permits to find the geometric structure of infinities contained in (1). To do so, one replaces $E_0(a_x, a_y, a_z)$ from (1) with $E_0^{(\delta)}(a_x, a_y, a_z)$ by introducing the cut-off function

$$f(\delta\omega_{nlp}) = \mathrm{e}^{-\delta\omega_{nlp}} \tag{3}$$

under the sign of summation in (1). After the repeated application of the Abel-Plana formula [10] to $E_0^{(\delta)}(a_x, a_y, a_z)$ one finds that there are three different types of divergent quantities in the limit $\delta \to 0$, I_1, I_2 and I_3 of order δ^{-4}, δ^{-3} and δ^{-2}, respectively. Then, the finite, renormalized, Casimir energy can be defined as

$$E_0^{\text{ren}}(a_x, a_y, a_z) = \lim_{\delta \to 0} \left[E_0^{(\delta)}(a_x, a_y, a_z) - I_1 - I_2 - I_3 \right]. \tag{4}$$

Here, I_k ($k = 1, 2, 3$) are the counter terms having the following geometrical structure:

$$I_1 = \frac{12\pi^2\hbar a_x a_y a_z}{c^3\delta^4}, \qquad I_2 = -\frac{\pi^2\hbar(a_x a_y + a_x a_z + a_y a_z)}{c^2\delta^3}, \qquad I_3 = \frac{\pi\hbar(a_x + a_y + a_z)}{8c\delta^2}. \tag{5}$$

A similar situation takes place for the electromagnetic field, where the renormalized Casimir energy, $E_{0,\text{em}}^{\text{ren}}$, also takes the form of (4) (with E_0 replaced for E_0^{em}) and

$$I_1^{\text{em}} = 2I_1, \qquad I_2^{\text{em}} = 0, \qquad I_3^{\text{em}} = -2I_3. \tag{6}$$

It is seen that in both cases the counter terms are proportional to the volume of the box $V = a_x a_y a_z$, to the area of box surface and to the sum of sides.

In the last few years the configuration of a rectangular box with so-called *movable* partition (piston) has attracted much attention [11–14]. This means that the piston can have any fixed position parallel to the two opposite faces of the box (the configuration where the piston is not fixed and may slide between the opposite faces is in fact a nonequilibrium case). Let the piston be parallel to the plane xy and have an equation $z = a_{z1} < a_z$. In this case our box is divided into the two boxes $a_x \times a_y \times a_{z1}$ and $a_x \times a_y \times (a_z - a_{z1})$. Calculating the sum of the regularized Casimir energies

$$E_0^{(\delta)}(a_x, a_y, a_{z1}) + E_0^{(\delta)}(a_x, a_y, a_z - a_{z1}), \tag{7}$$

one finds that the contribution from the singular terms of the form of (5) does not depend on the position of the piston a_{z1}. This leads to a finite force acting on the piston

$$F(a_x, a_y, a_z, a_{z1}) = -\frac{\partial}{\partial a_{z1}}\left[E_0^{(\delta)}(a_x, a_y, a_{z1}) + E_0^{(\delta)}(a_x, a_y, a_z - a_{z1})\right]. \tag{8}$$

This force is well defined and does not require the renormalization procedure (4).

In both scalar and electromagnetic cases the force acting on the piston attracts it to the nearest face of the box. On this ground the existence of the Casimir repulsion in cubes in the electromagnetic case was cosidered doubtful [12]. Specifically, it was claimed [12,13] that the definition of the pressure acting on a cube face requires elastic deformations of single bodies treated as perfect. The attraction (or repulsion for a piston with Neumann boundary conditions [15]) of a piston to the nearest face of the box does not, however, negate the Casimir repulsion for boxes without a piston that have some appropriate ratio of a_x, a_y and a_z. The point is that the cases with an empty space outside the box and that with another section of the larger box outside the piston are physically quite different. In the first case the vacuum energy outside the box does not depend on a_x, a_y and a_z and there is no force acting on the box from the outside. Whereas in the second case there is an extra section of the larger box outside the piston which gives rise to the additional force acting on it. In fact one need not admit elastic deformations to define a force and a pressure in static configurations. This is simply done using the principle of virtual work and virtual displacements through real forces [16,17]. In addition, from a thermodynamic point of view any equilibrium system can be characterized by the free energy (energy if the temperature is equal to zero) and the respective pressure [18]

$$P = -\left.\frac{\partial \mathcal{F}}{\partial V}\right|_{T=\text{const}}. \tag{9}$$

From this point of view it would be illogical to admit consideration of the force acting on a piston, but exclude from consideration forces acting on the faces of a box where this piston serves as a partition.

In this respect it seems important to provide a finite definition of the Casimir free energy in ideal metal rectangular boxes satisfying general physical requirements. The first calculations on this subject [9] resulted in a divergent free energy after removing the regularization. More recent results appear to be either infinite [19] or ambiguous [20]. Paper [21] reconsidered the derivation of the Casimir free energy in rectangular boxes using zeta functional regularization. However, the used formalism does not include all necessary subtractions.

The following definition of the Casimir free energy in rectangular boxes suggests itself [12,13,21]

$$\mathcal{F}_0 = E_0^{\text{ren}} + \Delta_T\mathcal{F}_0, \qquad \Delta_T\mathcal{F}_0 = k_BT\sum_{n,l,p=1}^{\infty}\ln\left(1-\mathrm{e}^{-\frac{\hbar\omega_{nlp}}{k_BT}}\right). \tag{10}$$

This expression is finite. However, it cannot be considered as physically satisfactory. The problem is that at high temperature the thermal correction (10) behaves as [22]

$$\Delta_T\mathcal{F}_0 = \alpha_1\frac{(k_BT)^4}{(\hbar c)^3} + \alpha_2\frac{(k_BT)^2}{(\hbar c)^2} + \alpha_3\frac{(k_BT)^2}{\hbar c} + \alpha_4 k_BT + \dots, \tag{11}$$

where $\alpha_1 = -V\pi^2/90$, $\alpha_{2,3} = \alpha_{2,3}(a_x, a_y, a_z)$ can be expressed in terms of the heat kernel coefficients and $\alpha_4 = \text{const}$. Then at high temperature $\Delta_T\mathcal{F}_0$ contains terms of quantum origin which increase with the increase of temperature. In the general case, these terms lead to respective forces acting on the box faces which increase with the increase of the sides of the box. Such paradoxical properties are physically unacceptable. Because of this it was suggested [23] to define the physical Casimir free energy of the box as

$$\mathcal{F} = E_0^{\text{ren}} + \Delta_T\mathcal{F}_0 - \alpha_1\frac{(k_BT)^4}{(\hbar c)^3} - \alpha_2\frac{(k_BT)^2}{(\hbar c)^2} - \alpha_3\frac{(k_BT)^2}{\hbar c}. \tag{12}$$

With this definition, the respective Casimir forces acting on the box faces go to zero when all the box sides a_x, a_y, a_z go to infinity in agreement with physical intuition.

The physical meaning of all three subtractions made on the right-hand side of (12) can be clearly understood. The first term is actually the contribution of the blackbody radiation in the volume of the box. This is seen from the fact that the free energy density of the blackbody radiation in empty space is given by

$$f_{bb} = k_BT\int\frac{d^3k}{(2\pi)^3}\ln\left(1-\mathrm{e}^{-\frac{\hbar c|k|}{k_BT}}\right) = -\frac{\pi^2(k_BT)^4}{90(\hbar c)^3}, \tag{13}$$

where for the electromagnetic case $f_{bb}^{\text{em}} = 2f_{bb}$.

For the scalar Casimir effect in a rectangular box with sides $a_x \times a_y \times a_z$ the asymptotic behavior of $\Delta_T\mathcal{F}_0$ at high T was investigated in [23] with the result

$$\alpha_2 = \frac{\zeta(3)}{4\pi}(a_xa_y + a_xa_z + a_ya_z), \qquad \alpha_3 = -\frac{\pi}{24}(a_x + a_y + a_z). \tag{14}$$

In the electromagnetic case the following values of these coefficients were obtained:

$$\alpha_2^{\text{em}} = 0, \qquad \alpha_3^{\text{em}} = \frac{\pi}{12}(a_x + a_y + a_z). \tag{15}$$

This demonstrates that the geometric structures of all three terms subtracted in (12) are precisely the same as the terms subtracted in (4) to make the Casimir energy finite at zero temperature. Because of this, the subtraction procedure in (12) can be interpreted as the additional (finite) renormalization of the same geometric parameters as were renormalized at zero temperature to make the Casimir energy of the box finite.

The simplest application of the final expression for the physical Casimir free energy (12) is the case of two plane parallel plates. It is easily seen that in this configuration $\alpha_2 = \alpha_3 = 0$ and one is left with only a subtraction of the free energy of the blackbody radiation in the volume between the plates $V = aS$, where S is the infinite plate area. This leads to the well known result [10,24] for the electromagnetic Casimir free energy per unit area of the plates

$$\mathcal{F}(a,T) = -\frac{\pi^2}{720a^3}\left\{1+\frac{45}{\pi^3}\sum_{l=1}^{\infty}\left[\frac{\coth(\pi l t)}{t^3 l^3}+\frac{\pi}{t^2 l^2 \sinh^2(\pi t l)}\right]-\frac{1}{t^4}\right\}, \tag{16}$$

where $t \equiv T_{\text{eff}}/T$, and the effective temperature is defined as $k_B T_{\text{eff}} = \hbar c/(2a)$. In particular, at $T \ll T_{\text{eff}}$ one obtains

$$\mathcal{F}(a,T) = -\frac{\pi^2}{720a^3}\left[1+\frac{45\zeta(3)}{\pi^3}\left(\frac{T}{T_{\text{eff}}}\right)^3-\left(\frac{T}{T_{\text{eff}}}\right)^4\right], \tag{17}$$

where the last contribution on the right-hand side originates from the subtraction of the blackbody radiation. We emphasize that only this term contributes to the thermal correction to the electromagnetic Casimir pressure at low temperatures (short separations)

$$P(a,T) = -\frac{\pi^2}{240a^4}\left[1+\frac{1}{3}\left(\frac{T}{T_{\text{eff}}}\right)^4\right]. \tag{18}$$

Equation (12) solves the long-standing problem on the calculation of the physical Casimir free energies and pressures in rectangular boxes of any size. A few examples for both the scalar and electromagnetic Casimir effect are considered in [23]. Here we present the computational results for the electromagnetic free energy in a cube and for the respective Casimir force

$$F_x(a,T) = a^2 P(a,T) = -\frac{1}{3}\frac{\partial\mathcal{F}(a,T)}{\partial a} \tag{19}$$

acting on the opposite cube faces.

In Fig. 1(a) we plot the electromagnetic Casimir free energy in a cube as a function of a at $T = 300\,\text{K}$ (solid line). In the same figure the Casimir energy at $T = 0$ is shown by the dashed line. As is seen in this figure, the electromagnetic Casimir free energy decreases with the increase of separation. At large separations $\mathcal{F}$ approaches a constant. In Fig. 1(b) the electromagnetic Casimir free energy is shown as a function of temperature for a cube with $a = 2\,\mu$m. The free

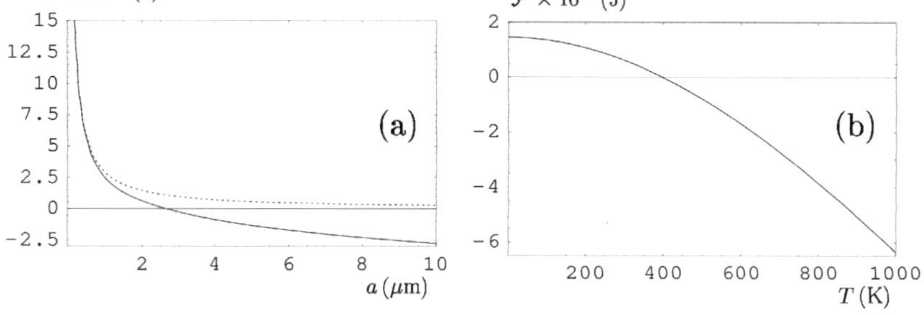

Figure 1. The electromagnetic Casimir free energy for a cube as a function of (a) size a at $T = 300\,\text{K}$ (solid line; the dashed line shows the energy at $T = 0$) and (b) temperature at $a = 2\,\mu$m.

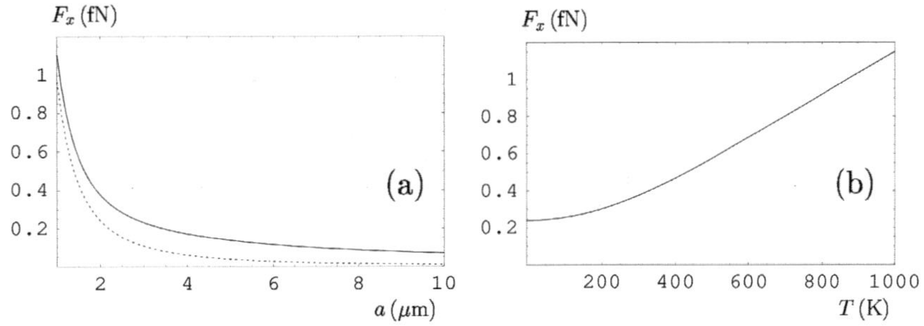

Figure 2. The electromagnetic Casimir force between the opposite faces of a cube as a function of (a) size a at $T = 300\,\mathrm{K}$ (solid line; the dashed line shows the force at $T = 0$) and (b) temperature at $a = 2\,\mu\mathrm{m}$.

energy decreases with the increase of T. At high temperatures $\mathcal{F}$ demonstrates the classical limit. The respective thermal electromagnetic Casimir force at $T = 300\,\mathrm{K}$, as a function of a, is shown in Fig. 2(a) by the solid line. It is positive (i.e., repulsive) for cubes of any size. Thus, thermal effects for cubes in the electromagnetic case increase the strength of the Casimir repulsion. The dashed line in Fig. 2(a) shows the electromagnetic Casimir force at $T = 0$ as a function of a. This force is given by

$$F_x(a) = \frac{0.09166}{3a^2}, \tag{20}$$

i.e., it is always repulsive. Fig. 2(b) demonstrates the electromagnetic Casimir force in a cube of size $a = 2\,\mu\mathrm{m}$ as a function of temperature. It is seen that the force increases with increasing temperature.

Note that the results presented differ from those found in [21] where the terms of order $(k_BT)^4$ and of lower orders in the Casimir free energy were obtained in the high-temperature regime. This is explained by the fact that the authors of [21] did not make subtractions of the contributions from the blackbody radiation and of the terms proportional to the box surface area and to the sum of its sides.

The thermal correction to the Casimir energy and force acting on a piston were investigated in [13] for the scalar field with Dirichlet or Neumann boundary conditions using the definition (10). The electromagnetic Casimir free energy and force acting on a piston were found in the case of ideal metal rectangular boxes and cavities with the general cross section [13]. In the limit of low temperatures the thermal correction to the Casimir force on a piston was shown to be exponentially small. In the case of medium temperature $a_x \ll \hbar c/(k_BT) \ll a_y, a_z$ the authors of [13] obtained terms of order $(k_BT)^4$ and of order $(k_BT)^2$ in the electromagnetic Casimir free energy. In the scalar Casimir free energy, a term of order $(k_BT)^3$ was also obtained. This results in the contribution to the force acting on a piston which increases with the increase of the temperature, depends on $\hbar$ and c and does not depend on the position of the piston. The scalar and electromagnetic thermal Casimir forces acting on a piston were also considered on the basis of equation (10) in [25].

The same results for the thermal correction to the Casimir force acting on a piston are obtained if the free energy is defined in accordance with equation (12). This is because the contribution of blackbody radiation to the energy of the entire box is equal to

$$-a_x a_y a_{z1}\, f_{bb} - a_x a_y (a_z - a_{z1}) f_{bb} = -a_x a_y a_z\, f_{bb}, \tag{21}$$

i.e., it does not depend on the position of the piston. This is also true for terms of order $(k_BT)^3$

and $(k_B T)^2$ which are proportional to the surface area of each section of the box and to the sum of its sides.

The above results were obtained for rectangular boxes with the Dirichlet boundary conditions (scalar case) and for ideal metal boxes (electromagnetic case). In the same way, as for zero temperature, the consideration of the thermal Casimir effect in rectangular boxes has to incorporate real material properties of the boundary surfaces. Till now this problem has not been conclusively solved.

3. Functional determinants and the justification of the proximity force approximation

The proximity force approximation [26] provides an important bridge between experiment and theory. Experimentally it is hard to use the configuration of two parallel plates. Because of this, most of experiments use the configuration of a sphere above a plate for which, even in the ideal metal case, the exact results for the electromagnetic Casimir force are not available. According to the proximity force approximation (PFA), the interaction energy between two curved surfaces Σ_1 and Σ_2 can be approximately calculated by replacing the small curved surface elements with respective plane plates. If the interaction energy between the opposite plane parallel elements is notated as $E(z)$ (where z is the separation distance), the interaction energy and force are approximately represented as

$$U(a) = \int_{\Sigma_1} E(z) d\sigma, \qquad F(a) = -\frac{\partial U(a)}{\partial a}. \tag{22}$$

For the configuration of an ideal metal sphere of radius R at a separation a above an ideal metal plane (22) results in

$$F^s_{\text{PFA}}(a) = 2\pi R E(a) = -\frac{\pi^3 \hbar c R}{360 a^3}. \tag{23}$$

For an ideal metal cylinder above an ideal metal plate the PFA leads to

$$F^c_{\text{PFA}}(a) = \frac{15\pi}{16}\sqrt{\frac{2R}{a}} E(a) = -\frac{\pi^3}{384\sqrt{2}}\sqrt{\frac{R}{a}}\frac{\hbar c}{a^3}. \tag{24}$$

Equations (22)–(24) are the approximate ones. They are applicable only at short separations between the surfaces. Thus, (23) and (24) work well only at $a \ll R$.

In many papers the PFA (22) is applied in a region where it is not applicable, for example at $a = R/2$. The obtained large deviations of the PFA result from the exact result are then considered as a "violation of the PFA". Such formulations are in fact misleading. The PFA gives only the main contribution to the force under some conditions. Specifically, it would be meaningless to calculate the integral in (22) up to higher orders in the related small parameter with the aim of obtaining a more exact result. What is really meaningful is the search of an exact analytical representation for the Casimir force in configurations where only the PFA result is so far available.

In the last few years the finite representation for the Casimir energy for two separated bodies A and B in terms of the functional determinants was obtained. In this representation the Casimir energy can be written in the form [27,28]

$$E(a) = \frac{1}{2\pi}\int_0^\infty d\xi \, \text{Tr} \ln(1 - \mathcal{T}^A \mathcal{G}^{(0)}_{\xi,AB} \mathcal{T}^B \mathcal{G}^{(0)}_{\xi,BA}) = \frac{1}{2\pi}\int_0^\infty d\xi \, \ln\det(1 - \mathcal{T}^A \mathcal{G}^{(0)}_{\xi,AB} \mathcal{T}^B \mathcal{G}^{(0)}_{\xi,BA}). \tag{25}$$

Here, $\mathcal{G}^{(0)}_{\xi,AB}$ is the operator for the free space Green function with the matrix elements $\langle \boldsymbol{r}|\mathcal{G}^{(0)}_{\xi,AB}|\boldsymbol{r}'\rangle$ where $\boldsymbol{r}$ belongs to the body A and $\boldsymbol{r}'$ to B. $\mathcal{T}^A$ ($\mathcal{T}^B$) is the operator of the T-matrix for a body A and B, respectively. The latter is widely used in light scattering theory, where it is the basic object for expressing the properties of the scatterers [29]. Using such a representation, in [28] the analytic results for the electromagnetic Casimir energy for an ideal metal cylinder above an ideal metal plane were obtained. Eventually, the result is expressed through the determinant of an infinite matrix with elements given in terms of the Bessel functions. The analytic asymptotic behavior of the exact Casimir energy at short separations was found in [30]. It results in the following expression for the Casimir force at $a \ll R$:

$$F^c(a,0) = F^c_{\rm PFA}(a)\left[1 - \frac{1}{5}\left(\frac{20}{\pi^2} - \frac{7}{12}\right)\frac{a}{R}\right]. \tag{26}$$

The PFA result (24) in this case matches with the first term on the right-hand side of (26).

Equation (26) is very important. It demonstrates that the relative error of the electromagnetic Casimir force between a cylinder and a plate calculated using the PFA is equal to $0.2886\,a/R$. Thus, for typical experimental parameters of $R = 100\,\mu$m and $a = 100\,$nm this error is approximately equal to only 0.03%.

For a sphere above a plate made of ideal metals the exact analytic solution in the electromagnetic case has not yet been obtained. The scalar Casimir energy for a sphere above a plate was found in [30,31]. The scalar Casimir energies for both a sphere and a cylinder above a plate have also been computed numerically using the wordline algorithms [32,33], but it was noted that the Casimir energies for the Dirichlet scalar field should not be taken as an estimate for those in the electromagnetic case. For an ideal metal sphere above an ideal metal plane a correction of order a/R beyond the PFA was computed numerically in [34] for $a/R \geq 0.075$ and in [35] for $a/R \geq 0.15$. In both cases the extrapolation of the obtained results to smaller a/R leads to a coefficient near a/R approximately equal to 1.4.

In addition, the validity of the PFA for a sphere above a plate has been estimated experimentally [36] and the error introduced from the use of this approximation was shown to be less than a/R. This is in disagreement with the extrapolations made in [34,35]. To solve this contradiction, it is desirable to find the analytical form of the first correction beyond the PFA for a sphere above a plane, like in (26) for the cylinder-plane configuration.

In fact the representation (25) provides a far-reaching generalization of the Lifshitz formula. From conceptual point of view it can be applied not only to ideal metals, but to real materials as well. The problem, however, is to find the matrix elements of the T-matrix operator which would take proper account of both geometric shape and material properties of the test bodies used in the experimental situation.

4. The experimental error and reliability of experiments

The concept of the experimental error is often confused with the theoretical error and with the measure of agreement between experiment and theory. However, when we deal with an *independent measurement*, the experimental error has nothing to do with any theory of the measured quantity. The independent measurement of the Casimir force or its gradient does not use any theory of the Casimir effect. Thus, the experiments [4,5,37–42] are independent in this respect. In other experiments (in [43], for instance) the measurement data are fitted to some theoretical expression for the Casimir force. Such kind of measurements are not independent and we do not consider them below.

Some papers arrive to theoretical conclusions which are inconsistent with the measurement data. This is sometimes surrounded by the statement that the measurements might be not as precise as indicated by the authors. It is our opinion that such statements made without an indication of any specific cause are inappropriate. Both random, $\Delta^{\rm rand}F^{\rm expt}(a)$, and systematic, $\Delta^{\rm syst}F^{\rm expt}(a)$, experimental errors in the Casimir force measurements are found using the

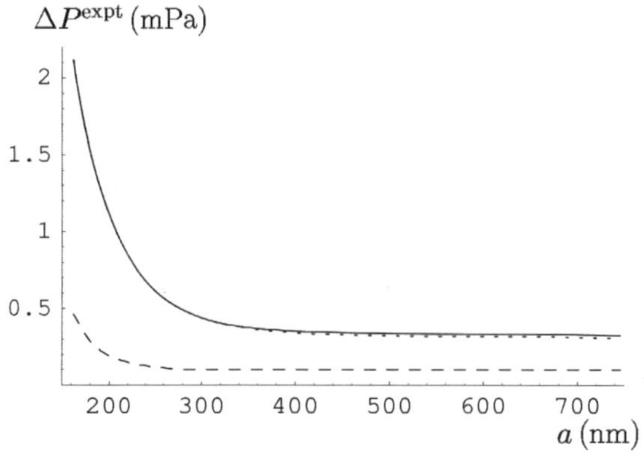

Figure 3. The total absolute experimental error of the Casimir pressure measurements [4,5] (the solid line), the random error (the long-dashed line), and the systematic error (the short-dashed line) are shown as functions of separation.

rigorous statistical procedures. They can be combined to find the total experimental error

$$\Delta^{\rm tot}F^{\rm expt}(a) = q_\beta(r)\left[\Delta^{\rm rand}F^{\rm expt}(a) + \Delta^{\rm syst}F^{\rm expt}(a)\right]. \qquad (27)$$

Here, $q_\beta(r)$ determined at $\beta = 0.95$ (i.e., at 95% confidence level) varies between 0.71 and 0.81 depending of the value of the quantity $r = \Delta^{\rm syst}F^{\rm expt}(a)/s_{\bar{F}}(a)$, where $s_{\bar{F}}(a)$ is the variance of the mean measured quantity [44]. In fact there is no arbitrariness in the determination of the total experimental error which is the ultimate characteristic of the precision of the measurements. The most valuable esperiments are marked by a negligible role of the random error. For such experiments

$$\Delta^{\rm tot}F^{\rm expt}(a) \approx \Delta^{\rm syst}F^{\rm expt}(a). \qquad (28)$$

For today there is only one indirect measurement of the Casimir pressure between Au coated plates by means of micromechanical torsional oscillator satisfying this condition [4,5]. The total experimental error in this measurement at shortest separations is as small as 0.2% of the measured Casimir pressure. We stress once again that this error is unrelated to much larger errors inherent to theoretical computations on the basis of the Lifshitz theory or to the measure of agreement between experiment and theory. This is just the resulting error with which the experimental data are taken. As an example, the total absolute experimental error in the experiment on measuring the Casimir pressure by means of a micromechanical torsional oscillator [4,5] is shown in Fig. 3 as a function of separation (the solid line). The long-dashed and short-dashed lines show the random and systematic errors, respectively. As a result, the relative total experimental error $\delta^{\rm tot}P^{\rm expt}(a) = \Delta^{\rm tot}P^{\rm expt}(a)/|P^{\rm expt}(a)|$ varies from 0.19% at $a = 162\,$nm to 0.9% at $a = 400\,$nm, and to 9.0% at $a = 746\,$nm.

Sometimes the experimental precision can be questioned if there are some doubts in the calibration procedures used. For example, the electrostatic calibration is of prime importance in the independnet measurements of the Casimir force. Specifically, it is usually carefully verified that the residual potential between the grounded test bodies does not depend on separation where the measurements of the electric force are performed. Recently it was claimed that the residual potential V_0 from the electrostatic calibration in the sphere-plate configuration is separation dependent [45]. The authors used an Au-coated sphere of 30.9 mm radius above an Au coated plate. On the basis of these measurements a reanalysis of the independence of V_0 on separation in the earlier measurements of the Casimir force by means of an atomic force microscope and a micromachined oscillator was invited. The results [45] are, however, not directly relevant to the earlier measurements. The point is that the radius of the sphere used in [45] is a factor of 300 larger than in the earlier precision measurements of the Casimir force. It

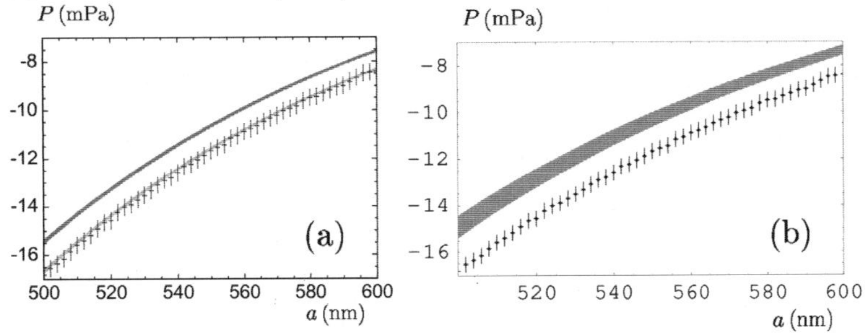

Figure 4. The crosses show the measured mean Casimir pressures together with the absolute errors in the separation and pressure as a function of the separation. (a) The theoretical Casimir pressures computed using the generalized plasma-like model and the optical data extrapolated by the Drude model are shown by the light-gray and dark-gray bands, respectively. (b) The theoretical Casimir pressures computed using different sets of optical data available in the literature versus separation are shown as the dark-gray band.

is well known that for large test bodies (i.e., large interaction areas) there are large variations of electric forces due to deviations of the mechanically polished and ground lens surface from perfect spherical shape [46].

5. Comparison between experiment and theory

Experiment is the supreme arbiter in physics. Because of this, the comparison between experiment and theory is a painful point for those theories that are found to be experimentally inconsistent. It happens that in such cases both the experimental data and the methods of comparison are questioned. The Casimir force is a strongly nonlinear function of the separation distance. As a consequence, such global characteristics of the agreement between experiment and theory as the root-mean-square deviation were found to be inadequate [47]. In the last few years two local methods on how to compare experiment with theory in the Casimir force measurements were elaborated and successfully applied. Within the first method [4,48,49], the experimental data are represented as crosses with arms determined by the total experimental errors in the measurement of separation and a related quantity (the force, the pressure or the frequency shift) determined at some chosen confidence level. In the same figure, one should plot the theoretical band whose width is equal to the total theoretical error determined at the same confidence as the experimental errors. The overlap (or its absence) of the experimental crosses and the theoretical band can be used to make a conclusion on the consistency or inconsistency between experiment and theory.

In Fig. 4 the first method of comparison between experiment and theory is illustrated on the measurement data by Decca et al. [4,5] discussed in Sec. 3. The light-gray band in Fig. 4(a) shows the theoretical results computed using the Lifshitz theory combined with the generalized plasma-like dielectric permittivity [50,51]

$$\varepsilon_{gp}(\mathrm{i}\xi) = \varepsilon(\mathrm{i}\xi) + \frac{\omega_p^2}{\xi^2}, \qquad \varepsilon(\mathrm{i}\xi) = 1 + \sum_{j=1}^{K} \frac{f_j}{\omega_j^2 + \xi^2 + \gamma_j \xi}. \tag{29}$$

Here, ω_p is the plasma frequency, $\omega_j \neq 0$ are the frequencies of the oscillators describing core electrons, f_j are the oscillator strengths and γ_j are the relaxation parameters. The dark-gray band in Fig. 4(a) is computed by the same Lifshitz theory using the tabulated optical data for Au [52] extrapolated to low frequencies by means of the Drude model [53–55]

$$\varepsilon_D(\mathrm{i}\xi) = 1 + \frac{\omega_p^2}{\xi(\xi+\gamma)} = 1 + \frac{4\pi\sigma(\mathrm{i}\xi)}{\xi}. \tag{30}$$

Here $\sigma(\mathrm{i}\xi)$ is the conductivity. It is connected with the dc conductivity by the equation

$$\sigma(\mathrm{i}\xi) = \frac{\sigma(0)}{1+\frac{\xi}{\gamma}}. \tag{31}$$

Note that the plasma frequency and the dc conductivity are expressed as [56]

$$\omega_p^2 = \frac{4\pi e^2 n}{m}, \qquad \sigma(0) = \mu\,|e|\,n, \tag{32}$$

where e and m are the charge and the mass of an electron, n is the charge carrier density and μ is their mobility. As is seen in Fig. 4(a), the experimental data shown as crosses (the experimental errors are determined at a 95% confidence level) are consistent with the theoretical approach using the generalized plasma-like permittivity. The Drude model approach is excluded at a 95% confidence level. In Fig. 4(b) the same experimental data are reproduced and compared with the Drude model approach using all sets of optical data available in the literature [57]. As is seen in Fig. 4(b), the use of optical data alternative to [52] makes the disagreement deeper between the experimental data and the Drude model approach. In Fig. 4(a,b) the comparison between experiment and theory is performed within the separation region from 500 to 600 nm. However, exactly the same conclusions follow over the entire measurement range in this experiment from 160 to 750 nm.

In the second method for the comparison of experiment and theory in the Casimir force measurements [40,48,58], the differences between the theoretical and mean experimental quantity, for instance, $P^{\text{theor}}(a) - \bar{P}^{\text{expt}}(a)$, are plotted as dots. In the same figure the borders of the confidence intervals $[-\Xi_P(a), \Xi_P(a)]$ for this difference at a chosen confidence level (usually 95%) are plotted as the function of separation. If no less than 95% of the dots representing the above differences belong to the confidence interval the theoretical approach is consistent with the data. Alternatively, if almost all the dots are outside the confidence interval, the theoretical approach is excluded by the data at a 95% confidence level. In Fig. 5 we illustrate the second method for the comparison of experiemnt with theory using the experimental data of the same measurements [4,5]. In Fig. 5(a) the theoretical approach using the generalized plasma-like permittivity (29) is compared with the data. It is seen that all dots are inside the

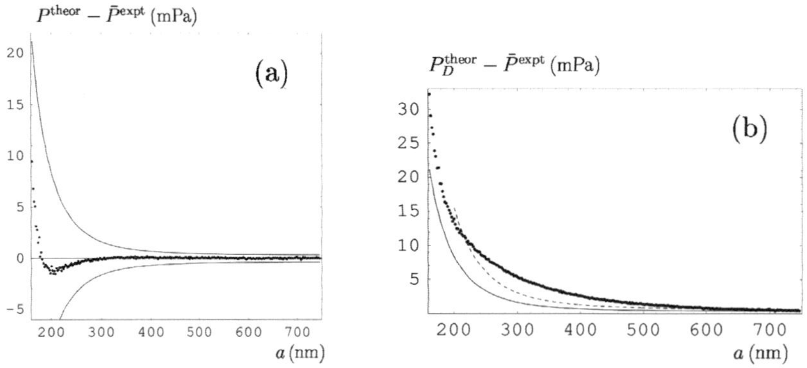

Figure 5. The differences of the theoretical and the mean experimental Casimir pressures between the two Au plates versus separation are shown as dots. The theoretical results are calculated using the Lifshitz theory at room temparature using (a) the generalized plasma-like model and (b) the Drude model approach. The solid lines indicate the boundaries of the 95% confidence intervals. The dashed line indicates the boundary of the 99.9% confidence intervals.

Journal of Physics: Conference Series **161** (2009) 012003 doi:10.1088/1742-6596/161/1/012003

error bars. Thus, this approach is consistent with the data. In Fig. 5(b) the same data are compared with the theoretical approach using the tabulated optical data extrapolated by the Drude model. The solid and dashed line represent the borders of 95% and 99.9% confidence intervals, respectively. As is seen in Fig. 5(b), the Drude model approach is experimentally excluded at a 95% confidence level within the entire measurement range from 160 to 750 nm. Within a more narrow measurement range from 210 to 620 nm the Drude model approach is excluded at a 99.9% confidence level.

If the theoretical approach is experimentally consistent [see Fig. 5(a)] the quantity $\Xi_P/|\bar{P}^{\rm expt}|$, determined at a 95% confidence level, can be used as the quantitative measure of agreement between experiment and theory. Thus, at $a = 162$ nm this measure is equal to 1.9%. It decreases to 1.4% at $a = 300$ nm and than gradually increases up to 9.7% at $a = 745$ nm. It is evident that at the shortest separation the agreement between experiment and theory is almost an order of magnitude worse than the total experimental error equal to only 0.19%. This is explained by large theoretical errors which dominate in the determination of Ξ_P at the shortest separations.

The above explanations aim to make absolutely clear that the calculation of errors and the comparison between experiment and theory is not an arbitrary, but a rigorously determined procedure. Recently, the measurement data of the experiment [4,5] was independently reanalyzed in [59] with the conclusion: "The data rule out the Drude approach..., while they are consistent with the plasma-model approach..."

6. Attempt to account for screening effects

Recently, the above discussed problems of the Drude model approach in application to real metals, and related problems arising for dielectric and semiconductor materials [60–64], motivated an attempt to modify the reflection coefficients in the Lifshitz formula by including the screening effects and diffusion currents [65,66]. The modified reflection coefficients for the transverse magnetic and transverse electric modes were obtained through use of Boltzmann transport equation which takes into account not only the standard drift current $\boldsymbol{j}$, but also the diffusion current $eD\nabla n$, where D is the diffusion coefficient and ∇n is the gradient of the charge carrier density [66]. The transverse magnetic coefficient takes the form

$$\tilde{r}_{\rm TM}(\mathrm{i}\xi,k_\perp) = \frac{\tilde{\varepsilon}(\mathrm{i}\xi)q - k - \frac{k_\perp^2}{\eta(\mathrm{i}\xi)}\frac{\tilde{\varepsilon}(\mathrm{i}\xi)-\varepsilon(\mathrm{i}\xi)}{\varepsilon(\mathrm{i}\xi)}}{\tilde{\varepsilon}(\mathrm{i}\xi)q + k + \frac{k_\perp^2}{\eta(\mathrm{i}\xi)}\frac{\tilde{\varepsilon}(\mathrm{i}\xi)-\varepsilon(\mathrm{i}\xi)}{\varepsilon(\mathrm{i}\xi)}}, \tag{33}$$

where $k_\perp$ is the projection of the wave vector in the plane of the plates, $\omega = \mathrm{i}\xi$ is the imaginary frequency and the following notations are introduced

$$q^2 = k_\perp^2 + \frac{\xi^2}{c^2}, \qquad k^2 = k_\perp^2 + \tilde{\varepsilon}(\mathrm{i}\xi)\frac{\xi^2}{c^2}, \qquad \tilde{\varepsilon}(\mathrm{i}\xi) = \varepsilon(\mathrm{i}\xi) + \frac{\omega_p^2}{\xi(\xi+\gamma)},$$
$$\eta(\mathrm{i}\xi) = \left[k_\perp^2 + \kappa^2\frac{\varepsilon(0)}{\varepsilon(\mathrm{i}\xi)}\,\frac{\tilde{\varepsilon}(\mathrm{i}\xi)}{\tilde{\varepsilon}(\mathrm{i}\xi)-\varepsilon(\mathrm{i}\xi)}\right]^{1/2}. \tag{34}$$

In this equation, $1/\kappa$ is the screening length and the dielectric permittivity of core electrons $\varepsilon(\mathrm{i}\xi)$ is defined in (29). The transverse electric coefficient is given by the standard expression

$$\tilde{r}_{\rm TE}(\mathrm{i}\xi,k_\perp) = \frac{q-k}{q+k}, \tag{35}$$

as is used in the Drude model approach.

The paper [66] claims the application of the above approach to intrinsic semiconductors only. It uses a specific Debye-Hückel expression for the screening length

$$\frac{1}{\kappa} = \frac{1}{\kappa_{\mathrm{DH}}} = R_{\mathrm{DH}} = \sqrt{\frac{\varepsilon(0) k_B T}{4\pi e^2 n}}. \tag{36}$$

This expression is applicable to particles obeying the Maxwell-Boltzmann statistics. It is obtained from the general representation for the screening length [67]

$$\frac{1}{\kappa} = R = \sqrt{\frac{\varepsilon(0) D}{4\pi\sigma(0)}} \tag{37}$$

if one uses the expression (32) for the dc conductivity and Einstein's relation [56,67]

$$\frac{D}{\mu} = \frac{k_B T}{|e|} \tag{38}$$

valid in the case of Maxwell-Boltzmann statistics. In the limiting case $\xi \to 0$ the reflection coefficient (33) coincides with that obtained in [65].

However, the application region of the reflection coefficients (33), (34) with the Debye-Hückel screening length (36) cannot be restricted to only intrinsic semiconductors. These coefficients should be applicable to all materials where the density of charge carriers is not too large so that they are described by Maxwell-Boltzmann statistics. This means that in the framework of the proposed approach it is legal to apply (33)–(36) to doped semiconductors with dopant concentration below critical and to solids with ionic conductivity etc.

Here, we consider the application of this approach to metallic plates. Metals and semiconductors of metallic type are characterized by rather high concentration of charge carriers which obey the quantum Fermi-Dirac statistics. The general transport equation, however, is equally applicable to classical and quantum systems. The only difference one should take into account is the type of statistics. Substituting Einstein's relation, valid in the case of Fermi-Dirac statistics [56,67]

$$\frac{D}{\mu} = \frac{2E_F}{3|e|}, \tag{39}$$

where $E_F = \hbar\omega_p$ is the Fermi energy, into (37), one arrives to the following expression for the Thomas-Fermi screening length [67]

$$\frac{1}{\kappa} = \frac{1}{\kappa_{\mathrm{TF}}} = R_{\mathrm{TF}} = \sqrt{\frac{\varepsilon(0) E_F}{6\pi e^2 n}}. \tag{40}$$

With this definition of the parameter κ, it is legal to apply equations (33)–(35) to metals.

Now we consider two thick metallic plates separated by a distance a at temperature T in thermal equilibrium. Under these conditions the Casimir free energy per unit area of the plates is given by the Lifshitz formula [68]. Let us assume that the reflection coefficients (33)–(35), (40) can be substituted into this formula. Then in terms of dimensionless variables $y = 2aq$, $\zeta = \xi/\omega_c \equiv 2a\xi/c$ one obtains

$$\tilde{\mathcal{F}}(a,T) = \frac{k_B T}{8\pi a^2} \sum_{l=0}^{\infty}{}' \int_{\zeta_l}^{\infty} y\,dy \left\{ \ln\left[1 - \tilde{r}_{\mathrm{TM}}^2(\mathrm{i}\zeta_l, y)\,\mathrm{e}^{-y}\right] + \ln\left[1 - \tilde{r}_{\mathrm{TE}}^2(\mathrm{i}\zeta_l, y)\,\mathrm{e}^{-y}\right] \right\}, \tag{41}$$

where $\zeta_l = 4\pi a k_B T l/(\hbar c)$ are the dimensionless Matsubara frequencies and a prime near the summation sign adds a multiple 1/2 to the term with $l = 0$. In terms of the dimensionless variables the reflection coefficient (33) takes the form

$$\tilde{r}_{\rm TM}({\rm i}\zeta, y) = \frac{\tilde{\varepsilon}y - [y^2 + (\tilde{\varepsilon}-1)\zeta^2]^{1/2} - \frac{(y^2-\zeta^2)(\tilde{\varepsilon}-\varepsilon)}{\tilde{\eta}\varepsilon}}{\tilde{\varepsilon}y + [y^2 + (\tilde{\varepsilon}-1)\zeta^2]^{1/2} + \frac{(y^2-\zeta^2)(\tilde{\varepsilon}-\varepsilon)}{\tilde{\eta}\varepsilon}}, \tag{42}$$

where

$$\tilde{\eta} = 2a\eta = \left[y^2 - \zeta^2 + \kappa_a^2 \frac{\varepsilon(0)\tilde{\varepsilon}}{\varepsilon(\tilde{\varepsilon}-\varepsilon)}\right]^{1/2}, \qquad \kappa_a \equiv 2a\kappa_{\rm TF}. \tag{43}$$

Note that all dielectric permittivities here are functions of ${\rm i}\omega_c\zeta$. Below we do not use the explicit expression for the reflection coefficient $\tilde{r}_{\rm TE}({\rm i}\zeta, y)$ because it coincides with the standard one, as defined in the Drude model approach, and considered in detail in [69].

Let us determine the behavior of the Casimir free energy (41) at low temperature. For all metals the screening length (40) is very small. As a result, at any reasonable separation distance between the plates, the dimensionless parameter κ_a defined in (43) is very large and the inverse quantity can be used as a small parameter

$$2a\kappa_{\rm TF} = \kappa_a \gg 1, \qquad \beta_a \equiv \frac{1}{\kappa_a} \ll 1. \tag{44}$$

Expanding the reflection coefficient (42) up to the first power of the parameter β_a one obtains

$$\tilde{r}_{\rm TM}({\rm i}\zeta, y) = r_{\rm TM}({\rm i}\zeta, y) - 2\beta_a Z + O(\beta_a^2), \tag{45}$$
$$Z \equiv \sqrt{\frac{\tilde{\varepsilon}(\tilde{\varepsilon}-\varepsilon)^3}{\varepsilon(0)\varepsilon}}\,\frac{y(y^2-\zeta^2)}{[\tilde{\varepsilon}y + \sqrt{y^2+(\tilde{\varepsilon}-1)\zeta^2}]^2},$$

where $r_{\rm TM}({\rm i}\zeta, y)$ is the standard TM reflection coefficient calculated with the dielectric permittivity $\tilde{\varepsilon}({\rm i}\omega_c\zeta)$ [it is given by (42) with the third term in both numerator and denominator omitted]. From (45) one arrives at

$$\ln\left[1 - \tilde{r}_{\rm TM}^2({\rm i}\zeta, y)\,{\rm e}^{-y}\right] = \ln\left[1 - r_{\rm TM}^2({\rm i}\zeta, y)\,{\rm e}^{-y}\right] + 4\beta_a \frac{r_{\rm TM}({\rm i}\zeta, y)\,Z}{{\rm e}^y - r_{\rm TM}^2({\rm i}\zeta, y)} + O(\beta_a^2). \tag{46}$$

Now we substitute (46) and the respective known expression for the TE contribution [69] into (41). Calculating the sum with the help of the Abel-Plana formula, we obtain in perfect analogy to [69]

$$\tilde{\mathcal{F}}(a,T) = \mathcal{F}_{gp}(a,T) - \frac{k_B T}{16\pi a^2}\int_0^\infty y\,dy\,\ln\left[1 - r_{{\rm TE},gp}^2(0,y)\,{\rm e}^{-y}\right] + \mathcal{F}^{(\gamma)}(a,T) + \beta_a \mathcal{F}^{(\beta)}(a,T), \tag{47}$$

where $\mathcal{F}^{(\gamma)}(a,T)$ is determined by equation (17) in [69]. It goes to zero together with its derivative with respect to temperature when $T \to 0$. The quantity $\mathcal{F}^{(\beta)}(a,T)$ originates from the second contribution on the right-hand side of (45). It is easily seen that $\mathcal{F}^{(\beta)}(a,T) = E^{(\beta)}(a) + O(T^3/T_{\rm eff}^3)$ at low T. The Casimir free energy $\mathcal{F}_{gp}(a,T)$ is defined by substituting the dielectric permittivity (29) of the generalized plasma-like model into the Lifshitz formula. It was found in [50,51] and the respective thermal correction was shown to be of order $(T/T_{\rm eff})^3$ when $T \to 0$. The TE reflection coefficient at zero frequency entering (47) is given by

$$r_{{\rm TE},gp}(0,y) = \frac{cy - \sqrt{4a^2\omega_p^2 + c^2y^2}}{cy + \sqrt{4a^2\omega_p^2 + c^2y^2}}. \tag{48}$$

As a result, calculating the Casimir entropy

$$\tilde{S}(a,T) = -\frac{\partial\tilde{\mathcal{F}}(a,T)}{\partial T} \qquad (49)$$

with the use of (47) and considering the limiting case of zero temperature, one arrives at

$$\tilde{S}(a,0) = \frac{k_B}{16\pi a^2}\int_0^\infty y\,dy\,\ln\left[1-\left(\frac{cy-\sqrt{4a^2\omega_p^2+c^2y^2}}{cy+\sqrt{4a^2\omega_p^2+c^2y^2}}\right)^2 \mathrm{e}^{-y}\right] < 0 \qquad (50)$$

in violation of the Nernst heat theorem. This result is obtained for metals with perfect crystal lattices. In the presence of impurities the Casimir entropy abruptly jumps to zero at $T < 10^{-3}\,$K [70].

Thus, the modified reflection coefficients taking the screening effects into account lead to a violation of the Nernst heat theorem for metals with perfect crystal lattices in the same way as the standard Drude model approach. Because of this, the theoretical approach using such reflection coefficients is thermodynamically inconsistent.

Now we briefly compare the theoretical predictions, following from the use of reflection coefficients $\tilde{r}_{\rm TM}$ and $\tilde{r}_{\rm TE}$, with the measurement data of the most precise experiment by means of micromachined torsional oscillator [4,5]. This experiment was already discussed in Sec. 5. In Fig. 6(a) the experimental data for the Casimir pressure between two Au plates are shown as crosses with the absolute errors determined at a 95% confidence level. The solid line presents the computational results for $P(a,T) = -\partial\mathcal{F}(a,T)/\partial a$ using the Lifshitz formula and the generalized plasma-like dielectric permittivity (29). The parameters of oscillators for Au were determined in [5] with high precision. The dashed line was computed using the Lifshitz formula for $P^{\rm mod}(a,T) = -\partial\tilde{\mathcal{F}}(a,T)/\partial a$ with the reflection coefficients $\tilde{r}_{\rm TM,TE}$ taking the screening effects into account. As is seen in Fig. 6(a), the theoretical approach taking into account the Thomas-Fermi screening length is experimentally excluded at a 95% confidence level over the separation region from 500 to 600 nm. The same conclusion follows within the entire measurement range from 160 to 750 nm.

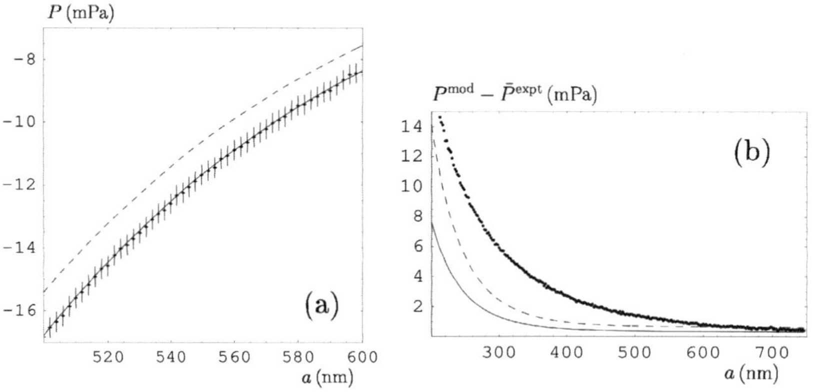

Figure 6. (a) The crosses show the measured mean Casimir pressures together with the absolute errors as a function of the separation. The theoretical Casimir pressures computed using the generalized plasma-like model and the approach including the screening effects are shown as solid and dashed lines, respectively. (b) Differences of the theoretical Casimir pressures computed with inclusion of the screening effects and the mean experimental Casimir pressures versus separation are shown as dots. The 95% and 99.9% confidence intervals are shown as the solid and dashed lines, respectively.

Fig. 6(a) illustrates the first method for the comparison between experiment and theory in Casimir force measurements discussed in Sec. 5. In Fig. 6(b) the second method for the comparison of experiment and theory is illustrated. Here, the differences between the theoretical Casimir pressures computed with inclusion of the screening effects and the mean experimental pressures are shown as dots. The solid line indicates the borders of 95% confidence intervals. Dots are outside the confidence interval $[-\Xi_P(a), \Xi_P(a)]$ over the entire measurement range from 160 to 750 nm. In the same figure, the dashed line shows the borders of 99.9% confidence intervals. As is seen in Fig. 6(b), dots are outside of this confidence interval within the separation region from 160 to 640 nm. Thus, within this region of separations the theoretical approach taking the screening effects into account [66] is experimentally excluded at a 99.9% confidence level.

The physical reasons why the inclusion of the screening effects into the Lifshitz theory is thermodynamically and experimentally inconsistent can be understood as follows. The Lifshitz theory is formulated for systems in thermal equilibrium. As was indicated in [71], the drift current of conduction electrons leads to heating of the crystal lattice. In this case, if the constant temperature is preserved, there must be a unidirectional flux of heat from the Casimir plates to the heat reservoir. The existence of such an interaction between a system and a heat reservoir is strictly prohibited in a state of thermal equilibrium [72] and is in contradiction with its definition [73]. According to this definition, in thermal equilibrium all irreversible processes connected with the dissipation of energy are terminated. Specifically, in thermal equilibrium any nonzero gradients of charge carrier density and any diffusion are impossible. Thus, the inclusion of the screening effects and diffusion currents into the Lifshitz theory is in violation of its applicability conditions.

7. Conclusions and discussion

In the above, we have discussed several problems at the interface between field-theoretical description of the Casimir effect and experiments on measuring the Casimir force. The consideration of the Casimir energies and forces in ideal metal rectangular boxes leads to the conclusion that even when using ideal models it is important to take into account some general physical requirements. Thus, it is not productive to use the free energy which leads to the Casimir forces of quantum nature which increase with increasing size of the box. It also seems thermodynamically inconsistent to claim that the Casimir force acting on a piston is a well defined quantity, whereas the forces acting on all other faces of the box are excluded from consideration. The reason is that if the free energy is defined correctly (see Sec. 2), there is a uniquely defined pressure on all faces of the box equal to the negative derivative of the free energy with respect to the box volume calculated at constant temperature.

An important tool for the comparison of experiment with theory is the proximity force approximation. In Sec. 3 we have discussed some inexact formulations which can be found in theoretical publications on this subject. We have also discussed recent achievements in quantum-field-theoretical approach to the calculation of the Casimir energies in terms of functional determinants and scattering matrices. This scientific direction has already obtained the first analytical results beyond the PFA. It is of great promise for many experimentally relevant applications of the theory.

In Secs. 4 and 5 we tried to add clarity to the widely discussed problems of the precision of experiments on the Casimir effect and the agreement between experiment and theory. It was stressed that the precision of some independent measurements can be much higher than of respective theoretical computations using the values of parameters which may not be known precisely enough. In such cases the agreement of experiment with theory can also be not as good as the precision of the measurements.

Finally, in Sec. 6 we have analyzed a recent theoretical approach to the thermal Casimir force taking into account the screening effects and diffusion currents. Using quantum Fermi-Dirac statistics and respective Thomas-Fermi screening length, we have applied this approach to calculate the Casimir free energy between two metallic plates. It was shown that the obtained free energy results in a violation of the Nernst heat theorem for metals with perfect crystal lattices. Thus, the approach under consideration is inconsistent with thermodynamics. The calculational results for the Casimir pressure in the configuration of two Au plates were compared with the results of the most precise experiment performed using a micromachined oscillator. It was shown that the theoretical predictions following from the inclusion of the screening effects are rejected by the experimental data at a 99.9% confidence level. The reason for the failure of this approach is the inclusion of irreversible diffusion processes violating thermal equilibrium which is the basic applicability condition of the Lifshitz theory.

Phenomenologically, the Lifshitz theory combined with the generalized plasma-like dielectric permittivity provides a description of dispersion forces between metallic test bodies which is in agreement with thermodynamics and consistent with all available experimental information. For now there is no other theoretical approach to the thermal Casimir force between metals which would satisfy the requirements of thermodynamics and be simultaneously consistent with all measurement data.

Acknowledgments

This work was supported by Deutsche Forschungsgemeinschaft, Grant No 436 RUS 113/789/0–4. The author is grateful to the Center of Theoretical Studies and Institute of Theoretical Physics, Leipzig University where this work was performed for kind hospitality.

[1] Casimir H B G 1948 *Proc. K. Ned. Akad. Wet.* **51** 793
[2] Sparnaay M J 1958 *Physica* **24** 751
[3] van Blockland P H G M and Overbeek J T G 1978 *J. Chem. Soc. Faraday Trans.* **74** 2637
[4] Decca R S, López D, Fischbach E, Klimchitskaya G L, Krause D E and Mostepanenko V M 2007 *Phys. Rev.* D **75** 077101
[5] Decca R S, López D, Fischbach E, Klimchitskaya G L, Krause D E and Mostepanenko V M 2007 *Eur. Phys. J.* C **51** 963
[6] Lukosz W 1971 *Physica* **56** 109
[7] Mamayev S G and Trunov N N 1979 *Teor. Matem. Fiz.* **38** 345 (*Theor. Math. Phys.* **38** 228)
[8] Mamayev S G and Trunov N N 1979 *Izv. Vuzov, Fizika* N9 51 (*Rus. Phys. J.* **22** 966)
[9] Ambjørn J and Wolfram S 1983 *Ann. Phys., N.Y.* **147** 1
[10] Bordag M, Mohideen U and Mostepanenko V M 2001 *Phys. Rep.* **353** 1
[11] Cavalcanti R M 2004 Phys. Rev. D **69** 065015
[12] Hertzberg M P, Jaffe R L, Kardar M and Scardicchio A 2005 *Phys. Rev. Lett.* **95** 250402
[13] Hertzberg M P, Jaffe R L, Kardar M and Scardicchio A 2007 *Phys. Rev.* D **76** 045016
[14] Edery A 2007 *Phys. Rev.* D **75** 105012
[15] Zhai X-H and Li X-Z 2007 *Phys. Rev.* D **76** 047704
[16] Charlton T M 1973 *Energy Principles in Theory of Structures* (Oxford: Oxford University Press)
[17] Langhaar H L 1989 *Energy Methods in Applied Mechanics* (Brooklyn: Krieger Press)
[18] Kubo R 1968 *Thermodynamics* (Amsterdam: North-Holland Publishing Company)
[19] Santos F C and Tort A 2000 *Phys. Lett.* B **482** 323
[20] Jáuregui R, Villarreal C and Hacyan S 2006 *Ann. Phys., NY* **321** 2156
[21] Lim S C and Teo L P 2007 *J. Phys. A: Math. Theor.* **40** 11645
[22] Dowker J S and Kennedy G 1978 *J. Phys. A: Math. Gen.* **11** 895
[23] Geyer B, Klimchitskaya G L and Mostepanenko V M 2008 *Europ. Phys. J.* C, **57** 823
[24] Brown L S and Maclay G J 1969 *Phys. Rev.* **184** 1272
[25] Lim S C and Teo L P 2008 Casimir piston at zero and finite temperature. *Preprint* arXiv:0808.0047
[26] Blocki J, Randrup J., Swiatecki W J and Tsang C F 1977 *Ann. Phys., NY* **105** 427.
[27] Kenneth O and Klich I 2006, *Phys. Rev. Lett.* **97** 0160401
[28] Emig T, Jaffe R L, Kardar M and Scardicchio A 2006 *Phys. Rev. Lett.* **96** 080403
[29] Bohren C F and Huffmann D R 1998 *Absorption and Scattering of Light by Small Particles* (New York: Wiley)
[30] Bordag M 2006 *Phys. Rev.* D **73** 125018

[31] Bulgac A, Magierski P and Wirzba A 2006 *Phys. Rev.* D **73** 025007
[32] Gies H and Klingmüller K 2006 *Phys. Rev. Lett.* **96** 220401
[33] Gies H and Klingmüller K 2006 *Phys. Rev.* D **74** 045002
[34] Emig T 2008 *J. Stat. Mech.* P04007
[35] Maia Neto P A, Lambrecht A and Reynaud S 2008 *Phys. Rev.* A **78** 012115
[36] Krause D E, Decca R S, López D and Fischbach E 2007 *Phys. Rev. Lett.* **98** 050403
[37] Harris B W, Chen F and Mohideen U 2000 *Phys. Rev.* A **62** 052109
[38] Chen F, Klimchitskaya G L, Mohideen U and Mostepanenko V M 2004 *Phys. Rev.* A **69** 022117
[39] Decca R S, Fischbach E, Klimchitskaya G L, Krause D E, López D and Mostepanenko V M 2003 *Phys. Rev.* D **68** 116003
[40] Decca R S, López D, Fischbach E, Klimchitskaya G L, Krause D E and Mostepanenko V M 2005 *Ann. Phys. NY* **318** 37
[41] Jourdan G, Lambrecht A, Comin F and Chevrier J 2009 *Europhys. Lett.* **85** 31001
[42] Chan H B, Bao Y, Zou J, Cirelli R A, Clemens F, Mansfield W M and Pai C S 2008 *Phys. Rev. Lett.* **101** 030401
[43] Bressi G, Carugno G, Onofrio R. and Ruoso G 2002 *Phys. Rev. Lett.* **88** 041804
[44] Rabinovich S G 2000 *Measurement Errors and Uncertainties. Theory and Practice* (New York: Springer-Verlag)
[45] Kim W J, Brown-Hayes M, Dalvit D A R, Brownell J H and Onofrio R 2008 *Phys. Rev.* A **78** 020101(R)
[46] Decca R S, Fischbach E, Klimchitskaya G L, Krause D E, López D, Mohideen U and Mostepanenko V M 2009 *Phys. Rev. A* **79** 026101
[47] Ederth T 2000 *Phys. Rev.* A **62** 062104
[48] Chen F, Mohideen U, Klimchitskaya G L and Mostepanenko V M 2006 *Phys. Rev.* A **74** 022103
[49] Obrecht J M, Wild R J, Antezza M, Pitaevskii L P, Stringari S and Cornell E A 2007 *Phys. Rev. Lett.* **98** 063201
[50] Geyer B, Klimchitskaya G L and Mostepanenko V M 2007 *J. Phys. A: Math. Theor.* **40** 13485
[51] Mostepanenko V M and Geyer B 2008 *J. Phys. A: Mat. Theor.* **41** 164014
[52] Palik E D (ed) 1985 *Handbook of Optical Constants of Solids* (New York: Academic)
[53] Boström M and Sernelius B E 2000 *Phys. Rev. Lett.* **84** 4757
[54] Brevik I, Aarseth J B, Høye J S and Milton K A 2005 *Phys. Rev.* E **71** 056101
[55] Milton K A 2004 *J. Phys. A: Math. Gen.* **37** R209
[56] Ashcroft N W and Mermin N D 1976 *Solid State Physics* (Philadelphia: Saunders College)
[57] Pirozhenko I, Lambrecht A and Svetovoy V B 2006 *New J. Phys.* **8** 238
[58] Chen F, Klimchitskaya G L, Mostepanenko V M and Mohideen U 2006 *Phys. Rev. Lett.* **97** 170402
[59] Bimonte G 2007 The thermal Casimir effect for rough metallic plates *Preprint* arXiv:0711.0278
[60] Geyer B, Klimchitskaya G L and Mostepanenko V M 2005 *Phys. Rev.* D **72** 085009
[61] Geyer B, Klimchitskaya G L and Mostepanenko V M 2006 *Int. J. Mod. Phys.* A **21** 5007
[62] Geyer B, Klimchitskaya G L and Mostepanenko V M 2008 *Ann. Phys. NY* **323** 291
[63] Klimchitskaya G L and Geyer B 2008 *J. Phys. A.: Mat. Theor.* **41** 164032
[64] Ellingsen S A, Brevik I, Høye J S and Milton K A 2008 *Phys. Rev.* E **78** 021117
[65] Pitaevskii L P 2008 *Phys. Rev. Lett.* **101** 163202
[66] Dalvit D A R and Lamoreaux S K 2008 *Phys. Rev. Lett.* **101** 163203
[67] Chazalviel J-N 1999 *Coulomb Screening of Mobile Charges: Applications to Material Science, Chemistry and Biology* (Boston: Birkhauser)
[68] Lifshitz E M 1956 *Zh. Eksp. Teor. Fiz.* **29** 94 (*Sov. Phys. JETP* **2** 73)
[69] Bezerra V B, Klimchitskaya G L, Mostepanenko V M and Romero C 2004 *Phys. Rev.* A **69** 022119
[70] Høye J S, Brevik I, Ellingsen S A and Aarseth J B 2007 *Phys. Rev.* E **75** 051127
[71] Geyer B, Klimchitskaya G L and Mostepanenko V M 2003 *Phys. Rev. A* **67** 062102
[72] Bryksin V V and Petrov M P 2008 *Fiz. Tverdogo Tela* **50** 222 (*Phys. Solid State* **50** 229)
[73] Kondepugi D and Prigogine I 1998 *Modern Thermodynamics* (New York: Wiley)

Journal of Physics: Conference Series **161** (2009) 012004 doi:10.1088/1742-6596/161/1/012004

On electrostatic and Casimir force measurements between conducting surfaces in a sphere-plane configuration

W J Kim[1*], M Brown-Hayes[1], D A R Dalvit[2], J H Brownell[1], and R Onofrio[3,1]

[1] Department of Physics and Astronomy, Dartmouth College, 6127 Wilder Laboratory, Hanover, NH 03755, USA

[2] Theoretical Division, MS B213, Los Alamos National Laboratory, Los Alamos, NM 87545, USA

[3] Dipartimento di Fisica "Galileo Galilei", Università di Padova, Via Marzolo 8, Padova 35131, Italy

E-mail: `onofrio@pd.infn.it` and `roberto.onofrio@dartmouth.edu`

Abstract. We report on measurements of forces acting between two conducting surfaces in a spherical-plane configuration in the 35 nm-1 μm separation range. The measurements are obtained by performing electrostatic calibrations followed by a residuals analysis after subtracting the electrostatic-dependent component. We find in all runs optimal fitting of the calibrations for exponents smaller than the one predicted by electrostatics for an ideal sphere-plane geometry. We also find that the external bias potential necessary to minimize the electrostatic contribution depends on the sphere-plane distance. In spite of these anomalies, by implementing a parametrization-dependent subtraction of the electrostatic contribution we have found evidence for short-distance attractive forces of magnitude comparable to the expected Casimir-Lifshitz force. We finally discuss the relevance of our findings in the more general context of Casimir-Lifshitz force measurements, with particular regard to the critical issues of the electrical and geometrical characterization of the involved surfaces.

PACS numbers: 12.20.Fv, 03.70.+k, 04.80.Cc, 11.10.Wx

1. Introduction

Casimir-Lifshitz forces [1, 2] provide an experimental window on the nature of quantum vacuum at the macroscopic level [3, 4, 5, 6, 7, 8, 9, 10, 11]. After pioneering attempts in the original parallel plate configuration [12], and in a variant of this configuration based upon a sphere and a plane [13], the claimed accuracy of recent experiments ranges from 15% in the parallel plane case [14] to 0.2-5% in the sphere-plane case [15, 16, 17, 18, 19]. In the case of the sphere-plane configuration, the Casimir force is typically evaluated by using the so-called proximity force approximation (PFA) [20, 21], introducing a theoretical uncertainty estimated to be in the 0.1 % range, and recently investigated experimentally [22].

A sphere-plane experiment has been performed, with a radius of curvature for the sphere of order of cm, at relatively large distances above 1μm [15]. In this case the largest deviation from the bare Casimir formula for an idealized configuration (perfect conductors, zero temperature) is expected to be the thermal contribution to the radiation pressure on the surfaces, still below the sensitivity of the apparatus at the largest explored distances. Three sphere-plane experiments have also been performed in a quite distinct regime, with smaller radius of curvature, of order 100 μm, and in the range below one micrometer [16, 17, 18, 23]. At distances less than 1 μm the correction to the Casimir force due to finite conductivity and surface roughness of the substrates cannot be neglected at the level of accuracy claimed in these three experiments. At least other five groups have recently performed Casimir force experiments in the sphere-plane configuration [24, 25, 26, 27, 28].

Mastering the Casimir force at the highest level of accuracy is essential to provide reliable limits to other macroscopic forces acting in the micrometer range, such as corrections to the Newtonian gravitational force independently predicted in many attempts to unify gravitation with the remaining fundamental interactions [29, 30, 31]. Although in the nanometer range the Casimir force loses its universal nature morphing into more specific, structure-dependent, molecular van der Waals forces, its study is sometimes considered of some relevance for designing nanomechanical structures and to investigate nonlinear effects, as first discussed in [32] and then experimentally demonstrated in [33, 34].

In this paper we report more extensively on measurements performed in a sphere-plane geometry exploring a novel range of parameters that interpolates between the two previous sets of sphere-plane measurements [35]. We use a gold-coated spherical lens with large radius of curvature, similar to the experiment performed by Steve Lamoreaux in Seattle, while at the same time exploring distances down to few tens of nanometers from the point of contact between the sphere and the plane, similar to more microscopic setups using microresonators. While we do achieve some evidence for the expected Casimir-Lifshitz force, our measurements suggest that a reanalysis of systematic effects in previous experiments will be beneficial to assess the accuracy with which this force has been measured so far, with particular regard to the role of the residual electrostatic forces [36, 37].

The paper is organized as follows: in Section II we describe the experimental apparatus and the basics of our measurement technique. In Section III we describe a procedure for electrostatic calibrations to determine various critical parameters such as the calibration factor, the absolute separation, and the minimizing potential related to the contact potential. Careful control over the system parameters allows us to search for a non-Coulombian contribution to the observed force signal and therefore to test this residual against various hypothesis such as a possible uncompensated electrical voltage, and the presence of the Casimir-Lifsthiz force, as described in Section IV. In Section V we critically assess systematic effects in our measurements and compare our findings to both previous short-range Casimir force measurements and long-range atomic force microscopy measurements.

2. Experimental set-up

The experimental setup is an upgrade of the one already described in a previous paper [38] for measuring the Casimir force in the cylinder-plane configuration as first theoretically discussed in [39]. Forces acting between a flat surface and a spherical surface are detected by measuring the shift in the mechanical oscillation frequency of a silicon cantilever that acts as the plane surface. The main components in this setup are the use of a phase-locked loop scheme [40] to drive the cantilever at its resonant frequency, the use of optical quality surfaces for both the resonator and the sphere, and a faster and more flexible data acquisition scheme.

A schematic of the apparatus is shown in Fig. 1. The rectangular cantilever, of length $L = (22.56 \pm 0.01)$ mm, width $w = (9.93 \pm 0.01)$ mm and thickness $t = (330 \pm 10)$ μm was laser-cut from a silicon wafer. Given the density of silicon, $\rho = 2.3 \times 10^3$ kg/m^3, the physical mass of the cantilever is $m_p = (1.72 \pm 0.05) \times 10^{-4}$ kg. The cantilever is sandwiched between two aluminum clamping structures, electrically isolated and thermally stabilized by a Peltier cooler.

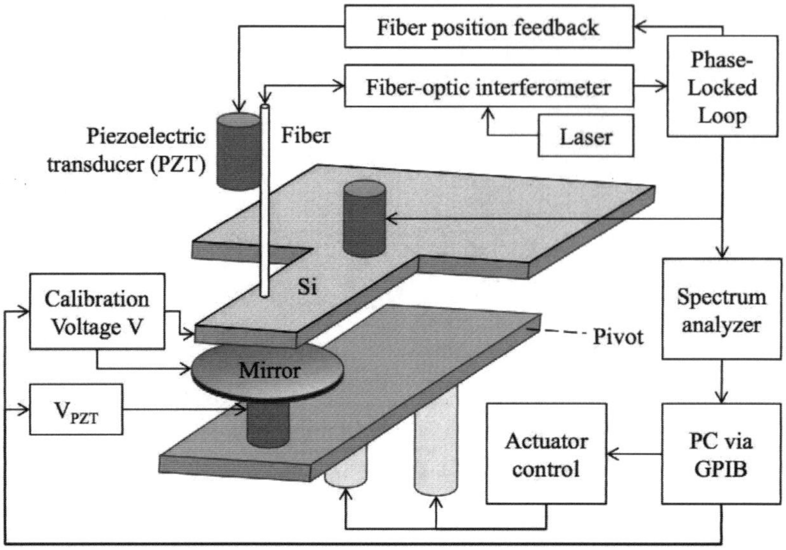

Figure 1. Schematic of the experimental setup and the data acquisition scheme. The silicon resonator is opposed by a spherical mirror, both Au coated. The latter can be moved along the vertical direction by driving two mechanical actuators for coarse approach, and with a piezoelectric actuator for fine control of the separation. The outside face of the resonator reflects light coming from a fiber optic interferometer, located 20-100 μm above it. The resonator is driven weakly at its resonant frequency with a piezoelectric actuator connected in a phase-locked loop (PLL) to the optical fiber.

This last feature is required since we have observed that the resonant frequency of the cantilever drifts as much as 2 Hz as the ambient temperature fluctuates about 1 K over the duration of a typical run of a few hours. The effect of the thermal drifts has also been mitigated off-line by a proper implementation of the data acquisition sequence, as described in Section III.

Below the cantilever a spherical mirror with radius of curvature $R = (30.9 \pm 0.15)$ mm and diameter $a = (8.00 \pm 0.25)$ mm is mounted on an aluminum frame connected to two motorized actuators allowing for coarse translational motion, plus an additional piezoelectric transducer for fine translational motion driven with a bias $V_{\rm PZT}$. The spherical lens is gold-coated by the manifacturing company with a thickness of 250 nm ($\pm$ 10 %) The surface of the cantilever facing the spherical lens is coated with a 200 nm layer of gold ($\pm$ 10 %) by thermal evaporation, with the rate of evaporation kept below 50 $\mathring{A}$/sec at a pressure of 3×10^{-6} mbar to ensure better conditions for homogeneous coating.

The predicted frequency of the fundamental flexural mode of the cantilever is $\nu_p = (0.162t/L^2)\sqrt{E/\rho}$ with E the Young's modulus of the material [41]. In our case this yields a resonant frequency around 894 Hz, within few percent from the measured frequency at the typical vacuum pressure of 1.6×10^{-4} Torr in the vacuum chamber. The stiffness k of the resonator can be estimated by assuming an effective mass for the resonator mode roughly equal to the physical mass, which yields $k = 1.036Ewt^3/L^3 \simeq 5.4 \times 10^3$ N/m. The stiffness is three or four orders of magnitude higher than the typical cantilevers used in atomic force microscopy, implying both advantages and drawbacks. A large stiffness allows us to reliably achieve small

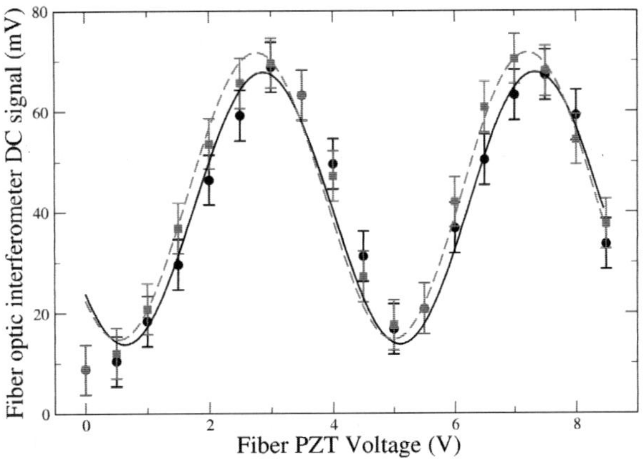

Figure 2. Calibration of the PZT actuator governing the sphere-plane separation. Four calibrations of the PZT actuator have been performed during the overall data taking, of which only two are reported here for simplicity, indicated with red-square and black-dot points, and their best fit with a sinusoidal function (red, dotted line, and black, continuous line, respectively). The average value of the actuation coefficient from the four data sets is $\beta = (87 \pm 2)$ nm/V.

gaps with little static deflection of the cantilever in the presence of strong electric fields for calibration purposes, at the expense of a lower force sensitivity. By decreasing the stiffness of the cantilever both the reachable minimum gap before snapping between the two surfaces occurs and the largest explorable gap for a given signal to noise ratio are increased.

The motion of the cantilever is detected by using a fiber optic interferometer [42] positioned a few microns above the upper cantilever face and fed by temperature-stabilized diode laser with an adjustable power in the 5-10 mW range at the wavelength of 781 nm. The output signal from the interferometer is filtered and amplified through a single reference mode lock-in amplifier and is fed back into the piezoelectric actuator driving the cantilever motion. The phase between the input and the output signals is properly chosen to maximize the vibration amplitude, typically around 30-40 degrees. This scheme is much faster and more efficient than the previous open-loop scheme as described in [38] in which white noise was used as the driving source without a feedback, and the complete FFT spectrum was acquired. The vibration frequency was measured by a counter with 10 mHz resolution, 100 times faster than in the open-loop scheme. The front side of the chamber has a large viewport allowing for visual inspection of the relevant components through an optical microscope, as shown in Fig. 10 of [38]. This also provides for a quick assessment of the optical fiber location, which can be manually adjusted within few hundred micrometers from the underlying resonator using a feedthrough micrometer.

The PZT was calibrated by using the same fiber optic interferometer with the cantilever removed and the spherical mirror replaced by a flat mirror. Assuming a linear relationship $\Delta x = \beta \Delta V_{\mathrm{PZT}}$, the DC interference amplitude versus the voltage applied to the PZT, as shown in Fig. 2, was fit with a sinusoidal function and the distance inferred from its period to be half of the 781 nm laser wavelength. The average of four separate calibration runs yields $\beta = (87 \pm 2)$ nm/V, and it was found to be independent of the voltage applied to the PZT, within the error, in the 0-100 V range. The sensitivity of the fiber optic interferometer is evaluated in the same configuration by driving the PZT with a monochromatic signal at various amplitudes, and looking the corresponding signal at the FFT spectrum analyzer. When the driving amplitude at a frequency of 1 KHz (still well below the maximum response frequency of the PZT of 69 KHz) is reduced at the level of a peak-to-peak signal of 0.25 mV, the measured amplitude at the FFT spectral analyzer becomes equal to the FFT broadband noise level of 10 $\mu V_{\rm rms}$ with an integration time of 10 s. Using the determined value of the actuation coefficient of the PZT, we estimate the minimum detectable signal with SNR $\simeq 1$ to be $\simeq 0.80\text{Å}/\sqrt{\rm Hz}$, similar to the one reported in [42].

For a generic distance-dependent force and for a small amplitude of the cantilever oscillation (2-3 nm as inferred from the interferometer signal) with respect to the average separation between the two surfaces x, the measured frequency of the resonator ν_m is shifted with respect to the proper frequency of the resonator ν_p in such a way that

$$\Delta\nu^2 = \nu_m^2 - \nu_p^2 = -\frac{1}{4\pi^2 m_{\rm eff}}\frac{\partial F(x)}{\partial x}, \tag{1}$$

where $m_{\rm eff}$ is the effective mass of the mode of oscillation of the resonator. In general, the total force acting on the resonator is the aggregate from independent sources and the square of the measured frequency ν_m of a cantilever can be identified as the sum of contributions from those sources,

$$\nu_m^2(x,V) = \nu_p^2 + \Delta\nu_e^2(x,V) + \Delta\nu_r^2(x), \tag{2}$$

where $\Delta\nu_e^2$ is the frequency shift due to electrical force gradients for instance due to an external bias voltage V, and $\Delta\nu_r^2$ is the frequency shift subject to force gradients of non-electrostatic nature, for instance the Casimir force, or random drifts of instrumental and environmental origin. In the next section we will discuss the response of the resonator in the case of electrostatic forces.

3. Electrostatic calibrations

We start with the electrostatic energy stored in a generic capacitance $C(x)$ biased by an external potential V, and under the presence of a contact potential $V_c(x)$ which in general may depend on the distance:

$$E_{el} = \frac{C(x)}{2}[V - V_c(x)]^2. \tag{3}$$

As recently pointed out by Lamoreaux [43], if the contact potential depends on distance the electrostatic force will include an additional term independent upon the external bias voltage. Since in our configuration the observable is the frequency shift, proportional to the second derivative of the electrostatic energy with respect to x, we have, after regrouping the various terms in powers of $V - V_c$:

$$E''_{el} = \frac{C''}{2}(V - V_c)^2 - [2C'V'_c + CV''_c](V - V_c) + CV'^2_c. \tag{4}$$

This can be regrouped again as $E''_{el} = A(V - V_c + B)^2 + D$ and, identifying the coefficients A, B, D to match the terms in Eq. 4, we obtain

$$A = \frac{C''}{2} \quad , \quad B = -\frac{2C'V'_c + CV''_c}{C''} \quad , \quad D = CV'^2_c - \frac{(2C'V'_c + CV''_c)^2}{2C''}. \tag{5}$$

This allows to write:

$$E''_{el} = \frac{C''}{2}(V - V_0)^2 + CV'^2_c - \frac{(2C'V'_c + CV''_c)^2}{2C''}, \tag{6}$$

where

$$V_0(x) = V_c(x) + \frac{2C'V_c' + CV_c''}{C''} \tag{7}$$

is the *minimizing* external potential, *i.e.* the external bias minimizing the electrostatic contribution to the frequency shift. It is clear that, even if the external bias voltage is chosen to minimize the electrostatic contribution, an irreducible term containing the first and second spatial derivatives of V_c in Eq. 7 still causes a frequency shift. The expected frequency shift due to Coulombian interactions is therefore:

$$\nu^2 = {\nu_0}^2 - \frac{C''}{8\pi^2 m_{\text{eff}}}(V - V_0)^2 + \frac{1}{4\pi^2 m_{\text{eff}}}\left[-CV_c'^2 + \frac{(2C'V_c' + CV_c'')^2}{2C''}\right]. \tag{8}$$

We identify three contributions on the RHS of Eq. 8: the first term is unrelated to external or internal voltages, the second term depends on the external voltage, and the last term is a Coulombian contribution only related to the spatial dependence of the contact potential. The electrostatic calibrations consist in the study of the response of the cantilever frequency to an external potential, therefore selecting the second contribution alone which allows to infer both the effective mass of the resonator and the minimizing potential V_0. This knowledge is then used to infer the contact potential assuming a specific form of the functional dependence on distance of V_0, which is in turn used to identify the third contribution in Eq. 8. The subtraction of the total Coulombian contribution allows then to evaluate the square frequency related to non-Coulombian forces and to random drifts due to background noise in the apparatus (see section IV) as first discussed in AFM measurements by [44] and in the specific context of Casimir measurements in [45].

Regarding the capacitance, for the sphere-plane configuration and for our choice of parameters, the proximity force approximation (PFA) holds and the capacitance is $C(x) = 2\pi\epsilon_0 R \ln(R/x)$. This formula in principle must be corrected for the finite spherical mirror diameter used in alternative to a full sphere, but this is a sub-leading PFA correction approximately equal to 0.1% for a typical separation of $d = 1\mu$m. Therefore, the measured frequency ν_m of the cantilever can be parameterized as

$$\nu_m^2(x) = \nu_0^2(x) - K_{\text{el}}(x)(V - V_0)^2, \tag{9}$$

where $\nu_0^2(x) = \nu_p^2 + \Delta\nu_e^2(x, V_0) + \Delta\nu_r^2(x)$, a parabola whose maximum is reached when the applied voltage at a given distance equals to the minimizing potential V_0. The parabola curvature $K_{\text{el}}(x) = \epsilon_0 R/4\pi m_{\text{eff}} x^2$ reflects the cantilever response to externally applied electric forces at a given distance x and allows to extract the effective mass of the resonator mode, and it can be measured by sweeping the applied bias over a bipolar range to discern the parabolic dependence of the frequency shift. The minimizing potential V_0 is found by looking at the horizontal displacement of the parabola. In Fig. 3 we show examples of electrostatic calibrations in which the square resonator frequency is plotted versus the bias voltage for different values of the gap separation between the sphere and the plane. In order to minimize the resonator frequency drifts due to fluctuations of ambient temperature, we measure the unbiased frequency before and after each measurement taken at a given bias voltage V. The average of the two unbiased frequencies is then used as the reference to evaluate the shift. We notice two remarkable features in Fig. 3. First, the peak value of the frequency is different in the four cases, reflecting the possible presence of a distance-dependent force combined with possible drifts in the value of the intrinsic frequency. Second, the location of the peak value of the resonator frequency in the parabolic fit does not occur at the same value of external voltage, indicating a distance-dependent minimizing potential.

The absolute distance between the sphere and the plane is not known with sufficient accuracy prior to the force measurement. In the experiment, the gap is varied by the voltage applied to

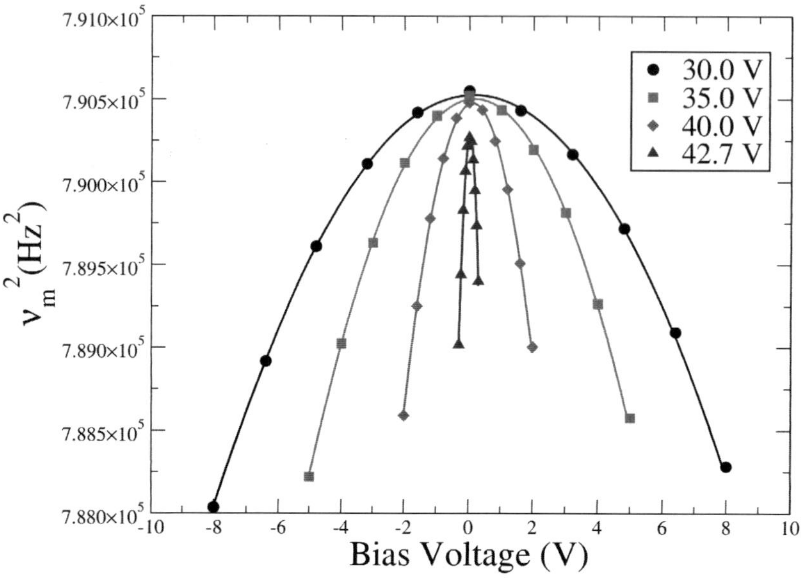

Figure 3. Parabolic dependence of the squared frequency ν_m^2 upon the external bias voltage V for a typical experimental run. The different curves correspond to different values of the sphere-plane separation (defined in the legend in terms of the voltage applied to the PZT used for approaching the sphere). The range of external bias voltages is adapted for each sphere-plane separation to maintain comparable values of the maximum detected frequency shift, of about 1 Hz. This corresponds to keep a constant value for the electric field in the gap. The best fit with a parabolic function is also shown, with the parabolic curves relative to the largest PZT voltages (corresponding to smaller gap separations) displaying largest curvatures. No significant dependence of the fitting parameters has been observed by changing the external bias voltage span.

the PZT ($V_{\rm PZT}$) and consequently $K_{\rm el}$ is a function of the relative distance (*i.e.*, of the applied $V_{\rm PZT}$). This requires an additional fitting parameter $V^0_{\rm PZT}$, which would cause contact and must be inferred from fitting the function

$$K_{\rm el}(V_{\rm PZT}) = \alpha(V^0_{\rm PZT} - V_{\rm PZT})^{-2}, \tag{10}$$

where α is a calibration factor containing the effective mass of the cantilever through $\alpha \equiv \epsilon_0 R/4\pi m_{\rm eff}\beta^2$. In Fig. 4 the curvature parameter $K_{\rm el}$ is plotted versus the piezoelectric actuator voltage $V_{\rm PZT}$ for four different experimental runs. Assuming the PZT actuation is linear, the absolute separation for a given $V_{\rm PZT}$ can be inferred in two ways: $x(V_{\rm PZT}) = \beta(V^0_{\rm PZT} - V_{\rm PZT})$ or $x(V_{\rm PZT}) = \beta[\alpha/K_{\rm el}(V_{\rm PZT})]^{1/2}$. Therefore, the absolute distance can be inferred either from the asymptotic limit $V^0_{\rm PZT}$ of the fit function or from the calibration factor α of the same function, indicating an interdependency of the two physical parameters appearing in Eq. (10). Note that quite large values of $K_{\rm el}$, of order $-2 \times 10^4 {\rm Hz}^2/{\rm V}^2$ in runs 1, 3, and 4, have been achieved. These large signals serve to accurately constrain the fit parameters owing to the high stiffness of the cantilever, which allows us to explore very small distances between the sphere and the cantilever without stiction.

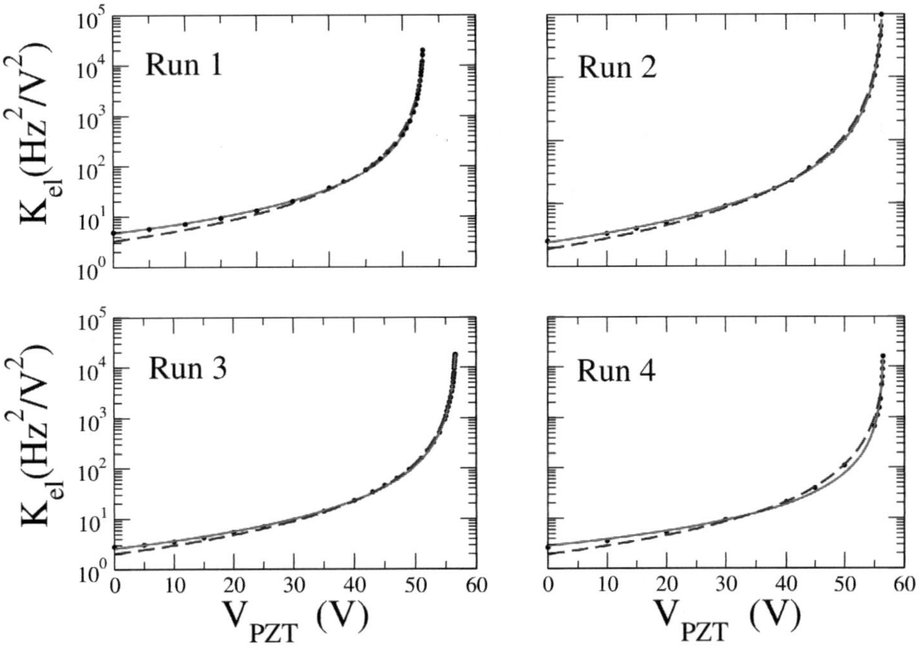

Figure 4. Electrostatic curvature coefficient versus the voltage applied to the PZT actuator for four different runs, and related fits with the expected fixed exponent $e = -2$ (blue dashed line) as in Eq. 10 and by allowing the exponent to be a free parameter (red continuous line). In run 1 the location of the surface of the sphere in proximity of the cantilever is different from the other runs, as the latter were taken after tilting the sphere by an angle equal to 5×10^{-2} radians.

Comparison of the two methods for determining $x(V_{PZT})$ in Fig. 5 (left plot) reveals disagreement when applying the inverse square law in Eq. 10. Direct measurement of the sphere-plane capacitance versus gap provides a third assessment of V^0_{PZT}. In Fig. 6 the capacitance versus PZT voltage is shown with the fitting curve corresponding to the capacitance between a sphere and a plane in the PFA approximation.

Another issue complicating the electrostatic calibrations is evident in Fig. 7 displaying V_0 inferred from Fig. 3 versus V_{PZT}. While the error bars are relatively large at large gap separations, they become smaller at smaller gaps, and two distinct dependencies are observed. In runs 1 and 2 the minimizing potential seems to depend linearly on the gap separation at small separations, while in the remaining runs it appears to tend towards a constant value. The significance of this dependence in the context of high-precision measurements of the Casimir force is discussed in section V.

The error in the curvature parameter K_{el} has been determined with dedicated runs in which conditions nominally leading to the same gap size are maintained, and the result is shown in Fig. 8. The overall time drift in K_{el}, if converted into a thermally induced change in gap size, corresponds to ± 200 nm. The average random uncertainty in K_{el} has been evaluated by taking the data of Fig. 8 and subtracting for each point the long term drift by means of a moving average with a defined time window. By choosing the average time window equal to 4 curvature measurements (each measurement taking 8-10 minutes depending on the number of points obtained in the frequency versus bias voltage measurement) we obtain a relative error

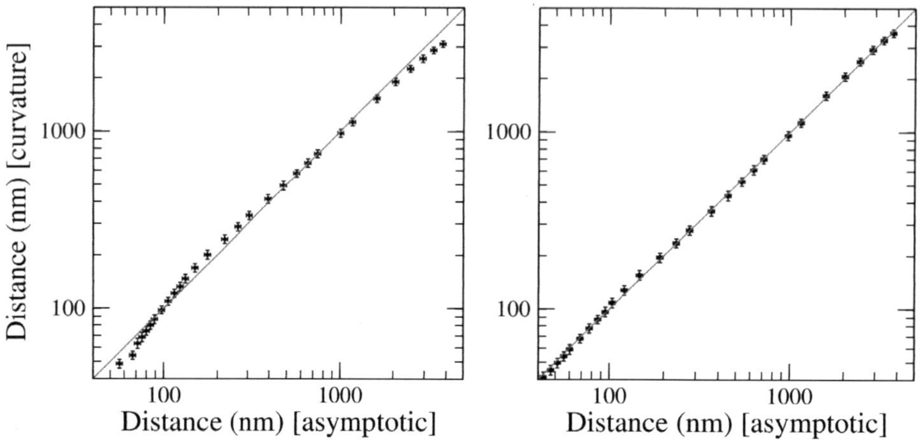

Figure 5. Plot of $x(V_{\rm PZT})$ converted into distance, as evaluated from the calibration factor α for the inverse square law, versus the same function evaluated from the asymptotic limit of the fit function. The left plot uses the expected electrostatic law with an exponent $e = -2$, while the right plot is obtained by optimizing the exponent in the curvature scaling with distance. Red lines are added to provide an eyeguide for the deviations from the expected consistency between the two determinations of the same quantity.

in $K_{\rm el}$ equal to about 4% error on top of the fitting uncertainty associated with the parabolic fit. This allows us to quantitatively compare fitting functions for the distance dependence of $K_{\rm el}$, and has resulted in another anomalous behavior. Indeed, the reduced χ^2 is near one when the exponent is fit in the -1.7 to -1.8 range, while it is at least one order of magnitude higher for fixed -2.0 exponent, as shown in Table 1. Adherence of our data to a strict power law from the farthest to the closest approach in all four cases implies the drift played little role during our data runs, except possibly in the case of Run 4. However, we find that the data clearly *fail* to follow the inverse square law dependence of the electrostatic coefficient upon the sphere-plane separation. More quantitatively, if the fitting exponent is left as a free parameter, our experimental data from four separate sequences follow a power law with exponents $-1.70 \pm 0.01, -1.77 \pm 0.02, -1.80 \pm 0.01, -1.54 \pm 0.02$, far from the expected value of -2. When substituted into Eq. 10, these exponent values produce better agreement between the two methods for determining $x(V_{\rm PZT})$, as shown in the right plot in Fig. 5.

The stability of the electrostatic result has been checked through repeating the data fit of $K_{\rm el}$ versus $V_{\rm PZT}$, starting with few points at the largest distances and by progressively including the data point corresponding to the closer distances. In this way the intrinsic instability of the fit will manifest itself through a systematic variation in the fit parameters subject to a choice of data points included. Figures 9 and 10 show the result of the stability test for Run 1-4, revealing dependence of the fits on the number of points considered. As we will discuss later in Section IV, this is limiting the accuracy of the determination of the Casimir force. In Table 1 we report the values of the parameters of the best fit of the electrostatic calibration for the four runs, leaving

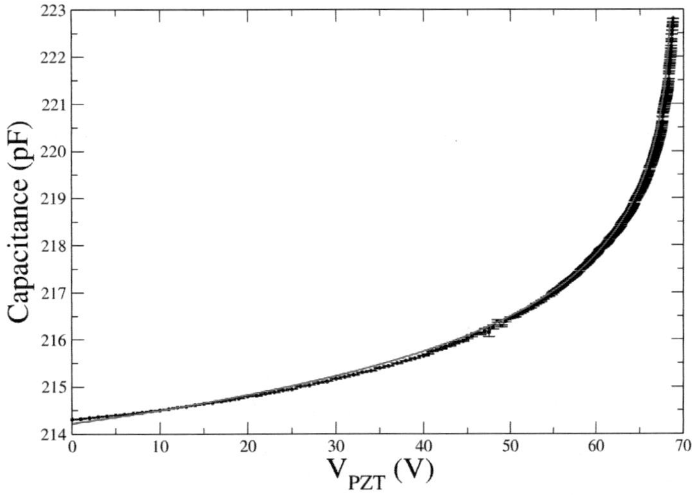

Figure 6. Capacitance versus PZT voltage data (black circles) and its best fit (red line). The fit is based on the PFA formula $C(d) = C_0 + A\ln[\beta(V^0_{\mathrm{PZT}} - V_{\mathrm{PZT}})]$, where the actuation coefficient $\beta = (87\pm2)$ nm/V. The best fit yields, for the coefficients C_0 and A, the values $C_0 = (193.9\pm0.2)$ pF and $A = -(1.757\pm0.002)$ pF, this last in agreement within two standard deviations with the less accurate theoretical expectation $A = -2\pi\epsilon_0 R = -(1.72\pm0.02)$ pF. The same fit provides a value for the asymptotic value of the PZT voltage corresponding to zero gap distance the value $V^0_{\mathrm{PZT}} = (69.31\pm0.01)$ Vm, equivalent to a distance of 47 ± 2 nm. The best fit corresponds to a reduced $\chi^2 = 2.9$.

the exponent free, and with the exponent frozen at the expected value of -2. In the latter case

	Run 1	Run 2	Run 3	Run 4
fixed exponent	-2	-2	-2	-2
α ($\mathrm{Hz^2V^{-2}m^2}$)	6200 ± 98	6197 ± 97	6701 ± 101	6438 ± 156
x_0 (nm)	64.4 ± 1.7	90.5 ± 2.6	62.6 ± 1.7	93.1 ± 2.6
χ^2/DOF	15.9	7.7	6.9	36.5
free exponent	-1.70 ± 0.01	-1.77 ± 0.02	-1.80 ± 0.01	-1.54 ± 0.02
α ($\mathrm{Hz^2V^{-2}m^e}$)	2805 ± 92	3021 ± 153	3732 ± 144	1415 ± 69
x_0 (nm)	29.6 ± 0.9	49.6 ± 1.7	35.7 ± 1.7	20.0 ± 1.7
χ^2/DOF	1.0	0.8	1.2	7.0

Table 1. Fitting parameters of the electrostatic calibrations, and values of the reduced χ^2, as the absolute χ^2 normalized by the number of degrees of freedom (DOF) in the fitting for a fixed exponent equal to 2 as expected from electrostatic considerations, and for a free exponent e. Note that the calibration factor α characterizes the sensitivity of our apparatus, yielding the effective mass in the case of fitting with the expected Coulomb exponent equal to 2. Variations of α from run to run are attributable to the rearrangement of the resonator clamping system, which strongly affects its effective mass. If the exponent is different from the expected value of 2, then the dimension of α changes and the effective mass cannot be properly evaluated.

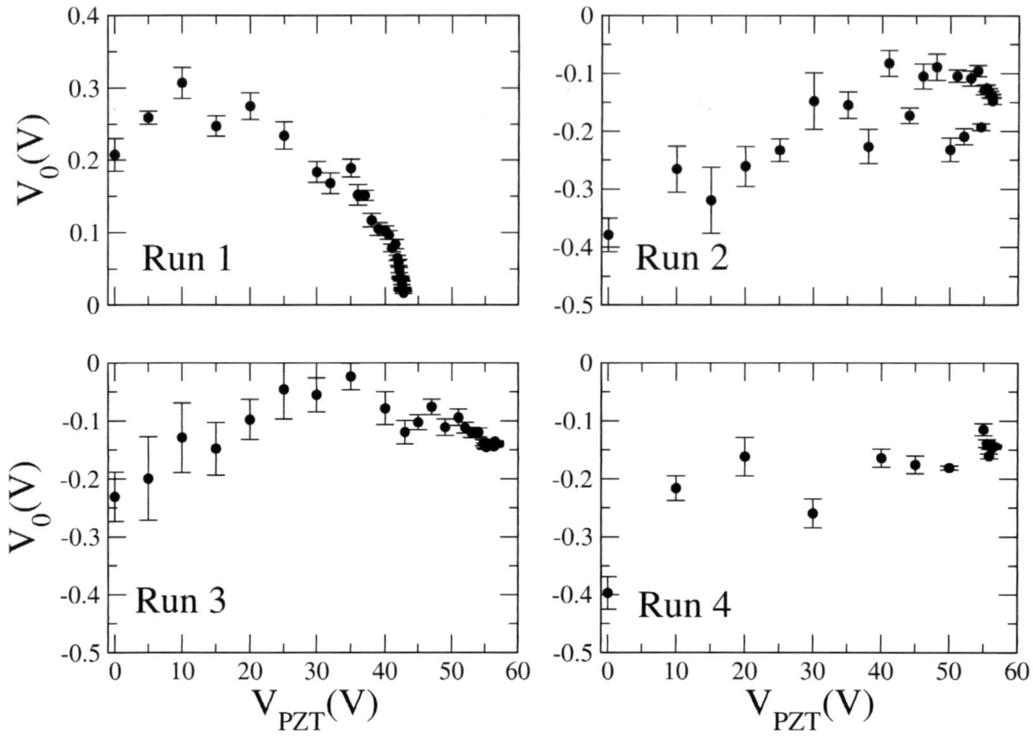

Figure 7. Plot of the minimizing potential versus $V_{\rm PZT}$ for the same four runs shown in Fig. 4. Two behaviors are evident at small separation (large $V_{\rm PZT}$): either an approximately constant potential (Run 3 and Run 4) or a potential linearly dependent upon the separation (Run 1 and, to a less extent, Run 2). Note that in run 1 V_0 is positive while for Runs 2-4 it is negative and converges to $V_0 \simeq$ -150 mV at the smallest separations.

one can infer the effective mass of the cantilever and this turns out to be larger by an order of magnitude than the physical mass, against the expectation to be a fraction of the physical mass [46, 47, 48]. Apart from the uncertainties related to the clamping of the resonator, this is another signal that the calibration fitting with a pure electrostatic contribution neglects some systematic effect.

The unexpected power law poses a significant limit on the validity of our electrostatic calibration. Among the possible systematic effects causing this deviation from the expected Coulombian behaviour, we have considered the following.

3.1. Static deflection of cantilever

The spring constant of our cantilever is extremely stiff (about 5400 N/m). Using Hooke's law, a deflection experienced by the cantilever due to an electrostatic force at 100 nm with an applied voltage of 100 mV is less than 0.2 Å. Hence, the static deflection should play little role.

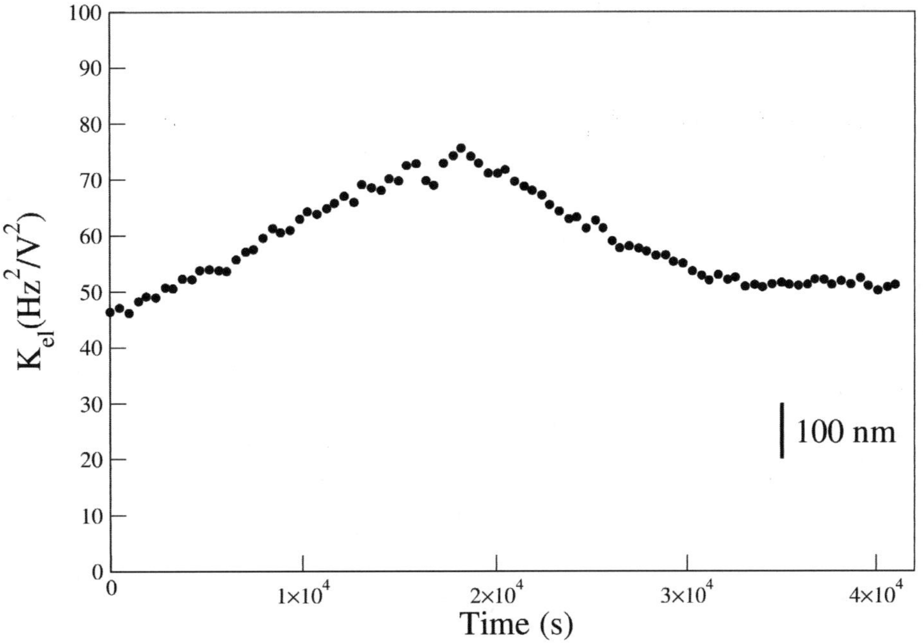

Figure 8. Long-term drift of the curvature coefficient $K_{\rm el}$ during a 12 hours test. The vertical bar represents the equivalent drift in terms of the change in the sphere-plane separation, in the conservative hypothesis that all the drift is attributable to the gap drift. Subtracting each point with a window average of 4 measurements one gets a relative error in $K_{\rm el}$ of 4 %. The sensitivity to the choice of the average window time has been also assessed, with relative errors of 2.9 %, 4.1 %, 4.4 % and 4.6 % for alternative choices of 2, 6, 8, and 10 measurements, respectively.

3.2. Thermal drift

Even though the temperature of the cantilever is actively stabilized by a Peltier cooler, the rest of the system is still subject to global thermal variation. In order to see this, we have measured $k_{\rm el}$ with respect to time at a nominally fixed distance. In the worst circumstance, the gap separation during the course of measurements can drift as much as 200 nm in *either* direction. Although such a drift could in principle affect the inferred exponent, a highly unlikely non-linear monotonic drift would be necessary to account for the consistently observed anomaly in each independent run.

3.3. Nonlinearity of the PZT actuation

The linearity of the PZT translation has been tested under a number of different circumstances. Notice that the translation intervals between the data points in each of the runs shown in are completely random. Yet, all of the runs obey a specific power law in all distances. The PZT was also independently calibrated by means of the fiber optic interferometer with a consistent, linear actuation coefficient factor $\beta = 87 \pm 2$ nm/V. Also, the presence of significant nonlinearities should manifest also as deviations from the PFA fitting of the sphere-plane capacitance versus the PZT voltage. As visible in Fig. 6, this class of measurements strongly constraint the PZT nonlinearity.

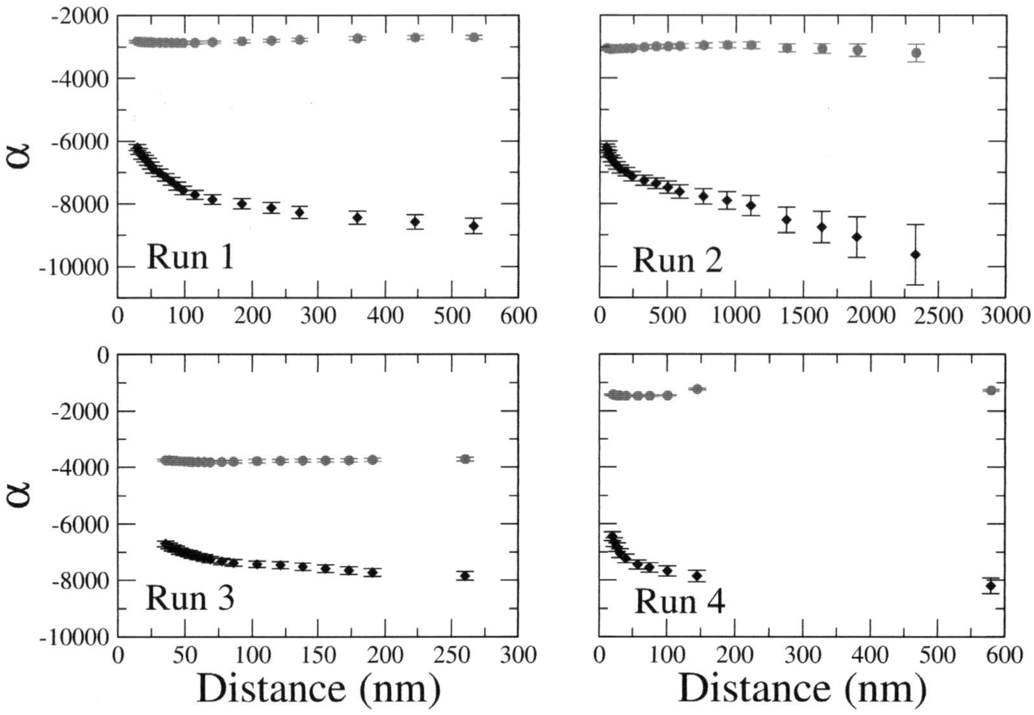

Figure 9. Test of stability of the fit of the electrostatic calibrations: shown is the determination from the fitting of the calibration factor α in the case of the fixed -2 exponent versus the sphere-plane distance for the fixed -2 exponent (in units of Hz^2 V^{-2} m^2) expected from electrostatics (diamond points, black), and for the best exponent left as a free parameter (circle points, red). The number of points used in the data fit includes up to the gap distance evaluated at that point. The best exponent points show convergence to a constant value at the closest distances in all four runs, unlike the electrostatic case with the fixed exponent.

3.4. Nonlinear oscillation of cantilever

The cantilever is driven at resonance in a phase-locked loop, a routine technique adopted by many groups [18, 25, 49, 50]. Higher order terms in the force expansion should produce higher harmonics of the drive frequency. Then the assumption that the frequency shift is simply proportional to the gradient of the external force could lead to erroneous assessments of $k_{\rm el}$, eventually affecting the exponent. We have not observed higher harmonics in the frequency spectrum of the resonator.

3.5. Deviation from geometrical ideality

The deviation from geometrical ideality and its influence on the local capacitances [51] could in principle play a role, both in having a different shape or a variable radius of curvature, and in the deviations from the PFA formula. In our case, even at the smallest explored distances, the PFA correction due to the roughness is estimated to be of 0.4% in the electrostatic calibration, well below our sensitivity. A more careful analysis beyond the PFA is possible taking into

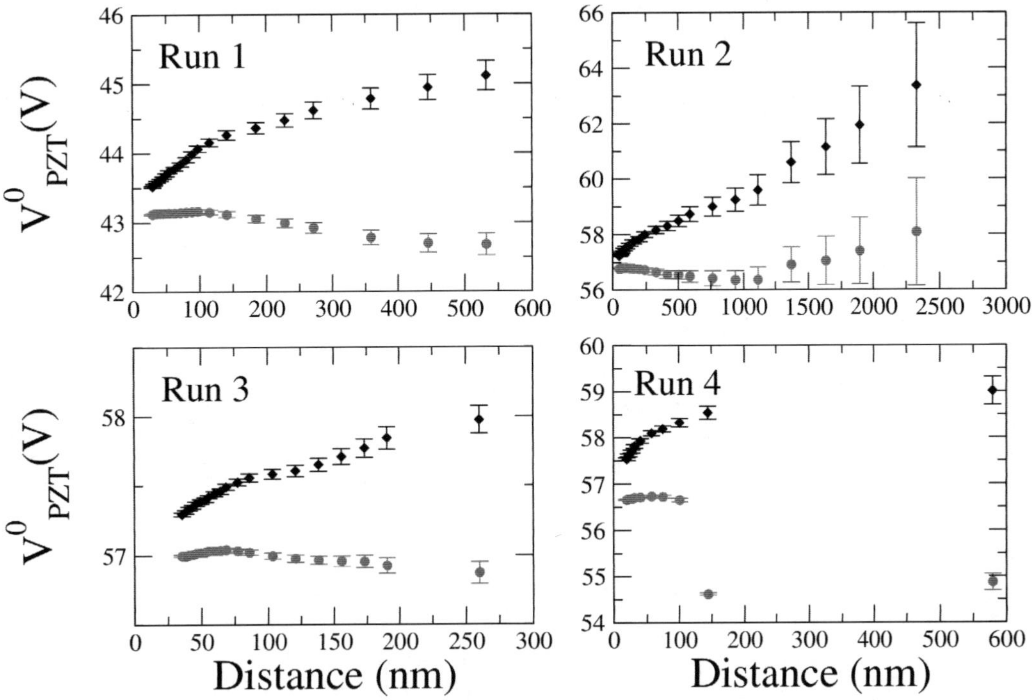

Figure 10. Test of stability of the fit of the electrostatic calibrations: shown is the determination from the fitting of the offset distance $x_0 = \beta V^0_{\rm PZT}$ versus the sphere- plane distance in the case of the fixed -2 exponent expected from electrostatics (diamond points, black), and for the best exponent left as a free parameter (circle points, red). Clearly, the trend associated with the fixed exponent displays a wide variation in the fitting parameter $V^0_{\rm PZT}$, while the overall behavior driven by the best exponent shows a degree of stability in all distances.

account the actual surface profiles. It is however worth to point out that at least the simple hypothesis of a non uniform radius of curvature of the spherical surface cannot take into account the anomalous scaling. In PFA, the force signal (for both electrostatic and Casimir) is directly proportional to the radius of curvature and the total force will be therefore obtained as an integral along the radii of curvature with no deformations to the inverse square law for the distance dependence. In [52] an interesting geometry has been discussed which could take into account the anomalous exponent we have observed, however the stringent agreement between the capacitance measurements and the PFA expression for a pure sphere-plane geometry rules out this hypothesis.

3.6. Surface roughness

Roughness corrections become important at small separations between the surfaces, both for electrostatic calibrations and for analyzing their residuals in the search for Casimir forces. Usually roughness corrections in the electrostatic calibration are disregarded as such calibrations are performed at large distances. Here we consider the simplest analysis based on the proximity force approximation. Let us assume that both surfaces have stochastic roughness with rms amplitudes $\langle h_s^2 \rangle$ and $\langle h_p^2 \rangle$ for the sphere and plane respectively. Apart from the usual PFA condition to treat the curvature effects due to the spherical lens ($d \ll R$), we further assume that the sphere-plane distance d is much smaller than the lateral roughness correlation length ξ on each surface (this is the condition for the applicability of PFA to roughness considerations, only valid for very smooth surfaces). Further assuming that the rms roughness amplitudes are the smallest lengthscales in the problem, the PFA second-order perturbation correction to the sphere-plane electrostatic force is

$$F_{\rm el}(x) = F_0(x)\left(1 + \frac{\langle h_s^2 \rangle}{x^2} + \frac{\langle h_p^2 \rangle}{x^2}\right), \tag{11}$$

where F_0 is the same force evaluated for smooth surfaces. Figure 11 shows typical images of the cantilever surface taken using AFM microscopy. We measured the surface profile for our surfaces and estimate the rms roughness of $\langle h_s^2 \rangle$= 4 nm^2 and $\langle h_p^2 \rangle$= 2.4 nm^2. The resulting deviation from ideality is below the estimated sensitivity of the apparatus, and an attempt to fit the electrostatic calibrations including the roughness at the PFA level as in Eq. 11 did result in significantly larger reduced χ^2. Roughness corrections to the electrostatic calibration potentially represent a systematic effect that could influence the residuals in the search of the Casimir contribution.

3.7. Patch potentials

Patch effects are expected to induce deviations from the Coulombian scaling with distance [37], and the fact that we have found in some calibrations a distance dependent minimizing potential could validate this hypothesis. However, this does not explain yet while the anomalous exponent is systematically observed in all runs, while only in some of them a manifest distance-dependent minimizing potential occurs.

4. Analysis of the residuals of the electrostatic calibrations and tests of the Casimir force

In the previous section we have found that the electrostatic calibrations performed in the entire measurement range show the presence of a distance-dependent minimizing potential (and therefore a distance-dependent contact potential) and an apparent deviation from the expectation for the sphere-plane capacitance dependence on distance. While the first systematic effect can be taken into account in the data analysis, the anomalous distance scaling of the electric force and the lack of its complete characterization prevents in principle to master the calibrations. This prevents the reliable determination of the relevant parameters of the system, for instance its effective mass, and therefore the derivation of the residuals containing the Casimir force signal. In the absence of a complete analysis, and waiting for further experimental and theoretical input on the anomalous exponent observed in our configuration, in this section we analyze the cantilever square frequency $\nu_0^2(x)$ obtained by subtracting to the electrostatic calibration data the Coulombian electrostatic contribution based on the fitting procedure, as first performed in [45]. This off-line procedure is necessary as we have evidenced a dependence on distance of V_0 in various runs, and therefore we cannot assume *a priori* a unique counterbias voltage for an on-line subtraction of the residual electrostatic force as typically performed in various experiments. A distance-dependent residual frequency shift implies the presence of a non-Coulombian force or drifts in the intrinsic resonator frequency or a combination of both. It is evident in Eq. 8 that the electrostatic residual, $\Delta\nu_e^2(x, V_0)$, is zero only if the contact potential is constant. This term must be evaluated by first inferring the contact potential V_c through numerical integration of the minimum condition V_0.

Based on Eq. 7 we have a second order differential equation for $V_c(x)$:

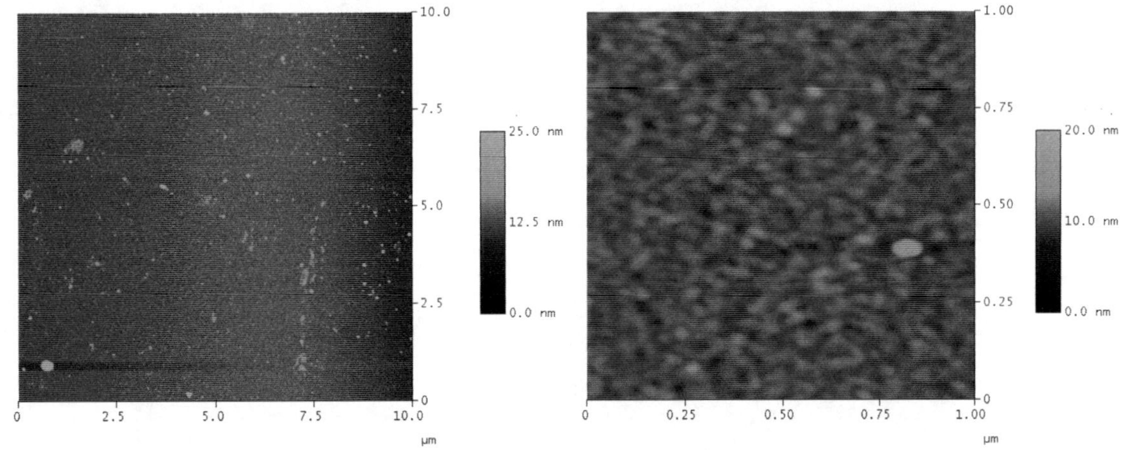

Figure 11. Atomic force microscopy images of the cantilever surfaces, of size 10 × 10 μm (left) and 1 × 1 μm (right). The overall rms roughness for the larger surface is 1.54 nm (with a peak in the profile due to dust of 4.78 nm) and for the smaller surface is 0.79 nm (with a peak in the profile of 2.2 nm).

$$\frac{C}{C''}\frac{d^2V_c(x)}{dx^2} + 2\frac{C'}{C''}\frac{dV_c}{dx} + V_c(x) = V_0(x), \tag{12}$$

where $V_0(x)$ is determined experimentally. For the sphere-plane case in the PFA this results in the following differential equation:

$$x^2 \ln(R/x)\frac{d^2V_c(x)}{dx^2} - 2x\frac{dV_c(x)}{dx} + V_c(x) = V_0(x). \tag{13}$$

to be solved numerically for instance using a Runge-Kutta integration method, once the experimental data on $V_0(x)$ are interpolated with an analytic function. In the absence of a model for the dependence on distance of V_0, different equivalent empirical functions can be used to fit the actual data. This obviously is a strong limitation to the predictive power on the residuals, but at a phenomenological level has to be considered as a conservative and realistic procedure to infer some physics beyond the Coulombian contribution. We use, as first guess function, an exponential one such as $V_0(x) = V_0 + \Delta V[1 - \exp(-x/\lambda)]$ with V_0 representing the potential when the two surfaces are in contact, $V_0 + \Delta V$ is the asymptotic value of the minimizing potential at large distances, and λ the characteristic lengthscale on which the minimizing potential varies. Alternatively, we may use the function $V_0(x) = V_{\log} + \Delta V_{\log} \ln(x/\Lambda)$, characterized by similar parameters $V_{\log}$, $\Delta V_{\log}$ and Λ, although in this case a defined asymptotic value is not available. In Fig. 12 we show the sensitivity of V_c obtained through integration of Eq. 13 to the choice of the function interpolating the V_0 distance dependence.

Another issue requiring careful attention regards the boundary conditions chosen to solve the second order differential equation for $V_c(x)$. At large distances $V_c(x)$ is expected to be constant and thus $V_c(x) = V_0(x)$. Also, the first derivative of $V_c(x)$ should be zero in order to satisfy the differential equation. Therefore we choose the boundary condition $V_c(x_n) = V_0(x_n); V_c'(x_n) = 0$,

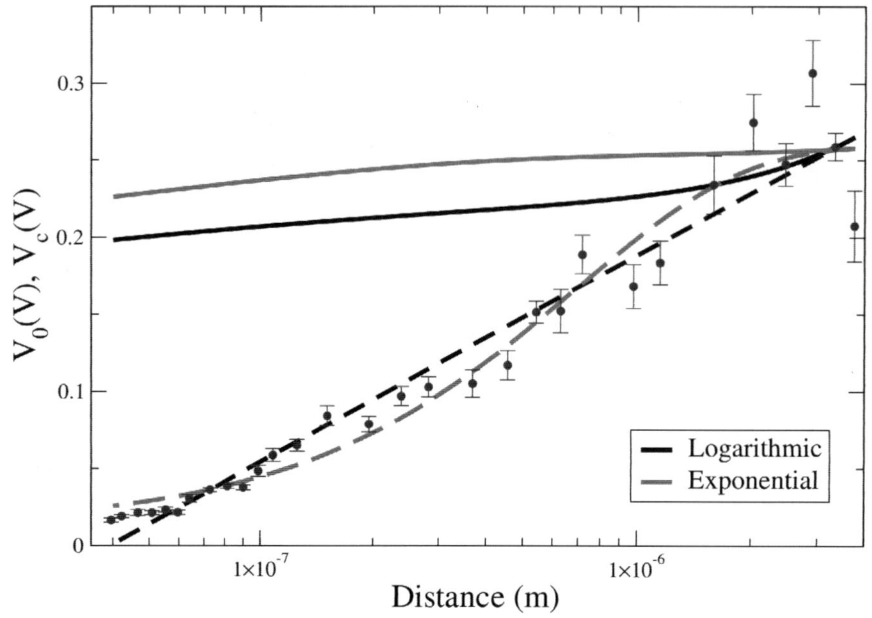

Figure 12. Data for the minimizing potential V_0 versus sphere-plane distance for run 1. The dashed curves represent the best fits of V_0 assuming an exponential dependence (red) and a logarithmic dependence (black). The best fit parameters are $V_0 = (0.011 \pm 0.007)$V, $\Delta V = (0.25 \pm 0.01)$V, and $\lambda = (703 \pm 93)$ nm in the case of the exponential function, and $V_{\log} = (0.07 \pm 1140)$ V, $\Delta V_{\log} = (0.058 \pm 0.003)$ V, and $\Lambda = (140.4 \pm 2.8)$ nm in the case of the logarithmic function. The continuous curves represent the corresponding plots of the contact potential V_c after solving the differential equation Eq. 13 as described in the text.

where x_n is the largest distance measured in the data set. In the case of the exponential fitting of $V_0(x)$, the curve becomes quickly flat as x increases and therefore the use of x_n at very large distances is a good approximation. For the logarithmic fitting this is more problematic, making the solution for $V_c(x)$ quite sensitive to the choice of x_n.

Based on this input, one can evaluate now the electrostatic contribution independent on the external bias potential, $\Delta\nu_e^2(x, V_0)$, and further subtract it from electrostatic contribution dependent on the external bias evaluate at the minimizing potential, $\nu_m^2(x, V = V_0)$ (*i.e.*, the maximum of each parabola in Fig. 3). The result is shown in Fig. 13, where the dashed and dotted-dashed curves are relative to the choice of the V_0 dependence on distance, as given in Fig. 12 (either exponential or logarithmic). We see that the plotted quantities $\nu_m^2(x, V = V_0) - \Delta\nu_e^2(x, V_0)$ indicate that there are further distance-dependent residuals $\nu_p^2 + \Delta\nu^2(x)$ of non-electrostatic nature. In this analysis it is critical to have the most accurate control on the dependence of V_0 on distance. In case of runs 2, 3, and 4 such a control is quite limited, and neither an exponential function nor a logarithmic function adequately describe its spatial dependence. In the remainder of this section we will concentrate on run 1.

Given the serious limitations of our electrostatic calibration, it is difficult to perform a rigorousanalysis of these extra residuals, as they could be Casimir-Lifshitz forces, patch potential forces,etc. In the following we will *assume* that these extra residuals are only due to the

Casimir-Lifshitz force, and evaluate the corresponding prediction for $\nu_p^2 + \Delta\nu_{\rm Cas}^2(x)$. As a first attempt, one can fit either the exponential (dashed line) or the logarithmic (dotted-dashed line) residuals in Fig. 13 with the plain Casimir formula for zero temperature and perfect metals. In the proximity force approximation,the sphere-plane Casimir force is $F_{\rm Cas}(x) = 2\pi R E_{\rm PP}(x)$, where $E_{\rm PP}(x)$ is the Casimir energy per unit area for the plane-plane configuration, $E_{\rm PP}(x) = -\hbar c\pi^2/720x^3$. Therefore, the frequency shift due to the Casimir force in the sphere-plane case is $\Delta\nu_{\rm Cas}^2(x) = -K_{\rm Cas}/x^4$, with $K_{\rm Cas} = \pi\hbar c R/480 m_{\rm eff}$. Fitting the above curves with $\nu_p^2 - K_{\rm Cas}/x^4$, we obtain $\nu_p^2 = (790440 \pm 19.3){\rm Hz}^2$, $K_{\rm Cas} = (5.5164 \pm 0.3390) \times 10^{-27}{\rm Hz}^2/{\rm m}^4$ for the dashed line (exponential formula for $V_0(x)$), and $\nu_p^2 = (790450 \pm 18.3){\rm Hz}^2$, $K_{\rm Cas} = (6.5795 \pm 0.3220) \times 10^{-27}{\rm Hz}^2/{\rm m}^4$ for the dashed-dotted line (logarithmic formula for $V_0(x)$). Using the values for the radius of curvature of the sphere of $R = 30.9$mm, and an effective mass of $m_{\rm eff} = 0.46$g, the obtained coefficients $K_{\rm Cas}$ are lower than theideal Casimir prediction($K_{\rm Cas,th} = 1.3 \times 10^{-26}{\rm Hz}^2/{\rm m}^4$)by about a factor 2, in line with the expectation that conductivity corrections could play a role.

In order to analyze possible conductivity (and temperature) corrections to the Casimir force we now use the Lifshitz formalism [53], written in terms of the frequency-dependent reflection coefficients of the two gold surfaces. The plane-plane free energy is

$$E_{\rm PP}(x) = \frac{k_B T}{2\pi x^2} \sum_p \sum_{m=0}^{\infty\,\prime} \int_{m\gamma}^{\infty} dy\, y^2 \, \log(1 - r_p^2 \, e^{-2y}). \tag{14}$$

Here p denotes the two possible polarizations (TE and TM), $\gamma = 2\pi k_B T x/\hbar c$, and the prime on the summation sign indicates that the $m = 0$ term is counted with half weight. The reflection amplitudes are given by the usual Fresnel formulas,

$$r_{\rm TM}(y, \xi_m) = \frac{s_m - \epsilon(i\xi_m)p_m}{s_m + \epsilon(i\xi_m)p_m} \quad ; \quad r_{\rm TE}(y, \xi_m) = -\frac{s_m - p_m}{s_m + p_m}, \tag{15}$$

where $\xi_m = 2\pi k_B T m/\hbar$ are the Matsubara frequencies, $p_m = y/m\gamma$, and $s_m = \sqrt{\epsilon(i\xi_m) - 1 + p_m^2}$. The dielectric permittivity at imaginary frequencies is evaluated using Kramers-Kronig relations with gold optical data from Palik [54], extrapolated to low frequencies using a Drude model with parameters: $\omega_p = 7.5$eV for the plasma frequency, and $\gamma_p = 0.061$eV for the plasma relaxation parameter. The resulting theoretical prediction (using once again PFA for the sphere-plane case) for $\nu_p^2 + \Delta\nu_{\rm Cas}^2(x)$ at $T = 300$K is plotted as the dotted line in Fig. 13, using as the free cantilever square frequency $\nu_p^2 = 7.9047767 \times 10^5$ Hz^2. It is apparent from the figure that the Casimir curve $\nu_p^2 + \Delta\nu_{\rm Cas}^2(x)$ approximately follows the behavior of the curves $\nu_m^2(x, V = V_0) - \Delta\nu_e^2(x, V_0)$ (that is, the electrostatic residuals *after* substraction of the distance-dependent contact potential term). Note that $\Delta\nu_e^2(x, V_0) > 0$, that means a positive frequency shift associated to a *repulsive* electrostatic residual (V-independent) force. Had we not taken this term into account, the Lifshitz theory curve would noticeably depart from the electrostatic residuals $\nu_m^2(x, V = V_0)$.

Finally, it is important to emphasize once again that, due to the anomalous exponentwe found in our electrostatic calibration, we cannot make any precision measurement claims on the Casimir force after the residual analysis. Nor does our data have enough sensitivity to detect corrections due to temperature effects or sample dependence. On the other hand, it is quite possible that part of the residuals are due not only to Casimir-like forcesbut also to other external voltage-independent forces, not taken into account in a simple Coulomb law, such as patch potentials.

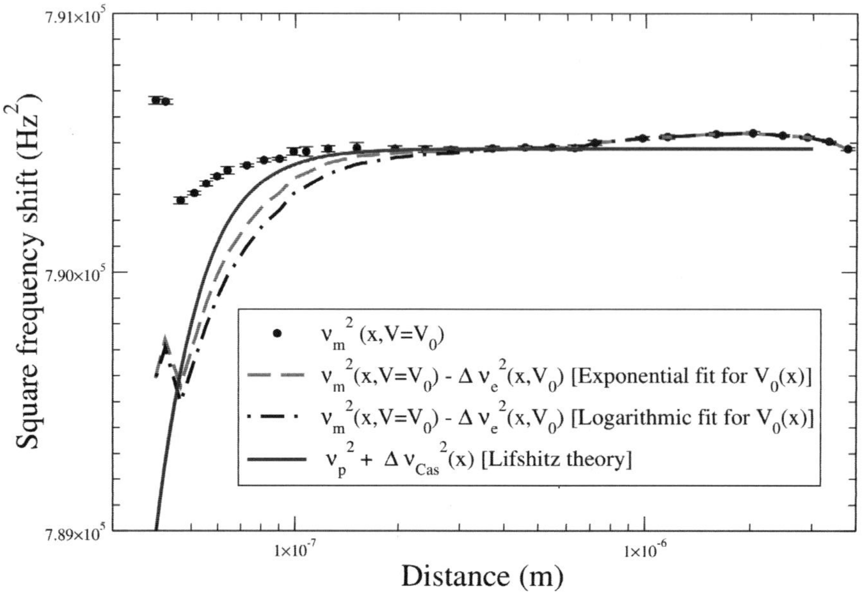

Figure 13. Plot of the electrostatic residuals versus distance. The dots correspond to the maximum value of each parabola in Fig. 3 as a function of distance, $\nu_m^2(x, V = V_0)$. The dashed (exponential fit for $V_0(x)$) and dotted-dashed (logarithmic fit for $V_0(x)$) lines result from substracting from $\nu_m^2(x, V = V_0)$ the extra term $\Delta\nu_e^2(x, V_0)$ due to the distance-dependence of the contact potential $V_c(x)$. The dotted line is the Lifshitz prediction for $\nu_p^2 + \Delta\nu_{\text{Cas}}^2(x)$. Parameters are an effective mass m_{eff} =0.46 g and a cantilever free oscillation frequency ν_p = 889.09Hz. Optical data of gold are from Palik [54], and Drude parameters are $\omega_p = 7.5$eV and$\gamma = 0.061$eV. Temperature is set to $T = 300$K.

5. Systematic effects and relationship to previous Casimir and AFM experiments

Our experiment spans between two different classes of experiments. In previous short-distance Casimir force experiments, the minimum explored distance is on the order of few hundred nanometers. This is due to the fact that typically, in order to increase the sensitivity at larger distances, the stiffness of the resonator is designed to be low. While this yields a stronger signal at large distances with respect to harder resonators, this also results in an earlier snapping of the resonator to the attracting surface. On the other hand, typical AFM experiments aimed at mapping the profile of a surface at subnanometer resolution (both depth and lateral) do not need to reach distances larger than 10-20 nm. We discuss here more in general the specific issues we have found that could affect at least in principle the analysis of previous measurements, or at least the assessment of their accuracy.

5.1. Distance dependence of the contact potential

Our finding of the dependence of the contact potential V_c upon the distance calls for a more careful study of previous experiments in which the contact potential was assessed at relatively larger gaps than the one used for looking at the Casimir force. In order to seek for similar regularities in previous Casimir force experiments, we have requested electrostatic calibration data kindly provided by various groups. The result of the analysis for the data of the Lucent

Laboratories group [17] working with spheres of much smaller radius of curvature (100 μm) is reported in the top-right plot in Fig. 14, showing a small but clearly visible linear correlation between V_0 and the sphere-plane distance. The data of the AFM group in Grenoble (with microspheres of radius 20 μm) in Fig. 14 bottom-left show also a strong dependence of V_0 at large distances, with a more complex behavior as evident from the change in polarity of the minimizing potential around 5 μm. In the case of the experiment performed at Indiana University-Purdue University Indianapolis (IUPUI), with a microsphere of radius (148.7 $\pm$ 0.2) μm, see bottom-right plot, no evident dependence on the distance can be assessed, and in that experiment a constant counterbias equal to the average value was assumed. A linear trend, with a systematic difference in the contact potential in the explored range of about 3 mV around random fluctuations of comparable size, seems also present in Fig. 4 of [55] using microspheres of radius 200 μm, although the coarse scale used for the vertical units makes harder a more careful and quantitative assessment.

Although it would be necessary to collect more data, one noticeable emerging pattern from the comparison of the experiments in Fig. 14 is that experiments with intermediate radii of curvature generally seem to provide a milder dependence of V_0 within a comparable distance range. It is plausible that spheres with very small radius of curvature as in the example from the Grenoble group are more vulnerable to local defects or impurities, while large radii of curvature (and therefore large active interaction surfaces) like in our case are sensitive to large scale variations in the patch potentials. It will be interesting to investigate, beside the obvious material composition, the geometry dependence of minimizing and contact potentials. An interesting study of the minimizing potential and its dependence upon distance and drift in time has been recently reported for the more macroscopic setup of a torsional balance [56].

To assess the impact of the distance-dependent minimizing potential on the precision of the Casimir force measurements, following [57], we evaluate the equivalent voltage necessary to mimic the ideal (zero temperature, perfect conductors) Casimir force at a given distance, and in the sphere-plane case we have

$$V_{\rm eq}(x) = \frac{\pi}{120}\left(\frac{\hbar c}{\epsilon_0}\right)^{1/2}\frac{1}{x}, \tag{16}$$

implying that the Casimir force can be simulated, at a distance of 1 μm, by just having 17.5 mV of uncompensated voltage between the two surfaces. For real materials the Casimir force is weaker, and can therefore be mimicked by an even smaller uncompensated voltage. If the minimizing potential V_0 is independent of the distance over the entire range of explored distances, it is legitimate to use an external counterbias to cancel its effect. Otherwise, a fixed counterbias will only cancel out the minimizing potential at a given distance and, depending on the slope of V_0 versus distance at smaller gaps, most if not all of the measured force could be due to the uncompensated potential. Some of the determinations of the contact potential in previous experiments have been performed at relatively large gaps with respect to those at which the Casimir forces observations are reported. In some cases, in particular for low stiffness microcantilevers, this is necessary because electrostatic forces at small distances will cause snapping or instabilities, unless very small bias voltages, limited by the voltage supply specifics, are applied. This suggests that a reanalysis of the data collected on Casimir forces so far will be beneficial in the light of these findings, also including a more careful scrutiny in the next round of measurements. Notice also that to have for instance a 0.1 % precision in the determination of the Casimir force the electrostatic background should be controlled with an accuracy better than 550 μV at 1 μm, which it is also equivalent to perform precision electrometry. In other words, the precision of 0.1 % is equivalent to be able to discern a minimum voltage of 550 μV applied to the gap, a very difficult task considering the presence of comparable

Journal of Physics: Conference Series **161** (2009) 012004 doi:10.1088/1742-6596/161/1/012004

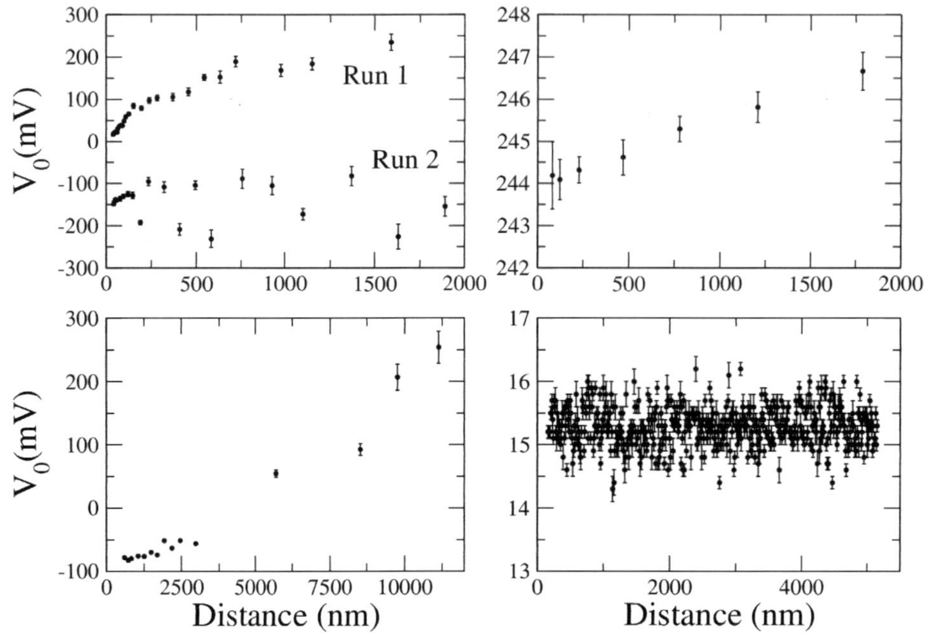

Figure 14. Minimizing voltage versus sphere-cantilever distance resulting for the data analysis of various experiments. Shown are data from run 1 and run 2 of our experiment (top-left, with a zero minimizing potential at the closest approach for run 1, and a non-zero value of $V_0 \simeq$-150 mV for the closest approach of run 2), data from [17] (top-right, courtesy of F. Capasso and H.B. Chan), results from the experiment of the Grenoble group (bottom-left, courtesy of J. Chevrier and G. Jourdan) , and the experiment at IUPUI (bottom-right, courtesy of R. Decca). The first three sets of data evidence a distance dependence of V_0 upon the sphere-plane separation, which could affect significantly the accuracy without taking into account this systematic effect, while this behaviour is absent in the bottom-right plot.

voltage drifts and patch potentials, although other groups have instead concluded that the estimated patch potential contributions are under control in their experiments [23, 58].

5.2. Thermal expansion

Even though the temperature of the cantilever clamping system is actively stabilized by a Peltier cooler, the rest of the system is still subject to global thermal variation as large as 1 K during the measurement runs. In order to understand the stability issues of our apparatus, we have studied the cantilever frequency for a time interval much longer (12 hours) than the typical duration of a run (of 3-4 hours) and we have show the temporal evolution of the curvature coefficient $K_{\rm el}$ at a given nominal distance in Fig. 8.

To understand the impact of thermal expansions let us consider an aluminum slab of L =1 cm with its coefficient of linear expansion $\alpha = 24 \times 10^{-6}$/K, situated underneath the cantilever. For a typical temperature change over a period of 12 hours of ΔT = 1K, the incremental expansion due to temperature is then $\Delta L = \alpha L \Delta T$ =240 nm, which is almost identical as the total distance variation shown in Fig. 8. This translates into a range of 5-10 nm of average variation in the

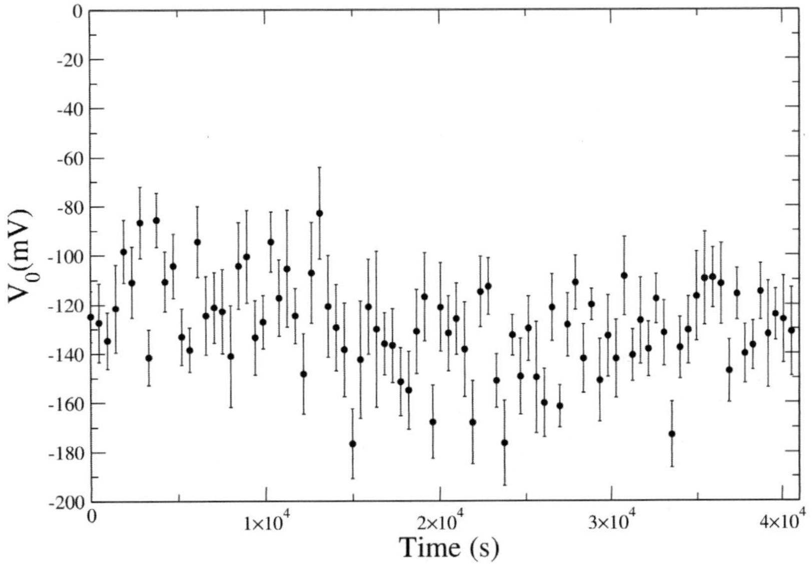

Figure 15. Time behaviour of V_0 in the same long run of Fig. 8. The apparent lack of correlation between this drift and the one presented in Fig. 8, which is mainly due to the drift in the sphere-plane gap distance, seems to imply that there are genuine time-dependent drifts of the minimizing potential even if the sphere-plane distance is kept constant, consistently with the finding in [56].

shot-to-shot electrostatic measurement, *i.e.* the change in the absolute gap separation during the acquisition of an individual data point. To evaluate the robustness of the fits, let us assume that the last position registered by the PZT right before hard contact is misplaced by 8 nm either closer or away from the surface, for instance due to thermal expansion. One underlying assumption held throughout our analysis has been the distance measurement is solely controlled by the action of the piezoelectric transducer, but it is evident at this point that the actual distances could be also affected by thermal expansion. In order to see how this small variation due to temperature change influences the final results of the fitting procedure, we have adjusted the last data point of the closest approach by 8 nm forward or backward in an attempt to mimic the effect of a thermal expansion hypothetically affecting only the last collected point.

The relocation of the single data point of the final distance has modified the entire set of fitting parameters as shown in Table 2. For the case of the calibration factor α, variations due to the modification of the single last data point were $14\%, 6.4\%, 18\%, 21\%$ for Run 1-4, respectively. Even larger changes in the values of the distance offset x_0 were found, up to 40 %.

The exponent e of the electrostatic power law, however, remains stable over the forward and backward displacements of the single data point, although its overall variation is evidently greater than the fitting uncertainties at a given position, demonstrating how vulnerable fitting parameters are to small changes in the absolute gap distance. It should be emphasized that only a single data point is readjusted in this analysis. Because the thermal expansion is present throughout the whole measurement at all distances, what has been shown here only represents a subset of many other possibilities.

Another source of drift has been identified in the minimizing potential, as evidenced in Fig.

15. This is consistent with recent reports of dedicated experiments studying the long term drift of the surface potential using torsional balances [56]. The presence of a minimizing potential slowly drifting in time implies severe limitations on the possibility to integrate for long time intervals precision force measurements.

5.3. Relationship between Casimir and AFM measurements

For the sphere-plane geometry, Casimir force experiments have used sphere radii much larger than typical in AFM apparatus ($R \simeq 100\mu$m versus 10-100 nm respectively) due to different motivations (mesoscopic effective surface versus higher image resolution, respectively). However, the physics ruling the instrument should be the same and one expects a smooth transition between the two regimes, and a close comparison between the accuracies reachable in these two regimes. Many of the systematic effects that can potentially mimic the Casimir force have been extensively studied in the AFM experiments [59, 44, 50], in particular for the investigation of adhesion and friction surface forces [60, 61, 62, 63], many years before Lamoreaux's landmark experiment [15]. Dürig *et al.* have studied metallic adhesion and short range forces using Scanning Tunneling Microscopy (STM) [64, 65], while a number of papers are devoted to the subject of radiative transfer [66], noncontact friction [67], and dissipative interaction between a resonant cantilever and various sample surfaces in the dynamic AFM mode [36, 68, 69, 70, 71, 72], still an active topic of research. It would be productive to bridge the short-range AFM measurements and Casimir force measurements that, after the pioneering measurements of the macroscopic van der Waals force [73], seem to have developed with somewhat divergent methodologies.

The Casimir force may be considered as the van der Waals force in the long-range, retarded regime and therefore the study of its nature may be seen as an extension of the general studies of the previous van der Waals force measurements with AFM techniques. From this perspective, the measurement of forces between macroscopic surfaces at small distances becomes a rich field of study, including also the possible role played by the meniscus force, double-layer force, capillary force, hydration force, hydrophobic force, and steric, depletion, and bridge forces, all ultimately of electromagnetic origin, yet different in distance scaling and strength [47]. Although not all

		Run 1	Run 2	Run 3	Run 4
F	$\alpha(\mathrm{Hz^2V^{-2}m^e})$	3194 ± 97	3200 ± 158	4265 ± 155	1749 ± 75
	x_0 (nm)	35.7 ± 0.1	53.9 ± 0.3	43.5 ± 0.2	28.7 ± 0.2
	e	-1.74 ± 0.01	-1.78 ± 0.02	-1.84 ± 0.01	-1.60 ± 0.01
	χ^2/DOF	1.6	0.5	2.2	7.0
	$\alpha(\mathrm{Hz^2V^{-2}m^e})$	2805 ± 92	3021 ± 153	3732 ± 144	1415 ± 69
	x_0 (nm)	29.6 ± 0.1	49.6 ± 0.2	35.7 ± 0.2	20.0 ± 0.2
	e	-1.70 ± 0.01	-1.77 ± 0.02	-1.80 ± 0.01	-1.54 ± 0.02
	χ^2/DOF	1.0	0.8	1.2	7.0
B	$\alpha(\mathrm{Hz^2V^{-2}m^e})$	2788 ± 97	3006 ± 162	3608 ± 149	1451 ± 78
	x_0(nm)	28.7 ± 0.2	48.7 ± 0.3	33.9 ± 0.2	20.0 ± 0.2
	e	-1.70 ± 0.01	-1.77 ± 0.02	-1.79 ± 0.01	-1.54 ± 0.02
	χ^2/DOF	2.8	1.6	1.7	11.5

Table 2. Stability test of the fitting parameters for the electrostatic calibrations. The middle rows present the fitting parameters α and x_0 for the electrostatic calibrations with a power law exponent e taken to be a free parameter, such that $K_{el} \propto d^e$. F and B in the first and third rows represent hypothetical situations in which the last data point right before hard contact were moved by 8 nm forward (F) and backward (B) from the original position, respectively.

of these forces may be relevant in a given experimental situation, one must carefully check the presence of each of these forces and must be able to distinguish them, on the top of residual electrostatic forces and Casimir forces in themselves. It is plausible that, especially at the level of accuracy reachable in current and future high precision measurements of the Casimir force at small distances, some of these forces should appear already at the level of residuals from the best fits.

A similar argument may be also applied to the electrostatic calibrations. The AFM community has already discussed issues such as friction, electrification, patch charges and potentials, while these have received little attention so far in the Casimir framework (apart from the estimates for the effect of the patch potentials discussed in [23, 58]), in spite of relevant analysis appearing in the recent literature [37, 74, 75]. Also, Burnham *et al.* [63] have evaluated the effect of patch charges by means of the image method, and compared with the previously observed long-range interaction of van der Waals force in the surface force microscopy. In [36], the force fluctuations between a cantilever tip ($R = 1\mu$m) and a surface (both gold-coated) are interpreted in terms of inhomogeneous electric fields due to the presence of atomic steps, adsorbates, hillocks, pits, and other defects. The ubiquity of these electric fields may significantly limit precision and accuracy of Casimir force measurements. It is therefore of great relevance to aim also at calibrations with other physical effects, such as through radiation pressure [26] and hydrodynamic forces in the case of the Casimir-Lifshitz force in the presence of liquid media [76], and to perform detailed comparisons among the various calibration techniques, especially to filter out issues specific to the electrostatic calibrations.

6. Conclusions

We have discussed electrostatic calibrations in a sphere-plane setup different from the previous ones for two features, namely a large radius of curvature of the sphere and a microresonator with large stiffness. This combination allows us to extend electrostatic calibrations to small gap distances in the presence of large spherical surfaces. We have evidenced anomalies in the expected scaling of the electrostatic force, and a distance dependent potential minimizing the electrostatic contribution. In the absence of a complete control of the underlying electrostatic physics, it is difficult to calibrate the apparatus at the extent to meaningfully discuss the residuals signal. By using the electrostatic expectations, we have found evidence of the Casimir-Lifshitz force at small separation in one out of the four runs for which optimal working conditions were found and a complete data analysis have been performed. While we believe that some systematic effects are being emphasized by using spheres with large radius of curvature and small sphere-plane distances, we draw two conclusions from our results which may be of more general interest. Firstly, we have shown that the determination at all distances of the contact potential V_0 is crucial, and its uncertainty can affect the entire data analysis procedure. This implies that electrostatic calibrations have to be performed in the entire range of explored distances, rather than being limited to larger separations for which the Casimir force is expected to be negligible. Secondly, we have discussed the stability of the determination of the fit parameters, like distance offset, minimizing potential, and effective mass, coming from the electrostatic calibrations. Various groups are now trying to confirm or rule out the anomalies we have observed in our experimental setup [77]. Apart from pointing out some limitations of the sphere-plane geometry, we believe that our discussion of systematic effects and data fitting robustness is beneficial for more rigorous data analysis in the next generation of experiments for any geometrical configuration, allowing to explore non-Coulombian forces, in the spirit of an experiment able to discover unknown physics rather than the demonstration of an *a priori* known effect [31].

*Present address: Department of Physics, Yale University, 217 Prospect Street, New Haven, CT 06520-8120, USA

Acknowledgments

We thank R.L. Johnson for skillful technical support, and Q. Wei for experimental assistance. M.B.H. acknowledges support from the Dartmouth Graduate Fellowship program and the NSF GAAN program. W.J.K. acknowledges useful discussions with S.K. Lamoreaux and A. Parsegian, while D.A.R.D. acknowledges interesting discussions with G. Carugno, H.B. Chan, R. Decca, S. de Man, S.K. Lamoreaux, U. Mohideen, J. Munday, and G. Ruoso. R.O. acknowledges partial support by the NSF through the Institute for Theoretical Atomic and Molecular Physics at Harvard University and the Smithsonian Astrophysical Observatory, and is grateful to F. Capasso and H.B. Chan for kind permission to use a set of their electrostatic calibration data from [17] for the analysis reported in Fig. 14, to J. Chevrier and G. Jourdan for useful discussions and for providing data reported in Fig. 14 prior to their publication, and to R. Decca for useful discussions and for providing data of his experiment reported in Fig. 14.

References

[1] Casimir H B G 1948 *Proc. K. Ned. Akad. Wet.* B **51** 793
[2] Lifshitz E M 1956 *Zh. Eksp. Teor. Fiz* **29** 94 [*Sov. Phys. JETP* **2** 73]
[3] Plunien G, Müller B, and Greiner W 1986 *Phys. Rep.* **134** 87
[4] Milonni P W 1994 The Quantum Vacuum (San Diego: Academic Press)
[5] Mostepanenko V M and Trunov N N 1997 The Casimir Effect and its Applications (London: Clarendon)
[6] Bordag M 1999 The Casimir Effect 50 Years Later (Singapore: World Scientific)
[7] Bordag M, Mohideen U, and Mostepanenko V M 2001 *Phys. Rep.* **353** 1
[8] Reynaud S, Lambrecht A, Genet C, and Jaekel M T 2001 *C. R. Acad. Sci. Paris* **IV-2** 1287
[9] Milton K A 2001 The Casimir Effect: Physical Manifestations of the Zero-Point Energy (Singapore: World Scientific)
[10] Milton K A 2004 *J. Phys. A: Math. Gen.* **37** R209
[11] Lamoreaux S K 2005 *Rep. Prog. Phys.* **68** 201
[12] Sparnaay M J 1958 *Physica* **24** 751
[13] van Blokland P H G M and Oveerbeek J T G 1978 *J. Chem. Soc. Faraday Trans. I* **74** 2637
[14] Bressi G, Carugno G, Onofrio R, and Ruoso G 2002 *Phys. Rev. Lett.* **88** 041804
[15] Lamoreaux S K 1997 *Phys. Rev. Lett.* **78** 5
[16] Mohideen U and Roy A 1998 *Phys. Rev. Lett.* **81** 4549; Harris B W, Chen F, and Mohideen U 2000 *Phys. Rev. A* **62** 052109
[17] Chan H B, Aksyuk V A, Kleiman R N, Bishop D J, and Capasso F 2001 *Science* **291** 1941
[18] Decca R S, López D, Fischbach E, and Krause D E 2003 *Phys. Rev. Lett.* **91** 050402; Decca R S *et al.* 2005 *Phys. Rev. Lett.* **94** 240401
[19] Decca R S, López D, Fischbach E, Klimchitskaya G, Krause DE, Mostepanenko VM 2007 *Phys. Rev. D* **75** 077101; 2007 *Eur. Phys. J. C* **51**, 963
[20] Derjaguin B V and Abrikosova I I 1957 *Sov. Phys. JETP* **3** 819; Derjaguin B V 1960 *Sci. Am.* **203** 47
[21] Blocki J, Randrup J, Swiatecki W J, and Tsang F 1977 *Ann. Phys.* **105** 427
[22] Krause D E, Decca R S, López D, and Fischbach E 2007 *Phys. Rev. Lett.* **98** 050403
[23] Decca R S, López D, Fischbach E, Klimchitskaya G L, Krause D E, and Mostepanenko V M 2005 *Annals of Physics* **318** 37
[24] van Zwol P J, Palasantzas G, and de Hosson J Th M 2007 *Appl. Phys. Lett.* **91** 144108; 2008 *Phys. Rev. B* **77** 035439; 2008 *Phys. Rev. B* **77** 075412; van Zwol P J, Palasantzas G, van de Schootbrugge M, de Hosson J Th M, and Craig V S J 2008 *Langmuir* **24** 7528
[25] Jourdan G, Lambrecht A, Comin F, and Chevrier J 2007 arXiv:0712.1767v1 [physics.gen-ph] 11 Dec 2007
[26] Petrov V, Hahn J, Petter J, Petrov M, and Tschudi T 2005 *Optics Lett.* **30** 3138; Petrov V, Petrov M, Bryksin V, Petter J, and Tschudi T 2006 *Optics Lett.* **31** 3167; Bryksin V and Petrov M 2007 *JETP Letters* **86** 368; Bryksin V and Petrov M 2008 *Phys. Solid State* **50** 229
[27] Ludwig T 2008 *J. Phys. A* **41** 164025
[28] Chan H B, Bao Y, Zou J, Cirelli R A, Klemens F, Mansfield W M, and Pai C S 2008 *Phys. Rev. Lett.* **101** 030401
[29] Fischbach E and Talmadge C L 1999 The Search for Non-Newtonian Gravity (New York: AIP/Springer-Verlag)
[30] Gundlach J H 2005 *New J. Phys.* **7** 205
[31] Onofrio R 2006 *New J. Phys.* **8** 237
[32] Serry F M, Walliser D, and Maclay G J 1995 *J. Microelectromech. Syst.* **4** 193

[33] Buks E and Roukes M L 2001 *Europhys. Lett.* **24** 220; 2001 *Phys. Rev. B* **63** 033402
[34] Chan H B, Aksyuk V A, Kleiman R N, Bishop D J, and Capasso F 2001 *Phys. Rev. Lett.* **87** 211801
[35] Kim W J, Brown-Hayes M, Dalvit D A R, Brownell J H, and Onofrio R 2008 *Phys. Rev. A* **78** 020101(R)
[36] Stipe B C, Mamin H J, Stowe T D, Kenny T W, and Rugar D 2001 *Phys. Rev. Lett.* **87** 096801
[37] Speake C C and Trenkel C 2003 *Phys. Rev. Lett.* **90** 160403
[38] Brown-Hayes M, Dalvit D A R, Mazzitelli F D, Kim W J, and Onofrio R 2005 *Phys. Rev. A* **72** 052102
[39] Dalvit D A R, Lombardo F C, Mazzitelli F D, and Onofrio R 2004 *Europhys. Lett.* **67** 517
[40] Albrecht T R, Grutter P, Horne D, and Rugar D 1991 *J. Appl. Phys.* **69** 668; Kim B I 2004 *Rev. Sci. Instr.* **75** 5035
[41] Timoshenko S, Young D H and Weaver W Jr. 1974 Vibration Problems in Engineering (New York: Wyley)
[42] Rugar D, Marmin H J, and Guethner P 1989 *Appl. Phys. Lett.* **55** 2588
[43] Lamoreaux S K 2008 E-Print arXiv:0808.0885v1 [quant-ph] 6 Aug 2008
[44] García R and Pérez R 2002 *Surface Sci. Rep.* **47** 197
[45] Iannuzzi D, Lisanti M, and Capasso F 2004 *Proc. Nat. Ac. Sci. USA* **101** 4019.
[46] Sader J E *et al.* 1995 *Rev. Sci. Instrum.* **66** 3789
[47] Cappella B and Dietler G 1999 *Surf. Sci. Rep.* **34** 1
[48] Bressi G, Carugno G, Galvani A, Onofrio R, Ruoso G, and Veronese F 2001 *Class. Quantum Grav.* **18** 3943
[49] Albrecht T R *et al.* 1991 J. Appl. Phys. **69** 668
[50] Giessibl F 2003 *Rev. Mod. Phys.* **75** 949
[51] Boyer L *et al.* 1994 J. Phys. D **27** 1504
[52] Decca RS, Fischbach E, Klimchitskaya GL, Krause DE, Lopez D, Mohideen U, and Mostepanenko VM 2008 e-print arXiv:0809.3576v1 [quant-ph]
[53] Lifshitz E M 1956 *Sov. Phys. - JETP* **2** 73; Lifshitz E M and Pitaevskii L P 1980 Statistical Physics, Part 2 (Oxford: Pergamon Press)
[54] Palik E D (editor) 1995 Handbook of Optical Constants of Solids (New York: Academic Press)
[55] Chiu H-C, Chang C-C, Castillo-Garza R, Chen F, and Mohideen U 2008 *J. Phys. A* **41** 164022
[56] Pollack S E, Schlamminger S, and Gundlach J H 2008 *Phys. Rev. Lett.* **101** 071101
[57] Onofrio R and Carugno G 1995 *Phys. Lett. A* **198** 365
[58] Chen F, Klimchitskaya, Mohideen U., and Mostepanenko V M 2004 *Phys. Rev. A* **69** 022117
[59] Binnig G, Quate C F, and Gerber Ch 1986 *Phys. Rev. Lett.* **56** 930
[60] Weaver J M R and Abraham D W 1991 *J. Vac. Sci. Techn. B* **9** 1559
[61] Terris B D, Stern J E, Rugar D, and Mamin H J 1989 *Phys. Rev. Lett.* **63** 2669
[62] Blackman G S, Mate C M, and Philpott M R 1990 *Phys. Rev. Lett.* **65** 2270
[63] Burnham N A, Colton R J, and Pollock H M 1992 *Phys. Rev. Lett.* **69** 144
[64] Dürig U, Züger O, and Pohl D W 1990 *Phys. Rev. Lett.* **65** 349
[65] Dürig U and Züger O 1994 *Phys. Rev. B* **50** 5008
[66] Hargreaves C M 1969 *Phys. Lett. A* **30** 491
[67] Volokitin A I and Persson B N 2003 *Phys. Rev. B* **68** 155420
[68] Dorofeyev I, Fuchs F, Gotsmann B, and Wenning G 1999 *Phys. Rev. B* **60** 9069
[69] Gotsmann B, Seidel C, Anczykowski B, and Fuchs H 1999 *Phys. Rev. B* **60** 11051
[70] Loppacher C *et al.* 2000 *Phys. Rev. B* **62** 13674
[71] Pfeiffer O, Nony L, Bennewitz R, Baratoff A, and Meyer E 2004 *Nanotechnology* **15** 101
[72] Volokitin A I, Persson B N J and Ueba H 2006 *Phys. Rev. B* **73** 165423
[73] Israelachvili J N and Tabor D 1972 *Proc. R. Soc. London, Series A, Math. and Phys. Sci.* **331** 4435
[74] Robertson M A *et al.* 2006 *Class. Quant. Grav.* **23** 2665
[75] Palacios-Lidon E and Colchero J 2006 *Nanotechnology* **17** 5491
[76] Munday J and Capasso F 2007 *Phys. Rev. A* **75** 060102(R)
[77] de Man S, Heeck K, and Iannuzzi D 2008 e-print arXiv:0809.3858v1 [quant-ph]; J. Chevrier (*private communication*); S.K. Lamoreaux (*private communication*); G. Palasantzas (*private communication*)

Journal of Physics: Conference Series **161** (2009) 012005 doi:10.1088/1742-6596/161/1/012005

Customized silicon cantilevers for Casimir force experiments using focused ion beam milling

R Castillo-Garza[1], Chia-Cheng Chang[1], Dong Yan[2] and U Mohideen[1,3]
[1] Department of Physics and Astronomy, University of California Riverside, Riverside, CA 92521, USA
[2] Center of NanoSciences and NanoEngineering, Bourns Hall 900 University Ave Riverside, CA 92521, USA

E-mail: umar.mohideen@ucr.edu

Abstract. Higher sensitivity cantilevers will lead to exploration of new phenomena in the Casimir effect. We have used focused ion beam milling to reduce the width of a commercial single crystal, rectangular-shaped silicon cantilevers with a massive Cr/Au-coated–hollow sphere attached at their free end. Theoretically these milled and modified cantilevers should have better Casimir force sensitivity than their non-milled counterparts. In this preliminary report however only 1 out of 4 modified cantilevers were found to have a higher force sensitivity. Future studies will be needed to determine the general applicability of focused ion beam milling for force sensitivity improvements in comparison to the complete nanofabrication of cantilevers.

1. Introduction

Many aspects of the Casimir effect [1] such as the temperature and shape dependence remain to be experimentally explored [2]. In the standard Casimir effect, the force between two ideal metal plates parallel to each other is given by:

$$F_{\min} = -\left(\frac{\pi^2 \hbar c}{240}\right)\frac{1}{d^4} \quad (1)$$

This effect plays an important role in many areas of fundamental and applied physics [2]. It has found applicatins in grand unification theories were non-Newtonian forces are predicted at distances where the Casimir force is currently being measured [2]. In fact, the Casimir force has already been used as a constraint for these predicted forces [2, 3]. Theorists have also predicted that the Casimir phenomenon is responsible for repulsive forces in non-parallel-plate geometries, such as an ideal spherical metal shell [4]. Technologically, the continuous miniaturization of micro electro mechanical systems (MEMS) has decreased their characteristic lengths to submicrometer distances. In this region, the Casimir force is the dominant forces for neutral devices and thus could be very detrimental to their

[3] To whom any correspondence should be addressed.

function due to the resultant friction, dissipation and adhesion issues [5]. The Casimir effect has also been measured using the MEMS technique [6].

In the last ten years, various experimental techniques have been utilized to precisely measure this effect in different materials, geometries and conditions [See Ref. 2 for a list of previous measurements]. Most experiments measure the Casimir force using the sphere-plate geometry because of the difficulty to precisely align surfaces in parallel-plate experiments [3,6,7] (note that there has been one recent measurement using parallel surfaces [8]). For the sphere-plate geometry, the proximity force approximation is used to obtain the Casimir force, and thus places ultimate limits on its precision. However, these limits are beyond the sensitivity of the measurements to date. We pioneered the technique of measuring the Casimir force using an atomic force microscope (AFM) in which the cantilever has a metal-coated polystyrene sphere attached to its free end [7]. These spheres are perfectly smooth, and have much less than 10^{-3} deviation from sphericity. We have used static AFM techniques to measure the mentioned force between the sphere and a metal coated plate [7]. The dynamical AFM (DAFM) technique which measures the force gradient between the sphere and substrate is expected to be more sensitive [9-13]. More sensitivity implies higher precision in the Casimir force measurements at the explored sphere to plate distances. In addition, the increased sensitivity will allow us to investigate separation distances usually inaccessible with the static AFM technique. Thus, it might be possible to investigate the thermal corrections to the Lifshitz theory of the Casimir effect which are a large fraction of the total force at sphere plate separations exceeding one micron.

Static AFM and DAFM techniques use cantilevers as probes for interaction forces. According to theoretical models, rectangular cantilevers which are thin, long, and narrow will produce highly sensitive force probes [9,10]. Many experimental groups have utilized this recipe to measure the spin of small ensembles of electrons or nuclei [12]. Researchers have also used these types of cantilevers to make noncontact friction measurements [11], and non-Newtonian force measurements [13]. These experiments were mostly done with cantilevers fabricated in a traditional nanofabrication facility. The force resolution obtained with these cantilevers surpasses by orders of magnitude the resolution currently obtained in Casimir force experiments [6-8]. However, these cantilevers need specialized steps such as critical point drying, etching, annealing etc. to achieve their sensitivity. These steps might not be compatible with the configuration of cantilevers necessary for Casimir force measurements. As mentioned above, to measure the Casimir force we use the sphere-plate geometry, where we attach a Cr/Au-coated sphere to a single crystal silicon (Si) cantilever, and measure the force between the sphere and a Cr/Au plate. The sensitive cantilevers, used by others [10-13] might not be suitable for our experiment, because the metal coated sphere is large and massive. The cantilevers might break or the internal stresses induced will decrease force sensitivity. They might also be difficult to align with our detection system or reduce the signal to noise ratio because of their excessive curvature on attachment of the sphere. A simpler and reliable way to achieve both the above unprecedented force sensitivities and meet the configuration requirements of Casimir force measurements, would be by sculpting the cantilever with a focused ion beam (FIB) milling technique [14]. By performing the sculpting after attachment of the sphere, we should be able avoid distortions of the cantilever due to the presence of the large sphere. Traditionally, the FIB technique has been used for precise alteration or manufacture of small structures [14]. Recently, researchers have also used them to modify larger devices such as cantilevers [14,15], utilizing a larger flux (current) of positive Gallium ions (Ga+) than the ones used for small structures. The larger currents lead to higher rates of etching. It appears that the FIB might be ideally suited for the fabrication of the specialized cantilevers necessary for Casimir force measurements.

In this article we compare and quantify the benefits of cantilever milling with respect to force sensitivity in the context of Casimir force experiments. In addition, we briefly discuss the role of the internal stresses induced when a massive sphere is attached to the cantilever's free end. In our case, the cantilever milling will primarily decrease its mass and its width. The intrinsic thermomechanical noise limits the cantilevers ultimate sensitivity [10], and provides us with the minimum force detectable:

$$F_{\min} = \left(\frac{2kk_B TB}{\pi Q f_0} \right)^{1/2} \tag{2}$$

where k is the spring constant of the cantilever, f_0 its resonant frequency and Q is the quality factor, k_B is the Boltzmann constant, B is the bandwidth, and T is the temperature.

Equation (2) is expressed in terms of the cantilever's dimensions and its intrinsic material parameters in the following way:

$$F_{\min} = 1.007 \left(\frac{wt^2}{lQ} \right)^{1/2} (E\rho)^{1/4} (k_B TB)^{1/2} \tag{3}$$

Equation (3) applies to rectangular cantilevers with thickness t, width w, and longitude l. E is the Young's modulus of the Si cantilever and ρ is its density. The following expressions are used in equation (3):

$$f_0 = \frac{0.507}{\pi} \left(\frac{E}{\rho} \right)^{1/2} \frac{t}{l^2} \tag{4}$$

and

$$k = 0.25E \frac{wt^3}{l^3} \tag{5}$$

The two previous equations are obtained for the first order resonance of an elastic beam fixed at a single end [16]. In our experiment, we have used FIB milling to reduce the width of a soft single crystal Si cantilever with a relatively massive metal coated sphere attached to its free end as shown in Figure 1. Using a network analyzer, we obtain the power spectral density of these modified cantilevers. From this power spectrum we measure their k, Q, and f_0. The values obtained permit us, in both cases, to calculate and compare the minimum force through the loss parameter, which is defined with the following expression:

$$\gamma = \frac{k}{Qf_0} \tag{6}$$

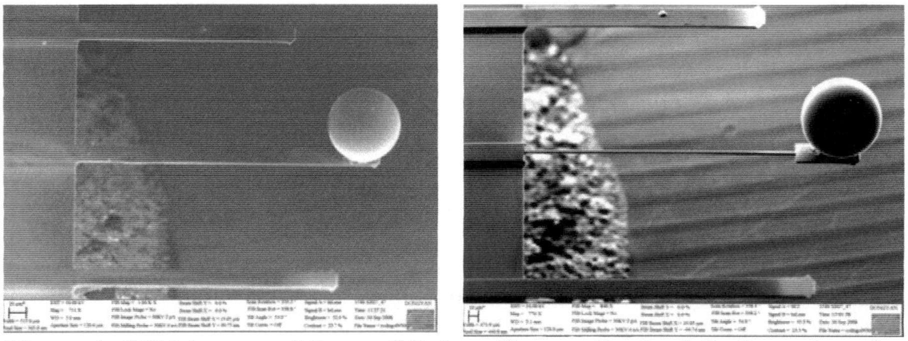

Figure 1. SEM images of the modified cantilevers. On the left, the rectangular cantilever with sphere attached before FIB milling. On the right is the same cantilever with sphere attached after FIB milling.

2. Instrument Design

We use a modified AFM capable of reaching ultra-high vacuum environments of 10^{-8} torr to measure the cantilevers intrinsic characteristics. The measurements reported here were taken at a pressure of about 10^{-6} Torr. The vacuum chamber that contains the AFM has been described previously [17]. This AFM apparatus utilizes a mix of commercial and lab fabricated components, e.g. a Nanoscope V controller for controlling the piezoelectric actuator, and an interferometer to measure the cantilever oscillations. The schematic diagram can be seen in figure 2. We have designed and built an AFM head capable of detecting the cantilever displacements through an all-fiber Rugar type interferometer [18]. This custom-built interferometer consists of a laser diode pigtail with a 1550 nm wavelength, an inline isolator, a 50X50 fiber coupler and two photodiodes. One photodiode, detects the interferometric signal, and is connected to a network analyzer. The other photodiode measures the laser power. Note that the top of the cantilever's free end and the end of the optical fiber form the interferometric cavity. Hence only one leg of the fiber coupler needs to be introduced into the vacuum chamber; the rest of the interferometer's components are outside the chamber and on an optical bench. Figure 3 has a schematic diagram of the all-fiber interferometer. It is worth mentioning that similar all-fiber interferometer techniques have been used to attain force sensitivities of the order of sub-attoNewtons [19].

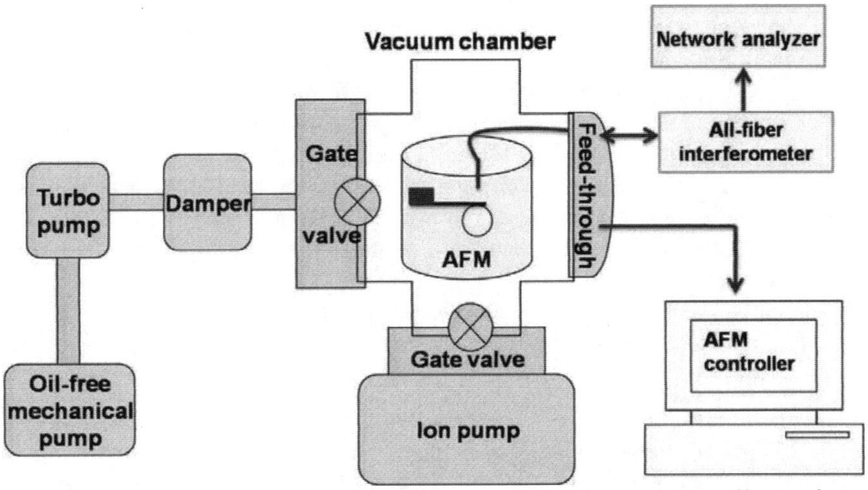

Figure 2. Schematic diagram of the complete setup used to measure the cantilever characteristics. The vacuum chamber and the all-fiber interferometer are on an optical bench.

3. Experimental procedure

Four Si cantilevers were first modified by attaching a Cr/Au coated sphere. Next, the resonance characteristics of the cantilevers with the spheres attached were measured in high-vacuum. Then the width was reduced by FIB milling and their resonance characteristics were measured again in the same high vacuum. The cantilever's power spectral density served two purposes: (i) obtain k at a temperature T and (ii) to obtain the Q. Assuming the cantilever behaves like a harmonic oscillator with one degree of freedom, we use the equipartition theorem to obtain the spring constant k of the cantilever, using the following expression [10]:

$$\frac{1}{2}k(x_{rms})^2 = \frac{1}{2}k_B T \tag{7}$$

Here, x_{rms} is the root mean square amplitude of the cantilever tip. For a temperature T we integrate the square of the cantilever power spectral density to obtain the $(x_{rms})^2$ value. Eq. (7) can be used to obtain the spring constant k. Note that x_{rms} is measured when the interferometric cavity is set at a quadrature point of the interference fringes. The quadrature point location, assures us that change in the interferometric signal due to a change in the cantilevers position with respect to the end of the optical fiber is linear. This is shown in the following expression

$$x_c = \frac{\lambda}{2\pi V_{pp}} V_{Det} \tag{8}$$

Here λ corresponds to the laser wavelength and V_{pp} is the peak-to-peak voltage corresponding to the amplitude of the cavity interferometer fringes. V_{Det} is the photodetector's signal and x_c is the relative cantilever displacement with respect to the end of the optical fibers. Equation (8) is obtained assuming a single reflection in the interferometric cavity and assuming a small cavity finesse. To modify the cavity length a piezoelectric actuator is attached to the cantilever holder. This piezoelectric allows us to keep the cavity length at a qudrature point. The calibration measurements used to achieve the quadrature point has been described elsewhere [17].

For one of the ends of the interferometric cavity we use a commercially available single crystal Si cantilevers with a k= 0.01 N/m to 0.1 N/m. The reported estimated dimensions for these cantilevers are: Length (l) = 350 µm, width (w) = 30 µm, and the thickness (t) = 1 µm [20]. We attached a hollow borosilicate sphere of around 100 nm diameter with silver (Ag) epoxy to the cantilever's free end. Using a thermal evaporator we coat the sphere with a 10 nm adhesion layer of Cr and subsequently with a 300 nm layer of Au. The cantilever with the sphere attached is shown in Figure 1. During metal coating, the modified cantilever is rotated for coating uniformity. Care is taken so that only the sphere and the cantilever's free end are coated. A more detailed coating process for a triangular cantilever can be found elsewhere [17]. The process for the rectangular Si cantilever is essentially the same. The particular choice of coating position serves two purposes: increases the cantilever's reflectivity at infrared frequencies, while maintaining the exceptional mechanical properties of the Si cantilever.

As soon as the modified cantilever has been coated it is set on the AFM and the vacuum chamber is evacuated. Calibration of the piezoelectric actuator is done. Next, using a spectrum analyzer we obtain the thermomechanical noise response of the cantilever. The data is averaged 10 times. This response is fit to a Lorentzian and analyzed to obtain the cantilever's spring constant k using eq. (7), its resonant frequency f_0 and its quality factor $Q = f_0/\Delta f$. The uncertainty in these 2 values is around 60 mHz. After this the cantilever with the sphere is taken out and the FIB milling is done. For FIB milling [21] a 4nA current of 30 keV accelerated Ga+ ions is used. The right panel in Figure 1 shows the modified cantilever after this procedure. Finally, the milled, modified cantilever is inserted into the vacuum chamber, the piezoelectric actuator is calibrated again, and the measurement of the cantilever's k, f_0 and Q is repeated.

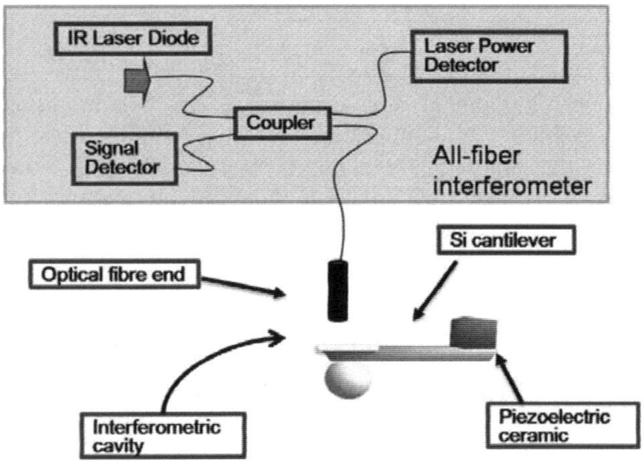

Figure 3. Schematic diagram of the experimental setup used to measure the cantilever's characteristics. The layout of the all-fiber interferometer is also shown.

4. Modified cantilever characteristics before and after the Ion milling process

The cantilever dimensions before and after ion milling are shown in table 1. A scanning electron microscope (SEM) was used to measure the cantilever dimensions as well as the radii of the coated hollow spheres. To compare and obtain accurate values of the dimensions we use a diffraction grating standard. The length of the cantilever in table 1 corresponds to the distance between the beams clamped-end to its tip. Although in theory the effective length of the cantilever should include radius of the attached sphere. Note that after the milling procedure two out of four cantilevers have two different widths. This is due to the nonuniformity of the ion milling process. The opposite ends, one near the tip close to the sphere and the other near the substrate are different and shown in Table 1. The consequence of these kinds of discrepancies will be important when comparing theoretical calculations of the intrinsic cantilever characteristics for these non-prismatic beams with the measured values for these cantilevers. One more effect that should be accounted for in these calculations is the pseudo trapezoidal shape of the cantilever cross-section after milling. This shape is produced because of the Gaussian shape of the focused ion beam. This could be avoided if the cantilever milling is done with a smaller ion current.

Table 1. Cantilever dimensions *before* and *after* FIB milling.

	Width (μm) *Before*	Width (μm) *After*	Length (μm)	Sphere diameter (μm)
Cantilever #4	35.3	3.8/2.4	296.6	80.5
Cantilever #3	35.8	3.0	300.8	83.0
Cantilever #2	31.6	4.1/2.7	299.7	77.2
Cantilever #1	33.0	3.2	291.3	84.0

Care was taken to pick cantilevers with similar characteristics. However intrinsic cantilever discrepancies, different masses of the attached spheres, and the amount of Ag epoxy used, have produced noticeable differences in their characteristics. The characteristic values for the four cantilevers: spring constant k, the resonant frequency f_0, and the quality factor Q are shown in table 2. To easily compare these measured values, table 2 is arranged, so the before and after values are next to each other. The most important values are the Q's. These values changed in a fairly unexpected fashion. Previous research has found that the internal structure of single crystal cantilevers is modified by the Ga+ ions during ion milling [15]; thereby affecting the Q. We believe that in our case this modification should have been minimal since we etched away most of the region interacting with the ion beam. However, the scattered Ga+ ions still lead to radiation damage in the silicon, leading to a decrease in the Q, just as we have observed in our cantilevers. As the Q measurement is done in vacuum, the viscous drag due to the air is negligible in our experiments and only the internal stresses of the cantilever determine the quality factor Q [22]. Nevertheless since the ratio of the volume to surface area changed with the milling process, surface effects might play a larger role on the measured Q for these cantilevers. For example the cantilever Q might have decreased because the cantilevers' surface was contaminated or because the cantilever surface was oxidized by the formation of SiO_2, similar to what other researchers have reported [22]. However, this alone cannot explain the decrease in Q observed here, which is much larger than that reported by other researchers for contamination and oxidation. Wet chemical etching of the oxide layer or contaminants cannot be done because of possible damage to the attached sphere. Not only do the cantilevers Q have valuable information; their k suggests that the modified cantilevers Young's modulus can be treated as those of regular unmodified ones. Using (5) and the cantilevers dimensions from table 1 the calculated spring constant

k is listed in table 2. The measured k values closely match the theoretical values for cantilevers without an attached sphere. This is true even though the values of the measured k have changed by an order of magnitude, from the change in the cantilever width.

Table 2. Cantilevers' intrinsic characteristics *before* and *after* FIB milling.

	f_0 (Hz) *Before*	f_0 (Hz) ***After***	k (N/m) *Before*	k (N/m) ***After***	Q *Before*	Q ***After***
Cantilever #4	5433.3	2269.2	0.3029	0.0311	15311.2	1684.0
Cantilever #3	5704.9	2265.4	0.1074	0.0022	11838.8	3982.8
Cantilever #2	5994.7	2520.8	0.0120	0.0063	11660.8	2290.7
Cantilever #1	5426.0	1984.0	0.0157	0.0022	18901.9	3715.8

In table 3 the cantilevers loss parameters and their minimum detectable force are compared before and after they went through the ion milling process. The minimum force values are calculated at $T =$ 300K using (2) and using the values of table 2. We have used the relative percentage change in equation (9) to emphasize the difference between the before and after values. For the case of the minimum force it leads to the following expression:

$$\text{Relative percentage change } \% = 100 \times \frac{F_{\min}^{Before} - F_{\min}^{After}}{F_{\min}^{Before}} \tag{9}$$

The results in table 3 show that only 1 out of 4 cantilevers present an increase in force sensitivity while the other 3 result in decreased values. The cantilevers minimum detectable force and the loss parameter depend on the values from table 2. Therefore a careful study of the values of this table should result in understanding why the minimum detectable force/loss parameter increased. However the data in table 3 is valuable because it demonstrates the combined effect of the values in table 2.

Table 3. Modified cantilevers minimum force (F_{min}) and loss parameter (γ) *before* and *after* FIB milling occurred. Columns 4 and 5 present the relative percentage change of these quantities due to the FIB process.

	F_{min} (femtoN) *Before*	F_{min} (femtoN) ***After***	Relative F_{min} percentage change (%)	γ *(nN/m)(Hz)$^{-1}$* *Before*	γ *(nN/m)(Hz)$^{-1}$* ***After***	Relative γ percentage change (%)
Cantilever #4	3.099	4.635	-49.6	3.64	8.12	-123.8
Cantilever #3	2.048	0.803	60.8	1.59	0.25	84.6
Cantilever #2	0.667	1.691	-153.4	0.17	1.08	-542.2
Cantilever #1	0.636	0.896	-40.8	0.15	0.30	-98.3

This data shows us the range of minimum force achieved by the modified cantilevers before milling. For all these cantilevers the minimum force already surpasses the sensitivity achieved in previous Casimir force experiments [6,7]. But if this milling technique on the cantilevers is successful the minimum detectible force can be substantially reduced with minimal nano processing. For instance:

Cantilever No. 3 increased its minimum force by 60% mainly due to the 2 order of magnitude decrease in the spring constant k. We believe that during milling not only is the cantilever's width reduced but also its thickness t was decreased. Consequently this cantilever's minimum force decreased. If future research is able to keep the milled cantilevers energy dissipation closer to the non-milled values, and the thickness and width reduction is the same, sensitivities of the order of tens of attonewtons could be achieved.

One of our objectives is to understand if the sphere and the Ag epoxy affect any of the mechanical properties of the unmodified single crystal Si cantilever. Since the cantilever mass, regardless of milling is comparable to the mass of the hollow sphere plus Ag epoxy, it is possible that the quality factor might have decreased because of the internal stresses introduced when the sphere is attached. In addition, we are studying the possibility that the theoretical model of a prismatic bar fixed at one end is no longer suitable for our milled cantilevers and that a better model with boundary conditions which resemble our modified cantilever with sphere attached will be more appropriate. In fact if a different theoretical model is used, the cantilevers theoretical characteristics might be different from the ones resulting from equations (4) and (5), changing the expected cantilever's minimum force. This is indeed the case for non-prismatic cantilevers [23].

To understand the origin of the internal stress in the cantilever, we annealed the milled cantilever. The result was promising; the cantilevers Q returned approximately to their previous non-milled value. However during the annealing process the cantilevers lost their sphere. The positive aspect of these results is that the cantilevers f_0 and k were close to the milled cantilevers values obtained previous to annealing with the hollow sphere attached to its free end. We believe that the reason for this is that most of the cantilevers added mass come from the Ag epoxy, which remained on the milled cantilever after the annealing process, i.e. the hollow metal coated sphere does not add much mass to the system. In any case further studies are currently being done to obtain a more accurate solution that better describes our modified cantilevers.

5. Conclusions

Reducing the width of single-crystal silicon cantilevers with a Cr/Au-coated-hollow sphere attached at their free-end leads to higher force sensitivities appropriate for better Casimir force detection. The FIB milling technique used to reduce the width seems to damage the material of the Si cantilevers and lead to a decrease in the cantilever quality factor. Annealing of the modified cantilevers suggests that this effect could be reversed. Future studies could allow us to attain the minimum detection forces of the order of tens of attonewtons for cantilevers with attached metal coated spheres required for sensitive Casimir force measurements.

Acknowledgments

The experimental section of this work was supported by NSF Grant No.PHY0653657 and the instrumental supplies along with the analysis by DOE Grant No. DE-FG02-04ER46131. The nanofabrication was supported by DOD/DMEA H94003-07-2-0703. R C-G is grateful for the financial support from UCMexus and CONACYT. We also appreciate the technical assistance of Dr. Dong Yan with the FIB milling technique.

References

[1] Casimir H B G 1948 *Poc. K. K. Akad. Wet.* **51** 793
[2] Bordag M, Mohideen U and Mostapenenko V M 2001 *Phys. Rep.* **353** 1 and references therein
Milton K A, 2001 *The Casimir Effect* (Singapore: World Scientific)
[3] Decca R S, Lopez D, Fischbach E, Klimchitskaya G L, Krause D E and Mostepanenko V M, 2007 Eur. Phys. J. C. 51, 963–75 and references therein
[4] Boyer T H 1968 *Phys. Rev.* **174** 1764–76
[5] Serry F M, Walliser D and Maclay J 1998 *J. Appl. Phys.* **84** 2501
Buks E and Roukes M L 2001 *Phys. Rev. B.* **63** 033402

[6] Chan H B, Aksyuk V A, Kleiman R N, Bishop D J and Capasso F 2001 *Science* **291** 1941
[7] Mohideen U and Roy A 1998 *Phys. Rev. Lett.* **81** 4549
Harris B W, Chen F and Mohideen U, 2000 *Phys. Rev.* A **62** 052109
Chen F, Mohideen U, Klimchitskaya G L and Mostepanenko V M 2002 *Phys. Rev. Lett.* **88** 101801
Chen F, Mohideen U, Klimchitskaya G L and Mostepanenko V M 2005 *Phys. Rev. A* **72** 020101 R
Chen F, Mohideen U, Klimchitskaya G L and Mostepanenko V M 2006 *Phys. Rev.* A **74** 022103
Chen F, Klimchitskaya G L, Mostepanenko V M and Mohideen U 2006 *Phys. Rev. Lett.* **97** 170402
Chen F, Mohideen U, Klimchitskaya G L and Mostepanenko V M 2007 *Opt. Exp.* **15** 4823
Chen F, Klimchitskaya G L, Mostepanenko V M and Mohideen U 2007 *Phys. Rev.* B **76** 035338
[8] Bressi G, Carugno G, Onofrio R and Ruoso G 2002 *Phys. Rev. Lett.* **88** 041804
[9] Giessibl F J 2003 *Rev. Mod. Phys.* **75** 949
[10] Stowe T D, Yasumura K, Kenny T W, Botkin D, Wago K and Rugar D 1997 *Appl. Phys. Lett.* **71** 288
Jenkins N E, DeFlores L P, Allen J, Ng T N, Garner S R, Kuehn S, Dawlaty J M and Marohn J A 2004 *J. Vac. Sci. Technol. B* **22**, 909
[11] Stipe B C, Mamin H J, Stowe T D, Kenny T W and Rugar D 2001 *Phys. Rev. Lett.* **87** 096801
Kuehn S, Loring R F and. Marohn J A 2006 *Phys. Rev. Lett.* **96** 156103
[12] Mamin H J, Poggio M, Degen C L and Rugar D 2007 *Nature Nanotech.* **2** 301-6
Budakian R, Mamin H J,. Chui B W and Rugar D 2005 *Science* **307** 408-411
[13] Chiaverini J, Smullin S J, Geraci A A, Weld D M and Kapitulnik A 2003 *Phys. Rev. Lett.* 90 151101
Smullin S J, Geraci A A, Weld D M, Chiaverini J, Holmes S and Kapitulnik A 2005 *Phys. Rev. D* 72 122001
[14] Tseng AA 2004 *J. Micromech. Microeng.* **14** R15–34
Maali A, Cohen-Bouhacin T, Jai C, Hurth C, Boisgard R, Aime J P, Mariolle D and Berin F 2006 *J. Appl. Phys.* **99** 024908
Hodges A R, Bussmann K M and Hoh J H 2001 *Rev. Sci. Instr.* **72** 3880
[15] Prewett P D, Anthony C J, Cheneler D and Ward M C L 2008 *Micro & Nano Letters* **3** 25–28
Antognozzi M, Ulcinas A, Picco L, Simpson S H, Heard P J, Szczelkun M D, Brenner B and Miles M J 2008 *Nanotechnology* **19** 384002
[16] Chen C J 2008 *Introduction to Scanning Tunneling Microscopy* (Oxford University Press, New York) pp 389-95
[17] Chiu H-C, Chang C-C, Castillo-Garza R, Chen F and Mohideen U 2008 *J. Phys. A: Math. Theor.* **41** 164022
[18] Rugar D, Mamin H J and Guethner P 1989 *Appl. Phys. Lett.* **55** 2588
[19] Mamin and Rugar D 2001 *Appl. Phys. Lett.* **79** 3358
[20] Mikromasch cantilevers Part No. CSC12/tipless/no Al
[21] Carl Zeiss Leo XB1540
[22] Yasumura K, Stowe T D, Chow E M, Pfafman T, Kenny T W, Stipe B C and Rugar D 2000 *J. Microelec. Syst.* **9** 117
[23] Fernando S, Austin M and Chaffey J 2007 *J. Phys. D: Appl. Phys.* **40** 7652-55

Journal of Physics: Conference Series **161** (2009) 012006 doi:10.1088/1742-6596/161/1/012006

An experimental apparatus for measuring the Casimir effect at large distances

P Antonini[1], G Bimonte[2], G Bressi[3], G Carugno[1], G Galeazzi[4,5], G Messineo[1,4] and G Ruoso[5]

[1] INFN sez di Padova, via Marzolo 8, 35131 Padova, Italy
[2] Dipartimento di Scienze Fisiche Università di Napoli Federico II Complesso Universitario MSA, Via Cintia, 80126 Napoli, Italy and INFN, Sezione di Napoli, Napoli, Italy
[3] INFN sez. di Pavia, via Bassi 6, 27100 Pavia, Italy
[4] Dipartimento di Fisica, Università di Padova, via Marzolo 8, 35131 Padova, Italy
[5] INFN Lab. Naz. Legnaro, viale dell'Università 2, 35020 Legnaro (Pd), Italy

E-mail: `Giuseppe.Ruoso@lnl.infn.it`

Abstract. An experimental set-up for the measurement of the Casimir effect at separations larger than a few microns is presented. The apparatus is based on a mechanical resonator and uses a homodyne detection technique to sense the Casimir force in the plane-parallel configuration. First measurements in the 3-10 micrometer range show an unexpected large force probably due to patch effects.

1. Introduction and motivation

In quantum electrodynamics the description of the vacuum is modified by the presence of boundary conditions: the introduction of conducting surfaces changes the ground state of the electromagnetic field in a way that depends on the allowed mode frequencies. A variation of the position of the surfaces results in a change of the mode frequencies and thus in an energy change. This corresponds to a net force acting on the surfaces, and is known as Casimir effect [1, 2, 3]. The attractive force between two parallel and perfectly conducting metal plates is

$$F = \frac{\pi^2 \hbar c S}{240\, d^4}, \tag{1}$$

where c is the speed of light, $\hbar$ the reduced Planck constant, S the surface of the plates, and d their separation.

After the pioneering works of Sparnaay [4] and van Blokland and Overbeek [5], the experimental study of the Casimir effect has received a strong impulse in the last decade, following the first precision measurement conducted by Lamoreaux [6]. Several experiments followed these measurements, all but one using a plane-sphere geometry: Mohideen *et al.* [7], Decca *et al.* [8], Ederth [9] and Chan *et al.* [10]. The group of Bressi *et al.* [11] was able to perform a measurement in the plane parallel geometry originally proposed by Casimir. With this geometry the experimental difficulties are bigger, and this explains why the precision is worse compared to other configurations. More on the subject can be found on this proceedings.

The Casimir formula (1) is valid only in an ideal situation, and has to be corrected to take into account for example the finite conductivity of the metal plates and their roughness. The experimental study of these corrections, that are not negligible only at very small distances ($\leq$1 μm), was possible up to now only with the plane-sphere configuration due to its better precision. These measurements were performed for separations smaller than 1 micrometer, while for larger distances the Casimir force was below the sensitivity of the experiments.

An important correction to the Casimir formula is due to the presence of thermal photons: the original calculation has been made at zero temperature but experiments are normally performed at room temperature. Due to the form of the black body spectrum, thermal corrections start to be significative only for separations larger than a few microns, and become the dominant part for distances larger than

$$\lambda_T = \frac{2\pi c}{\omega_t} = \frac{\hbar c}{k_B T}, \tag{2}$$

corresponding to 7 μm at 300 K [12]. Since the separation has to be rather large, the most promising geometry for observing the thermal contributions to the Casimir force is the plane parallel one, where the possibility to use relatively wide surfaces results in measurable forces. A plane - cylinder geometry has also been recently proposed [13, 14].

The effect of thermal radiation in a Casimir like experiment has recently been measured by the group of E. A. Cornell [15]: in this case a Casimir-Polder configuration has been used [16], studying the force between a bulk object and a gas-phase atom [17].

In this paper we present a set-up based on the plane-parallel geometry where we aim to measure the Casimir force for distances of several microns, thus entering in the regime of strong thermal corrections. Our apparatus is based on an oscillating plate whose motion is detected using a Michelson type interferometer. This mechanical resonator is made of aluminized silicon and its motion is excited at low frequency by means of a movable source plate. A homodyne detection scheme is employed, allowing for very long integration times necessary to measure the small motions of the resonator induced by the weak force. Care has been taken in the design of the apparatus to reduce background noises and systematic effects. Vibrations are in fact driving the resonator and are thus limiting the sensitivity and increasing the integration time. The most important systematic contribution comes from the presence of voltage biases between the plates even when they are shortcut. These voltages must be counterbiased with high precision otherwise they will hide the Casimir force.

The exact calculation of the finite temperature corrections to the Casimir force is still an open question: there are at least two models, both are predicting a sizable change in the force for separations above 1 - 2 microns. Without the necessity to enter into this dispute, we quote as an example the results of Bordag *et al.* [3]: in the plane-parallel configuration, at a temperature T = 300 K and a separation of 5 μm, the correction to the standard force is about 50%, thus a precision of the order of 10-20% in the measurement of the force would then be sufficient to separate this contribution.

2. Experimental set-up

A detailed description of the apparatus and of the detection technique has been given in [18], here only a brief summary will be given.

2.1. The apparatus

The plane-parallel configuration is the most convenient geometry for the measurement of the force at large separations, providing a possibly measurable signal even at distances of the order of a few microns. In our set-up two flat parallel surfaces are fronting each other. One surface is called the source and the other the resonator. The distance between the two can be changed by means of a piezoelectric actuator which holds the source. The source can exert a force on the resonator, through an electrostatic field or the Casimir effect. The resonator is free to move around its equilibrium position, and by monitoring changes of its position using a laser interferometer it is possible to gather information on the forces acting on it.

The resonator is made of silicon crystal and is essentially a plate 1 cm × 1 cm × 50 μm connected to a frame through four parallel wires 1 cm long and 50 × 50 μm^2 in section. Plate, wires and frame are carved from the same crystal which was initially coated with 0.4 μm of aluminium. We measured its resonance frequency to be ν_r = 125 Hz with a quality factor around 2000.

The source is a 1.2 cm × 1.5 cm, 500 μm thick aluminium block, with its surface polished with diamond at the optical level. Measurement of the overall flatness showed a peak to valley distortion less than 200 nm. The source is glued to a glass slab, which is in turn glued to a piezoelectric actuator (PZT1) providing a controllable translational motion of the source itself. The PZT is clamped to a 6-axis stepper-motor-controlled positioning stage (Thorlabs, 6-Axis NanoMax Nanopositioner), which is used to control the parallelism between the two plates. This arrangement is shown in figure 1.

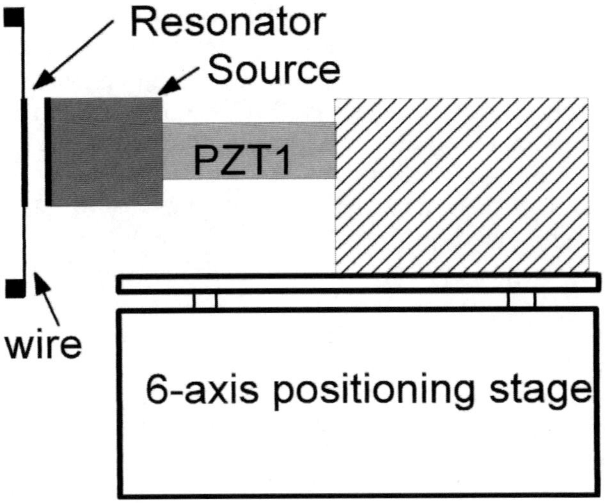

Figure 1. Side view showing the resonator, the source and the positioning system used for the parallelization.

The source, the resonator and the positioning stage are in a vacuum chamber, at a pressure of about 3 × 10^{-6} mbar. The vacuum is provided by a vibration-free ion pump. The whole experiment lies inside a clean room (class 1000), with thermal stabilization at the level of ΔT = 0.1°, but showing a long term drift of the order of half a degree per day. The motion of the resonator is detected by means of a Michelson type interferometer set-up with a 1 mW amplitude stabilized He-Ne laser at 633 nm. The interference signal is collected by a photodiode, whose voltage readout V_M converts into distance by using:

$$\delta x = \frac{\lambda}{2\pi} \frac{V_M}{V_{fr}}. \quad (3)$$

where $\lambda = 633$ nm and V_{fr} is the voltage corresponding to a interference fringe, in our case normally $V_{fr} = 2.3$ V.

In order to achieve the maximum sensitivity of the resonator, particular care has been taken to minimize its free motion. The whole apparatus is mounted on a passive low frequency damper and on a floating optical table. As explained next, we want to minimize the low frequency noise, below a few Hz: the passive damper (Minus-k Technologies 250 BM 6) is a mechanical low pass filter with a cut-off frequency at 0.5 Hz in the transmissibility curve. The clean room also represents a quiet environment to avoid acoustical noise: all the electronics lie in a separate control room. Moreover, a special metallic enclosure has been constructed around the apparatus. This enclosure has walls equipped with copper tubes through which temperature stabilized water flows, thus allowing very good long term stability of the entire system and thus eliminating the temperature drifts of the clean room. We discovered that temperature stability is crucial in order to perform repeated measurements over several days. In our case a stability of 0.1 degree of the experimental apparatus has been obtained for a time span of several days.

2.2. Detection technique

The set-up presented in this paper is based on an amplitude homodyne detection technique, already described in [19].

A static detection scheme is not sensitive enough, because it is affected by low-frequency drifts due to thermal expansion, electronic instabilities, drift of the laser frequency. For this reason we use the dynamical homodyne detection technique: we move the source with a periodical movement in the direction of the resonator. The Casimir force acting between the two surfaces makes the resonator move, and its motion is detected using the interferometer. If the chosen frequency is well below the resonator proper frequency, a fixed phase difference will be present between source motion and resonator motion, thus allowing for vector averaging of the interferometer signal.

If between the source and the resonator there is a spatially dependent force $F(d)$, then the equation of motion of the equivalent harmonic oscillator is

$$m\ddot{x}_r(t) = -m\omega_r^2 x_r(t) + F(d), \tag{4}$$

where $x_r(t)$ is the position of the resonator, $\omega_r = 2\pi\nu_r$ is its angular proper frequency and m its mass.

We can decompose the separation d between source and resonator into two components, namely a fixed distance d_0 and a time-dependent component $x_s(t)$, due to the movement of the source, that is $d = d_0 + x_s(t)$ ($|x_s/d_0| \ll 1$). The free motion of the resonator is considered much smaller and is neglected. In the case of a force of the form $F(d) = C/d^n$, a periodic modulation of the position of the source $x_s = x_s^0 \cos\omega_s t$ (with $\omega_s \ll \omega_r$), results in (at first order):

$$F(t) = \frac{C}{d_0^n}\left(1 - n\frac{x_s^0}{d_0}\cos\omega_s t\right). \tag{5}$$

The oscillation of the resonator will be at the same frequency, and its amplitude together with the value of the spring constant of the resonator allows a calculation of the force parameters. In the Fourier spectrum of the solution $x_r(t)$ of equation (4) there is a peak at the frequency $\nu_s = \omega_s/2\pi$, whose amplitude is:

$$A_s^0 = n\frac{Cx_s^0}{m\omega_r^2 d_0^{n+1}}. \tag{6}$$

An overall calibration of the system is obtainable using controllable electrostatic forces. In fact, a constant voltage V between the resonator and the source will induce a force

$$F_V(d) = -\frac{1}{2}\epsilon_0 S\left(\frac{V}{d}\right)^2, \tag{7}$$

Journal of Physics: Conference Series **161** (2009) 012006 doi:10.1088/1742-6596/161/1/012006

It is then possible to compare the amplitude induced by the Casimir force A_C and the one obtained by the voltage calibration at fixed bias A_V:

$$A_C = \frac{1}{60}\frac{\pi^2 \hbar c S}{m\omega_r^2 d_0^5} x_s^0 \tag{8}$$

$$A_V = \frac{\epsilon_0 S V^2}{m\omega_r^2 d_0^3} x_s^0. \tag{9}$$

The calibrations are useful to infer the parameters common to both expressions, without relying on their direct determination.

2.3. Stray effects

In order to measure the Casimir force, one must be able to eliminate or compensate other forces acting between the two plates. In particular, forces of electrostatic nature can be present. The origin of these forces can be divided into two different categories: residual bias voltage and patch potentials.

Even when two conductors are shortened together, a residual bias V_0 could be present between them due to contact potential. A counterbias $-V_0$ eliminates the effect of this residual potential, but one has to be sure that its value is not changing with time and gap separation. A careful study must be performed. Counterbias accuracy must be kept at a few mV level in order to disentangle the contribution of the Casimir force.

It is well known that local changes in the crystallographic directions exposed at the surface of a clean polycrystalline metal, as well as surface contaminations, give rise to a point-dependent surface density $m(\vec{x})$ of dipoles on the surface of a metal [20]. The presence of such an inhomogeneous double layer of charges results in a spatially varying electrostatic potential $V(\vec{x})$ at the metal surface, relative to its interior, known as "patch" potential. This zero-mean electrostatic component cannot be eliminated by a counterbias, and gives rise to a force whose distance dependance is related to the spatial distribution of the patches. In order to calculate the force of attraction between the two plates, it is convenient to describe the patches with a characteristic length scale λ_L, which satisfies the condition:

$$d \ll \lambda_L \ll L \tag{10}$$

where d is the gap separation between the plates having lateral dimension L. The patch potentials $V(\vec{x})$ can be split as:

$$V(\vec{x}) = V^{(L)}(\vec{x}) + V^{(S)}(\vec{x}) \tag{11}$$

In this equation, $V^{(L)}(\vec{x})$ represents the *long wavelength* component, that does not vary appreciably under displacements smaller than λ_L along the surface, while $V^{(S)}(\vec{x})$ represents the *short wavelength* component. Physically, $V^{(L)}(\vec{x})$ can be thought to arise from the charges of a few isolated impurities distributed over the surface of the plates, while $V^{(S)}(\vec{x})$ describes the patch potentials of the microscopic crystallites forming the surfaces. The corresponding variances σ_L^2 and σ_S^2 of the two components give rise to a force with a characteristic distance dependence. Following [20] it is possible to obtain a force of the form:

$$F(d) = \epsilon_0 L^2 \left(\frac{\sigma_L^2}{2d^2} + \frac{2\sigma_S^2}{k_{\max}^2 - k_{\min}^2} \int_{k_{\min}}^{k_{\max}} dk \frac{k^3}{\sinh^2(kd)} \right) \tag{12}$$

where the wave numbers $k_{\max,\min} = 2\pi/\lambda_{\min,\max}$ are related to the maximum and minimum sizes $\lambda_{\max,\min}$ of the crystallites.

3. Experimental results

The homodyne detection technique has been optimized looking for the modulation frequency of the source position providing the best signal to noise ratio. Long noise spectra are taken at the beginning of each measurement session, as a result the source is normally modulated at frequencies between 7 Hz and 14 Hz. This frequency range, well below the resonator proper frequency, ensures a complete decoupling between source motion and resonator itself due to mechanical and electrical pick-ups, as it was checked by a very long run with the plates kept at a separation of about 50 microns. Most of the measurements have been performed with 45 nm amplitude for the source motion. The resonator motion noise was normally at the level of $3 \cdot 10^{-11}$ m/$\sqrt{\text{Hz}}$, but this figure was stable only at night, degrading by a factor of 2 during daytime.

By imposing an external bias voltage V_g to the gap, it is then possible to calibrate the apparatus. This is done in order to obtain the value of the residual bias voltage V_0, the effective resonator mechanical response and the absolute gap width separation. Again, for a detailed description of these steps, we refer to a previous paper [18], while presenting here the most recent results.

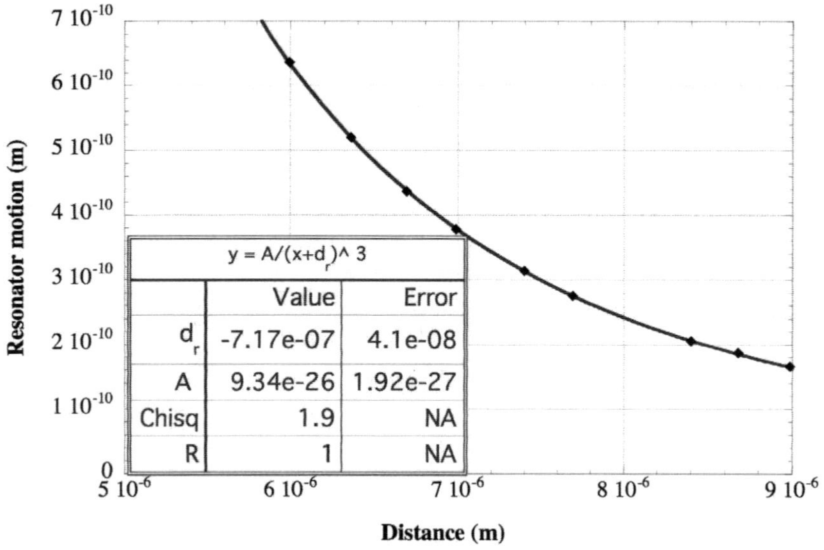

Figure 2. Measured signal versus distance when imposing an external bias voltage in the gap. The solid line is the best fist with a $1/d^3$ function. The experimental error for each measurement is smaller than the marker size.

By applying different values of an external bias voltage in the gap at fixed distance between the plates, one obtain the residual bias V_0. The resulting value was $V_0 = 244.7 \pm 0.8$ mV, with a stability of 1-2 mV over a few days period. For longer time scales changes up to 10 mV have been observed: the reason for these changes is not yet clear to us, a possibility is the temperature of the apparatus. The value of V_0 has also been measured for different distances, and no change has been observed.

The measurement of the resonator motion versus gap separation, while keeping the external bias V_g constant, provides the reference distance d_r and also a cross check of the resonator mechanical response. The absolute gap separation is obtained by adding to the reference distance d_r the displacement of the source given by the transducer PZT1.

Figure 2 shows a typical calibration response. The important parameter is the error on d_0, which is 40 nm. The coefficient A differs from the expected one by about 5 %.

Casimir-type measurement are performed when $V_g = -V_0$, thus having a net zero voltage between the plates. Figure 3 shows the results obtained in 4 different measurement runs, performed in different days as indicated. The abscissa of the figure is the real gap separation already corrected using d_r. The ordinate shows the force derivative, i.e. the pulsating term of equation (5). Data are fitted with a function of the form A/d^n, where A and n are free parameters and d is the gap separation. Results of the fits are rather astonishing: for n, values are in the range 3.5-5, and A results to be way too large for being due to the Casimir force. Imposing a Casimir type dependence ($n = 5$), the experimental coefficient A is about 100 times larger than expected.

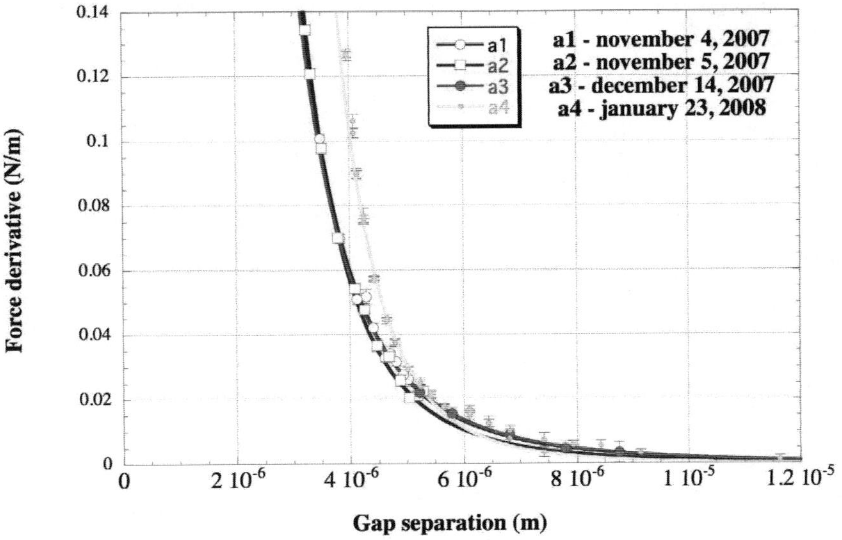

Figure 3. Summary of the measurements performed in the Casimir like configuration, i.e. a net zero voltage in the gap. Four different sets have been taken, corresponding to different measurement days. The abscissa is the absolute distance between the plates, as obtained by adding to the reference value d_r the displacement of PZT1.

It is then important to try to understand the reasons for these results. The presence of an uncompensated bias voltage is excluded since a too large value would be necessary, much larger than the error on the determination of V_0. The influences of AC voltages on the gap have been eliminated by using filters along the connecting cables and replacing the cables themselves. Also the effect of the residual pressure was studied by changing the amount of gas in the chamber: no variation has been seen. A more detailed analysis concerns the presence of patch effects.

4. Patch effects

In order to study the presence of patches in the surface of the plates, a Kelvin probe analysis [21] has been done using plates similar to those mounted on the apparatus. Measurements of the work function across the samples were made by scanning the surfaces with a tip having an approximate diameter of 2 mm, with steps of 317.5 μm. The data (see Figure 4) showed that the work function exhibits significant fluctuations over distances of the order of one millimeter, and therefore this should be regarded as a long wavelength component of the patch potential in a force measurement involving separations of the order of a few microns. The variances of the work functions were found to be in the range 10 - 30 mV.

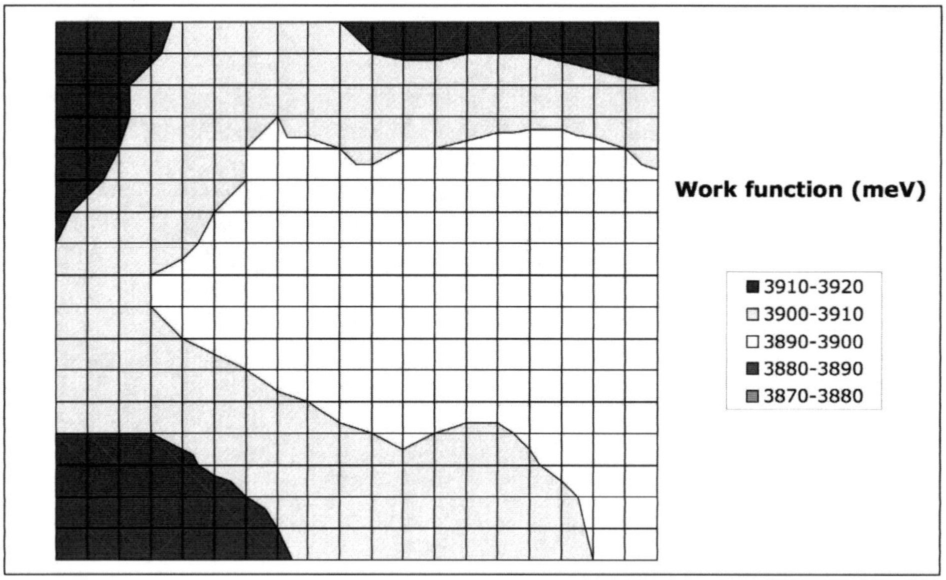

Figure 4. Kelvin probe analysis of a sample of bulk aluminium of the same type of the one used in the apparatus for measuring the Casimir effect. The measuring tip has a diameter of about 2 mm, each little square has a 0.35 mm side.

Another analysis that we performed was the X ray diffractometry, in order to determine the grain size of the microscopic crystallites forming the surfaces. This analysis gave a 0.1 μm size for the grains forming the aluminium bulk, and a 0.03 μm size for the aluminium coating on the silicon resonator. This analysis gives no information on the amplitude and variance of the work function associated with the crystallites, which should form the short wavelength component of the patch related force.

The three sets of data a1, a2, a3 of Figure 3 have been fitted together with the derivative $F'(d)$, with respect to d, of the force $F(d)$ of equation (12), in order to extract from the fit the parameters related to the short and long wavelength components. The result is showed in Figure 5, the best fit is obtained with $\lambda_{\rm min} = 3\,\mu$m, $\sigma_S = 91.3$ mV, $\sigma_L = 51.6$ mV. As it is clear from the figure, the fit is good, and the probable values for σ_S and σ_L appear both very reasonable. The favored value for σ_L is perhaps a bit larger than what expected from the Kelvin probe measurement. However, it should be considered that the tip used is very thick (2 mm), and therefore their data actually represent smoothed measurements of the work function, over areas of 4 mm^2. The smoothing may have led to underestimate the real fluctuation of the work function at scales of, say, 200-300 microns.

On the other side, the value $\lambda_{\rm min} = 3\,\mu$m is indicating that the short wavelength component is not the one seen by the X ray diffractometry. There could be some other patches, for example again due to impurities, with micron size areas, that it is not possible to see with an independent surface analysis. This leaves still some open question on the understanding of our data.

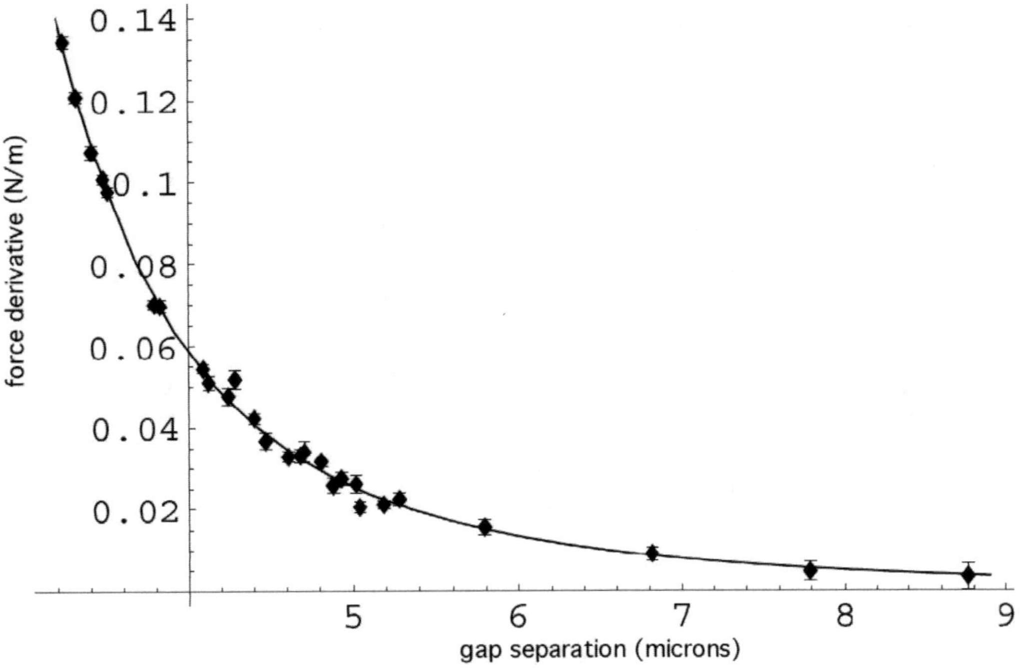

Figure 5. Fitting of the sets a1, a2, a3 of Figure 3. The solid line is the best fit with the derivative of the force (12), for $L = 1$ cm, $\lambda_{\rm min} = 3\,\mu$m, $\sigma_S = 91.3$ mV, $\sigma_L = 51.6$ mV.

5. Conclusions

We have set up an apparatus to measure the Casimir effect between parallel plates at separation in the range up to 5-6 microns. The possibility to achieve good parallelization and small gaps, down to 3 μm, using 1 cm^2 surfaces has been accomplished. The effort to measure the Casimir force is not yet successful, having encountered a much larger force of instrumental origin. Possible explanation of this force could be related to patch effects, but a clear identification of the patch potentials was not possible. It has to be noted that the use of aluminium might have enhanced the presence of patches: for this reason another set up is under construction with gold plated surfaces, which should reduce the effect of patches. If this problem will persist also using gold, this will put a serious obstacle in the measurement of the Casimir effect at a large separation.

Acknowledgments
We are grateful to the high skill of E. Berto (Università di Padova) and F. Calaon (INFN Padova). They helped with the control of the vacuum set-up and the construction of mechanical components. G. C. thanks his father for the help he provided to the construction of the experimental set-up, tests and improvements during the years.

Bibliography

[1] Casimir H B G *Proc. K. Ned. Akad. Wet.* **51**, 793 (1948).
[2] Lifshitz E M *Sov. Phys. JETP* **2**, 73 (1956).
[3] Bordag M, Mohideen U, and Mostepanenko V M *Phys. Rep.* **353**, 1 (2001).
[4] Sparnaay M J *Physica (Utrecht)* **24**, 751 (1958).
[5] van Blokland P H G M, and Overbeek J T G *J. Chem. Soc. Faraday Trans. I* **74**, 2637 (1978).
[6] Lamoreaux S K *Phys. Rev. Lett.* **78**, 5 (1997).
[7] Mohideen U and Roy A *Phys. Rev. Lett.* **81**, 4549 (1998).
[8] Decca R S, Fischbach E, Klimchitskaya G L, Krause D E, López D, and Mostepanenko V M *Phys. Rev. D* **68**, 116003 (2000).
[9] Ederth T *Phys. Rev. A* **62**, 062104 (2000).
[10] Chan H B, Aksyuk V A, Kleiman R N, Bishop D J, and Capasso F *Science* **291**, 1941 (2001).
[11] Bressi G, Carugno G, Onofrio R, and Ruoso G *Phys. Rev. Lett.* **88**, 041804 (2002).
[12] Genet C, Lambrecht A, and Reynaud S *Phys. Rev. A* **62**, 012110 (2000).
[13] Brown-Hayes M, Dalvit D A R, Mazzitelli F D, Kim W J, and Onofrio R *Phys. Rev. A* **72**, 052102 (2005).
[14] Hertzberg M P, Jaffe R L, Kadar M, and Scardicchio A *Phys. Rev. Lett.* **95**, 250402 (2005).
[15] Obrecht J M, Wild R J, Antezza M, Pitaevskii L P, Stringari S, and Cornell E A *Phys. Rev. Lett.* **98**, 063201 (2007) .
[16] Casimir H B G, and Polder D *Phys. Rev.* **73**, 360 (1948).
[17] Antezza M, Pitaevskii L P, Stringari S, and Svetovoy V B *Phys. Rev. Lett.* **97**, 223203 (2006).
[18] Antonini P, Bressi G, Carugno G, Galeazzi G, Messineo G, and Ruoso G *New J. Phys.* **8**, 239 (2006).
[19] Bressi G, Carugno G, Galvani A, Onofrio R, Ruoso G, and Veronese F *Class. Quantum Grav.* **18**, 3943 (2001).
[20] Speake C C, and Trenkel C *Phys. Rev. Lett.* **90**, 160403 (2003).
[21] Measurements were performed by KP Technology, Burn Street, Wick, Caithness, KW1 5EH, Scotland

Journal of Physics: Conference Series **161** (2009) 012007 doi:10.1088/1742-6596/161/1/012007

Size quantization effects in thin film Casimir interaction

A Benassi[1,2] and C Calandra[2]

[1] CNR/INFM-National Research Center on nanoStructures and bioSystems at Surfaces (S3), Via Campi 213/A, I-41100 Modena, Italy

[2] Dipartimento di Fisica, Università di Modena e Reggio Emilia, Via Campi 213/A, I-41100 Modena, Italy

E-mail: benassi.andrea@unimore.it

Abstract. We investigate the role of size quantization in the vacuum force between metallic films of nanometric thickness. The force is calculated by the Lifshitz formula with the film dielectric tensor derived from the one-electron energies and wavefunctions under the assumption of a constant potential inside the film and a uniform distribution of the positive ion charge. The results show that quantization effects tend to reduce the force with respect to the continuum plasma model. The reduction is more significant at low electron densities and for film size of the order of few nanometers and persists for separation distances up to $10-50$ nm. Comparison with previous work indicates that the softening of the boundary potential is important in determining the amount of the reduction. The calculations are extended to treat Drude intraband absorption. It is shown that the inclusion of relaxation time enhances the size quantization effects in the force calculations.

1. Introduction

Thin metal films of nanometric size can have physical properties different from their bulk counterparts due to quantum size effects (QSE) caused by the valence electron confinement in the direction normal to the film surfaces [1, 2, 3]. Since thin films are among the basic components of modern nanodevices an increasing attention has been addressed to the quantum vacuum fluctuations forces (van der Waals and Casimir forces) between them. Theoretical determinations of these forces have been based mainly on a continuum description of the material dielectric properties, neglecting modifications of the electronic structures due to the boundaries, i.e. under the assumption that the film dielectric properties be the same of the bulk material [4, 5, 6, 7, 8]. This assumption is not valid for small systems with low electron density and/or large surface to volume ratio. In such cases a significant size dependence of the basic properties, like the Fermi energy and the density of electron states, arises as a consequence of the electron confinement [9].

In a previous paper [10] we have shown that the inclusion of QSE in the plasma model of a free electron metal leads to modifications of the force intensity between nanometric size films with respect to the bulk plasma model which range from several to few percent depending upon the film electron density and the separation distance. The calculated forces show quantum size oscillations and are less intense compared to those determined with the bulk plasma model. The calculations were performed using the *particle in a box* model [11] (hereafter indicated as PBM)

in which independent electrons are confined along the direction normal to the film surfaces by hard walls and behave as a two dimensional gas parallel to the surfaces. Such model represents a simplified picture of the one electron potential in the film.
In this paper we improve our description of the film dielectric properties along two lines of development: first we introduce a finite well model for the one-electron potential along the surface normal, second we include intraband absorption by introducing the relaxation time in a manner that allows to keep number conservation. As to the first issue, we notice that the use of a soft confining potential like a finite well, besides being closer to the real shape of the one-electron potential as determined by first principles calculations [12, 13, 14, 15], allows for a better treatment of QSE for films of small thickness, since it does not introduce a priori a distance within which electrons are to be confined and leaves the Fermi energy free to oscillate with the film size [9, 16], a feature that is not present in the PBM model, where the Fermi energy is kept equal to the bulk value. The inclusion of Drude absorption in vacuum force calculations has been a subject of considerable debate, mainly in relation with finite temperature corrections [17, 18, 19, 20, 21, 22]. Very accurate measurements have been reported that shows that the simple plasma model of bulk dielectric can give better agreement with the experiments than the model with finite relaxation times [23]. Our results refer to the $T = 0°$K case where difficulties do not seem to be present [24]. Our purpose is to understand how QSE may affect the force between thin films when intraband absorption is included. In this paper we introduce such corrections and we show their importance in the calculation of the intensity of the force between metallic films.

2. Theoretical framework

We consider two identical metal films of thickness D with plane boundaries separated by a distance ℓ (see figure 1 (a)). D represents the extension, along the z-direction normal to the surface, of the positive ion distribution, which is supposed to be uniform with the same density of the bulk system. By extending previous results for isotropic slabs to the case of films with anisotropic dielectric tensor we can write the expression of the force per unit area as [25, 26, 27, 28]:

$$F = -\frac{\hbar}{2\pi^2}\int_0^\infty k\,dk \int_0^\infty d\omega\,\gamma(\omega)\left[\frac{Q_{TM}(i\omega)^2}{1-Q_{TM}(i\omega)^2} + + \frac{Q_{TE}(i\omega)^2}{1-Q_{TE}(i\omega)^2}\right] \tag{1}$$

$$Q_{TM} = \frac{\rho_{TM}(1-e^{-2\gamma_{TM}D})}{1-\rho_{TM}^2 e^{-2\gamma_{TM}D}}e^{-\gamma\ell} \qquad Q_{TE} = \frac{\rho_{TE}(1-e^{-2\gamma_{TE}D})}{1-\rho_{TE}^2 e^{-2\gamma_{TE}D}}e^{-\gamma\ell} \tag{2}$$

$$\rho_{TM} = \frac{\gamma_{TM}(\omega)-\gamma\epsilon_{xx}(\omega)}{\gamma_{TM}(\omega)+\gamma\epsilon_{xx}(\omega)} \qquad \rho_{TE} = \frac{\gamma_{TE}(\omega)-\gamma(\omega)}{\gamma_{TE}(\omega)+\gamma(\omega)} \tag{3}$$

$$\gamma_{TE}(\omega) = \sqrt{k^2 - \frac{\omega^2}{c^2}\epsilon_{xx}(\omega)} \qquad \gamma_{TM}(\omega) = \sqrt{\left(\frac{k^2}{\epsilon_{zz}(\omega)} - \frac{\omega^2}{c^2}\right)\epsilon_{xx}(\omega)} \tag{4}$$

$$\gamma(\omega) = \sqrt{k^2 - \frac{\omega^2}{c^2}} \tag{5}$$

Here ϵ_{xx} and ϵ_{zz} are the diagonal components of the dielectric tensor along the planar directions and along the surface normal respectively. We assume $\epsilon_{xx} = \epsilon_{yy}$ and the off diagonal components to be zero. This assumption is consistent with the two dimensional gas behavior of the electrons parallel to the surface. The anisotropy of the dielectric tensor is a consequence of the finite extension of the film along the z-direction and it is the main feature introduced by the size

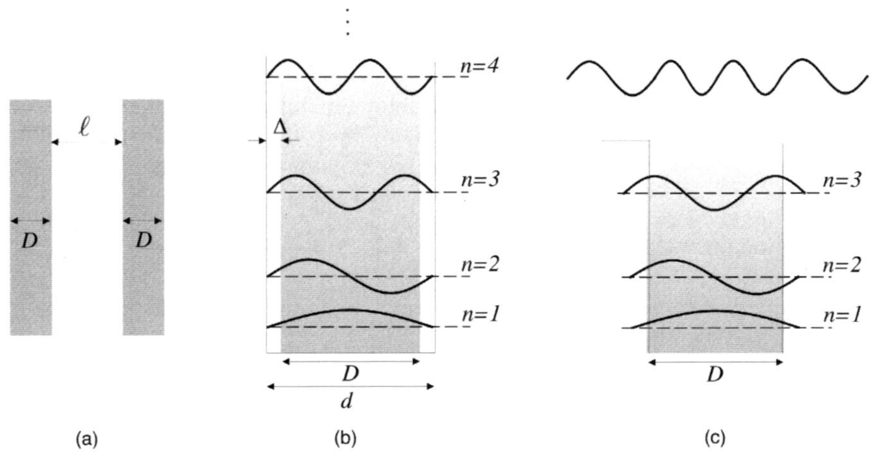

Figure 1. (a) Sketch of the two identical interacting films. (b) Infinitely deep quantum well with artificial spill-out. (c) Finite quantum well with natural spill-out.

quantization. To calculate the force one needs the expression of the dielectric tensor, which has to be derived from the film electronic structure. We assume an independent particle model with the one electron potential $V(z)$. The electron energies are given by

$$E_{\mathbf{k}_\parallel,n} = \frac{\hbar^2}{2m}\mathbf{k}_\parallel^2 + E_n \tag{6}$$

i.e. they are described by the continuous quantum number $\mathbf{k}_\parallel$ and by the discrete sub-band index n coming from the quantization of the perpendicular wavevector, m being the electron mass. The corresponding wavefunctions are given by

$$\psi_{\mathbf{k}_\parallel,n}(\mathbf{r}_\parallel, z) = \frac{1}{\sqrt{A}} e^{i\mathbf{k}_\parallel\cdot\mathbf{r}_\parallel}\phi_n(z) \tag{7}$$

here A is the surface area, $\mathbf{k}_\parallel$ and $\mathbf{r}_\parallel$ are two dimensional wavevectors parallel to the surface, and $\phi_n(z)$ is supposed to be independently normalized. The functions $\phi_n(z)$ are the solutions of the equation

$$\left\{\frac{\hbar^2}{2m}\frac{\partial^2}{\partial z^2} + V(z)\right\}\phi_n(z) = E_n\phi_n(z) \tag{8}$$

with the proper boundary conditions. The Fermi energy E_F is obtained through the *aufbau* procedure i.e. by arranging the eigenvalues in ascending numerical order and counting until the number of states needed to accommodate all the electrons in the film is reached. This procedure leads to a Fermi energy that depends upon the film size and is generally different from the bulk value [16]. This can be understood by noting that, to ensure charge neutrality, the number of electrons and the number of ions per unit area have to be equal. The electron density $n(z)$ can be simply obtained from the wavefunctions:

$$n(z) = \frac{1}{2\pi}\sum_{m=1}^{m_0}\left(\frac{2mE_F}{\hbar^2} - E_n\right)|\phi_m(z)|^2 \tag{9}$$

where m_0 is the label of the last occupied state, while the ion density is simply given by $n_0 = k_{F_B}^3/3\pi^2$, where k_{F_B} is the bulk Fermi wavevector. By integrating the densities along the z axis and imposing that both give the same number of charges per unit area, one gets the relation

$$\frac{1}{2\pi}\sum_{m=1}^{m_0}\left(\frac{2mE_F}{\hbar^2}-E_n\right)=n_0D \tag{10}$$

which, for finite D values, is generally not satisfied if one replaces E_F with its bulk counterpart $E_{F_B}=\hbar^2k_{F_B}^2/2m$. In the case of the PBM model this equation is not satisfied, since one assumes that the Fermi level is the same as in the bulk. To obtain charge neutrality one has to impose the additional condition that the electron density be confined on a length d larger than D. This artificially introduces the electronic charges spill-out but has the consequence that the average electronic density is lower than the ion density [9, 10, 29, 30].

For the purpose of the present study we assume the potential to be a finite well $V(z)=-V_0$ inside the film and zero outside. For such finite well model (FWM) the energies of the bound states can be written as

$$E_{\mathbf{k}_{\|},n}=\frac{\hbar^2}{2m}\mathbf{k}_{\|}^2+\frac{\hbar^2}{2m}k_{zn}^2-V_0 \tag{11}$$

where k_{zn} are the quantized transverse wavevectors. They are obtained from the equation giving the condition for the existence of bound states in a quantum well [31]:

$$k_{zn}=\frac{n\pi}{D}-\frac{2}{D}sin^{-1}\left(\frac{k_{zn}}{k_0}\right) \tag{12}$$

with $k_0=\sqrt{2mV_0}/\hbar$. Notice that the first term at the second member is the value of the transverse wavevector for an infinite well model of size D and the second term goes to zero as V_0 goes to infinity. This implies that, for given film size and number of electrons, the Fermi energy referred to the well bottom is higher in the infinite well model. Notice that when V_0 goes to infinity one does not recover the PBM, since the Fermi energy is varied with respect to the bulk value in order to satisfy the charge neutrality condition (10). Figure 1 (b) and (c) illustrate the difference: in the PBM the electronic charge density is confined within a distance $d=D+\Delta$, larger than the size of the ionic charge distribution, to allow for the electronic charge spill-out and to ensure global neutrality for $E_F=E_{F_B}$. In the FWM the charge spill-out results naturally from the behaviour of the single particle states while the charge neutrality is achieved by varying the Fermi energy with respect to the bulk value. Obviously in the limit of infinitely deep well (hereafter indicated as IWM) the electronic charge turns out to be entirely localized within the length D and the Fermi energy is strongly increased compared to its bulk value.

Once the electron energies and wavefunctions have been obtained, one can calculate the dielectric tensor from the expression [11]:

$$\begin{aligned}\epsilon_{\alpha\alpha}(\omega) &= 1-\frac{\omega_P^2}{\omega^2}-\frac{8\pi e^2}{Adm^2\omega^2}\sum_{\mathbf{k}_{\|},n}\sum_{\mathbf{k}'_{\|},n'}f(E_{\mathbf{k}_{\|},\mathbf{n}})(E_{\mathbf{k}_{\|},\mathbf{n}}-E_{\mathbf{k}'_{\|},\mathbf{n}'})\times\\ &\times\frac{|\langle\psi_{\mathbf{k}_{\|},n}|\hat{p_\alpha}|\psi_{\mathbf{k}'_{\|},n'}\rangle|^2}{(E_{\mathbf{k}_{\|},\mathbf{n}}-E_{\mathbf{k}'_{\|},\mathbf{n}'})^2-\hbar^2\omega^2}\end{aligned} \tag{13}$$

here $\alpha=x,y,z$ labels the cartesian component of the tensor, $\hat{p}_\alpha$ indicates the component of the electron linear momentum, $\omega_P=\Omega_Pn(D)/n_0$ is the plasma frequency of the quantized electron gas ($\Omega_P=\sqrt{4\pi e^2N_0/m}$ is the free electron plasma frequency and $n(D)$ is the average electron density of the film) and $f(E_{\mathbf{k}_{\|},\mathbf{n}})$ is the occupation factor of the $(\mathbf{k}_{\|},\mathbf{n})$ state. In the IWM and the FWM $\omega_P=\Omega_P$, while in PBM $n(D)$ is smaller than n_0, since the electronic charge is distributed over a larger distance than the ionic charge (see Fig.1 (b)). The off-diagonal components are equal to zero. This expression differs from the plasma model dielectric function adopted in previous studies in that: (i) it has a tensor character with $\epsilon_{xx}=\epsilon_{yy}\neq\epsilon_{zz}$, (ii) through the double sum in the second member it accounts for transitions between lateral subbands, whose probability amplitude is expressed by the momentum matrix element between the one electron wavefunctions (7). It can be easily shown that these transitions do not affect the lateral components of the dielectric tensor, which are given by the simple expression of the plasma dielectric function

$$\epsilon_{xx}(\omega) = \epsilon_{yy}(\omega) = 1 - \frac{\omega_P^2}{\omega^2} \tag{14}$$

because the momentum matrix element for x and y component vanishes.

3. Results for finite well potential

In this section we present the results of calculations for finite well potentials. We take *Al*, *Ag* and *Cs* corresponding to a radius per electron in Bohr units r_s/a_0 equal to 2.07, 3.02 and 5.62 respectively, in order to illustrate QSE at different densities and potential depths. The value of the potential depth is obtained by summing the metal work function W with the calculated Fermi energy. Figure 2 reports the calculated Fermi energies as a function of the product between the Fermi wavevector and the film thickness. This allows to better point out the oscillations and the cusps arising from the crossing of the Fermi energy by new subbands upon varying the film size. The figure shows the V_0 value appropriate to the bulk and to the large D limit. We give in the same figure the results obtained by assuming an infinitely deep potential (IWM). The comparison allows to illustrate the effects of the potential softening. In agreement with previously published results [12, 16, 33] we find that:

- E_F is systematically larger than E_{F_B} and goes to the bulk value as D goes to infinity. The difference is more pronounced and the bulk limit is achieved at larger size in the low density systems, as it is clearly shown by the comparison between *Cs* and *Al* curves;
- As expected from the discussion of the previous section, the softening of the potential leads to less pronounced deviations from the bulk values. Because of the stronger electron confinement, the IWM has a larger Fermi energy that the FWM;
- The cusps correspond to integer values of half of the Fermi wavelength in the IWM case. This feature is only approximately verified for the FWM.

As a consequence of size quantization, the z-component of the dielectric tensor is expected to go to a finite value $\epsilon_{zz}(0)$ as the frequency goes to zero. This value increases proportionally to D^2 in the large size limit [11]. In Figure 3 we plot the quantity $\epsilon_{zz}(0)/D^2$ for the three cases under study. Again the results show the cusps due to the filling of new subbands as D increases [10]. The asymptotic limit is obtained for $k_F D/\pi$ of the order of $5 - 6$ in the three cases. Significant differences appear in the large D behaviour when the FWM results are compared with those from the IWM model: $\epsilon_{zz}(0)$ is larger for finite wells of small size, while it is smaller at high D values. The convergence to the asymptotic limit is considerably slower for the infinite well, specially in the low density metals. This behaviour reflects the differences in the distribution of the eigenvalues of equation (8). For the infinite well there are infinite bound states whose energy scales like n^2, see equation (12), and the separation between two successive levels increases linearly with n. Such behaviour is not present in the FWM, for which equation (8) has a finite number of eigenvalues corresponding to bound states and a continuum spectrum at positive energies. The PBM curve takes values closer to the FWM than to the IWM. This is primarily a consequence of the fact that PBM allows for electron charge spill-out, while in IWM the electron distribution is confined within D i.e. it has the same size of the positive charge.

In the following we show the changes in the force caused by the size quantization with respect to the results obtained by using the isotropic continuum plasma model, where the z-component of the film dielectric tensor is equal to the planar components i.e. is given by equation (14). We use the symbol F_Q to indicate the force per unit area calculated for the quantized film, while

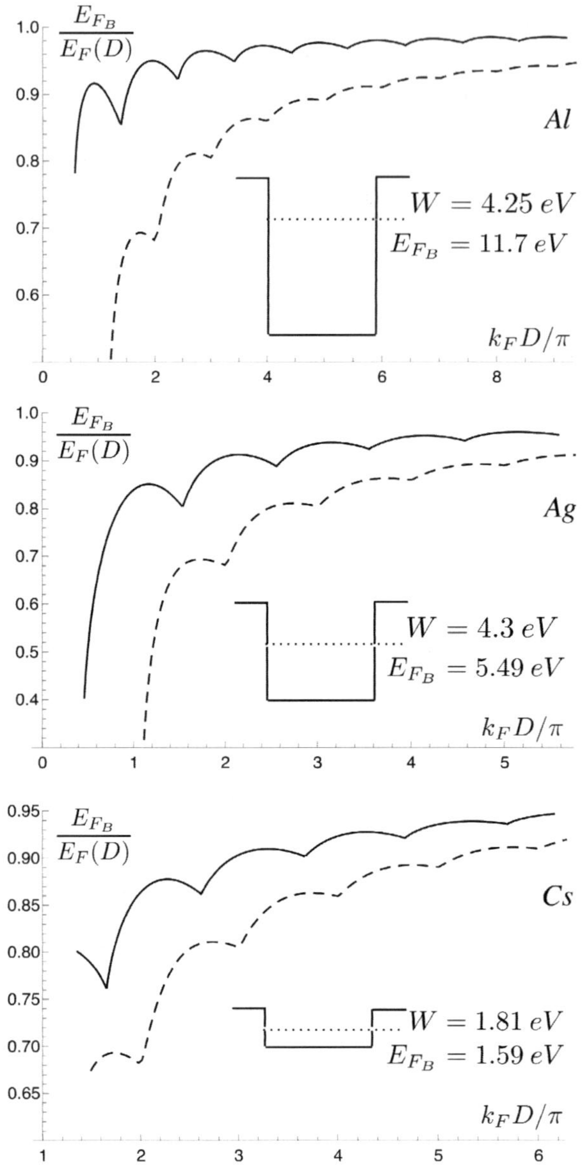

Figure 2. Fermi energy normalized to its bulk value for Al, Ag and Cs using the IWM (dashed line) and the FWM (continuous line). The bulk Fermi energies and the work functions have been taken from ref. [32].

F_P is the force per unit area calculated in the isotropic plasma model. To better illustrate the results, in figures 4 and 5 we plot the quantity

$$\delta_P = \frac{F_P - F_Q}{F_P} \tag{15}$$

as a function of the separation distance ℓ for films of 1 and 5 nm thickness respectively. In each figure we display the results for the three cases under study and we compare the finite well with the IWM at the same density. This allows us to point out the modifications caused by the potential softening. We also show the curves appropriate to the PBM. In agreement with our

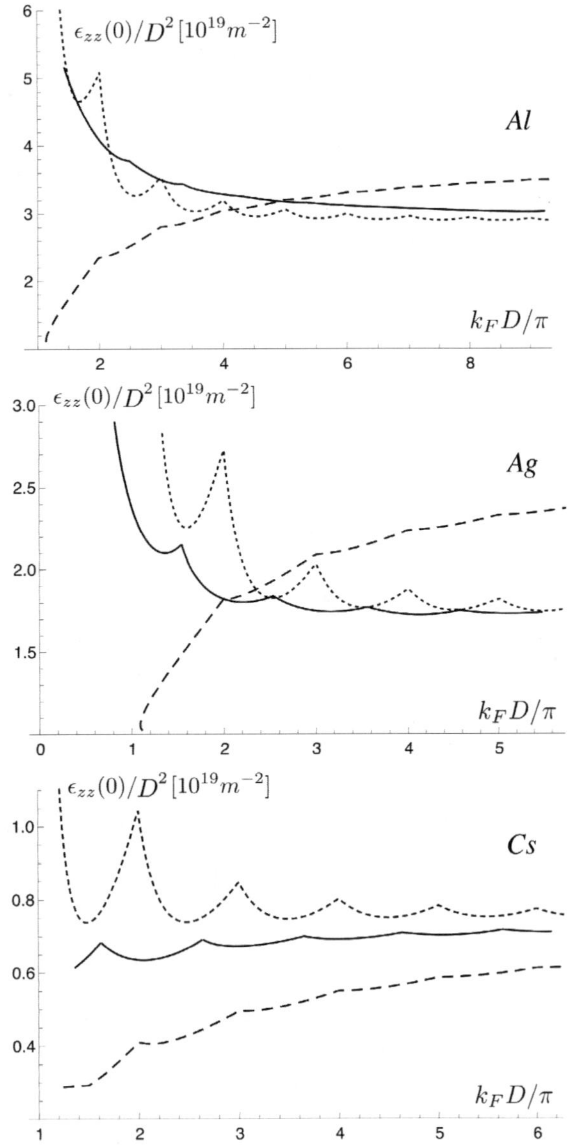

Figure 3. Static value of the zz component of the dielectric tensor for Al, Ag and Cs using the IWM (dashed lines), the PBM (dotted lines) and the FWM (continuous lines).

previous findings we observe that

- QSE tend to reduce the intensity of the force;
- the reduction is more significant at low density (Cs) than at high density (Al);
- it may be considerably higher than 10% for 1 nm thickness and reduce to few per cent at 5 nm;
- it can be appreciable over a distance interval up to $10 - 50$ nm.

The most important conclusion that can be drawn from the figures is that the potential shape is important and can lead to a substantial modifications of the quantum size effects both at small and at large distances. The models which confine the electronic charge tend to overestimate the

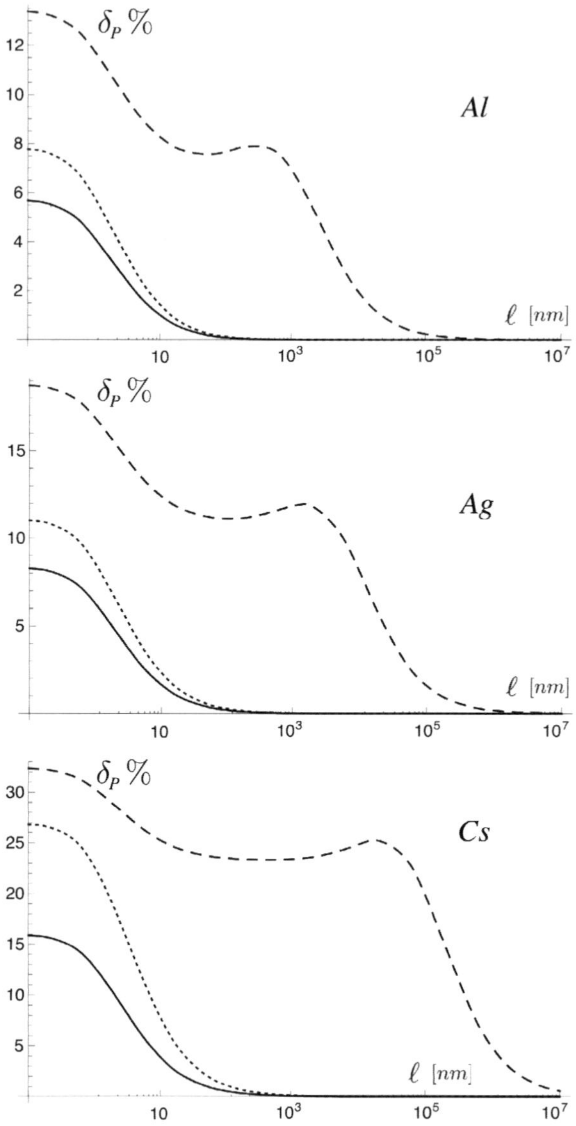

Figure 4. Relative percentual difference for the force between two identical films of thickness $D = 1$ nm as a function of the films separation ℓ, for Al, Ag and Cs. Dashed lines have been obtained using the PBM, dotted lines have been obtained using the IWM and the continuous lines represents the finite well results.

force reduction induced by size quantization. The curves for the PBM show large force reduction (greater δ_P values) over a wide interval of distances. On passing to the IWM case one notice that the removal of the constriction that the Fermi energy be equal to the bulk value, still keeping an infinitely deep potential, leads to smaller δ_P values and to a more rapid decay of the curves at large distances. Reducing the well depth to finite values has a similar effect: it causes a general decrease of δ_P and a narrowing of the distance interval over which QSE are appreciable. This also implies that any increase of the confining potential depth at fixed ion density leads to higher δ_P values and to more significant QSE. The large values taken by δ_P in the PBM case do not arise from the charge confinement only, since, as pointed out before, the constraint on the charge

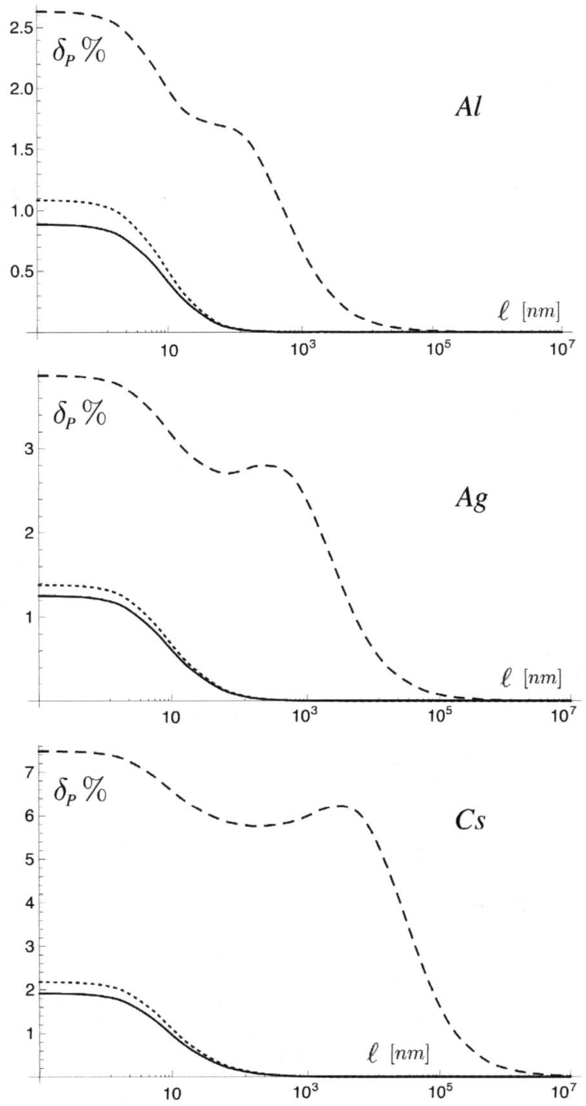

Figure 5. Relative percentual difference for the force between two identical films of thickness $D = 5$ nm as a function of the films separation ℓ, for Al, Ag and Cs. Dashed lines have been obtained using the PBM, dotted lines have been obtained using the IWM and the continuous lines represents the finite well results.

distribution is weaker that in the IWM. To a large extent they are a consequence of the plasma frequency normalization caused by the decrease in the average electron charge density that it is necessary in order to achieve global charge neutrality [10, 29, 30]. In the isotropic plasma model one takes $\omega_P = \Omega_P$. In the PBM this value is obtained only at large film thickness. Neglecting this normalization i.e. taking the free electron plasma frequency in the parallel components of the dielectric tensor (but not in ϵ_{zz}) would lead to δ_P values closer to the well potential models.

4. Intraband absorption effects

To introduce intraband absorption we have to modify the dielectric tensor in a way that allows to include relaxation time effects in the parallel components and to recover the Drude behaviour in the large D limit. This cannot be done by simply introducing an imaginary part of the frequency ω, since this violates the continuity equation locally [34, 35]. The appropriate recipe is to replace into equation (13) ω^2 with $\omega(\omega + i2\pi/\tau)$, where τ is the relaxation time. For the parallel components this leads to the Drude dielectric function

$$\epsilon(\omega) = 1 + \frac{\omega_P^2}{\omega(\omega + i\gamma)} \tag{16}$$

where $\gamma = 2\pi/\tau$. The results of the calculation of the force per unit area with the modified dielectric tensor are displayed in Figures 6 and 7 for 1 nm and 5 nm films. We report the quantity

$$\delta_D(\gamma) = \frac{F_D(\gamma) - F_{QD}(\gamma)}{F_D(\gamma)} \tag{17}$$

where F_D is the force calculated using the bulk Drude model with a given relaxation time and F_{QD} in the one obtained by the calculation with the same relaxation time and with size quantization included. For $\tau \to \infty$ we recover the plasma model so that $\delta_P = \delta_D(0)$. The calculations have been performed assuming a finite potential well and for two values of the relaxation frequency γ. We have taken values that approximately correspond to those reported for the metals under consideration [32]. The figures show a comparison with the curves obtained by the plasma model with QSE. It is clear from these results that the main effect of the inclusion of intraband absorption is to increase δ i.e. to increase the difference with respect to the calculations with the bulk dielectric function. The smaller is the relaxation time the larger is the reduction of the force. The effect is qualitatively the same in the three metals under study, but it depends upon the well depth and the film size. The influence of the shape of the potential is illustrated in Fig.8 showing the curves of Ag films at a given relaxation frequency for the various models. Again it should be noticed that the FWM gives the lower δ_D values. The PBM results show large δ_D values over a very wide interval of distances. To a large extent this behaviour has to be imputed to the renormalization of the plasma frequency. Figure 9 shows typical curves of δ_D as a function of the film thickness for different values of the relaxation frequencies at a given separation distance of 5 nm. As expected δ_D decreases with D, but the slope at large thicknesses (D of the order of $10 - 50$ nm) depends significantly upon the relaxation frequency.

5. Conclusions and lines of development

We have presented a rather complete set of theoretical results illustrating the possible role of size quantization effects in the electromagnetic vacuum force between very thin films and showing how the determination of these effects depends upon the description of the film electronic structure and upon the inclusion of intraband effects. However we want to point out that our analysis is still a mere indication of the corrections to the simple picture that assumes the same dielectric function for films and bulk solids. Although the basic features of size quantization (confinement of the electronic charge, anisotropy of the dielectric tensor, presence of inter-subbands transitions in the dielectric function) are already present in the models we have studied, there is room for substantial improvements before an accurate comparison with experimental data, like those obtained by Lisanti et al. for Pd films [36], can be done. A more detailed treatment should include (i) band structure effects, (ii) non-locality of the dielectric response and (iii) non-local treatment of the reflectivity. A self consistent first principles calculation of the inverse dielectric matrix for a slab of the appropriate size and symmetry, from which a macroscopic dielectric function can be derived with band structure and non-local effects included,

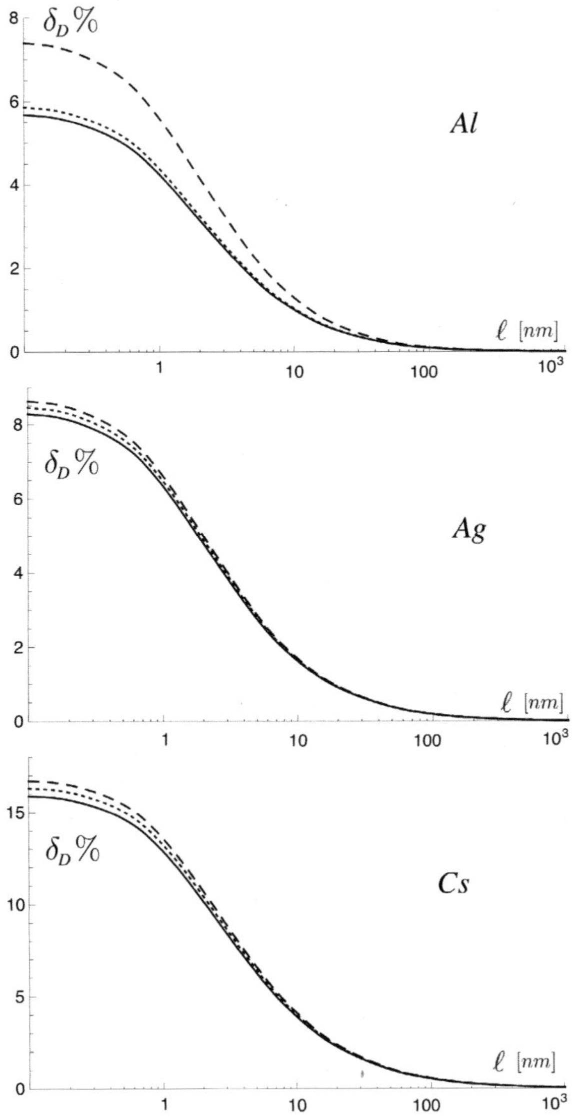

Figure 6. Relative percentual difference for the force between two identical films of thickness $D = 1$ nm as a function of the films separation ℓ, for Al, Ag and Cs. Continuous lines represent the results for $\gamma = 0$. Dotted lines have been obtained using $\gamma = 5 \times 10^{13}$ rad/s for Ag and Cs and $\gamma = 10^{14}$ rad/s for Al. Dashed lines have been obtained using $\gamma = 10^{14}$ rad/s for Ag and Cs and $\gamma = 10^{15}$ rad/s for Al.

could provide the appropriate treatment of the first two issues [37, 38]. Corrections to the Fresnel optics, along the lines indicated by several authors [39, 40, 41], can lead to important modifications of the reflectivity even in the case of a free electron gas film. We are currently investigating these matters and the results will be presented elsewhere. Still when comparing theory with experiments for thin films one should consider the fact that measured relaxation frequency turn out to depend upon the film size and morphology [42, 43, 44, 45, 46]. This is largely due to the so called classical size effects arising from the scattering of the electrons at the film boundaries. In view of the sensitivity of QSE to the value of the relaxation time, comparison

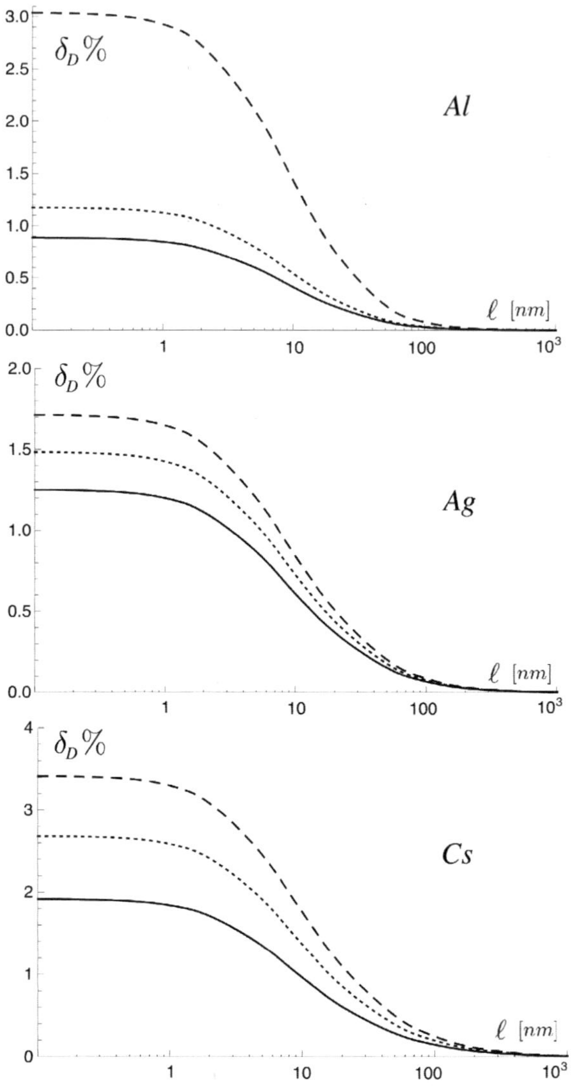

Figure 7. Relative percentual difference for the force between two identical films of thickness $D = 5$ nm as a function of the films separation ℓ, for Al, Ag and Cs. Continuous lines represent the results for $\gamma = 0$. Dotted lines have been obtained using $\gamma = 5 \times 10^{13}$ rad/s for Ag and Cs and $\gamma = 10^{14}$ rad/s for Al. Dashed lines have been obtained using $\gamma = 10^{14}$ rad/s for Ag and Cs and $\gamma = 10^{15}$ rad/s for Al.

with experimental data may be possible only if a realistic estimate of the modifications in the relaxation times caused by surface scattering is available.

Acknowledgments

AB thanks *CINECA Consorzio Interuniversitario* (`www.cineca.it`) for funding his Ph.D. fellowship.

Reference

[1] T-C Chiang. *Surf. Sci. Reports*, 39:181, 2000.

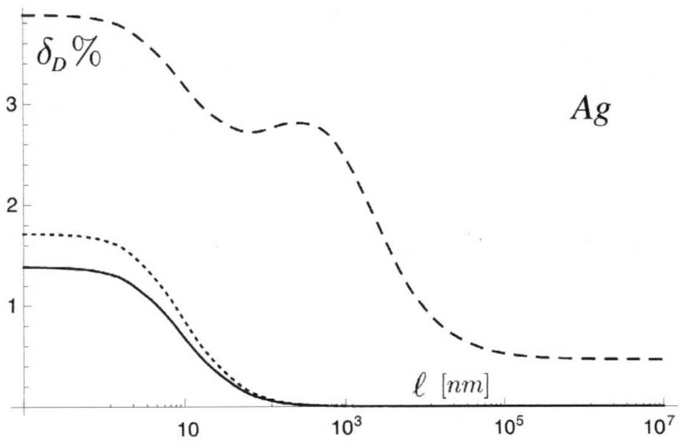

Figure 8. Relative percentual difference for the force between two identical *Ag* films of thickness $D = 5$ nm as a function of the films separation ℓ, $\gamma = 10^{14}$ rad/s. Continuous lines represent the FWM, dotted line the IWM and dashed line the PBM.

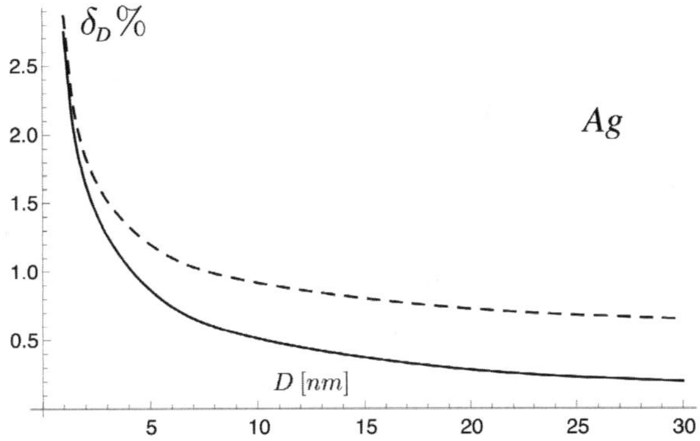

Figure 9. Relative percentual difference for the force between two identical *Ag* films separated by a distance $\ell = 5$ nm as a function of the films thickness D. Continuous line represents the result for $\gamma = 0$ whereas the dashed line has been obtained with $\gamma = 10^{14}$ rad/s.

[2] M C Tringides, M Jalochowski, and E Bauer. *Phys. Today*, 60:50, 2007.
[3] Jin-Feng Jia, Shao-Chun Li, Yan-Feng Zhang, and Qi-Kun Xue. *J.Phys.Soc.Japan*, 76:082001, 2007.
[4] R Esquivel-Sirvent and V B Svetovoy. *Phys. Rev. B*, 72:045443, 2005.
[5] A Benassi and C Calandra. *J.Phys.A: Math.Theor.*, 40:13453, 2007.
[6] A Lambrecht, I Pirozhenko, L Duraffourg, and Ph Andreucci. *Europhys. Lett.*, 77:44006, 2007. Erratum: *Europhys. Lett.* 81:19901, 2008.
[7] I Pirozhenko and A Lambrecht. *Phys.Rev.A*, 77:013811, 2008.
[8] Z Lenac and M S Tomas. *Phys. Rev. A*, 78:023834, 2008.
[9] J P Rogers III, P H Cutler, T R Feuchtwang, N Miskovski, and A A Lucas. *Surf. Sci.*, 141:61, 1984.
[10] A Benassi and C Calandra. *Europhys. Lett.*, 82:61002, 2008.
[11] D M Wood and N W Aschcroft. *Phys.Rev.B*, 25:6255, 1982.
[12] K F Schulter. *Surf.Sci.*, 55:427, 1976.
[13] P J Feibelman. *Phys. Rev. B*, 27:1991, 1983.
[14] S Ciraci and I P Batra. *Phys. Rev. B*, 33:4294, 1985.
[15] E Ogando, N Zabala, E V Chulkov, and M J Puska. *Phys.Rev.B*, 71:205401, 2005.
[16] J P Rogers III, P H Cutler, T R Feuchtwang, and A A Lucas. *Surf. Sci.*, 181:436, 1987.
[17] M Boström and B E Sernelius. *Phys. Rev. Lett.*, 84:4757, 2000.
[18] V B Bezerra, G L Klimchitskaya, V M Mostepanenko, and C Romero. *Phys. Rev. A*, 69:022119, 2004.
[19] I Brevik, J B Aarseth, J S Høye, and K A Milton. *Phys.Rev.E*, 71:056101, 2005.
[20] V B Bezerra, R S Decca, E Fischbach, B Geyer, G L Klimchitskaya, D E Krause, D Lopez, V M Mostepanenko,

and C Romero. *Phys. Rev. E*, 73:028101, 2006.
[21] J S Høye, I Brevik, J B Aarseth, and K A Milton. *J.Phys.A.: Math. Gen.*, 39:6031, 2006.
[22] I Brevik, S A Ellingsen, and K A Milton. *New J. Phys.*, 8:236, 2006.
[23] R S Decca, D Lopez, E Fischbach, G L Klimchitskaya, D E Krause, and V M Mostepanenko. *Ann.Phys.*, 318:37, 2005.
[24] S K Lamoreaux. *Rep. Prog. Phys.*, 68:201, 2005.
[25] E M Lifshitz. *Sov. Phys. JEPT*, 2:73, 1956.
[26] I E Dzyaloshinskii, E M Lifshitz, and L P Pitaevskii. *Adv. Phys.*, 10:165, 1958.
[27] F Zhou and L Spruch. *Phys. Rev. A*, 52:297, 1995.
[28] M Bordag, U Mohideen, and V M Mostepanenko. *Phys. Rep.*, 353:1, 2001.
[29] P Czoschke, Hong Hawoong, L Basile, and T C Chiang. *Phys.Rev.B*, 72:035305, 2005.
[30] P Czoschke, Hong Hawoong, L Basile, and T C Chiang. *Phys.Rev.B*, 72:075402, 2005.
[31] A S Davydov. *Quantum Mechanics.* Pergamon Press, 1976.
[32] N W Ashcroft and N D Mermin. *Solid State Physics.* Cornell University, Philadelphia, 1976.
[33] B E Sernelius, K F Berggren, M Tomak, and C McFadden. *J. Phys. C: Solid State Phys.*, 18:225, 1985.
[34] N D Mermin. *Phys. Rev. B*, 1:1019, 1970.
[35] P Garik and N W Ashcroft. *Phys.Rev.B*, 21:391, 1980.
[36] M Lisanti, D Iannuzzi, and F Capasso. *Proc. Nat. Ac. Sci. USA*, 102:11989, 2005.
[37] J L Li, J Chun, N S Wingreen, R Car, I A Aksay, and D A Saville. *Phys. Rev. B*, 71:235412, 2005.
[38] B E Sernelius. *J. Phys. A: Math. Gen.*, 39:6741, 2006.
[39] P Appel. *Physica Scr.*, 24:795, 1981.
[40] P J Feibelman. *Prog.Surf.Sci.*, 12:287, 1982.
[41] K Kempa and W L Scaich. *Phys. Rev. B*, 34:547, 1986.
[42] M Jalochowski, H Knoppe, G Lilienkamp, and E Bauer. *Phys. Rev. B*, 46:4693, 1992.
[43] G Fahsold, A Bertel, O Krauth, N Maggi, and A Pucci. *Phys. Rev. B*, 61:1408, 2000.
[44] J Sotelo, J Ederth, and G Niklasson. *Phys. Rev. B*, 67:195106, 2003.
[45] G K Pribil, B Johs, and N J Ianno. *Thin Solid Films*, 455-456:443, 2004.
[46] H Hoffmann and J Vancea. *Thin Solid Films*, 85:147, 1981.

60 Years of the Casimir Effect IOP Publishing
Journal of Physics: Conference Series **161** (2009) 012008 doi:10.1088/1742-6596/161/1/012008

Lateral Casimir forces on parallel plates and concentric cylinders with corrugations

Inés Cavero-Peláez

Laboratoire Kastler Brossel, Université Pierre et Marie Curie, ENS, CNRS, Campus Jussieu, University Paris 6, Case 74, F-75252 Paris, cedex 05, France.

E-mail: caverо@spectro.jussieu.fr

Kimball A Milton, Prachi Parashar and K V Shajesh

Oklahoma Center for High Energy Physics and Homer L. Dodge Department of Physics and Astronomy, University of Oklahoma, Norman, OK 73019, USA.

E-mail: milton@nhn.ou.edu, prachi@nhn.ou.edu, shajesh@nhn.ou.edu

Abstract. In this paper we are giving a quantitative description of two different configurations for noncontact gears. We consider the solutions from a perturbative calculation for two semitransparent parallel plates and concentric cylinders both with corrugations on the inner surfaces. In the case of corrugated parallel plates we discuss results from first- and second-order perturbation calculation in the corrugation amplitudes and we will concentrate on the first-order perturbation for the case of the corrugated concentric cylinders (the second order calculation is under study), both for the weak and strong couplings. We compare the perturbative results with the results from the proximity force approximation (PFA) and an exact weak coupling calculation.

1. Introduction.

When two parallel plates with corrugations are situated close to each other, besides the usual Casimir attraction that takes place between them [1], there exists another contribution to the attraction that depends on the corrugations of the surface of the plates and therefore, on the non-normal coordinates (in the case of the parallel plates this coordinate is along the parallel surfaces where the corrugations are imposed, and in the case of the concentric cylinders this coordinate is the azimuthal angle with origin at the center of the cylinders). This new term gives rise to a lateral force or torque that could, in principle, make one plate slide with respect to the other or one cylinder rotate with respect to the other.

This is of course of both theoretical and experimental interest. Some nanoscale mechanical devices based on the lateral Casimir force between different geometries have been proposed in [2–4]. Roy and Mohideen [5] studied experimentally the Casimir interaction between corrugated surfaces. They calculated the lateral Casimir force between a plate and a large sphere both with the same small sinusoidal corrugations. The theoretical treatment of this was analised in terms of the PFA [6] however its validity has been questioned and consequently debated by others. Several perturbative QED calculations have been carried out [7–9] but unfortunately do not settle the issue of the PFA validity because the error keeping only the leading order may

Journal of Physics: Conference Series **161** (2009) 012008 doi:10.1088/1742-6596/161/1/012008

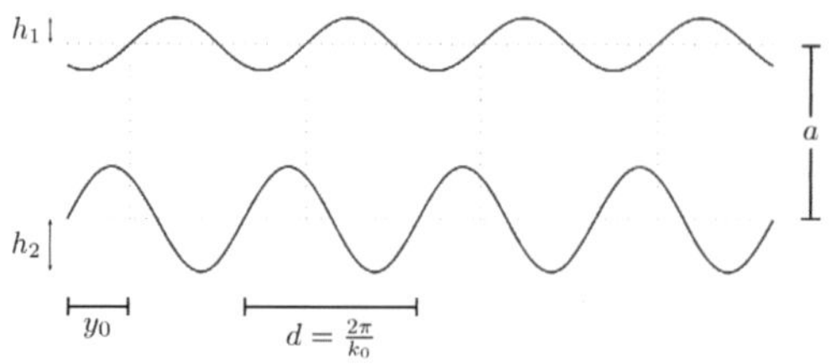

Figure 1. Parallel plates with sinusoidal corrugations.

be high. It is believed that the next-to-leading-order calculation will be able to throw light on the issue. Here we discuss the result of such calculation for the scalar fields hoping that an analogous calculation for the electromagnetic case will follow. We base our discussion on a calculation details of which can be found in [10–12].

We divide the paper in two relevant parts. The first one is devoted to corrugated parallel plates. We show and discuss the perturbative results for two limiting cases, strong and weak-coupling. The weak-coupling limit is also compared with an exact calculation and the PFA solution. The second part of the paper shows the results for corrugated concentric cylinders. We give perturbative results for both the Dirichlet and the weak-coupling limits. We extrapolate these results to the case of very large radii of the cylinders while keeping constant the distance between them. This situation corresponds to parallel plates.

2. Parallel plates with corrugations.

Let's consider two semitransparent plates lying on the x-y plane and placed parallel to each other. The plates have corrugations on the facing surfaces along the y-axes. We consider translational invariance on the x-axes and we work at a given frequency. The average distance between the parallel slabs is a and we model the corrugations by the functions $h_1(y)$ and $h_2(y)$. These are defined such that at any point the inequality $a \geq h_1(y) - h_2(y)$ holds.

For the cases studied here we choose sinusoidal corrugations, $h_i(y) = h_i \sin(k_i y)$. Each plate is described by a potential of the form,

$$V_i(z, y) = \lambda_i \, \delta(z - a_i - h_i(y)), \tag{1}$$

where $i = 1, 2$ are labels that identify the individual plates. In this way we can consider our system as two semitransparent parallel plates making up the background and described by the potentials $V_i^{(0)}(z) = \lambda_i \delta(z - a_i)$ for $i = 1, 2$. On the top of it we have the corrugations $h_i(y)$ which can be considered to be a perturbation of the background (see figure 1).

Let's make one of the plates shift a distance y_0 with respect to the other one.

$$h_1(y) = h_1 \sin[k_0(y + y_0)], \tag{2a}$$
$$h_2(y) = h_2 \sin[k_0 y], \tag{2b}$$

where $k_0 = 2\pi/d$ is the wavenumber corresponding to the corrugation wavelength d [1]. This gives rise to a force in the lateral direction, the so-called lateral Casimir force,

$$F_{\text{Lat}} = -\frac{\partial E}{\partial y_0}. \tag{3}$$

[1] The same wavelength is assumed in both plates because it is only in this case that we find a leading order contribution to the lateral force.

Since the corrugations can be considered a perturbation over the background $V_i^{(0)}(z)$, we can easily extract their contribution by calculating the deviation of the total potential $V_i(z,y)$ from that of the background,

$$\Delta V_i(z,y) = V_i(z,y) - V_i^{(0)} = \lambda_i\, \delta(z - a_i - h_i(y)) - \lambda_i \delta(z - a_i)$$
$$= \sum_{n=1}^{\infty} \frac{[-h_i(y)]^n}{n!} \frac{\partial^n}{\partial z^n} V_i^{(0)}(z) = \sum_{n=1}^{\infty} V_i^{(n)}(z,y). \tag{4}$$

The key formula to calculate the Casimir energy is [13–15],

$$\Delta E = \frac{i}{2\tau} \operatorname{Tr} \ln \mathrm{G}\mathrm{G}^{(0)^{-1}}, \tag{5}$$

where the Green's function G satisfies a differential equation with a potential function that describes the corrugations in both semitransparent plates,

$$\Big[- \partial^2 + V_1 + V_2 \Big] G = 1, \tag{6}$$

and the corresponding Green's function associated with the background obeys the differential equation

$$\Big[- \partial^2 + V_1^{(0)} + V_2^{(0)} \Big] G^{(0)} = 1, \tag{7}$$

where the corrugations are not present. However, since only the term containing the perturbative potential in both plates play a role in determining the lateral Casimir force, we can reduce the above more general formula (5) to

$$E_{12} = -\frac{i}{2\tau} \operatorname{Tr} \ln \Big[1 - \mathrm{G}_1 \Delta \mathrm{V}_1 \mathrm{G}_2 \Delta \mathrm{V}_2 \Big], \tag{8}$$

where now G_i refers to the Green's function when only one of the plates has corrugations on it. It obeys the equation

$$\Big[-\partial^2 + V_1^{(0)} + V_2^{(0)} + \Delta V_i \Big] G_i = 1. \tag{9}$$

It is worth noticing that only the corrugations from plate i enter this equation.

Expanding the above formula (8) in a second- and fourth-order perturbative calculation in the corrugation amplitudes, gives expressions for the leading- and next-to-leading-order lateral Casimir energy. The general formulas turn out to be very long and complicated and they will not be shown here (details of the calculation as well as final formulas can be found in [10, 12]); here we focus on the solutions for the Dirichlet and weak coupling limits.

2.1. Dirichlet limit.

Here we consider the strong coupling limit, where $a\lambda_{1,2} \gg 1$, both in the leading and next-to-leading orders.

2.1.1. Leading order. Expanding (8) to second order gives us the first non-zero contribution to the interaction energy,

$$\frac{E_{12}^{(2)}}{L_x L_y} = \cos(k_0 y_0) \frac{\pi^2}{240\, a^3} \frac{h_1}{a} \frac{h_2}{a} A_D^{(1,1)}(k_0 a), \tag{10}$$

Journal of Physics: Conference Series **161** (2009) 012008 doi:10.1088/1742-6596/161/1/012008

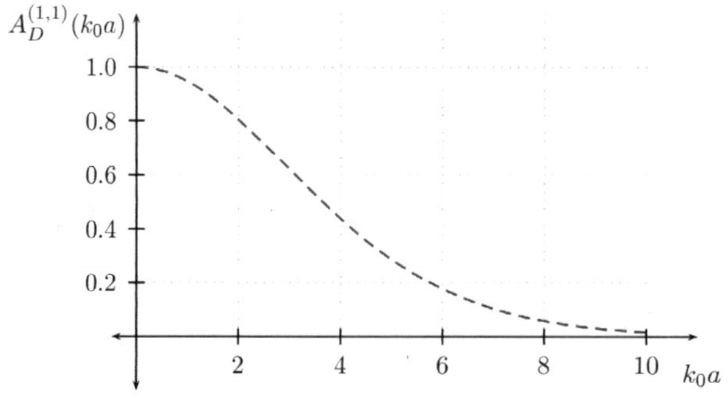

Figure 2. Dirichlet limit: Plot of $A_D^{(1,1)}(k_0a)$ versus k_0a.

where the function $A_D^{(1,1)}(k_0a)$ has the form

$$A_D^{(1,1)}(t_0) = \frac{15}{\pi^4}\int_0^\infty \bar{s}\, d\bar{s} \int_{-\infty}^{\infty} dt \frac{s}{\sinh s} \frac{s_+}{\sinh s_+}, \tag{11}$$

and it is normalized such that $A_D^{(1,1)}(0) = 1$, which becomes a convenience when we compare the results with those obtained by the proximity force approximation. We have used the notation $s^2 = \bar{s}^2 + t^2$, $s_\pm^2 = \bar{s}^2 + (t \pm t_0)^2$, and $t_0 = k_0a$. This expression can be numerically evaluated and has been plotted in figure 2.

The lateral Casimir force can be calculated using (3) which yields

$$F_{\text{Lat,D}}^{(2)} = 2\, k_0a\, \sin(k_0y_0) \left| F_{\text{Cas,D}}^{(0)} \right| \frac{h_1}{a}\frac{h_2}{a}\, A_D^{(1,1)}(k_0a), \tag{12}$$

where $|F_{\text{Cas,D}}^{(0)}|$ is the magnitude of the normal scalar Casimir force between two uncorrugated parallel Dirichlet plates given as

$$\frac{F_{\text{Cas,D}}^{(0)}}{L_xL_y} = -\frac{\pi^2}{480}\frac{1}{a^4}. \tag{13}$$

2.1.2. Next-to-leading order. The third-order term in the perturbation expansion gives zero contribution in the particular case when the wavelength is the same in both plates. If we do not restrict ourselves to this condition, the third order would give a finite contribution to the interacting energy becoming, in this case, the leading contribution itself[2]. Since we are assuming $k_1 = k_2 = k_0$, we expand (8) up to fourth order to get the next-to-leading-order contribution to E_{12}. In the strong-coupling limit this gives us

$$\frac{E_{12}^{(4)}}{L_xL_y} = \frac{\pi^2}{240\,a^3}\frac{h_1}{a}\frac{h_2}{a}\frac{15}{4}\Bigg[\cos(k_0y_0)\left\{ \frac{h_1^2}{a^2} A_D^{(3,1)}(k_0a) + \frac{h_2^2}{a^2} A_D^{(1,3)}(k_0a) \right\} - \cos(2k_0y_0)\,\frac{1}{2}\frac{h_1}{a}\frac{h_2}{a}\, A_D^{(2,2)}(k_0a) \Bigg], \tag{14}$$

[2] because in such a case the second order expansion exactly cancels.

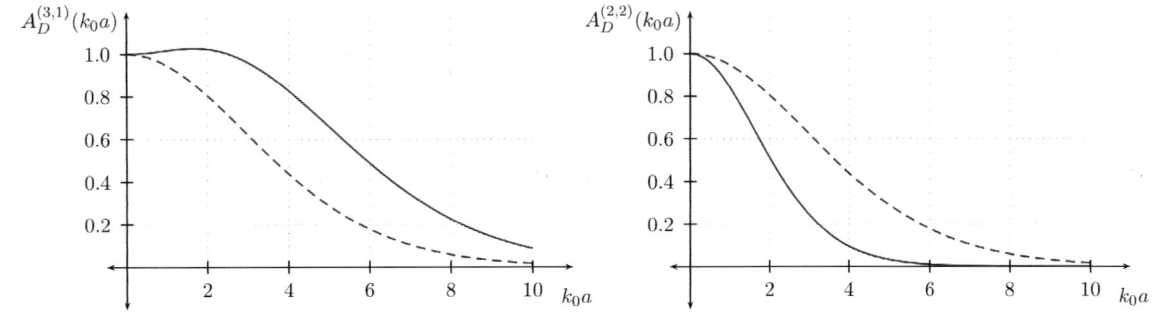

Figure 3. Dirichlet limit: Plot of $A_D^{(3,1)}(k_0a)$ and $A_D^{(2,2)}(k_0a)$ versus k_0a. The dashed curve represents $A_D^{(1,1)}(k_0a)$ which is plotted here for reference.

where we have introduced the functions,

$$\begin{aligned} A_D^{(3,1)}(t_0) &= A_D^{(1,3)}(t_0) \\ &= \frac{1}{2\pi^4}\int_0^\infty \bar{s}\,d\bar{s}\int_{-\infty}^{\infty} dt \frac{s}{\sinh s}\frac{s_+}{\sinh s_+}\left[4\frac{s}{\tanh s}\frac{s_-}{\tanh s_-} + 2\frac{s}{\tanh s}\frac{s_+}{\tanh s_+} - s^2 - s_-^2\right], \end{aligned} \tag{15a}$$

$$A_D^{(2,2)}(t_0) = \frac{1}{\pi^4}\int_0^\infty \bar{s}\,d\bar{s}\int_{-\infty}^{\infty} dt\left[\frac{s^2}{\sinh^2 s}\frac{s_-^2}{\sinh^2 s_-} + 2\frac{s^2}{\tanh^2 s}\frac{s_+}{\sinh s_+}\frac{s_-}{\sinh s_-}\right]. \tag{15b}$$

These functions can be calculated numerically and they are plotted in figure 3 for a later discussion.

The next-to-leading-order contribution to the lateral Casimir force in the Dirichlet limit reads

$$\begin{aligned} F_{\text{Lat,D}}^{(4)} &= 2\,k_0a\,\sin(k_0y_0)\left|F_{\text{Cas,D}}^{(0)}\right|\frac{h_1}{a}\frac{h_2}{a}\frac{15}{4}\left[\left(\frac{h_1^2}{a^2}+\frac{h_2^2}{a^2}\right)A_D^{(3,1)}(k_0a)\right. \\ &\quad \left. -2\cos(k_0y_0)\frac{h_1}{a}\frac{h_2}{a}A_D^{(2,2)}(k_0a)\right]. \end{aligned} \tag{16}$$

2.1.3. Discussion. In order to compare our results, we are going to confine ourselves to the case where the amplitudes of the corrugations in both plates are the same, $h_1 = h_2 = h$. Also, by observing equations (12) and (16) we notice that there is a common factor in all of them. It involves the normal Casimir force, as well as terms dependent on the geometry of the corrugations and on the initial shift y_0 between the plates. By removing this common factor, what is left is just a dimensionless weight factor. We can therefore normalize the above expressions to the common factor and compare the resulting dimensionless term that we obtain in this manner from each of the mentioned formulas. In this spirit we define the weight factors as

$$\mathcal{F}_D\left(k_0a, \frac{h}{a}, k_0y_0\right) = \frac{F_{\text{Lat,D}}\left(k_0a, \frac{h}{a}, k_0y_0\right)}{2\left|F_{\text{Cas,D}}^{(0)}\right|\frac{h^2}{a^2}\,k_0a\,\sin(k_0y_0)}. \tag{17}$$

We shall use superscripts $[m, n]$ on these dimensionless quantities to denote the order in the power series expansion to which the lateral force (not $\mathcal{F}$) has been calculated. In particular m

will signify the order in the parameter $k_0 a$, and n will denote the order in h/a. The perturbative results are complete in the $k_0 a$ dependence and thus will be denoted by $\mathcal{F}^{[\infty,n]}$. In these notations our results can be summarized as (see equations (12) and (16))

$$\mathcal{F}_D^{[\infty,4]} = A_D^{(1,1)}(k_0 a) + \frac{15}{2}\frac{h^2}{a^2}\Big[A_D^{(3,1)}(k_0 a) - \cos(k_0 y_0)\, A_D^{(2,2)}(k_0 a)\Big], \tag{18}$$

where $\mathcal{F}_D^{[\infty,4]}$ means weight factor of the lateral Casimir force up to the fourth order. That is, it includes both leading and next-to-leading orders.

We can in fact look at how much the next-to-leading order contributes to the force compared to the leading order. The left plots in figure 4 show the second-order contribution to the force with a dashed line and the fourth-order contribution with a solid line for different values of the angle and constant values of $k_0 h$. We see that for a fixed offset $k_0 y_0$ the fourth order contribution becomes more important for high $k_0 h$ that means, for bigger corrugations. We show the plot for $k_0 y_0 = 0$ to illustrate this fact more clearly because in this case the weight factor does not diverge for small $k_0 a$. However, at that value of the offset, the lateral force is zero as it can be noticed straightforwardly from (12) and (16). For any other value of the offset, there are points where the graph diverges. This is because as $k_0 h$ approaches $2k_0 a$, the expansion is not longer valid. It corresponds to the case when the distance between the plates is smaller than the sum of the corrugations on both plates.

On the right hand side of the same figure we plot the fractional correction of the next-to-leading-order contribution relative to the leading-order. It shows that at any given offset $k_0 y_0$ and $k_0 h$ the fractional error is higher for small separations and big corrugations give higher next-to-leading-order relative errors.

In the absence of an exact answer, for the Dirichlet case, for comparison is not possible to extract the precise error in the perturbative results. In the next section we shall evaluate the corresponding result in the weak coupling limit and evaluate the lateral Casimir force non-perturbatively. We show that the error in the lateral force, for the weak coupling limit, after the next-to-leading order has been included is small for $k_0 h \ll 1$. Presuming that the same would hold for the Dirichlet case, we can expect the error to be sufficiently small to use the fourth-order results for comparison with experimental data.

2.2. Weak coupling limit.

We consider now $a\lambda_{1,2}$ to be very small and we study the second and fourth order in the perturbation series.

2.2.1. Leading order. In this case the expansion of (8) leads to an expression that can be evaluated analytically

$$E_{12,\text{W}}^{(2)} = \cos(k_0 y_0)\left|E_{\text{Cas,W}}^{(0)}\right| \frac{h_1}{a}\frac{h_2}{a}\, A_W^{(2)}(k_0 a), \tag{19}$$

where we have introduced the function

$$A_W^{(2)}(t_0) = \frac{t_0^3}{2}\frac{\partial^2}{\partial t_0^2}\left[\frac{1}{t_0}\, e^{-t_0}\right] = e^{-t_0}\sum_{m=0}^{2}\frac{t_0^m}{m!} = \frac{e_2(t_0)}{e^{t_0}}, \tag{20}$$

plotted in figure 5. We evaluate the lateral Casimir force to this perturbation order as

$$F_{\text{Lat,W}}^{(2)} = k_0 a\, \sin(k_0 y_0)\left|F_{\text{Cas,W}}^{(0)}\right| \frac{h_1}{a}\frac{h_2}{a}\, A_W^{(2)}(k_0 a). \tag{21}$$

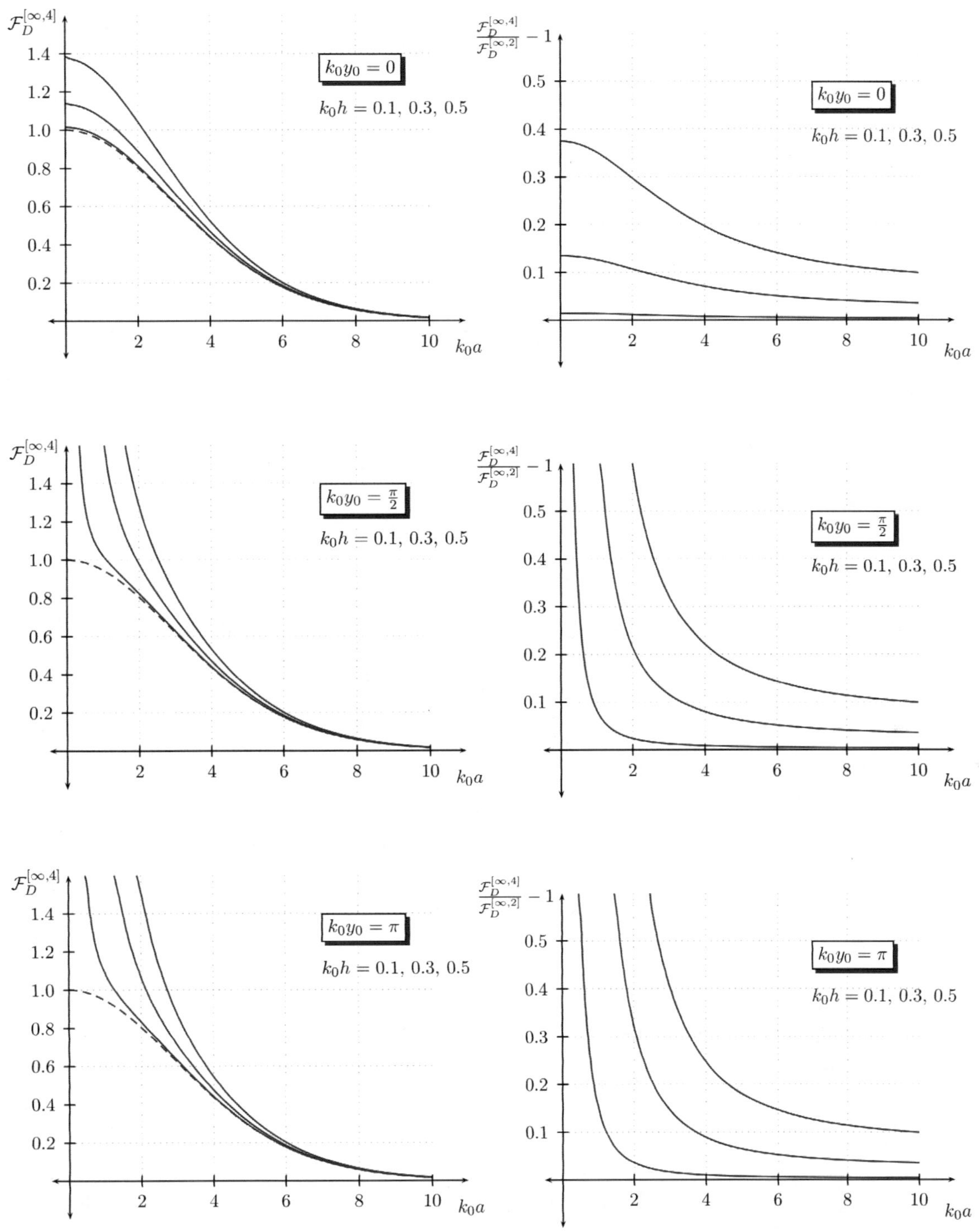

Figure 4. Dirichlet limit: The plots on the left show $\mathcal{F}_D^{[\infty,4]}$ versus k_0a for various values of k_0y_0. The dashed curve represents $\mathcal{F}_D^{[\infty,2]}$ which is plotted here for reference. In each plot the higher values of k_0h deviate more from the reference. The plots on the right show the fractional correction in the next-to-leading-order contribution. In each plot the higher values of k_0h have larger corrections.

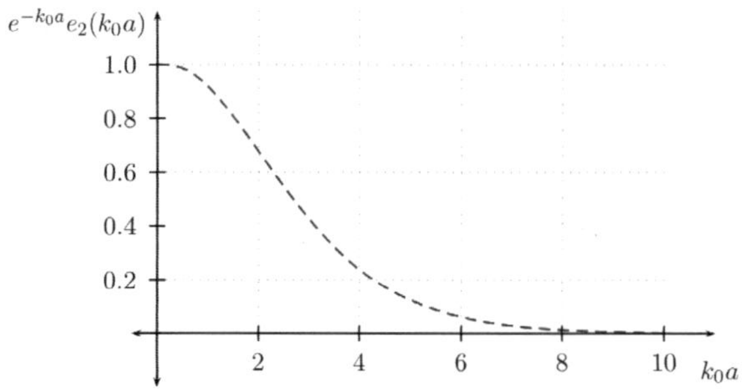

Figure 5. Weak coupling limit: Plot of $A_W^{(2)}(k_0a) = e^{-k_0a}e_2(k_0a)$ versus k_0a.

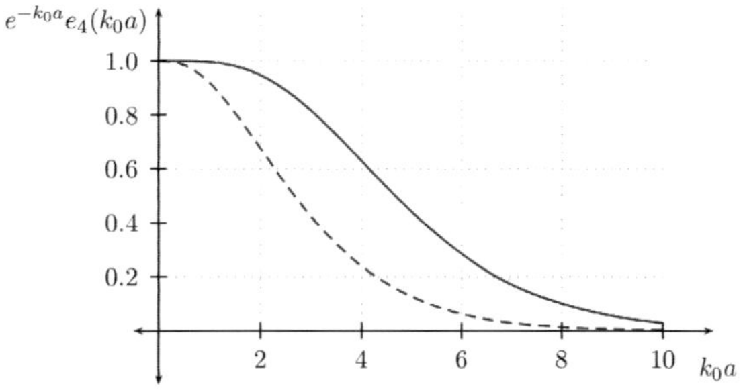

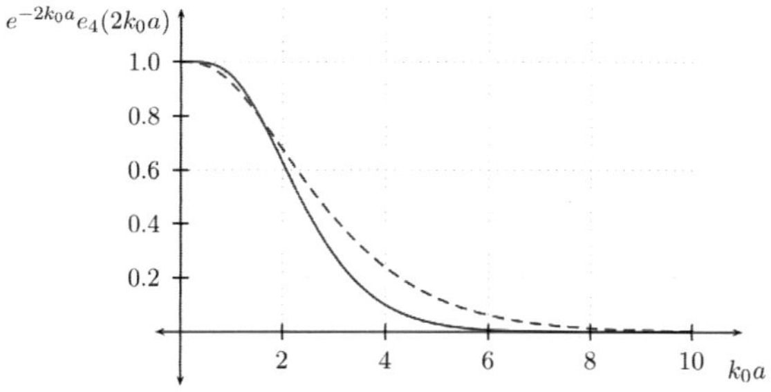

Figure 6. Weak coupling limit: Plot of $A_W^{(4)}(k_0a)$ and $A_W^{(4)}(2k_0a)$ versus k_0a. The dashed curve represents $A_W^{(2)}(k_0a)$ which is plotted here for reference.

2.2.2. Next-to-leading order. We can write the total fourth-order contribution to the interaction energy in the presence of corrugations in the weak limit to be

$$E_{12,W}^{(4)} = \left|E_{\text{Cas,W}}^{(0)}\right| \frac{h_1}{a}\frac{h_2}{a}\frac{3}{2}\left[\cos(k_0y_0)\left(\frac{h_1^2}{a^2}+\frac{h_2^2}{a^2}\right)A_W^{(4)}(k_0a) - \cos(2k_0y_0)\frac{1}{2}\frac{h_1}{a}\frac{h_2}{a}A_W^{(4)}(2k_0a)\right], \tag{22}$$

where in contrast to the result in the Dirichlet limit we observe the appearance of the same function, though with different arguments, as coefficients. This function has been suitably defined as

$$A_W^{(4)}(t_0) = \frac{t_0^5}{4!}\frac{\partial^4}{\partial t_0^4}\left[\frac{1}{t_0}e^{-t_0}\right] = e^{-t_0}\sum_{m=0}^{4}\frac{t_0^m}{m!} = \frac{e_4(t_0)}{e^{t_0}}, \tag{23}$$

so that it equals unity at $t_0 = 0$. These are plotted in figure 6. Using (22) in (3) we find that the fourth-order contribution to the lateral Casimir force in the weak coupling limit equals

$$F_{\text{Lat,W}}^{(4)} = k_0a\,\sin(k_0y_0)\left|F_{\text{Cas,W}}^{(0)}\right| \frac{h_1}{a}\frac{h_2}{a}\frac{3}{2}\left[\left(\frac{h_1^2}{a^2}+\frac{h_2^2}{a^2}\right)A_W^{(4)}(k_0a) - 2\cos(k_0y_0)\frac{h_1}{a}\frac{h_2}{a}A_W^{(4)}(2k_0a)\right]. \tag{24}$$

2.2.3. Proximity force approximation. It is interesting to find this result since, as we shall see later, the weak-coupling limit can be solved exactly. That will allow us to compare both the perturbative and the proximity force approximation results to the exact solution. For sufficiently small distances a (in comparison to $d = 2\pi\, k_0^{-1}$) we can treat the plates to be built out of small sections for which the energy is approximately that of the parallel plate geometry. This is the so-called proximity force approximation. To compute the force under this approximation we consider the distance between the plates to be

$$a(y) = a + h_2 \sin[k_0 y] - h_1 \sin[k_0(y + y_0)], \tag{25}$$

so that we can integrate the resulting energy after substituting the above in the expression for the Casimir energy between two plates in the weak coupling limit.

Therefore, in the PFA, the lateral Casimir force in the weak coupling limit reads,

$$F_{\text{Lat,W}}^{\text{PFA}} = k_0 a \sin(k_0 y_0) \left| F_{\text{Cas,W}}^{(0)} \right| \frac{h_1}{a} \frac{h_2}{a} \frac{1}{(1 - \frac{r^2}{a^2})^{\frac{3}{2}}}, \tag{26}$$

for $|h_1| + |h_2| < a$.

2.2.4. Nonperturbative result. It was shown in [13] that exact results for the Casimir force, in the weak coupling limit, can be achieved for specific geometries. This is being extended for a class of geometries, for the scalar case, in [16], and for dielectrics in electromagnetism [17]. In our case the solution of the exact weak coupling reads

$$\frac{F_{\text{Lat,W}}}{L_x L_y} = -\frac{\lambda_1 \lambda_2}{32\pi^2 a^2} \frac{h_1}{a} \frac{h_2}{a} (k_0 a)^4 \frac{1}{\pi} \int_{-\infty}^{\infty} \frac{dt}{t} \,\text{Re}\left[\frac{\sin(t + k_0 y_0)}{[(t + i k_0 a)^2 + \{k_0 r(t)\}^2]^{\frac{3}{2}}} \right], \tag{27}$$

for $|h_1| + |h_2| < a$.

2.2.5. Discussion. Plots analogous to the ones showed in figure 4 can be drawn for the case of weak-coupling. However, since these have the same physical content as those for the Dirichlet case we do not show them here. Instead we are going to compare the perturbation expansion and the PFA to the exact weak-coupling solution.

In order to do that, we follow the notation in (18) and normalize the force for the weak-coupling as

$$\mathcal{F}_W \left(k_0 a, \frac{h}{a}, k_0 y_0 \right) = \frac{F_{\text{Lat,W}} \left(k_0 a, \frac{h}{a}, k_0 y_0 \right)}{\left| F_{\text{Cas,W}}^{(0)} \right| \frac{h^2}{a^2} k_0 a \sin(k_0 y_0)}. \tag{28}$$

The entire expression up to fourth order in the perturbation series in the weak coupling reads

$$\mathcal{F}_W^{[\infty,4]} = \frac{e_2(k_0 a)}{e^{k_0 a}} + 3\frac{h^2}{a^2} \left[\frac{e_4(k_0 a)}{e^{k_0 a}} - \cos(k_0 y_0) \frac{e_4(2k_0 a)}{e^{2k_0 a}} \right]. \tag{29}$$

The result in the proximity force approximation should be valid to the first order in $k_0 a$ and any order in h/a. Thus $\mathcal{F}^{[1,\infty]}$ will represent the PFA result in our notation. In the weak coupling this is

$$\mathcal{F}_W^{[1,\infty]} = \frac{1}{(1 - \frac{r^2}{a^2})^{\frac{3}{2}}}, \tag{30}$$

and for the exact calculation from (27) gives

$$\mathcal{F}_W^{[\infty,\infty]} = -\frac{(k_0 a)^3}{\sin(k_0 y_0)} \frac{1}{\pi} \int_{-\infty}^{\infty} \frac{dt}{t} \,\text{Re}\left[\frac{\sin(t + k_0 y_0)}{[(t + i k_0 a)^2 + \{k_0 r(t)\}^2]^{\frac{3}{2}}} \right], \quad \text{for} \quad 2\text{h} < \text{a}. \tag{31}$$

Journal of Physics: Conference Series **161** (2009) 012008 doi:10.1088/1742-6596/161/1/012008

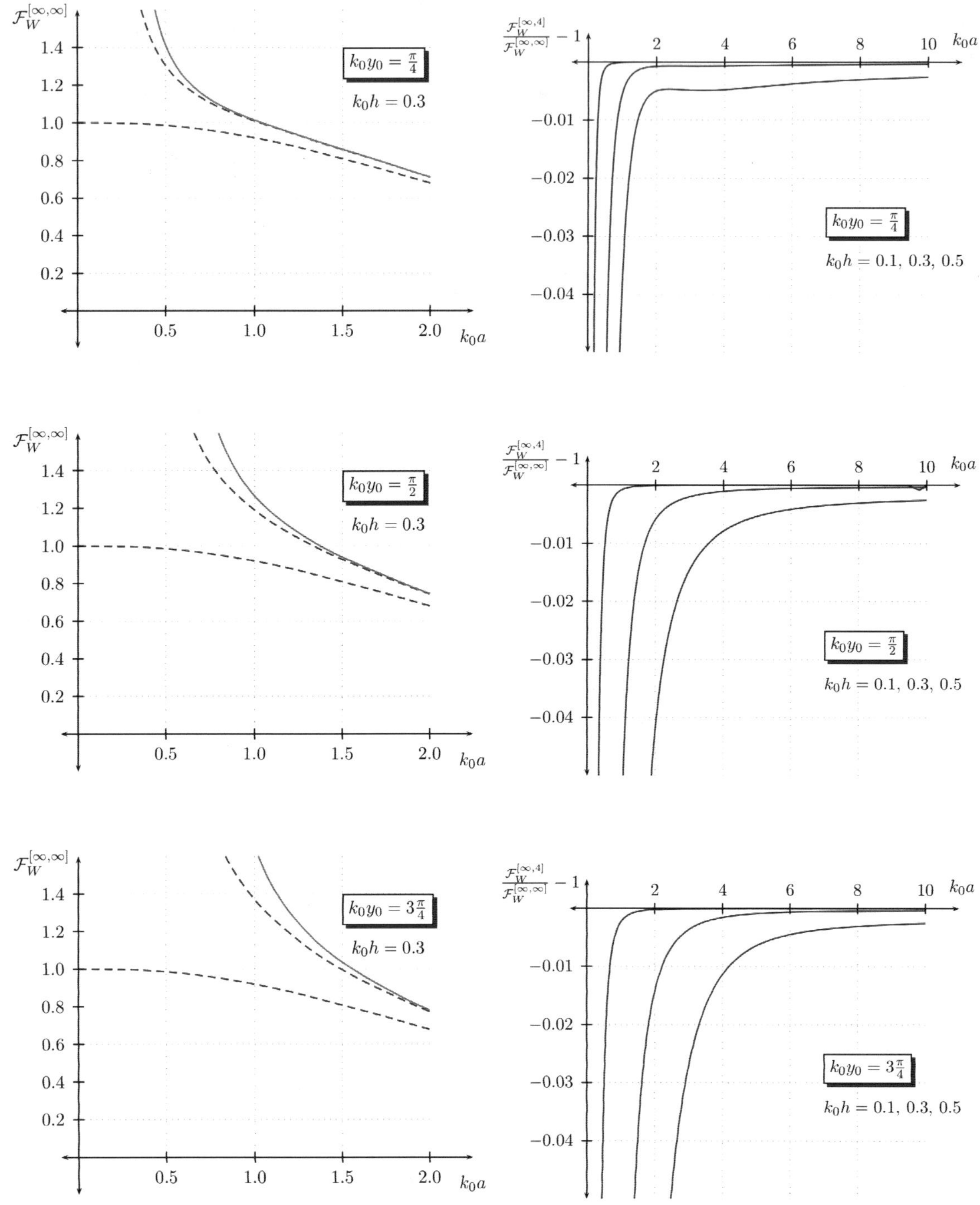

Figure 7. Weak coupling limit: The plots on the left show $\mathcal{F}_D^{[\infty,\infty]}$ versus k_0a for various values of k_0y_0 at $k_0h = 0.3$. The dashed curves represents $\mathcal{F}_D^{[\infty,2]}$ and $\mathcal{F}_D^{[\infty,4]}$ which is plotted here for reference. In each plot the exact value is greater than the value estimated by $\mathcal{F}_D^{[\infty,4]}$. The plots on the right show the fractional error in the perturbative result. In each plot the higher values of k_0h have larger errors.

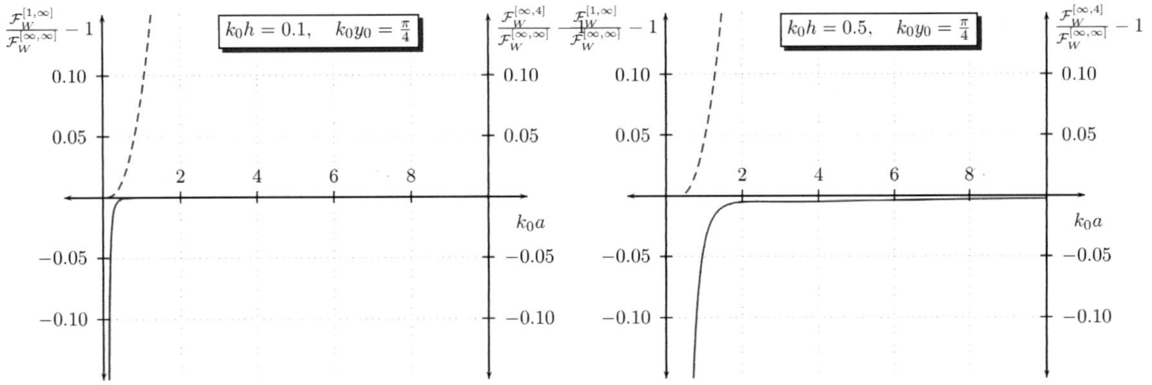

Figure 8. Weak coupling limit: Fractional error in PFA is plotted as the dashed curve. The corresponding error in the perturbative result is plotted as the bold curve and is described by the axis on the right.

The above non-perturbative expression for the lateral Casimir force on corrugated plates (in the weak-coupling limit), has been plotted in figure 7. The perturbative results are plotted as dashed curves in the same figure. We observe that the perturbative results, when the next-to-leading order is included, compares with the exact result remarkably well for $k_0h \ll 1$ keeping the condition that $2h < a$. The fractional error in the lateral force when the next-to-leading order contribution is included is displayed on the right column in figure 7. For a fixed offset, the error gets larger for higher values of k_0h. For k_0h small, we get very good approximation in the whole range of k_0a.

We can also compare the exact result with the one obtained from the PFA. In figure 8 we plot the fractional error in the PFA result, for fixed k_0h, and compare it to the fractional error in the perturbative results. We note that PFA is a good approximation for $k_0a \ll 1$ and perturbative results are valid for $k_0h \ll 1$. Both the PFA and the perturbative analysis are restricted to $2h < a$.

We earlier noted that the PFA limit is obtained by taking the limit $k_0a \to 0$ while keeping the ratio h/a fixed. In figure 9 we plot the lateral force in the PFA limit and compare it with the exact result for various values of k_0a. We note that the error in the PFA is less than 1% for $k_0a \ll 1$ for arbitrary h/a. For $k_0y_0 = \pi/4$, we observe that the PFA is a very good approximation for $h \sim a$ which satisfies $2h\sin(k_0y_0/2) < a$. After viewing the plots for various values of k_0y_0 we note that in general the PFA is a good approximation for $h \sim a$ and further beyond for offsets $k_0y_0 < \pi/4$. It is, in fact, plausible that the PFA holds for large amplitude corrugations for small offsets because the corrugations fit together like fingers in a glove.

The experiments are highly grounded on calculations based on the proximity force approximation. There has been some misunderstanding regarding the range of validity of the PFA versus numerical calculations [9, 18, 19]. As we have shown here the ranges of validity of both solutions may not overlap.

3. Concentric cylinders with corrugations.

In this section we show the numerical calculation for the leading-order Casimir torque between two concentric corrugated cylinders. The cylinders have average radii a_1 and a_2 and the mean distance between them is $a = a_2 - a_1$ like it is shown in figure 10. The set-up is equivalent to the

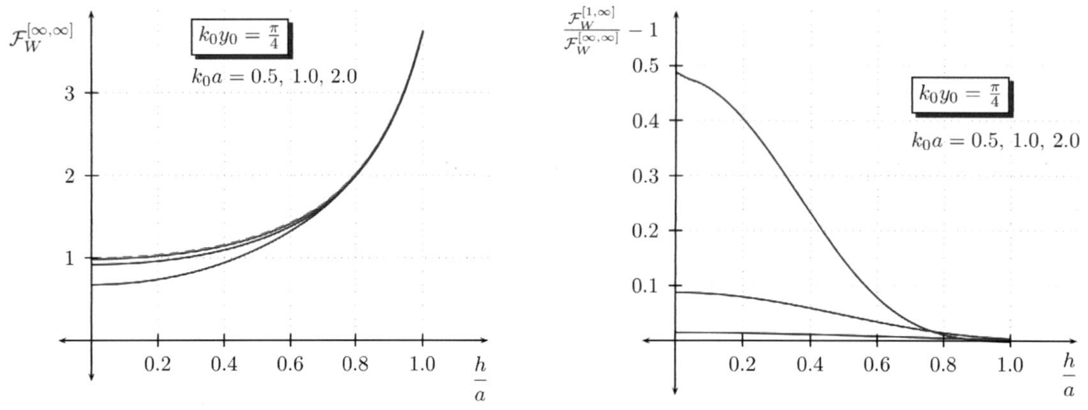

Figure 9. Weak coupling limit: On the left the PFA is shown as the dashed curve. The exact results for higher values of $k_0 a$ are seen to deviate from the dashed curve. On the right the fractional error in PFA is shown to increase with higher values of $k_0 a$.

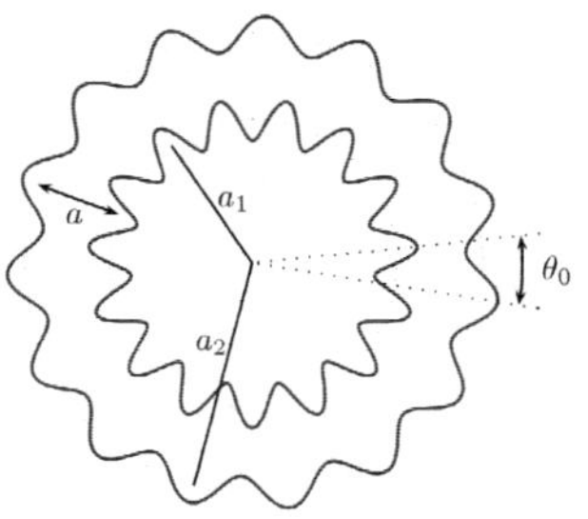

Figure 10. Non-contact gears: Concentric corrugated cylinders with the same corrugation frequency, $\nu = 15$, on each cylinder. θ_0 is the angular shift between the gears.

one in the previous case. The cylinders are described by a potential of the same kind as before,

$$V_i(r, \theta) = \lambda_i \, \delta(r - a_i - h_i(\theta)), \tag{32}$$

where $i = 1, 2$ identify the individual cylinders and $h_i(\theta)$ are the functions describing the corrugations. For an initial angular shift θ_0 as shown in figure 10, and assuming sinusoidal corrugations, we have

$$h_1(\theta) = h_1 \sin[\nu(\theta + \theta_0)], \tag{33a}$$
$$h_2(\theta) = h_2 \sin[\nu\theta], \tag{33b}$$

where $h_{1,2}$ are the corrugation amplitudes and ν is the frequency associated with the corrugations.

In the next sections we will show the results for the strong and weak coupling limit up to second order perturbation. The Casimir torque is calculated by taking the derivative of the

Journal of Physics: Conference Series **161** (2009) 012008 doi:10.1088/1742-6596/161/1/012008

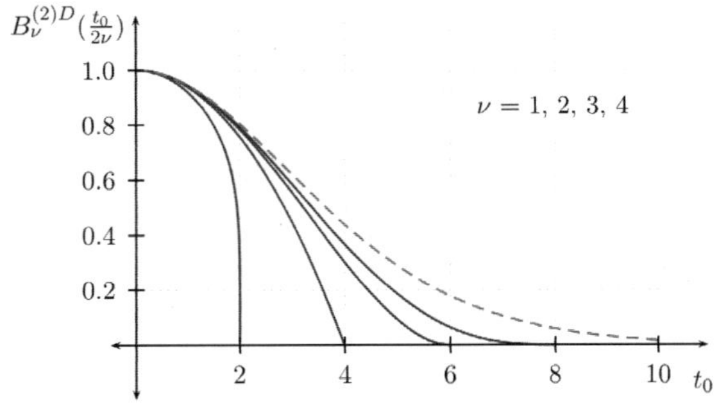

Figure 11. Dirichlet limit: Plots of $B_\nu^{(2)D}(\frac{t_0}{2\nu})$ versus t_0, for $t_0 < 2\nu$ and fixed ν. The dashed curve is the corresponding plot for corrugated plates which is approached by the corrugated cylinders for larger values of ν.

interaction energy with respect to the shifted angle,

$$\mathcal{T} = -\frac{\partial E_{12}}{\partial \theta_0}. \tag{34}$$

Details of the calculation can be found in reference [11, 12]. In the last section we discuss the results.

3.1. Dirichlet limit.
In the Dirichlet limit the interaction energy can be expressed in the form

$$\frac{E_{12}^{(2)}}{2\pi R\, L_z} = \cos(\nu\theta_0)\, \frac{\pi^2}{240\, a^3}\, \frac{h_1}{a}\frac{h_2}{a}\, B_\nu^{(2)D}(\alpha), \tag{35}$$

where we have divided by a factor of $2\pi R$, which is the mean circumference. The function $B_\nu^{(2)D}$ is defined as

$$B_\nu^{(2)D}(\alpha) = \frac{15}{\pi^4} \sum_{m=-\infty}^{\infty} 8\alpha^3 \int_0^\infty x\, dx \frac{4\,\alpha^2}{(1-\alpha^2)}\frac{1}{D_m(\alpha;x)}\frac{1}{D_{m+\nu}(\alpha;x)}, \tag{36}$$

where $\alpha = \frac{a}{2R}$, $R = \frac{a_1+a_2}{2}$ and the functions D_m are given by

$$D_m(\alpha;x) = I_m(x[1+\alpha])K_m(x[1-\alpha]) - I_m(x[1-\alpha])K_m(x[1+\alpha]), \tag{37}$$

Using (34) we evaluate the Casimir torque to be

$$\frac{\mathcal{T}^{(2)D}}{2\pi R\, L_z} = \nu\sin(\nu\theta_0)\, \frac{\pi^2}{240\, a^3}\, \frac{h_1}{a}\frac{h_2}{a}\, B_\nu^{(2)D}(\alpha), \tag{38}$$

where $B_\nu^{(2)D}(\alpha)$ behaves as in figure 11.

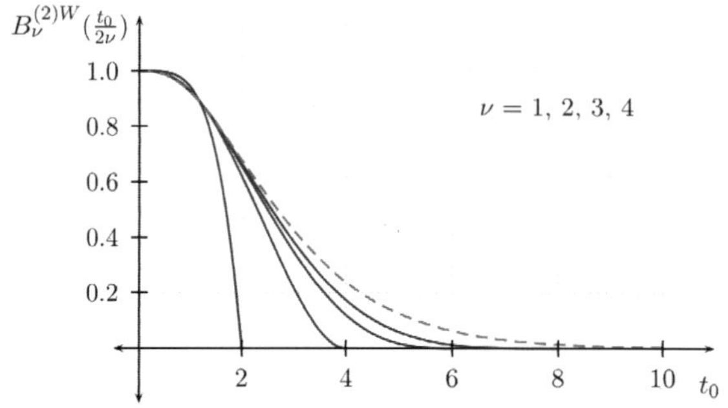

Figure 12. Weak coupling limit: Plots of $B_\nu^{(2)W}(\frac{t_0}{2\nu})$ versus t_0, for $t_0 < 2\nu$ and fixed ν. The dashed curve is the corresponding plot for corrugated plates which is approached by the corrugated cylinders for larger values of ν.

3.2. Weak coupling limit.

According to reference [11] the interaction energy in the limit $a\lambda_i \ll 1$ becomes,

$$\frac{E_{12}^{(2)W}}{2\pi R\, L_z} = \cos(\nu\theta_0)\,\frac{\lambda_1\lambda_2}{32\pi^2\, a}\,\frac{h_1}{a}\frac{h_2}{a}\,B_\nu^{(2)W}(\alpha), \tag{39}$$

where we have defined the function

$$B_\nu^{(2)W}(\alpha) = -\frac{\alpha^3}{2}\frac{\partial}{\partial\alpha}\left[\frac{1}{\alpha^2}\left(\frac{1-\alpha}{1+\alpha}\right)^\nu (1-\alpha^2)(1+2\alpha\nu+\alpha^2)\right]. \tag{40}$$

The Casimir torque per unit area, for the weak coupling limit, can thus be evaluated to be

$$\frac{\mathcal{T}^{(2)W}}{2\pi R\, L_z} = \nu\sin(\nu\theta_0)\,\frac{\lambda_1\lambda_2}{32\pi^2\, a}\,\frac{h_1}{a}\frac{h_2}{a}\,B_\nu^{(2)W}(\alpha). \tag{41}$$

The function $B_\nu^{(2)W}(\alpha)$ is plotted in figure 12.

3.3. Discussion.

We can compare our results with the results from the corrugated parallel plates. In the limiting case when the two cylinders have very large radii but the distance between them is kept constant we can consider that we have parallel plates in a region of small variations in the angle θ_0. This corresponds to $m, n \to \infty$ such that m/R is finite and $\alpha \to 0$.

In figure 11 we plot the function $B_\nu^{(2)D}(\alpha)$, where $t_0 = 2\alpha\nu$. The dashed curve corresponds to the parallel plates. We see that the plot approaches the result for parallel plates as we increase ν. Notice that α cannot get values bigger than 1 since that case corresponds to the inner radius equal to 0. When $\alpha \to 0$ we reach the PFA limit.

The same conclusions can be drawn from figure 12 for the weak coupling case.

In general we see that our results for the Casimir torque reproduce the results for the lateral force on corrugated parallel plates in the limit of large radius and small corrugation wavelengths.

Acknowledgments
We thank the US National Science Foundation, grant number PHY-0554826, and the US Department of Energy, grant number DE-FG-04ER41305, for partial support of the research reported here. ICP would like to thank the French National Research Agency (ANR) for support through Carnot funding. We show our gratitude to Victor Dodonov for his hospitality and his great organization of this meeting on "60 Years of the Casimir Effect." We also like to thank the International Center for Condensed Matter Physics at the University of Brasilia where the meeting was held.

References
[1] Casimir H B G 1948 *Proc. Kon. Ned. Akad. Wetensch.* **51** 793
[2] Ashourvan A, Miri M and Golestanian R 2007 *Phys. Rev. Lett.* **98** 140801
[3] Lombardo F C, Mazzitelli F D and Villar P I 2008 *J. Phys. A: Math. Theor.* **41** 164009
[4] Miri M, and Golestanian R 2008 *Appl. Phy. Lett.* **92**, 113103 (*Preprint* 0802.4144 [quant-ph])
[5] Roy A and Mohedeen U 2002 *Phys. Rev. A***82** 4380
[6] Chen F, Mohideen U, Klimchitskaya G L and Mostepanenko V M 2002 *Phys. Rev. Lett.* **88** 101801
[7] Emig T, Hanke A, Golestanian R and Kardar M 2001 Phys. Rev. Lett. **87** 260402
[8] Emig T, Hanke A, Golestanian R and Kardar M 2003 *Phys. Rev. A* **67**, 022114
[9] Rodrigues R B, Maia Neto P A, Lambrecht A and Reynaud S 2006 *Phys. Rev. Lett.* **96**, 100402
[10] Cavero-Peláez I, Milton K A, Parashar P and Shajesh K V 2008 *Phys. Rev. D* **78** 065018 (*Preprint* 0805.2776 [hep-th])
[11] Cavero-Peláez I, Milton K A, Parashar P and Shajesh K V 2008 *Phys. Rev. D* **78** 065019 (*Preprint* 0805.2777 [hep-th])
[12] Cavero-Peláez I, Milton K A, Parashar P and Shajesh K V (To be published in the proceedings of *Alexander Friedman International Seminar on Gravitation and Cosmology*, Satellite Symposium Devoted to 60 years of the Casimir Effect.)
[13] Milton K A and Wagner J 2008 *J. Phys. A* **41**, 155402 (*Preprint* 0712.3811 [hep-th])
[14] Kenneth O and Klich I 2006 *Phys. Rev. Lett.* **97**, 160401
[15] Emig T, Graham N, Jaffe R L and Kardar M 2008 *Phys. Rev. D* **77**, 025005 (*Preprint* 0710.3084 [cond-mat.stat-mech])
[16] Milton K A and Wagner J 2008 *Phys. Rev. D* **77**, 045005 (*Preprint* 0711.0774 [hep-th])
[17] Milton K A, Parashar P and Wagner J *Phys. Rev. Lett.*, in press *Preprint* 0806.2880 [hep-th])
[18] Chen F, Mohideen U, Klimchitskaya G L, and Mostepanenko V M 2007 *Phys. Rev. Lett.* **98**, 068901
[19] Rodrigues R. B., Maia Neto P. A., Lambrecht A., and Reynaud S 2007 *Phys. Rev. Lett.* **98**, 068902

Journal of Physics: Conference Series **161** (2009) 012009 doi:10.1088/1742-6596/161/1/012009

Computation of Casimir forces for dielectrics or intrinsic semiconductors based on the Boltzmann transport equation

Diego A R Dalvit[1] **and Steve K Lamoreaux**[2]

[1] Theoretical Division, MS B213, Los Alamos National Laboratory, Los Alamos, NM 87545, USA

[2] Yale University, Department of Physics, P.O. Box 208120, New Haven, CT 06520-8120, USA

E-mail: dalvit@lanl.gov

Abstract. The interaction between drifting carriers and traveling electromagnetic waves is considered within the context of the classical Boltzmann transport equation to compute the Casimir-Lifshitz force between media with small density of charge carriers, including dielectrics and intrinsic semiconductors. We expand upon our previous work (Phys. Rev. Lett. 2008, in press; arXiv:0805.1676) and derive in some detail the frequency-dependent reflection amplitudes in this theory and compute the corresponding Casimir free energy for a parallel plate configuration. We critically discuss the the issue of verification of the Nernst theorem of thermodynamics in Casimir physics, and explicity show that our theory satisfies that theorem. Finally, we show how the theory of drifting carriers connects to previous computations of Casimir forces using spatial dispersion for the material boundaries.

PACS number: 42.50.Ct, 12.20.-m, 78.20.-i

1. Introduction

Quantum vacuum forces acting between dielectric planar surfaces or between an atom and a dielectric semi-space were computed long ago by Lifshitz [1] in terms of the complex frequency-dependent dielectric permittivity $\bar{\epsilon}(\omega)$ of the material boundaries. In this original formulation for ideal dielectrics, $\bar{\epsilon}(\omega)$ does not include contributions from current carriers, and as such can be called the "bare" permittivity. Extensions of the Lifshitz theory to media with large free charge carrier density, such as metals or highly doped semiconductors, are typically done by adding a frequency-dependent conduction term, computed from the optical data of the material and extrapolated to low frequencies by different theoretical models (e.g., a Drude-like term $i4\pi\sigma_0/\omega$, where σ_0 is the dc Drude conductivity). The finite temperature Casimir-Lifshitz force is extremely sensitive to the optical response of the materials at low frequency, and therefore different theoretical extrapolations have resulted in conflicting conclusions about the nature of the Casimir force between metals and/or highly doped semiconductors.

Systems with small density of current carriers, such as insulators or intrinsic-semiconductors, were recently considered by Pitaevskii [2] and by us [3]. In [2] the thermal Lifshitz force between an atom and a conductor with low charge density was computed in terms of the Green function formalism, taking into account the penetration of the static component of the fluctuating EM field into the conductor. This approach is quasi-static, appropriate for the large distance regime of the thermal Lifshitz atom-surface interaction. The relevant (longitudinal) Green function, expressed in terms of an auxiliary static potential field, can be computed assuming that the gas of carriers in nondegenerate. The static potential satisfies the equation $(\nabla^2 - \kappa^2)\varphi = 0$, where $\kappa^2 = 4\pi e^2 n_0/\bar{\epsilon}_0 k_B T$. Here $-e$ is the electron charge, $\bar{\epsilon}_0$ is the static bare dielectric constant of the medium (which does not take into account the contribution from current carriers), and n_0 is the (uniform) carrier density. Note that $\kappa = 1/R_D$ is the inverse of the Debye-Hückel screening radius R_D. For good metals the Debye radius is very small (on the order of inter-atomic distances), while for insulators and intrinsic semiconductors it is much larger (on the order of microns or more).

In [3] we have extended Pitaevskii's calculation beyond the quasi-static limit and proposed a theory for the Casimir interaction taking into account Debye screening and carrier drift based on the classical Boltzmann equation. Rather than computing the force with the Green function formalism, we use the form of the Lifshitz formula written in terms of frequency-dependent reflection amplitudes $r^p_{\mathbf{k},j}(w)$ of the j-th material boundary. Here p denotes the polarization of incoming waves (transverse electric TE or transverse magnetic TM). For simplicity, we will assume that the material is such that there is no mixing of polarizations upon reflection (the more general case can be treated replacing the reflection amplitudes by 2×2 reflection matrices). The projection on the plane of the interface of the linear momentum of incoming waves is denoted by $\mathbf{k}$. The Casimir-Lifshitz pressure between two plane semi-spaces separated by a gap of length d is

$$P(d) = 2k_B T \sum_{n=0}^{\infty}{}' \int \frac{d^2\mathbf{k}}{(2\pi)^2} \sqrt{k^2 + \xi^2/c^2} \sum_p \frac{r_1^p r_2^p e^{-2d\sqrt{k^2+\xi^2/c^2}}}{1 - r_1^p r_2^p e^{-2d\sqrt{k^2+\xi^2/c^2}}}. \tag{1}$$

The prime in the sum over n means that the zero frequency $n = 0$ term has to be multiplied by a $1/2$ factor, and all reflection coefficients are evaluated at imaginary frequencies $\omega = i\xi_n$, where $\xi_n = 2\pi n k_B T/\hbar$ are the Matsubara frequencies.

2. From Boltzmann transport equation to reflection amplitudes

In order to compute the appropriate frequency-dependent reflection coefficients for materials with small density of carriers we will consider that the EM field interacts with the gas of drifting carriers, and that these can be modeled as a continuum nondegenerate system. Under these conditions, it is reasonable to model the carriers with the classical Boltzmann transport equation coupled to Maxwell's equations for the electromagnetic field [5, 6].

For a dielectric the carriers are charged particles (electrons or ions) hopping from site to site of the crystalline array. For an intrinsic semiconductor, the density of carriers and hole is equal, but their dynamics are different; however, in this work we treat them as dynamically equivalent, which doubles the charge density. Assuming that there is no external applied field on the material, and all fields have a time dependency of the form $e^{-i\omega t}$, Maxwell's equations take the form

$$\nabla \times \mathbf{E} = i\mu_0\omega\mathbf{H}, \quad \nabla \times \mathbf{H} = -i\bar{\epsilon}(\omega)\omega\mathbf{E} + \mathbf{J}, \quad \nabla \cdot \mathbf{E} = -\frac{en}{\bar{\epsilon}(\omega)}. \tag{2}$$

Here n is the carrier density, μ_0 is the permeability of vacuum, and $\mathbf{J} = -en\mathbf{v}$ is the carrier current, where $\mathbf{v}$ is the mean velocity of carriers. The charge transport in the system is described by the classical Boltzmann equation,

$$\left(\frac{\partial}{\partial t} + \mathbf{v}\cdot\nabla\right)\mathbf{v} = -\frac{e}{m}\mathbf{E} - \frac{v_T^2}{n}\nabla n - \frac{\mathbf{v}}{\tau}, \tag{3}$$

where m is the effective mass of charge carriers, $v_T = \sqrt{k_B T/m}$ is their mean thermal velocity, and τ is the carrier relaxation time. Now we linearize Eq.(3) with respect to the ac fields.

Writing the charge density as $n \to n_0 + n(\mathbf{r})e^{-i\omega t}$, the current is $\mathbf{J} = -en_0\mathbf{v}$ (after discarding the term $n(\mathbf{r})\mathbf{v}e^{-2i\omega t}$). Since by assumption there is no external field applied, there is no net motion of charges, $\mathbf{v}_0 = 0$, so the term $\mathbf{v} \cdot \nabla\mathbf{v}$ in Eq.(3) behaves as $e^{-2i\omega t}$, and can also be discarded. Finally, the linearized Boltzmann equation is

$$\left(-i\omega + \frac{1}{\tau}\right)\mathbf{v} = -\frac{e}{m}\mathbf{E} - \frac{v_T^2}{n_0}\nabla n. \tag{4}$$

Inserting $n = -\bar{\epsilon}\nabla \cdot \mathbf{E}/e$ into this equation, we can solve for $\mathbf{v}$:

$$\mathbf{v} = \frac{\tau}{1 - i\omega\tau}\left[-\frac{e}{m}\mathbf{E} + \frac{v_T^2\bar{\epsilon}}{en_0}\nabla \cdot (\nabla \cdot \mathbf{E}\right]. \tag{5}$$

Plugging this expression into Maxwell's equations we derive the fundamental equation for the electric field inside the material,

$$\left[\nabla^2 + \mu_0\bar{\epsilon}(\omega)\omega^2\left(1 + i\frac{\omega_c}{\omega(1 - i\omega\tau)}\right)\right]\mathbf{E} = \left[1 + i\mu_0\bar{\epsilon}(\omega)\frac{\omega D}{1 - i\omega\tau}\right]\nabla \cdot (\nabla \cdot \mathbf{E}). \tag{6}$$

Here $\omega_c = 4\pi en_0\mu/\bar{\omega}$, $\mu = e\tau/m$ is the mobility of carriers, and $D = v_T^2\tau$ is the diffusion constant. Note that the frequency-dependent ratio $\omega_c/D = 4\pi e^2 n_0/\bar{\epsilon}(\omega)k_{\rm B}T$ coincides with $\kappa^2 = 1/R_{\rm D}^2$ in the quasi-static limit.

As we will show below, Eq. (6) allows TM and TE solutions, so that there is no cross-polarization upon reflection on the material, and we can safely use reflection amplitudes rather than reflection matrices in the Lifshitz formula. Let us assume that the material occupies the semi-space region $z < 0$ and the region $z > 0$ is vacuum.

2.1. TM modes

For transverse magnetic modes $e_y = 0$, so that the electric field is

$$\mathbf{E}(\mathbf{r}) = [e_x(z)\hat{\mathbf{x}} + e_z(z)\hat{\mathbf{z}}]e^{ikx}, \tag{7}$$

where, from now on, we are omitting the phase factors $e^{-i\omega t}$. Substituting this into Eq. (6) we obtain two coupled second-order differential equations for e_x and e_z:

$$\left[\partial_z^2 + \mu_0\bar{\epsilon}(\omega)\omega\left(1 + i\frac{\tilde{\omega}_c}{\omega} + i\frac{\tilde{D}k^2}{\omega}\right)\right]e_x = ik[1 + i\mu_0\bar{\epsilon}(\omega)\omega\tilde{D}]\partial_z e_z,$$

$$\left[-i\mu_0\bar{\epsilon}(\omega)\omega\tilde{D}\partial_z^2 - k^2 + \mu_0\bar{\epsilon}(\omega)\omega^2\left(1 + i\frac{\tilde{\omega}_c}{\omega}\right)\right]e_z = ik[1 + i\mu_0\bar{\epsilon}(\omega)\omega\tilde{D}]\partial_z e_x,$$

where $\tilde{\omega}_c \equiv \omega_c/(1 - i\omega\tau)$ and $\tilde{D} \equiv D/(1 - i\omega\tau)$. It is possible to combine these two coupled equations into two uncoupled fourth-order differential equations. To this end one takes the ∂_z^2 derivative of the first equation above, which results in terms proportional to $\partial_z^4 e_x$, $\partial_z^2 e_x$, and $\partial_z^3 e_z$. This last term $\partial_z^3 e_z$ can be obtained from taking the ∂_z derivative of the second equation above, which results in terms proportional to $\partial_z^2 e_x$ and e_x. Putting all together, one can derive the following fourth-order differential equation for e_x:

$$(\partial_z^2 - \eta_T^2)\,(\partial_z^2 - \eta_L^2)e_x = 0, \tag{8}$$

Journal of Physics: Conference Series **161** (2009) 012009 doi:10.1088/1742-6596/161/1/012009

where

$$\eta_T^2(\omega) = k^2 - \mu_0\bar{\epsilon}(\omega)\omega^2\left(1 + i\frac{\tilde{\omega}_c}{\omega}\right), \tag{9}$$

$$\eta_L^2(\omega) = k^2 - i\frac{\omega}{\tilde{D}}\left(1 + i\frac{\tilde{\omega}}{\omega}\right). \tag{10}$$

In a similar fashion one obtains the following equation for e_z:

$$(\partial_z^2 - \eta_T^2)\,(\partial_z^2 - \eta_L^2)e_z = 0. \tag{11}$$

The solutions of these equations that vanish for $z \to -\infty$ are $e_x(z) = A_T e^{\eta_T z} + A_L e^{\eta_L z}$ and $e_z(z) = A'_T e^{\eta_T z} + A'_L e^{\eta_L z}$, where we assume $\text{Re}\eta_T > 0$ and $\text{Re}\eta_L > 0$. The amplitudes are related as $A'_L = -i\eta_L A_L/k$ and $A'_T = -ikA_T/\eta_T$. Given this TM electric field, the associated TM magnetic field can be readily computed, $\mathbf{H} = i\hat{\mathbf{y}}A_T e^{\eta_T z}e^{ikx}(k^2 - \eta_T^2)/\mu_0\omega\eta_T$.

Now we compute the TM reflection amplitude, imposing the boundary conditions on the $z = 0$ interface. These conditions are E_x, H_y, and $\bar{\epsilon}E_z$ continuous (the continuity of B_z is automatically satisfied for TM modes). On the vacuum side ($z > 0$) the condition $\nabla \cdot \mathbf{E} = 0$ implies $ike_x + \partial_z e_z = 0$, so that the fields incident on the interface from the $z > 0$ side are $\mathbf{E}^{\text{in}} = E_0[-(k_z/k)\hat{\mathbf{x}} + \hat{\mathbf{z}}]e^{ik_z z}e^{ikx}$ and $\mathbf{H}^{\text{in}} = -E_0\omega\hat{\mathbf{y}}e^{ik_z z}e^{ikx}/\mu_0 kc^2$, where we have used that in vacuum $k^2 + k_z^2 = \omega^2/c^2$. The reflected fields are $\mathbf{E}^{\text{r}} = rE_0[+(k_z/k)\hat{\mathbf{x}} + \hat{\mathbf{z}}]e^{-ik_z z}e^{ikx}$ and $\mathbf{H}^{\text{r}} = -rE_0\omega\hat{\mathbf{y}}e^{-ik_z z}e^{ikx}/\mu_0 kc^2$, where r is the reflection amplitude. The transmitted fields into the material ($z < 0$) are $\mathbf{E}^{\text{t}} = \left[(A_T e^{\eta_T z} + A_L e^{\eta_L z})\,\hat{\mathbf{x}} + \left(-\frac{ik}{\eta_T}A_T e^{\eta_T z} - \frac{i\eta_L}{k}A_L e^{\eta_L z}\right)\hat{\mathbf{z}}\right]e^{ikx}$ and $\mathbf{H}^{\text{t}} = i\hat{\mathbf{y}}(k^2 - \eta_T^2)A_T e^{\eta_T z}e^{ikx}/\mu_0\omega\eta_T$. Imposing the boundary conditions, and after some straightforward algebra, the reflection amplitude can be written as $r(\omega) = (1-\alpha)/(1+\alpha)$, where $\alpha = \frac{k^2}{i\eta_L k_z}\left[\frac{1}{\bar{\epsilon}(\omega)} - \frac{\omega^2/c^2}{k^2-\eta_T^2} + \frac{\eta_L\eta_T\omega^2/c^2}{k^2(k^2-\eta_T^2)}\right]$. Expressed along imaginary frequencies $\omega = i\xi$, the TM reflection amplitude is

$$r_{\mathbf{k}}^{\text{TM}}(i\xi) = \frac{\bar{\epsilon}(i\xi)\sqrt{k^2 + \xi^2/c^2} - \chi}{\bar{\epsilon}(i\xi)\sqrt{k^2 + \xi^2/c^2} + \chi}, \tag{12}$$

where

$$\chi = \frac{1}{\eta_L}\left[k^2 + \bar{\epsilon}(i\xi)\frac{\xi^2}{c^2}\frac{\eta_L\eta_T - k^2}{\eta_T^2 - k^2}\right]. \tag{13}$$

Along imaginary frequencies, η_L and η_T take the form:

$$\eta_L(i\xi) = \sqrt{k^2 + \frac{4\pi e^2 n_0}{\bar{\epsilon}(i\xi)k_B T} + \frac{\xi(1+\xi\tau)}{v_T^2\tau}}, \tag{14}$$

$$\begin{aligned}\eta_T(i\xi) &= \sqrt{k^2 + \bar{\epsilon}(i\xi)\frac{\xi^2}{c^2}\left(1 + \frac{4\pi e^2 n_0\tau}{m\bar{\epsilon}(i\xi)\xi(1+\xi\tau)}\right)} \\ &= \sqrt{k^2 + [\bar{\epsilon}(i\xi) + 4\pi\sigma(i\xi)/\xi]\xi^2/c^2},\end{aligned} \tag{15}$$

where $\sigma(i\xi) = \sigma_0/(1+\xi\tau)$ and $\sigma_0 = e^2 n_0\tau/m$ are the ac and dc Drude conductivities, respectively. Therefore, Eq. (12) gives a modified Fresnel TM coefficient due to the presence of Debye-Hückel screening and charge drift in the material.

2.2. TE modes

For transverse electric modes $e_z = 0$, so that the electric field is

$$\mathbf{E}(\mathbf{r}) = [e_x(z)\hat{\mathbf{x}} + e_y(z)\hat{\mathbf{y}}]e^{ikx}. \tag{16}$$

Substituting this into Eq. (6) we obtain two second-order differential equations for e_x and e_z:

$$\left[\partial_z^2 + \mu_0\bar{\epsilon}(\omega)\omega^2\left(1 + i\frac{\tilde{\omega}_c}{\omega} + i\frac{\tilde{D}k^2}{\omega}\right)\right] e_x = 0,$$

$$\left[\partial_z^2 - k^2 + \mu_0\bar{\epsilon}(\omega)\omega^2\left(1 + i\frac{\tilde{\omega}_c}{\omega}\right)\right] e_y = ik[1 + i\mu_0\bar{\epsilon}(\omega)\omega\tilde{D}]\partial_z e_x.$$

The solution to the first equation is $e_x(z) = Ae^{\beta z}$, where A is a constant and $\beta^2 = -i\mu_0\bar{\epsilon}(\omega)\omega\tilde{D}\eta_L^2$ (we assume $\mathrm{Re}\beta > 0$). Plugging this solution into the second equation we obtain $e_y(z) = Be^{\eta_T z} + Ce^{\beta z}$, where $C = ikA\beta(1 + i\mu_0\bar{\epsilon}(\omega)\omega\tilde{D})/(\beta^2 - \eta_T^2)$. Given this TE electric field, the associated TE magnetic field is $\mathbf{H} = (1/i\mu_0\omega)[-(B\eta_T e^{\eta_T z} + C\beta e^{\beta z})\hat{\mathbf{x}} + A\beta e^{\beta z}\hat{\mathbf{y}} + ik(Be^{\eta_T z} + Ce^{\beta z})\hat{\mathbf{z}}]e^{ikx}$.

Now we compute the TE reflection amplitude imposing the boundary conditions on the interface (the continuity of $\bar{\epsilon}E_z$ is automatically satisfied for TE modes). On the vacuum side we have $e_x = 0$, so that the incident fields are $\mathbf{E}^{\rm in} = E_0 e^{ik_z z}e^{ikx}\hat{\mathbf{y}}$ and $\mathbf{H}^{\rm in} = E_0(-k_z\hat{\mathbf{x}} + k\hat{\mathbf{z}})e^{ik_z z}e^{ikx}/\mu_0\omega$, and the reflected fields are $\mathbf{E}^{\rm r} = rE_0 e^{-ik_z z}e^{ikx}\hat{\mathbf{y}}$ and $\mathbf{H}^{\rm r} = rE_0(+k_z\hat{\mathbf{x}} + k\hat{\mathbf{z}})e^{-ik_z z}e^{ikx}/\mu_0\omega$. The transmitted fields into the material are given above. A simple calculation leads to the expression of the reflection amplitude $r = (ik_z - \eta_T)/(ik_z + \eta_T)$. Upon performing the rotation $\omega \to i\xi$, we get

$$r_{\mathbf{k}}^{\rm TE}(i\xi) = \frac{\sqrt{k^2 + \xi^2/c^2} - \eta_T}{\sqrt{k^2 + \xi^2/c^2} + \eta_T}. \tag{17}$$

Using Eq. (15) we see that $r^{\rm TE}$ is the usual Fresnel TE reflection coefficient with a dielectric permittivity equal to the sum of the "bare" one and the ac Drude (conduction) permittivity, $\epsilon(i\xi) = \bar{\epsilon}(i\xi) + 4\pi\sigma(i\xi)/\xi$.

As discussed in detail in [3], these modified TE and TM reflection coefficients have appropriate limiting behaviors. In the quasi-static limit ($\xi \to 0$) they coincide with the ones derived in [2] for conductors with small density of carriers in the large distance (low frequency) regime, namely $r_{\mathbf{k}}^{\rm TE}(0) = 0$ (in the static limit the TE polarized field is a pure magnetic field, which fully penetrates the nonmagnetic material) and $r_{\mathbf{k}}^{\rm TM}(0) = (\bar{\epsilon}_0 q - k)/(\bar{\epsilon}_0 q + k)$, with $q = \sqrt{k^2 + \kappa^2}$ (in the static limit $r^{\rm TM}$ interpolates between a good conductor and an ideal dielectric). On the other hand, for any frequency $\xi \geq 0$, and in the limit of ideal dielectrics (small free charge density and small effective thermal velocity), we recover the usual Fresnel equations written in terms of the bare permittivity $\bar{\epsilon}(\omega)$.

3. Influence of drifting carriers in the Casimir-Lifshitz free energy

We now study the implications of our theory in the computation of the Casimir-Lifshitz free energy

$$\frac{E}{A} = k_{\rm B}T\sum_p\sum_{n=0}^{\infty\prime}\int\frac{d^2\mathbf{k}}{(2\pi)^2}\ln[1 - r_{\mathbf{k},1}^p(i\xi_n)r_{\mathbf{k},2}^p(i\xi_n)e^{-2d\sqrt{k^2+\xi_n^2/c^2}}], \tag{18}$$

between two identical planar semi-spaces with small density of charge carriers, such as intrinsic semiconductor media. As examples, we consider the cases of pure germanium and pure silicon. The reflection coefficients Eqs.(12,17) that enter into this equation depend on temperature explicitly through the Matsubara frequencies and implicitly through the optical and conductivity parameters, which we proceed to quote.

For intrinsic Ge, the bare permittivity can be approximately fitted with a Sellmeier-type expression

$$\bar{\epsilon}(i\xi) = \bar{\epsilon}_\infty + \omega_0^2\frac{\bar{\epsilon}_0 - \bar{\epsilon}_\infty}{\xi^2 + \omega_0^2}, \tag{19}$$

where $\bar{\epsilon}_0 \approx 16.2$, $\bar{\epsilon}_\infty \approx 1.1$, and $\omega_0 \approx 5.0 \times 10^{15}$rad/sec at $T \approx 300$K. The temperature dependence of the permittivity has been measured in the $20-300$K range at wavelengths $1.9-5.5\mu$m [7], and shown to be very weak. Therefore, in this paper we assume that the permittivity is approximately constant as a function of temperature, and given by the above Sellmeier fitting function. The intrinsic carrier density varies with temperature as

$$n_0(T) = \sqrt{n_c n_v} e^{-\frac{E_g}{2k_B T}}, \tag{20}$$

where n_c and n_v are the effective density of states in the conduction and valence band, respectively. These depend on temperature as $n_c(T) = 1.98 \times 10^{15} T^{3/2}\text{cm}^{-3}$ and $n_v(T) = 9.6 \times 10^{14} T^{3/2}\text{cm}^{-3}$ (temperature is measured in degrees K). The band gap energy also depends on temperature, $E_g(T) = 0.742 - 4.8 \times 10^{-4} T^2/(T+235)$eV. The effective mass of conductivity is $m = 0.12 m_e$, where m_e is the free electron mass [8]. The relaxation time depends on temperature as $\tau(T) = \tau_0 + \tau_1 e^{C_1(T/300)^2 + C_2(T/300)}$ [9], where $\tau_0 = 0.26$ps, $\tau_1 = 1.49$ps, $C_1 = -0.434$, and $C_2 = 1.322$. At $T = 300$K one has $E_g = 0.66$eV, $n_c = 1.0 \times 10^{19}\text{cm}^{-3}$, $n_v = 5.0 \times 10^{18}\text{cm}^{-3}$, and $\tau = 3.9$ps.

For intrinsic Si, $\bar{\epsilon}_0 \approx 11.87$, $\bar{\epsilon}_\infty \approx 1.035$, and $\omega_0 \approx 6.6 \times 10^{15}$rad/sec at $T \approx 300$K. The temperature dependence of the permittivity has been measured in the $20-300$K range at wavelengths $1.1-5.6\mu$m [7], and also shown to be very weak. The effective density of states in the conduction and valence bands are $n_c(T) = 6.2 \times 10^{15} T^{3/2}\text{cm}^{-3}$ and $n_v(T) = 3.5 \times 10^{15} T^{3/2}\text{cm}^{-3}$ respectively. The band gap energy is $E_g(T) = 1.17 - 4.73 \times 10^{-4} T^2/(T+636)$eV. The effective mass of conductivity is $m = 0.26 m_e$ [8]. The relaxation time parameters are $\tau_0 = 1.0$ps, $\tau_1 = -0.538$ps, $C_1 = 0.0015$, and $C_2 = -0.09$ [9]. At $T = 300$K one has $E_g = 1.12$eV, $n_c = 3.2 \times 10^{19}\text{cm}^{-3}$, $n_v = 1.8 \times 10^{19}\text{cm}^{-3}$, and $\tau = 0.5$ps.

In Fig 1 we plot the Casimir-Lifshitz free energy between two identical planar intrinsic semiconducting (Ge and Si) semi-spaces as a function of the distance between them. We use our theory of Casimir forces with account of Debye-Hückel screening and charge drift to compute the reflection coefficient Eqs. (12,17) and compare these predictions with the simple model in which the reflection coefficients are given by the usual Fresnel formulas in which the permittivity of the materials $\epsilon(i\xi)$ is computed by adding to the bare permittivity $\bar{\epsilon}(i\xi)$ a dc conductivity term $4\pi\sigma_0/\xi$. In both models we normalize the free energies to the free energy computed using the standard Lifshitz theory using the bare permittivity only. For intrinsic Ge and intrinsic Si, $\tilde{\omega}_c$ and $\tilde{D}/\xi$ are both very small in the relevant range of frequencies for the Lifshitz formula. Therefore only the $n = 0$ TM mode is significantly modified by the screening and charge drift effects, $r_{\mathbf{k}}^{\text{TE}}(0) = 0$ and $r_{\mathbf{k}}^{\text{TM}}(0) = (\bar{\epsilon}(0)q - k)/(\bar{\epsilon}(0)q + k)$, with $q = \sqrt{k^2 + \kappa^2}$, as in [2]. In all other $n \geq 1$ terms in Eq. (18) the reflection coefficients can be replaced by the standard Fresnel expressions in terms of the bare permittivity $\bar{\epsilon}(i\xi)$. Since for intrinsic carrier density for Ge ($\approx 10^{13}\text{cm}^{-3}$) is much larger than that for Si ($\approx 10^{10}\text{cm}^{-3}$), the Debye radius is of Ge, $R_D = 1/\kappa = 0.68\mu$m is much smaller than that of Si, $R_D = 24\mu$m. As follows from Fig. 1, the effect of Debye screening and drifting carriers (denoted as "drift" in Fig. 1) becomes important for distances much larger than the Debye radius, so that this effect is more likely to be detected in Ge than in Si. In the latter case, for distances $d > R_D$, the Casimir force is too weak, at such a large distance, to be measured by any current or proposed experimental technique. From Fig. 1 we also note that when $d \gg R_D$ the plates appear as perfect conductors for the TM $n = 0$ mode, while in the case of the additive term $(\bar{\epsilon}(i\xi) + 4\pi\sigma_0/\xi$, denoted as "cond" in the figure), the plates appear as perfect conductors for the TM $n = 0$ mode at distances of the order of $\lambda_T = \hbar c/k_B T$ ($\approx 7\mu$m at $T = 300$K), independent of the material properties.

4. On the satisfaction of Nernst theorem as a prerequisite for a Casimir theory

Calculations of finite temperature Casimir-Lifshitz forces between media with large charge density, such as metals and highly doped semiconductors, have resulted in a heated debate on the

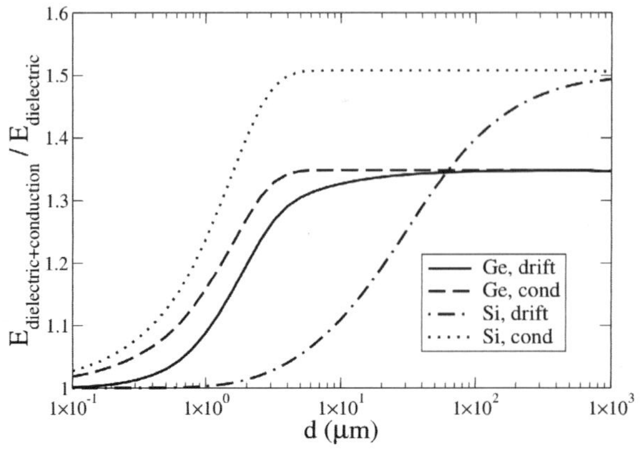

Figure 1. Casimir-Lifshitz free energy at $T = 300\text{K}$ for intrinsic Ge and Si taking into account charge drift Eqs. (12,17) (curves denoted by "drift") or an additive dc conductivity term $4\pi\sigma_0/\xi$ to the bare permittivity (curves denoted by "cond"). For Ge, $\sigma_0 = 1/(43\Omega\ \text{cm})$, and for Si, $\sigma_0 = 1/(2.3 \times 10^5\Omega\ \text{cm})$. The free energies are normalized to those computed using only the bare permittivity in the usual Lifshitz theory.

adequate way to describe the optical properties of such systems within the Lifshitz formalism. Different phenomenological ways to extrapolate optical data to low frequencies, either with a Drude model including dissipation or with a plasma model setting dissipation to zero from the start, result in completely different predictions for the force at finite temperature [10].

It has been suggested that the Nernst theorem of thermodynamics serves as a way to accept or discard conductivity models when applied to the computation of the Casimir-Lishitz entropy $S(T) = -\partial E(T)/\partial T$. The Nernst theorem states that the entropy of a physical system of N particles in thermal equilibrium at zero temperature is a well-defined constant, determined only by the degeneracy Ω_N of the ground state of the system, that is, $S(T = 0) = k_B \ln \Omega_N$. For systems with non-degenerate ground states $\Omega_N = 1$ (e.g., a perfect crystal lattice), the entropy should vanish at zero temperature. This is not the case for a large class of systems, including spin networks and glasses, that can have a large collection of degenerate ground states (with degeneracy Ω_N depending on the total number of particles), so that $S(T = 0) > 0$ in such systems. In some textbooks [11] it is further required as part of Nernst theorem that the degeneracy Ω_N be independent of any varying parameters of the system (such as pressure, volume, field intensities, etc). This has been used by some authors to discard Casimir theories which lead to a zero-temperature entropy that depends on the separation d between the Casimir plates. The requirement of independence of $S(T = 0)$ on volume is at odds with the fact that entropy is an extensive quantity, and should grow with system size [12]. It is not clear to us that one can simply discard a model of conductivity for the Casimir plates based on the fact that the zero temperature entropy depends on the distance between plates. We believe that this issue requires further study. Of course, if for a given theoretical model for the Casimir plates $S(T = 0) < 0$, then such model violates Nernst theorem.

In the remainder of this section we explicitly prove that our theory for Casimir forces with intrinsic semiconductor media is compatible with Nernst theorem of thermodynamics, resulting in $S(T = 0) = 0$, as for systems with a non-degenerate ground state ($\Omega_N = 1$). We will closely follow the approach in [13]. We start by expressing the Casimir free energy as

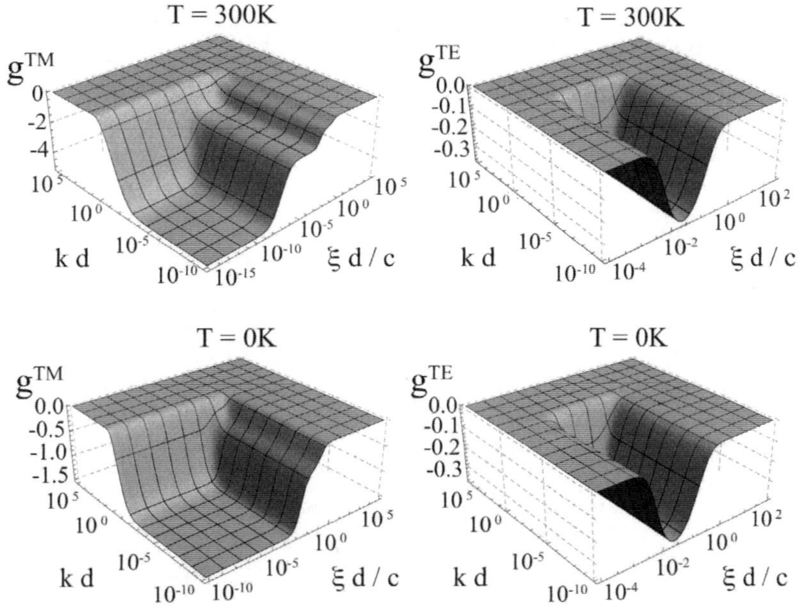

Figure 2. Behavior of the functions $g^p(i\xi, \mathbf{k})$ used to compute the Casimir-Lifshitz free energy and entropy for semiconductor materials with account of drifting carriers. The reflections coefficients are given by (12) and (17), parameters are for intrinsic Ge (see text), and the distance is set to $d = 1\mu$m. The variation with temperature (in the range $T = 0 - 300$K) of the TE function is not perceptible on the scale of the figure. The corresponding functions without account of Debye screening and carrier drift correspond to the $T = 0$K plots in this figure.

$E = \frac{\hbar}{2\pi} \sum_{p,\mathbf{k}} \sum_{n=0}^{\infty'} \theta g^p(in\theta, \mathbf{k}; \theta)$, where

$$g^p(\omega, \mathbf{k}; \theta) = \ln[1 - r^p_{\mathbf{k},1}(\omega, \theta)\, r^p_{\mathbf{k},2}(\omega, \theta) e^{-2d\sqrt{k^2 - \omega^2/c^2}}], \tag{21}$$

and $\theta = 2\pi k_{\rm B} T/\hbar$. Note that we have allowed for an implicit dependence of the reflection coefficients on temperature. The Casimir-Lifshitz entropy is $S = -(2\pi/\hbar)\partial E/\partial\theta$. In Fig. 2 we plot the behaviour of $g^p(i\xi, \mathbf{k}; \theta)$ as a function of the imaginary frequency $\omega = i\xi$ and as a function of $k = |\mathbf{k}|$ for TM and TE polarizations for different temperatures (the corresponding reflection amplitudes are obtained from Eqs. (12,17)). Let us consider the TE and TM contributions to the entropy separately.

For TE modes, since the reflection coefficient (17) depends implicitly on temperature only through $\tilde{\omega}_c$, which becomes exponentially small as low temperatures because the carrier density vanishes as $T \to 0$, it is possible to show that the $\theta \to 0$ and $\omega \to 0$ limits of $r^{\rm TE}_{\mathbf{k}}(\omega, \theta)$ commute, that $g^p(\omega, \mathbf{k}; \theta)$ is analytic in the upper-half complex ω plane, and that the sum over n and the derivative with respect to θ in the expression for the entropy can be interchanged. Therefore, the contribution of TE modes to the entropy is [13]

$$S^{\rm TE}(T) = -\sum_{\mathbf{k}} \sum_{n=0}^{\infty'} [g^{\rm TE}(in\theta, \mathbf{k}; \theta) + in\theta g^{\rm TE}_{\omega}(in\theta, \mathbf{k}; \theta) + \theta g^{\rm TE}_{\theta}(in\theta, \mathbf{k}; \theta)]. \tag{22}$$

Here we have defined $g^{\rm TE}_{\omega} \equiv \partial_\omega g^{\rm TE}(\omega, \mathbf{k}; \theta)$ and $g^{\rm TM}_{\theta} \equiv \partial_\theta g^{\rm TM}(\omega, \mathbf{k}; \theta)$. Using the analytical properties of the function $g^{\rm TE}$ it is possible to write an expansion of the first two terms in

(22) in powers of temperature, resulting in $(g_\xi^{\rm TE}(0,\mathbf{k})/6)\tau + (5g_{\xi^2}^{\rm TE}(0,\mathbf{k})/12)\tau^2 + \ldots$, where $g_\xi^{\rm TE}(0,\mathbf{k}) = \lim_{\xi\to 0}\partial g^{\rm TE}(i\xi,\mathbf{k};0)/\partial\xi$ and $g_{\xi^2}^{\rm TE}(0,\mathbf{k}) = \lim_{\xi\to 0}\partial^2 g(i\xi,\mathbf{k};0)/\partial\xi^2$ [13]. Thus, the first two terms in (22) give a vanishing entropy at $T=0$, and imply a low-temperature behaviour of the entropy proportional to T^2, since $g_\xi^{\rm TE}(0,\mathbf{k}) = 0$ and $g_{\xi^2}^{\rm TE}(0,\mathbf{k}) = -e^{-2kd}\overline{\epsilon}_0^{-2}\omega_c^2/8k^4 < 0$ (see TE plots in Fig. 2). The last term in (22) is proportional to $\partial_\theta\tilde{\omega}_c$, which is exponentially small at low temperatures. Therefore, the full TE contribution to the entropy vanishes at zero temperature, namely $S^{\rm TE}(0) = 0$.

For TM modes, the reflection coefficient (12) depends implicitly on temperature both through $\tilde{\omega}_c$ and $\tilde{D}$ in a complicated fashion. Contrary to the TE case, the $\theta\to 0$ and $\omega\to 0$ limits of $r^{\rm TM}(\omega,\mathbf{k};\theta)$ do not commute, and therefore it is not possible to write the contribution of TM modes to the entropy in the simple form (22). The $n\geq 1$ and $n=0$ terms have to be treated separately. This can be done by defining a new function $\tilde{g}^{\rm TM}(in\theta,\mathbf{k})$ which is identical to $g^{\rm TM}(in\theta,\mathbf{k};\theta)$ for $n\geq 1$, and for $n=0$ it is defined as $\tilde{g}^{\rm TM}(0,\mathbf{k}) \equiv \lim_{\theta\to 0} g^{\rm TM}(\alpha\theta,\mathbf{k};\theta)$, where we approach zero along the path $\omega=\alpha\theta$ [13]. The TM contribution to the entropy is

$$S^{\rm TM}(T) = -\sum_{\mathbf{k}}\left\{\frac{g^{\rm TM}(0,\mathbf{k};\theta)-\tilde{g}^{\rm TM}(0,\mathbf{k})}{2}\right. \\ \left. + \sum_{n=0}^{\infty\,\prime}[\tilde{g}^{\rm TM}(in\theta,\mathbf{k};\theta) + in\theta\tilde{g}_\omega^{\rm TM}(in\theta,\mathbf{k};\theta) + \theta\tilde{g}_\theta^{\rm TM}(in\theta,\mathbf{k};\theta)]\right\}. \qquad (23)$$

As in the TE case, the first two terms in the sum $\sum_{n=0}^{\infty\,\prime}$ vanish in the zero temperature limit $\theta\to 0$. Their first non-vanishing contribution to the entropy is linear in T, since $g_\xi^{\rm TM}(0,\mathbf{k}) > 0$ (see plots TM in Fig. 2). On the other hand, in the low-temperature limit the third term in the sum over n in (23) is,

$$\sum_{n=0}^{\infty\,\prime}\theta\tilde{g}_\theta^{\rm TM}(in\theta,\mathbf{k};\theta) \stackrel{\theta\to 0}{\longrightarrow} \partial_\theta\omega_c(\theta)\int_0^\infty d\xi\frac{\partial\tilde{g}^{\rm TM}(i\xi,\mathbf{k};\theta)}{\partial\omega_c} + \partial_\theta D(\theta)\int_0^\infty d\xi\frac{\partial\tilde{g}^{\rm TM}(i\xi,\mathbf{k};\theta)}{\partial D}. \qquad (24)$$

The first term in (24) is zero at $T=0$ due to the exponential decay of ω_c at low temperatures. Although $\partial_\theta D$ does not vanish at $T=0$ (since $\tau(T)$ goes to a non-zero constant at $T=0$ and then $D(T)\propto T$ at low temperatures), the second term in (24) is also zero at $T=0$ because the integrand is exponentially small. Finally, in the limit $\theta\to 0$, the first line in (23) vanishes since $g^{\rm TM}(0,\mathbf{k};0) = \tilde{g}^{\rm TM}(0,\mathbf{k})$ by definition. Therefore, the TM contribution to the zero temperature entropy is $S^{\rm TM}(T=0)=0$. We conclude that our theory for Casimir-Lifshitz forces in systems with low density of carriers (intrinsic semiconductors, dielectrics, etc) with account of Debye screening and charge drift is in agreement with Nernst theorem.

5. Connection between this theory of drifting carriers and spatial dispersion

As we have already mentioned, the static limit of the reflection coefficients (12,17) obtained by us using the Boltmann transport approach coincide with those previously derived in [2], where it was noted that the same static reflection amplitudes can be interpreted in terms of spatial dispersion. Indeed, using as the static permittivity tensor $\boldsymbol{\epsilon} = {\rm diag}(\epsilon^\perp,\epsilon^\perp,\epsilon^\parallel)$ with the transverse permittivity $\epsilon^\perp = \overline{\epsilon}_0$, and the longitudinal permittivity depending on wavevector as $\epsilon^\parallel(k) = \overline{\epsilon}_0[1+1/(kR_{\rm D})^2]$, and computing the reflection coefficients for an anisotropic (uniaxial) material [14], it is straightforward to recover the static versions of (12,17).

Although the original Lifshitz paper [1] for dispersion forces between bodies separated by vacuum did not allow for spatial dispersion, it can certainly be generalized to include nonlocal dielectric response in those setups, for example by averaging the vacuum Maxwell stress tensor and calculating field strengths via the retarded Green tensor of the field, that should include spatial dispersion when it is important [2, 15]. As noted in [2], problems arise when the bodies are separated by a liquid instead of vacuum.

Instead of proceeding via the Green function method, here we use the scattering formalism generalized to spatial dispersion to compute the force between plates separated by vacuum. The Casimir pressure between the plates is given by the same Eq. (1), and the effects of spatial dispersion are incorporated by appropriately writing the reflection amplitudes $r^p_{\mathbf{k}}(\omega)$ in terms of the permittivity tensor $\boldsymbol{\epsilon}(\omega, k)$ [16, 17]. In [16] the permittivity tensor is computed in the random phase approximation (Lindhard dielectric function) including dissipation [18]. The reflection amplitudes are written as

$$r^p_{\mathbf{k}}(\omega) = \frac{H^p(k,\omega) - 1}{H^p(k,\omega) + 1}, \tag{25}$$

where the TM and TE H-functions are (our H, h functions correspond to the G, g functions in [16])

$$H^{\mathrm{TM}}(k,\omega) = \frac{k}{\gamma^0}\tilde{h}_a(k,\omega) - \frac{(\omega/c)^2}{(\gamma^0)^2}\tilde{h}_b(k,\omega) + \frac{k(k-\gamma^0)}{(\gamma^0)^2}\tilde{h}_c(k,\omega) + 1, \tag{26}$$

$$H^{\mathrm{TE}}(k,\omega) = \tilde{h}(k,\omega) + 1, \tag{27}$$

where

$$\begin{aligned} h_a(k,\omega) &= 2k\int_{-\infty}^{\infty}\frac{dq_z}{2\pi}\frac{1}{q^2\epsilon^{\parallel}(q,\omega)}, \\ h_b(k,\omega) &= 2\gamma^0(k,\omega)\int_{-\infty}^{\infty}\frac{dq_z}{2\pi}\frac{1}{q^2 - \epsilon^{\perp}(q,\omega)(\omega/c)^2}, \\ h_c(k,\omega) &= \frac{2(\omega/c)^2 k\gamma^0(k,\omega)}{k-\gamma^0(k,\omega)}\int_{-\infty}^{\infty}\frac{dq_z}{2\pi}\frac{1}{q^2[q^2 - \epsilon^{\perp}(q,\omega)(\omega/c)^2]}. \end{aligned} \tag{28}$$

Here $\gamma^0(k,\omega) = \sqrt{k^2 - (\omega/c)^2}$, $\mathbf{q} = (\mathbf{k}, q_z)$, and the tilde above an h-function means the h-function minus the same function except that the dielectric function is set to unity [16]. In the limit of neglegible spatial dispersion ($\epsilon^{\parallel}$ and $\epsilon^{\perp}$ independent of q) one gets the usual Fresnel expressions for the reflection coefficients.

We now show that our reflection coefficients can be linked to spatial dispersion even beyond the static limit, connecting in this way our approach based on Boltzmann transport equation with spatial nonlocality. We closely follow the approach of [16]. As in the static case, we assume that the nonlocal permittivity depends solely on k, being independent of q_z. This allows us to straightfowardly compute the integrals in (28). From the H^{TE} function we obtain

$$\frac{H^{\mathrm{TE}}(k,i\xi) - 1}{H^{\mathrm{TE}}(k,i\xi) + 1} = \frac{\sqrt{k^2 + \xi^2/c^2} - \sqrt{k^2 + \epsilon^{\perp}(k,i\xi)\xi^2/c^2}}{\sqrt{k^2 + \xi^2/c^2} + \sqrt{k^2 + \epsilon^{\perp}(k,i\xi)\xi^2/c^2}}. \tag{29}$$

We recover our TE reflection coefficient (17) for a transverse dielectric function independent of k,

$$\epsilon^{\perp}(k,i\xi) = \overline{\epsilon}(i\xi)\left[1 + \frac{\omega_c}{\xi(1+\xi\tau)}\right]. \tag{30}$$

For H^{TM} we obtain

$$\begin{aligned} H^{\mathrm{TM}}(k,i\xi) &= 1 + \frac{k}{\gamma^0}\left[\frac{1}{\epsilon^{\parallel}(k,\omega)} - 1\right] + \frac{\xi^2/c^2}{\gamma^0}\left[\frac{1}{\eta_T} - \frac{1}{\gamma^0}\right] \\ &- \frac{k\xi^2/c^2}{\gamma^0}\left[\frac{1}{k\eta_T - \eta_T^2} - \frac{1}{k\gamma^0 - (\gamma^0)^2}\right], \end{aligned} \tag{31}$$

where, after rotation $\omega \to i\xi$, $\gamma^0 = \sqrt{k^2 + \xi^2/c^2}$ and $\eta_T = \sqrt{k^2 + \epsilon^{\perp}(k, i\xi)\xi^2/c^2}$ from (17) and (30). Equating $(H^{\rm TM}-1)/(H^{\rm TM}+1)$ to the expression (12) for the TM reflection coefficient, one obtains $H^{\rm TM}(k, i\xi) = \bar{\epsilon}(i\xi)\gamma^0/\chi$. Using this expression in (31) one can derive the longitudinal permittivity

$$\epsilon^{\|}(k, i\xi) = \frac{k}{\gamma^0} \tag{32}$$

$$\times \left[\frac{\bar{\epsilon}(i\xi)\gamma^0}{\chi} - 1 + \frac{k}{\gamma^0} + \frac{\xi^2/c^2}{\gamma^0}\left(\frac{1}{\eta_T} - \frac{1}{\gamma^0}\right) + \frac{k\xi^2/c^2}{\gamma^0}\left(\frac{1}{k\eta_T - \eta_T^2} - \frac{1}{k\gamma^0 - (\gamma^0)^2}\right)\right]^{-1}.$$

As follows from the above considerations, our theory for Casimir forces with media with low density of charge carriers can be directly connected to spatial dispersion, as mentioned by us in [3].

6. Conclusions

In this paper we have expanded on our previous work [3] to compute Casimir-Lifshitz forces between bodies with low density of charge carriers (intrinsic semiconductors, dielectrics, disordered systems, etc) taking into account Debye-Hückel screening and charge drift. Our approach is based on the classical Boltzmann transport equation, and is applicable to non-degenerate systems with an energy gap. We have shown how the finite conductivity of such systems modifies the Casimir-Lifshitz force between such materials, and made numerical predictions for the force using germanium and silicon plates. Our theory can be seen as a special case of spatial dispersion, and provides a simple way to take into account nonlocal effects in terms of readily available material properties. We have explicitly shown that our theory is compatible with Nernst theorem of thermodynamics. This is in agreement with previous work [16] that demonstrated that spatial dispersion resolves the issues with Nernst theorem. Work related to our approach, totally based on nonlocal dielectric responses, recently appeared [19], extending our analysis to degenerate systems.

Acknowledgments

We acknowledge correspondence with C. Henkel, F. Intravaia, L.P. Pitaveskii, and discussions with S.A. Ellingsen, G.L. Klimchitskaya, V.M. Mostepanenko, and F.S.S. Rosa. D.A.R.D. is grateful to Victor Dodonov and the other organizers of the workshop on "60 Years of the Casimir Effect". He also acknowledges the support of the U.S. Department of Energy through the LANL/LDRD program for this work.

References

[1] Lifshitz E M 1956 *Zh. Eksp. Teor. Fiz* **29**, 94 [*Sov. Phys. JETP* **2** 73]
[2] Pitaevskii L P 2008 *Phys. Rev. Lett.* in press; arXiv:0801.0656
[3] Dalvit D A R and Lamoreaux S K 2008 *Phys. Rev. Lett.* in press; arXiv:0805.1676
[4] Landau L D and Lifshitz E M 1980 *Statistical Physics, Part 1* (Pergamon Press, Oxford)
[5] Sumi M 1967 *Japanese Journal of Applied Physics* **6** 688
[6] Thiennot J 1972 *Le Journal de Physique* **33** 219
[7] Frey B J, Levinton D B, and Madison T J 2006 *Proceedings of SPIE* **6273 II**; arXiv:physics/0606168
[8] http://www.ioffe.ru/SVA/NSM
[9] http://www.iue.tuwien.ac.at/phd/palankovski/node51.html
[10] See, for example, Bezerra V B, Klimchitskaya G L, Mostepanenko V M and Romero C 2004 *Phys. Rev. A* **69**, 022119; Høye J S, Brevik I, Ellingsen S A and Aarseth J B 2007 *Phys. Rev. E* **75** 051127, and references therein.
[11] See, for example, Rumer Yu B and Ryvkin M Sh 1980 *Thermodynamics, Statistical Physics, and Kinetics* (Mir Publishers, Moscow)
[12] A more precise definition of the Nernst theorem is that in the thermodynamic limit ($N \to \infty$) the entropy per particle $S/N = k_{\rm B}\ln(\Omega_N)/N$ should verify $\lim_{N\to\infty}\ln(\Omega_N)/N \to 0$ at $T = 0$, limiting the possible

number of ground states for a many-body system (see, for example, Kardar M 2007 *Statistical Physics of Particles* (Cambridge University Press, Cambridge)). Note that as long as the ground-state degeneracy does not grow exponentially with the system size, then the entropy per particle does vanish at zero temperature in the thermodynamic limit.

[13] Intravaia F and Henkel C 2008 *J. Phys. A: Math. Theor.* **41** 164018

[14] See, for example, Rosa F S S, Dalvit D A R, and Milonni P W 2008 *Phys. Rev. A* **78** 032117

[15] Ginzburg V L and Barash Yu S 1975 *Sov. Phys. Ups.* **18** 931

[16] Sernelius B E 2005 *Phys. Rev. B* **71** 235114; 2006 *J. Phys. A:Math. Gen.* **39** 6741

[17] Esquivel-Sirvent R, Villarreal C, Mochán W L, Contreras-Reyes A M, and Svetovoy V B 2006 *J. Phys. A:Math. Gen.* **39** 6323

[18] Kliewer K L and Fuchs R 1969 *Phys. Rev.* **181** 552

[19] Svetovoy V B, arXiv:0809.3901v1.

Journal of Physics: Conference Series **161** (2009) 012010 doi:10.1088/1742-6596/161/1/012010

Low temperature Casimir-Lifshitz free energy and entropy: the case of poor conductors

Simen Ådnøy Ellingsen[1], Iver Brevik[1], Johan S. Høye[2], Kimball A Milton[3]

[1]Department of Energy and Process Engineering, Norwegian University of Science and Technology, N-7491 Trondheim, Norway

[2]Department of Physics, Norwegian University of Science and Technology, N-7491 Trondheim, Norway

[3]Oklahoma Center for High Energy Physics and Department of Physics and Astronomy, The University of Oklahoma, Norman, OK 73019, USA

E-mail: simen.a.ellingsen@ntnu.no

Abstract. The controversy concerning the temperature correction to the Casimir force has been ongoing for almost a decade with no view to a solution and has recently been extended to include semiconducting materials. We review some theoretical aspects of formal violations of Nernst's heat theorem in the context of Casimir Lifshitz thermodynamics and the role of the exponent of the leading term of the dielectric permittivity with respect to imaginary frequency. A general formalism for calculating the temperature corrections to free energy at low temperatures is developed for systems which do not exhibit such anomalies, and the low temperature behaviour of the free energy in a gap between half-spaces of poorly conducting materials modelled with a Drude type permittivity is calculated.

1. Introduction

The Casimir force [1], once merely a theoretical curiosity, is becoming the center of widespread attention in the wake of rapid developments in microtechnology. The enormous experimental progress made over the last decade towards accurately measuring this force [2–14] has created the need to calculate the Casimir force with high accuracy in realistic settings, taking into account such effects as material optical properties, surface roughness and geometry effects. Reviews of recent progress include [15–17].

It was realised quite early that an ambiguity existed as to the interpretation of Lifshitz' formula [18] for the Casimir attraction between dielectric half-spaces: when describing an ideal metal by taking the permittivity to infinity, different results were obtained at finite temperatures depending on the way the limit was taken. The ambiguity was originally sidestepped by prescription [19] and re-examined only much later by Boström and Sernelius [20] who concluded that due to the finite relaxation time of conduction electrons in a metal the transverse electric (TE) reflection coefficient of a metal-vacuum interface must vanish in the zero frequency limit, contrary to Casimir's ideal metal approximation in which reflection coefficients are set to unity at all energies. This was further supported by another study by Høye et al. [21]. For finite temperatures this vanishing of the TE zero frequency reflection coefficient leads to a prediction of a relatively large reduction of the Casimir force between metal plates at finite temperature, up to 15% at 300K. The Boström-Sernelius analysis was opposed on thermodynamical grounds for violating Nernst's theorem (the third law of thermodynamics) which states that a nondegenerate system must have zero entropy at zero temperature [22]. Moreover it was concluded that the series of high accuracy experiments at Purdue exclude the thermal correction predicted by the theory in which the TE zero frequency mode does not contribute [11, 13]. The debate is summarised in recent reviews [23, 24].

Recently an analogous ambiguity in the Lifshitz formalism was brought up for the case of semiconductors [25–28]. A formal violation of Nernst's theorem is once again the difficulty, this time due to discontinuous behaviour in the transverse magnetic (TM) reflection coefficient, whose value in the limit of zero frequency depends intimitely on the way the small density of conducting electrons in semiconducting materials at finite temperatures are taken into account.

However, we do not expect such formal violations of Nernst's theorem stemming from the mathematical subtleties of the Lifshitz formula to have implications for the physics of the problem. In [21] it was concluded that on physical grounds, no TE zero mode should be present for real metals, and recently a quantum statistical mechanical treatment came to the same result [29]. For semiconductors, earlier statistical mechanical analyses by Jancovici and Šamaj [30, 31] and by Buenzli and Martin [32] for ionic systems are of interest. What is found is that the effective separation between the plates increases as twice the ionic shielding length, which implies a non-local behaviour of the dielectric function. This increase in effective separation also means that the ionic contribution to reflectivity vanishes with vanishing ionic concentration. The results of refs. [30–32] are restricted to ionic systems, but in the high temperature (classical) limit they recover the ideal metal result corresponding to no TE zero mode.

Several others have also developed non-local approaches to the problem of Lifshitz theory for poor conductors, notably Pitaevskii [33], Lamoreaux and Dalvit [34,35], Esquivel [36] and most recently Svetovoy [37]. Also some these have not gone without objections [38–42]. We restrict ourselves to local permittivity models in the present effort.

2. Formal violations of Nernst's heat theorem; general theory

The Lifshitz formula expresses the Casimir free energy between parallel surfaces described by polarisation specific reflection coefficients r_q where $q \in \{p, s\}$ is the polarisation (assuming specular reflection and no coupling between p and s modes). For simplicity we shall assume the surfaces to be identical in the following, in which case the Casimir free energy at temperature T reads

$$\mathcal{F}(a) = \frac{T}{2\pi} \sum_{m=0}^{\infty}{}' \int_{\zeta_m}^{\infty} \mathrm{d}\kappa\kappa \sum_{q=p,s} \ln(1 - r_q^2 \mathrm{e}^{-2\kappa a}) \tag{1}$$

wherein a is the plate separation, p, s denotes TM and TE polarisations respectively, and $\mathrm{i}\zeta_m$ are the (imaginary) Matsubara frequencies so that $\zeta_m = 2\pi m T$. As conventional, the prime on the summation mark signifies that the $m = 0$ term be taken with half weight. We will be using natural units $k_\mathrm{B} = \hbar = c = 1$ throughout. Henceforth we will frequently omit the subscript m on ζ_m and the various quantities depending on it. The integral in (1) is over all transverse momenta $\mathbf{k}_\perp$ of the field ($\perp$ denotes a direction parallel to the surfaces) and the substitution $\kappa^2 = \mathbf{k}_\perp^2 + \zeta^2$ has been made.

In the case where the interfaces are between vacuum and a half-space made of dielectric material, the Fresnel reflection coefficients read

$$r_s = \frac{\kappa - \tilde{\kappa}}{\kappa + \tilde{\kappa}}; \quad r_p = \frac{\varepsilon\kappa - \tilde{\kappa}}{\varepsilon\kappa + \tilde{\kappa}}; \quad \tilde{\kappa} \equiv \sqrt{\kappa^2 + \zeta^2(\varepsilon - 1)}. \tag{2}$$

Here $\varepsilon = \varepsilon(\mathrm{i}\zeta)$ denotes the dielectric permittivity relative to vacuum.

It is straightforward to verify that the values of r_q in the limit $\zeta \to 0$ depend on the leading exponent of $\varepsilon(\mathrm{i}\zeta)$ as this limit is approached. For materials with mobile charges, models of the permittivity will typically diverge in the zero frequency limit, whereas that of a pure dielectric isolator reaches a finite value, $\lim_{\zeta\to 0}\varepsilon(\mathrm{i}\zeta) = \bar{\epsilon}$. Assuming $\varepsilon(\mathrm{i}\zeta) \sim (\zeta/\tilde{\omega})^\lambda$ ($\tilde{\omega}$ is a constant) as $\zeta \to 0$, one readily obtains the limits arrayed in table 1 in which

$$\tilde{r}_s(\kappa) = -\frac{\kappa^2}{\tilde{\omega}^2}\left(\sqrt{1+\frac{\tilde{\omega}^2}{\kappa^2}}-1\right)^2 \le 0. \tag{3}$$

λ	r_s	r_p
0	0	$\frac{\bar{\epsilon}-1}{\bar{\epsilon}+1}$
-1	0	1
-2	$\tilde{r}_s(\kappa)$	1
< -2	-1	1

Table 1. Values of $r_q(\mathrm{i}\zeta \to 0, \kappa)$ for different exponents λ.

The model permittivities which cause formal violation of Nernst's theorem have the common trait that the exponent λ takes one value at all finite temperatures which changes abruptly at exactly $T = 0$. A general treatment demonstrates that such temperature dependence is necessary in order for a formal violation of the theorem to occur [43,44] as we will briefly explain.

An example is the application of a Drude model to describe the dielectric response of an infinitely large and perfectly pure metal lattice, for which $\varepsilon(\mathrm{i}\zeta)$ is modelled as

$$\varepsilon(\mathrm{i}\zeta) = 1 + \frac{\omega_\mathrm{p}^2}{\zeta[\zeta+\nu(T)]}. \tag{4}$$

For a real metal sample of finite size, $\nu(T)$ reaches a nonzero value at zero temperature due to electron scattering on boundaries, impurities and imperfections, and $\lambda = -1$ for all temperatures. In this case entropy vanishes in the zero temperature limit as it should [45,46]. In a perfect lattice of infinite size, however, electron relaxation is solely due to scattering on thermal phonons, so that $\nu \sim T^5$ as $T \to 0$. Thus λ changes from -1 to -2 at $T = 0$, making r_s jump discontinuously from zero to a finite value as seen in table 1. Clearly $\tilde{\omega} = \omega_\mathrm{p}$ in (3) in the case $\lambda = -2$.

Another example is the semiconductor whose conductivity vanishes as a function of T. If a Drude model is used to model the permittivity of such a material,

$$\varepsilon(\mathrm{i}\zeta) = 1 + \frac{\bar{\epsilon}-1}{1+\zeta^2/\omega_0^2} + \frac{4\pi\sigma(T)}{\zeta}, \tag{5}$$

a formal violation occurs when $\sigma(T)$ vanishes at exactly $T = 0$. In this case $\lambda = -1$ at all T until absolute zero, where it skips to $\lambda = 0$ and the magnitude of r_p jumps discontinuously.

Let

$$\lim_{\zeta\to 0} r_q = R_q(\kappa;\lambda) \tag{6}$$

as tabulated (note that apart from the $\lambda = -2$ s-mode, R_q is independent of κ). It can be shown that when reflection coefficients jump discontinuously at $\zeta = 0$, the free energy obtains a term linear in temperature equal to the difference of the $m = 0$ terms of (1) as obtained with the two zero-frequency reflection coefficients respectively. If the leading ζ exponent of ε changes from λ_1 to λ_2 at exactly $T=0$, therefore, it leads to a residual entropy $\mathcal{S}=-\partial\mathcal{F}/\partial T$ at zero temperature

$$\mathcal{S}_{\lambda_1\to\lambda_2}=\frac{1}{4\pi}\sum_{q=p,s}\int_0^\infty \mathrm{d}\kappa\kappa\ln\frac{1-R_q^2(\kappa;\lambda_2)\mathrm{e}^{-2\kappa a}}{1-R_q^2(\kappa;\lambda_1)\mathrm{e}^{-2\kappa a}}. \tag{7}$$

In particular, when $R_q(\kappa;\lambda_1)=R_q(\lambda_1)$ and $R_q(\kappa;\lambda_2)=R_q(\lambda_2)$ one may use the relation

$$\int_0^\infty \mathrm{d}\kappa\kappa\ln[1-R^2\mathrm{e}^{-2\kappa a}]=-\frac{1}{4a^2}\mathrm{Li}_3(R^2) \tag{8}$$

where $\mathrm{Li}_n(x)$ is the nth order polylogarithmic function

$$\mathrm{Li}_n(x)=\sum_{l=1}^\infty\frac{x^l}{l^n} \tag{9}$$

to write

$$\mathcal{S}_{\lambda_1\to\lambda_2}=\frac{1}{16\pi a^2}\sum_{q=p,s}\left\{\mathrm{Li}_3[R_q^2(\lambda_1)]-\mathrm{Li}_3[R_q^2(\lambda_2)]\right\}. \tag{10}$$

In the particular cases of Drude modelled metals and semiconductors discussed above it follows immediately from table 1 and Eqs. (7) and (10) that, respectively,

$$\mathcal{S}_{-1\to-2}=\frac{1}{4\pi}\int_0^\infty \mathrm{d}\kappa\kappa\ln[1-\tilde{r}_s^2(\kappa)\mathrm{e}^{-2\kappa a}];\qquad\text{(metals)} \tag{11a}$$

$$\mathcal{S}_{-1\to0}=\frac{1}{16\pi a^2}\left\{\zeta(3)-\mathrm{Li}_3[R_p^2(0)]\right\}\qquad\text{(semiconductors)} \tag{11b}$$

where $R_p(0)=(\bar{\epsilon}-1)/(\bar{\epsilon}+1)$ and where we have used $\mathrm{Li}_n(1)=\zeta(n)$, the Riemann zeta function. These two exponent transitions are those which come into play for metals and dielectrics respectively, or more precisely, upon plugging a Drude-type permittivity model with vanishing $\nu(T)$ or $\sigma(T)$ into the Lifshitz formula and extrapolating to zero temperature. Other exponent transitions, naturally, would give other zero point entropy expressions.

Note that by letting $\tilde{\omega}\to\infty$ in (3) so that $\tilde{r}_s(\kappa)\to-1$, the entropy (11a) becomes

$$\mathcal{S}_{-1\to-2}\stackrel{\tilde{\omega}\to\infty}{\longrightarrow}-\frac{\zeta(3)}{16\pi a^2}, \tag{12}$$

which is the well known result for the so-called modified ideal metal model obtained by this procedure [21].

3. Free energy temperature correction for poor Drude conductor

The model for the conductivity we will be studying in the following is assumed not to depend on T within a finite range of temperatures including $T=0$, in which case it is clear from the above that Nernst's theorem will be satisfied. While the consideration of such a model cannot resolve such anomalies as reported in the previous section, it is nonetheless useful to establish benchmark results in various models within the Lifshitz formalism which exhibits very nontrivial behaviour in the joint limit of zero temperature and frequency ζ.

We consider a semiconductor modelled by a Drude type permittivity such as (5), but where we assume σ to be constant within a finite range of low temperatures including $T=0$ (the case $\sigma=0$ was worked out in [25]). A more detailed treatment of this model may be found in [47].

The method used is to note that $\mathcal{F}(a,T)$ depends on T only through the prefactor T and the Matsubara frequencies $\zeta_m = 2\pi mT$, so we may write the free energy (1) on the form[1]

$$\mathcal{F} = T\sum_{q=p,s} f_q(a) \sum_{m=0}^{\infty}{}' g(\mu) \tag{13}$$

where the function $f_q(a)$ is a convenient prefactor and we use the shorthand notation $\mu \equiv mt$ where $t = T/T_0$ is a dimensionless rescaled temperature to be defined in Eq. (18) below.

In the limit $T = 0$ the sum becomes an integral, and we are interested in the difference between sum and integral, which may be found from the Euler-Maclaurin formula. It turns out $g(\mu)$ is not analytical at $m = 0$ so it is necessary to start the sum at $m = 1$ (or a higher value), writing

$$\begin{aligned}\tilde{\Gamma} \equiv \left[\sum_{m=0}^{\infty}{}' - \int_0^\infty \mathrm{d}m\right] g(mt) &= \frac{1}{2}g(0) - \int_0^1 g(mt)\mathrm{d}m + \frac{1}{2}g(t) - \sum_{k=1}^{\infty}\frac{B_{2k}}{(2k)!}g^{(2k-1)}(t) \\ &= \frac{1}{2}g(0) - \int_0^1 g(mt)\mathrm{d}m + \frac{1}{2}g(t) - \frac{1}{12}g'(t) + \frac{1}{720}g'''(t) - \dots. \end{aligned} \tag{14}$$

Since the correction terms in (14) are evaluated at $m = 1$ and we are considering small T, we may choose $T_0 \gg T$ so that $\mu \ll 1$. We anticipate that when expanded for small μ, the function $g(\mu)$ is of the form

$$g(\mu) \sim c_0 + c_1\mu + c_{\frac{3}{2}}\mu^{\frac{3}{2}} + c_{2l}\mu^2\ln(\mu) + c_2\mu^2 + c_3\mu^3 + ..., \quad \mu \to 0 \tag{15}$$

Upon insertion into (14) one finds that the c_0 and c_2 terms do not contribute. Terms of integer powers of μ are determined by a finite number of terms in the series (14), but for the terms containing logarithms or half-integer powers every term contributes. The series obtained is asymptotic but a meaningful value may nonetheless be assigned to all terms of (15) by defining the series by Borel summation[2] or zeta regularisation. The result for polarisation mode $q = p, s$ is [47]

$$\Delta\mathcal{F}_q = Tf_q(a)\tilde{\Gamma}_q = Tf_q(a)\left[-\frac{c_1}{12}t + \zeta(-\tfrac{3}{2})c_{\frac{3}{2}}t^{\frac{3}{2}} + \frac{\zeta(3)c_{2l}}{4\pi^2}t^2 + \frac{c_3}{120}t^3 + ...\right]_q \tag{16}$$

giving terms proportional to respectively $T^2, T^{\frac{5}{2}}, T^3$ and T^4. It is understood that the coefficients c_1 through c_3 are polarisation mode specific. We shall content ourselves with expanding free energy to order T^3 in the present paper.

Equation (16) effectively reduces the problem to one of determining the expansion coefficients of Eq. (15). This task is still not trivial, however, and we consider only the "intermediate asymptotic" region in which conductivity is very small compared to inverse separation but much greater than temperature

$$T \ll 4\pi\sigma \ll \frac{1}{a}. \tag{17}$$

It is convenient now to define $T_0 = 2\sigma$, that is

$$t = \frac{T}{T_0} \equiv \frac{\zeta_1}{4\pi\sigma} = \frac{2\pi T}{4\pi\sigma}. \tag{18}$$

[1] Note that this convention differs sightly from that of [47].

[2] See appendix of [47]

Assumption (17) ensures that $t \ll 1$. The frequency which enters into (14) is $\zeta_1 \sim T \ll \sigma$, so $\varepsilon(\mathrm{i}\zeta)$ simplifies to

$$\varepsilon = \bar{\epsilon} + \frac{1}{\mu} \tag{19}$$

We will consider the p and s modes individually in the following.

4. The TM mode

The TM mode expression for Casimir Lifshitz free energy exhibits highly nontrivial behaviour near zero temperature. To simplify matters we note from (17) that the quantity

$$\alpha \equiv 2a(4\pi\sigma) \tag{20}$$

obeys $\alpha \ll 1$. We will determine free energy corrections perturbatively in powers of α in order to obtain analytical results for the coefficients in (15).

Substituting the integration variable

$$x = 2\kappa a = \frac{\kappa\alpha\mu}{\zeta} \tag{21}$$

the TM free energy may be written

$$\mathcal{F}_p = \frac{(4\pi\sigma)^3 t}{4\pi^2\alpha^2}\sum_{m=0}^{\infty}{}' \left\{\int_{\alpha\mu}^{\infty} \mathrm{d}x x \ln(1 - r_p^2 \mathrm{e}^{-x})\right\} \equiv \frac{(4\pi\sigma)^3 t}{4\pi^2\alpha^2}\sum_{m=0}^{\infty}{}' g_p(m) \tag{22}$$

where $g_p(m)$ is now chosen to be the expression within the curly braces. To leading order in α one finds

$$\ln(1 - r_p^2\mathrm{e}^{-x}) = \ln(1 - A_\mu \mathrm{e}^{-x}) + \mathcal{O}(\alpha^2); \tag{23a}$$

$$A_\mu \equiv \left(\frac{1 + (\bar{\epsilon} - 1)\mu}{1 + (\bar{\epsilon} + 1)\mu}\right)^2. \tag{23b}$$

With the reflection squared coefficient now a constant with respect to x the integral in (22) can be solved explicitly and expansion in μ as in (15) yields the coefficients c_1 and c_{2l} to leading order in α as simply [47]

$$c_1 = \frac{2\pi^2}{3}; \quad c_{2l} = 8 \tag{24}$$

and with (16) the low temperature correction to free energy is to leading order in α in terms of T and σ reads:

$$\Delta\mathcal{F}_p = -\frac{\pi^2 T^2}{72(4\pi\sigma)a^2} + \frac{\zeta(3)T^3}{\pi(4\pi\sigma)^2 a^2}. \tag{25}$$

An examination shows that the correction to this result is approximately a factor α^2 smaller as anticipated.

5. The TE mode

A similar procedure is performed for the s (or TE) mode. With the substitution

$$x = \frac{\kappa}{\zeta\sqrt{\varepsilon(i\zeta) - 1}} = \frac{\kappa\mu}{\chi\zeta}; \quad \chi \equiv \sqrt{\mu + (\bar{\epsilon} - 1)\mu^2}, \tag{26}$$

the TE term of the free energy (1) may be written

$$\mathcal{F}_s = \frac{(4\pi\sigma)^3 t}{4\pi^2}\sum_{m=0}^{\infty}{}' \left\{\chi^2\int_{\mu/\chi}^{\infty} \mathrm{d}x\, x \ln(1 - r_s^2 \mathrm{e}^{-\alpha\chi x})\right\} = \frac{(4\pi\sigma)^3 t}{4\pi^2}\sum_{m=0}^{\infty}{}' g_s(m). \tag{27}$$

Again $g_s(m)$ is defined as the expression between the curly braces and in the new variables the squared reflection coefficient reads

$$r_s^2(x) = (\sqrt{x^2+1} - x)^4. \tag{28}$$

Expanding the logarithm of (27) to linear order in α,

$$\ln(1 - r_s^2\mathrm{e}^{-\alpha\chi x}) = \ln(1 - r_s^2) + \frac{\alpha\chi x r_s^2}{1 - r_s^2} + \mathcal{O}(\alpha^2) \tag{29}$$

allows us to determine the temperature corrections for these orders of α. Both terms on the right hand side of (29) give integrals over x which are explicitly solvable, and subsequent expansions in powers of μ give the coefficients

$$c_1 = -\frac{1}{4}(2\ln 2 - 1); \quad c_{2l} = -\frac{1}{4}; \quad c_{\frac{3}{2}} = \frac{\alpha}{12}. \tag{30}$$

Clearly $c_{\frac{3}{2}} \ll c_1, c_{2l}$, being of linear order in α. The temperature corrections for the TE mode thus read

$$\Delta\mathcal{F}_s = \frac{(4\pi\sigma)T^2}{48}(2\ln 2 - 1) + \frac{1}{6}\sqrt{2\pi}\zeta(-\tfrac{3}{2})a(4\pi\sigma)^{\frac{3}{2}}T^{\frac{5}{2}} - \frac{\zeta(3)T^3}{8\pi} + \mathcal{O}(T^{\frac{7}{2}}). \tag{31}$$

This result is in fact in perfect agreement with that obtained for Drude metals [45] but includes one more order. Thus the concordance implies that a further expansion in α would only yield corrections of higher orders in temperatures, and that for the case of the TE mode the expansion in α was not essential for obtaining this result. Eq. (31) is therefore valid also when α is not small, as is the case for a good conductor.

Notably the temperature corrections are independent of $\bar{\epsilon}$ to order T^3 for both polarisations and only enters in higher orders, an insight not included in [45].

6. Numerical verification of results

The final results (25) and (31) have been checked numerically to verify their correctness [47]. A plot of the theoretical correction (25) compared to a direct numerical calculation of the free energy is provided in figure 1.

A much more sensitive test is provided by the quantity

$$R_p = \frac{\Delta\mathcal{F}_p^{\mathrm{th}} - \Delta\mathcal{F}_p^{\mathrm{num}}}{\Delta\mathcal{F}_p^{\mathrm{th}}} \tag{32}$$

where $\Delta\mathcal{F}_p^{\mathrm{th}}$ is the theoretically predicted free energy correction (25) and $\Delta\mathcal{F}_p^{\mathrm{num}}$ is that found by direct numerical calculation. The data used for calculation are (in SI units) $a = 1000\mathrm{nm}$, $\sigma^{\mathrm{SI}}/\epsilon_0 = 10^{12}\mathrm{s}^{-1}$, $\bar{\epsilon} = 11.66$ and $\omega_0 = 8\cdot 10^{15}\mathrm{s}^{-1}$.

We have found that $\Delta\mathcal{F}_p^{\mathrm{th}}$ is of the form $\Delta\mathcal{F}_p^{\mathrm{th}} = -CT^2(1 - C_1T)$ and assume $\Delta\mathcal{F}_p^{\mathrm{th}}$ to have the expansion

$$\Delta\mathcal{F}_p^{\mathrm{num}} = -DT^2(1 - D_1T + D_2T^2 + \dots), \tag{33}$$

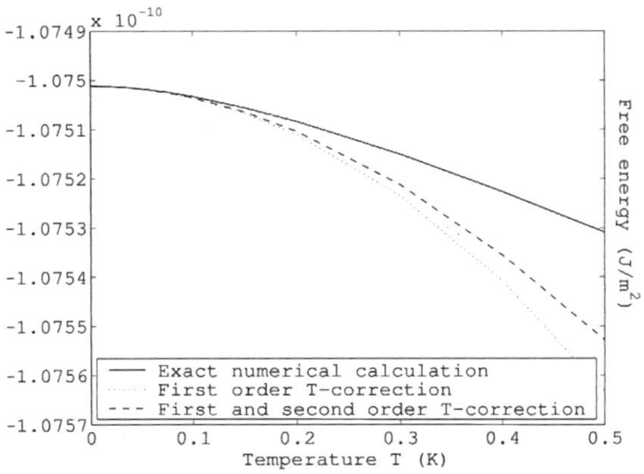

Figure 1. Free energy $\mathcal{F}_p$ calculated using direct numerical calculation and the theoretical correction (25) shifted to coincide with the numerical result at $T = 0$.

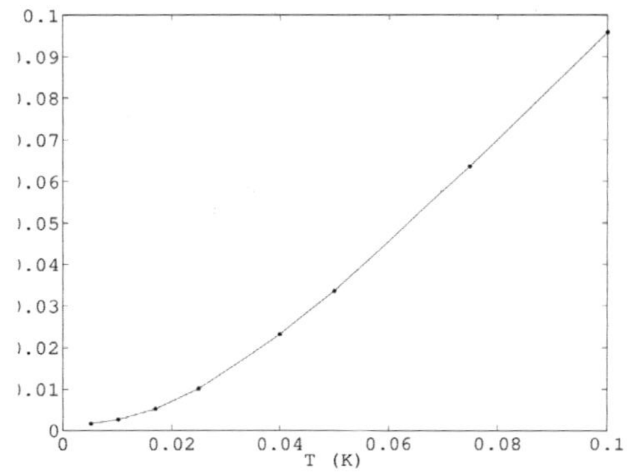

Figure 2. Plot of the quantity R from Eq. (32).

which predicts the following expansion for R:

$$R = \frac{C - D}{C} - \frac{D}{C}(C_1 - D_1)T - \frac{D}{C}[D_2 + C_1(C_1 - D_1)]T^2 + \ldots. \tag{34}$$

In the special case where $C = D$ and $C_1 = D_1$, this becomes

$$R = -D_2T^2 + \mathcal{O}(T^3). \tag{35}$$

Thus if the coefficient C is incorrect R would not converge to 0 at $T = 0$, and an incorrect C_1 would show as a linear behaviour at small temperatures. None of these effects are perceptible in the figure, demonstrating that the corrections to (25) are small as predicted.

A similar high precision check of the result for the s mode (31) was not possible with these numerical parameter values because the correction relative to the free energy at zero temperature is extremely small, of order 10^{-9} at 1K. The correctness of the terms proportional to T^2 and $T^{\frac{5}{2}}$ was however thoroughly verified for a good conductor in [45]. The term $\propto T^3$ is numerically elusive because while requiring very high accuracy for verification when $\alpha \ll 1$, it is completely dominated by other terms for good conductors. A comparison of a direct numerical calculation with the prediction (31) to order T^2 in figure 3 reveals that the difference between these graphs, plotted in figure 4, is in the same order of magnitude as the T^3 term in (31), while the term proportional to $T^{\frac{5}{2}}$ is too small to be visible at this level.

7. Conclusions

We have reviewed the theory of formal violations of Nernst's heat theorem emphasising the way such a formal violation can only occur when the leading order behaviour of $\varepsilon(\mathrm{i}\zeta)$ with respect to ζ undergoes a discontinuous change at exactly $T = 0$. Such apparent problems with the Lifshitz formalism occur when the double limit where T and ζ are both taken to zero is not unique and depends intimately on the exact way in which a material's dielectric (and, in general, magnetic) response is modelled in this limit. As a general remark, Nernst's theorem concerns zero temperature only, and it is not *a priori* clear that one can simply extrapolate between these two very different temperatures and use a result at one temperature to draw conclusions at another. In particular, if a system behaves essentially different at $T = 0$ than

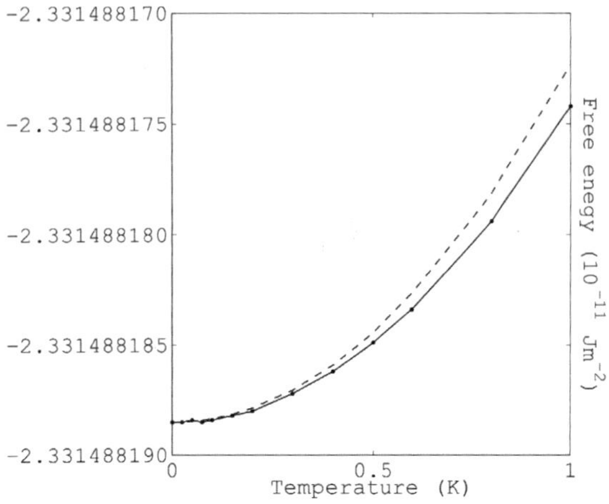

Figure 3. Free energy $\mathcal{F}_s$ calculated using direct numerical calculation and the theoretical correction (31) to order T^2 shifted to coincide with the numerical result at $T = 0$.

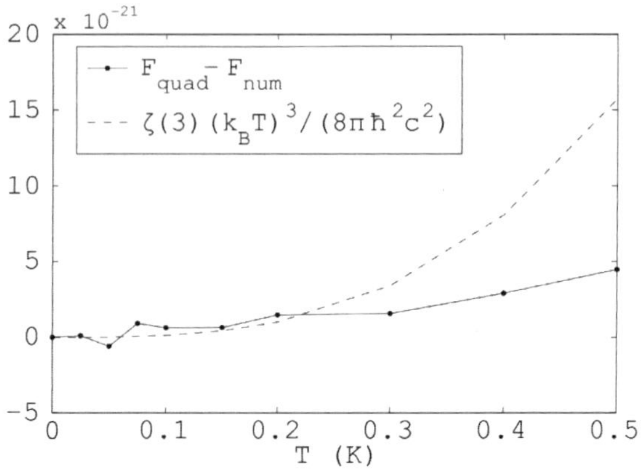

Figure 4. The difference between the graphs in figure 3 plotted against the T^3 term of (31).

at room temperature, a formal violation of Nernst's theorem by extrapolation is not necessarily worrisome. The results obtained complement those found in the case of zero conductivity in [25] and generalise TE mode calculations for Drude metals in [45].

Using a Drude type model to describe a poor conductor whose conductivity stays finite at zero temperature we establish the low temperature corrections to Casimir Lifshitz free energy between two identical half-spaces separated by vacuum. As modelled, both TE and TM modes exhibit a quadratic temperature behaviour at low temperatures.

References

[1] Casimir H B G 1948 *Proc. Kon. Ned. Akad. Wetensch.* **51** 793
[2] Lamoreaux S K 1997 *Phys. Rev. Lett.* **87** 5
[3] Mohideen U and A. Roy A 1998 *Phys. Rev. Lett.* **81** 4549
[4] Roy A, Lin C-Y, and Mohideen U 1999 *Phys. Rev. D* **60** 111101(R)
[5] Harris B W, Chen F, and Mohideen U 2000 *Phys. Rev. A* **62** 052109
[6] Ederth T 2000 *Phys. Rev. A* **62** 062104
[7] Chan H B et al. 2001 *Phys. Rev. Lett.* **87** 211801; 2001 *Science* **291** 1941
[8] Chen F, Mohideen U, Klimchitskaya G L, and Mostepanenko V M 2002 *Phys. Rev. Lett.* **88** 101801
[9] Bressi G, Carugno G, Onofrio R, and Ruoso G 2002 *Phys. Rev. Lett.* **88** 041804
[10] D. Iannuzzi, M. Lisanti and F. Capasso, Proc. Natl. Acad. Sci. USA **101** 4019 (2004)
[11] Decca R S et al. 2005 *Ann. Phys. (N.Y.)* **318** 37
[12] M. Lisanti, D. Iannuzzi and F. Capasso, Proc. Natl. Acad. Sci. USA **102** 11989 (2005)
[13] Bezerra V B et al. 2006 *Phys. Rev. E* **73** 028101
[14] Chen F, Klimchitskaya G L, Mostepanenko V M, and Mohideen U 2007 *Phys. Rev. B* **76** 035338
[15] Milton K A 2004 *J. Phys. A: Math. Gen.* **37** R209
[16] Buhmann S Y and Welsch D-G 2007 *Prog. Quantum Electron.* **31** 51
[17] Capasso F, Munday J N, Iannuzzi D and Chan H B 2007 *IEEE J. Selec. Top. Quant. Electron.* **13** 400
[18] Lifshitz E M 1955 *Zh. Eksp. Teor. Fiz.* **29** 94
[19] Schwinger J, DeRaad Jr. L L, and Milton K A 1978 *Ann. Phys. (N.Y.)* **115** 1
[20] Boström M and Sernelius B E 2000 *Phys. Rev. Lett.* **84** 4757
[21] Høye J S, Brevik I, Aarseth J B, and Milton K A 2003 *Phys. Rev. E* **67** 056116
[22] Klimchitskaya G L and Mostepanenko V M 2001 *Phys. Rev. A* **63** 062108
[23] Brevik I, Ellingsen S A and Milton K A 2006 *New J. Phys.* **8** 236
[24] Klimchitskaya G L and Mostepanenko V M 2006 *Comtemp. Phys.* **47** 131

[25] Geyer B, Klimchitskaya G L, and Mostepanenko V M 2005 *Phys. Rev. D* **72** 085009
[26] Klimchitskaya G L, Geyer B, and Mostepanenko V M 2006 *J. Phys. A: Math. Gen.* **39** 6495
[27] Geyer B, Klimchitskaya G L, and Mostepanenko V M 2006 *Int. J. Mod. Phys. A* **21** 5007
[28] Klimchitskaya G L and Geyer B 2008 *J. Phys. A: Math. Theor.* **41** 164014
[29] Buenzli P R and Martin Ph A 2008 *Phys. Rev. E* **77** 011114
[30] Jancovici B and Šamaj L 2004 *J. Stat. Mech.* P08006
[31] Jancovici B and Šamaj L 2005 *Europhys. Lett.* **72** 35
[32] Buenzli P R and Martin Ph A 2005 *Europhys. Lett.* **72** 42
[33] Pitaevskii L P 2008 *Phys. Rev. Lett* **101** 163202
[34] Lamoreaux S K 2008 *Preprint* quant-ph:0801.1283
[35] Dalvit D A R and Lamoreaux S K 2008 *Preprint* quant-ph:0805.1676
[36] Esquivel-Sirvent R 2008 *Phys. Rev. A* **77** 042107
[37] Svetovoy V B 2008 *Phys. Rev. Lett.* **101** 163603
[38] Klimchitskaya G L, Mohideen U and Mostepanenko V M 2008 *J. Phys. A* **41** 432001
[39] Decca R S et al. 2008 *Preprint* quant-ph:0803.4247
[40] Geyer B, Klimchitskaya G L, Mohideen U and Mostepanenko 2008 *Preprint* quant-ph:0810.3243
[41] Decca R S et al. 2008 *Preprint* quant-ph:0810.3244
[42] Klimchitskaya G L, Mohideen U and Mostepanenko 2008 *Preprint* cond-mat:0810.3247
[43] Ellingsen S A 2008 *Phys. Rev. E* **78** 021120
[44] Intravaia F and Henkel C 2008 *J. Phys. A: Math. Theor.* **41** 164018
[45] Høye J S, Brevik I, Ellingsen S A, and Aarseth J B, 2007 *Phys. Rev. E* **75** 051127
[46] Brevik I, Ellingsen S A, Høye J S, and Milton K A 2008 *J.Phys. A: Math. Theor.* **41** 164017
[47] Ellingsen S A, Brevik I, Høye J S, and Milton K A 2008 *Phys. Rev. E* **78** 021117

The Casimir frequency spectrum: can it be observed?

Simen Ådnøy Ellingsen

Department of Energy and Process Engineering, Norwegian University of Science and Technology, N-7491 Trondheim, Norway

E-mail: simen.a.ellingsen@ntnu.no

Abstract. The frequency spectrum of the Casimir force between real materials is studied with a view to assess possible ways of by which its dramatic oscillatory behaviour might be observed. Two simple attempts are made in which a model material is perturbed so as to become semitransparent in a band of frequencies. It is found that the real frequency formalism of the Casimir Lifshitz force is extremely volatile upon manipulations of optical properties and produces nonsensical results when extreme care is not taken to ensure the physicality of all perturbations, whereas the imaginary frequency formalism is highly robust and behaves well even under unphysical manipulations. The indication is thus that the general physical requirements of response functions preclude the observation of the frequency spectrum.

1. Introduction

The dependence of the Casimir force on the frequency dependent dielectric response of materials has long been an issue of intense research and at times a source of controversy. Casimir's original calculation of the fluctuation induced attraction between parallel plates assumed ideal conductors, perfectly reflecting at all field energies [1], and was generalised by Lifshitz a few years later [2] to materials of arbitrary frequency dependent dielectric response. The Lifshitz formula has since come to be recognised as a special case of a much more general formalism of multiple scattering [3, 4], yet while much progress has been made in research on the Casimir effect between more general bodies, the role of the inclusion of material dispersion properties in the simple two-plate geometry is still imperfectly understood. This has been most clearly demonstrated in the long ongoing disagreement over the thermal behaviour of the Casimir force and free energy (e.g. [5, 6] and references therein), a controversy centered in essence on the exact optical properties of materials at very low frequencies.

Upon incorporating the dispersive properties of materials, the expression for the total Casimir attraction between objects must necessarily include an integral over all frequencies of the electromagnetic field. Since all real materials obey the laws of causality it is possible [7] to shift the integral from the real to the positive imaginary frequency axis, which for the purposes of calculation is most often preferable. While the integrand along the imaginary frequency axis will typically be nicely peaked and exponentially decaying at high frequencies and therefore well suited for numerical evaluation, the real-frequency integrand is the direct opposite. Ford and others investigated the real-frequency spectrum of the Casimir force some time ago and showed how it is wildly oscillating, with the area under one oscillation peak of the frequency integrand much greater than the measurable Casimir force itself [8–10].

Formally, thus, the Casimir effect appears to emerge as the sum of an alternating series of very large and almost exactly canceling contributions. This inspired the suggestions that the optical properties of materials may be tuned so as to produce a Casimir force of desired magnitude and even sign [10, 11]. A handful of efforts have since elaborated this prospect [12–15], while an analysis by Genet and co-workers concluded that due to general restrictions of causality and analyticity such possibilities could not be realised [16].

Recently, the interpretation of an experiment by Iannuzzi and co-workers [17] was revisited in light of the considerations of the frequency spectrum of the Casimir force [18]. The experiment measured the Casimir attraction between so-called hydrogen switchable mirrors (HSMs). A HSM is a material which is a good metallic reflector in its as-deposited state but becomes transparent in the visible frequency range upon placement in a hydrogen rich atmosphere. While the reflection properties of the HSM are quite drastically different in the two states, the group was unable to detect a significant change in the Casimir attraction from one atmosphere to the other at a precision in the order of 15%. At face value this negative result seems at odds with the predictions of Ford, yet an alternative calculation of the effects of such a change of reflectivity indicates that the expected effects on the force might indeed be rather modest. A more careful analysis was made in [19].

In the following the recent analysis on the effect of perturbations of optical properties on the real frequency axis is reviewed and extended.

2. Frequency spectrum of the Casimir Lifshitz force

The frequency spectrum of Casimir interactions may be taken from the Lifshitz formula for real frequencies or worked out on more general grounds from the stress tensor operator of the quantised electromagnetic field. The Lifshitz formula in the form most useful for this purpose reads [2] (zero temperature is assumed henceforth)

$$P(a) = -\frac{1}{2\pi^2}\Re\mathrm{e}\int_0^\infty \mathrm{d}\omega\omega^3\int_C \mathrm{d}pp^2\sum_{\sigma=s,p}\frac{r_\sigma^2\exp(2\mathrm{i}p\omega a)}{1-r_\sigma^2\exp(2\mathrm{i}p\omega a)}. \tag{1}$$

Here $P(a)$ is the Casimir pressure between the parallel plates separated by a distance a where the minus sign signifies an attractive force. Natural units $\hbar = c = k_\mathrm{B} = 1$ are used throughout. ω is the real-valued frequency and the integral over the Lifshitz variable $p = \sqrt{(\mathbf{k}_\perp/\omega)^2 - 1}$ (physically the sum over all momentum components parallel to the surfaces) runs from 1 to 0 covering the propagating part of the spectrum, and thence to $\mathrm{i}\infty$ summing up all evanescent field contributions.

The frequency spectrum $P_\omega(a)$ is simply defined as

$$P(a) = \int_0^\infty \mathrm{d}\omega P_\omega(a) = \int_0^\infty \mathrm{d}\omega \sum_{\sigma=s,p} P_{\omega,\sigma}(a). \tag{2}$$

This spectrum in fact equals that derived by Ford with a more fundamental procedure [8, 10]. With the special modelling assumption that reflection coefficients are constant with respect to $\mathbf{k}_\perp$ the p-integral (1) may be solved explicitly by use of the polylogarithmic function

$$\mathrm{Li}_n(x) = \sum_{k=1}^\infty \frac{x^k}{k^n} \tag{3}$$

which obeys the useful recursion relation (A and K are constants)

$$\int \mathrm{d}y\mathrm{Li}_n(Ae^{\eta y}) = \frac{1}{\eta}\mathrm{Li}_{n+1}(Ae^{\eta y}) + K \tag{4}$$

to yield the following frequency spectrum

$$P_\omega(a;\{r_\sigma\}) = \frac{-1}{16\pi^2 a^3}\sum_{\sigma=s,p}\left[-\xi^2\Im\mathrm{mLi}_1(r_\sigma^2 e^{\mathrm{i}\xi}) - 2\xi\Re\mathrm{eLi}_2(r_\sigma^2 e^{\mathrm{i}\xi}) + 2\Im\mathrm{mLi}_3(r_\sigma^2 e^{\mathrm{i}\xi})\right] \tag{5}$$

where the shorthand dimensionless quantity $\xi = 2\omega a$ is introduced. The spectrum (5) is plotted in figure 1 assuming r_σ are also constant with respect to frequency.

When the latter assumption is made, the frequency integral is explicitly solvable and the Casimir zero temperature pressure is found as

$$P(a,\{r_\sigma\}) = -\frac{3}{16\pi^2 a^4}\sum_{\sigma=p,s}\Re\mathrm{eLi}_4(r_\sigma^2). \tag{6}$$

Casimir's result $P_\mathrm{C}(a) = -\pi^2/(240a^4)$ is regained by inserting $r_p^2 = r_s^2 = 1$ and $\mathrm{Li}_4(1) = \zeta(4) = \pi^4/90$. Exactly the same treatment starting from the Lifshitz formula for the zero temperature free energy rather than pressure yields

$$\mathscr{F}(a,\{r_\sigma\},T=0) = -\frac{1}{16\pi^2 a^3}\sum_{\sigma=p,s}\Re\mathrm{eLi}_4(r_\sigma^2) \tag{7}$$

which again simplifies trivially to Casimir's result in the limit $r_\sigma = 1$.

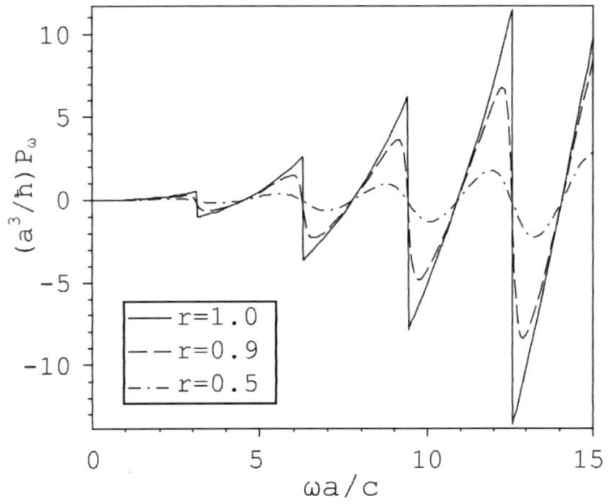

Figure 1. The frequency spectrum (5) normalised by a^{-3} for different (constant) values of $r_\sigma = r$, assumed equal for both polarisations.

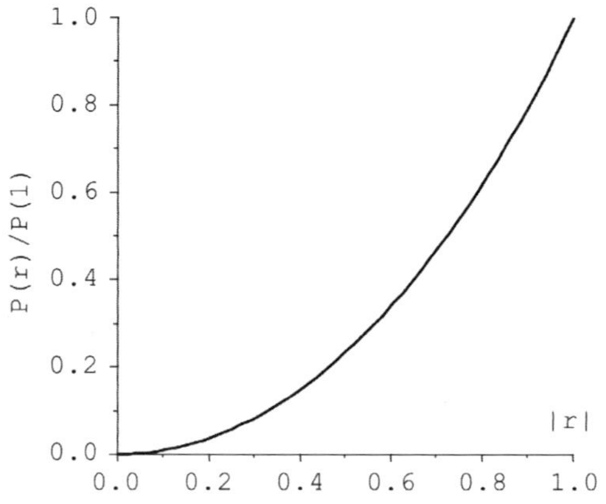

Figure 2. Casimir pressure (6) as function of constant reflection coefficient r, normalised by the ideal conductor limit $P_\mathrm{C}(a) = -\pi^2/(240a^4)$.

3. Frequency spectrum with real materials

When more realistic material properties are employed, the frequency spectrum of the Casimir force is no longer as regular as that shown in figure 1, but shares the property of wild oscillation of increasing amplitude throughout the frequency range in which optical data are typically available. Given an explicit expression of reflection coefficients as functions of frequency and transverse momentum, it is straightforward to calculate and plot the frequency spectrum, as shown in figure 3. In these calculations the permittivity function for gold is used, represented by, respectively, the simple plasma model $\epsilon(\omega) = 1 - \omega_p^2/\omega^2$ with $\omega_p = 9.0$eV and interpolation of experimental optical data tabulated in [20]. The graphs share the trait of large oscillations which make them unsuitable for most numerical purposes. It is notable, however, that the two spectra are radically different even though a calculation of the total Casimir force by the standard method of rotating the integral onto the imaginary frequency axis gives almost identical results whether the plasma model or the corresponding data set from [20] is used.

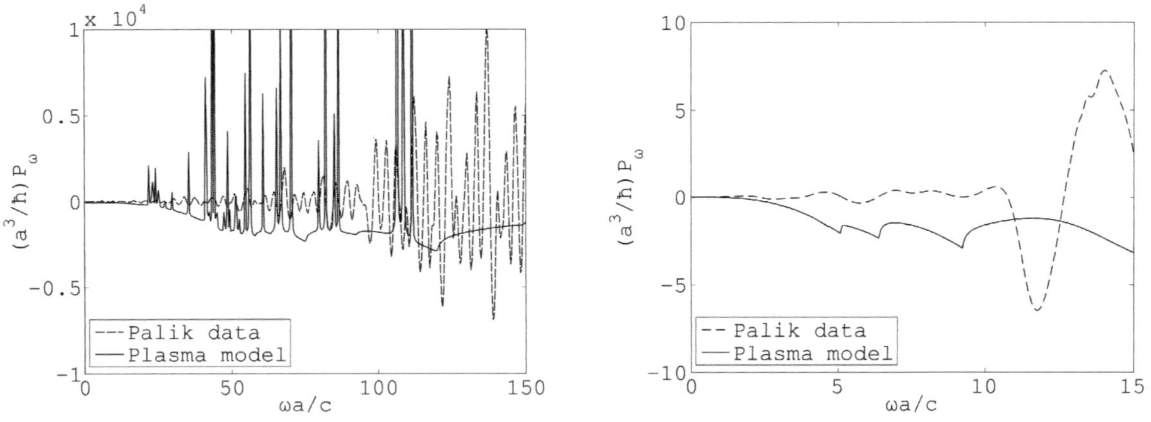

Figure 3. Numerically calculated frequency spectrum using repectively data from [20] and the plasma model, $\epsilon(\omega) = 1 - \omega_p^2/\omega^2$. Note that the spectrum is now a function of a and ω individually, not only their product; $a = 100$nm is used in the figure.

The striking nature of these spectra naturally raises the question whether or not such giant oscillations could be observable, and, if so, might even be used for experimental or technological purposes as proposed in previous publications. On purely qualitative grounds one may conclude from the extreme difference between the graphs in figure 3 that if observation of such a spectrum were indeed possible, quantitative predictions cannot be made using simple models such as the plasma or Drude model (the latter produces a spectrum reminiscent of that of the plasma model in figure 3 with a greater number of sharp peaks), and the direct use of extremely accurate optical data would be required.

4. Effects of a band perturbation of reflectivity

As figure 3 shows clearly, due to its highly oscillating nature the real frequency formalism is not very helpful for calculating the absolute Casimir force [21]. One may think, however, that it could nonetheless be fruitful for calculating a *change* of optical properties which was limited to a small interval of frequencies, in which case an integral might not need to span the entire frequency axis.

The experiment described above in which HSMs were used for the measurement of Casimir forces would seem a natural candidate for the use of a real-frequency formalism. Upon hydrogenation of the HSM, the material became transparent in the visible frequency range whereas the optical properties in the infrared and ultraviolet remained largely unchanged. The authors of [17, 19] propose a rough model for the description of the alteration of reflectivity in which the metallic permittivity of the mirror in the as-deposited state is "switched off" at $\omega_1 \approx 7.5 \cdot 10^{14}$rad/s and "on" again at $\omega_2 \approx 9.4 \cdot 10^{15}$rad/s. That is, $\epsilon(\omega) = 1 + 4\pi\chi(\omega) \longrightarrow \tilde{\epsilon}(\omega) = 1 + 4\pi\tilde{\chi}(\omega)$ where

$$\tilde{\chi}(\omega) = \chi(\omega)[1 - \theta(\omega - \omega_1)\theta(\omega_2 - \omega)]. \tag{8}$$

Here $\theta(x)$ is the unit step function. In fact the transition alters the properties at all frequencies to some extent [19], but as a first approximation one might expect that the "off/on" model would give at least an idea of the magnitude of the effect.

Using the formalism of imaginary frequencies, the effect of such an "off/on" model can be calculated by integration over the imaginary part of $\epsilon(\omega)$ in the relevant Kramers-Kronig relation. Using as an example the Drude model,

$$\epsilon(\omega) = 1 - \frac{\omega_{\mathrm{p}}^2}{\omega(\omega + \mathrm{i}\nu)}, \tag{9}$$

with ν being the electronic relaxation frequency (for gold $\nu = 35\mathrm{meV}$ is often used [22]), one may readily calculate the correction $\epsilon(\mathrm{i}\zeta) \to \epsilon(\mathrm{i}\zeta) - \Delta\epsilon(\mathrm{i}\zeta)$ explicitly as

$$\begin{aligned} \Delta\epsilon(\mathrm{i}\zeta) &= \frac{2}{\pi}\int_0^\infty \frac{\mathrm{d}\omega\omega[\epsilon''(\omega) - \tilde{\epsilon}''(\omega)]}{\omega^2+\zeta^2} = \frac{2}{\pi}\int_{\omega_1}^{\omega_2} \frac{\mathrm{d}\omega\omega\epsilon''(\omega)}{\omega^2+\zeta^2} \\ &= \frac{\omega_p^2}{\zeta^2-\nu^2}\frac{2}{\pi}\left[\arctan\frac{\nu\Delta\omega}{\nu^2+\omega_1\omega_2} - \frac{\nu}{\zeta}\arctan\frac{\zeta\Delta\omega}{\zeta^2+\omega_1\omega_2}\right] \end{aligned} \tag{10}$$

with $\Delta\omega = \omega_2 - \omega_1$. Here the real and imaginary parts of $\epsilon(\omega)$ are denoted with a single and double prime respectively. For moderate values of of $\Delta\omega$ this gives only a relatively small contribution to $\epsilon(\mathrm{i}\zeta)$, on the level of a few percent for the values of ω_1 and ω_2 mentioned, and the consequences for the pressure would be unobservable at the precision of the experiment [17], in line with the negative conclusion of that paper. The "off/on" model is appealing as a first approach due to its simplicity and intuitiveness and when applied this way provides a very simple qualitative explanation of the effects of the introduction of a transparency window.

Using the real-frequency spectrum directly, however, gives very different results. The change in the force is calculated by subtracting a band of frequencies from the spectrum plotted in figure 3. To avoid possible troubles stemming from cutting the spectrum off sharply, an envelope function is used,

$$\varphi(\omega) = 1 - \frac{\Delta}{\pi}\{\arctan[sa(\omega-\omega_1)] + \arctan[sa(\omega_2-\omega)]\}, \tag{11}$$

so that $\epsilon(\omega) \to 1 + 4\pi\chi(\omega)\varphi(\omega)$. The parameter $\Delta \in [0,1]$ is the relative reduction of $\epsilon(\omega)$ in the band and s determines the smoothness of the edges. The limit $s \to \infty$ giving the unit step behaviour of (8) [18]. A numerical integration over a sufficiently large frequency integral around $\Delta\omega$ with and without $\varphi(\omega)$ yields a rough prediction of the change in the force, which is plotted in figure 4. Clearly the change found this way is absurdly large and, counterintuitively, has opposite sign whether the Drude or plasma model is employed. At separation $a = 100\mathrm{nm}$ as used in this calculation, the absolute Casimir pressure between ideally conducting parallel plates is approximately 13Pa, shown as a grey band in the figure for scale. It is immediately clear that a simple "off/on" model such as this is not compatible with direct real frequency calculation.

5. Causal perturbation on narrow frequency band

For all its intuitive appeal, the "off/on" model, even in its smooth form (11) is not very realistic and suffers from problems which could give rise to the dramatic failure when applied on the real frequency axis. The sharp edges of the frequency band initially employed violate conditions of continuity and analyticity, but the indication is that this is not important since the introduction of smooth boundaries through the envelope function $\varphi(\omega)$ makes for only qualitative adjustments of the results.

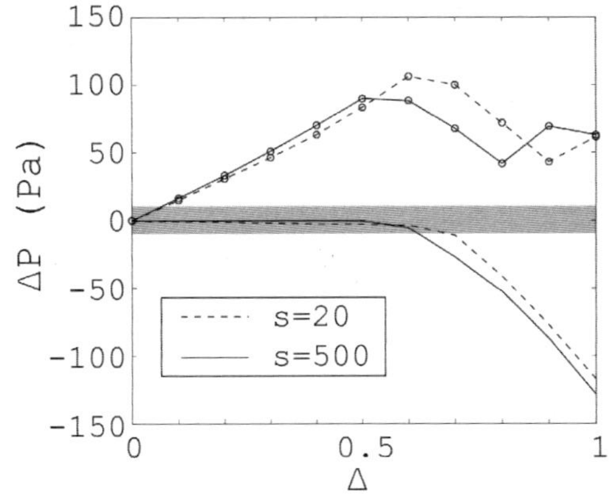

Figure 4. Numerically calculated change in the force based on an integral over a band of frequencies of the spectra using respectively the Drude model (circles) and plasma model (no circles). The grey band shows the range $\pm P_C(a)$, the absolute Casimir force between ideal conductors.

More serious is the fact that the way the transparency band is calculated on the real axis does not adhere to requirements of causality, in particular, that $\epsilon(\omega)$ fulfils the Kramers-Kronig relations. In the following an investigation is made of the feasibility of using real-frequency calculation for the assessment of the effect of a narrow-band perturbation of $\epsilon(\omega)$ in a way so that (12) and the so-called f-sum rule are obeyed.

It is a standard procedure in the study of the optical properties of materials to model the absorption spectrum $\epsilon''(\omega)$ of the material and work out $\epsilon'(\omega)$ based on this using

$$\epsilon'(\omega) = 1 + \frac{2}{\pi}\mathscr{P}\int_0^\infty d\xi \frac{\xi\epsilon''(\xi)}{\xi^2 - \omega^2}. \tag{12}$$

Likewise, $\epsilon(i\zeta)$ is calculated using the corresponding Kramers Kronig relation (10). Remarkably informative and useful information about different classes of materials can be extracted from very simple models [23]. This provides motivation to regard the consequences of perturbing $\epsilon''(\omega)$ in a causal way.

The perturbation in the dissipation function is modelled in the following way[1]:

$$\epsilon''(\omega) \to \tilde{\epsilon}''(\omega) = \epsilon''(\omega)[(1+\eta) - \Delta\cdot\theta(\omega-\omega_0)\theta(\omega_0+\delta\omega-\omega)], \quad \omega \geq 0 \tag{13}$$

that is, ϵ'' is approximately unchanged everywhere except in a small frequency band at ω_0 of width $\delta\omega$ in which dissipation is reduced by a constant quantity Δ. The small parameter η has the same sign as Δ and is introduced because the imaginary part of the permittivity must satisfy the f-sum rule (e.g. [24]):

$$\int_0^\infty d\omega\omega\epsilon''(\omega) = \text{const.} = \frac{\pi}{2}\sum_\alpha \frac{4\pi\aleph_\alpha q_\alpha^2}{m_\alpha^*} \equiv \frac{\pi}{2}\Omega_p^2 \tag{14}$$

where sum is over the types of particles, α, causing dissipation, characterised by number density $\aleph_\alpha$, charge q_α and effective mass m_α^*. When dissipation is due to conduction electrons only, $\Omega_p = \omega_p$ as in the Drude and plasma models. Insisting that (14) be satisfied both before and after the perturbation implies

$$\eta = \frac{2}{\pi}\frac{\Delta}{\Omega_p^2}\int_{\delta\omega} d\omega\omega\epsilon''(\omega) \approx \frac{2\Delta}{\pi}\frac{\omega_0}{\Omega_p^2}\epsilon''(\omega_0)\delta\omega \tag{15}$$

[1] Causality requires that the perturbation is an odd function of ω, but only positive frequencies are considered here.

where $\int_{\delta\omega} \mathrm{d}\omega$ denotes integration from ω_0 to $\omega_0 + \delta\omega$ and the latter form is true when $\delta\omega$ is small compared to all other frequency scales. Assuming $\delta\omega$ small implies that $\eta \ll 1$.

To keep the considerations a little more concrete, consider the special case of the Drude model (9), for which η becomes:

$$\eta_{\mathrm{D}} = \frac{2}{\pi}\Delta \cdot \arctan\left(\frac{\nu \cdot \delta\omega}{\nu^2 + \omega_0(\omega_0 + \delta\omega)}\right) \approx \frac{2\Delta}{\pi}\frac{\nu \cdot \delta\omega}{\nu^2 + \omega_0^2}. \tag{16}$$

The perturbed real part $\tilde{\epsilon}'(\omega)$ now follows from (12)

$$\tilde{\epsilon}'(\omega) = 1 + \frac{2}{\pi}\mathscr{P}\left\{(1+\eta)\int_0^\infty \mathrm{d}\xi - \Delta\int_{\delta\omega}\mathrm{d}\xi\right\}\frac{\xi\epsilon''(\xi)}{\xi^2 - \omega^2}$$

which gives, to linear order in $\delta\omega$ or η, the change in $\epsilon'(\omega)$

$$\delta\epsilon'(\omega) \equiv \tilde{\epsilon}'(\omega) - \epsilon'(\omega) \approx \eta\left[\epsilon'(\omega) - 1 - \frac{\Omega_{\mathrm{p}}^2}{\omega_0^2 - \omega^2}\right] \tag{17}$$

where (15) was used. The case of the Drude model gives

$$\delta\epsilon'_{\mathrm{D}}(\omega) = -\eta_{\mathrm{D}}\frac{\omega_{\mathrm{p}}^2(\omega_0^2 + \nu^2)}{(\omega^2 + \nu^2)(\omega_0^2 - \omega^2)}. \tag{18}$$

In the same fashion, inserting (13) into the corresponding Kramers-Kronig integral for imaginary argument, (10), yields to linear order in η

$$\delta\epsilon(\mathrm{i}\zeta) \equiv \tilde{\epsilon}(\mathrm{i}\zeta) - \epsilon(\mathrm{i}\zeta) \approx \eta\left[\epsilon(\mathrm{i}\zeta) - 1 - \frac{\Omega_{\mathrm{p}}^2}{\omega_0^2 + \zeta^2}\right]. \tag{19}$$

The important difference between (17) and (19) is that while $\delta\epsilon(\mathrm{i}\zeta) \ll \epsilon(\mathrm{i}\zeta)$ over the entire imaginary frequency axis, the perturbation $\delta\epsilon'(\omega)$ grows large close to its pole at ω_0. While the perturbation (13) is negligible when calculated for imaginary frequencies, thus, it may be worth taking a closer look at what happens in the real frequency setting, equation (1).

The permittivity $\epsilon(\omega)$ for a Casimir cavity of two dielectric half-spaces (assumed to be made of the same material for simplicity) enters into the Lifshitz formula through the Fresnel reflection coefficients

$$r_s = \frac{p - s}{p + s}; \quad r_p = \frac{\epsilon p - s}{\epsilon p + s} \tag{20}$$

where $s \equiv \sqrt{p^2 + \epsilon - 1}$. The reflection coefficients stay approximately unchanged by $\delta\epsilon'$ except when ω is in the neighbourhood of ω_0 where $\tilde{\epsilon}'(\omega)$ becomes dominated by the otherwise negligible correction $\delta\epsilon'$. Here, thus, $\tilde{s} \sim \sqrt{\delta\epsilon'(\omega)}$ and $\tilde{s}$ therefore rises towards $+\infty$, skips to $+\mathrm{i}\infty$ and decreases quickly thence to its unperturbed value once more. The effect for r_p and r_s is, roughly, that over a frequency interval of order $\delta\omega$ near ω_0 they rise to unity and return to their unperturbed value once more. Since $\delta\omega$ is assumed smaller than all other frequency scales involved, all other quantities can be seen as approximately constant over this interval from which it follows that the change in Casimir pressure from (1) behaves as

$$\delta P(a) \sim \delta\omega \cdot P_{\omega_0}(a; \{r_\sigma\})|_{r_s = r_p = 1} \tag{21}$$

where $P_{\omega_0}(a; \{r_\sigma\})$ is the discontinuous spectrum function (5) at $\omega = \omega_0$ taken with unity reflection coefficients just as worked out by Ford [10] and shown in figure 1.

The result (21) is again counterintuitive in that it depends sensitively on the frequency ω_0 and could be either positive or negative depending on the quantity $\omega_0 a$. Indeed the spectrum $P_{\omega_0}(a)|_{r_\sigma=1}$ is discontinous at $\omega_0 a = n\pi,\ n \in \mathbb{N}$, meaning δP would make a leap were $\omega_0 a$ to be shifted slightly past such a value, e.g. by changing the separation slightly close to $a = n\pi/\omega_0$.

In conclusion it seemes that the fact that the perturbation is made causal by use of (12) and made to fulfil the f-sum rule (14) does not in itself salvage the apparent paradox that a calculation using the real frequency integral yields apparently nonsensical results while a calculation along the imaginary axis appears robust. If the perturbation $\Delta\epsilon''(\omega)$ fulfilled all the requirements of analyticity and causality and were taken into account by an integral (12) spanning the entire positive frequency axis, the real-frequency and imaginary frequency formalisms must necessarily give the same results. The observation of the vast fluctuations of the Casimir frequency spectrum, however, requires that access be somehow gained to finite intervals of the spectrum, and although a somewhat coarse analysis, the above considerations strongly indicate a pessimistic conclusion with regards to the feasibility of such an enterprise.

6. Conclusions and outlook

The frequency spectrum of the Casimir force between real materials is studied with a view to assess whether it may be possible to observe the dramatic oscillatory behaviour of the real frequency Casimir energy spectrum. A generalisation of Ford's result for ideal conductors [10] to a model of constant subunity reflection coefficients reveals a smoothening of the spectrum, but the oscillatory behaviour remains unchanged.

Upon inserting more realistic optical data for real materials, the frequency spectra obtain an even more wild and erratic behaviour. Observation of the dependence on real frequencies requires that access be somehow gained to finite intervals of real frequencies as opposed to the absolute force itself, which dependes only on the integral over all frequencies.

Two simple attempts are made to investigate the question whether a perturbation of the optical properties of materials which is restricted to a finite band of real frequencies could reveal a way to observe the large oscillations described. These have not been concerned with whether and when such perturbations may be made in practice, but have focussed on a problem of theoretical nature which occurs upon attempting to calculate the effects of such perturbations on the Casimir force: the predictions are radically different whether it is performed using the real frequency or imaginary frequency formulation of the Lifshitz formula. This paradox was first reported in [18].

In the first and coarsest approach the materials involved are assumed to be made transparent in a band of frequencies but remain unchanged outside this band. At a qualitative level this mimics the situation created in a recent experiment by Iannuzzi and co-workers [17] in which a hydrogen-switchable mirror is used which becomes transparent in the optical region upon hydrogenation. Using an imaginary time formalism the change in the force predicted is quite modest, in accordance with the experiment, whereas an exclusion of a part of the *real* frequency axis leads to nonsensical predictions of enormous amplitude and apparently arbitrary sign.

In a second attempt the perturbation is made in the imaginary part of permittivity $\epsilon(\omega)$ over a very small frequency range and causality is ensured by the invocation of the Kramers Kronig relations to calculate ϵ for imaginary frequencies and the real part of ϵ for real frequencies. Again a real frequency calculation reveals highly counterintuitive results whereas the imaginary frequency calculation appears reasonable in sign and magnitude and in accordance with intuition.

It is clear that the real frequency formalism of the Casimir Lifshitz force is very volatile upon manipulations of optical properties whereas the imaginary frequency formalism is robust and behaves well even with permittivities violating criteria of analyticity and causality. While it may be shown that the two formalisms must yield the same result for measurable quantities when the entire frequency integral is evaluated, both attempts to access the dependency on a finite frequency band made herein have yielded very different results in the two formalisms, of which there are good reasons to believe the result in the imaginary frequency domain to be the physical one. While this study is far from exhaustive, it indicates that the peculiar real-frequency spectrum of the Casimir force is not observable.

Acknowledgements
The author thanks Professor Victor Dodonov and the rest of the organising team for hosting such an excellent conference in Brasilia. In the endeavours reported above I have benefited from discussions with professor Bo E. Sernelius.

References
[1] Casimir H B G 1948 *Proc. K. Ned. Akad. Wet.* **60** 793
[2] Lifshitz E M 1955 *Zh. Eksp. Teor. Fiz.* **29** 94 [1956 *Sov. Phys. JETP* **2** 73]
[3] Lambrecht A, Maia Neto P A and Reynaud S 2006 *New J. Phys.* **8** 243
[4] Klich I 2008 Contribution to this volume.
[5] Brevik I, Ellingsen S A, and Milton K A 2006 *New J. Phys.* **8** 236
[6] Geyer B and Klimchitskaya G L 2008 *J. Phys. A* **41** 164032
[7] Landau L D and Lifshitz E M 1980 *Statistical Physics* Pt. 1, 3rd ed. (Amsterdam: Elsevier Butterworth-Heinemann) §123
[8] Ford L H 1988 *Phys. Rev. D* **38** 528
[9] Hacyan S, Jáuregui R, Soto F, and Villareal C 1990 *J. Phys. A* **23** 2401
[10] Ford L H 1993 *Phys. Rev. A* **48** 2962
[11] Iacopini E 1993 *Phys. Rev. A* **48** 129
[12] Ford L H 1998 *Phys. Rev. A* **58** 4279
[13] Sopova V and Ford L H 2004 *Phys. Rev. A* **70** 062109
[14] Lang A S I D 2005 *J. Math. Phys.* **46** 102105
[15] Ford L H 2007 *Int. J. Theor. Phys.* **46** 2218
[16] Genet C, Lambrecht A, and Reynaud S 2003 *Phys. Rev. A* **67** 043811
[17] Iannuzzi D, Lisanti M and Capasso F 2004 *Proc. Natl. Acad. Sci. U.S.A.* **101** 4019
[18] Ellingsen S A 2008 *Europhys. Lett.* **82** 53001
[19] de Man S and Iannuzzi D 2006 *New J. Phys.* **8** 235
[20] Palik E D (ed) 1995 *Handbook of Optical Constants of Solids* (New York: Academic Press)
[21] Rodriguez A, Ibanescu M, Iannuzzi D, Joannopoulos J D, and Johnson S G 2007 *Phys. Rev. A* **76** 032106
[22] Lambrecht A and Reynaud S 2000 *Eur. Phys. J. D* **8** 309
[23] Sernelius B E 2001 *Surface Modes in Physics* (Berlin: Wiley-VCH) Ch. 2
[24] Peiponen K-E, Vartiainen E M, and Asakura T 1999 *Dispersion, Complex Analysis and Optical Spectroscopy* (Berlin: Springer) Ch. 6

Journal of Physics: Conference Series **161** (2009) 012012 doi:10.1088/1742-6596/161/1/012012

Frequency-dependent Drude damping in Casimir force calculations

R Esquivel-Sirvent

Instituto de Física, Universidad Nacional Aut'onoma de México, Apdo. Postal 20-364, México D.F. 01000, México

E-mail: raul@fisica.unam.mx

Abstract.
The Casimir force is calculated between Au thin films that are described by a Drude model with a frequency dependent damping function. The model parameters are obtained from available experimental data for Au thin films. Two cases are considered; annealed and nonannealed films that have a different damping function. Compared with the calculations using a Drude model with a constant damping parameter, we observe changes in the Casimir force of a few percent. This behavior is only observed in films of no more than 300 $\AA$ thick.

1. Introduction

The advent of precise and systematic Casimir force experiments since the late 90 [1, 2, 3, 4, 5, 6, 7, 8] has prompted an intense research on the role of the dielectric properties of the involved materials. Although, the Lifshitz theory [9] explicitly requires the dielectric function of the materials, an important issue is which one is the correct dielectric function that is consistent in describing the optical properties of the materials and the measurements of the Casimir force.

The first approach is to assume a plasma model for the dielectric function [10] or the more realistic Drude model, that has been extensively used when extrapolating to low frequencies tabulated data. Although it may be thought that the problem of using a dielectric function is straight forward, controversial results have been reported, in particular in relation to the use of the Drude model in finite temperature calculations of the Casimir force .

The use of of the Drude model in Lifshitz theory seems to violate Nernst heat theorem, while the plasma model presents no problem at all, but is not realistic in the representation of the dielectric properties of metals. The plethora of papers and comments shows that the issue is far from settled [11, 12, 13, 14, 15, 16, 17].

Even without considering finite-temperature effects, the choice of the dielectric function can change the calculations of the Casimir force. For example, for Au samples the Drude parameters extracted from tabulated data vary depending on the sample. The variations on the Drude parameters have important implications in the Casimir force calculations since difference of up to 5% are obtained[18]. A similar result was obtained by Svetovoy et al. where [19] different Au samples were prepared under similar conditions with thicknesses ranging from 120 nm to 400 nm. From measured ellipsometry data it was verified that the plasma frequency varies from 6.8 eV to 8.4 eV for this set of particular samples, changing the Casimir force a few percent. The conclusions of these works show that there is not a standard plasma frequency or damping parameter for Au, it is sample dependent and in situ measurements are needed. Experimentally, the effect of thin films on the Casimir force was shown experimentally by Iannuzzi [22, 21] who demonstrated that the Casimir attraction between a metallic plate and a metal coated sphere depended on the thickness of the coating.

The reduction of size can significantly change the physical parameters of a system. In the case of thin films, as the thickness of the film approaches the mean free path, the conductivity show a sharp decrease in its values. This was shown experimentally by Kastle [23] with Au films whose thickness varied from 2 nm to 70 nm. Indeed, a conductor-insulator transition is observed as a function of film thickness in Au [24]. As a function of film thickness, the Casimir force decreases with decreasing film thickness until a critical thickness is reached after which the Casimir force increases even with decreasing film thickness[25].

To further the discussion about the possible factors that influence Casimir force calculations, in this paper we introduce a frequency dependent damping $\gamma(\omega)$ in the Drude model. This model describes the dielectric properties of thin films and changes if the film has been annealed.

2. Frequency dependent damping

The classical Drude model the local dielectric function is given by

$$\epsilon(\omega) = 1 - \frac{\omega_p^2}{\omega(\omega + i\gamma)}, \tag{1}$$

where ω is the frequency, ω_p the plasma frequency and γ the damping parameter that is constant for a fixed temperature.

Measurements of the dielectric properties of Au thin films by M. L. Theye [26] from reflectance and transmittance data showed a deviation from the bulk Drude behavior of Au. The deviations from the Drude model were explained by introducing a frequency dependent relaxation time due to electron-phonon and electron-ion interactions of the form:

$$\gamma = \gamma_0 + A\omega^2. \tag{2}$$

However, a more precise correction to the Drude model was introduced by Nagel [27] to include the frequency dependence of the damping parameter. It was observed that sample preparation was relevant in the optical behavior of the material, since the measured data was different for annealed and nonannealed samples. An explanation of the frequency dependent relaxation time with the sample and the role that annealing plays in the optical properties was given by Nagel [27] using a classical two carrier model. The model assumes that in a thin film sample there are two regions labeled a and b. One where the electrons see a perfect lattice, inside crystallites and a second highly disordered region between the crystallites. The electrons respond differently in each of these regions and have a different damping rate, say γ_a and γ_b, and a different plasma frequency ω_{pa} and ω_{pb}. Ignoring local field corrections in the determination of the optical response, the two carrier model yields an effective damping parameter given by

$$\gamma_{eff} = \gamma_a \left[1 + \frac{\omega_{pb}}{\omega_{pa}}\left(\frac{\omega^2+\gamma_a^2}{\omega^2+\gamma_b^2}\right)\right]^{-1} + \gamma_b \left[1 + \frac{\omega_{pa}}{\omega_{pb}}\left(\frac{\omega^2+\gamma_b^2}{\omega^2+\gamma_a^2}\right)\right]^{-1}, \tag{3}$$

where $\omega_{pa,b} = 4\pi N_{a,b}e^2/m*_{a,b}$. This last expression is general and the behavior observed by Theyé Eq. (2) is obtained if $\omega\tau_a >> 1$ and $\omega\tau_b << 1$. Equation (3) assumes that the effective masses in both regions are the same, thus $N_b/N_a = \omega_{pb}/\omega_{pa}$. Thus the thin film can be modeled by a Drude dielectric function with an effective damping constant.

In this paper we will use the parameters considered by Nagel [27] that fit the experimental data of Theyé [26] for an annealed and a nonanneald Au film. The parameters are shown in Table 1.

Table 1. Parameters for the two films used in our calculations, after [27].

film	N_b/N_a	$\gamma_a \times 10^{14}s^{-1}$	$\gamma_b \times 10^{14}s^{-1}$
annealed	0.0077	0.93	25
nonannealed	0.058	1.18	25

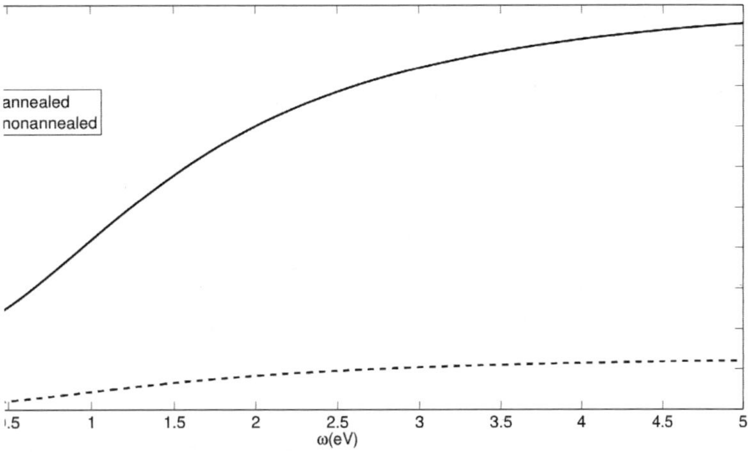

Figure 1. Effective damping as a function of frequency using Eq. (3) for an annealed sample (dashed line) and a nonannealed sample (solid line).

In Figure 1, we plot the effective damping γ_{eff} Eq. (3) as a function of frequency for the samples considered by Nagel. As expected, annealing the film will reduce the number of impurities and the damping constant should be smaller.

To study the effect that a frequency dependent damping has on the Casimir force, we consider two plates separated a distance L with a thin film deposited on their surface. The films are described by the parameters given in Table. 1 We calculate the reduction factor define as the Casimir force calculated using Lifshitz formula F divided by the Casimir force between perfect conductors F_0; this is $\eta = F/F_0$. This is,

$$\eta = \frac{120L^4}{\pi^4} \int_0^\infty QdQ \int_{q>0} dk \frac{k^2}{q} (G^s + G^p), \tag{4}$$

where $G_s = (r_{1s}^{-1} r_{2s}^{-1} \exp(2kL) - 1)^{-1}$ and $G_p = (r_{1p}^{-1} r_{2p}^{-1} \exp(2kL) - 1)^{-1}$. In these expressions, the factors $r_{p,s}$ are the reflectivities for p or s polarized light , Q is the wavevector component along the plates, $q = \omega/c$ and $k = \sqrt{q^2 + Q^2}$.

In Figure (2) we show the reduction factor for Au samples described by a classical Drude (η_c) model where the damping is constant, and the Drude model with a frequency dependent damping parameter for an annealed (η_a) and nonannealed samples (η_n). For the classical Drude model the parameters are $\omega_p = 9eV$ and $\gamma = 0.02eV$. The difference between the annealed and nonannealed sample is small. The reduction factor increases as compared to the annealed and nonannealed samples. The difference in reduction factor can better be seen by computing the

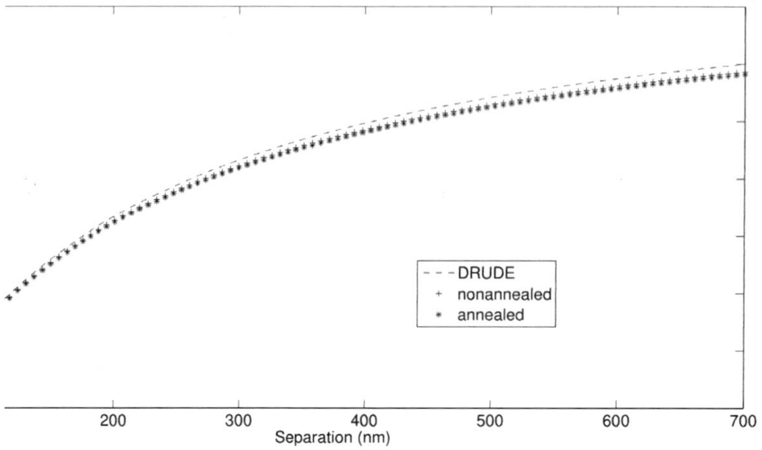

Figure 2. Reduction factor as a function of separation for the classical Drude model with a constant damping parameter (solid line), the annealed Au (*) sample and the nonannealed Au sample (+).

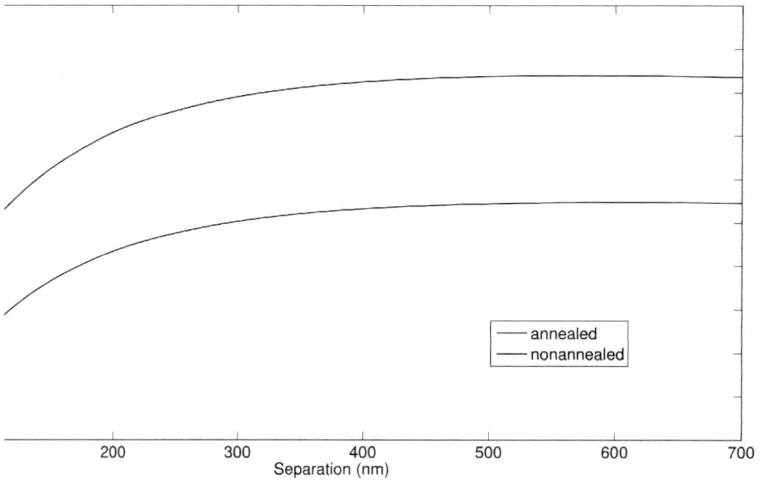

Figure 3. Percent difference Δ between the reduction factors calculated in Fig.(2).

percent difference $\Delta = 100|\eta_{a,n} - \eta_c|/\eta_c$ as a function of the separation between the plates. At large separations, the percent difference is at most of $\sim 2.2\%$ for the nonannealed sample and $\sim 1.6\%$ for the annealed film.

3. Conclusions

The goal of high-precission measurements at small separations of the Casimir force, requires that the dielectric function of the materials are well known. In this paper we have considered the effects of the frequency dependent damping when Au films of a few hundreds of Angstroms are considered. Besides, the effect of sample preparation such as annealing the film also changes the Casimir force. Although the changes are of less than 3% at large separations, we wish to point out that this is an example where using a simple Drude model, for example, to extrapolate to low frequencies tabulated data, is not granted unless thick films are used. This is important if high precision measurements are seek. Furthermore, preparing a sample in ultra high vacuum can change the optical properties of the Au films [28], showing the need for in-situ determination of the dielectric properties of the samples used in Casimir experiments.

Finally, the relevance of the frequency dependent damping parameter is in the finite-temperature effects, since now the damping will also depend on the Matsubara frequencies. This will be consider in a future paper.

Acknowledgements: Partial support from DGAPA-UNAM project No. 113208 .

References

[1] Lamoreaux S K 1997 *Phys. Rev. Lett.* **78** 5; 1998 *Phys. Rev. Lett.* **81** 5475
[2] Mohideen U and Roy A 1998 *Phys. Rev. Lett.* **81** 4549 ; Roy A , Lin C Y and Mohideen U 1999 *Phys. Rev.* D **60**, 111101(R) ; Harris B W , Chen F and Mohideen U 2000 *Phys. Rev.* A **62**, 052109
[3] Ederth T (2000) *Phys. Rev.* A **62** 062104
[4] Chan H B, Aksyuk V A , Kleiman R N, Bishop D J and F. Capasso 2001 *Science* **291** 1941; 2001 *Phys. Rev. Lett.* **87** 211801
[5] Bressi G, Carugno G , Onofrio R, and Ruoso G, *Phys. Rev. Lett.* **88** 041804 (2002)
[6] Decca R S, López D, Fischbach E, and Krause D E 2003 *Phys. Rev. Lett.* **91** 050402 ; Decca R S, Fischbach E, Klimchitskaya G L, Krause D E, López D, and Mostepanenko V M 2003 *Phys. Rev.* D **68** 116003
[7] van Zwol P J, Palasantzas G, De Hosson J Th M 2008 *Phys. Rev.* B **77** 075412
[8] Chan H B, Bao Y, Zou J Cirelli R A Klemens F Mansfield W M and Pai S C 2008 *Phys. Rev. Lett* **101** 030401
[9] E. M. Lifshitz 1956 *Zh. Eksp. Teor. Fiz.* **29** 94 [1956 *Sov. Phys. JETP* **2** 73].
[10] Lambrecht A, Reynaud S 2000 *Eur. Phys. J.* D **8** 309
[11] J. S Hoye, I. Brevik, S. E. Ellingsen and J. B. Aarseth 2007 Phys. Rev. E **75**, 051127
[12] Bimonte G 2007 *New J. Phys.* **9** 281
[13] Ellingsen S A 2008 *Phys. Rev.* E **78** 021120
[14] Mostepanenko V M and Geyer B 2008 J. Phys. A: Math. Thor. **41** 16014
[15] Lamoreaux S K 2008 Possible resolution of the Casimir force finite temperature correction "controversies" *Preprint* arXiv:0801.1283v1 [quant-ph]
[16] Decca R S, Fischbach E, Geyer B, Klimchitskaya G L, Krause D E, Lopez D, Mohideen U, Mostepanenko V M 2008 Comment on " Possible resolution of the Casimir force finite temperature correction "controversies"" *Preprint* arXiv:0803.4247v1 [quant-ph]
[17] Ellingsen A., Brevik I, Hoye J S , Milton K A (2008) Low temperature Casimir-Lifshitz free energy and entropy: the case of poor conductors *Preprint* arXiv:0809.0763v1 [quant-ph]
[18] Pirozhenko I, Lambrecht A and Svetovoy V B 2006 *New Journal of Physics* **8** 238
[19] Svetovoy V B, van Zwol P J, Palasantzas G, De Hosson J Th M 2008 *Phys. Rev.* B **77** 035439
[20] Pirozhenko I and Lambrecht A 2008 *Phys. Rev.* A **77**, 013811
[21] Iannuzzi D, Lisanti M, Munday J N and Capasso F 2006 *J. Phys. A: Math. Gen.* **39** 6445
[22] Lisanti M, Iannuzzi D and Capasso F 2005 *Proc. Nat. Acad. Sci.* **102** 11989
[23] Kastle G, Boyen H G , Schroder A , Plettl A and Ziemann P 2004 **70**, 165414
[24] Walther M , Cooke D G , Sherstan C, Hajar M, Freeman M R and Hegmann F A 2007 *Phys. Rev.* B **76** 125408
[25] Esquivel-Sirvent R. 2008 *Phys. Rev* A **77** 042107
[26] Theye M L 1970 *Phys. Rev.* B **2** 3060
[27] Nagel S R and Schnatterly E 1974 *Phys. Rev.* B **9** 1299
[28] Bennett H E, Bennett J M 1966 *Optical Properties and Electronic Structure of Metals and Alloys* ed F Abelés and L Erwin (Amsterdam: North Holland) p 175

Journal of Physics: Conference Series **161** (2009) 012013 doi:10.1088/1742-6596/161/1/012013

On thermal Casimir force between real metals

Bodo Geyer
Center of Theoretical Studies and Institute for Theoretical Physics, Leipzig University, D-04009, Leipzig, Germany

E-mail: Bodo.Geyer@itp.uni-leipzig.de

Abstract. The physical reasons why the Drude dielectric function is not compatible with the Lifshitz formula, as opposed to the generalized plasma-like permittivity, are presented. Essentially, the problem is connected with the applicability conditions of the Lifshitz theory. It is shown that the Lifshitz theory combined with the generalized plasma-like permittivity is thermodynamically consistent.

1. Introduction: Casimir free energy and different models of dielectric permittivity

The Casimir effect [1], arising due to the alteration of the spectrum of electromagnetic zero-point oscillations by material boundaries, is one of the most important subjects of interdisciplinary interest. Recent trends go toward complex experimental settings for real materials, including applications to nanotechnology, and their consistent theoretical explanation on the framework of quantum electrodynamics (see, e.g., monographs [2, 3], proceedings [4] and more recent papers [5, 6, 7]) appropriately adapted to real material bodies.

The basic theory of the van der Waals and Casimir forces between dielectric materials in terms of their frequency-dependent dielectric permittivities $\varepsilon(\omega)$ was proposed by Lifshitz [8] for the most simple configuration (two parallel plates/semispaces) and ideal surfaces. In the Lifshitz theory the *free energy* of the fluctuating field between two *infinite, electrically neutral* plane parallel plates of thickness d at temperature T in *thermal equilibrium* is given by [8, 9]

$$\mathcal{F}(a,T)=\frac{k_BT}{2\pi}\sum_{l=0}^{\infty}\Big(1-\tfrac{1}{2}\delta_{0l}\Big)\int_0^{\infty}k_\perp dk_\perp\Big\{\ln\Big[1-r_{\rm TM}^2(\xi_l,k_\perp)e^{-2aq_l}\Big]+\ln\Big[1-r_{\rm TE}^2(\xi_l,k_\perp)e^{-2aq_l}\Big]\Big\}. \quad (1)$$

Here, a is the separation distance between the plates, k_B is the Boltzmann constant, $\xi_l = 2\pi k_BTl/\hbar$ are the Matsubara frequencies defined for any $l=0,\,1,\,2,\,\ldots$, and $k_\perp=|\boldsymbol{k}_\perp|$ is the magnitude of the wave vector projection onto the plane of the plates. The reflection coefficients for the two independent polarizations of the electromagnetic field (transverse magnetic, TM, and transverse electric, TE) are expressed [10] in terms of the frequency-dependent dielectric permittivity, $\varepsilon(\omega)$, along the imaginary frequency axis:

$$r_{\rm TM}(\xi_l,k_\perp)=\frac{\varepsilon_l^2q_l^2-k_l^2}{\varepsilon_l^2q_l^2+k_l^2+2q_lk_l\varepsilon_l\coth(k_ld)},\qquad r_{\rm TE}(\xi_l,k_\perp)=\frac{k_l^2-q_l^2}{q_l^2+k_l^2+2q_lk_l\coth(k_ld)}, \quad (2)$$

where

$$q_l=\sqrt{k_\perp^2+\xi_l^2/c^2},\quad k_l=\sqrt{k_\perp^2+\varepsilon_l\xi_l^2/c^2},\quad \varepsilon_l=\varepsilon({\rm i}\xi_l). \quad (3)$$

The Lifshitz theory remains the basis also for real metals and dielectrics and more complicated experimental settings, usually ignoring the finite extension of the material probes under appropriate conditions. However, depending on theoretical model for their dielectric permittivity this theory meets problems and paradoxes.

Especially, the application of the *Drude model* for metals,

$$\varepsilon_D(\omega) = 1 - \frac{\omega_p^2}{\omega(\omega + \mathrm{i}\gamma)}, \tag{4}$$

where ω_p is the plasma frequency and γ is the relaxation parameter, led to conflict with Nernst's heat theorem for plates with perfect crystal lattice [11, 12], see also [13, 14, 15, 16]. In addition, recent precision measurements of the Casimir force [17, 18, 19, 20] excluded the Drude model at a 99.9% confidence level. Also the application of the Lifshitz theory to dielectrics or semiconductors with not too high density of free charge carriers results in a violation of the Nernst heat theorem if the nonzero dc-conductivity at zero frequency is taken into account [21, 22, 23, 24]. Its inclusion also leads to a contradiction with experiment [25, 26, 27].

On the other hand the usual, non-dissipative *plasma model*

$$\varepsilon_p(\omega) = 1 - \frac{\omega_p^2}{\omega^2}, \tag{5}$$

resulting from (4) for $\gamma = 0$, is in agreement with thermodynamics and longer separation experiments [17, 18, 19, 20], but is in contradiction with the experiment performed at short separations [28, 29]. Also the Leontovich surface impedance approach [35] was found to be in agreement with the longer separation experiments just mentioned but, unfortunately, is simply not applicable at shorter separations.

Recently, in order to overcome these difficulties, for real metals the *generalized plasma-like dielectric permittivity* has been proposed [30] as follows

$$\varepsilon_{gp}(\omega) = 1 - \frac{\omega_p^2}{\omega^2} + \sum_{j=1}^{K} \frac{f_j}{\omega_j^2 - \omega^2 - \mathrm{i}g_j\omega}, \tag{6}$$

where $\omega_j \neq 0$ are the resonant frequencies of oscillators describing the core electrons, g_j are the respective relaxation parameters, f_j are the oscillator strengths and K is the number of oscillators; the values of oscillator parameters for different materials can be found in [31]. That model includes dissipation processes due to the interband transitions of core electrons but disregards dissipation due to scattering processes of free electrons. As was shown in [30], the Lifshitz formula combined with the generalized plasma-like dielectric permittivity is consistent with both short and long separation experiments. It also exactly satisfies the Kramers-Kronig relations [30]. In [32] it has been shown that this model also satisfies the Nernst heat theorem and, therefore, is compatible with thermodynamics.

Here, following the lines of Ref. [32] and adopting the point of view of Ref. [33], we first show why the Drude dielectric function is incompatible with the Lifshitz formula. Then, we derive an analytic asymptotic expression for the Casimir free energy in the limit of low temperatures. Finally, we show the validity of the Nernst heat theorem in the Lifshitz theory combined with the generalized plasma-like dielectric permittivity.

2. Incompatibility of Drude model with the Lifshitz formula for metallic plates

Despite originally derived for dielectric plates of infinite area, Eq. (1) is correctly used [9, 34] also for both dielectric and ideal metal plates of finite area S under the condition $a \ll \sqrt{S}$. However, when plates of real metal are described by the Drude dielectric function (4), the appearance of a drift current of conduction electrons leads to a crucially new physical situation.

To gain a better understanding of this statement, let us derive Eq. (4) starting from Maxwell equations in an unbounded nonmagnetic metallic medium,

$$\begin{aligned} \mathrm{rot}\boldsymbol{B} &= \frac{1}{c}\frac{\partial \boldsymbol{E}}{\partial t} + \frac{4\pi}{c}\sigma_0 \boldsymbol{E}, & \mathrm{div}\boldsymbol{B} = 0, \\ \mathrm{rot}\boldsymbol{E} &= -\frac{1}{c}\frac{\partial \boldsymbol{B}}{\partial t}, & \mathrm{div}\boldsymbol{E} = 0. \end{aligned} \tag{7}$$

Here, the electric current density $\boldsymbol{j} = \sigma_0 \boldsymbol{E}$ is induced in a metal under the influence of external sources, and σ_0 is the conductivity at zero frequency. Physically the demand that the medium is unbounded means that it should be much larger than the extension of the wave fronts of electromagnetic waves coming from external sources (i.e., of zero-point oscillations and thermal photons).

Solving Eqs. (7) by monochromatic waves, $\boldsymbol{E} = \mathrm{Re}\left[\boldsymbol{E}_0(\boldsymbol{r})e^{-\mathrm{i}\omega t}\right]$ and $\boldsymbol{B} = \mathrm{Re}\left[\boldsymbol{B}_0(\boldsymbol{r})e^{-\mathrm{i}\omega t}\right]$ with $\boldsymbol{E}_0(\boldsymbol{r})$ and $\boldsymbol{B}_0(\boldsymbol{r})$ satisfying the well known wave equations, the wave vector $\boldsymbol{k}$ obeys,

$$k^2 = \frac{\omega^2}{c^2} + \mathrm{i}\frac{4\pi\sigma_0\omega}{c^2} \equiv \frac{\varepsilon_n(\omega)\,\omega^2}{c^2}\,. \tag{8}$$

Here, the dielectric permittivity of the *normal skin effect* is introduced being defined as

$$\varepsilon_n(\omega) = 1 + \mathrm{i}\frac{4\pi\sigma_0}{\omega} \qquad \text{with} \qquad \sigma_0 = \omega_p^2/(4\pi\gamma)\,. \tag{9}$$

This equation is applicable at not too high frequencies (the region of the normal skin effect) where the relation $\boldsymbol{j} = \sigma_0 \boldsymbol{E}$ is valid.

In order to extend the applicability of (9) to higher frequencies, up to the plasma frequency, the following replacement is made,

$$\sigma_0 \;\rightarrow\; \sigma_D(\omega) = \frac{\sigma_0\left(1 + \frac{\mathrm{i}\omega}{\gamma}\right)}{1 + \frac{\omega^2}{\gamma^2}}\,, \tag{10}$$

herewith obtaining the (complex) conductivity of the Drude model; namely, substituting (10) in (9) we recover the dielectric permittivity (4).

Obviously, in the limit $\omega \ll \gamma$, in both the numerator and the denominator of (10), unity dominates over ω/γ and ω^2/γ^2 leading back to the *real conductivity* of conduction electrons σ_0. This converts the dielectric permittivity of the Drude model (4) in the dielectric permittivity of the normal skin effect (9). Contrary, at sufficiently high frequencies $\gamma \ll \omega < \omega_p$ (the region of infrared optics) one can neglect unity as compared to ω/γ and ω^2/γ^2. Then (4) and (10) lead to the plasma model with dielectric permittivity (5) and *purely imaginary conductivity*,

$$\sigma_P(\omega) = \mathrm{i}\,\frac{\sigma_0\gamma}{\omega}\,. \tag{11}$$

The total current in the framework of the Drude model (4) is given by

$$\begin{aligned} \boldsymbol{j}_{\mathrm{tot}}(\boldsymbol{r},t) &= \mathrm{Re}\left[-\frac{\mathrm{i}\omega}{4\pi}\varepsilon_D(\omega)\boldsymbol{E}_0(\boldsymbol{r})e^{-\mathrm{i}\omega t}\right] \\ &= \frac{\omega}{4\pi}\left(1 - \frac{\omega_p^2}{\omega^2+\gamma^2}\right)\mathrm{Im}\left[\boldsymbol{E}_0(\boldsymbol{r})e^{-\mathrm{i}\omega t}\right] + \frac{\sigma_0\gamma^2}{\omega^2+\gamma^2}\mathrm{Re}\left[\boldsymbol{E}_0(\boldsymbol{r})e^{-\mathrm{i}\omega t}\right]. \end{aligned} \tag{12}$$

The first term on the right-hand side of this equation has the meaning of the *displacement current*, whereas the second term is proportional to the physical electric field $\boldsymbol{E} = \boldsymbol{E}(\boldsymbol{r}, t)$ and describes the *real current of conduction electrons*. Under the condition $\gamma \ll \omega < \omega_p$, i.e., in the region of infrared optics, the first term on the right-hand side of (12), i.e., the displacement current of the plasma model with pure imaginary conductivity dominates. Under the opposite condition $\omega \ll \gamma$, i.e., in the region of the normal skin effect, the second term, i.e., the real physical current of conduction electrons dominates.

By the foregoing discussion we found that the Drude dielectric function (4) is obtained from the Maxwell equations with, besides of the displacement current, also a real drift current of conducting electrons initiated by the external electric field $\boldsymbol{E}$. That current unavoidably leads to a Joule heating of the Casimir plates and an irreversible process driving the system out of thermal equilibrium [35]. In order to held the temperature constant it is necessary to admit an unidirectional flux of heat from the system to the heat bath [36]. However, this is prohibited by the definition of thermal equilibrium which requires not only that $T = \text{const}$, but also that all irreversible processes of dissipation of energy into heat have been terminated [37, 38]. As a consequence, this means that the substitution of the Drude dielectric permittivity into the Lifshitz formula violates at least one of the suppositions of its derivation and, therefore, may lead to a violation of Nernst's theorem and to contradictions with experimental data.

On the other hand, the dielectric permittivity of the plasma model is connected with only the displacement current which does not lead to Joule's heating and, therefore, no violation of thermodynamics should appear. Consequently, the free electron plasma model in combination with the Lifshitz formula satisfies the Nernst heat theorem [11, 12]. However, since it is in disagreement with the measurement of the Casimir force in short separation experiments it should be extended in such a way that additional dispersion forces do not lead to real drift currents of conducting electrons. This was achieved by the approach using the generalized plasma-like permittivity [30].

3. Thermodynamic test for the generalized plasma-like dielectric permittivity

Now, we demonstrate that the generalized plasma-like permittivity (6), which is in agreement with all experiments performed up to date also satisfies the requirements of thermodynamics. The generalized plasma-like dielectric permittivity takes into account the interband transitions of core electrons but does not include relaxation of the free conduction electrons. The latter are described by an oscillator with zero resonant frequency, $\omega_0 = 0$, which is explicitly included in by the second term of (6) with $g_0 = 0$ and $f_0 = \omega_p^2$. Thus, similar to the usual plasma model, the generalized plasma-like permittivity admits only the displacement current and does not allow for a heat transfer. Because of this, the generalized plasma-like permittivity is compatible with the Lifshitz formula. Reviewing shortly the proof of Ref. [32] we show that the Lifshitz formula combined with the generalized plasma-like permittivity is in agreement with Nernst's heat theorem and thus withstands the thermodynamic test.

To find the asymptotic behavior of the Casimir free energy and entropy at low temperature, we first present equations (1), (2) and (6) in terms of the following dimensionless parameters:

$$\begin{aligned} &\tilde{\omega}_p = \frac{\omega_p}{\omega_c} \equiv \frac{1}{\alpha}, \qquad \zeta_l = \frac{\xi_l}{\omega_c} \equiv \tau l, \qquad y = \sqrt{4a^2k_\perp^2 + \zeta_l^2}, \\ &\gamma_j = \frac{\omega_c^2}{\omega_j^2}, \qquad \delta_j = \frac{\omega_c g_j}{\omega_j^2}, \qquad C_j = \frac{f_j}{\omega_j^2}, \end{aligned} \tag{13}$$

where $\omega_c \equiv c/(2a)$ is the so-called characteristic frequency of the Casimir effect and the parameter τ is represented as $\tau = 2\pi T/T_{\text{eff}}$ with $k_B T_{\text{eff}} \equiv \hbar c/2a$; here, T_{eff} is the so-called effective temperature which, for example, at separation $a = 1\,\mu$m is equal $T_{\text{eff}} \approx 1145\,$K.

In terms of these new variables, the Lifshitz formula (1) takes the form

$$\mathcal{F}(a,T) = \frac{\hbar c \tau}{32\pi^2 a^3} \sum_{l=0}^{\infty} \left(1 - \frac{1}{2}\,\delta_{0l}\right) \int_{\zeta_l}^{\infty} dy\, f(\zeta_l, y) \qquad \text{with} \tag{14}$$

$$f(\zeta_l, y) = y \left\{ \ln\left[1 - r_{\rm TM}^2(\zeta_l, y) e^{-y}\right] + \ln\left[1 - r_{\rm TE}^2(\zeta_l, y) e^{-y}\right] \right\}. \tag{15}$$

The reflection coefficients (2) are given by

$$r_{\rm TM}(\zeta_l, y) = \frac{(\varepsilon_l^2 - 1)(y^2 - \zeta_l^2)}{(\varepsilon_l + 1)y^2 + (\varepsilon_l - 1)\zeta_l^2 + 2\varepsilon_l y h_l(y) \coth\left[(d/2a)h_l(y)\right]},$$

$$r_{\rm TE}(\zeta_l, y) = \frac{(\varepsilon_l - 1)\zeta_l^2}{2y^2 + (\varepsilon_l - 1)\zeta_l^2 + 2y h_l(y) \coth\left[(d/2a)h_l(y)\right]}, \tag{16}$$

where

$$h_l(y) = \sqrt{y^2 + (\varepsilon_l - 1)\zeta_l^2}\,. \tag{17}$$

The generalized plasma-like dielectric permittivity along the imaginary frequency axis can be presented as

$$\varepsilon_l = \varepsilon({\rm i}\zeta_l) = 1 + \frac{1}{\alpha^2 \zeta_l^2} + \sum_{j=1}^{K} \frac{C_j}{1 + \gamma_j \zeta_l^2 + \delta_j \zeta_l}. \tag{18}$$

Using the Abel-Plana formula [2, 9]

$$\sum_{l=0}^{\infty} \left(1 - \frac{1}{2}\delta_{0l}\right) F(l) = \int_0^{\infty} F(t)dt + {\rm i} \int_0^{\infty} dt \frac{F({\rm i}t) - F(-{\rm i}t)}{e^{2\pi t} - 1}, \tag{19}$$

where $F(z)$ is an analytic function in the right half of the complex plane, we can rearrange (14) to the form

$$\mathcal{F}(a,T) = E(a) + \Delta\mathcal{F}(a,t)\,. \tag{20}$$

Here, the energy of the Casimir interaction at zero temperature is given by

$$E(a) = \frac{\hbar c}{32\pi^2 a^3} \int_0^{\infty} d\zeta \int_{\zeta}^{\infty} f(\zeta, y) dy\,, \tag{21}$$

and the thermal correction to the Casimir energy is expressed as follows:

$$\Delta\mathcal{F}(a,T) = \frac{{\rm i}\hbar c \tau}{32\pi^2 a^3} \int_0^{\infty} dt \frac{F({\rm i}\tau t) - F(-{\rm i}\tau t)}{e^{2\pi t} - 1}, \tag{22}$$

where

$$F(x) = \int_x^{\infty} dy\, f(x,y)\,. \tag{23}$$

The behavior of the thermal correction (22), (23) at low temperature will be the subject of our further consideration.

Perturbation expansion can be performed in analogy to papers [11, 12, 39, 40]. At first, we expand the reflection coefficients $r_{\rm TM}(\zeta, y)$ and $r_{\rm TE}(\zeta, y)$ in powers of the parameter α preserving all powers up to the fourth inclusive. Since α can be identically presented as $\alpha = \lambda_p/(4\pi a)$, where λ_p is the plasma wavelength, this means that $\alpha \ll 1$ at all separation distances between the plates larger than λ_p. Furthermore, an expansion at low τ is performed up to fifth order.

As is seen from (15), it is more convenient to expand the logarithmic functions containing the reflection coefficients being multiplied by the variable y. The details of these expansion are contained in Ref. [32] and will not repeated here.

It is remarkable that these expansions effectively do not depend on d, the thickness of the plates, contained in (16). This is because asymptotically, when α goes to zero, the factor containing the thickness d behaves as $\coth\left[(d/2a)h_l(y)\right] \approx 1+2\exp\left[-d/(a\alpha)\right]$. Thus, this factor could only contribute exponentially small terms in these expansions providing the plate thickness d is much larger than the penetration depth of electromagnetic oscillations into the metal [recall that $2a\alpha = \lambda_p/(2\pi)$]. Under this condition the perturbation expansions are common for two semispaces and for two plates of finite thickness.

In [32] it was shown that up to second power in α there appear no contributions of the core electrons, i.e., that contributions coincide with those of the free electron plasma model [39, 40],

$$\Delta\mathcal{F}_p(a,T) = -\frac{\hbar\, c\, \tau^3}{32\pi^2\, a^3}\left\{\frac{\zeta(3)}{4\pi^2} - \frac{\tau}{360} + \alpha\left(\frac{\zeta(3)}{\pi^2} - \frac{\tau}{45}\right) - \alpha^2\frac{\zeta(5)}{2\pi^4}\tau^2\right\}, \tag{24}$$

where $\zeta(z)$ is the Riemann zeta function. As was shown in [40], the terms in (24) of order α^0 and α do not contain corrections of order τ^n with $n \geq 5$; they contain only exponentially small corrections of order $\exp(-2\pi/\tau)$.

Now, let us consider the contributions from the core electrons which are contained only in the terms of order α^3 and α^4. Their contribution to the thermal correction is as follows:

$$\Delta\mathcal{F}_g(a,T) = \frac{\hbar c\,\alpha^3\tau^4}{32\pi^2 a^3}\left\{\left[\frac{\zeta(3)}{60}\sum_{j=1}^{K} C_j\delta_j + \tau\frac{3\zeta(5)}{2\pi^4}\left(\sum_{j=1}^{K} C_j + 2\right)\right] - \alpha\left[\frac{4\zeta(3)}{15}\sum_{j=1}^{K} C_j\delta_j + \tau\frac{6\zeta(5)}{\pi^4}\right]\right\}. \tag{25}$$

The total Casimir free energy computed using the generalized plasma-like permittivity can be now found from (20), (24) and (25) as follows:

$$\mathcal{F}(a,T) = E(a) + \Delta\mathcal{F}_p(a,T) + \Delta\mathcal{F}_g(a,T). \tag{26}$$

Taking into account the definition of τ, it can be represented in the form

$$\begin{aligned}\mathcal{F}(a,T) \;=\; & E(a) - \frac{\hbar c\zeta(3)}{16\pi a^3}\left(\frac{T}{T_{\rm eff}}\right)^3\left\{1 + 4\alpha\right. \\ & -\frac{\pi^3}{45\zeta(3)}\frac{T}{T_{\rm eff}}\left(1 + 8\alpha + 6\zeta(3)\alpha^3\sum_{j=1}^{K} C_j\delta_j - 96\zeta(3)\alpha^4\sum_{j=1}^{K} C_j\delta_j\right) \\ & \left. -\frac{8\zeta(5)}{\zeta(3)}\left(\frac{T}{T_{\rm eff}}\right)^2\alpha^2\left[1 + 3\alpha\left(\sum_{j=1}^{K} C_j + 2\right) - 12\alpha^2\right]\right\}.\end{aligned} \tag{27}$$

Here one can see that the free energy calculated using the generalized plasma-like permittivity contains the correction of order $(T/T_{\rm eff})^4$ not only in the terms of order α^0 and α (as in the usual plasma model) but also in the third and fourth order expansion terms in α. In the usual plasma model the terms of order α^3 and α^4 contain the thermal corrections only of order of $(T/T_{\rm eff})^5$ and higher [39].

To estimate the relative role of the additional terms arising due to the use of the generalized plasma-like permittivity, one can use the parameters of oscillator terms in (18) for Au [20]. This results in

$$\sum_{j=1}^{6} C_j = 6.3175, \qquad \sum_{j=1}^{6} C_j\delta_j = \left\{\begin{array}{ll} 0.272, & a = 200\,\text{nm}, \\ 0.109, & a = 500\,\text{nm}. \end{array}\right. \tag{28}$$

From (27) it is easy to find the asymptotic behavior of the Casimir entropy

$$S(a,T) = -\frac{\partial \mathcal{F}(a,T)}{\partial T} \tag{29}$$

at low temperature. Without performing the simple differentiation it is immediatly seen from (27) that the entropy goes to zero (and remains positive) when the temperature vanishes,

$$S(a,T) \to 0 \quad \text{when} \quad T \to 0\,. \tag{30}$$

This means that the Nernst heat theorem is satisfied and the Lifshitz theory combined with the generalized plasma-like dielectric permittivity withstands the thermodynamic test.

4. Conclusions and discussion

In the foregoing we have discussed the physical reasons why the Drude dielectric function is not applicable in the case of finite plates made of real metal. It was shown that for the validity of the Drude model a nonzero current of conduction electrons exists, which leads to a Joule's heating of the Casimir plates. As a consequence, one of the suppositions for the derivation of the Lishitz formula, namely, thermal equilibrium is violated. Of course, any dielectric permittivity being related to a drift current of conducting electrons is ruled out by that consideration.

On the contrary, not only the free electron plasma dielectric permittivity but also the generalized plasma-like permittivity leads to only a displacement current and does not result to a dissipation of energy into heat. From this consideration it becomes clear why the generalized plasma-like permittivity, which does not include the relaxation processes of conduction electrons, is consistent with all available measurements of the Casimir force at both short and large separations.

To conclusively establish the applicability of the generalized plasma-like permittivity in the theory of the thermal Casimir force between real metals we reviewed the thermodynamic test of this model. Studying the asymptotic behavior of both the Casimir free energy and Casimir entropy at low temperature it is shown that the latter is positive and takes zero value at zero temperature. Thus, the Nernst heat theorem is satisfied. Of course, when the oscillator parameters describing the core electrons go to zero, the newly obtained expressions for the Casimir free energy and entropy go into the ones found for the usual plasma model.

From this we conclude that no problems with the Lifshitz theory occur if it is applied under observation of its region of applicability. Similar conclusion is made in Ref. [33] agreeing "with the fact that the conductivity processes in real materials connected with the drift current of conduction electrons violate the applicability conditions of the Lifshitz theory and, thus, are not described by this theory." Since these processes, however, exist in reality, one must ask if they are to "be taken into account in some future general theory of dispersion forces."

References

[1] Casimir H B G 1948 *Proc. K. Ned. Akad. Wet.* **51** 793
[2] Mostepanenko V M and Trunov N N 1997 *The Casimir Effect and its Applications* (Oxford: Clarendon Press)
[3] Milton K A 2001 *The Casimir Effect* (Singapore: World Scientific)
[4] Elizalde E and Odintsov S D (eds) 2006 Papers presented at the 7th Workshop on Quantum Field Theory Under the Influence of External Conditions, *J. Phys. A: Mat. Gen.* **39** 6109
[5] Emig T, Jaffe R L, Kardar M and Scardicchio A 2006 *Phys. Rev. Lett.* **96** 080403
[6] Emig T 2007 *Phys. Rev. Lett.* **98** 160801
[7] Ashourvan A, Miri M and Golestanian R 2007 *Phys. Rev. Lett.* **98** 140801
[8] Lifshitz E M 1956 *Sov. Phys. JETP* **2** 73
[9] Bordag M, Mohideen U and Mostepanenko V M 2001 *Phys. Rep.* **353** 1
[10] Blagov E V, Klimchitskaya G L and Mostepanenko V M 2005 *Phys. Rev.* B **71** 235401
[11] Bezerra V B, Klimchitskaya G L and Mostepanenko V M 2002 *Phys. Rev.* A **66** 062112

[12] Bezerra V B, Klimchitskaya G L, Mostepanenko V M and Romero C 2004 *Phys. Rev.* A **69** 022119
[13] Bezerra V B, Decca R S, Fischbach E, Geyer B, Klimchitskaya G L, Krause D E, López D,
[14] Mostepanenko V M, Bezerra V B, Decca R S, Fischbach E, Geyer B, Klimchitskaya G L, Krause D E, López D and Romero C 2006 *J. Phys. A: Mat. Gen.* **39** 6589
[15] Milton K A 2004 *J. Phys. A: Mat. Gen.* **37** R209
[16] Høye J S, Brevik I, Aarseth J B and Milton K A 2006 *J. Phys. A: Mat. Gen.* **39** 6031
[17] Decca R S, Fischbach E, Klimchitskaya G L, Krause D E, López D and Mostepanenko V M 2003 *Phys. Rev.* D **68**, 116003
[18] Decca R S, López D, Fischbach E, Klimchitskaya G L, Krause D E and Mostepanenko V M 2005 *Ann. Phys. NY* **318** 37
[19] Decca R S, López D, Fischbach E, Klimchitskaya G L, Krause D E and Mostepanenko V M 2007 *Phys. Rev* D **75** 077101
[20] Decca R S, López D, Fischbach E, Klimchitskaya G L, Krause D E and Mostepanenko V M 2007 *Eur. Phys. J* C **51** 963
[21] Geyer B, Klimchitskaya G L and Mostepanenko V M 2005 *Phys. Rev.* D **72** 085009
[22] Geyer B, Klimchitskaya G L and Mostepanenko V M 2006 *Int. J. Mod. Phys.* A **21** 5007
[23] Klimchitskaya G L, Geyer B and Mostepanenko V M 2006 *J. Phys. A.: Mat. Gen.* **39** 6495
[24] Geyer B, Klimchitskaya G L and Mostepanenko V M 2008 *Ann. Phys. NY*, **323** 291
[25] Chen F, Klimchitskaya G L, Mostepanenko V M and Mohideen U 2007 *Optics Express* **15** 4823
[26] Chen F, Klimchitskaya G L, Mostepanenko V M and Mohideen U 2007 *Phys. Rev.* B **76** 035338
[27] Klimchitskaya G L and Mostepanenko V M 2008 *J. Phys. A.: Mat. Theor.* **41** 312002
[28] Harris B W, Chen F and Mohideen U 2000 *Phys. Rev.* A **62** 052109
[29] Chen F, Klimchitskaya G L, Mohideen U and Mostepanenko V M 2004 *Phys. Rev.* A **69** 022117
[30] Klimchitskaya G L, Mohideen U and Mostepanenko V M 2007 *J. Phys. A.: Mat. Theor.* **40** F339
[31] Parsegian V A 2005 *Van der Waals forces: A Handbook for Biologists, Chemists, Engineers, and Physicists* (Cambridge: Cambridge University Press)
[32] Geyer B, Klimchitskaya G L and Mostepanenko V M 2007 *J. Phys.* A: *Math. Theor.* **40** 13485
[33] Klimchitskaya G L, *Problems and paradoxes of the Lifshitz theory*, these proceedings
[34] Gies H and Klingmüller K 2006 *Phys. Rev. Lett.* **97** 220405
[35] Geyer B, Klimchitskaya G L and Mostepanenko V M 2003 *Phys. Rev.* A **67** 062102
[36] Bryksin V V and Petrov M P, 2008 *Fzi. Tverdogo Tela* **50** 222, (*Phys. Solid State* **50** 229)
[37] Rumer Yu B and Ryvkin M S 1980 *Thermodynamics, Statistical Physics, and Kinetics* (Moscow)
[38] Kondepui D and Prigogine I 1998 *Modern Thermodynamics* (New York)
[39] Bordag M, Geyer B, Klimchitskaya G L and Mostepanenko V M 2000 *Phys. Rev. Lett.* **85** 503
[40] Geyer B, Klimchitskaya G L and Mostepanenko V M 2001 *Int. J. Mod. Phys.* A **16** 3291

New geometries in the Casimir effect: dielectric gratings

Astrid Lambrecht[1] **and Valery N Marachevsky**[1,2]

[1] Laboratoire Kastler Brossel, CNRS, ENS, UPMC, Campus Jussieu case 74, 75252 Paris, France

[2] V. A. Fock Institute of Physics, St. Petersburg State University, 198504 St. Petersburg, Russia

E-mail: `lambrecht@spectro.jussieu.fr`

E-mail: `maraval@mail.ru`

Abstract. An exact solution for the Casimir force between two arbitrary dielectric gratings with the same period d is presented. The Casimir energy for two dielectric gratings or periodic dielectrics is expressed in terms of Rayleigh coefficients. The theory is applied to calculate the Casimir force in several cases of interest.

1. Introduction

In 1948 H. B. G. Casimir showed that two electrically neutral, perfectly conducting plates, placed parallel in vacuum, modify the vacuum energy density with respect to the unperturbed vacuum[1]. The vacuum energy density varies with the separation between the mirrors and leads to the Casimir force, which scales with the inverse of the fourth power of the mirrors separation.

The Casimir force is highly versatile. Changing materials and shape of the boundaries can modify its strength and even its sign. For example the role of surface plasmons to tailor the Casimir force [2, 3, 4, 5] and the possible interest of metamaterials [6, 7, 8, 9] in order to produce repulsive forces have been studied during the last years. The modification of the Casimir force between a silicon membrane and a gold coated sphere due to the illumination of the silicon membrane with laser light was demonstrated recently [10]. Modifying the Casimir force could potentially be useful in the design and control of micro- and nanomachines.

The availability of experimental set-ups that allow accurate measurements of surface forces between macroscopic objects at submicron separations has recently stimulated a renewed interest in the Casimir effect. While the material dependence of the Casimir force has been thoroughly studied between two plane mirrors (see e.g. [11, 12, 13, 14]), for most other geometries exact calculations exist only for perfectly reflecting boundaries (see e.g. [15]). If finite conductivity effects are taken into account, the shape dependence of the Casimir force is usually treated using the proximity force approximation (PFA) which amounts to summing up contributions at different distances as if they were independent. PFA is known to be valid if the typical length scale of geometric deviation from the plane-plane configuration is much larger than the distance between the two bodies. However, for periodic metallic or dielectric gratings such as manufactured in nanotechnology PFA is expected to fail as all important length scales are of the same order.

Journal of Physics: Conference Series **161** (2009) 012014 doi:10.1088/1742-6596/161/1/012014

The first measurement of the Casimir force between an aluminium grating with small sinusoidal corrugations and an aluminium sphere was performed by Roy and Mohideen [16]. In a recent paper[17], Chan et al. present the first measurement of the Casimir force between a rectangular silicon grating of high aspect ratio and a gold sphere and demonstrate the violation of the PFA in this geometry. Corresponding calculations taking into account the periodic structure beyond PFA, but only for perfect mirrors[18], turn out to lead to a too large deviation from PFA[17].

In this paper we present a novel exact theory of the Casimir effect between gratings of arbitrary periodic structure, where we take explicitly into account the (arbitrary) dielectric permittivity of the material. First we describe the Rayleigh basis and outline the way to calculate the Rayleigh coefficients for an arbitrary grating. Then we present the derivation of the Casimir energy in terms of Rayleigh coefficients for two dielectric gratings or periodic dielectrics with the same period d. This derivation generalizes the idea of the argument principle method in the Casimir effect, which was used for plane geometries before (see e.g.[19, 20]). We insert the reflection matrices for two separated gratings into the argument principle explicitly and express the formula for the Casimir energy in terms of them. Our general derivation can be applied to various Casimir systems. We apply our formulation to the situation of two rectangular silicon gratings and show that our calculation yields deviation of the real force from the PFA prediction up to 24 percents. We also performed calculations corresponding to the measurement by Chan et al. allowing therefore a first quantitative theory-experiment comparison. The result taking into account the finite conductivity gives a smaller deviation of the exact force from the PFA prediction than the calculation for perfect mirrors.

2. General method

We consider two periodic dielectric gratings of arbitrary form separated by a vacuum slit. The special case of lamellar (or rectangular) gratings is depicted in Fig.1. The geometrical parameters of the grating are the corrugation depth a, the period d and the gap d_1. The gaps of both gratings are separated by a distance L. For simplicity we will suppose the space between the two gratings to be filled with vacuum with $\epsilon = \mu = 1$.

The physical problem is time and z invariant, so electric and magnetic fields can be written in the form:

$$E_i(x,y,z,t) = E_i(x,y)\exp(ik_z z - i\omega t), \quad (1)$$
$$H_i(x,y,z,t) = H_i(x,y)\exp(ik_z z - i\omega t). \quad (2)$$

Let us suppose for the moment that the upper grating is absent. We consider a generalized conical diffraction problem on the lower grating. The longitudinal components of the electromagnetic field outside the corrugated dielectric region ($y > a$) may be written by making use of a generalization of the Rayleigh expansion for an incident monochromatic wave:

$$E_z(x,y) = I_p^{(e)}\exp(i\alpha_p x - i\beta_p^{(1)}y) + \sum_{n=-\infty}^{+\infty} R_{np}^{(e)}\exp(i\alpha_n x + i\beta_n^{(1)}y), \quad (3)$$
$$H_z(x,y) = I_p^{(h)}\exp(i\alpha_p x - i\beta_p^{(1)}y) + \sum_{n=-\infty}^{+\infty} R_{np}^{(h)}\exp(i\alpha_n x + i\beta_n^{(1)}y), \quad (4)$$
$$\alpha_p = k_x + 2\pi p/d, \quad \beta_p^{(1)2} = \omega^2 - k_z^2 - \alpha_p^2, \quad (5)$$
$$\alpha_n = k_x + 2\pi n/d, \quad \beta_n^{(1)2} = \omega^2 - k_z^2 - \alpha_n^2 \quad (6)$$

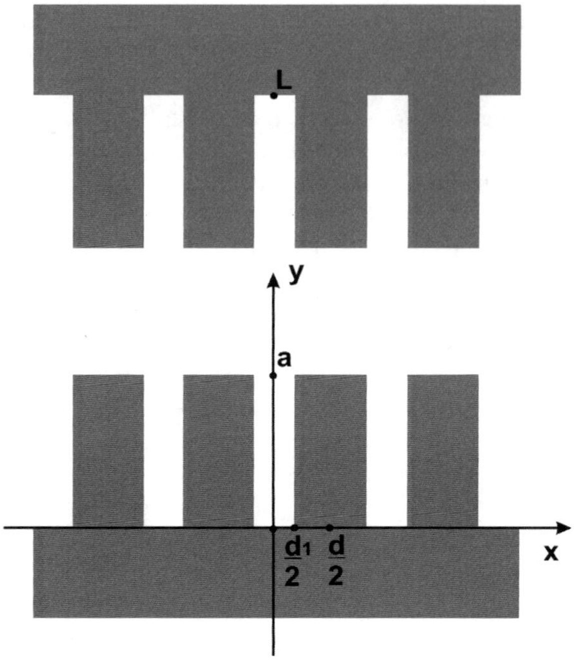

Figure 1. Rectangular gratings geometry.

with an integer p, the sums are performed over all integers n. All other field components can be expressed via the longitudinal components E_z, H_z. This solution is valid outside any periodic one-dimensional structure. Note that the expansions should satisfy the quasi-periodicity conditions:

$$E_z(x+d,y) = e^{ik_x d} E_z(x,y), \tag{7}$$
$$H_z(x+d,y) = e^{ik_x d} H_z(x,y). \tag{8}$$

We now have to determine the coefficients $R_{np}^{(e)}, R_{np}^{(h)}$ for a specific periodic geometry profile. For this purpose it is convenient to rewrite the Maxwell equations inside the corrugation region $0 < y < a$ in the form of first order differential equations, $\frac{\partial A}{\partial y} = MA$, where M is a square matrix of dimension $8N+4$, $A^T = (E_z, E_x, H_z, H_x)$ and $2N+1$ is the number of Rayleigh coefficients considered in every Rayleigh expansion. For a rectangular dielectric grating the matrix M is a constant matrix. At $y = 0$ the solution has to satisfy the expansions

$$E_z(x,y) = \sum_{n=-\infty}^{+\infty} T_{np}^{(e)} \exp(i\alpha_n x - i\beta_n^{(2)} y), \tag{9}$$
$$H_z(x,y) = \sum_{n=-\infty}^{+\infty} T_{np}^{(h)} \exp(i\alpha_n x - i\beta_n^{(2)} y), \tag{10}$$
$$\beta_n^{(2)2} = \epsilon\mu\omega^2 - k_z^2 - \alpha_n^2, \tag{11}$$

which are valid for $y \leq 0$. We then determine the unknown Rayleigh coefficients by matching the solution of equations $\frac{\partial A}{\partial y} = MA$ inside the corrugation region with Rayleigh expansions (3),(4) at $y = a$ and expansions (9),(10) at $y = 0$. Everywhere in the calculations we assumed $\mu = 1$.

The fields E_z and H_z are not decoupled for $k_z \neq 0$. This is why the reflection matrix R_1 for a reflection from a lower grating can be defined as follows:

$$R_1(\omega) = \begin{pmatrix} R^{(e)}_{n_1 q_1}(I^{(e)}_p = \delta_{pq_1}, I^{(h)}_p = 0) & R^{(e)}_{n_2 q_2}(I^{(e)}_p = 0, I^{(h)}_p = \delta_{pq_2}) \\ R^{(h)}_{n_3 q_3}(I^{(e)}_p = \delta_{pq_3}, I^{(h)}_p = 0) & R^{(h)}_{n_4 q_4}(I^{(e)}_p = 0, I^{(h)}_p = \delta_{pq_4}) \end{pmatrix}. \tag{12}$$

Performing a change of variables $y = -y' + L$, $x = x' - s \quad (s < d)$ in (3), (4), it is possible to obtain the reflection matrix R_{2up} for the reflection of an upward wave from a grating with the same profile turned upside-down, displaced from the lower grating by $\Delta x = s$, $\Delta y = L$. Note that for the upper grating in Fig.1 the special case $s = 0$ is depicted.

Up to now we considered a diffraction problem on a single grating. In [21] the Casimir energy between two bodies, the diffraction properties of which can be described by a scattering matrix, has been derived in plane geometries on the basis of canonical quantization. Roughness corrections were derived on the basis of a scattering approach in [22, 23]. The path integral method was used to obtain multipole expansion of the Casimir energy between the two compact objects [24], exact results in spherical geometries [24, 25] were also derived.

We outline a novel derivation here, which can be applied to various Casimir systems. To obtain the Casimir energy we need to determine the eigenfrequencies of all stationary solutions of the generalized diffraction problem of subsequent diffraction of the electromagnetic field on two periodic dielectrics separated by a gap-gap distance L. These eigenfrequencies can be summed up by making use of an argument principle, which states:

$$\frac{1}{2\pi i} \oint \phi(\omega) \frac{d}{d\omega} \ln f(\omega) d\omega = \sum \phi(\omega_0) - \sum \phi(\omega_\infty), \tag{13}$$

where ω_0 are zeroes and ω_∞ are poles of the function $f(\omega)$ inside the contour of integration, degenerate eigenvalues are summed over according to their multiplicities (see e.g. [19, 20]). For the Casimir energy we have $\phi(\omega) = \hbar\omega/2$. The equation for eigenfrequencies of the corresponding problem of classical electrodynamics is $f(\omega) = 0$.

Consider first the plane-plane geometry when two dielectric parallel slabs (slab 1: $y < 0$, slab 2: $y > L$) are separated by a vacuum slit ($0 < y < L$). In this case TE and TM modes do not couple. The equation for TE eigenfrequencies is:

$$f(\omega) = 1 - r_{1TE}(\omega) r_{2TEup}(\omega) = 0. \tag{14}$$

Here $r_{1TE}(\omega)$ is the reflection coefficient of a downward plane wave which reflects on a dielectric surface of slab 1 at $y = 0$, while $r_{2TEup}(\omega)$ is the reflection coefficient of an upward plane wave which reflects on a dielectric surface of slab 2 at $y = L$. One can deduce from Maxwell equations that $r_{2TEup}(\omega) = r_{2TE}(\omega)\exp(2ik_yL)$ ($r_{2TE}(\omega)$ is a reflection coefficient of a downward TE plane wave which reflects on a dielectric slab 2 now located at $y < 0$). From (14) and the analogous equation for TM modes one immediately obtains the Lifshitz formula [11] by making use of the argument principle (13).

For two gratings or periodic dielectrics separated by a vacuum slit one has to consider a reflection of downward and upward waves from a unit cell $0 < k_x < 2\pi/d$. Due to the structure of the surface, TE and TM modes do not decouple anymore, but they are coupled by the diffraction process. The equation for normal modes states:

$$R_1(\omega_i) R_{2up}(\omega_i) \psi_i = \psi_i, \tag{15}$$

where ψ_i is an eigenvector describing the normal mode with a frequency ω_i. Instead of equation (14) one obtains from (15) the following condition for eigenfrequencies:

$$\det(I - R_1(\omega) R_{2up}(\omega)) = 0. \tag{16}$$

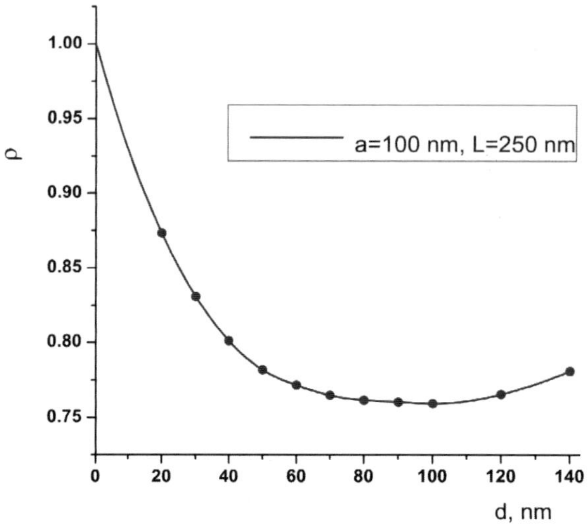

Figure 2. Casimir force normalized by its PFA value for two Si gratings with $a = 100$nm and $d_1 = \frac{d}{2}$ as a function of d at a fixed distance $L = 250$nm.

For every k_x, k_z the solution of (16) yields possible eigenfrequencies ω_i of the solutions of Maxwell equations that should be substituted into the definition of the Casimir energy $E = \sum_i \hbar\omega_i/2$. These solutions should tend to zero for $y \to \pm\infty$. The summation over the eigenfrequencies is performed by making use of the argument principle (13), which yields the Casimir energy of two parallel dielectrics on a "unit cell" of period d and unit length in z direction:

$$E = \frac{\hbar c\, d}{(2\pi)^3} \int_0^{+\infty} d\omega \int_{-\infty}^{+\infty} dk_z \int_0^{\frac{2\pi}{d}} dk_x \quad \ln\det\Big(I - R_1(i\omega)R_{2up}(i\omega)\Big), \tag{17}$$

c is the speed of light in vacuum. This is an exact expression valid for two arbitrary gratings or periodic dielectrics with the same period d separated by a vacuum slit. It can be applied to calculate the Casimir energy of any parallel gratings or periodic dielectrics made of a material described by a dielectric function.

Consider the particular case $s = 0$, which is depicted in Fig.1. From the derivation sketched above it follows that

$$R_{2up}(i\omega) = K(i\omega)R_2(i\omega)K(i\omega), \tag{18}$$

where $K(i\omega)$ is a diagonal $2(2N+1)$ matrix of the form:

$$K(i\omega) = \begin{pmatrix} G & 0 \\ 0 & G \end{pmatrix}, \tag{19}$$

with matrix elements $e^{-L\sqrt{\omega^2+k_z^2+(k_x+\frac{2\pi m}{d})^2}}$ $(m = -N \dots N)$ on a main diagonal of a matrix G. Note that in all Rayleigh expansions the Fourier basis is taken symmetrically around $m = 0$.

3. Rectangular gratings

We have numerically calculated the exact Casimir force for two rectangular gratings at zero temperature in the geometry of Fig.1 for silicon for different values of d, $d1 = d/2$ and $a = 100$ nm

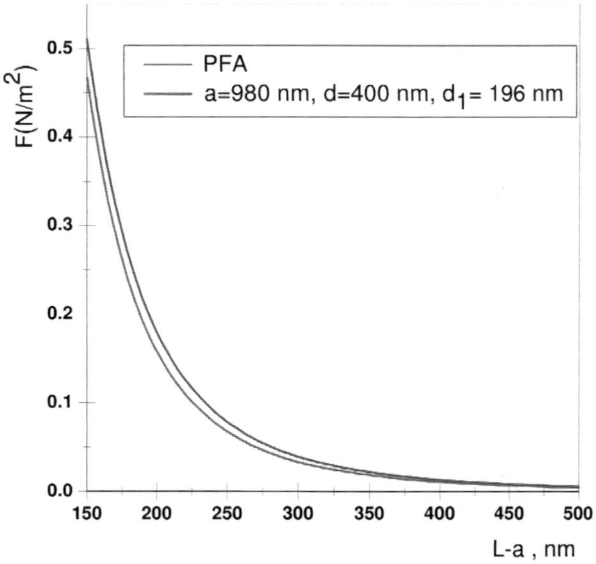

Figure 3. Casimir force between a Si grating and a gold plate as a function of distance.

by making use of the formulas (17, 18, 19) and a Drude-Lorentz model for a dielectric permittivity of an intrinsic silicon obtained in [14]. We compare our exact results of the Casimir force for different values of d to the PFA results. Calculated with the proximity force approximation, the Casimir force between the two gratings is just the geometric sum of two contributions corresponding to the Casimir force between two plates F_{PP} at distances L and $L-2a$, that is $F_{\rm PFA} = \frac{1}{2}(F_{PP}(L) + F_{PP}(L-2a))$. In particular it is independent of the corrugation period d. To assess quantitatively the validity of PFA, we plot the dimensionless quantity $\rho = \frac{F}{F_{\rm PFA}}$. The ratio is presented on Fig.2. Exact and PFA results differ for silicon by up to 24 percents for a corrugation period of 100nm and the PFA violation could thus be demonstrated experimentally. We recover the PFA result in two limiting cases, for a vanishing corrugation period and for very large corrugation periods. In between the exact result for the Casimir force is always *smaller* than the PFA prediction, in contrast to calculations for perfect conductors, where the resulting force is always larger than the PFA prediction.

We will now apply our method to the recent experiment by Chan et al.[17], who measured the Casimir force gradient between a silicon grating with nanostructured trenches and a gold sphere of radius $R = 50\mu$m. The force gradient F'_{PS} between a sphere of radius R and a plate can be expressed via the force F_{PP} in the plane-plane configuration as $F'_{PS} = 2\pi R F_{PP}$. This is why we show in figure 3 the zero temperature result for the absolute force values evaluated for a grating with the experimental parameters $a = 980$nm, $d = 400$nm, $d_1 = 196$nm placed in front of a gold plate (we used a plasma model with a plasma frequency $\omega_p = 9eV$ for gold [14] and a Drude-Lorentz model for intrinsic silicon [14]).

From our calculation we obtain a force $F_{PP} = 0.51\text{N/m}^2$ for a plate separation of 150nm. With the experimental parameters this leads to a prediction for the Casimir force gradient of $F' = 160.8, 56.4, 24.6$ pN/μm at respectively $L - a = 150, 200, 250$ nm. The absolute values of the force are thus in good agreement with the measured values depicted in Fig.3c of [17].

Now we present ratios of our results for the force to the predictions of PFA for two different gratings. Figure 4 shows ρ as a function of $L-a$ for two gratings corresponding to the experiment with $a = 980$nm, $d = 400$nm, $d_1 = 196$nm (green line) and $a = 1070$nm, $d = 1000$nm,

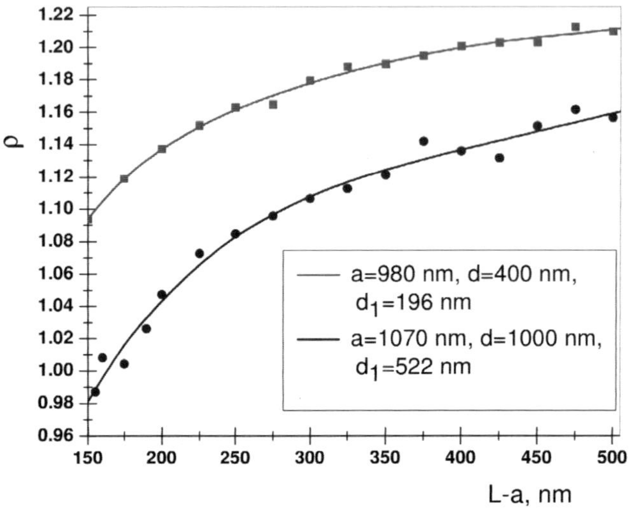

Figure 4. Casimir force normalized by the PFA value between a Si grating and a gold plate as a function of distance for two different gratings. Solid curves are calculated by making use of the least square method from the theoretical points on the figure.

$d_1 = 522$nm (blue line). One can compare our results with the experimental points and the fit shown in Figure 3d of [17], there is a reasonable agreement between the theoretical results and experimental data. Possible improvements in future may include calculations with more detailed models of dielectric permittivities of silicon and gold performed at finite temperature.

The fact that the perfect conductor model fails might be due to the influence of surface plasmons, as the grating affects their dispersion relation. Surface plasmons contribute essentially and at all distances to the Casimir force [2, 3, 4, 5], the Casimir force thus has to change considerably when structured surfaces are considered. These changes are not visible in a perfect conductor model which ignores the existence of surface plasmons.

Acknowledgments

Valery Marachevsky is grateful to professor Victor Dodonov and all organizers of the conference "60 years of the Casimir effect" in Brasilia, June 23-27, 2008, for an excellent organization of the conference and partial support. V.M. is grateful to colleagues for interesting and valuable discussions and comments during the conference. The authors would like to thank Serge Reynaud for helpful discussions. This work was supported by the French National Research Agency (ANR) through grant n ANR-06-NANO-062 - MONACO project. V.M. acknowledges financial support from French-Brazilian collaboration CAPES-COFECUB. V.M. also acknowledges financial support from grants RNP 2.1.1.1112, RFBR $07-01-00692-a$, SS.5538.2006.2.

References

[1] Casimir H B G 1948 *Proc. K. Ned. Akad. Wet.* **51** 793
[2] Barton G 1979 *Rep. Prog. Phys.* **42** 65
[3] Intravaia F and Lambrecht A 2005 *Phys. Rev. Lett.* **94** 110404
[4] Henkel C, Joulain K, Mulet J Ph and Greffet J J 2004 *Phys. Rev. A* **69** 023808

[5] Bordag M 2005 *J. Phys. A* **39** 6173
[6] Henkel C and Joulain K 2005 *EPL* **9**(6) 929
[7] Leonhardt U and Philbin T G 2007 *New J. Phys.* **9** 254
[8] Pirozhenko I G and Lambrecht A 2008 *J. Phys. A: Math.Theor.* **41** 164015
[9] Rosa F S, Dalvit D A and Milonni P W 2008 *Phys. Rev. Lett.* **100** 183602
[10] Chen F, Klimchitskaya G L, Mostepanenko V M and Mohideen U 2007 *Phys. Rev. B* **76** 035338
[11] Lifshitz E M 1956 *Soviet Phys. JETP* **2** 73
[12] Lambrecht A and Reynaud S 2000 *Eur. Phys. J. D* **8** 309
[13] Bezerra V B, Klimchitskaya G L and Mostepanenko V M 2000 *Phys. Rev. A* **62** 014102
[14] Pirozhenko I and Lambrecht A 2008 *Phys. Rev. A* **77** 013811
[15] Marachevsky V N 2008 *J.Phys.A: Math.Theor.* **41** 164007 ; Marachevsky V N 2007 *Phys.Rev.D* **75** 085019
[16] Roy A and Mohideen U 1999 *Phys.Rev.Lett.* **82** 4380
[17] Chan H B, Bao Y, Zou J, Cirelli R A, Klemens F, Mansfield W M and Pai C S 2008 *Phys.Rev.Lett.* **101** 030401
[18] Büscher R and Emig T 2004 *Phys. Rev. A* **69** 062101
[19] Schram K 1973 *Phys.Lett.A* **43** 282
[20] Barash Yu S and Ginzburg V L 1975 *Sov. Phys. Usp.* **18** 305
[21] Genet C, Lambrecht A and Reynaud S 2003 *Phys. Rev. A* **67** 043811
[22] Maia Neto P A, Lambrecht A and Reynaud S 2005 *Phys. Rev. A* **72** 012115
[23] Lambrecht A, Maia Neto P A and Reynaud S 2006 *New J. Phys.* **8** 243
[24] Emig T, Graham N, Jaffe R L and Kardar M 2007 *Phys. Rev. Lett.* **99** 170403
[25] Maia Neto P A, Lambrecht A and Reynaud S 2008 *Phys.Rev.A* **78** 012115

On the accuracy of the PFA: analogies between Casimir and electrostatic forces

Francisco D Mazzitelli, Fernando C Lombardo
Departamento de Física *Juan José Giambiagi*, Facultad de Ciencias Exactas y Naturales, UBA; Ciudad Universitaria, Pabellón I, 1428 Buenos Aires, Argentina

Paula I Villar
Departamento de Física *Juan José Giambiagi*, Facultad de Ciencias Exactas y Naturales, UBA; Ciudad Universitaria, Pabellón I, 1428 Buenos Aires, Argentina and Computer Applications on Science and Engineering Department, Barcelona Supercomputing Center (BSC), 29, Jordi Girona 08034 Barcelona, Spain

Abstract. We present an overview of the validity of the Proximity Force Approximation (PFA) in the calculation of Casimir forces between perfect conductors for different geometries, with particular emphasis for the configuration of a cylinder in front of a plane. In all cases we compare the exact numerical results with those of PFA, and with asymptotic expansions that include the next to leading order corrections. We also discuss the similarities and differences between the results for Casimir and electrostatic forces.

1. Introduction

The experimental and theoretical activity in the analysis of the Casimir effect is, nowadays, extremely intense. After 60 years, there are several high precision experiments and theoretical calculations for a variety of geometries. In the last years, there has been a remarkable progress in this field. On the experimental side, the new generation of experiments started about ten years ago [1]. The precision achieved, much larger than that of the first generation of experiments [2], triggered a lot of theoretical activity. While there were exact calculations for single cylindrical [3] and spherical [4], perfectly conducting shells, the calculation of the interaction of two different bodies, beyond the original two parallel plates, started about eight years ago. Since then, various theoretical techniques have been developed in order to understand the geometric dependence of the Casimir force. These include the use of the argument theorem to perform explicitly the sum over modes [5, 6, 7, 8], semiclassical and optical approximations [9], methods based on functional integrals [10] and scattering theory [11]. Many of these approaches have a common root in the multiple scattering theory developed in the seventies [12] (see also [13] for an updated review and applications to semitransparent bodies), and the evolution in the computational power allowed a precise numerical evaluation that involves, in general, the computation of determinants of infinite matrices. There are also full numerical approaches, as the worldline numerics [14], that has been applied to scalar fields satisfying Dirichlet boundary conditions, or finite difference methods that evaluate the Casimir energy from the two point function of the electromagnetic field [15]. As a consequence of this theoretical activity, we now have exact results for a variety of geometries that involve perfectly conducting shells: cylinder and sphere in front of a plane [16, 17, 18], eccentric cylinders [7, 8], two spheres [17], surfaces with periodic corrugations [19], Casimir pistons [20], etc. Some of these methods also apply to the case of imperfect mirrors, that we will not consider here.

For more than fifty years, the interaction between different bodies was computed mainly using the so called proximity force approximation (PFA) [21]. This approximation, expected to be valid as long as the interacting surfaces are smooth and very close, uses the original Casimir expression for the energy per unit area for parallel plates separated by a distance d

$$E_{\rm pp}(d) = -\frac{\pi^2}{720 d^3}, \tag{1}$$

and approximates the interaction between two conducting surfaces that form a curved gap of variable width z by

$$E_{\rm PFA} = \int_\Sigma d\sigma\, E_{\rm pp}(z). \tag{2}$$

It is clear that this formula does not take into account the non-parallelism of the surfaces. Moreover, the result will depend on the particular surface Σ chosen to perform the integral. However, these corrections are expected to be small for low-curvature, very close surfaces.

Until the development of the theoretical methods described above , the accuracy of the PFA was not assesed, simply because PFA is an uncontrolled approximation, and there were no exact calculations to compare with. On general grounds, denoting by $\mathcal{L}$ a typical length associated to the curvature of one of the surfaces (assumed much smaller than the curvature of the second surface) and by d the minimum distance between surfaces, one expects

$$E_{12} = E_{\rm PFA}\left\{1 + \Gamma\frac{\mathcal{L}}{d} + O\left[\left(\frac{\mathcal{L}}{d}\right)^2\right]\right\}, \tag{3}$$

where Γ is a constant, whose numerical value fixes the accuracy of the PFA in each particular geometry (one can write similar expressions for geometries that involve two surfaces of similar curvature). As we will see, the situation is a bit more complex, since the corrections to PFA may contain non-analytic corrections as $\left(\frac{\mathcal{L}}{d}\right)^n \ln\left(\frac{\mathcal{L}}{d}\right)$.

In this paper, we will present an overview of the accuracy of the PFA for the case of perfectly conducting shells with different geometries: concentric cylinders (Section 2), concentric spheres (Section 3), a cylinder in front of a plane (Section 4), and a sphere in front of a plane (Section 5). In all cases, we will compare the exact numerical results with the PFA, and obtain the numerical value of the constant Γ, which fixes the magnitude of the next to leading order (NTLO) correction. Moreover, we will also present, for each geometry, analogous comparisons for the electrostatic energy. These are, of course, trivial textbook examples. However, we think that the computation of the electrostatic energy using PFA is an interesting pedagogical exercise that illustrates the accuracy of the approximation for a different interaction, based on the result for the electrostatic energy contained between two parallel plates at a potential difference V

$$U_{\rm pp} = \frac{\epsilon_0 A V^2}{2d}. \tag{4}$$

Moreover, as we will also point out in our final remarks (Section 6), analogies with classical electromagnetism may be useful to suggest and/or to understand new effects in Casimir physics.

Some of the results presented here have been previously obtained by the authors and collaborators (concentric cylinders [6], cylinder in front of a plane [22]). The exact formula for the Casimir energy in the concentric-spheres geometry has been derived in Ref.[5, 23]. However, a numerical analysis and a discussion of the relevant limiting situations (in particular the proximity limit) has not been considered before. Therefore, in Section 4 we describe with some detail the derivation of analytic results in the small and large distance limits, along with numerical computations. For the sake of completeness, we also describe briefly the results for the sphere-plane configuration obtained by other authors [17, 18] (Section 5).

2. Concentric Cylinders

Let us first consider two concentric cylinders of length L, with radii a and b, respectively (with $L \gg a, b$ to neglect border effects). The exact formula for the Casimir interaction energy is given by [5, 6]

$$E_{12}^{\rm cc} = \frac{L}{4\pi a^2}\int_0^\infty d\beta\, \beta \ln M^{\rm cc}(\beta), \tag{5}$$

where

$$M^{\rm cc}(\beta) = \prod_n \left[1 - \frac{I_n(\beta)K_n(\alpha\beta)}{I_n(\alpha\beta)K_n(\beta)}\right]\left[1 - \frac{I_n'(\beta)K_n'(\alpha\beta)}{I_n'(\alpha\beta)K_n'(\beta)}\right], \tag{6}$$

where $\alpha = b/a$. The first factor corresponds to Dirichlet (TM) modes and the second one to Neumann (TE) modes. The concentric-cylinders configuration is interesting from a theoretical point of view, since it can be used to test analytic and numerical methods. It also has potential implications for the physics of nanotubes [7, 24]. This result can also be derived as a particular case from the general formula for eccentric cylinders [7, 8].

The short distance limit $\alpha - 1 \ll 1$ has already been analyzed for this case [6], and involves the summation over all values of n, that can be performed after using the uniform expansion for Bessel functions (in the next section, we will present a similar calculation for concentric spheres). As expected, the resulting value is equal to the one obtained via the proximity approximation, namely

$$E_{\rm PFA}^{\rm cc} = -\frac{\pi^3 L}{360a^2}\frac{1}{(\alpha-1)^3}. \tag{7}$$

When obtaining the PFA for a given configuration, the result is in general ambiguous, since it depends on the choice of the surface Σ (Eq.(2)). Eq.(7) corresponds to the energy per unit area for parallel plates times the area of the *inner* cylinder. In this case one could also choose, for instance, the area of the *outer* cylinder, which results in an extra factor of α that modifies the NTLO correction. The intermediate choice of the *geometric mean* of the areas gives

$$E_{\rm PFA}^{\rm cc} = -\frac{\pi^3 L}{360a^2}\frac{\alpha^{1/2}}{(\alpha-1)^3}, \tag{8}$$

and reproduces the result that is obtained using a semiclassical approximation based on periodic orbit theory [6].

In the opposite limit ($\alpha \gg 1$), it can be shown that to leading order only the TM $n = 0$ mode contributes to the interaction energy, and that the energy decreases logarithmically with the ratio $\alpha = b/a$,

$$E_{12}^{\rm cc} \approx -\frac{1.26L}{8\pi b^2 \ln\alpha}. \tag{9}$$

It is worth to stress that, while for small values of α both TM and TE modes contribute with the same weight to the interaction energy, the TM modes dominate in the large α limit.

In previous works, we have evaluated the analytic corrections to the PFA given in Eq.(7). Due to the simplicity of this configuration, it is possible to obtain not only the next to leading order, but also the next to next to leading contribution [22, 25]. The Casimir energy, beyond the proximity approximation, can be written as [22, 25]

$$E_{12}^{\rm cc} \approx -\frac{\pi^3 L}{360a^2(\alpha-1)^3}\left\{1+\frac{1}{2}(\alpha-1)-(\frac{1}{10}+\frac{2}{\pi^2})(\alpha-1)^2+...\right\}. \tag{10}$$

In the expression above, the first term inside the parenthesis corresponds to the proximity approximation contribution in Eq.(7), while the second and third terms are the first and second order corrections respectively. It is important to stress here that both TM and TE modes contribute with the same weight to the energy up to the next to leading order, but it is not the case in the second order correction [22, 25]. It is also remarkable that the PFA based on the geometric mean of the areas given in Eq.(8) reproduces the exact result not only to leading order but also to the NTLO. In Refs.[22, 25] we have shown that PFA can be used as a useful tool in order to improve the numerical evaluation at very small distances, and we have used this improvement in order to check numerically the non linear correction to PFA described in Eq.(10).

Let us now consider the electrostatic analogue for this configuration. It is trivial to evaluate the exact expression for the electrostatic interaction energy, which can be written as

$$U_{12}^{\rm cc} = \frac{\pi\epsilon_0 V^2}{\ln\alpha}, \tag{11}$$

where V is the difference between the electrostatic potential of the inner and outer cylinders.

The proximity approximation for the electrostatic interaction energy can be computed from the result of two parallel plates Eq.(4), and it is given by

$$U_{\rm PFA}^{\rm cc} \approx \frac{\pi\epsilon_0 V^2}{\alpha-1}, \tag{12}$$

where we have used the area of the inner cylinder. Taking the ratio between the exact and PFA results , it is possible to read the next to leading correction, which is given by

$$\frac{U_{12}^{\rm cc}}{U_{\rm PFA}^{\rm cc}} \approx 1+\frac{1}{2}(\alpha-1). \tag{13}$$

Remarkably, the NTLO correction has the same numerical factor both for the electrostatic interaction energy and for the Casimir interaction energy shown in Eq.(10). Related to this, the calculation of $U_{\rm PFA}^{\rm cc}$ using the geometric mean of the areas also reproduces the exact result $U_{12}^{\rm cc}$ including the NTLO.

There is an additional analogy between the calculations of the electrostatic and Casimir energies: in the large distance limit $\alpha \gg 1$, both energies vanish only logarithmically as the radius of the inner cylinder tends to zero (see Eqs.(9) and (11)).

3. Concentric spheres

Let us now consider two concentric spherical shells of radii a and b respectively, with $\alpha = b/a > 1$. The Casimir interaction energy can be computed using a procedure similar to that of the concentric cylinders. The exact energy is given by [5, 23]

$$E_{12}^{\rm cs} = \frac{1}{\pi a}\sum_{l\geq 1}\nu\int_0^\infty dy\,\ln[(1-F_\nu^{\rm TE})(1-F_\nu^{\rm TM})]\,, \tag{14}$$

where

$$F_\nu^{\rm TE} = \frac{I_\nu(y)K_\nu(\alpha y)}{I_\nu(\alpha y)K_\nu(y)}\,, \tag{15}$$

$$F_\nu^{\rm TM} = \frac{(I_\nu(y) + 2yI'_\nu(y))(K_\nu(\alpha y) + 2\alpha y K'_\nu(\alpha y))}{(I_\nu(\alpha y) + 2\alpha y I'_\nu(\alpha y))(K_\nu(y) + 2yK'_\nu(y))}\ , \tag{16}$$

and $\nu = l + 1/2$.

As far as we know, this energy has not been studied in detail before, so we analyze the opposite limits $\alpha \to 1$ and $\alpha \to \infty$. In order to obtain an analytic expression in the proximity limit $\alpha \to 1$, it is useful to perform the change of variables $y \to \nu y$ in the integral appearing in Eq.(14), so we can use the uniform expansion for the Bessel functions. For example we have

$$\frac{K_\nu(\nu\alpha y)}{K_\nu(\nu y)} = \frac{\sqrt{1+y^2}}{\sqrt{1+\alpha^2y^2}}\frac{(1-\frac{u(t_\alpha)}{\nu})}{(1-\frac{u(t_1)}{\nu})}e^{\nu[\eta(\alpha y)-\eta(y)]}, \tag{17}$$

where

$$\eta(y) = \sqrt{1+y^2} + \ln\frac{y}{1+\sqrt{1+y^2}}\ ;\ u(t) = \frac{3t-5t^3}{24}\ ;\ t_\alpha = \frac{1}{\sqrt{1+\alpha^2y^2}}, \tag{18}$$

and similar expressions for the functions I_ν, I'_ν, and K'_ν. Inserting these asymptotic expansions in Eqs.(15) and (16), one can show that

$$F_\nu^{\rm TE} \simeq F_\nu^{\rm TM} \simeq e^{-2\nu\Delta\eta(y)}(1 + O(\frac{\alpha-1}{\nu})), \tag{19}$$

where

$$\Delta\eta(y) = \eta(\alpha y) - \eta(y) \simeq (\alpha-1)\sqrt{1+y^2} - \frac{(\alpha-1)^2}{2\sqrt{1+y^2}}\ . \tag{20}$$

The term proportional to $(\alpha-1)/\nu$ will not contribute to the leading and NTLO, so we will neglect it in what follows.

Using these expressions, we can write the interaction energy as

$$E_{12}^{\rm cs} \simeq -\frac{2}{\pi a}\int_0^\infty dy \sum_{k\geq 1}\frac{1}{k}\sum_{l\geq 1}\nu^2 e^{-2\nu k\Delta\eta(y)}. \tag{21}$$

The sum over l can be easily computed and gives

$$\begin{aligned}\textstyle\sum_{l\geq 1}\nu^2 e^{-2\nu k\Delta\eta(y)} &= \frac{1}{4k^3\Delta\eta^3} + O(\Delta\eta)\\ &= \frac{1}{4k^3(\alpha-1)^3(1+y^2)^{3/2}}\left(1 + \frac{3(\alpha-1)}{1+y^2} + O((\alpha-1)^2)\right).\end{aligned} \tag{22}$$

Inserting Eq.(22) into Eq.(21), computing first the sum over k and then the remaining integral we finally obtain

$$E_{12}^{\rm cs} = E_{\rm PFA}^{\rm cs}\left\{1 + (\alpha-1) + O((\alpha-1)^2)\right\}. \tag{23}$$

Here

$$E_{\rm PFA}^{\rm cs} = -\frac{\pi^2}{720(b-a)^3}4\pi a^2 = -\frac{\pi^3}{180a(\alpha-1)^3}, \tag{24}$$

is the Casimir energy computed with the PFA using the area of the inner surface. We have confirmed the analytic approximation given in Eq.(23) through a numerical evaluation of the exact energy given Eq.(14). The results are shown in Fig. 1.

There are some interesting properties, similar to those of the previous section, that are worth noticing. On the one hand, TE and TM modes give the same contribution to both the leading

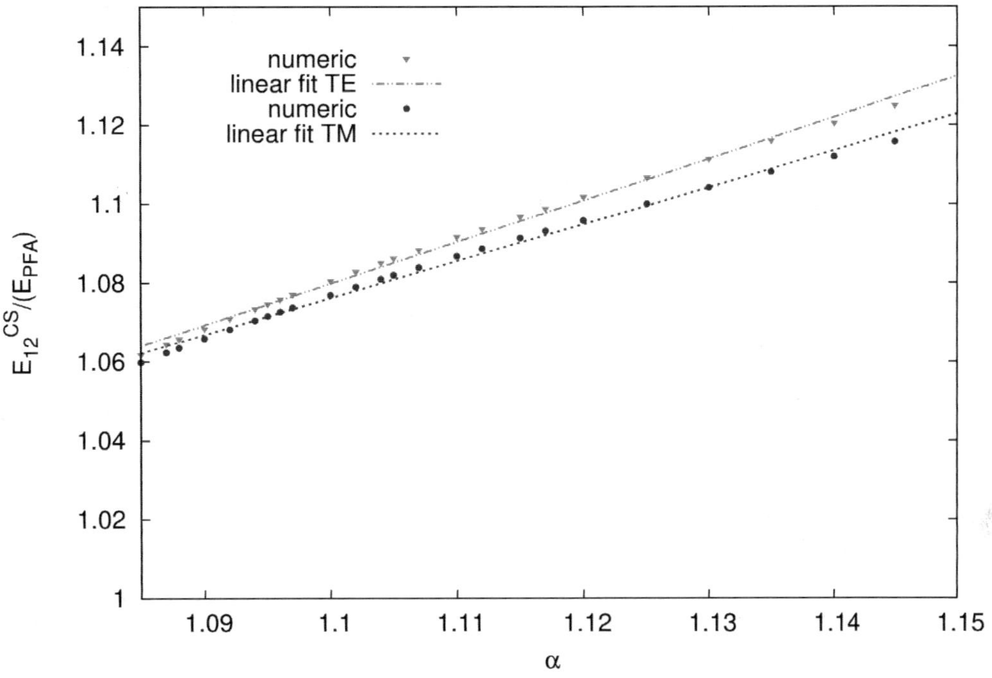

Figure 1. Numerical evaluation of the Casimir interaction energy for the configuration of concentric spheres, near the proximity limit. A simple fit $f(x) = a + bx$ of the numerical data gives, for the TM-modes $a = 0.97$, $b = 1.09$, and $a = 0.98$, $b = 0.97$ for TE-modes

and next to leading orders. On the other hand, if the PFA approximation is computed with the geometric mean area of the inner and outer spheres, the resulting expression

$$E^{\rm cs}_{\rm PFA} = -\frac{\pi^3\alpha}{180a(\alpha-1)^3} = -\frac{\pi^3}{180a(\alpha-1)^3}\left\{1+(\alpha-1)\right\}, \qquad (25)$$

reproduces not only the leading term of the exact interaction energy but also the NTLO.

One can also study the opposite limit, in which $a \ll b$. In this case, the Casimir interaction energy is dominated by the $l = 1$ modes. Keeping only this contribution in the exact expression Eq.(14), and after the change of variables $\alpha y = x$ we obtain

$$E^{\rm cs,TE}_{12} \approx \frac{3}{2\pi a\alpha}\int_0^\infty dx\,\ln\left(1-\frac{K_{3/2}(x)I_{3/2}(x/\alpha)}{I_{3/2}(x)K_{3/2}(x\alpha)}\right). \qquad (26)$$

Expanding the logarithm up to the leading order in $1/\alpha$ we get

$$E^{\rm cs,TE}_{12} \approx -\frac{1}{\pi^2 a\alpha^4}\int_0^\infty dx\,x^3\frac{K_{3/2}(x)}{I_{3/2}(x)} \approx -\frac{0.745}{a\alpha^4}. \qquad (27)$$

A similar analysis can be carried out for the TM modes. The result is

$$E^{\rm cs,TM}_{12} \approx -\frac{1.011}{a\alpha^4}. \qquad (28)$$

It is interesting to note that, unlike the case of concentric cylinders, both TE and TM modes contribute with the same order of magnitude to the Casimir energy in the large α limit. Moreover

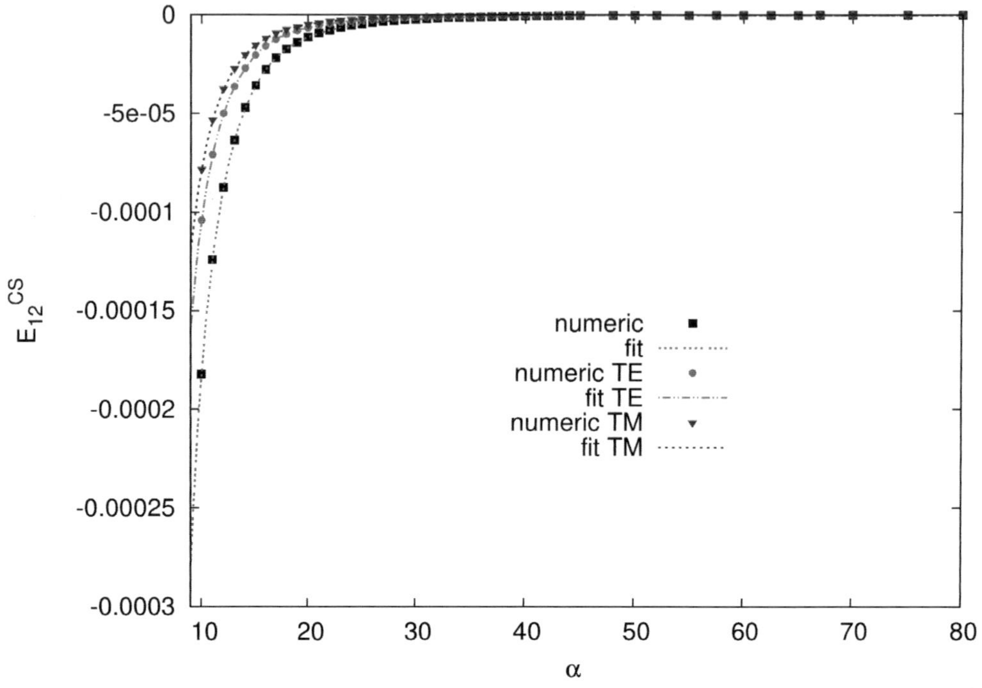

Figure 2. Large α behaviour of the Casimir interaction energy for concentric spheres. A simple fit of the form $f(x) = a/x^b$ gives $a = -1.05$, $b = -4.01$ for TE modes, and $a = -0.82$, $b = -4.02$ for TM modes.

the interaction energy vanishes as a^3 as the radius of the sphere tends to zero. We have checked these analytic results with numerical evaluations of the exact formula, as shown in Fig.2. A fit of the form $f(x) = a/x^b$ gives $a = -1.05$, $b = -4.01$ for TM modes, and $a = -0.82$, $b = -4.02$ for TE modes. Moreover, performing a fit with $g(x) = a/x^4$, we obtain $a = -1.03$ and $a = -0.78$ for TM and TE modes respectively.

Let us now compare the PFA in Casimir physics with the textbook electrostatic example. The electrostatic energy contained between the spherical shells is given by

$$U_{12}^{\rm cs} = 2\pi\epsilon_0 V^2 \frac{ba}{b-a}, \tag{29}$$

where V is the potential difference. A trivial application of the PFA, based on the inner sphere, gives

$$U_{\rm PFA}^{\rm cs} = 2\pi\epsilon_0 V^2 \frac{a^2}{b-a}, \tag{30}$$

so

$$U_{12}^{\rm cs} = U_{\rm PFA}^{\rm cs}\frac{b}{a} = U_{\rm PFA}^{\rm cs}\left\{1 + (\alpha - 1)\right\}. \tag{31}$$

We see that, as for the case of concentric cylinders, the next to leading order correction to PFA has the same numerical coefficient in electrostatic and Casimir energies. Moreover, in this case, the choice of the geometric mean area gives the exact result for the electrostatic energy. As we will see in the next sections, these are peculiarities of the geometries considered so far.

Journal of Physics: Conference Series **161** (2009) 012015 doi:10.1088/1742-6596/161/1/012015

4. A cylinder in front of a plane

We consider now a perfectly conducting cylinder of length L and radius a (with $L \gg a$ to neglect border effects). The cylinder is parallel to a perfectly conducting planar surface of area $A \gg a^2$, and the minimum distance between the two surfaces is denoted by d. This configuration is of experimental interest: being intermediate between the sphere-plane and the plane-plane geometries, it can shed some light on the longstanding controversy about thermal corrections to the Casimir force. Keeping the two plates parallel has proved very difficult, while the sphere and plate configuration avoids this problem, the force is not extensive. In the case of the cylinder-plane configuration, it is easier to hold the cylinder parallel and the force results extensive in its length. There is an ongoing experiment to measure the Casimir force for this configuration [26].

The Casimir energy for this configuration was first evaluated in the PFA in Ref.[27]. The exact formula has been derived in Refs.[16, 28], and has the same structure than Eq.(5), where

$$\begin{aligned} E_{12}^{\rm cp} &= \frac{L}{4\pi a^2}\int_0^\infty d\beta\ \beta\ \left[\ln(M^{\rm TE}(\beta)) + \ln(M^{\rm TM}(\beta))\right] \\ &= E_{12}^{\rm TE} + E_{12}^{\rm TM}, \end{aligned} \tag{32}$$

where $M^{\rm TM}(\beta) = \det[\delta_{np} - A_{n,p}^{\rm TM,CP}]$ and $M^{\rm TE}(\beta) = \det[\delta_{np} - A_{n,p}^{\rm TE,CP}]$. Here β is a dimensionless integration variable and n, p are arbitrary integers. The matrix elements are given by [16, 28]

$$A_{n,p}^{\rm TE,CP} = -\frac{I_n'(\beta)}{K_n'(\beta)} K_{n+p}(2\beta H/a), \tag{33}$$

and

$$A_{n,p}^{\rm TM,CP} = \frac{I_n(\beta)}{K_n(\beta)} K_{n+p}(2\beta H/a). \tag{34}$$

Note that the evaluation of the Casimir energy for this configuration involves the computation of the determinant of an infinite, non-diagonal matrix. Once more, the exact formula can be derived from the general formula for eccentric cylinders [8].

In the following we will numerically evaluate the cylinder-plane Casimir interaction energy for small distances, in order to discuss the leading correction to the PFA. In Figs.3, 4, and 5 we present the Casimir interaction energy for the cylinder-plane configuration. For the runs, we used a matrix of dimension (101,101) to reach the proximity limit ($d \to 0$). It must be mentioned that for smaller values of d, we need to increase the dimension of the A matrix and the integration range of β in Eq.(32). This fact becomes our major limitation to reach yet smaller values of d.

This problem has been considered from an analytical point of view in Ref.[28]. Using the uniform expansions for the Bessel functions appearing in the matrix elements $A_{n,p}^{\rm TE,CP}$ and $A_{n,p}^{\rm TE,CP}$, and after complex calculations, it can be shown that, in the proximity limit:

$$E_{12}^{\rm cp,TM} = -\frac{1}{2\pi}\sqrt{\frac{a}{d^5}}\frac{3\zeta(4)}{32\sqrt{2}}\left(1 + 0.1944\frac{d}{a} + ...\right), \tag{35}$$

$$E_{12}^{\rm cp,TE} = -\frac{1}{2\pi}\sqrt{\frac{a}{d^5}}\frac{3\zeta(4)}{32\sqrt{2}}\left(1 - 1.1565\frac{d}{a} + ...\right), \tag{36}$$

where we have written separately the contributions of TM and TE modes.

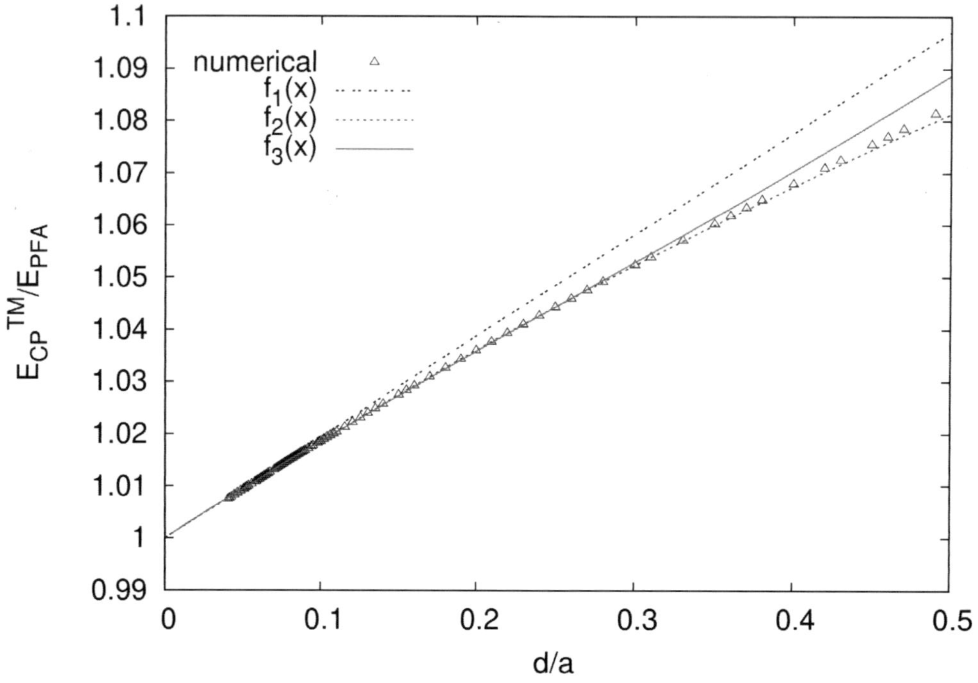

Figure 3. Numerical result for the TM modes for the cylinder-plane configuration, and the corresponding fits presented in Table 1. A simple linear fit $f(x) = a + bx$ of the numerical data in the interval $0.04 \leq d/a \leq 0.07$ gives $a = 0.9999$ and $b = 0.1900$. The theoretical values are $a = 1$ and $b = 0.1944$.

d/a	$f_1(x) = 1 + bx$	$f_2(x) = 1 + b * x + c * x^2$	$f_3(x) = 1 + b * x + c * x^2 * \log(x)$
$[0.04 : 0.15]$	$b = 0.1864$	$b = 0.1922, c = -0.0601$	$b = 0.1961, c = 0.0438$
$[0.04 : 0.20]$	$b = 0.1849$	$b = 0.1923, c = -0.0613$	$b = 0.1983, c = 0.0540$
$[0.04 : 0.25]$	$b = 0.1829$	$b = 0.1922, c = -0.0601$	$b = 0.2003, c = 0.0634$
$[0.04 : 0.30]$	$b = 0.1811$	$b = 0.1920, c = -0.0586$	$b = 0.2022, c = 0.0716$
$[0.04 : 0.35]$	$b = 0.1794$	$b = 0.1918, c = -0.0572$	$b = 0.2045, c = 0.0810$
$[0.04 : 0.40]$	$b = 0.1771$	$b = 0.1914, c = -0.0549$	$b = 0.2076, c = 0.0935$

Table 1. Different fits for the numerical results of Fig. 3 (TM modes). We fix $f_i(0) = 1$ since the numerical data agree this value with high precision.

We will discuss the first order corrections to PFA for TM and TE modes separately. In Fig.3, we show our numerical results for the TM modes. The fit of the numerical results depends of course on the interval chosen for d/a. There is an obvious compromise: on the one hand, as already mentioned, we cannot consider very small values for d/a because of numerical limitations. On the other hand, the expansion in powers of d/a are expected to be valid only for $d/a \ll 1$. In any case, as can be seen from Table 1, the different fits for the numerical results are stable, and confirm both the PFA to leading and next to leading orders. Indeed, the results are fully compatible with the analytic results given in Eq.(35), considering both linear and quadratic fits of the numerical results. Moreover, a simple linear fit in a smaller range of d/a gives $a = 0.9999$ and $b = 0.1900$ and already reproduces the analytical results [28] with high accuracy (see also numerical findings in [29]).

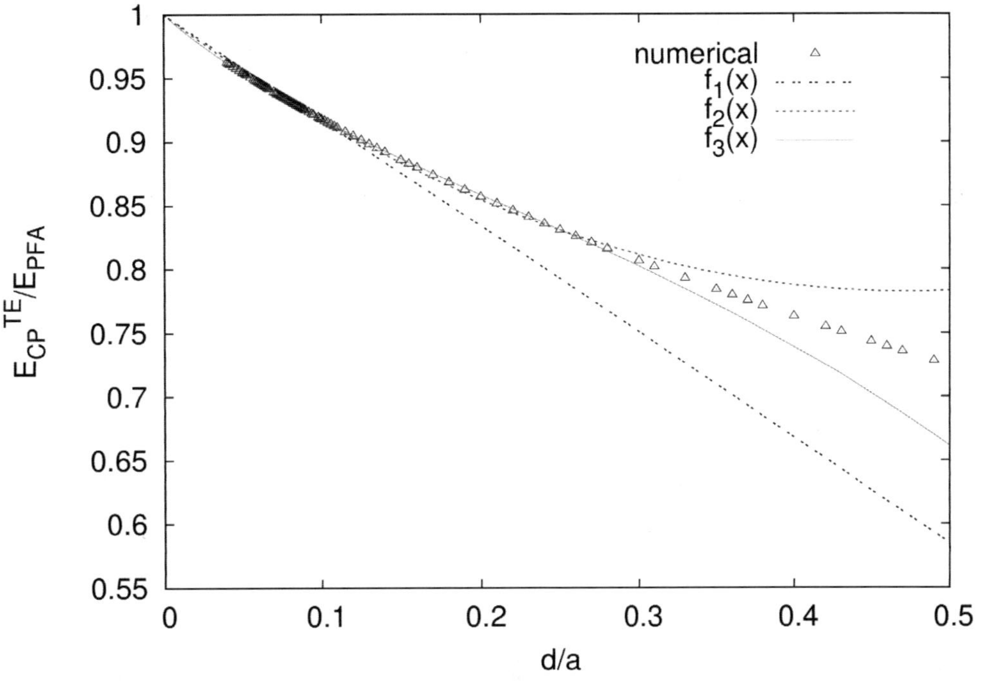

Figure 4. Numerical result for the TE modes for the cylinder-plane configuration, and the corresponding fits presented in Table 2. A simple linear fit $f(x) = a + bx$ of the numerical data in the interval $0.04 \leq d/a \leq 0.07$ gives $a = 0.9940$ and $b = -0.7808$. The theoretical values are $a = 1$ and $b = -1.1565$.

d/a	$f_1(x) = 1 + bx$	$f_2(x) = 1 + b * x + c * x^2$	$f_3(x) = 1 + b * x + c * x^2 * \log(x)$
$[0.04 : 0.15]$	$b = -0.8301$	$b = -0.9704, c = 1.4499$	$b = -1.0711, c = -1.0852$
$[0.04 : 0.20]$	$b = -0.8013$	$b = -0.9509, c = 1.2326$	$b = -1.0772, c = -1.1141$
$[0.04 : 0.25]$	$b = -0.7683$	$b = -0.9349, c = 1.0794$	$b = -1.0890, c = -1.1674$
$[0.04 : 0.30]$	$b = -0.7399$	$b = -0.9222, c = 0.9772$	$b = -1.1037, c = -1.2306$
$[0.04 : 0.35]$	$b = -0.7158$	$b = -0.9091, c = 0.8879$	$b = -1.1232, c = -1.3115$
$[0.04 : 0.40]$	$b = -0.6851$	$b = -0.8943, c = 0.7999$	$b = -1.1534, c = -1.4360$

Table 2. Different fits for the numerical results of Fig. 4 (TE modes). We fix $f_i(0) = 1$ since the numerical data agree this value with high precision.

In Fig.4, we show our results for the Neumann modes, and we include in Table 2 different fits of the numerical data. In this case, the value obtained for the linear correction to PFA depends strongly on the assumption about the next non trivial correction. This is not surprising: as we cannot consider extremely small values for d/a, the non linear corrections may have a non negligible contribution in the intervals chosen for the fits. For example, a simple linear fit gives $a = 0.994$ and $b = -0.7808$ which does not coincide with the result in Eq.(36). However, based on the discussion about the slower convergence of the Neumann corrections presented in Ref.[28], we have allowed the possibility of non linear corrections proportional to $(d/a)^2 \ln(d/a)$ in our fits. Remarkably, when this non linear corrections are taken into account, the coefficient of the linear correction gets closer to the analytic prediction in Eq.(36), that we reproduce with an error less than 7%. Note that, as can be seen in Fig.3, this is not the case for TM modes, since

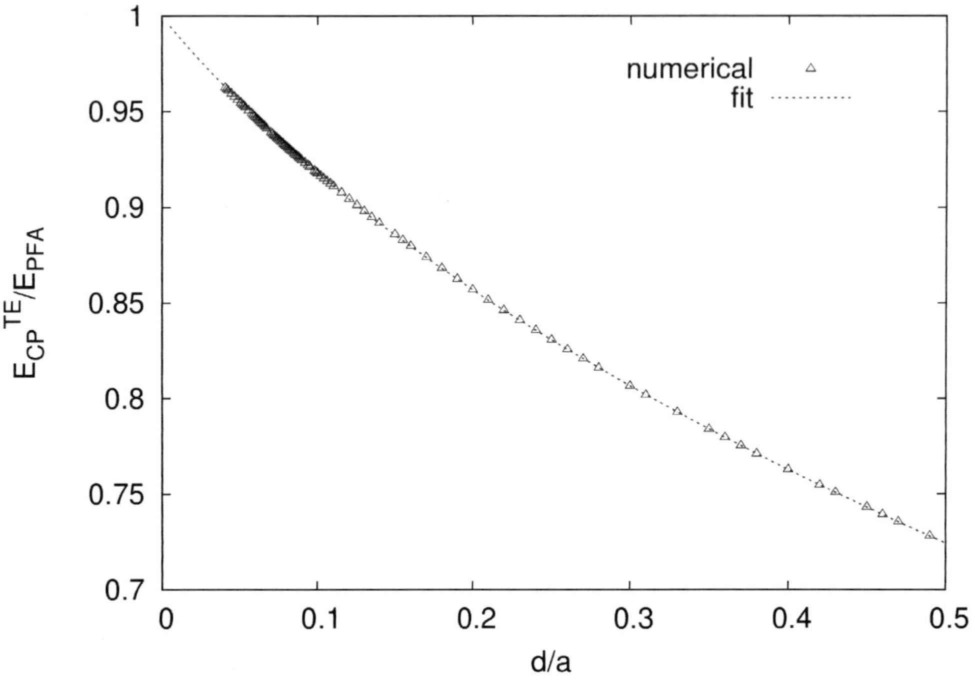

Figure 5. A numerical fit of the results for the TE modes including cubic corrections $f(x) = 1 + bx + cx^2 \log x + dx^3$. The coefficients are $b = -1.0478$, $c = -0.9485$, and $d = 0.6708$.

the best fit of the numerical data contains a quadratic term without a logarithm. In Fig.5 we show a fit of the numerical data for TM modes that includes a cubic correction $(d/a)^3$. With this additional term, the fit reproduces the numerical data up to $d/a = 0.5$.

To summarize our results, the fits of the numerical data clearly confirm the analytic prediction for the TM modes, and suggest that the next non trivial correction for the TE modes is not quadratic but proportional to $(d/a)^2 \ln(d/a)$.

As in the previous sections, we consider the electrostatic interaction. Let us assume that the conducting cylinder is kept at a fixed electrostatic potential V, while the planar surface is grounded. For this geometry, the exact electrostatic interaction energy given by

$$U_{12}^{\rm cp} = \frac{\pi L \epsilon_0 V^2}{\operatorname{arccosh}\left(1 + \frac{d}{a}\right)}. \tag{37}$$

In the limit of $d/a \ll 1$ it is simple to show that the electrostatic energy reduces to the PFA result

$$U_{\rm PFA}^{\rm cp} = \frac{L\pi\epsilon_0 V^2}{\sqrt{2}} \sqrt{\frac{a}{d}}. \tag{38}$$

As it was done before, we can compare the exact electrostatic energy (expanded in powers of d/a) with the PFA result, and extract from it the next to leading correction, i.e.,

$$\frac{U_{12}^{\rm cp}}{U_{\rm PFA}^{\rm cp}} = 1 + \frac{1}{12}\frac{d}{a}. \tag{39}$$

As in the previous examples, the exact electrostatic result shows a linear NTLO correction to PFA. However, the numerical value of the linear correction is different from that of Casimir energy.

Finally, we also point out that, in the large distance limit $d \gg a$, the electrostatic interaction becomes

$$U_{12}^{\rm cp} = \frac{\pi L \epsilon_0 V^2}{\ln(\frac{d}{a})}. \tag{40}$$

As for the Casimir energy [16], it vanishes logarithmically with as $a \to 0$.

5. A sphere in front of a plane

The sphere-plane geometry is, up to now, the most important geometry that have been used to measure precisely the Casimir forces. From the theoretical point of view, the evaluation of the Casimir energy in the electromagnetic case has been performed very recently in Refs. [17, 18], while the evaluation for scalar fields has been previously reported in Ref.[11]. See also [30] for asymptotic expansions in the scalar field case near the proximity limit.

For the sake of completeness, we quote here the results obtained in Refs.[17, 18] regarding the behaviour of the Casimir energy for this configuration. Denoting by a the radius of the sphere, and by d the minimum distance between the plane and the sphere, numerical fits in both references give

$$E_{12}^{\rm sp} \simeq E_{\rm PFA}^{\rm sp} \left\{1 - 1.4\frac{d}{a}\right\}. \tag{41}$$

The next to leading order correction is again linear, as in the previous cases. Both fits were performed by assuming that the next to NTLO is quadratic in d/a.

There is still no analytic prediction for the NTLO correction in the electromagnetic case. However, one can compare the results of the numerical calculations [17] and the asymptotic expansions in the scalar case [30]. Although the scalar results for TE and TM modes do not reproduce the electromagnetic result (this geometry does not allow this decomposition), there is an interesting similarity with the results described in the previous section. The theoretical asymptotic expansions, for scalar fields satisfying Dirichlet and Neumann boundary conditions read, respectively,

$$\begin{aligned} E_{12}^{\rm sp,D} &\simeq E_{\rm PFA}^{\rm sp} \left\{1 + \frac{1}{3}\frac{d}{a}\right\}, \\ E_{12}^{\rm sp,N} &\simeq E_{\rm PFA}^{\rm sp} \left\{1 + (\frac{1}{3} - \frac{10}{\pi^2})\frac{d}{a}\right\}. \end{aligned} \tag{42}$$

The numerical fits for the scalar case [17] give 0.33 and −2.43, for the Dirichlet and Neumann case, respectively.

While the agreement for Dirichlet modes is remarkable, for Neumann modes there is a strong discrepancy. So, based on the discussion for the cylinder-plane geometry, one can argue that also in this case the second order corrections could contain logarithmic factors.

The electrostatic problem can also be solved exactly, and it is relevant for the initial calibration in the measurements of the Casimir force. If the potential difference between the plane and the sphere is V, the electrostatic energy is given by [31]

$$U_{12}^{\rm sp} = 2\pi\epsilon_0 a V^2 \sinh\beta \sum_{n\geq 1} \frac{1}{\sinh(n\beta)}, \tag{43}$$

where $\cosh\beta = 1 + d/a$. In order to obtain an analytic expression in the limit $\beta \to 0$ we write

$$S \equiv \sum_{n\geq 1} \frac{1}{\sinh(n\beta)} = S - \int_1^\infty \frac{dn}{\sinh(n\beta)} + \frac{1}{\beta}\ln(\coth\beta)$$
$$= \frac{\gamma}{\beta} + \frac{1}{\beta}\ln(\coth\beta) + O(\beta), \tag{44}$$

where $\gamma = 0.5772$. Replacing this expression into Eq.(43) and expanding the result for small d/a we obtain

$$U_{12}^{\rm sp} = U_{\rm PFA}^{\rm sp}\left\{1 + \frac{1}{3}\frac{d}{a} + O\left(\frac{d/a}{\ln(d/a)}\right)\right\}, \tag{45}$$

where

$$U_{\rm PFA}^{\rm sp} = -\pi\epsilon_0 a V^2 \ln(2d/a) \; . \tag{46}$$

In Eq.(45) we omited an irrelevant constant term. It is interesting to remark that the next to NTLO correction in the electrostatic force is not quadratic but proportional to $d/(a\log(d/a))$.

6. Final remarks

We have presented a brief review of the calculations of the Casimir energy for different geometries involving perfect conductors, paying particular attention to the NTLO corrections to the PFA. In all cases considered, the first corrections to the PFA are linear, with a coefficient of order one. So, generically, the PFA results agree with the exact energies within 1% when $\mathcal{L}/d < 10^{-2}$. The situation for the next to NTLO is more complex. For concentric cylinders this correction is quadratic [22, 25], and it can be shown that this is also the case for concentric spheres. However, in the cylinder-plane configuration additional logarithmic factors could arise [22, 28]. This is probably also the case for a sphere in front of a plane.

We have compared validity of the PFA for the Casimir interaction energy with the same approximation in electrostatic examples, in each geometry considered. In all cases, the general result is also valid: the NTLO corrections to PFA are always linear. Moreover, for concentric spheres and cylinders, the next to leading order corrections to PFA have the same numerical coefficients for electrostatic and Casimir energies, and can be obtained from the PFA using a particular area, i.e. the geometric mean of the areas of both surfaces. This is certainly a property of this particular geometries, in which the distance between surfaces is constant and the normal to both surfaces are parallel at each point.

There is another property of the Casimir interaction energy that has its counterpart in electrostatics. In the case of concentric cylinders or cylinder-plane geometries, the Casimir energy vanishes only logarithmically as the radius $a \to 0$. This is also the case for the analogous electrostatic problems. Once more, this is a property of geometries involving cylinders, and the situation is different for geometries involving spheres, as we have shown in Section 3.

The analogies between the Casimir energy and the electrostatic energy could be useful to suggest and/or understand the behaviour of the vacuum forces in different situations. Let us consider, instead of perfect conductors, the case of surfaces that separate media with different electromagnetic properties. For example, consider three media described by different dielectric constants $\epsilon_1 > \epsilon_2 > \epsilon_3$, separated by flat surfaces. It is a simple exercise to show that, even if the interfaces have free electric charges of different sign, the interaction between them may be repulsive, due to the polarization of the media. This suggests that the same situation may happen for the vacuum fluctuations, and this is indeed the case, as can be easily shown using Lifshitz formula [32]. Similar electrostatic effects arise for all the geometries considered here. Therefore, based on this analogy, one can argue that repulsive Casimir forces can take place in all of these geometries, if the boundaries become interfaces between different media. This should be valid even beyond the obvious situation in which one uses the PFA starting from Lifshitz formula. There is a concrete example that has been recently analyzed, the repulsive interaction between eccentric cylinders [33]. According to the electrostatic analogy this property should be valid as long as the radii of the cylinders and the dielectric constants satisfy certain relations. It would be interesting to check if this is also the case for the Casimir interaction. Work on this issue is in progress.

Acknowledgments

We would like to thank V. Dodonov for his hospitality and organization of the meeting *"60 years of Casimir effect"*. We also thank Mariano Vazquez for useful computational help. This work was supported by UBA, Conicet, and ANPCyT Argentina.

References

[1] Lamoreaux S K, Phys. Rev. Lett. **78**, 5 (1997);
Mohideen U and Roy A, Phys. Rev. Lett. **81**, 4549 (1998);
Bressi G, Carugno G, Onofrio R and Ruoso G, Phys. Rev. Lett. **88**, 041804 (2002);
Harris B W, Chen F and Mohideen U, Phys. Rev. A **62**, 052109 (2000);
Ederth T, Phys. Rev. A **62**, 062104 (2000);
Chan H B, Aksyuk V A, Kleiman R N, Bishop D J and Capasso F, Science 291, 1941 (2001);
Chan H B, Aksyuk V A, Kleiman R N, Bishop D J and Capasso F, Phys. Rev. Lett. **87**, 211801 (2001);
Iannuzzi D, Gelfand I, Lisanti M and Capasso F, Proc. Nat. Ac. Sci. USA 101, 4019 (2004);
Decca R S, Lopez D, Fischbach E and Krause D E, Phys. Rev. Lett. **91**, 050402 (2003);
Decca R S et al., Phys. Rev. Lett. **94**, 240401 (2005); Decca R S et al., Annals of Physics **318**, 37 (2005).

[2] Sparnaay M J, *The historical background of the Casimir effect Physics in the Making: Essays on Developments in 20th Century Physics*: in Honour of H.B.G. Casimir on the occasion of his 80th Birthday ed A Sarlemijn and M J Sparnaay (Amsterdam: North-Holland) (1989).

[3] DeRaad L L, Jr and Milton K A, Ann. Phys. NY, **136**, 229 (1981).

[4] Boyer T H, Phys. Rev. **174**, 1764 (1968).

[5] Saharian A A, ICTP Report No. IC/2000/14, e-print hep-th/0002339.

[6] Mazzitelli F D, Sánchez M J, Scoccola N N and von Stecher J, Phys. Rev. A **67**, 013807 (2003).

[7] Dalvit D A R, Lombardo F C, Mazzitelli F D and Onofrio R, Phys. Rev. A **74**, 020101 (2006).

[8] Mazzitelli F D, Dalvit D A R and Lombardo F C, New Journal of Physics **8**, 240 (2006).

[9] Schaden M and Spruch L, Phys. Rev. A **58**, 935 (1998);Phys. Rev. Lett. **84**, 459 (2000).

[10] Golestanian R and Kardar M, Phys. Rev. A **58**, 1713 (1998);
Emig T, Hanke A and Kardar M, Phys. Rev. Lett. **87**, 269492 (2001);
Emig T and Buscher R, Nucl. Phys. B **696**, 468 (2004);
Emig T, Jaffe R L, Kardar M and Scardicchio A, Phys. Rev. Lett. 96 (2006) 080403;
Bordag M, Phys. Rev. D **73**, 025007 (2006).

[11] Bulgac A, Magierski P and Wirzba A, Phys. Rev. D 73, 025007 (2006);
Wirzba A, Bulgac A and Magierski P, J. Phys. A 39 (2006) 6815.

[12] Balian R and Duplantier B, Ann. Phys. NY, 112, 165 (1978);Ann. Phys. NY, 104, 300 (1977).

[13] Milton K and Wagner J, J. Phys. A: Math. Theor. 41 (2008) 155402.

[14] Gies H, Langfeld K and Moyaerts L, JHEP **06**, 018 (2003).

[15] Rodriguez A, Ibanescu M, Iannuzzi D, Capasso F, Joannopoulos J D and Johnson S G, Phys. Rev. Lett. **99**, 080401 (2007).

[16] Emig T, Jaffe R J, Kardar M and Scardicchio A, Phys. Rev. Lett. **96**, 080403 (2006).
See also Rahi S J, Rodriguez A W, Emig T, Jaffe R L, Johnson S G and Kardar M, Phys. Rev. A **77**, 030101(R) (2008);
Rahi S J, Emig T, Jaffe R L and Kardar M, Phys. Rev. A **78**, 012104 (2008).

[17] Emig T, J. Stat. Mech., P04007 (2008).

[18] Maia Neto P A, Lambrecht A and Reynaud S, Phys. Rev. A **78**, 012115 (2008).

[19] Emig T, Hanke A, Golestanian R and Kardar M, Phys. Rev. Lett. **87**, 260402 (2001).

[20] Cavalcanti R M, Phys. Rev. D **69**, 065015 (2004);
Hertzberg M P, Jaffe R L, Kardar M and Scardicchio A, Phys. Rev. D **76**, 045016 (2007);
Marachevsky V N, Phys. Rev. D **75**, 085019 (2007).

[21] Derjaguin B V and Abrikosova I I, Sov. Phys. JETP **3**, 819 (1957);
Derjaguin B V, Sci. Am. **203**, 47 (1960).

[22] Lombardo F C, Mazzitelli F D and Villar P I, Phys. Rev. D **78**, 085009 (2008).
[23] Saharian A A, Phys. Rev. D **63**, 125007 (2001).
[24] Blagov E V, Klimchitskaya G L and Mostepanenko V M, Phys. Rev. B **71**, 235401 (2005).
[25] Lombardo F C, Mazzitelli F D and Villar P I, J. Phys. A **41**, 164009 (2008).
[26] Brown-Hayes M, Dalvit D A R, Mazzitelli F D, Kim W J and Onofrio R, Phys. Rev. A **72**, 052102 (2005).
[27] Dalvit D A R, Lombardo F C, Mazzitelli F D and Onofrio R, Europhys. Lett. 68, 517 (2004).
[28] Bordag M, Phys. Rev. D **73**, 025007 (2006).
[29] Gies H and Klingmuller K, Phys. Rev. Lett. **96**, 220401 (2006).
[30] Bordag M and Nikolaev V, J. Phys. A: Math. Theor. **41**, 164002.
[31] Smythe W R, *Static and Dynamic Electricity*, McGraw-Hill, New York (1968).
[32] Lifshitz E M, Zh. Eksp. Teor. Fiz. **29**, 94 (1955) [Sov. Phys. JETP **2**, 73 (1956)].
[33] Rodriguez A W, Munday J N, Joannopoulos J D, Capasso F, Dalvit D A R and Johnson S G, Phys. Rev. Lett. **101**, 190404 (2008).

Journal of Physics: Conference Series **161** (2009) 012016 doi:10.1088/1742-6596/161/1/012016

Test of the Proximity Force Approximation

Bo E Sernelius and C E Román-Velázquez

Department of Physics, Chemistry, and Biology, Linköping University, SE-58183 Linköping, Sweden

E-mail: bos@ifm.liu.se

Abstract. We study the geometrical corrections to the simple Proximity Force Approximation (PFA) for the non-retarded Casimir force. We extend traditional PFA in two ways: We take the whole surfaces of the objects facing each other into account, not just the curvatures at the point of closest distance; we take the thickness of the coating of coated objects into account in the formalism. We present analytical and numerical results for a sphere above a substrate, for a spherical shell above a substrate, and for two interacting spheres. We compare the results to those from a multi-polar expansion method, a method based on a more solid foundation.

1. Introduction

J. D. van der Waals found empirically in 1873 that there is an attractive force between non-polar atoms. He studied the equation of state for real gases and found that there had to be a correction added to the pressure in the ideal-gas law. This indicated an attractive force between the gas atoms. It took a long time, until 1930, before there was an explanation to this force. London [1] gave the explanation in terms of fluctuations in the electron density within the atoms (fluctuating dipoles). In an alternative description [2] one may, instead of discussing the particles, focus on the electromagnetic fields. The force may be expressed as a result of changes in the zero-point energy of the electromagnetic normal modes of the system. There are modes associated with the atoms and modes associated with the vacuum. Casimir studied a more pure and idealized system consisting of two perfectly reflecting metal plates. In this geometry there are no modes associated with the plates themselves; there are only vacuum modes. The presence of the plates changes the vacuum modes and their zero-point energy. He published his findings in a classical paper [3] in 1948, the Casimir force was born, 60 years ago. Casimir's paper is one of the most important papers in the history of physics since it demonstrates that the boundary conditions of a system may change its zero-point energy and hence its properties. In the case of objects made from real materials both the modes associated with the objects and the vacuum modes contribute to the force [2]. The first type dominates in the van der Waals region which is for smaller separations. For larger separations between the objects the vacuum modes dominate and the result is the Casimir force. In the idealized case of perfect metals there is no van der Waals range. In 1997 Lamoreaux [4] performed the first modern high-precision measurements of the Casimir force. The accuracy was good enough to make the direct comparison with theory feasible. This spurred a burst of renewed interest in the Casimir effect.

On the theory side it may be complicated to calculate the force between objects of general shape. There are exceptions, though. The van der Waals and Casimir forces between half spaces is feasible to calculate. For finite objects at large distances one may use multipolar expansions

[5]. However, for small separations one needs to keep more and more terms in the expansion the smaller the distance; one reaches a limit when the method is no longer feasible to use.

In this work we study an approximation that comes handy in this situation. The Proximity Force Approximation (PFA). It was first used already in 1934 [6] in connection with coagulation of aerosols. It is a very powerful and widely used approximation for the interaction at short distances between two objects. Lamoreaux' experiment, discussed above, is interpreted within this approximation. It has also been used in the determination of surface roughness effects [7]. It is difficult to make a strict estimate of how good the approximation is since it is based on a very shaky foundation but it has gained a wide-spread acceptance in recent years.

The basic idea of the approximation is that the interaction potential between the objects is an average interaction energy between parallel planar interfaces,

$$V(z) = \int_S dS E_p(w), \tag{1}$$

where $E_p(w)$ is the interaction energy per unit area between planar interfaces a distance w apart. The variable z is the closest distance between the two objects. The surface S is not uniquely defined and the choice of S is in the general case not a trivial choice to make. In the examples we discuss here the choice is more obvious. For objects above a substrate we may choose S to be that part of a flat surface, parallel to the substrate, that is covered by the projection of the object; w is the distance at dS between the substrate and the object along the normal to S. In the case of two interacting objects we suggest that S is that part of the planar surface, perpendicular to the resulting force, where the projections of the two objects overlap; w is the distance at dS between the two objects along the normal to S.

Some attempts have been made to show its validity in different cases. However, the lack of a general and rigorous proof still remains, as well as estimates of the level of accuracy at all distances. In the present work we follow an alternative procedure. We extend the traditional PFA by taking the geometry into account in an optimal way. We produce both analytical and numerical results. The numerical results we compare to those from the multi-polar calculation method [5]. In this article we limit the results to those for a sphere above a substrate, for a spherical shell above a substrate, and for two interacting spheres. A more exhaustive study of the PFA results for a wide range of geometries and configurations will appear in a forthcoming publication [8].

We use extensions to the traditional PFA as discussed in Sec. 2. In Sec. 3 we show how the calculation of the interaction energy per unit area in planar structures are performed. These results are used as input in our PFA calculations. In Sec. 4 we give a general expression for the interaction energy for half spaces, cylinders and spheres. Sec. 5 is devoted to spherical objects, Finally, summary and conclusions are found in Sec. 6.

2. The Proximity Force Approximation in this work

Traditionally in the PFA of the interaction between two objects one only takes into consideration the surface of each object that is facing the other object. The function $E_p(w)$, in Eq. (1), is the energy per unit area for two half spaces, made up from the materials of the two objects, separated by the distance w. If the objects are immersed in an ambient medium the gap between the two half spaces is also filled with this medium. With this treatment the backsides of the objects have no effects at all. One gets, e.g., the same result for two spheres as for two halfspheres. We know that the normal modes contributing to the interaction may extend through the objects and continue on the other side. Thus, the backside may have important effects on the results. One may extend the treatment by using $E_p(w)$ from a planar structure with four interfaces instead of two. The distance between the interfaces surrounding the object material is the local thickness of the object. We will not consider this in the present work, but address another limitation that

Journal of Physics: Conference Series **161** (2009) 012016 doi:10.1088/1742-6596/161/1/012016

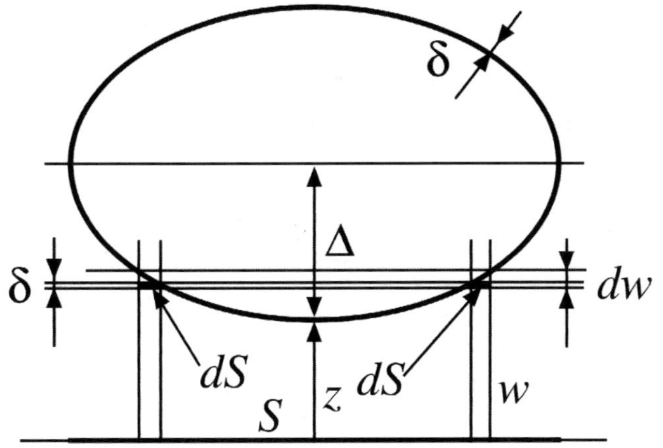

Figure 1. Coated oblate spheroid above a substrate used to illustrate the parameters discussed in the text.

is more severe. Two coated spheres, e.g., are in PFA treated as solid spheres. We will extend the treatment by considering a planar structure with four interfaces. In the most general case the two coatings are of different materials and of different thickness. We let $E_p(w,\delta)$ be the energy per unit area for two coated half spaces, made up from the materials of the two objects, separated the distance w. The thickness of the coating is denoted by δ. We may express the potential as

$$V(z) = \int_S dS E_p(w,\delta) = \int_z^{z+\Delta} dw \underbrace{\frac{dS}{dw}}_{g(w-z)} E_p(w,\delta), \tag{2}$$

and find the force as

$$F(z) = -\frac{dV}{dz} = -g(\Delta) E_p(z+\Delta,\delta) + g(0) E_p(z,\delta) - \int_z^{z+\Delta} dw \frac{dg(w-z)}{dz} E_p(w,\delta). \tag{3}$$

In many cases $g(\Delta)$ vanishes, like in the illustrating example in Fig. 1. Here we have a coated spheroid above a substrate. The surface S is defined through the projection of the object onto the substrate. When $g(\Delta)$ vanishes we have

$$F(z) = g(0) E_p(z,\delta)\left[1 + \frac{1}{g(0) E_p(z,\delta)} \int_z^{z+\Delta} dw \frac{d^2S}{dw^2} E_p(w,\delta)\right], \tag{4}$$

where the first part is what one usually means with PFA. The remaining part, within brackets, is a correction factor depending on the geometry. This factor is often dropped without any motivation at all or with the argument that the resulting error is of the same order of magnitude as the error in PFA itself. Often $E_p(w,\delta)$ has a known power dependence in w over the whole integration interval. Then we may often find an analytical expression for the correction factor. In Fig. 1 we study a coated oblate spheroid above an uncoated substrate. The lowest horizontal line is the substrate boundary. The thicker part of this line indicates the surface S, which is defined by the projection of the spheroid on the substrate. The integration variable w varies from z, the closest distance, to $z+\Delta$. The region between w and $w+dw$ picks out a surface on the object whose projection on the substrate is a ring of area dS. This surface contributes, in the PFA spirit, to the energy with dS times the energy per unit area, $E_p(w,\delta)$, of the planar configuration defined by the substrate and the film of thickness δ, indicated in the figure. If the spheroid is not an empty shell, the half space above the film should be filled by the material of the spheroid.

To summarize, in this work we extend the traditional PFA in two ways; we retain the correction factor of Eq. (4); we extend PFA to include the finite coat thickness of coated objects. The effects of these extensions are demonstrated and comparisons are made to numerical results from more accurate calculations. Throughout the text we refer to results from traditional PFA as PFA results and to extended or corrected PFA results as full PFA results.

Before we proceed with the various objects we derive, in next section, the interaction energy in those planar structures we need for the PFA calculations.

3. Planar structures

A general expression for the interaction energy per unit area in a planar system where only one of the distances, w, between neighboring interfaces is allowed to vary is [2]

$$\begin{aligned} E_p(w) &= \frac{\hbar}{2}\int_{-\infty}^{\infty}\frac{d\omega}{2\pi}\int\frac{d^2k}{(2\pi)^2}\left\{\ln\left[f(\mathbf{k},i\omega,w)\right]-\ln\left[f(\mathbf{k},i\omega,\infty)\right]\right\} \\ &= \frac{\hbar}{2}\int_{-\infty}^{\infty}\frac{d\omega}{2\pi}\int\frac{d^2k}{(2\pi)^2}\ln\left[\frac{f(\mathbf{k},i\omega,w)}{f(\mathbf{k},i\omega,\infty)}\right], \end{aligned} \tag{5}$$

where the reference energy is set to when w is infinite. The variable $\mathbf{k}$ is the two dimensional wave vector in the plane of the interfaces. The function $f(\mathbf{k},\omega,w)$ is the function in the condition for having electromagnetic normal modes in the planar system,

$$f(\mathbf{k},\omega,w)=0. \tag{6}$$

In the general case there are two of these functions, one for TE modes and one for TM modes. Here we limit the treatment to the non-retarded limit. One of the effects from neglecting retardation is that the TE modes are absent. In the case of a coated object above a coated substrate or of two coated objects we need four interfaces between five regions. For a structure of the type 1|2|3|4|5 where medium 3 has the variable thickness w we have

$$\frac{f(\mathbf{k},\omega,w)}{f(\mathbf{k},\omega,\infty)}=1+e^{-2kw}\frac{r_{23}r_{34}+e^{-2kd_2}r_{12}r_{34}+e^{-2kd_4}r_{23}r_{45}+e^{-2kd_2}e^{-2kd_4}r_{12}r_{45}}{1+e^{-2kd_2}r_{12}r_{23}+e^{-2kd_4}r_{34}r_{45}+e^{-2kd_2}e^{-2kd_4}r_{12}r_{23}r_{34}r_{45}}, \tag{7}$$

where $k=|\mathbf{k}|$ and

$$r_{ij}=\frac{\varepsilon_j(\omega)-\varepsilon_i(\omega)}{\varepsilon_j(\omega)+\varepsilon_i(\omega)}. \tag{8}$$

The function $\varepsilon_i(\omega)$ and d_i are the dielectric function and thickness, respectively, of region i. In the case of a coated object above a substrate or of two objects where only one is coated we need three interfaces between four regions. For a structure of the type 1|3|4|5 where medium 3 has the variable thickness w we let $r_{12}=0$ and get

$$\frac{f(\mathbf{k},\omega,w)}{f(\mathbf{k},\omega,\infty)}=1+e^{-2kw}\frac{r_{13}r_{34}+e^{-2kd_4}r_{13}r_{45}}{1+e^{-2kd_4}r_{34}r_{45}}. \tag{9}$$

In the case of an uncoated object above an uncoated substrate or of two uncoated objects we need two interfaces between three regions. For a structure of the type 1|3|5 where medium 3 has the variable thickness w we let $r_{45}=0$ and get

$$\frac{f(\mathbf{k},\omega,w)}{f(\mathbf{k},\omega,\infty)}=1+e^{-2kw}r_{13}r_{35}. \tag{10}$$

For this last structure the integration over momentum in Eq. (5) may be performed and results in an infinite series.

$$\begin{aligned}\mathrm{E_p}(w) &= \frac{\hbar}{8\pi^2}\int\limits_{-\infty}^{\infty} d\omega \int\limits_{0}^{\infty} dk k \ln\left[1 - e^{-2kw} r_{13} r_{53}\right] = \frac{\hbar}{32\pi^2 w^2}\int\limits_{-\infty}^{\infty} d\omega \int\limits_{0}^{\infty} dk k \ln\left[1 - e^{-k} r_{13} r_{53}\right] \\ &= -\frac{\hbar}{32\pi^2 w^2}\int\limits_{-\infty}^{\infty} d\omega \sum_{l=1}^{\infty}\frac{(r_{13} r_{53})^l}{l^3} = -\frac{\hbar}{32\pi^2 w^2}\sum_{l=1}^{\infty}\frac{\langle\omega_l\rangle}{l^3},\end{aligned} \tag{11}$$

where the characteristic dielectric integrals $\langle\omega_l\rangle$ are

$$\langle\omega_l\rangle = \int\limits_{-\infty}^{\infty} d\omega \left[\frac{\varepsilon_1(\omega) - \varepsilon_3(\omega)}{\varepsilon_1(\omega) + \varepsilon_3(\omega)}\frac{\varepsilon_5(\omega) - \varepsilon_3(\omega)}{\varepsilon_5(\omega) + \varepsilon_3(\omega)}\right]^l. \tag{12}$$

This result is obtained after a variable substitution, series expansion of the logarithm, followed by the integration over momentum.

4. General expression for half spaces, cylinders and spheres

For solid objects in neglect of the geometrical correction in Eq. (4) one may find the following general expression for half spaces, cylinders and spheres: [2, 9]

$$E(z) = -\frac{\hbar}{32\pi^2 z^{(1+n/2)}}\Gamma(1+n/2)\left[\frac{2\pi R_1 R_2}{R_1 + R_2}\right]^{1-n/2}\sum_{l=1}^{\infty}\frac{\langle\omega_l\rangle}{l^3}, \tag{13}$$

where $n = 0$ for spheres, $n = 1$ for cylinders, and $n = 2$ for half spaces. The result is for the interaction energy in case of spheres, the interaction energy per unit length for cylinders and the interaction energy per unit area for half spaces. The variable z is the closest distance between the objects. Thus, for two half spaces we have

$$E(z) = -\frac{\hbar}{32\pi^2 z^2}\sum_{l=1}^{\infty}\frac{\langle\omega_l\rangle}{l^3} = E_p(z), \tag{14}$$

for two cylinders of radii R_1 and R_2

$$E(z) = -\frac{\hbar}{32\pi^2 z^{3/2}}\Gamma(3/2)\left[\frac{2\pi R_1 R_2}{R_1 + R_2}\right]^{1/2}\sum_{l=1}^{\infty}\frac{\langle\omega_l\rangle}{l^3} = \Gamma(3/2)\sqrt{\frac{2\pi R_1 R_2 z}{R_1 + R_2}}E_p(z), \tag{15}$$

and for two spheres of radii R_1 and R_2 the result is

$$E(z) = -\frac{\hbar}{32\pi^2 z}\left[\frac{2\pi R_1 R_2}{R_1 + R_2}\right]\sum_{l=1}^{\infty}\frac{\langle\omega_l\rangle}{l^3} = \left[\frac{2\pi R_1 R_2 z}{R_1 + R_2}\right]E_p(z). \tag{16}$$

To get the result for a cylinder of radius R above a substrate we let R_2 go to infinity and replace R_1 with R in Eq. (15). This results in

$$E(z) = -\frac{\hbar}{32\pi^2 z^{3/2}}\Gamma(3/2)\sqrt{2\pi R}\sum_{l=1}^{\infty}\frac{\langle\omega_l\rangle}{l^3} = \Gamma(3/2)\sqrt{2\pi R z}E_p(z). \tag{17}$$

To get the result for a sphere of radius R above a substrate we let R_2 go to infinity and replace R_1 with R in Eq. (16). This results in

$$E(z) = -\frac{\hbar}{32\pi^2 z}2\pi R\sum_{l=1}^{\infty}\frac{\langle\omega_l\rangle}{l^3} = 2\pi R z E_p(z). \tag{18}$$

This is as far as we get with the general formula. Now, we continue in next and the following sections with the geometrical corrections for spherical objects.

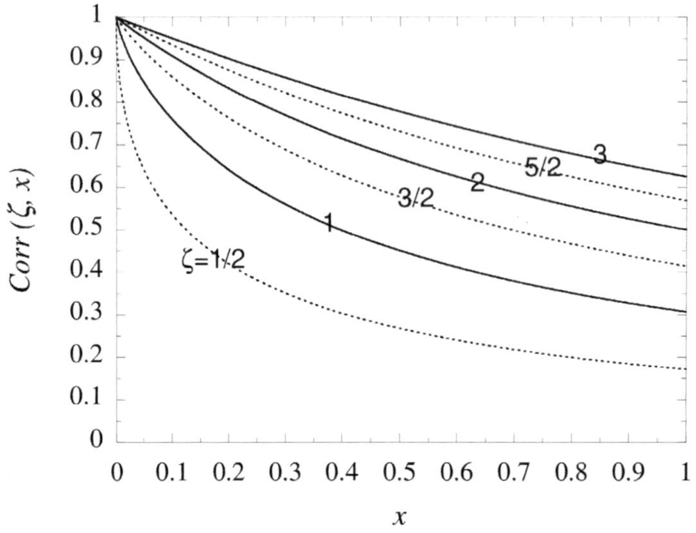

Figure 2. The correction factor $Corr(\zeta, x)$ for a set of ζ values.

5. Spherical objects

Spherical objects are often used in experiments. The advantage is that one avoids the problem of alignment. This is the case both for a sphere above a substrate and for two interacting spheres. In Lamoreaux's [4] classical measurement of the force between two gold plates one actually measured the force between a sphere and a planar surface.

5.1. Sphere-substrate interaction

For a sphere above a substrate the parameters entering Eq. (4) are

$$\Delta = R; \quad g(x) = 2\pi (R - x); \quad g(0) = 2\pi R; \quad \frac{d^2 S}{dw^2} = -2\pi, \tag{19}$$

and this results in

$$F(z) = 2\pi R E_p(z,\delta) \left[1 - \frac{1}{R E_p(z,\delta)} \int\limits_z^{z+R} dw E_p(w,\delta) \right], \tag{20}$$

where we have included the possibility for the sphere and/or the substrate to have a coating of thickness δ. If $E_p(w,\delta)$ varies as

$$E_p(w,\delta) = -C/w^{\zeta} \tag{21}$$

in the whole integration interval we find the correction factor, $Corr(\zeta, x)$ on analytical form

$$Corr(\zeta, x) = \begin{cases} 1 - \left[x - x^{\zeta}/(1+x)^{\zeta-1} \right] / (\zeta - 1); & \zeta \neq 1 \\ 1 - x \ln[(1+x)/x]; & \zeta = 1 \end{cases} \tag{22}$$

where $x = z/R$. The correction factor is illustrated for some ζ values in Fig. 2. Eq. (20) is valid both in the non-retarded and retarded separation regions. The PFA is only good for small x-values but if R is big enough one may still be in the retarded region. For uncoated sphere and substrate $\zeta = 2$ ($\zeta = 3$) in the non-retarded (retarded) region. If either of the sphere and substrate or both are coated the separation dependence is in general more complex. If a metallic coating is thin enough there is a separation range where $\zeta = 5/2$.

With the proper scaling one may produce universal figures, i.e., figures that are identical for all values of R. From Eqs. (5) and (7) follows that $E_p(w,\delta) = f(w/R, \delta/R)/w^2$. From Eq. (20) follows that if one plots $R \times F$ as a function of z/R the results look identical as long as the coat thickness is the same fraction of R. This means that there is no need to repeat the calculations for different sphere radii. In Fig. 3 we show R times the non-retarded force on a gold sphere above a gold substrate as function of normalized separation z/R. Throughout we use the dielectric function of gold given in Ref. [10]. The differences between the curves

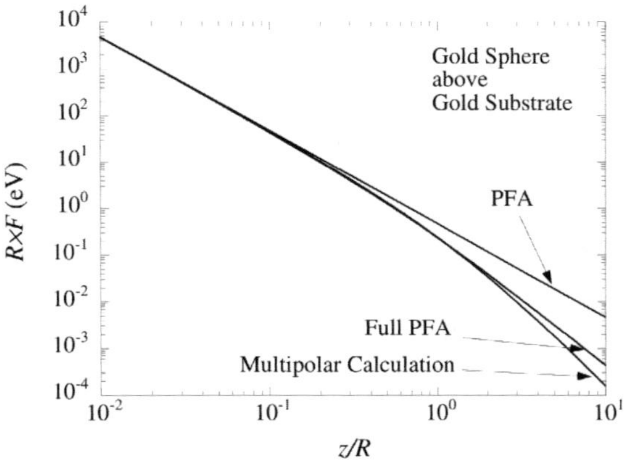

Figure 3. R times the non-retarded force on a gold sphere above a gold substrate as function of z/R for the multipolar result, the PFA result and the geometry corrected or full PFA result.

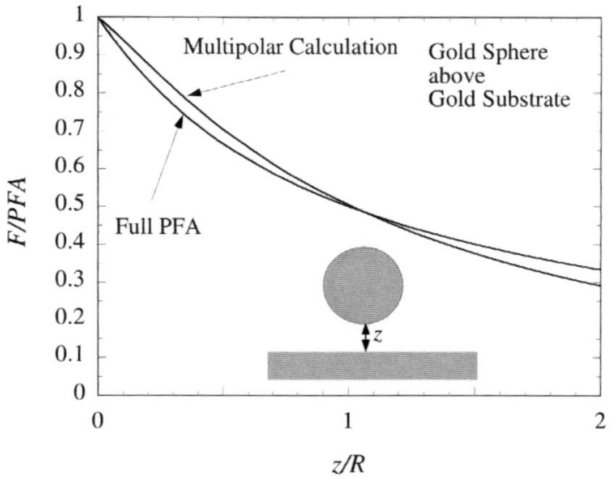

Figure 4. The non-retarded force on a gold sphere above a gold substrate relative the PFA result as function of z/R for the multipolar result and the full PFA result.

are not clearly seen in a figure like this. In Fig. 4 we show the result from the multipolar calculation and the geometry corrected or full PFA result relative the PFA result. In Fig. 5 we show R times the non-retarded force on a spherical gold shell above a gold substrate as function of normalized separation z/R. The thick solid curve and dashed curve are the PFA and full PFA curves, respectively, for a solid gold sphere and are the same as in Fig. 3; the solid curve with open circles is the result from the multipolar calculation for a shell of a thickness of one percent of the radius; the filled circles are our extended PFA result for the gold shell; the solid curve with triangles is our extended and full PFA result for the gold shell. We note that the full result, from the multipolar calculation, follows the PFA result for a solid gold sphere for distances smaller than approximately the coat thickness. Then for larger separations there is a region where it follows rather closely the 2D PFA result obtained by replacing the gold film with a 2D (two dimensional) metallic sheet with 2D electron density given as the projection of the 3D electron density of the film. These results are derived in Refs. [11, 12],

$$R \times F = 2\pi R^2 E_p(z) \approx 0.1556\sqrt{n\hbar^2 e^2 / m_e}\sqrt{\delta/R}\Big/(z/R)^{5/2}, \tag{23}$$

where n is the conduction electron density of gold.

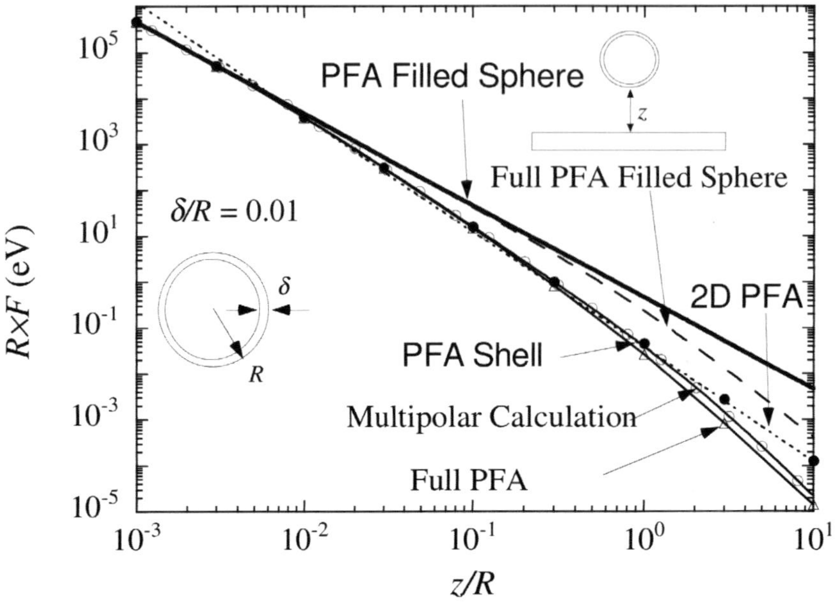

Figure 5. R times the non-retarded force on a spherical gold shell above a gold substrate as function of z/R. The curves are for the multipolar result (thin solid curve with open circles), the PFA result for a solid sphere (thick solid curve), the full PFA result for a solid sphere (dashed curve), the two-dimensional PFA result (dotted curve), the PFA result for the shell (solid circles) and the full PFA for the shell (triangles).

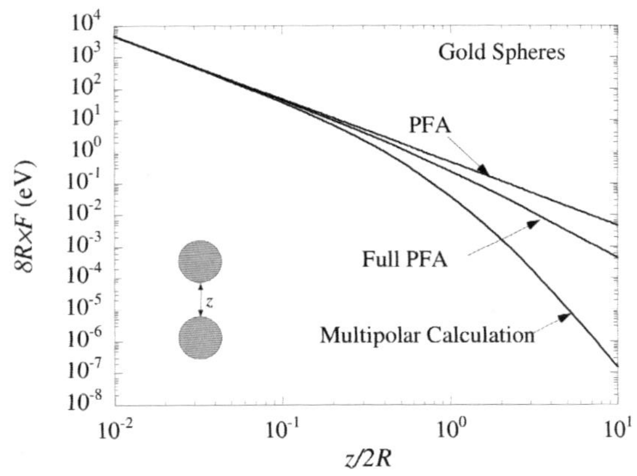

Figure 6. $8R$ times the non-retarded force between two gold spheres as function of $z/2R$ for the multipolar result, the PFA result and the full PFA result.

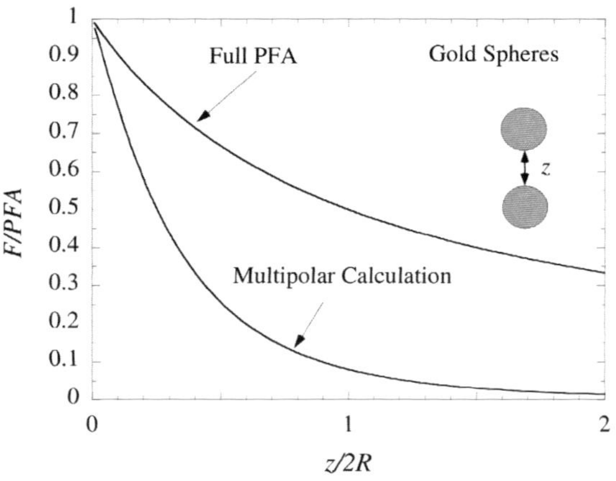

Figure 7. The non-retarded force between two equal size gold spheres relative the PFA result as function of $z/2R$ for the multipolar result and the full PFA result.

5.2. Sphere-sphere interaction

For two spheres of equal size the parameters entering Eq. (4) are

$$\Delta = 2R; \quad g(x) = \pi (R - x/2); \quad g(0) = \pi R; \quad \frac{d^2 S}{dw^2} = -\pi/2, \tag{24}$$

and this results in

$$F(z) = \pi R E_p(z,\delta)\left[1 - \frac{1}{2RE_p(z,\delta)}\int\limits_z^{z+2R} dw E_p(w,\delta)\right]. \tag{25}$$

If $E_p(w,\delta)$ varies as in Eq. (21) in the whole integration interval we again find the correction factor, $Corr(\zeta, x)$, on analytical form and it is identical to the expression in Eq. (22) but now $x = z/2R$. In Fig. 6 we show $8R$ times the non-retarded force between two gold spheres of equal size as function of normalized separation $z/2R$. Plotted in this way the PFA and full PFA curves are identical to the corresponding ones in Fig. 3; the result from the multipolar calculation is not. In Fig. 7 we show the result from the multipolar calculation and the full PFA result relative the PFA result. We find that the agreement is not as good as in the case of a sphere above a substrate.

6. Summary and conclusions

In this work we have performed a critical test of the validity of the Proximity Force Approximation. This was done by comparisons with the results from a calculation method of more solid foundation, the multi-pole expansion method. Calculations were limited to a sphere above a substrate, a spherical shell above a substrate, and two interacting spheres. The results from a more extensive study will appear elsewhere [8].

We extended the traditional PFA in two ways; we did not just keep the small separation asymptote but took the geometrical shapes of the objects fully into account; we took the finite thickness of the coatings and free standing shells into account. For the three geometries studied in this work we found that the full PFA worked best for the shell above a substrate, second best for the solid sphere above a substrate, and third best for two interacting spheres.

In Fig. 1 we see that there are surface elements with high inclination angles. It would be tempting to try to improve and go beyond PFA by using element planes of variable inclination angle. However, the improvement is rudimentary. The angular dependence decreases rapidly with the ratio between the element distance to the substrate and the element-length.

Throughout this work we have presented universal figures, independent of the system scale. A word of caution is in place. All results are limited to the non-retarded range. This means that the range of validity in each figure depends on the scaling parameter. At the present status of the available measurement techniques one requires big enough objects, in the order of tens of micrometers, in close proximity, in the range of hundreds of nanometers. In a majority of cases this means that one is in the non-retarded limit and that the full PFA should be a good approximation.

References

[1] London F 1930 *Z. Phys. Chem.* B**11** 222; 1930 *Z. Phys.* **63** 245
[2] Sernelius Bo E 2001 *Surface Modes in Physics* (Berlin: Wiley-VCH)
[3] Casimir H B G 1948 *Proc. Kon. Akad. Wet.* **51** 793
[4] Lamoreaux S K 1997 *Phys. Rev. Lett.* **78** 5
[5] Noguez C, Román-Velázquez C E, Esquivel-Sirvent R, and Villarreal C 2004 *Europhys. Lett.* **67** 191
[6] Derjaguin B 1934 *Kolloid Z.* **69** 155
[7] Genet C, Lambrecht A, Maia Neto P, and Reynaud S 2003 *Eurphys. Lett.* **62** 484
[8] Sernelius Bo E and Román-Velázquez C E, t o be published
[9] Langbein D 1974 *Theory of the Van der Waals Attraction* (New York: Springer) in the series *Springer Tracts in Modern Physics*
[10] Lambrecht A and Reynaud S 2000 *Eurphys. J.* D**8** 309
[11] Boström M and Sernelius Bo E 2000 *Phys. Rev.* B**61** 2204
[12] Sernelius Bo E and Björk P 1998 *Phys. Rev.* B**57** 6592

Journal of Physics: Conference Series **161** (2009) 012017 doi:10.1088/1742-6596/161/1/012017

Casimir pressure on a thin metal slab

M S Tomaš[1] and Z Lenac[2]
[1]Rudjer Bošković Institute , P. O. Box 180, 10002 Zagreb, Croatia
[2]Department of Physics, University of Rijeka, 51000 Rijeka, Croatia
E-mail: tomas@thphys.irb.hr

Abstract. We consider the vacuum-field pressure on boundaries of a metal slab in the middle of a cavity with perfectly reflecting mirrors adopting the plasma model for the metal and paying special attention to the surface plasmon polariton contribution to the pressure. We demonstrate that, with increasing cavity length, the pressure on a thin ($d \ll \lambda_P$) slab in this system decreases from the Casimir pressure $F_C = -\pi^2\hbar c/240d^4$ at zero slab-mirror distances to the non-retarded force per unit area $F_{\rm nr} = 1.19(d/\lambda_P)F_C$ in the case of an isolated slab. In the first case the pressure is entirely due to the photonic modes propagating through the metal whereas in the second case it is entirely due to the (nonretarded) surface plasmon modes supported by the free-standing thin slab. In either case the pressure decreases with the slab thickness. These considerations indicate that the vacuum-field pressure on a thin metal layer (and its modal structure) can be in a symmetric cavity significantly influenced when changing the cavity length.

1. Introduction

The Casimir effect most commonly refers to the existence of forces between neutral macroscopic bodies due to the change of the zero-point energy of the electromagnetic field in confined space [1, 2, 3, 4]. Soon upon its prediction for two perfectly conducting plates in vacuum [5], the theory of the Casimir effect was extended to more realistic systems consisting of two dielectrics separated by a vacuum gap [6] and by a gap filled by a medium [7, 8, 9] and, more recently, to systems involving more layers [10] including general dielectric and magnetodielectric multilayers [11, 12, 13, 14]. Evidently, when the gap between two stacks of layers is filled by a medium, as in multilayers, the vacuum-field force (per unit area) on the stacks can also be regarded as the pressure on the medium between them. Moreover, since the vacuum-field fluctuations are always present, the pressure on the medium persists even in the absence of other layers, that is, even in the case of a free standing single material slab. Consequently, as pointed out recently [15], in addition to the traditional Casimir force due to the presence of other layers, every material layer in a multilayered systems experiences a vacuum-field pressure on its boundaries. Changes caused by this pressure (e.g. change of the layer thickness) can be taken as an alternative signature of the Casimir effect and it is therefore of fundamental interest to explore it in more details. From the practical point of view, however, particularly interesting systems in this respect are those involving metallic plates as metal components are often met in micromechanical (MEMS) and nanomechanical (NEMS) devices [16, 17].

Vacuum-field pressure on a metal slab has already been addressed (to some extent) by Dzyaloshinskii *et al.* [7] when discussing the Casimir force between two dielectric media separated by a metal layer. The same system has also recently been considered by Imry [18] who pointed out strong dependence of the zero-point radiation pressure on a metal film on properties of the surrounding media. Very recently, Benassi and Calandra [15] used the Lifshitz formula to calculate the pressure on a metal slab and explore its dependence on the properties of the slab as well as on the distance and properties of a nearby (metallic) substrate. By combining a Lifshitz-like formula and the mode summation method, in our previous work [19] we have explored the effect of surrounding media as well as of a cavity on the pressure on a metal slab paying special attention to the contribution of the surface polariton (SP) modes to the pressure. In the present work, we consider in more details the effect of a cavity on the pressure on surfaces of a metal slab in its center and derive several new results concerning this pressure. Since the ordinary Casimir force on the slab vanishes in symmetric configurations, consideration of this system is a very convenient way to explore the pressure on the slab surfaces. As in aforementioned works, we adopt free-electron (plasma) model to describe the metal and, for simplicity, assume perfectly reflecting cavity mirrors. Accordingly, our aim here is to establish trends of the pressure with the system parameters and calculate its limiting values rather than to discuss it for a realistic system.

The paper is organized as follows. In Section II, we briefly adapt the theory of the ordinary Casimir force on a slab in a cavity [11, 14, 20] to include also the vacuum-field pressure on the slab surfaces and derive a Lifshitz-like formula for this pressure (see also [15]). In Sections III-V, we use this formula to discuss the pressure on surfaces of a free standing metal slab, of a metal layer between perfect mirrors and of a metal slab in an ideal cavity, respectively. Our conclusions are summarized in Section VI.

2. Preliminaries

Consider a dielectric slab inserted in a planar cavity, as depicted in Fig. 1. The total force (per unit area) acting on the slab consists of the pressure $F = f_s$ on the slab boundaries and the slab-mirror interaction force $F' = f_2 - f_1$ [15], where according to the theory of the Casimir

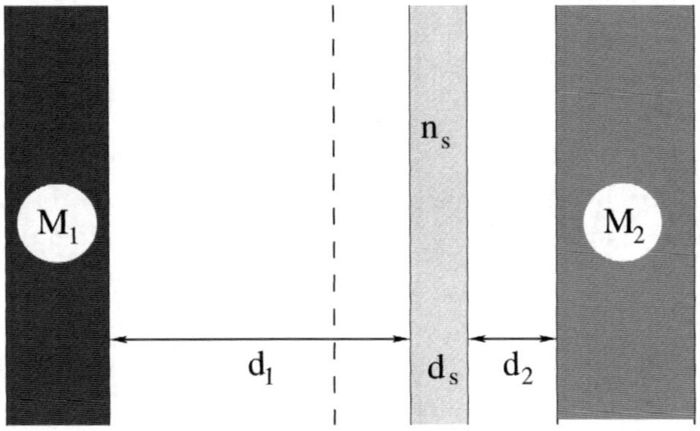

Figure 1. System considered schematically.

force in multilayers [11]

$$f_j = -\frac{\hbar}{2\pi^2}\int_0^\infty d\xi \int_0^\infty dk k \kappa_j \sum_{q=p,s} \frac{1 - D_{qj}(i\xi, k)}{D_{qj}(i\xi, k)}, \quad D_{qj}(i\xi, k) = 1 - r^q_{j-} r^q_{j+} e^{-2\kappa_j d_j}. \qquad (1)$$

Here $\kappa_j(i\xi, k) = \sqrt{\varepsilon_j(i\xi)\xi^2/c^2 + k^2}$ is the perpendicular wave vector at the imaginary frequency in the jth layer and $r^q_{j\pm}(i\xi, k)$ are the reflection coefficients of the right and left stack of layers bounding the layer.

Considering first the pressure on the slab, we have

$$F = -\frac{\hbar}{2\pi^2}\int_0^\infty d\xi \int_0^\infty dk k\kappa_s \sum_{q=p,s} \frac{r^q_{s-}r^q_{s+}e^{-2\kappa_s d_s}}{1 - r^q_{s-}r^q_{s+}e^{-2\kappa_s d_s}}, \tag{2}$$

where

$$r^q_{s-(+)}(i\xi,k) = \frac{-\rho^q + R^q_{1(2)}e^{-2\kappa d_{1(2)}}}{1 - \rho^q R^q_{1(2)}e^{-2\kappa d_{1(2)}}} \tag{3}$$

are reflection coefficients for the waves reflected within the slab. Here $\kappa(i\xi,k) \equiv \kappa_1 = \kappa_2 = \sqrt{\xi^2/c^2 + k^2}$ is the perpendicular wave vector in the cavity,

$$\rho^p(i\xi,k) = \frac{\varepsilon_s\kappa - \kappa_s}{\varepsilon_s\kappa + \kappa_s}, \qquad \rho^s(i\xi,k) = \frac{\kappa - \kappa_s}{\kappa + \kappa_s}, \tag{4}$$

are the *vacuum-slab* reflection coefficients and $R^q_{1(2)}(i\xi,k)$ are those of the mirrors. According to Eqs. (2) and (3), f_s can be rewritten as

$$F = -\frac{\hbar}{2\pi^2}\int_0^\infty d\xi \int_0^\infty dk k\kappa_s \sum_{q=p,s} \frac{(\rho^q - R^q_1 e^{-2\kappa d_1})(\rho^q - R^q_2 e^{-2\kappa d_2})e^{-2\kappa_s d_s}}{\tilde{D}_{qs}}, \tag{5}$$

where

$$\begin{aligned}\tilde{D}_{qs}(i\xi,k) &= 1 - \rho^{q2}e^{-2\kappa_s d_s} - \rho^q(1 - e^{-2\kappa_s d_s})(R^q_1 e^{-2\kappa d_1} + R^q_2 e^{-2\kappa d_2}) \\ &\quad + (\rho^{q2} - e^{-2\kappa_s d_s})R^q_1 R^q_2 e^{-2\kappa(d_1+d_2)}. \end{aligned} \tag{6}$$

The traditional force per unit area on the slab due to the presence of the mirrors $F' = f_2 - f_1$ is found similarly using [11]

$$r^q_{1-(2+)}(i\xi,k) = R^q_{1(2)}, \qquad r^q_{1+(2-)}(i\xi,k) = r^q + \frac{t^{q2}R^q_{2(1)}e^{-2\kappa d_{2(1)}}}{1 - r^q R^q_{2(1)}e^{-2\kappa d_{2(1)}}}, \tag{7}$$

where $r^q(i\xi,k)$ and $t^q(i\xi,k)$ are the Fresnel coefficients for the (whole) slab. We find that F' is given by Eq. (5) provided that the nominators in that equation are replaced by [11]

$$\rho^q(1 - e^{-2\kappa_s d_s})(R^q_2 e^{-2\kappa d_2} - R^q_1 e^{-2\kappa d_1}) \tag{8}$$

and that $\kappa_s \longrightarrow \kappa$ in front of the sum. Clearly, in contrast to F, the pressure F' vanishes when the slab is in the center ($d_1 = d_2$) of a symmetric ($R^q_1 = R^q_2$) cavity.

Frequencies $\omega^q_n(k)$ of SP and other bound modes supported by the system are given as solutions of

$$D_{qs}(\omega,k) \equiv 1 - r^q_{s-}r^q_{s+}e^{-2\alpha_s d_s} = 0, \quad \alpha_s(\omega,k) \equiv \kappa_s(-i\omega,k) = \sqrt{k^2 - \varepsilon_s\omega^2/c^2}, \tag{9}$$

[or, equivalently, $\tilde{D}_{qs}(\omega,k) = 0$] in the corresponding part of the (ω,k)-plane. Separate contribution of each SP mode to the Casimir force can therefore be obtained by using the real frequency counterpart of Eq. (5) and calculating the corresponding residuum. Alternatively, since this equation is (for a lossless system) compatible with the standard definition of the Casimir energy (with respect to the slab)

$$E_s = A\int \frac{d^2\mathbf{k}}{(2\pi)^2} \sum_{q=p,s}\sum_n \frac{\hbar}{2}\left[\omega^q_n(k,d_s) - \omega^q_n(k,d_s\to\infty)\right], \tag{10}$$

we may calculate the surface contribution to the pressure using directly

$$F_S = -\frac{1}{A}\frac{\partial E_s^{\rm SP}}{\partial d_s} = -\frac{\hbar}{4\pi}\int_0^\infty dkk \sum_\sigma \frac{\partial \omega_\sigma(k)}{\partial d_s}, \tag{11}$$

where σ enumerates SP modes. In the following, we use this method to perform a modal analysis of the pressure on a thin metal slab in a symmetric cavity. Clearly, in such a configuration, frequencies of SP (and other) modes are found as solutions of

$$r_s^q(\omega,k)e^{-\alpha_s d_s} = \pm 1, \quad r_s^q(\omega,k) = \frac{-\rho^q + R^q e^{-2\alpha d_s}}{1-\rho^q R^q e^{-2\alpha d_s}}, \quad \alpha(\omega,k) = \sqrt{k^2-\omega^2/c^2}, \tag{12}$$

and the modes are further characterized by an index $\nu = \pm$ describing their symmetry with respect to the central plane of the system.

3. Free-standing metal slab

Let us start with the case of a free-standing metallic slab. Adopting the electron plasma model, the slab is described by the dielectric function

$$\varepsilon_s(i\xi) = 1 + \frac{\omega_P^2}{\xi^2}, \tag{13}$$

where ω_P is the metallic plasma frequency. Letting $R^q_{1(2)} = 0$ Eq. (5), the pressure on the slab is then given by

$$F = -\frac{\hbar}{2\pi^2}\int_0^\infty d\xi \int_0^\infty dkk\kappa_s \sum_{q=p,s} \frac{[\rho^q(i\xi,k)]^2 e^{-2\kappa_s d_s}}{1-[\rho^q(i\xi,k)]^2 e^{-2\kappa_s d_s}}, \tag{14}$$

with $\rho^q(i\xi,k)$ given by Eqs. (4) and (13). Note that this formula differs from the corresponding formula for the standard Casimir force (per unit area) between two metal half-spaces only in the (explicit) presence of κ_s instead of κ.

For a thin, $d_s \ll c/\omega_p$, slab the main contribution to F comes from large wave vectors. Accordingly, we may let $\kappa_s \simeq \kappa \simeq k$ and consequently $\rho^p \simeq \rho^p_{\rm nr} = (\varepsilon_s - 1)/(\varepsilon_s - 1)$ and $\rho^s \simeq \rho^s_{\rm nr} = 0$ in Eq. (14). Thus, the pressure on the slab is in this nonretarded (quasistatic) approximation equal to the pressure on two identical semi-infinite metals separated by a thin vacuum gap. Introducing $x = \xi/\omega_P$ and $t = 2kd$ as the integration variables, in this way we obtain

$$F_{\rm nr} = -\frac{\hbar\omega_P}{16\pi^2 d^3}\int_0^\infty dx \int_0^\infty dt t^2 \frac{e^{-t}}{(2x^2+1)^2 - e^{-t}} = -0.00781\frac{\hbar\omega_P}{d^3}. \tag{15}$$

It is easy to see that $F_{\rm nr}$ is entirely due to the two surface plasmon modes supported by the slab (or two semi-infinite free-electron metals) [18, 24, 25, 26, 27]. Indeed, from the nonretarded limit of Eq. (12)

$$\frac{\varepsilon_s(\omega)-1}{\varepsilon_s(\omega)+1}e^{-kd_s} = \mp 1$$

we find familiar frequencies of surface plasmons

$$\omega_\pm(k) = \frac{\omega_P}{\sqrt{2}}\sqrt{1 \pm {\rm e}^{-kd_s}}, \tag{16}$$

so that the integrand in Eq. (3) becomes:

$$-k\,\frac{\partial\omega_\pm(k)}{\partial d_s} = \pm\frac{\omega_P}{2\sqrt{2}}\frac{k^2 e^{-kd_s}}{\sqrt{1 \pm e^{-kd_s}}}.$$

Using this in Eq. (3), the total (nonretarded) contribution to the pressure of the two plasmon modes is easily calculated to be equal to $F_{\rm nr}$ given above as a result of compensation between the pressing ($F_-^{\rm SP} = 7.83F_{\rm nr}$) and relaxing ($F_+^{\rm SP} = -6.83F_{\rm nr}$) contributions from the ω_- and ω_+ mode, respectively.

To estimate the pressure on a thick slab, we proceed in the standard way [6, 7] and introduce in Eq. (14) the variable p instead of k by letting $\kappa_s = \sqrt{\varepsilon_s(i\xi)}\xi p/c$. With $x = \xi/\omega_P$, this gives

$$F \simeq -\frac{\hbar c k_P^4}{2\pi^2}\int_0^\infty dx(1+x^2)^{3/2}\int_1^\infty dpp^2\sum_{q=p,s}\frac{[\rho^q(ix\omega_P,p)]^2 e^{-2p\sqrt{1+x^2}k_Pd_s}}{1-[\rho^q(ix\omega_P,p)]^2 e^{-2p\sqrt{1+x^2}k_Pd_s}}, \qquad (17)$$

where $k_P = \omega_P/c = 2\pi/\lambda_P$ and

$$\rho^p(ix\omega_P,p) = \frac{(1+x^2)s - x^2p}{(1+x^2)s + x^2p}, \quad \rho^s(ix\omega_P,p) = \frac{s-p}{s+p}, \quad s = \sqrt{p^2 - 1 + \frac{x^2}{1+x^2}}. \qquad (18)$$

Since for $k_Pd_s \gg 1$ the main contribution to the integral comes from small-x region, we can let $x = 0$ everywhere except in the exponents where we use $\sqrt{1+x^2} \simeq 1 + x^2/2$. Retaining only the leading terms, in this way we obtain

$$\begin{aligned} F &\simeq -\frac{\hbar c k_P^4}{4\pi\sqrt{\pi k_P d_s}}\int_1^\infty dpp^{3/2}e^{-2pk_pd_s}\left[1 + \left(\frac{\sqrt{p^2-1}-p}{\sqrt{p^2-1}+p}\right)^2\right] \\ &\simeq -\frac{\hbar c k_P^4}{4}\frac{e^{-2k_pd_s}}{\sqrt{(\pi k_P d_s)^3}}, \qquad k_Pd_s \gg 1, \end{aligned} \qquad (19)$$

where the final result follows upon a partial integration. Accordingly, as already noted by Dzyaloshinskii *et al* [7], in the plasma model for the metal the pressure on the slab surfaces exponentially decreases with d_s.

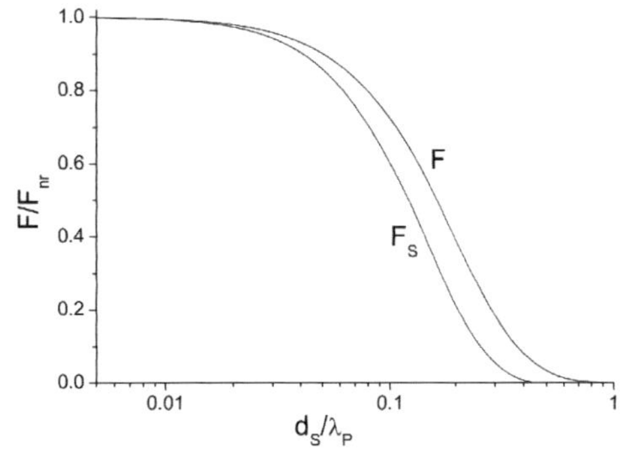

Figure 2. Vacuum-field pressure on a free standing metal slab relative to its nonretarded value as a function of the slab thickness. Lower line gives the surface polariton contribution to the total vacuum-field pressure.

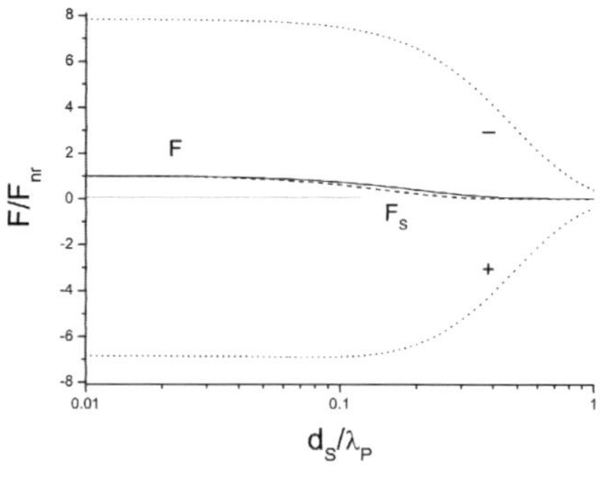

Figure 3. Surface polariton contribution (dashed line) to the total vacuum-field pressure (full line) on a free standing metal slab. Separate contributions $F_\pm^{\rm SP}$ of the two surface polariton modes are presented by dotted lines

At intermediate slab thicknesses, the pressure must be calculated using the exact result Eq. (14). The result of such a calculation are presented in Figs. 2 and 3 where we have plotted the pressure on a free standing slab relative to its nonretarded value $F_{\rm nr}$ [Eq. (15)] and the surface polariton contribution to it F_S, respectively. As seen, owing to the field retardation, the true Casimir pressure F deviates significantly from $F_{\rm nr}$ for slab thicknesses $d_s/\lambda_P > 0.01$ and can be in this region rather well approximated by the contribution of two SP modes supported by the slab F_S. Of course, at even larger slab thicknesses $(d_s/\lambda_P > 1)$, F_S vanishes owing to the decoupling of SP modes at two slab surfaces. The same happens to F since, as shown above, it attenuates exponentially for large thicknesses of the slab.

4. Metal slab between perfect mirrors

Next we consider the other limiting case namely that of a metal layer sandwiched between perfect mirrors. Such a situation is described by Eqs. (2) and (3) when letting $R_i^q = \delta_{qp} - \delta_{qs}$ and $d_i = 0$. From Eq. (3) it then follows that also $r_{s\pm}^q = \delta_{qp} - \delta_{qs}$ and we have

$$F = -\frac{\hbar}{\pi^2}\int_0^\infty d\xi \int_0^\infty dkk\kappa_s \frac{e^{-2\kappa_s d_s}}{1-e^{-2\kappa_s d_s}}. \tag{20}$$

This formula can be significantly simplified by exploiting the fact that the integral over k vanishes when $\xi \to \infty$. Following Schaden *et al.* [21], we partially integrate over ξ and simultaneously introduce κ_s as the integration variable in the integral over k. This gives

$$\begin{aligned} F &= \frac{\hbar}{2\pi^2}\int_0^\infty d\xi\xi\frac{d}{d\xi}\int_{\sqrt{\varepsilon_s(i\xi)}\xi/c}^\infty d\kappa_s\kappa_s^2 \sum_{q=p,s}\frac{e^{-2\kappa_s d_s}}{1-e^{-2\kappa_s d_s}} \\ &= -\frac{\hbar}{\pi^2 c^3}\int_0^\infty d\xi\xi^3\sqrt{\varepsilon_s(i\xi)}\frac{e^{-2\sqrt{\varepsilon_s(i\xi)}\xi d_s/c}}{1-e^{-2\sqrt{\varepsilon_s(i\xi)}\xi d_s/c}}, \end{aligned} \tag{21}$$

where, in the second step, we have noted that for the free-electron dielectric function $(d/d\xi)\sqrt{\varepsilon_s(i\xi)}\xi = 1/\sqrt{\varepsilon_s(i\xi)}$. Finally, making the substitution $y = \sqrt{\varepsilon_s(i\xi)}\xi/\omega_P$ and expanding the integrand, we obtain for the pressure on the slab surfaces

$$F = -\frac{\hbar c k_P^4}{\pi^2}\sum_{n=1}^\infty \frac{d^2}{da_n^2}\frac{K_1(a_n)}{a_n}, \qquad a_n = 2nk_P d_s, \tag{22}$$

where

$$K_1(a) = a\int_1^\infty dy\sqrt{y^2-1}e^{-ay} = \int_1^\infty dy\frac{y}{\sqrt{y^2-1}}e^{-ay} \tag{23}$$

is recognized as modified Bessel function of the second kind [22, 23].

To find pressure on a thin slab we use

$$\frac{d^2}{da^2}\frac{K_1(a)}{a} = \frac{6}{a^4} - \frac{1}{2a^2} + \mathcal{O}(a^0), \qquad a \ll 1.$$

This gives

$$f_s = F_C\left[1 - \frac{5}{\pi^2}(k_P d_s)^2 + \mathcal{O}[(k_P d_s)^4]\right], \qquad F_C = -\frac{\pi^2\hbar c}{240 d_s^4}, \tag{24}$$

so that the pressure on a thin metal layer is, to the leading order, given by famous Casimir result. Evidently, it is due to the modes propagating through the layer. Indeed, as follows from Eq. (12), this system supports only modes with frequencies $(\alpha_s = -in\pi/d_s)$

$$\omega_n^q(k) = c\sqrt{k_P^2 + k^2 + n^2\pi^2/d_s^2},$$

where n is an integer. For $k_P \ll d_s^{-1}$, these mode frequencies lead to the same Casimir energy as in the original Casimir configuration. The pressure on a thick slab can be obtained using

$$\frac{d^2}{da^2}\frac{K_1(a)}{a} = e^{-a}\sqrt{\frac{\pi}{2a^3}}\left[1+\frac{27}{8a}+\mathcal{O}(a^{-2})\right], \qquad a \gg 1$$

and is, to the leading order, again given by Eq. (19). These considerations are illustrated by the black full line in Fig. 4 giving the pressure F (relative to the Casimir pressure F_C) on a free-electron metal layer between ideal mirrors as a function of $k_P d_s$.

5. Metal slab in an ideal cavity

The pressure on a metal slab in the center of a cavity with perfectly reflecting mirrors is given by Eqs. (2) and (3) [or Eqs. (5)] with $R_i^q = \delta_{qp} - \delta_{qs}$ and $d_1 = d_2 \equiv d$ and can only be calculated numerically. Its behaviour with the slab thickness and the mirror-slab distance is illustrated in Figs. 4 and 5.

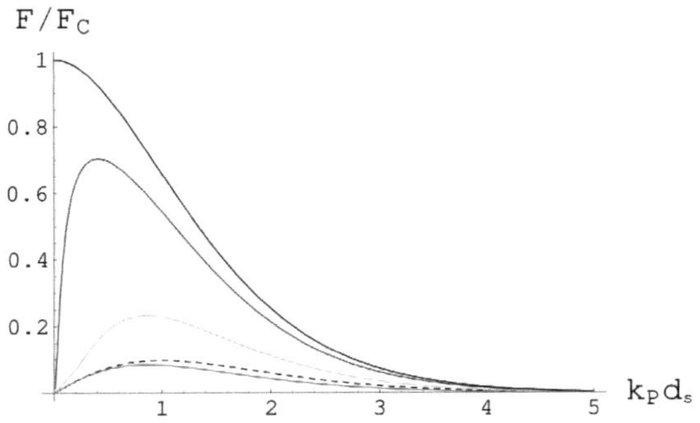

Figure 4. Relative vacuum-filed pressure on a metal slab in the center of an ideal cavity. The distance of the slab from the cavity mirrors is $k_p d = 0$ (black full line), $k_p d = 0.01$ (blue line), $k_p d = 0.1$ (green line), $k_p d = 1$ (red line) and $k_p d = \infty$ (black dashed line).

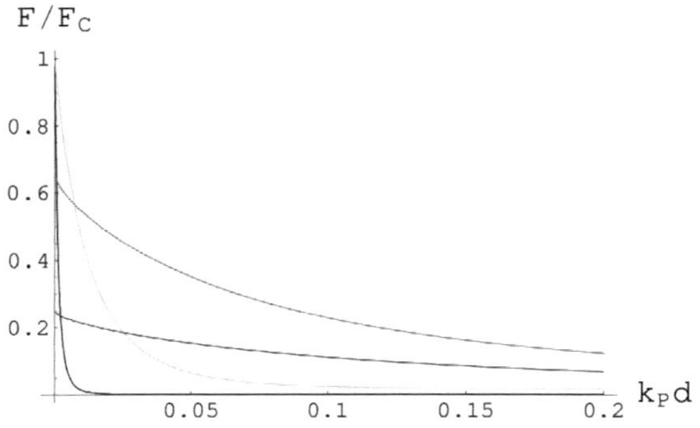

Figure 5. Relative vacuum-filed pressure on a metal slab in the center of an ideal cavity as a function of its distance from cavity mirrors. The thickness of the slab is $k_p d_s = 0.01$ (black line), $k_p d_s = 0.1$ (green line), $k_p d_s = 1$ (red line) and $k_p d_s = 2$ (blue line).

As seen, cavity strongly affects the pressure on a thin ($k_P d_s < 0.1$) slab; one observes its strong drop from F_C for non zero mirror-slab distances. As shown above, with increasing d the pressure on a thin slab decreases to $F_{\rm nr} = 0.19 k_P d_s F_C$. We also recall that, whereas for $d = 0$ the pressure is due to modes propagating through the metal, at large mirror-slab separations it is due solely to surface plasmon (evanescent) modes. This can be understood when realizing that nearby mirrors effectively damp surface modes since their field cannot accommodate to the perfect mirror boundary condition. Indeed, according to Eq. (12) (with $q = p$) the dispersion relations of SP modes in the present configuration can be rewritten as

$$\varepsilon(\omega) = -\frac{\alpha_s}{\alpha}\frac{\tanh^{\pm 1}\alpha_s d_s}{\tanh \alpha d}. \tag{25}$$

Accordingly, when $d \to 0$ frequencies of SP modes tend to zero. Thus when changing the cavity length, in addition to changing its magnitude, one effectively modify the mode spectrum of the pressure on the slab.

As expected, the effect of the cavity on the pressure on a thick ($k_P d_s > 1$) slab is less pronounced and diminish with the slab thickness owing to the attenuation of the vacuum-field fluctuations with d_s. We also observe from the red and black dashed curve in Fig. 4 that for systems with mirror-slab distances $k_P d \geq 1$ the cavity effect on the pressure is very small and the pressure on the slab differs very little (on this scale) from the pressure on a free standing slab. Thus, the effect of the cavity on the pressure on a slab in its center is largest for systems with the mirror-slab distances $k_P d < 1$, as illustrated in Fig. 5. Since λ_P for (noble) metals is of the order of 10^2nm, this corresponds to the mirror-slab distances less than (say) 20nm.

6. Summary

In this work we have considered and performed a modal analysis of the vacuum-field pressure on surfaces of a metal slab in the center of a cavity adopting the plasma model for the metal and assuming ideally reflecting cavity mirrors. We have confirmed analytically previous results for the pressure on a thin and a thick free-standing slab and derived an exact formula for the pressure on a metallic layer between perfect mirrors. According to these results, the pressure on a thin slab decreases with increasing slab-mirror distances from the well-known Casimir pressure F_C to the quasistatic pressure $F_{\rm nr} = 1.19(d/\lambda_P)F_C$. In the first case it is entirely due to the photonic modes propagating through the metal and in the second case it is entirely due to the surface plasmon modes of the free-standing slab. We recall that the pressure on metallic plates in the standard Casimir configuration behaves with decreasing their separation precisely in the same way: at large separations between the plates it is equal to F_C and is due to the photonic modes [5, 6] whereas at small separations between the plates it is equal to $F_{\rm nr}$ and is due to the surface plasmon modes supported by the plates [24, 25, 26, 27]. This similarity of the pressures in two configurations is lost in the case of a thick slab as the pressure on its surfaces exponentially decreases for large slab thicknesses. These considerations demonstrate that a cavity may influence the vacuum-field pressure on a thin metal slab (and its modal structure) to a considerable extent.

Acknowledgments

This work was supported by the Ministry of Science, Education and Sport of the Republic of Croatia under contract No. 098-1191458-2870.

Referencess

[1] Millonni P W 1994 *The Quantum Vacuum. An Introduction to Quantum Electrodynamics* (San Diego: Academic Press)
[2] Bordag M, Mohiden U and Mostepanenko V M 2001 *Phys. Rep.* **353** 1
[3] Milton K A 2004 *J. Phys. A: Math. Gen.* **37** R209
[4] Lamoreaux S K 2005 *Rep. Prog. Phys.* **68** 201
[5] Casimir H B G 1948 *Proc. Kon. Ned. Akad. Wet.* **51** 793
[6] Lifshitz E M 1955 *Zh. Eksp. Teor. Fiz.* **29** 94 [1956 *Sov. Phys. JETP* **2** 73]
[7] Dzyaloshinskii I E, Lifshitz E M and Pitaevskii L P 1961 *Adv. Phys.* **10** 165
[8] Abrikosov A A, Gorkov L P and Dzyaloshinski I E 1963 *Methods of Quantum Field Theory in Statistical Physics* (Englewood Cliffs NJ: Prentice-Hall) Ch 6
[9] Schwinger J, DeRaad L L Jr. and Milton K A 1978 *Ann. Phys. (N.Y.)* **115** 1
[10] Zhou F and Spruch L 1995 it Phys. Rev. A **52** 297
[11] Tomaš M S 2002 *Phys. Rev. A* **66** 052103
[12] Raabe C and Welsch D-G 2003 *Phys. Rev. A* **68** 033810
[13] Tomaš M S 2005 *Phys. Lett. A* **342** 381
[14] Raabe C and Welsch D-G 2005 *Phys. Rev. A* **71** 013814
[15] Benassi A and Calandra C 2007 *J. Phys. A: Math. Theor.* **40** 13453
[16] Maclay G J 2006 *arXiv:physics/0608196*
[17] Gusso A and Delben G J 2008 *J. Phys. D: Appl. Phys.* **41** 175405
[18] Imry Y 2005 *Phys. Rev. Lett.* **95** 080404

[19] Tomaš M S and Lenac Z 2008 *Phys. Rev.* A **57** 1108
[20] Ellingsen S A and Brevik I 2007 *J. Phys. A: Math. Theor.* **40** 3643
[21] Schaden M, Spruch L and Zhou F 1998 *Phys. Rev.* A **78** 023834
[22] Gradshteyn I S and Ryzhik I M 1994 *Table of Integrals, Series, and Products* ed. Jeffrey A (New York: Academic Press) p 363
[23] http://functions.wolfram.com/Bessel-TypeFunctions/BesselK
[24] van Kampen N G, Nijboer B R A and Schram K 1968 *Phys. Lett.* **26A** 307
[25] Genet G, Intravaia F, Lambrecht A and Reynaud S 2004 *Ann. Fond. L. de Broglie* **29** 311
[26] Henkel C, Joulain K, Mulet J-P and Greffet J-J 2004 *Phys. Rev.* A **67** 023808
[27] Bordag M 2006 *J. Phys.* A **39** 6173

Journal of Physics: Conference Series **161** (2009) 012018 doi:10.1088/1742-6596/161/1/012018

Ultraviolet divergences, repulsive forces and a spherical plasma shell

M Bordag

Institute for Theoretical Physics, Leipzig University, Vor dem Hospitaltore 1, D-04103 Leipzig

E-mail: Michael.Bordag@itp.uni-leipzig.de

Abstract. We discuss the vacuum energy of the electromagnetic field interacting with a spherical plasma shell together with a model for the classical motion of the shell. We discuss the ultraviolet divergences in terms of the heat kernel coefficients. Using these, we carry out the renormalization by redefining the parameters of the classical model. It turns out that this is possible and that the resulting model has a vacuum energy which changes sign in dependence on the parameters of the plasma shell. In the limit of the plasma shell becoming an ideal conductor the vacuum energy found by Boyer in 1968 is reproduced.

1. Introduction

In general, the repulsive forces found for the Casimir energy of a conducting sphere and a conducting cube remain a still not finally understood problem. It was a quite big surprise for the community when Boyer in 1968 [1] found the Casimir effect for the conducting sphere to be repulsive. Not only that this invalidated Casimir's electron model [2], this result was to some extend also counterintuitive. The background for that is in the relation between Casimir and van der Waals forces. From the one hand side, the Casimir forces follow from the zero point energy of the electromagnetic field confined between material bodies and on the other hand side the same formula emerges as a limiting case for becoming ideal conductors from the Lifshitz formula for the same material bodies. It is common understanding that this limiting case where the surfaces become ideal conductors is just the bridge between quantum field theory with its vacuum energy and the dispersion forces in many particle theory represented by the Lifshitz formula. These forces generalize the interaction between two molecules, which is given by the Casimir-Polder potential and which is always attractive. For this attraction there exists even a very general reason; it is a second order perturbation to a ground state energy which in quantum mechanics is always negative. Because of this deep connection Casimir's assumption that the attractive force he found for parallel planes should keep its sign when considering a sphere instead, was so very natural. For basically the same reasons the repulsive character of the force for the sphere was so unexpected.

During the last years with the development of nanotechnology, the discussion of repulsive forces became actual again. The reason is in the stickyness of the attractive forces. Because their increase at small separations any configuration of closely nearby located nanoobjects is potentially instable with respect to these bodies sticking together. This is a limitation in a number of applications. The question was discussed whether it is possible to use the repulsive force on a sphere for reducing the stickyness. Here one must remember that this repulsive

Journal of Physics: Conference Series **161** (2009) 012018 doi:10.1088/1742-6596/161/1/012018

force was found for an ideal configuration, namely an infinitely thin conducting spherical shell. This thinness is necessary because otherwise the ultraviolet divergences do not go away. In fact these cancel each other when taking the vacuum energies from inside the sphere and from outside together. But this compensation does not work for any finite thickness. In this sense the repulsive force must be considered a pure theoretical phenomenon. In this connection it must also be mentioned that recently in [3] quite general arguments were given in favor of the statement that the forces between separated bodies areattractive.

There is another situation with a repulsive force which makes the puzzle even worse. This is a dielectric ball. In general, this is nothing else than a macroscopic body characterized by a permittivity and one should expect the vacuum energy of the electromagnetic field in its presence to be finite. This is, however, not the case, except for the dilute approximation [4]. This means that the permittivity ϵ of the ball is close to its vacuum value such that $\epsilon - 1 = 4\pi N\alpha$, where N is the density and α is the static polarizability of the molecules, is small. Then the expansion of the vacuum energy up to second order in this small parameter is called the dilute approximation. There are two remarkable facts. The first is that in this approximation the vacuum energy is always finite in zetafunctional regularization or can be made finite in a unique manner. This was found for the first time in the beginning of the 80ies and mathematically confirmed in the end of the 90ies (see [4] and the citations therein). The second fact is that this vacuum energy can be calculated in two ways. The first way is the standard one by calculating the zero point energy of the electromagnetic field in the presence of the ball, see [5]. The second one is to sum up the pairwise Casimir-Polder potential

$$V(r_1, r_2) = -\frac{23}{(4\pi)^3 N^2} \frac{(\epsilon_1 - 1)^2}{|r_1 - r_2|^7}, \tag{1}$$

where r_1 and r_2 are the locations of these molecules, over all pairs of molecules constituting the ball. This was done in [6] carrying out the integrations in

$$E = -\frac{23}{8\pi} \frac{(\epsilon_1 - 1)^2}{(4\pi)^2} \int_V dr_1 \int_V dr_2 \, \frac{1}{|r_1 - r_2|^\gamma}. \tag{2}$$

Here the power in the denominator was changed to γ in order to make the integrations convergent. This is a regularization and to make the integrals convergent one needs to have $\gamma < 3$. In the end one has to perform the continuation to $\gamma = 7$. One arrives at

$$E = -\frac{23}{2^{2+\gamma}\pi} (\epsilon_1 - 1)^2 \frac{\Gamma\left(2 - \frac{\gamma}{2}\right)}{\Gamma\left(4 - \frac{\gamma}{2}\right)(3 - \gamma)} \frac{1}{R^{\gamma - 6}}. \tag{3}$$

This expression has a unique analytic continuation to $\gamma = 7$ where it takes a finite value and one comes the the known result

$$E_0 = \frac{23}{384\pi R} (c_1 - c_2)^2 + O\left[(c_1 - c_2)^3\right]. \tag{4}$$

As seen it is repulsive despite the fact that the pairwise potential is attractive. A deeper physical understanding of this fact is missing. Of course one can trace back mathematically the sign change, but the answer that it happens during an analytic continuation which is necessary to do in order to work with correct defined quantities, is not especially illuminating.

Besides the puzzle with the sign, another motivation for the present talk is to contribute to the discussion of the renormalization of vacuum energy in the presence of boundaries or singular background fields in application to the Casimir effect.

As it was pointed out in [7], there is a gap between the two well understood situations. These are, on the one side, the Casimir force between distinct objects which is always finite and, on the other side, the vacuum energy in smooth background fields which can be renormalized by standard methods of quantum field theory. In between these two, the situation is not finally settled. The mentioned paper, it was questioned whether boundaries can be incorporated at all into a well posed renormalization program. For instance, it was argued that the process of making the background field concentrated on a surface is not physical.

The aim of the present contribution is to discuss an example of a background field concentrated on a surface having both, a well posed renormalization procedure for the vacuum energy and a meaningful physical interpretation. As model we take a spherical plasma shell interacting with the electromagnetic field and we allow for a classical vibrational motion of the shell. The investigation of the plasma shell model was pioneered by Barton [8] and it is aimed to describe the π-electrons in a C_{60}-molecule.

2. The plasma shell model and its renormalization

In the plasma shell model the electrons are described by an electrically charged fluid whose motion is confined the shell. Further, the model contains an immobile, overall electrically neutralizing background aimed to describe the carbon atoms and the remaining electrons. The fluid is allowed a non-relativistic motion. Of course, this model is a quite crude simplification, especially because the motion of the electrons should rather follow a relativistic dispersion relation [9, 10]. On the other hand side it appears to be physically meaningful and should therefore result in physically meaningful results for the vacuum energy. For instance, it should allow for a treatment of the vacuum fluctuation of the electromagnetic field coupled to the plasma shell.

The interaction of the plasma shell with the electromagnetic field results in matching conditions on the electromagnetic field across the shell as shown in [8] (and earlier, for a plane sheet, in [11]). These conditions do not depend on the state of the excitations of the fluid. The vacuum energy can be calculated from the fluctuations of the electromagnetic field whereas the fluctuations of the fluid must not be taken into account as shown in [12] (or vice verse). In this setup, the polarizations of the electromagnetic field separate into TE- and TM-modes. For the corresponding amplitudes the following matching conditions hold ([8], [13]),

$$\begin{aligned} \lim_{r\to R+0} f_{l,m}(kr) - \lim_{r\to R-0} f_{l,m}(kr) &= 0, \\ \lim_{r\to R+0} (r f_{l,m}(k,r))' - \lim_{r\to R-0} (r f_{l,m}(k,r))' &= \Omega R f_{l,m}(kR), \end{aligned} \tag{5}$$

for the TE-mode, and

$$\begin{aligned} \lim_{r\to R+0} (r g_{l,m}(k,r))' - \lim_{r\to R-0} (r g_{l,m}(k,r))' &= 0, \\ \lim_{r\to R+0} g_{l,m}(kr) - \lim_{r\to R-0} g_{l,m}(kr) &= -\frac{\Omega}{k^2 R}(R g_{l,m}(k,R))', \end{aligned} \tag{6}$$

for the TM mode. The parameter

$$\Omega = \frac{4\pi n e^2}{mc^2} \tag{7}$$

carries information on the properties of the fluid like its density n and mass m.

Considered as a scalar problem, the matching conditions (5) of the TE mode are equivalent to a delta function potential $\Omega\delta(r-R)$ in the wave equation and the conditions (6) of the TM mode loosely speaking correspond to the derivative of a delta function. A difference is that in the scalar problems the zeroth orbital momentum, $l=0$, or s-wave contribution is present whereas in the electromagnetic case it is absent and the sum over l in the vacuum energy starts from $l=1$. In the limit $\Omega\to\infty$ which is formally the ideal conductor limit the boundary conditions

(5) and (6) became Dirichlet boundary conditions for TE polarization and Neumann for TM polarization.

We extend this model by allowing for radial vibrations (breathing mode) of the plasma shell. In C_{60} these are determined by the elastic forces acting between the carbon atoms. Without going here in any detail we describe these vibrations phenomenologically by a Hamilton function

$$H_{\rm class} = \frac{p^2}{2m} + \frac{m}{2}\,\omega_b^2\,(R-R_0)^2 + E_{\rm rest} \tag{8}$$

with a momentum $p = m\dot{R}$. Here m is the mass of the shell, ω_b is the frequency of the breathing mode, R_0 is the radius at rest and $E_{\rm rest}$ is the energy which is required to bring the pieces of the shell apart, i.e., it is some kind of ionization energy.

Now we consider a system consisting of the classical motion of the shell as described by $H_{\rm class}$ and the vacuum energy $E_{\rm vac}$ of the electromagnetic field interacting with the shell by means of the matching conditions (5) and (6). We assume the classical motion adiabatically slow such that the vacuum energy can be taken as a function of the momentarily radius of the shell, $E_{\rm vac} = E_{\rm vac}(R)$, and we neglect the backreaction of the electromagnetic field on the shell. Under these assumptions the energy of the classical system, $E_{\rm class}(R) = H_{\rm class}$, and the vacuum energy add up to the total energy of the considered system,

$$E_{\rm tot} = E_{\rm class}(R) + E_{\rm vac}(R). \tag{9}$$

Next we consider the ultraviolet divergences of the vacuum energy. These are given in general terms by the heat kernel coefficients a_n (we use the notations of [14] and we can define a 'divergent part' of the vacuum energy which is, as known, not uniquely defined. It depends on the kind of regularization one has to introduce. For instance, in zeta functional regularization, the regularized vacuum energy reads

$$E_{\rm vac}(s) = \frac{\mu^{2s}}{2}\sum_n \omega_n^{1-2s}, \tag{10}$$

where μ is an arbitrary parameter with the dimension of a mass and with a frequency damping function it is,

$$E_{\rm vac}(\delta) = \frac{1}{2}\sum_n \omega_n\, e^{-\delta\omega_n}, \tag{11}$$

where ω_n are the frequencies of the quantum fluctuations of the electromagnetic field. In our problem the spectrum is continuous, but for the moment it is more instructive to keep the notations of a discrete spectrum. In zeta functional regularization, the divergent part reads

$$E_{\rm vac}^{\rm div}(s) = -\frac{a_2}{32\pi^2}\left(\frac{1}{s} + \ln\mu^2\right), \tag{12}$$

where we used the notations of [14] in which the heat kernel expansion reads

$$K(t) \sim \frac{1}{(4\pi t)^{3/2}}\left(a_0 + a_{\frac{1}{2}}\sqrt{t} + a_1\, t + \dots\right). \tag{13}$$

In the scheme with the frequency damping we have

$$E_{\rm vac}^{\rm div}(\delta) = \frac{3a_0}{2\pi^2}\frac{1}{\delta^4} + \frac{a_{1/2}}{4\pi^{3/2}}\frac{1}{\delta^3} + \frac{a_1}{8\pi^2}\frac{1}{\delta^2} + \frac{a_2}{16\pi^2}\ln\delta. \tag{14}$$

The regularizations are removed by $s \to 0$ resp. $\delta \to 0$. These formulas follow, for example, from section 3.4 in [14] for $m = 0$.

The idea of the renormalization is to have in the classical energy $E_{\rm class}$ parameters which can be changed in a way to absorb $E_{\rm vac}^{\rm div}$. In the considered model such parameters are the mass m of the shell, the frequency ω_b of the breathing mode, the radius at rest R_0, and the energy $E_{\rm rest}$. Now, whether this is possible, is a matter of the dependence of the heat kernel coefficients, especially of a_2, on the radius R which is the dynamical variable of the classical system. In the considered, very simple model we have only a polynomial dependence on R up to second order in (8). Since we assumed adiabaticity for the motion of the shell we do not have a time dependence in a_2 so that it cannot contain $\dot{R}$. Hence, the kinetic energy remains unchanged and, together with it, the mass m. Only the remaining parameters, ω_b, R_0 and $E_{\rm rest}$ can be used to accommodate the divergent part. In fact, this turns out to be sufficient for the considered model. As it will be seen below, the heat kernel coefficients $a_0, \ldots, a_2$ entering the divergent part, depend on the radius polynomial and at most quadratically. In this way this model is renormalizable.

It should be mentioned that this scheme follows closely the corresponding one in quantum field theory with $E_{\rm vac}^{\rm div}$ in place of the counterterms. Also the interpretation of the renormalization is similar. Namely, we argue that the vacuum energy in fact cannot be switched off and what we observe are parameters like, for example in QED, electron mass and charge, after renormalization.

Within this scheme of renormalization, the specific form of the heat kernel coefficients is insignificant. One has to bother of its dependence on R in order to fit into the freedom of redefining the parameters in $E_{\rm class}$. If this is the case, one may define a renormalized vacuum energy by means of

$$E_{\rm vac}^{\rm ren} = \lim_{s\to 0} \left(E_{\rm vac}(s) - E_{\rm vac}^{\rm div}(s) \right) \tag{15}$$

(and the same with δ in place of s) and one has now to consider

$$E_{\rm tot} = E_{\rm class} + E_{\rm vac}^{\rm ren} \tag{16}$$

in place of (9). In this way, the question on how to remove the ultraviolet divergences is answered.

It remains, however, the question about the uniqueness of his procedure which comes in from the parameter μ in the zeta functional scheme or from the possibility of a redefinition $\delta \to c\delta$, where c is an arbitrary constant, in the other scheme.

In the case of QED at this place one imposes conditions on the analog of $E_{\rm vac}^{\rm ren}$ in a way, that the mass and the charge take the values one observes experimentally.

In our case a similar scheme is conceivable too. A different scheme, suggested in [4], using the large mass expansion to fix the ambiguity does not work here since the electromagnetic field is massless. A way out could be to look for a minimum of the total energy, $E_{\rm tot}$, (9), which however would imply to take the model (8) seriously. This is not the aim of the present paper. Instead, as a normalization condition we demand that in the limit of the plasma frequency $\Omega \to \infty$, where the matching conditions (5) and (6) turn into that of an ideal conductor, we shall recover the vacuum energy of a conducting spherical shell, i.e., just the quantity which was first calculated by Boyer in [1]. Indeed, as we will see in the next section, this is possible using the freedom of a finite renormalization.

3. The heat kernel coefficients and the renormalized vacuum energy

The electromagnetic field interacting with the plasma shell is defined in the whole space and it has a continuous spectrum. In that case the vacuum energy, after the subtraction of the contribution of the empty space, can be represented in the form (see Eq.(3.43) in [14])

$$E_0(s) = -\frac{\cos \pi s}{\pi}\mu^{2s}\sum_{l=1}^{\infty}\nu\int_0^{\infty} dk\, k^{1-2s}\frac{\partial}{\partial k}\ln f_l(ik) \tag{17}$$

with $\nu = l + 1/2$. The arbitrary parameter μ has the dimension of a mass and $f_l(k)$ is the Jost function of the corresponding scattering problem. With argument rotated to the imaginary axis these read

$$\begin{aligned} f_l^{\rm TE}(ik) &= 1 + \frac{\Omega}{k}\, s_l(kR)e_l(kR), \\ f_l^{\rm TM}(ik) &= k^2\left(1 - \frac{\Omega}{k}\, s_l'(kR)e_l'(kR)\right), \end{aligned} \tag{18}$$

where we used the modified Riccati-Bessel functions

$$s_l(x) = \sqrt{\frac{\pi x}{2}}\, I_{l+1/2}(x), \qquad e_l(x) = \sqrt{\frac{2x}{\pi}}\, K_{l+1/2}(x). \tag{19}$$

By these formulas, the heat kernel coefficients, including that for double poles, which are related to the vacuum energy by means of

$$2\mu^{-2s}(4\pi)^{3/2}\Gamma\left(s-\frac{1}{2}\right)E_0(s) = \sum_{k\geq 0}\frac{a_{k/2}}{s-2+\frac{k}{2}} + \sum_{k\geq 3}\frac{a'_{k/2}}{\left(s-2+\frac{k}{2}\right)^2} + \dots, \tag{20}$$

can be calculated. For the TE mode this repeats earlier calculations, for the TM mode is was done for the first time in [13]. The result is shown in the table. It must be mentioned that for the TM-mode double poles in s appear which were included in Eq. (20) and the corresponding coefficients are marked with a prime. These result in logarithmic terms in the heat kernel asymptotics. In fact, in our case these appear starting with $k = 5/2$. Therefore these do not affect the renormalization.

$k/2$	TE-mode	TM-mode
0	0	0
1/2	0	$8\pi^{3/2}R^2$
1	$-4\pi\Omega R^2$	$-\frac{4\pi}{3}\Omega R^2$
3/2	$\pi^{3/2}\Omega^2R^2$	$-\frac{10}{3}\pi^{3/2}$
2	$-\frac{2}{3}\pi\Omega^3R^2 + 4\pi\Omega$	$-4\pi\Omega + \frac{2\pi}{15}\Omega^3R^2$

Table 1. The first few heat kernel coefficients for the plasma shell as defined in Eq. (20).

Using these heat kernel coefficients, the renormalization program discussed in the preceding section was carried out in [13]. The vacuum energy can be divided into asymptotic and numeric parts. The asymptotic parts were found to be

$$\begin{aligned} E_{\rm vac}^{\rm TE,\,as}(s) &= -\frac{a_2^{\rm TE}}{32\pi^2}\left[\frac{1}{s} - 2\ln\frac{\Omega}{2\mu}\right] + \frac{\Omega^3R^2}{72\pi} + \frac{\Omega}{180\pi} + E_{\rm vac}^{\rm TE,\,an} + O(s), \\ E_{\rm vac}^{\rm TM,\,as}(s) &= -\frac{a_2^{\rm TM}}{32\pi^2}\left[\frac{1}{s} - 2\ln\frac{\Omega}{2\mu}\right] + \frac{7\Omega^3R^2}{1800\pi} - \frac{29\Omega}{36\pi} + E_{\rm vac}^{\rm TM,\,an} + O(s). \end{aligned} \tag{21}$$

These expressions are sums of a divergent part (it is proportional to the heat kernel coefficient a_2 as expected), two terms growing with Ω and an 'analytical' part which has a finite limit for $s \to 0$.

All quantities entering $E_{\rm vac}^{\rm TE,\,an}$ and $E_{\rm vac}^{\rm TM,\,an}$ are defined in [13]. The analytical parts have a limit for $\Omega \to \infty$,

$$\begin{aligned} \lim_{\Omega\to\infty} E_{\rm vac}^{\rm TE,\,an} &= \frac{17}{128R}, \\ \lim_{\Omega\to\infty} E_{\rm vac}^{\rm TM,\,an} &= -\frac{11}{128R} \end{aligned} \tag{22}$$

which matches the corresponding expressions known from the conducting sphere.

With Eqs. (21) and the property (22) we have all information we need to complete the renormalization. We remind the discussion the preceding section that all terms which are proportional to R, R^2 or which do not depend on R can be removed by a redefinition of the parameters in the classical part. In (21) this concerns all except the last ones, $E_{\rm vac}^{\rm TE,\,an}$ and $E_{\rm vac}^{\rm TM,\,an}$. That means, that we not only can remove the contribution proportional to a_2, but also the terms growing with Ω. For this reasons we define the renormalized vacuum energies by

$$\begin{aligned} E_{\rm vac}^{\rm TE,\,ren} &= E_{\rm vac}^{\rm TE,\,num} + E_{\rm vac}^{\rm TE,\,an}, \\ E_{\rm vac}^{\rm TM,\,ren} &= E_{\rm vac}^{\rm TM,\,num} + E_{\rm vac}^{\rm TM,\,an}. \end{aligned} \tag{23}$$

With these formulas we completed the model consisting of the classical energy and the vacuum energy which is the sum of the two contributions in (23). The main merit of this vacuum energy is that it turns for $\Omega \to \infty$ into the ideal conductor limit. Using the formulas for the Jost functions and also the formulas given in [13], it is possible to evaluate this vacuum energy numerically. The results are shown in the figures 1 and 2 for the dimensionless function $\mathcal{E}$ defined by

$$E_{\rm vac}^{\rm ren} = \frac{\mathcal{E}(\Omega R)}{R} \tag{24}$$

as functions of their arguments $x = \Omega R$.

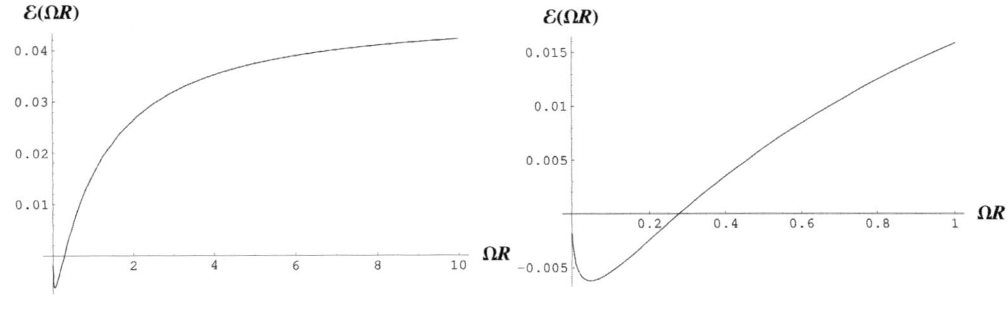

Figure 1. The function $\mathcal{E}(\Omega R) = R E_{\rm vac}^{\rm ren}$ plotted as function of Ω. For large Ω it tends to the ideal conductor limit, $\lim_{\Omega R\to\infty} = 0.0046$ (left panel). For small Ω (right panel) it takes negative values and decreases as $\mathcal{E}(\Omega R) \sim -0.0589\sqrt{\Omega R}$.

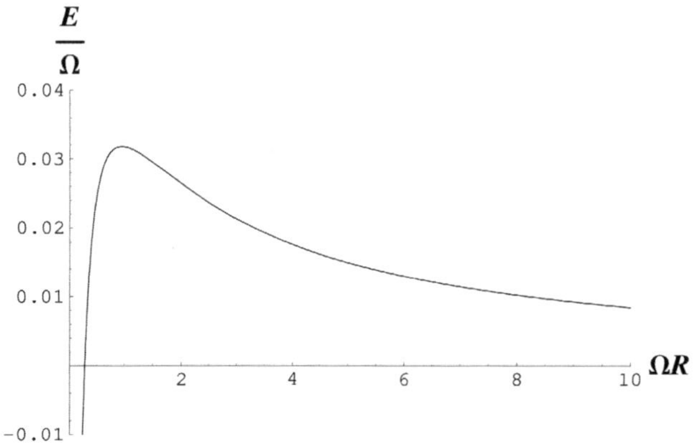

Figure 2. The renormalized vacuum energy divided by Ω, $E^{\rm ren}_{\rm vac}/\Omega$, as a function of R. For large R it approaches the ideal conductor limit and for small radii, $R \leq \Omega^{-1}$, it becomes attractive.

4. Conclusion

In the foregoing sections we considered the vacuum energy of the electromagnetic field interacting with a spherical plasma shell. We considered the heat kernel coefficients for both polarizations. It turned out that the vacuum energy in zeta functional regularization and, with it the corresponding zeta function, have double poles. This implies that the corresponding spectral problem is more complicated. On the other hand, at least based on the calculations carried out in this paper, there is nothing which would diminish the reasonableness of this model.

A basic concern of this paper is to construct a model allowing for a physically meaningful interpretation of the renormalization. We considered with the breathing mode of the shell the simplest model for the classical motion of the shell. It turned out that this model is able to accommodate all renormalizations which we were like to carry out. These are the removal of the pole in $s = 0$, i.e., of the ultraviolet divergence, and the removal of all contributions growing together with the plasma frequency Ω. It should be mentioned that this includes also the removal of the arbitrary constant μ which came in with the regularization. The nontrivial statement which allowed for doing so is that the dependence on the radius R of all these contributions is polynomial not exceeding R^2.

It should be mentioned that the same construction which we made in this paper for the plasma shell model interacting with the electromagnetic field can be done along the same lines also with a shell carrying a delta function and interacting with a scalar field which could be considered as a kind of acoustic field. All calculations for such a system were done in literature, the heat kernel coefficients were found in [4] and the renormalized vacuum energy was calculated in [15]. However, these fragments were not assembled into a consistent renormalization scheme.

It would be interesting to investigate the question whether the presented here renormalization procedure can be carried out also for more general deformations of the shell. In principle, most ingredients for such a calculation are available. Especially, the heat kernel coefficients for the TE modes can be taken from [16]. It would remain to calculate the coefficients for the TM modes.

The renormalized vacuum energy within our model is represented in the Figures 1 and 2. It smoothly interpolates between the limiting cases of large and small plasma frequency Ω. In the limit of $\Omega \to \infty$, where we recover the known result for the conducting sphere, the energy is positive, whereas for small Ω it becomes negative. This is in accordance with expectations coming from the weak coupling limit where the forces should become like van der Waals forces which are attractive. It must be mentioned, however, that a simple perturbative expansion for small Ω would fail since the behavior for $\Omega \to 0$ is nonanalytic.

Concerning the arbitrariness of the normalization procedure we would like to mention that the removal of the contributions growing together with Ω can be considered as a normalization condition. The motivation of such a condition is twofold. First of all the fact must be emphasized that in this way a reasonable condition exists. This is nontrivial having in mind, for example, the situation with the dielectric ball which shows beyond the dilute approximation inevitably a divergence [4], even if dispersion is included [17] and no normalization condition was found so far. A second motivation comes from the physical sense of the limit $\Omega \to \infty$. A plasma shell becomes for growing Ω more and more like a conductor. If one starts for finite Ω from assuming that a finite vacuum energy makes physically sense, then an unbounded increase with increasing Ω would be unphysical. Hence a finite limit must be reached. From the calculations carried out in this paper, especially, from Eqs.(22), it turns out that this is just the ideal conductor limit. It should be underlined that there is no way to get any different finite value since all ambiguities in (21), especially the logarithmic term which is proportional to a_2, grow with powers of Ω. There is a further reason for our choice of the normalization condition. It follows from dimensional considerations suggesting the general form (24) of the vacuum energy. From these it follows, that a function $\mathcal{E}(\Omega R)$ growing with Ω would result in a vacuum energy which does not vanish for $R \to \infty$ (at fixed Ω). This is clearly unphysical.

In this way, our normalization condition ensures the uniqueness of the renormalized vacuum energy and makes this model physically meaningful. In addition, the gap between the renormalization procedure in quantum field theory in smooth background fields and the removal of divergences of the Casimir energy in the background of boundaries, as suggested for example in [18] (section 6.5), is narrowed. At once, in this way, the much discussed vacuum energy of a conducting spherical shell now appears as a limiting case of a slightly more physical model.

Acknowledgments

This contribution is based on the paper [13] performed together with N. Khusnutdinov.
I would like to thank the organizers of the International Workshop *60 years of Casimir effect* for their kind hospitality and the excellent conditions.

References

[1] T H Boyer. Quantum Electromagnetic Zero-Point Energy of a Conducting Spherical Shell and Casimir Model for a Charged Particle. *Phys. Rev.*, 174:1764, 1968.
[2] H B G Casimir. Quantum electrodynamics - introductory remarks. *Physica*, 19(7):846–849, 1953.
[3] O Kenneth and I Klich. Opposites attract: A theorem about the Casimir force. *Physical Review Letters*, 97(16), 2006.
[4] M Bordag, K Kirsten and D Vassilevich. On the ground state energy for a penetrable sphere and for a dielectric ball. *Phys. Rev.*, D59:085011, 1999.
[5] G Lambiase, G Scarpetta and V V Nesterenko. Exact value of the vacuum electromagnetic energy of a dilute dielectric ball in the mode summation method. *Mod. Phys. Lett.*, A16:1983–1995, 2001.
[6] Kimball A Milton and Y Jack Ng. Observability of the bulk Casimir effect: Can the dynamical Casimir effect be relevant to sonoluminescence? *Phys. Rev.*, E57:5504–5510, 1998.
[7] N Graham et al. The Dirichlet Casimir problem. *Nucl. Phys.*, B677:379–404, 2004.
[8] G Barton. Casimir energies of spherical plasma shells. *J. Phys. A: Math. Gen.*, 37(3):1011–1049, 2004.
[9] S A Maksimenko and G Ya Slepyan. Electrodynamics of Carbon Nanotubes. *J. Comm. Tech. Electron.*, 47:235–252, 2002.
[10] Mikhail I Katsnelson. Graphene: carbon in two dimensions. 2006. arXiv:cond-mat/0612534.
[11] G Barton. Perturbative Casimir energies of dispersive spheres, cubes and cylinders. *J. Phys. A: Math. Gen.*, 34(19):4083–4114, 2001.
[12] M Bordag. On the interaction of a charge with a thin plasma sheet. *Phys.Rev.D*, 76:065011, 2007.

Journal of Physics: Conference Series **161** (2009) 012018 doi:10.1088/1742-6596/161/1/012018

[13] M Bordag and N Khusnutdinov. On the vacuum energy of a spherical plasma shell. *Phys.Rev.D*, 77:085026, 2008.
[14] M Bordag, U Mohideen and V M Mostepanenko. New developments in the Casimir effect. *Phys. Rep.*, 353:1–205, 2001.
[15] Marco Scandurra. The ground state energy of a massive scalar field in the background of a semi-transparent spherical shell. *J. Phys. A: Math. Gen.*, 32:5679–5691, 1999.
[16] D V Vassilevich. Heat kernel expansion: User's manual. *Phys. Rept.*, 388:279–360, 2003.
[17] M Bordag and K Kirsten. Heat kernel coefficients and divergencies of the Casimir energy for the dispersive sphere. *Int. J. Mod. Phys. A*, 17:813–819, 2002.
[18] Steve Blau, Matt Visser and Andreas Wipf. Zeta Functions and the Casimir Energy. *Nucl. Phys.*, B310:163, 1988.

Journal of Physics: Conference Series **161** (2009) 012019 doi:10.1088/1742-6596/161/1/012019

A remembrance of Hendrik Casimir in the 60th anniversary of his discovery, with some basic considerations on the Casimir Effect

Emilio Elizalde

Consejo Superior de Investigaciones Científicas
Instituto de Ciencias del Espacio & Institut d'Estudis Espacials de Catalunya (IEEC/CSIC)
Campus UAB, Facultat de Ciències, Torre C5-Parell-2a planta
08193 Bellaterra (Barcelona) Spain

E-mail: elizalde@ieec.uab.es

Abstract. When the number and importance of the applications of the Casimir effect are flourishing, and on the occasion of the 60th anniversary of his beautiful discovery, as a tribute to the memory of Hendrik Brugt Gerhard Casimir I discuss here some fundamental issues related with the effect that need to be recalled from time to time, as well as on some of my personal impressions of Prof. Casimir. This article may also serve as an easy introduction for the non-specialist willing to learn something about the quantum vacuum.

1. Introduction: Hendrik Casimir

Though many of the things I will say in what follows may be well known to some of the readers, I cannot avoid recalling them here in order to put into context the different points I will address in the following sections. Let me here start with a short biographic introduction. Hendrik Brugt Gerhard Casimir was born in Den Haag (Holland) on July 15, 1909. Starting in 1928 he studied Physics at the University of Leiden, where he finished his PhD in 1931. In his PhD Thesis he dealt, from the point of view of Quantum Mechanics, with the problem of a rotating rigid body, as well as with the theory of the rotation group as applied to molecules. During this period Casimir spend a short time in Copenhagen, with Niels Bohr. I was really impressed to read in Casimir's autobiography the intensity of the effect that produced on him the first encounter with the great genius. He tells there the story that, in order to be able to convince his parents in Holland that Bohr, the professor he was visiting, was a really important person, he told them to write a letter to him [their son] in Denmark just putting on the envelope as address the following: Hendrik Casimir c/o Niels Bohr, Denmark, and nothing else. They did so and, in fact, the letter punctually traveled from Holland to Casimir's hands, in very few days. It is nice to recall that in those times the important physicists were recognized as great personalities, as are now movie stars or rap singers. They had wide recognition in their own countries, and some of them in the whole world. Physics itself was very highly regarded too.

After finishing his PhD, Casimir worked as an assistant of Wolfgang Pauli in Zürich. In 1938 he became a professor of physics in Leiden. During this time, he studied some problems on heat and electricity conduction and he contributed to projects aimed at reaching the lowest temperatures known at that time, of the order of a few millikelvin. A low temperature lab was (as is also today) fundamental in order to carry out interesting research in material science, with a wide spectrum of aims. In 1942, during the second World War, Casimir moved to the Research Laboratories of Philips Incorporation in Eindhoven, Holland. In his new workplace que continued to be an active scientist, and in 1945 he wrote a paper, that became famous, on Onsager's principle of microscopic irreversibility. As soon as 1946, Casimir already became co-director of Philips' Research Laboratories, and a member of the Board of Directors of the Philips Incorporation in 1956, where he remained until 1972. In spite of having been working for the industry for most of his professional life, Casimir continues to be one of the more prominent theoretical physicists that have come out of his country, Holland that, let us remember, had not long ago a very important success with the 1999 Nobel Prizes obtained by Martinus Veltman and Gerardus 't Hooft.

Aside from the effect that now carries his name, as well as its relation with de van der Waals forces [1] (which I will discuss more extensively in what follows), Casimir published other important contributions to science, as to the theory of Lie groups, to hyperfine structure and the calculus of quadripolar models, the physics of low temperatures, magnetism and the thermodynamics of superconductors, and the already mentioned contribution to Onsager's theory. Casimir played also a role in the creation of the European Physical Society, of which he became president in 1972. In 1979 he was one of the main speakers in the celebration of the 25th Anniversary of CERN, the European Center for Nuclear Research, in Geneva (Switzerland). In September 1998, when he was almost 90, in which was finally to be the very last of his scientific activities, he gave the opening lecture under the title *"Some remarks on the history of the so called Casimir effect"*, in the 4th Meeting of the series QFEXT, that took place in Leipzig (Germany). He passed away on May 4, 2000, in Heeze (Holland).

I keep in my office, as a treasure, his self-biographical book *"Haphazard Reality: Half a Century of Science"*, that he kindly send to me together with a handwritten letter, in 1999, shortly before his dead. In the letter he gave an answer to my invitation to come to Barcelona and visit our Institute, that I had extended to him personally during the Leipzig meeting. Regretfully, the visit could never take place.

2. A few Historical Notes on the Casimir Effect

It is, at the very least, quite shocking that in the above named biography, Casimir makes almost no mention of the very important effect that bears his name [1]. He just writes on passing, in page 247, that his phenomenon *"is mentioned sometimes under the name Casimir effect"*, that *"it has been found experimentally"*, and that *"their generalizations have been seen to have certain theoretical importance"*. He describes, in a different paragraph, how in 1951, while both of them were participating in a conference in Heidelberg, he tried very hard to make understandable to the prominent physicist Wolfgang Pauli his arguments about the physical reality of the effect he had discovered, and how Pauli was unable to grasp them. All of Casimir's efforts, from one perspective, from another, were useless since Pauli kept building one objection after another until the point when, having Pauli realized that all of his counterarguments had no effect on the enthusiastic Casimir, Pauli called him, repeatedly, *"Stehaufmanderl"*, which is a very strong form of 'stubborn', and makes literal reference to a toy, very common in Germany at the time, with the form of a clown of wood that keeps all the way its vertical position, owing to its having a very heavy spherical base made of steel that does not allow it to fall down, even if you try it by any means! May this anecdote serve as a convincing example which makes perfectly clear that understanding the essence of the Casimir effect is far from easy. Some of the problems associated with this phenomenon still remain nowadays, as may have become clear to the reader from some of the contributions to these Proceedings.

During about ten years, starting towards the end of the 80's, Barcelona was surely the most active place on earth working on (the theoretical aspects mainly) of the Casimir effect: Rolf Tarrach, Enric Verdaguer, Sonia Pabàn, August Romeo, Sergio Leseduarte, Klaus Kirsten, and the author of this paper himself, with some punctual contributions of Leipzig colleagues too, published a considerable number of papers, some of which have become, with the passage of years, fundamental and much loved references on the subject (judged from the several hundreds of citations they have collected since then). Several PhD thesis came out of this activity, not to speak of the researchers that got important promotions, to finish as professors inside and outside of Spain, thanks to this activity. Needless to say, it was a period of my scientific life of which I am very proud. Much more recently, in 2005, I had the pleasure and the honor to organize, in CosmoCaixa, Barcelona, the seventh meeting, QFEXT'05, of the already mentioned series on "Quantum Field Theory under the Influence of External Conditions". Casimir, for one, had participated in the fourth of these meetings, which took place in Leipzig, as previously mentioned.

3. On Heisenberg's Indeterminacy (or Uncertainty) Principle

Quantum Mechanics (QM), later extended to the Quantum Theory of Fields (QFT), constituted one of the great scientific revolutions of the XXth Century. Some of the fundamental principles in which it is based, as well as the consequences that follow from the axioms of these theories, in particular, the ones related with the collapse of the wave function and with the theory (and practice) of the quantum measurement, have not started to be well understood and successfully used until the first years of the XXIst Century. The situation will most probably improve in the years to come. The effect we are here dealing with refers precisely to one of these basic principles, and the very concepts associated to the same were the object of lively discussions among some of the founders of quantum physics (and some of its detractors, too).

The Casimir effect is usually 'defined' as "any measurable consequence of the quantum fluctuations of the vacuum state of a quantum system, as they are revealed by the imposition of boundary conditions, a specific topology, a background field, etc." In Heisenberg's version a quantum system has associated an energy operator, the system's Hamiltonian, which has a spectrum of eigenvalues and corresponding eigenvectors. The state of minimal energy (eigenvalue) is called the "vacuum state". It is important to understand that the quantum system is here, so to say, the 'universal domain' the playground or battlefield, in the same way as the 'universal set' and the 'probability space' one builds up in probability theory. In this way a classical similarity is closely established —in fact quantum mechanics has much to do with probability theory (it had even more directly to do in the case of the original formulation by E. Schrödinger, that was unsuccessful, as the reader will surely know). To sum up, the vacuum state of a quantum system is that of minimal energy, what can be ideally represented as the one of maximal 'quietness' an 'ocean of tranquility' in the most strict sense possible within the given quantum system. In order to exit from the vacuum state, the system needs a supply of energy. In the absence of this, how can there possibly be fluctuations of this vacuum state? The quantum fluctuations of the vacuum state are allowed by Heisenberg's uncertainty (or indeterminacy) principle, which is one of the pillars of the quantum world. In particular, one of its consequences is that a statement saying that "the energy of a quantum system in a given state has, at given time, a specific value'" has no strict sense. It is because of this reason that it turns out to be impossible to prevent the possibility of the vacuum state to fluctuate, the fluctuations having the order of magnitude given by Heisenberg's principle, that is, of Planck's fundamental constant. In this way, creation and anihilation of virtual pairs of a particle and an antiparticle occur, in a way such that all characteristics of the vacuum of our quantum system, as its charge, spin, etc., are kept unaltered during such processes. We should here stress the enormous importance of Heisenberg's principle. Let us recall, for instance, that thirty years ago now, that is, also thirty years after Casimir's discovery, Hawking formulated his nowadays very famous theory on the quantum radiation of black holes taking as a fundamental point in his argument the uncertainty principle.

Let us now just do a further step, by choosing a specific model. It is very common in Quantum Mechanics to use the harmonic oscillator model in order to clarify any argumentation. This is also true, to some extent, in Classical Mechanics, but in the quantum paradigm it turns out to be almost unavoidable. The quantization of the simple harmonic oscillator leads to the result that the energy eigenvalues are given by: $E_n = (n + 1/2)\hbar\omega$, $n = 0, 1, 2, 3, ...$, where $\hbar$ is Planck's constant (over 2π) and ω the frequency of oscillation. The state of minimal energy corresponds to $n = 0$ and the vacuum to vacuum transition of the system turns into: $\langle\emptyset|H|\emptyset\rangle = (\hbar/2)\sum_\omega \omega$, where the sum must be interpreted in a general sense as a 'summation' over all possible frequencies: it turns into an infinite sum (a mathematical series) if the spectrum is discrete (as, for instance, when the quantum system is confined to a box, or if it is bound to have periodic boundary conditions); but it is an integral in the case of a continuous spectrum (as for an unbounded system); or, more generally, it is a multidimensional combination of both cases as, for instance, when the system is confined in one or more directions but unbounded in other directions. This is just a technical issue, of very little relevance in fact, but it turns out that some scholars not working on the subject but who got interested in it (this happens more and more often recently) get already stuck at this point and keep asking why we put a simple summation sign at that place.

What is indeed important is to ask the question (as already did so many physicists in the last century, as Albert Einstein, Wolfgang Pauli or Niels Bohr): has this 'vacuum energy', that is, the energy associated to the vacuum to vacuum transition and which is given in terms of the half-Planck-quantum, any physical meaning? That is, could it be seen in the lab? Can it have any observable consequences? In fact these question were not difficult to answer within the standard formalism of Quantum Field Theory. Continuing with the study of the simple harmonic oscillator in terms of creation and anihilation operators, it is a standard procedure in QFT to take the normal ordering prescription, which fixes as equal to zero the vacuum energy, by postulating that all creation operators are to be put to the right, while all anihilation operators should go to the left. Of course this is a prescription, and one could equally well take a different one, resulting in a different value for the vacuum energy. All this hints already to the answer of the questions above: within QFT the vacuum energy is an arbitrary quantity (it just means fixing an arbitrary origin of energies) and cannot be measured experimentally. This seems to be generally accepted by everybody now (see, however [2]). The situation radically changes when general relativity is involved: the vacuum itself gravitates and the vacuum fluctuation energy fulfills the equivalence principle [3, 4]), and here indeed it has an absolute meaning as a physically measurable quantity. Anyway, for a long time gravity considerations did not come into play and vacuum fluctuations remained a useless idea[1], until Casimir —who studied the problem with care under the suggestion of Bohr— came up with his brilliant proposal to compare two situations, one involving the vacuum alone and the other being a slightly modified situation, where just some boundary conditions were imposed on the vacuum field. But this is a very well known story to any potential readers of this article, so I will stop this description here.

In any case some word must still be uttered about an issue that is very important in respect to the specific calculations needed to produce a physically meaningful answer. Aside from the conceptual considerations before, it is clear from the very beginning that a regularization procedure will be needed all the time when dealing with this subject. Indeed, the very simple case of the harmonic oscillator already leads to an infinite and divergent sum. A part of the problem is in fact physically clear and natural. On putting some boundary conditions and comparing the two situations above, an infinite number of the modes appearing in the vacuum

[1] The contribution of vacuum fluctuations to the Lamb shift must not be forgotten either.

sum go away when computing the difference, because they are compatible with the boundary conditions and appear on both terms (so they disappear when taking the difference). However, in a real generic situation this does not mean that the resulting value is already finite —an even if it is, it often happens that an additional, so-called finite renormalization is needed! This is indeed an elusive point, not fully appreciated in the literature, and thus the origin of different erroneous results. I cannot discuss this issue in more detail here, but the reader is warned that it is of paramount importance in this subject and makes it, in a way, quite far from trivial from a technical point of view.

Back to a historical context, the appearance of divergences in QFT lead some important physicists, as Albert Einstein or Paul Dirac, to reject it as a wrong theory (the right one not having been found yet). Now, a number of colleagues (as the already mentioned 't Hooft, Veltman, and many others) have developed a consistent renormalization theory that leads to extremely precise results that can be confronted with laboratory measurements to more than 14 digits in some cases. This cannot be mere coincidence. However, when dealing with the vacuum energy fluctuations in a cosmological context involving general relativity these same kind of calculations lead to an enormous value of the vacuum energy density which is in flagrant contradiction with what we observe. Again, this discrepancy has led some physicists (as Asim Barut and others) to the conviction that maybe vacuum fluctuations do not exist at all, and thus this big problem (ordinarily known as the cosmological constant problem) just disappears! From a fundamental viewpoint this basically leads to a new start in Quantum Mechanics and QFT: a non-linear theory where the second quantization is not need. The reader should be informed about this possibility, that has produced some striking results (as a god approximation in the order of magnitude to the Lamb shift and other quantum field effects), but it is a fact that this alternative does not have much support in the scientific community nowadays. On the other hand, the number of more standard approaches to solve the cosmological constant problem [5] are flourishing (just search for 'cosmological constant' in the ArXives).

4. The Casimir Effect and the Fulling-Davies Theory

The 1948 paper by H.B.G. Casimir [1] was the beginning of a whole branch of research which aims at answering, both from a fundamental and from a practical perspective, very profound questions on the vacuum structure of Quantum Field Theory. The interest of the subject is certified by the impressive number of papers published in recent years (see, e.g., [6]). To wit, fluctuations of the quantum vacuum appear practically everywhere, the key issue being to ascertain if these contributions are relevant or not for the physical (chemical or biological) process in question. They have been argued to be irrelevant for sonoluminiscence (a 10^{-5} contribution), but they are very important for accurate calculations in laser cavities, wetting phenomena of alkali compounds by Helium-3, the sticking of small plates in MEM and NEM devices, and so on [6].

When the plates move quickly, as with a high-frequency vibration, one has the dynamical Casimir effect, a phenomenon studied by S. Fulling and P. Davies in 1976 [7]. Actually, moving mirrors further modify the structure of the quantum vacuum, what manifests in the creation and annihilation of particles. Once the mirrors return to rest, a number of the produced particles may still remain which can be interpreted as radiated particles. This flux has been calculated in several situations by different procedures (averaging over fast oscillations multiple scale analysis, with the rotating wave approximation, numerical techniques, etc.). For a single, perfectly reflecting mirror, the number of produced particles as well as their energy diverge while the mirror moves. Several renormalization prescriptions had been devised to obtain a well-defined energy, however, for some trajectories this finite energy turned not to be a positive quantity and could not be identified with the energy of the produced particles (see e.g. [7]), a situation that was resolved with our approach [8], which relies upon two rather simple ingredients: (i) proper use of a Hamiltonian method and (ii) the consideration of partially transmitting mirrors, which

Journal of Physics: Conference Series **161** (2009) 012019 doi:10.1088/1742-6596/161/1/012019

become transparent at sufficiently high frequencies. Realistic mirrors will always satisfy this condition and the result we have proven, in this way, is both that the number of created particles is finite and also that their energy is always positive for the whole trajectory corresponding to the mirrors' displacement. We have also calculated, in [8], the radiation-reaction force that acts on the mirrors owing to the emission and absorption of particles, and which is related with the field's energy through the energy conservation law, so that the energy of the field at any time t is equal, with the opposite sign, to the work performed by the reaction force up to time t. Such force is usually split into two parts: a dissipative force whose work equals minus the energy of the particles that remain, and a reactive force vanishing when the mirrors return to rest. We have shown for the first time that the radiation-reaction force calculated from the Hamiltonian approach for partially transmitting mirrors satisfies, all the way the energy conservation law and can thus naturally account for the creation of positive energy particles. Also, the dissipative part we obtain agrees with the one calculated by other methods, as using the Heisenberg picture or other effective Hamiltonians. To repeat, this is basically the result of a proper, physically meaningful renormalization prescription.

4.1. On the Nature of Quantum Vacuum Fluctuations

A fact that may be closely related with the considerations above, is that our universe appears to be spatially flat and to possess a non-vanishing cosmological constant (cc), as introduced by Einstein and with the same sign. Actually, for elementary particle physicists the cc constitutes a great embarrassment, calculations of it being off by the famous over-one-hundred orders of magnitude which one gets by setting a plausible cut-off of the order of the Planck length. The cc has to do with cosmology—through Einstein's equations and the FRW universe obtained from them—and also with the local structure of elementary particle physics, as the stress-energy density of the vacuum. Being precise, by the equivalence principle (wee more about that below) the vacuum expectation value of the stress-energy tensor $\langle T_{\mu\nu}\rangle \equiv -\mathcal{E} g_{\mu\nu}$ appears on the rhs of Einstein's equations:

$$R_{\mu\nu} - \frac{1}{2} g_{\mu\nu} R = -8\pi G(\tilde{T}_{\mu\nu} - \mathcal{E} g_{\mu\nu}). \tag{1}$$

It affects cosmology: $\tilde{T}_{\mu\nu}$ contains excitations above the vacuum, and is thus equivalent to a cosmological constant $\Lambda = 8\pi G\mathcal{E}$. Recent observations yield [9]

$$\Lambda_{\text{obs}} = (2.14 \pm 0.13 \times 10^{-3}\ \text{eV})^4 \quad \sim 4.32 \times 10^{-9}\ \text{erg/cm}^3$$

This idea is old and goes back to Zel'dovich [10], who already stated that the cc gets contributions from the zero point fluctuations

$$E_0 = \frac{\hbar c}{2} \sum_n \omega_n, \qquad \omega = k^2 + m^2/\hbar^2, \quad k = 2\pi/\Lambda. \tag{2}$$

Evaluating in a box and putting a cut-off at maximum k_{max} corresponding to reliable QFT physics (e.g., the Planck energy)

$$\rho \sim \frac{\hbar\, k^4_{\text{Planck}}}{16\pi^2} \sim 10^{123} \rho_{\text{obs}}. \tag{3}$$

The issue of the cc has got renewed thrust from the observational evidence of an acceleration in the expansion of our Universe, initially reported by two different groups [11]. As a consequence, many theoreticians have urged to try to explain this fact, and also to try to reproduce the precise value of the cc coming from these observations, but it is more difficult to explain why the cc is so small but non-zero, than to build theoretical models where it exactly vanishes [5].

Before going on, it is interesting to observe that the Casimir effect dates back from the very same year, 1948, as the discovery of renormalized QED. Curiously enough, however, Feynman is reported to have argued that vacuum fluctuations could not be affected by gravity (for a nice account of this point, with historical remarks, see [12]). Thus, a very basic issue is if one is right in writing the equations above and thus assuming that the vacuum energy satisfies the equivalence principle of GR. In other words, how the renormalized Casimir energy of a pair of plates couples to gravity? The answer to this question is less straightforward than one might suspect. In fact very recently disparate answers have appeared in the literature: forces that depend on the orientation of the Casimir apparatus wrt the gravitational field of the earth, etc. [3, 4] Rigorous evidence seemed still to be lacking that the vacuum energy should be taken seriously in Gravity.

Essentially there are two ways to proceed with the calculations. A gauge-invariant procedure: as the energy-momentum tensor of the physical system must be conserved, one needs to include a physical mechanism holding the plates apart against the Casimir force, what leads in practice to a very complicated model-dependent calculation [3]. The alternative, and more practical, way is to find a physically natural coordinate system, more realistic than any other. A reasonable one, used for these purposes, is the Fermi coordinate system, a general-relativistic extrapolation of the concept of an inertial coordinate frame. This has been used in [3], a paper initially criticized in [4] (again, the reason one gets different answers in different coordinate systems is that the starting point is not gauge-invariant). However, in the end agreement seems to have been reached (at least in Fermi and Rindler coordinates), after renormalizing the mass of the plates, that the Casimir energy contributes with a gravitational mass in accordance with the equivalence principle. The end word is that, although a rigorous gauge-invariant proof is still lacking, the calculations performed until now hint towards the positive answer.

5. A Cosmological Imprint of the Casimir Effect?

Having clarified the important point above, in this section, rather than trying to understand the fine-tuned cancelation of the enormous contributions mentioned at the *local* level, we will elaborate on a quite simple idea (but, for the same reason, of potentially far reaching consequences), related with the *global* topology of the universe [13] and in connection with the possibility that fields of small mass pervading the universe most probably exist. They are indeed ubiquitous in inflationary models, quintessence theories, etc. We do not aim at solving the old cc problem, but just at nailing down some contributions to it which may be of the order of magnitude corresponding to the recent observations, as given above. We address here, so to say, the 'perturbative part' of the *new* cc problem [14]: we assume the existence of a quantum field background extending through the universe and calculate the contribution to the cc of the Casimir energy density for this field coming from some typical BCs. Ultraviolet contributions will be safely set to zero invoking some (to be fixed) mechanism of a fundamental theory.

Another hypothesis is the existence of both large and small dimensions (the total number of large spatial coordinates being always three), some of which may be compactified, so that the global topology of the universe will play an important role. There is an extensive literature both on what is the global topology of (spatial sections) of the universe and also on possible contribution of the Casimir effect as a source of some sort of cosmic energy, e.g., as in the case of the creation of a neutron star. Arguments favor different topologies, as a compact hyperbolic manifold, for the spatial section, that could have clear observational consequences. At a second stage it will have sense to consider all possibilities concerning the nature of the fields, the different models for the topology of the universe, and the possible BCs, with their effects on the sign of the force too.

Zeta function regularization techniques are used to this end with impressive success [15]. From previous experience [16], we know that the range of orders of magnitude of the vacuum energy density for the most common possibilities is not so widespread, and may only differ by at most a couple of digits. This has allowed us, both for the sake of simplicity and universality, to deal with simplified situations [17] as, to start, a space-time like $\mathbb{R}^{d+1} \times \mathbb{T}^p \times \mathbb{T}^q$, or $\mathbb{R}^{d+1} \times \mathbb{T}^p \times \mathbb{S}^q, \ldots$, which are very plausible models for the topology [13]. Here, $d \geq 0$ stands for a possible number of non-compactified dimensions. Recall the physical contribution to the vacuum or zero-point energy $\langle 0|H|0\rangle$ is obtained by subtracting the vacuum energy corresponding to the situation with the only change that compactification is absent (in practice this is done by conveniently sending the compactification radii to infinity). It is its *difference* $E_C = \langle 0|H|0\rangle|_R - \langle 0|H|0\rangle|_{R\to\infty}$ (R a typical compactification length) what gives rise to the additional Casimir energy E_C contribution to the cc coming from the BC. Renormalization is to be carried out on the spectral parameter μ: in terms of the spectrum $\{\lambda_n\}$ of H, $\langle 0|H|0\rangle = \frac{1}{2}\sum_n \lambda_n/\mu$, (the sum may involve several continuum and several discrete indices), and μ is needed to render the eigenvalues dimensionless [17]. The physical vacuum energy density in this case, where the contribution of a scalar field ϕ living in a partly compactified spatial section of the universe is considered, with

$$S = \frac{1}{2}\int d^4x\,\sqrt{-g}\,\left[g^{\mu\nu}\partial_\mu\phi\partial_\nu\phi + (m^2+\xi R)\phi^2\right], \tag{4}$$

will be denoted by ρ_ϕ (this is just the contribution to ρ_V coming from this field, there might be other, in general), $\rho_\phi = \frac{1}{2}\sum_n \lambda_n/\mu = \frac{1}{2}\sum_{\mathbf{k}} \frac{1}{\mu}\left(k^2+M^2\right)^{1/2}$, where the sum $\sum_{\mathbf{k}}$ is a generalized one and $M^2 = m^2 + \xi R$ an effective mass term [17].

An important issue is the specific *sign* of the resulting force. For scalar fields and the usual compactifications or BCs it seems not possible to get the right sign corresponding to the accelerated expansion of the universe. However, in worldbrane models and others, involving supergravitons and fermion fields, we have been able to prove that the appropriate sign can be obtained under particular but quite natural conditions. Specifically, in the case of the torus topology we have obtained [18] that the topological contributions to the effective potential have in fact a fixed sign, which depends on the BC one imposes. It is negative for periodic fermionic fields in a a supersymmetric theory, and positive for anti-periodic fields. Thus, topology provides a mechanism which, in a most natural way, permits to have a positive cc in a multi-supergraviton model with anti-periodic fermions [18]. Moreover, the value of the cc is regulated by the corresponding size of the torus (as is easy to see in the scalar case above). One can most naturally use in this case the minimum number, $N = 3$, of copies of bosons and fermions, and show that the order of magnitude of the observational values for the cc can be reproduced invoking quite natural assumptions.

6. Gravity Equations as Equations of State

Also in relation with the cosmological constant, different models have been proposed (T. Padmanabhan et al. [19], D. Blas, J. Garriga [20], etc.) where it appears under the form of an *integration constant*. This can be done, in particular, by modifying the gravity metric in a way that gives rise to the so-called unimodular gravity, but there are also other possibilities. In a seminal paper, in 1995, Ted Jacobson went even further, in a certain way, and obtained Einstein's equations from local thermodynamics arguments only [21]. He did this by way of generalizing black hole thermodynamics to space-time thermodynamics as seen by a local observer. This strongly suggest, in a fundamental context, that Einstein's equations are probably to be viewed as a sort of equation of state. As a consequence, Jacobson went further to suggest that EEs should probably not be taken as basic for quantizing gravity. More recently, C. Eling, R. Guedens, and T. Jacobson [22] extended these considerations to *polynomial* $f(R)$

gravity, with the important difference that the comparison only worked with *non-equilibrium* thermodynamics.

Modified gravity models, as $f(R)$ gravity, constitute a very important dynamical alternative to standard ΛCDM cosmology, in that they have the capability to describe the current accelerated expansion of our Universe (dark energy epoch) together with the initial de Sitter phase and inflation, and even some galaxy rotation curves for the dark (and ordinary) matter [23]. One could imagine that Jacobson's derivations can be generalized to cover all these more complicated theories of gravity that are extensively used nowadays. Indeed, we have been able to extend Jacobson's arguments (in a non-trivial way) to $f(R)$ gravities, being f a general function not just of the Ricci scalar but also of any covariant derivatives of the Ricci scalar of arbitrary order.

Moreover, we have obtained the corresponding field equations as an equation of state of *local* space-time thermodynamics [24]. This could be done by using (i) Iyer and Wald's more general definition of dynamic BH entropy [25] together with (ii) the concept of an effective Newton constant for graviton exchange (effective propagator) of Brustein et. al. [26]. After our work was issued, Wu et. al. [27] have obtained a direct extension of our results to Brans-Dicke and scalar-tensor gravities. Some other extensions of our arguments have also been issued [28].

Acknowledgments

Parts of this work were done in collaboration with G. Cognola, J. Haro, S. Nojiri, S.D. Odintsov, P.J. Silva and S. Zerbini. This work was supported in part by MEC (Spain), project FIS2006-02842, and by AGAUR (Generalitat de Catalunya), contract 2005SGR-00790 and grant DGR2008BE1-00180. Part of the research was performed while the author was on leave at Department of Physics and Astronomy, Dartmouth College, 6127 Wilder Laboratory, Hanover, NH 03755, USA.

[1] Casimir H B G, Proc. K. Ned. Akad. Wettensch. **60**, 793 (1948);
Casimir H B G and Polder D, Phys. Rev. **73**, 360 (1948).
[2] Volovik G E, Int. J. Mod. Phys. **D15**, 1987 (2006).
[3] Fulling S A, Milton K A, Parashar P, Romeo A, Shajesh K V and Wagner J, Phys. Rev. **D76**, 025004 (2007) [hep-th/0702091].
[4] Bimonte G, Calloni E, and Rosa L, Phys. Rev. **D77**, 044026 (2008) [arXiv:0707.1077].
[5] Weinberg S, Phys. Rev. **D61** 103505 (2000).
[6] Elizalde E and Odintsov S D, Eds., *Quantum Field Theory under the Influence of External Conditions*, J. Phys. A: Math. Gen. **39** (2006).
[7] Fulling S A and Davies P C W, Proc. Roy. Soc. Lond. **A348**, 393 (1976).
[8] Haro J and Elizalde E, Phys. Rev. Lett. **97**, 130401 (2006).
[9] Tegmark M *et al.* [SDSS Collaboration], Phys. Rev. **D69**, 103501 (2004);
Eisenstein D J et al., Astrophys. J. **633**, 560 (2005);
Adelman-McCarthy J K et al., Astrophys. J. Suppl. **162**, 38 (2006).
[10] Zeldovich Ya B, Soviet Physics Uspekhi **11**, 382 (1968) [Uspekhi Fiz. Nauk, **95**, 209 (1968)].
[11] Perlmutter S et. al. [Supernova Cosmology Project Collaboration], Astrophys. J. **517**, 565 (1999);
Riess A G et. al. [Hi-Z Supernova Team Collaboration], Astron. Journ. **116**, 1009 (1998).
[12] Woodward J F, Found. Phys. **31**, 819 (2001).
[13] Blanloeil V and Roukema B F, Eds., *Cosmological Topology in Paris 1998*, astro-ph/0010170. See also the entire Vol. 15 of Classical and Quantum Gravity (1998).
[14] Burgess C P et. al., hep-th/0606020;
Padmanabhan T, Int. J. Mod. Phys. **D15**, 1659 (2006).
[15] Elizalde E, Odintsov S D, Romeo A, Bytsenko A A and Zerbini S, *Zeta regularization techniques with applications* (World Scientific, Singapore, 1994);
Elizalde E, *Ten Physical Applications of Spectral Zeta Functions* (Springer-Verlag, Berlin, 1995);
Kirsten K, *Spectral functions in mathematics and physics* (Chapman & Hall, London, 2001);
Bordag M, Elizalde E and Kirsten K, J. Math. Phys. **37**, 895 (1996);J. Phys. **A31**, 1743 (1998);
Bordag M, Elizalde E, Kirsten K and Leseduarte S, Phys. Rev. **D56**, 4896 (1997);
Elizalde E, Vanzo L and Zerbini S, Commun. Math. Phys. **194**, 613 (1998);

Elizalde E, J. Comput. Appl. Math. **118**, 125 (2000);J. Phys. **A22**, 931 (1989);
Elizalde E, Bordag M and Kirsten K, J. Phys. A **31**, 1743 (1998);
Elizalde E and Romeo A, Phys. Rev. **D40**, 436 (1989); J. Math. Phys. **30**, 1133 (1989).

[16] Elizalde E, Nuovo Cim. **104B**, 685 (1989).

[17] Elizalde E, J. Phys. **A40**, 6647 (2007); J. Phys. **A39**, 6299 (2006); J. Phys. **A36**, L567 (2003); Phys. Lett. **B516**, 143 (2001).

[18] Cognola G, Elizalde E and Zerbini S, Phys. Lett. **B624**, 70 (2005);
Cognola G, Elizalde E, Nojiri S, Odintsov S D and Zerbini S, Mod. Phys. Lett. **A19**, 1435 (2004).

[19] Paranjape A, Sarkar S, and Padmanabhan T, Phys. Rev. **D74** (2006) 104015.

[20] Blas D, Deffayet C and Garriga J, Phys. Rev. **D76**, 104036 (2007);
Alvarez E, Blas D, Garriga J and Verdaguer E, Nucl. Phys. **B756**, 148 (2006).

[21] Jacobson T, Phys. Rev. Lett. **75** (1995) 1260.

[22] Eling C, Guedens R and Jacobson T, Phys. Rev. Lett. **96** (2006) 121301.

[23] Cognola G, Elizalde E, Nojiri S, Odintsov S D, Sebastiani L and Zerbini S, Phys. Rev. **D77** (2008) 046009;
Capozziello S, Cardone V F and Troisi A, Phys. Rev. **D71** (2005) 043503;
Elizalde E, Nojiri S and Odintsov S D, Phys. Rev. **D70**, 043539 (2004);
Elizalde E, Nojiri S, Odintsov S D and Wang P, Phys. Rev. **D71** 103504 (2005);
Capozziello S, Carloni S and Troisi A, Res. Dev. Astron. Astrophys. **1** (2003) 625;
Cognola G, Elizalde E, Nojiri S, Odintsov S D and Zerbini S, JCAP **0502** (2005) 010; Phys. Rev. **D73**, 084007 (2006).

[24] Elizalde E and Silva P J, "F(R) gravity equation of state," arXiv:0804.3721 [hep-th], Phys. Rev. **D78[RC]**, 061501 (2008).

[25] Iyer V and Wald R M, Phys. Rev. **D50** (1994) 846; Wald R M, Phys. Rev. **D48** (1993) 3427.

[26] Brustein R, Gorbonos D and Hadad M, arXiv:0712.3206.

[27] Wu S-F, Yang G-H and Zhang P M, arXiv:0805.4044.

[28] Zhu T, Ren Ji-R and Mo S-F, arXiv:0805.1162 [gr-qc];
Eling C, arXiv:0806.3165 [hep-th];
Cai R-G, Cao L-M and Ya-Peng Hu, arXiv:0807.1232 [hep-th] and arXiv:0809.1554 [hep-th].

Casimir effect: The TGTG formula

Israel Klich[1,2] **and Oded Kenneth**[3]

(1) Kavli Institute for Theoretical Physics, University of California Santa Barbara, CA 93106
(2) Department of Physics, University of Virginia, ,Charlottesville, VA 22904
(3) Department of Physics, Technion, Haifa 32000 Israel

E-mail: `klich@caltech.edu`

Abstract. The representation of the Casimir energy in terms of Lippmann-Schwinger operators in the TGTG formula is a convenient way to address both quantitative and qualitative aspects of the Casimir force. Since this form of the energy is already explicitly regularized it may be used in order to prove statements about the force. In addition, combined with approximations of the T operators involved, it allows for convergent asymptotic expressions for the force between bodies.

Keywords: *Casimir , Quantum Fluctuations , Van der Waals , Zero-Point Energy*

1. Introduction

In this talk we discuss various aspects of a representation of the Casimir energy in terms of T operators, which is adequate whenever one is interested in the interaction of two bodies.

The Casimir force [1] explores the interplay between a quantum field and external "classical" like objects such as boundary conditions, background dielectric bodies or space-time metric. While the classical objects modify the behavior of the field due to their presence, the field, in turn, acts on the objects by exerting forces. Much work has been devoted to understanding the effect, as it appears in varied branches of physics: from condensed matter (interaction between surfaces in fluids) to gravitation and cosmology.

The original method used by Casimir, that of mode summation, has led to a large body of work on the effect in simple geometries, where the modes may be computed exactly.

The scattering approach to Casimir physics has proved very useful in 1D. Indeed, many of the calculations of Casimir interaction between bodies are based on scattering theory, as the photon spectrum in an open geometry is continuous and its description requires scattering.

Here, we explore a scattering approach to the Casimir effect in higher dimensions. The approach is based on analysis of a determinant formula for Casimir interactions obtained in [2, 12], and may be viewed as a generalization of previous formulas, especially related to scattering, such as the Lifshitz formula [3], and the results of Balian and Duplantier [4]. Within this approach, the Casimir energy is encoded in the determinant of the operator $1 - T_A G_0 T_B G_0$ where T_A, T_B are Lippmann-Schwinger T operators associated with bodies A and B and G_0 is the photons Green's function; we shall therefore refer to the formula as the TGTG formula.

In [2] it was shown how general results regarding the direction of the force between bodies related by reflection can be obtained from the *TGTG* formula. For example, the sign problem of interaction between two hemispheres was resolved. This result was subsequently extended

Journal of Physics: Conference Series **161** (2009) 012020 doi:10.1088/1742-6596/161/1/012020

to a large class of interacting fields possessing the "reflection positivity" property [5] (See also [6], where use is made of reflection positivity arguments to infer attraction between vortices and anti-vortices in a frustrated XY model). In [7] an alternative derivation of the formula was presented.

We start with a derivation of the determinant formula by showing how to derive the T operator of a pair of perturbations, once the T operator of each perturbation by itself is known. We then illustrate how one obtains the appropriate formula in the vector (electromagnetic) case. Next, we will consider the special case of a body placed next to a perfect mirror, and the dilute limit, dealing with very weak dielectrics by expanding round $\epsilon = 1$.

The mathematical validity of the present approach to calculations of Casimir forces is then addressed. In particular we show that the formula is given in terms of $\log\det(1 + A)$. Here the sum $\sum_i |\lambda_i| < \infty$ is absolutely convergent, where λ_i are the eigenvalues of A, and so $\log\det(1+A)$ is mathematically well defined, thereby explicitly showing that the usual infinities plaguing Casimir calculations have been completely accounted for.

A natural setting for computations using the TGTG formula is by writing it in a basis of partial wave expansions respecting symmetries of bodies as much as possible. In 1D where only two modes (left and right movers) exist at each ω the formula leads to a known closed form formula for the Casimir energy in terms of reflection coefficient (see, e.g. [8, 9]).

Spherical wave expansions are convenient when considering the Casimir interaction between spherical bodies. We demonstrate this by computing the force between two spheres at all distances, thereby generalizing the approach of [10] to spheres beyond Dirichlet boundary conditions, and going beyond the proximity force approximation.

2. The TGTG formula

Let us start by briefly recalling the Lippman-Schwinger operators from scattering theory. Most of the derivation is standard and may be skipped by readers interested only in new results. However, we point out that our approach where the T operators of combined scatterers are utilized seems new.

The T operator was introduced by Lippmann and Schwinger in their celebrated paper on scattering theory [11]. As an application of their formalism, they have computed the scattering properties of a neutron by a proton bound in an inert molecule. T appears in the Lippmann-Schwinger equation as follows: Assume that a solution ϕ_{in} of a free wave equation, without a potential $H_0(\omega)\phi_{in} = 0$ is known. Here H_0 corresponds to a free wave equation, for example:

$$H_0 = \nabla^2 + \omega^2$$

if we are dealing with a scalar field, or

$$H_0 = \nabla \times \nabla \times + \omega^2 \mathbf{I}$$

for the wave equation of the vector potential A in the radiation gauge $A_0 = 0$.

We are interested in the modes of the system in the presence of some local perturbation. Such a perturbation will be represented by a potential V (In the electromagnetic case, when the magnetic response is not considered V corresponds to $\omega^2(\epsilon(x) - 1)$), and so we are interested in the equation $(H_0 + V)\psi = 0$. Note that all the operators above depend on ω, which we omit from the notation to streamline the arguments.

A solution ψ of the eigenvalue equation $(H_0 + V)\psi = 0$ having ϕ_{in} as the incoming part is now constructed. Formally, this is done by looking for a solution of:

$$\psi = \phi - G_0 V \psi,$$

which is the Lippmann-Schwinger equation.

It follows that

$$\psi = (I + G_0 V)^{-1}\phi = (I - G_0 V(I + G_0 V)^{-1})\phi = (I - G_0 T)\phi$$

where

$$T = V(I + G_0 V)^{-1}, \tag{1}$$

is called the transition matrix, or the Lippmann-Schwinger operator.

A typical example is choosing ϕ to be a plane wave solution of the scalar equation, and so the scattered wave-function is given by

$$\psi_k = e^{ikx} - \int \mathrm{d}k' G_0(k)\langle k|T|k'\rangle e^{ik'x}. \tag{2}$$

To find the interaction between two bodies, we consider the T operator of two perturbations V_A, V_B. We compute the joint transition matrix for both perturbations $T_{A\bigcup B}$, and show that the part independent of "self energy" is exactly (11).

Writing, formally, $V = G_0^{-1} - G^{-1}$ we have:

$$1 - G_0(V_1 + V_2) = 1 - G_0(2G_0^{-1} - G_1^{-1} - G_2^{-1}) = G_0 G_1^{-1} + G_0 G_2^{-1} - 1 \tag{3}$$

so that:

$$\frac{1}{1-G_0(V_1+V_2)} = \frac{1}{G_0G_1^{-1}+G_0G_2^{-1}-1} = \frac{1}{G_1^{-1}+G_2^{-1}-G_0^{-1}}G_0^{-1} = G_1\frac{1}{G_2+G_1-G_2G_0^{-1}G_1}G_2G_0^{-1}, \tag{4}$$

where we used $A^{-1} = BB^{-1}A^{-1}C^{-1}C = B(CAB)^{-1}C$.

Using the identity:

$$G = G_0 - G_0 T G_0 \tag{5}$$

as $G_i = G_0 - G_0 T_i G_0$ (with $i = A, B$), together with the definition of T (1) we find:

$$\frac{1}{1+G_0(V_A+V_B)} = (1 - G_0 T_A)\frac{1}{1-G_0T_BG_0T_A}(1 - G_0 T_B) \tag{6}$$

and so the joint T operator of a pair of perturbations may be factored as:

$$T_{A\bigcup B} = (V_A + V_B)\frac{1}{1+G_0(V_A+V_B)} = (V_A + V_B)(1 - G_0 T_A)\frac{1}{1-G_0T_BG_0T_A}(1 - G_0 T_B) \tag{7}$$

This expression represents a *re-summation* of the Born series for T, given by:

$$T = (V_A + V_B) - (V_A + V_B)G_0(V_A + V_B) + (V_A + V_B)G_0 V G_0(V_A + V_B) + ... \tag{8}$$

in a way which identifies the terms which are responsible for directly mixing the A and B perturbations, namely $1 - G_0 T_B G_0 T_A$, as well as the terms which will be responsible for repeated scattering from the same perturbation, which are responsible for self energy: $(1 - G_0 T_A)$.

The expression for $T_{A\bigcup B}$ can be used to find the ground state energy of the system. It is well known that the change of the density of states in the system due to introduction of the perturbations is given by[1]:

$$\delta\rho(E) = \frac{1}{\pi}\mathrm{Im}\partial_E \,\mathrm{Tr}\log T(E) \tag{10}$$

We now may use this formula in two ways:

For the Casimir effect, we compute the energy by integrating over the density of states and summing the allowed modes of the electromagnetic field, giving the result

$$E_C(a) = \int_0^\infty \tfrac{\mathrm{d}\omega}{2\pi} \log\det(1 - T_A G_0 T_B G_0). \tag{11}$$

From a more general point of view, we consider the change in the energy of the ground state of a many body system. In this case one needs to compute the many-body T operator, and try to identify how the pole (or cut) associated with the ground state in the density of states shifts.

Dirichlet boundary conditions

In many cases, and indeed in the original presentation by Casimir, one is interested in sharp boundary conditions, such as Dirichlet or Neumann. Sharp boundary conditions result in singular energy density at the surface, as field modes are required to vanish for all momentum scales. Typically, the local energy density diverges as the inverse fourth power of the distance from the boundary [13].

It is important to point out that the above considerations also describe the conducting case with minor changes. Following [14], assume conducting boundary conditions are given over a surface Σ, parameterized by internal coordinate u and by the embedding in $\mathbb{R}^3$ given by $\mathbf{x}(u)$. One may describe a simple metal by taking $\chi(i\omega) = \frac{\omega_p^2}{4\pi\omega^2}$ on Σ, and letting Σ have a thickness of a few skin depths l/ω_p, $l \sim \mathcal{O}(1)$, here ω_p is the plasma frequency (proportional to the effective electron density in the metal). In the limit of large ω_p one retains the same expression as (11), with the following substitutions:

$$E_C(a) = \tag{12}$$
$$-\tfrac{1}{4\pi}\int_0^\infty \mathrm{d}\omega (\log\det(1 - \mathcal{M}_{BA}\tfrac{1}{1+\mathcal{M}_A}\mathcal{M}_{AB}\tfrac{1}{1+\mathcal{M}_B})$$

where in the Dirichlet case $\mathcal{M}$ is given by:

$$M^{(D)}(u, u'; \omega) = l\omega_p \sqrt{g(u)} G_0(\mathbf{x}(u), \mathbf{x}(u'))\sqrt{g(u')} \tag{13}$$

and acting on the surfaces Σ (See also [18]).

3. Analytical properties of the $T_A G_0 T_B G_0$-operator:

Casimir physics is largely involved with understanding and separating cutoff dependent self energies from universal energies. Often various regularization methods are employed in the calculation. The aim of this section is to show that the form (11) for the energy is already

[1] A related form is known as the Krein formula:

$$\delta\rho(E) = \frac{1}{\pi}\mathrm{Im}\partial_E \log\det S(E) \tag{9}$$

where S is the scattering matrix.

fully regularized. We do this in a rigorous fashion, showing that the object $\log\det(1 - T_A G_{0AB} T_B G_{0BA})$ is mathematically well defined and finite.

In practical terms this means that replacing the infinite dimensional matrix of $1 - T_A G_{0AB} T_B G_{0BA}$ by its upper-left $n \times n$ block with n large enough and calculating the resulting ordinary determinant, gives an arbitrarily good approximation to a (finite) quantity which we call $\det(1 - T_A G_{0AB} T_B G_{0BA})$. The main mathematical notions and theorems which we use here, are briefly reviewed in section 11.

As remarked already in the introduction it is well known that $\det(1 - M)$ is well defined whenever M is a *trace class* operator (definition 11.4, denoted *t.c.*). We would like to show that for a large class of situations (including a pair of disjoint finite bodies A, B, separated by a finite distance) the operator $T_A G_{0AB} T_B G_{0BA} : H_A \to H_A$ is trace class in the continuum limit , and so prove that indeed the expression (11) is finite and well defined.

Indeed by theorem 11.5 the mere fact that $G_0(x, y)$ is a smooth function for $x \neq y$ is sufficient to guarantee that for any pair of compact volumes $A, B \in \mathbb{R}^3$ at finite mutual distance the operator G_{0AB} is trace class. To deduce that $T_A G_{0AB} T_B G_{0BA}$ is trace class (and by similar argument also $1 - G_0 \mathcal{J} T_A$ appearing in eq (26)) it is then enough (proposition 11.6) to make sure $T_{A,B}(i\omega)$ are bounded (definition 11.2).

In the context of dielectric interaction it is particularly easy to show that $T(i\omega)$ is bounded. In physical systems at equilibrium, it follows from causality properties of the dielectric function [15], that $\chi(i\omega, x) \geq 0$. We then have the following

Lemma 3.1. *For $\chi(i\omega, x) > 0$, the T operators are positive and bounded.*

Proof: Since $G_0, \chi > 0$ (definition 11.3) one may write $T = \sqrt{\chi}\frac{\omega^2}{1+\omega^2\sqrt{\chi}G_0\sqrt{\chi}}\sqrt{\chi}$ from which it is seen that $T > 0$ and that in the operator norm $||T|| \leq \omega^2||\chi||$ □.
In fact, this holds also for nonlocal χ as long as $f(x) \to \int_A \chi(i\omega, x, x')f(x')\mathrm{d}x'$ is a bounded positive operator $H_A \to H_A$. In the context of a more general type of interactions which may not be positive, one needs to use some assumption on the stability of the system to guarantee that $T(i\omega)$ is bounded. We do not elaborate on this here.

An alternative approach to proving the trace class property of $T_A G_{0AB} T_B G_{0BA}$ is based on the notion of a Hilbert Schmidt operator (definition 11.7, also denoted *H.S.*). Here the frequently used strategy in operator analysis is to use the following fact: if $U \in H.S.$ and $V \in H.S.$, then $UV \in t.c.$. The advantage of this approach is that it is very easy to check if an operator is Hilbert-Schmidt. Since the Hilbert-Schmidt norm is $||A||^2_{H.S.} = \mathrm{Tr}(A^\dagger A)$, one may evaluate it directly, (e.g. by computing $\int |A(x, x')|^2$).

Theorem 3.2. *For any two bodies A, B such that $\int_{A\times B} \mathrm{d}x\mathrm{d}y|G_0(x, y)|^2 < \infty$, $T_A G_{0AB} T_B G_{0BA}$ is trace class.*

Proof: First we show that $T_A G_{0AB}$ and $T_B G_{0BA}$ are Hilbert Schmidt operators. This can be verified in the following way. We have just seen that T_A, T_B are bounded operators. Now note that G_{0AB} is Hilbert Schmidt, since,

$$||G_{0AB}||^2_{H.S} = \int_{A\times B} \mathrm{d}x\mathrm{d}y|G_{0AB}(x, y)|^2, \tag{14}$$

which is finite under the condition above. Now the inequality $||T_A G_{0AB}||_{H.S} \leq ||T_A||||G_{0AB}||_{H.S}$ implies that $T_A G_{0AB}$ is Hilbert-Schmidt. Finally using $U, V \in H.S. \Rightarrow UV \in t.c.$ we see that $T_A G_{0AB} T_B G_{0BA} \in t.c.$ □

Corollary 3.3. *For any finite bodies A, B, such that distance$(A, B) > 0$, and any Green's function which is finite away from the diagonal, $T_A G_0 T_B G_0 \in t.c.$*

Example 3.4. *For the scalar field discussed above, $G_0(x,y) = \frac{e^{-\omega|x-y|}}{4\pi|x-y|}$, the condition is satisfied. In the same way it is satisfied for the electromagnetic field (one has to take into account also matrix indices but these discrete indices do not change the finiteness of the integrals)*

Remark 3.5. *The ω integration in (11), is convergent. To see this note that G_0 decays exponentially with ω, therefore $||G_0||_{H.S.}$ decays exponentially, also the $||T||$'s do not grow more then quadratically in ω.*

In the EM case one may also worry due to the factor $\frac{1}{\omega^2}$ appearing in $\mathcal{D}_{0ij}(x,y) = (\delta_{i,j} - \frac{1}{\omega^2}\nabla_i^{(x)}\nabla_j^{(y)})G_0(x,y)$ about convergence for $\omega \sim 0$. This factor, however, gets cancelled since $||T|| \le \omega^2||\chi||$ as shown in lemma 3.1.

One may also show that G_{0AB} are t.c. themselves by using H.S. properties. The bodies are assumed not to touch, thus we can choose a C_0^∞ (compactly supported and infinitely smooth) function f_A, such that $P_A f_A = P_A$, and $P_B f_A = 0$ where P_A, P_B are the projections on $L^2(A), L^2(B)$ (i.e. $f_A(x) = 1$ for $x \in A$, and it then smoothly goes to 0, before reaching body B) .

Writing:

$$\begin{gathered} G_{0AB} = L_1 L_2 \\ L_1 = P_A \frac{1}{(p^2+\omega^2)^\alpha} \;\; ; \;\; L_2 = (p^2+\omega^2)^\alpha f_A G_0 P_B, \end{gathered} \tag{15}$$

we see that if $2\alpha > d$,

$$\begin{gathered} ||L_1||^2_{H.S.} = \mathrm{Tr}(P_A \tfrac{1}{(p^2+\omega^2)^\alpha})(P_A \tfrac{1}{(p^2+\omega^2)^\alpha})^\dagger = \\ Vol(A) \int \mathrm{d}^d p |\tfrac{1}{(p^2+\omega^2)^{2\alpha}}| < \infty \end{gathered} \tag{16}$$

and so L_1 is Hilbert Schmidt. Next, we check that $L_2 \in H.S.$. To see this last point, note that

$$\begin{gathered} < x|L_2|x' > = < x|(p^2+\omega^2)^\alpha f_A G_0 P_B|x' > = \\ (-\triangle_x + \omega^2)^\alpha f_A(x) G_0(x-x') P_B(x') \end{gathered} \tag{17}$$

Since $G_0(x-x')$ is smooth away from $x = x'$, where the expression is anyway zero because $f_A P_B = 0$, and since $\langle x|L_2|x'\rangle$ has compact support we see that $||L_2||^2_{H.S.} = \int \mathrm{d}x \mathrm{d}x' |L_2|^2 < \infty$. Thus, G_{0AB} can be written as a product of two H.S. operators, and as such is trace class.

Finally, we have that

Theorem 3.6. (**Eigenvalues of TGTG**) *For $\chi > 0$, all eigenvalues λ of the (compact) operator $T_A G_{0AB} T_B G_{0BA}$ appearing in (11) satisfy $1 > \lambda \ge 0$.*

Proof: We will use repeatedly that for bounded operators X, Y the nonzero eigenvalues of XY and YX are the same. Note first that $G_0, \chi \ge 0$ (as operators) implies

$$\mathrm{spec}(\chi G_0) \setminus \{0\} = \mathrm{spec}(\sqrt{G_0}\chi\sqrt{G_0}) \setminus \{0\} \subset [0, \infty). \tag{18}$$

Writing $T_\alpha G_0 = 1 - \frac{1}{1+\omega^2 \chi_\alpha G_0}$ as an operator on $L^2(\mathbb{R}^3)$ it is then clear that its spectrum lies in [0,1). The same conclusion then applies to the operator $\sqrt{G_0}T_\alpha\sqrt{G_0}$ but since it is hermitian one concludes also $||\sqrt{G_0}T_\alpha\sqrt{G_0}|| < 1$ from which it follows $||\sqrt{G_0}T_A G_0 T_B \sqrt{G_0}|| < 1$ and hence $\lambda < 1$. Similarly $\sqrt{G_0}T_\alpha\sqrt{G_0} \ge 0$ imply $\lambda \ge 0$ □

4. The Electromagnetic Case

Here we follow the approach of [15]. The statistical properties of the electromagnetic field in a medium are described by the appropriate photonic Green's function. The electromagnetic fields are derived from the electromagnetic potentials A^α, $\alpha = 0, .., 3$. (It is convenient to work in the gauge $A^0 = 0$.) The retarded Green's function $\mathcal{D}_{ik}$ is defined by:

$$\mathcal{D}^R_{ik}(X_1, X_2) = \tag{19}$$
$$\begin{cases} \langle A_i(X_1)A_k(X_2) - A_k(X_2)A_i(X_1)\rangle & t_1 < t_2 \\ 0 & \text{otherwise} \end{cases}$$

where X_1, X_2 are 4-vectors $(X_1^0, ..X_1^3)$ and $k, i = 1, ..3$. The angular brackets denote averaging with respect to the Gibbs distribution.

The interaction of the electromagnetic field with a classical current $\mathbf{J}$ put in the medium is given by:

$$V = -\frac{1}{c}\int \mathbf{J}\cdot\mathbf{A}.$$

Kubo's formula allows us to treat this interaction within linear response. By Kubo's formula the mean value $\overline{\mathbf{A_i}}$ in presence of a current $\mathbf{J}$ satisfies:

$$\overline{\mathbf{A_i}}(\mathbf{r})_\omega = -\frac{1}{\hbar c}\int \mathcal{D}^R_{ik}(\omega; \mathbf{r}, \mathbf{r}')\mathbf{J}_k(\mathbf{r}')_\omega \mathrm{d}^3\mathbf{r}', \tag{20}$$

where

$$\mathcal{D}^R_{ik}(\omega; \mathbf{r}, \mathbf{r}') = \int_0^\infty e^{i\omega t}\mathcal{D}^R_{ik}(t; \mathbf{r}, \mathbf{r}')\mathrm{d}t \tag{21}$$

The function $\mathcal{D}$ is sometimes referred to as the generalized susceptibility of the system [15].

From Maxwell's equations it follows that in a medium with a given permittivity tensor ϵ_{ij}, permeability tensor μ_{ij}, and current $\mathbf{J}$, the vector potential A_i satisfies:

$$(\nabla \times (\mu^{-1}\nabla\times) - \frac{\omega^2}{c^2}\epsilon)\overline{\mathbf{A}} = \frac{4\pi}{c}\overrightarrow{\mathcal{J}}_\omega \tag{22}$$

Substituting (20) in (22), we see that $\mathcal{D}$ is a Green's function for the equation:

$$\nabla \times \mu^{-1}\nabla \times \mathcal{D} - \frac{\omega^2}{c^2}\epsilon\mathcal{D} = -4\pi\hbar \mathrm{I}\delta(\mathbf{r} - \mathbf{r}') \tag{23}$$

where I is the 3-dimensional unit matrix.

To get the energy we now simply use the TGTG formula (11) with the scalar propagator G_0 replaced by the vector propagator $\mathcal{D}_0$ everywhere (including in the definition of the T operators) (for details see [12]). Here the explicit expression for $\mathcal{D}_0$ is

$$\mathcal{D}_{0ij}(k, i\omega) = -\frac{4\pi\hbar}{k^2 + \omega^2 c^{-2}}(\delta_{ij} + \frac{c^2 k_i k_j}{\omega^2}) \tag{24}$$

5. Dielectric in front of a mirror

A somewhat simplified, but useful in practice, version of our formula is obtained in the case of a body placed close to a mirror. Consider the body A to the left of a Dirichlet mirror B located at $x_n = a/2$. It is well known (using the image method) that the effect of the Dirichlet mirror is to replace the free propagator G_0 by

$$G_B(\mathbf{x}, \mathbf{x}') = G_0(\mathbf{x}, \mathbf{x}') - G_0(\mathbf{x}, J(\mathbf{x}')) \tag{25}$$

Journal of Physics: Conference Series **161** (2009) 012020 doi:10.1088/1742-6596/161/1/012020

where $J(\mathbf{x}_{\|}, x_{\perp}) = (\mathbf{x}_{\|}, a - x_{\perp})$ denotes reflection through the mirror plane. This may be written as $G_B - G_0 = -G_0\mathcal{J}$ where $\mathcal{J}$ is the operator defined by $\mathcal{J}\psi(\mathbf{x}) = \psi(J(\mathbf{x}))$. Noting the standard relation (5) $G_B = G_0 - G_0 T_B G_0$ between the Green function in the presence of scatterer B to its T matrix one concludes $G_0 T_B G_0 = G_0\mathcal{J}$ which when substituted in (11) gives

$$E_C(a) = \int_0^{\infty} \frac{\mathrm{d}\omega}{2\pi} \log\det(1 - G_0\mathcal{J}T_A) \tag{26}$$

An alternative (though closely related) approach is to note the energy it costs to place a body A near a mirror B is

$$E_C = \int_0^{\infty} \frac{\mathrm{d}\omega}{2\pi} \log\det{}_\Lambda(1 + \omega^2 \chi_A(\mathbf{x}, i\omega) G_B(\mathbf{x}, \mathbf{x}')).$$

Subtracting the energy $E_C = \int_0^{\infty} \frac{\mathrm{d}\omega}{2\pi} \log\det_\Lambda(1 + \omega^2 \chi_A(\mathbf{x}, i\omega) G_0(\mathbf{x}, \mathbf{x}'))$ it costs to put A in vacuum then gives the Casimir interaction energy. Using the relation

$$\begin{aligned}(1 + G_B V_A)/(1 + G_0 V_A) &= \\ 1 + (G_B - G_0)T_A &= 1 - G_0\mathcal{J}T_A\end{aligned} \tag{27}$$

leads again to (26).

Yet another way of obtaining the same result is by substituting $\chi_B = \lambda\delta(x_n - a/2)$ in the definition of T_B and doing the algebra. One then finds

$$\begin{aligned}G_0 T_B G_0 &= \\ \int \frac{d^2 k_{\perp}}{(2\pi)^2} e^{ik_{\perp}(x-x')_{\perp}} &\frac{\lambda\omega^2}{2q(\lambda\omega^2 + 2q)} e^{-q|x_n| - q|x'_n|}\Big|_{q=\sqrt{\omega^2 + k_{\perp}^2}},\end{aligned} \tag{28}$$

which in the limit $\lambda \to \infty$ reduces, as expected, to the expression $G_0\mathcal{J}$ obtained through the image method.

We now address the case of a Neumann mirror. Note that the Green function in the presence of a Neumann mirror is $G = G_0 + G_0\mathcal{J}$. By repeating the arguments above we find that the Casimir interaction between an object A and a Neumann mirror is given by a similar formula to (26), involving the determinant $\det(1 + G_0\mathcal{J}T_A)$. We remark that while the Dirichlet mirror may be considered as the limit $\lambda \to \infty$ of a dielectric having e.g. $\chi_B = \lambda\delta(x_n - a/2)$ (or in more realistic model $\chi_B = \lambda\theta(x_n - a/2)$) it is hard to find a simple analog $\chi_B(x)$ that would lead in a similar limit to a Neumann mirror.

A similar treatment is applicable in the more physically relevant EM case. The boundary conditions $E_{\|} = 0$ may be enforced by requiring the vector potential to satisfy $\mathcal{J}A = -A$ where $\mathcal{J}$ is defined to act on vectors as $\mathcal{J}A(x) = (A_{\|}(J(x)), -A_{\perp}(J(x)))$ (Here $A_{\|}, A_{\perp}$ denote the components of A parallel and normal to the mirror surface. The temporal component is considered as a parallel component though in practice we usually choose a gauge where it vanishes.)

The EM Casimir interaction between a dielectric and a mirror is then given by a formula similar to (26) with $G_0, \mathcal{J}$ replaced by the EM propagator $\mathcal{D}_0$ and the vectorial $\mathcal{J}$ defined above.

It is interesting to consider also an ideal permeable mirror (having $\mu \to \infty, \epsilon = 1$). This corresponds to the boundary condition $B_{\|} = 0$ which may be enforced by requiring the vector potential to satisfy $\mathcal{J}A = +A$. Thus the Casimir interaction of body A with such a mirror will be given by an expression involving the determinant $\det(1 + T_A\mathcal{D}_0\mathcal{J})$.

6. Separable Potentials

Here we show a simple example where everything can be computed immediately in the TGTG approach. The example is based on so called "separable potentials". Such potentials arise in variety of situations in physics and in mathematics, and were first introduced for Casimir type problems by Jaffe and Williamson [21]. They correspond to "rank 1" perturbations, and can be written as $V = |f\rangle\langle f|$ for some function f, or in x space notation, the perturbation V has the kernel $V(x, x') = f(x)f(x')$.

For such potentials, one may readily calculate the T operators, which turn out to be also of rank 1, and so the interaction energy is easily calculated. We compute T:

$$T = \frac{1}{1+VG_0}V = \sqrt{V}\frac{1}{1+\sqrt{V}G_0\sqrt{V}}\sqrt{V} = \frac{1}{1+\langle f|G_0|f\rangle}|f\rangle\langle f|, \tag{29}$$

and so we have

$$\log\det(1 - T_A G_0 T_B G_0) = \log(1 - \frac{1}{1+\langle f_A|G_0|f_A\rangle}\frac{1}{1+\langle f_B|G_0|f_B\rangle}|\langle f_B|G_0|f_A\rangle|^2) \tag{30}$$

Since the (positive) terms $\frac{1}{1+\langle f_A|G_0|f_A\rangle}\frac{1}{1+\langle f_B|G_0|f_B\rangle}$ do not depend on the distance, to find the direction of the force it is enough to consider $|\langle f_B|G_0|f_A\rangle|$.

Let us show that the force can be repulsive when both V_A and V_B are positive perturbations. We take the following functions: $f_A = \alpha_1\delta(x+a) + \alpha_2\delta(x+a+1)$ and the potential in B is defined by $f_B = \beta_1\delta(x-a) + \beta_2\delta(x-a-1)$; here we assume that these points all lie on a particular line, which is the x direction.

We choose $\omega = 1$ for simplicity and compute:

$$\langle f_A|G_0|f_B\rangle = \alpha_1\beta_1\frac{e^{-2a}}{2a} + (\alpha_1\beta_2 + \alpha_2\beta_1)\frac{e^{-2a-1}}{2a+1} + \alpha_2\beta_2\frac{e^{-2a-2}}{2a+2}. \tag{31}$$

Upon choosing: $\alpha_1 = 1, \alpha_2 = -20, \beta_1 = 10, \beta_2 = -1$, and distance $a = 0.15$ we get

$$\partial_a|\langle f_A|G_0|f_B\rangle|^2 \sim 2200 > 0,$$

implying that (30) gives a repulsive contribution for the energy at this distance and frequency. It is possible to extend this example so that even integration over frequencies is still repulsive. This will be reported elsewhere.

7. Dilute limit

In the following sections we consider strategies of using the $TGTG$ formula in actual calculations. A particularly simple case is when χ is small, which is commonly referred to as the "dilute" case (and sometimes as "low contrast"). Here we briefly sketch how to best use the formula in this limit. As shown in Theorem 3.6, one always has $||TGTG|| < 1$, therefore we may expand the $\log\det(1 - ..)$ expression (11) in powers:

$$E_C = -\int\frac{d\omega}{2\pi}\sum\frac{1}{m}\operatorname{Tr}(T_A G_0 T_B G_0)^m \tag{32}$$

In the dilute limit $\chi_\alpha < 1$, so one may also substitute the expansion

$$T_\alpha = -\sum_{n=0}^{\infty}(-\omega^2\chi_\alpha G_0)^n\omega^2\chi_\alpha \tag{33}$$

in (32) and compute the involved integrals to desired order. This expansion is the continuum equivalent to summation of two body forces, and is equal to the Born series appearing, for example, in [15].

Journal of Physics: Conference Series **161** (2009) 012020 doi:10.1088/1742-6596/161/1/012020

8. Scattering Approach

As remarked above, the operator $T_A G_0 T_B G_0$ appearing in our formula is closely related to scattering data. The purpose of this section is to clarify this relation and make it more explicit. In order to keep better touch with conventions used in scattering theory, we usually avoid in this section using Wick rotation and thus we work in Lorentzian rather than Euclidean space, with a real rather then imaginary frequency and with the Feynman rather than the Euclidean propagator G_0.

As mentioned above, the arguments of G_0 in (11) never coincide, implying that when $G_0(x_a, x_b)$ is considered as a function of x_b alone it is a solution of the (homogeneous) free wave equation. Thus one may expand $G_0(x_a, x_b)$ in the form $\sum \mathcal{C}_{\alpha\beta}\phi^*_\alpha(x_a)\phi_\beta(x_b)$ where $\{\phi_\alpha(x_a)\}, \{\phi_\beta(x_b)\}$ are some sets of free wave solutions of energy ω. There is of course great freedom in choosing the sets $\{\phi_\alpha(x_a)\}, \{\phi_\beta(x_b)\}$. In practice one would choose these in a way that makes subsequent calculations easier. As mentioned earlier we consider $T_A G_0 T_B G_0$ as acting only on the volume of object A, therefore these considerations also apply to the propagator on the right of this expression.

The Lippmann-Schwinger operator $T(\omega)$ is related to the S-matrix by [2]

$$S = 1 - 2\pi i\delta(\omega^2 - \omega'^2)T(\omega). \tag{34}$$

Therefore $T(\omega)$ has the property that its matrix element $\langle\alpha|T|\beta\rangle$ between a pair of free states α, β having energy ω is equal to the corresponding matrix element of the transition matrix. Since the operator T_B in $T_A G_0 T_B G_0$ is sandwiched between a pair of free Feynman propagators corresponding to energy ω, we may identify it with the corresponding transition matrix. Due to the cyclicity of the determinant $\det(1 - T_A G_0 T_B G_0)$ the same is true of T_A.

Substituting the expansion $G_0(x_a, x_b) = \sum \mathcal{C}_{\alpha\beta}\phi^*_\alpha(x_a)\phi_\beta(x_b)$ we arrive at

$$T_A G_0 T_B G_0 = \sum_{\alpha\alpha'\beta\beta'} T_A|\alpha\rangle \mathcal{C}_{\alpha\beta}\langle\beta|T_B|\beta'\rangle \mathcal{C}_{\alpha'\beta'}\langle\alpha'|$$

The Casimir interaction will then be given explicitly by

$$E = \int_0^\infty \frac{d\omega}{2\pi} \log\det(1 - K(i\omega)). \tag{35}$$

Here $K_{\alpha''\alpha'} = (T_A)_{\alpha''\alpha}\mathcal{C}_{\alpha\beta}(T_B)_{\beta\beta'}\mathcal{C}_{\alpha'\beta'}$.

9. Partial waves expansion

In the following sections, we consider strategies of using the representation (35) by restricting the K matrix to a finite subspace which gives the dominant contribution to the force. Indeed, in many cases of interest only a few partial waves are significantly scattered, the best example for this is when objects are far apart, and from a large distance look point like. At this limit one expects significant contribution only from s-wave scattering. In the more general case, K may be approximated by a finite dimensional matrix corresponding to several partial waves. In order to see how this works in practice we consider below a few simple cases.

[2] Most standard textbooks discuss the non-relativistic case and therefore include a factor $\delta(\omega - \omega')$ instead of $\delta(\omega^2 - \omega'^2)$. Writing the delta function in terms of momentum the two cases reduce to the same expression: $\delta(k^2 - k'^2)$

One dimensional systems
A particularly simple case occurs when the system is one-dimensional. Consider, e.g., a scalar field in 1D. All states of energy ω are then spanned by two modes: left and right movers $|L\rangle, |R\rangle = \frac{1}{\sqrt{2\pi}} e^{\pm i\omega x}$. Hence in this case the determinant Eq (11) can easily be calculated exactly. To see how this is done we write the Feynman propagator explicitly as

$$G_0 = \int_{-\infty}^{\infty} \frac{dk}{2\pi} \frac{e^{ikx}}{\omega^2 - k^2 + i0} = -\frac{i}{2\omega} e^{i\omega|x|} \tag{36}$$

We consider a pair of scatterers A, B such that A is to the left of B. This immediately implies that we have $x_a < x_b$ and therefore

$$G_{0BA}(x_b, x_a) = -\frac{i}{2\omega} e^{i\omega(x_b - x_a)} = \frac{-2i\pi}{2\omega} |R\rangle\langle R| \tag{37}$$

Similarly we also have $G_{0AB} = \frac{-2i\pi}{2\omega}|L\rangle\langle L|$. Using this we see that the operator K in (35) turns into the c-number

$$K = (\frac{-2i\pi}{2\omega})^2 \langle R|T_A|L\rangle\langle L|T_B|R\rangle = \tilde{r}_A(\omega) r_B(\omega). \tag{38}$$

Here r_B ($\tilde{r}_A$) is the reflection coefficient for a wave hitting scatterer B from the left (A from the right) to be reflected back. Note that the normalization of T implied by Eq (34) is responsible to the cancelling of the factor $\frac{-2i\pi}{2\omega}$. (Had we used relativistic normalization for $|L, R\rangle$ the factor 2ω would not have appeared.) We thus conclude

$$\det(1 - T_A G_0 T_B G_0) = 1 - \tilde{r}_A(\omega) r_B(\omega).$$

The tilde on r_A serves to remind us that it is the reflection coefficient from the right side of A.

We remark that $\tilde{r}_A(\omega) r_B(\omega)$ depends implicitly on the distance between A, B through the (phase) dependence of r_A, r_B on the scatterers locations. To make this explicit, note that moving a scatterer a distance a affects the reflection coefficients as $r \to e^{-2ia\omega} r, \tilde{r} \to e^{2ia\omega}\tilde{r}$.

Moving the scatterers a distance a apart therefore result in

$$\det(1 - T_A G_0 T_B G_0) \to (1 - e^{2ia\omega} \tilde{r}_A(\omega) r_B(\omega)).$$

Substituting in (35) we obtain the familiar formula for 1d Casimir interaction between scatterers [8, 9, 16].

Multi-component field in 1d
The considerations used above for a single scalar field in one dimension extend to a situation where $\phi = (\phi_1, \phi_2, ...\phi_n)$ is an n component field. In this case the reflection coefficients $r_{A,B}$ turn into $n \times n$ matrices and one finds $\det(1 - T_A G_0 T_B G_0) = \det(1 - \tilde{r}_A(\omega) r_B(\omega))$ where the determinant on the right is of a usual $n \times n$ matrix.

Plane wave expansion.
In physical three dimensional space there are many different possible ways to expand the propagator $G_0(x_a, x_b) = \sum C_{\alpha\beta} \phi^*_\alpha(x_a) \phi_\beta(x_b)$ in terms of free wave solutions $\{\phi_\alpha(x_a)\}, \{\phi_\beta(x_b)\}$. In the next section we describe the expansion in spherical waves (which is probably the most useful expansion), and we demonstrate its use to calculating the Casimir force between compact object. However for the sake of simplicity we first describe here a plane wave expansion which is the immediate generalization of eq(37). A simple heuristic way to arrive at this generalization

is to formally think of the field ϕ in three dimensions as one dimensional field having infinitely many components labelled by its transverse momenta. Indeed, such point of view has been successfully used in describing transport in quasi 1D conductors in mesoscopic physics, whereby each transverse component corresponds to a scattering channel (see for example [17]). This suggests splitting $\vec{k}$ into its z-component k_z and its transverse components $k_{\|} = (k_x, k_y)$. The 3d propagator may then be written as:

$$G_0 = -\int \frac{d^2k_{\|}}{(2\pi)^2} \frac{ie^{i|z|k_z} e^{ik_{\|}x_{\|}}}{2k_z}\Big|_{k_z=\sqrt{\omega^2-k_{\|}^2+i0}}$$

Here $\sqrt{\omega^2 - k_{\|}^2 + i0}$ may be either real and positive (for $\omega^2 > k_{\|}^2$) or pure imaginary (for $\omega^2 < k_{\|}^2$) in which case the $i0$ prescription implies that it must be chosen on the positive imaginary axis. Assuming that A is located to the left of B along the z-axis it follows that

$$T_A G_0 T_B G_0 = \int \frac{dk_x dk_y dq_x dq_y}{(2\pi)^4} T_A|(q_x, q_y, -q_z)\rangle \times \\ \frac{1}{2q_z}\langle(q_x, q_y, -q_z)|T_B|(k_x, k_y, k_z)\rangle \frac{1}{2k_z}\langle(k_x, k_y, k_z)|, \quad (39)$$

where $q_z = \sqrt{\omega^2 - q_x^2 - q_y^2 + i0}$ and $k_z = \sqrt{\omega^2 - k_x^2 - k_y^2 + i0}$.

When considering only the terms satisfying $\omega^2 > q_x^2 + q_y^2, k_x^2 + k_y^2$, eq. (39) indeed looks like a straightforward generalization of the 1d result. However as this expression shows, to get the correct result one must also include the contribution of evanescent waves ($q_{\|}^2 > \omega^2$). Upon Wick rotation, however, the distinction between ordinary and evanescent waves disappears. It may also be noted that (since in general $q_z \neq k_z$) the variation of the $\langle(q_x, q_y, -q_z)|T_B|(k_x, k_y, k_z)\rangle$ matrix elements upon moving B along the z-axis is considerably more complicated then in the 1d case.

The above representation may be helpful in problems where the scatterers A, B have exact or approximate planar geometry (e.g. corrugated plates). Though the theorem guaranteeing finite trace does not apply for infinite plates one may show that dividing by the plate area leads to finite result. We remark that actual calculation of the determinant requires discretizing $k_{\|}$ which corresponds to assuming large but finite plates. Alternatively, one may use eq. (32) with continuous $k_{\|}$.

10. Spherical waves expansion

When describing interaction between two compact bodies, often it is convenient to represent the transition matrices T in a spherical wave basis. To do so, we choose two points P_A, P_B inside bodies A, B respectively. We parameterize the points of body A by the radius vector $\vec{r} = \vec{r_A}$ measured from the point P_A and the points of B by the radius vector $\vec{r'} = \vec{r_B}$ measured from the point P_B . The vector connecting P_A and P_B will be denoted by $\vec{a}$ (Fig.1). In the scalar case, the free spherical waves centered at P_A, P_B are given by

$$|(lm)_{A,B}\rangle = \sqrt{\frac{2\omega^2}{\pi}} j_l(\omega r_{A,B}) Y_{lm}(\hat{r}_{A,B}) \quad (40)$$

with the normalization $\langle\omega' l' m'|\omega l m\rangle = \delta_{ll'}\delta_{mm'}\delta(\omega - \omega')$.

To use (35), the scalar 3d Green function $G_0 = -\frac{e^{i\omega r}}{4\pi r}$, is expanded in terms of the spherical harmonic functions centered around P_A and those centered around P_B.

$$G_\omega = \sum_{lm;l'm'} |(lm)_B\rangle \mathcal{C}_{lm;l'm'} \langle(l'm')_A| \quad (41)$$

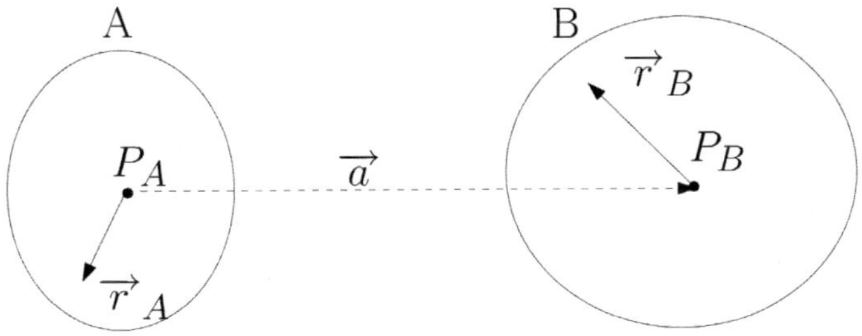

Figure 1. Coordinate system used for the partial wave approach

where

$$\mathcal{C}_{lm;l'm'}(\omega) = -\tfrac{i\pi}{2\omega}\sum_{l'',m''} C\begin{pmatrix} l & l' & l'' \\ m & m' & m'' \end{pmatrix} i^{l''+l'-l} h^{(1)}_{l''}(\omega a) Y_{l''m''}(\hat{a}), \tag{42}$$

Y_{lm} are spherical harmonics, j_l, h_l are spherical Bessel and Hankel functions, and the coefficients $C\begin{pmatrix} l & l' & l'' \\ m & m' & m'' \end{pmatrix}$ have known expressions in terms of the $3j$ symbol or as an integral of spherical functions:

$$C\begin{pmatrix} l & l' & l'' \\ m & m' & m'' \end{pmatrix} = 4\pi \int d\Omega Y_{lm} Y^*_{l'm'} Y^*_{l''m''} = (-1)^m \sqrt{4\pi(2l+1)(2l'+1)(2l''+1)} \times \begin{pmatrix} l & l' & l'' \\ 0 & 0 & 0 \end{pmatrix} \begin{pmatrix} l & l' & l'' \\ m & -m' & -m'' \end{pmatrix} \tag{43}$$

In actual computations it is often more convenient to use the Wick-rotated expression. This may be expressed as $\mathcal{C}_{lm;l'm'}(i\omega) = -\frac{\pi}{2\omega} i^{l'-l} g_{lm;l'm'}$. where the coefficients

$$g_{lm;l'm'} = \sum_{l'',m''} C\begin{pmatrix} l & l' & l'' \\ m & m' & m'' \end{pmatrix} \sqrt{\tfrac{2}{\pi\omega a}} K_{l''+\frac{1}{2}}(\omega a) Y_{l''m''}(\hat{a}), \tag{44}$$

are real. Equations(42,44) may be somewhat simplified by choosing the z-axis along $\hat{a}$.

The above expansion of G_ω allows expressing $T_A G_0 T_B G_0$ in terms of matrix elements $\langle l'm'|T|lm\rangle$ of the transition matrices of the two scatterers. The Casimir interaction may then be written as in (35) where

$$K_{lm;l'm'} = (-1)^{l_1+l_2}(T_A)_{lm;l_1m_1}\mathcal{C}_{l_1m_1;l_2m_2}(T_B)_{l_2m_2;l_3m_3}\mathcal{C}_{l_3m_3;l'm'}. \tag{45}$$

Here $\mathcal{C}_{lm;l'm'}$ are given by (42) or (44), summation over $l_1, m_1, l_2, m_2, l_3, m_3$ is implied and we note that the extra sign resulted from $\mathcal{C}_{lm;l'm'}(-\hat{a}) \equiv (-1)^{l+l'}\mathcal{C}_{lm;l'm'}(\hat{a}) = \mathcal{C}_{l'm';lm}(\hat{a})$.

If we assume that only waves having $l \le l_0$ are significantly scattered then K will turn into a finite $(l_0+1)^2 \times (l_0+1)^2$ matrix (since the dimension of the subspace $l \le l_0$ is $\sum_{l=0}^{l_0}(2l+1) = (l_0+1)^2$). We stress that this argument *does not* require us to assume spherical symmetry of the scatterers.

When A, B are very far apart the interaction between them is governed by waves of very low frequency and therefore also low l. At this limit the leading contribution comes from the s-wave scattering transition matrix element $\langle l = 0|T_{A,B}|l = 0\rangle \simeq 2\omega^2\lambda_{A,B}/\pi$ where λ is the scattering length.

The matrix K then reduces to the scalar $K = -\omega^2\lambda_A\lambda_B\left(h_0^{(1)}(\omega a)\right)^2 = 4\pi\frac{\lambda_A\lambda_B}{a^2}e^{2ia\omega}$. Doing the integral (35) one arrives at

$$E_C = -\frac{\lambda_A\lambda_B}{a^3}.$$

This limit corresponds to the scalar version of the well known Casimir-Polder interaction. Our formalism however allows calculating corrections to it up to any desirable finite order in $\frac{1}{a}$. For example for two Dirichlet spheres of radii R_1, R_2 at distance a between their centers the expansion gives:

$$E = \begin{array}{c} -\frac{R_1R_2}{4\pi a^3} - \frac{R_1R_2(R_1+R_2)}{8\pi a^4} \\ -\frac{R_1R_2(34R_1^2+9R_1R_2+34R_2^2)}{48\pi a^5} \\ -\frac{R_1R_2(R_1+R_2)(2R_1^2+21R_1R_2+2R_2^2)}{36\pi a^6} + \ldots \end{array} \qquad (46)$$

Acknowledgment It is a pleasure to thank the organizers of the International Workshop "60 years of Casimir Effect" and the International Center for Condensed Matter Physics at the University of Brasilia, and in particular Victor Dodonov for this successful conference.

11. Appendix: Some properties of (infinite dimensional) operators

Here we recall some mathematical notions that we have used in describing the trace class properties of (11).

Definition 11.1. *For an operator $B : H \to H$, the operator norm of $||B||$ is defined as* $||B|| = \sup_{\psi\in H,\psi\neq 0}\frac{|<\psi|B|\psi>|}{<\psi|\psi>}$

Definition 11.2. *An operator B is bounded if $||B|| < \infty$*

Definition 11.3. *An operator $A : H \to H$ is called a* **positive operator** *(denoted $A > 0$) iff $\langle\psi|A|\psi\rangle \geq 0$ for every $\psi \in H$.*

This implies that A is hermitian and its spectrum nonnegative. If $A : H \to H$ is a positive operator then there exist a unique positive operator $B : H \to H$ satisfying $A = B^2$. B is called the **square root** of A and denoted $\sqrt{A}$.

Definition 11.4. *An operator $A : H_1 \to H_2$ is called* **trace class** *(and denoted $A \in t.c.$ or $A \in \mathcal{J}_1$) iff $\sum ||A\psi_n|| < \infty$ where $\{\psi_n\}_{n=1}^{\infty}$ is some orthonormal basis of H_1. It can be shown that this condition does not depend on the choice of the orthonormal basis. (Note that the definition makes sense even when $H_1 \neq H_2$.)*

If $A : H \to H$ is trace class then for any orthonormal basis $\{\psi_n\}_{n=1}^{\infty}$ of H the sum $\sum\langle\psi_n|A|\psi_n\rangle$ converges to the same (finite) value which is denoted $tr(A)$ and called the trace of A. One then also has $tr(A) = \sum\lambda_n$ where $\{\lambda_n\}$ are the eigenvalues of A (Lidski's theorem)

If $A : H \to H$ is trace class then the determinant $\det(1 + A)$ may also be rigorously defined and one has $\det(1 + A) = \prod(1 + \lambda_n)$.

The following theorem may be proved using the well known fact that the Fourier coefficients of a smooth $K(x, y)$ decay faster than any power. (Note that these coefficients also serve as the matrix elements with respect to Fourier basis of the operator defined by K.)

Journal of Physics: Conference Series **161** (2009) 012020 doi:10.1088/1742-6596/161/1/012020

Theorem 11.5. *Consider an operator* $A : L^2(D_1) \to L^2(D_2)$ *where* D_1, D_2 *are some domains in* $\mathbb{R}^n$ *which is given explicitly as an integral* $A\psi(x) = \int_{D_1} K(x,y)\psi(y)dy$. *A sufficient condition for* A *to be trace class is that* D_1, D_2 *are compact and* $K(x,y)$ *is smooth in a neighborhood of* $D_1 \times D_2$.

Proposition 11.6. *If* A *is trace class and* B *bounded then* AB *and* BA *are also trace class and* $Tr(|AB|), Tr(|BA|) \leq ||B|| Tr(|A|)$.

Definition 11.7. M *is a Hilbert Schmidt operator (denoted* $M \in H.S.$ *or* $M \in J_2$*) if* $||M||^2_{H.S.} \equiv \mathrm{Tr}\, M^\dagger M < \infty$

In particular we mention that the product of two Hilbert Schmidt operators always gives a trace class operator.

References

[1] Casimir H B G, Proc. Koninkl. Ned. Akad. Wet. **51**, 793 (1948).
[2] Kenneth O and Klich I, Phys. Rev. Lett. 97, 160401 (2006).
[3] Lifshitz E M, Sov. Phys. JETP 2, 73 (1956)
[4] Balian R and Duplantier B, Ann. Phys.N.Y. 112, 165, (1978).
[5] Bachas C P, quant-ph/0611082
[6] Nussinov Z, cond-mat/0107339 (See appendix A and footnote [35]).
[7] Emig T, Graham N, Jaffe R L and Kardar M, quant-ph/07071862
[8] Kenneth O, preprint hep-th/9912102.
[9] Genet C, Lambrecht A and Reynaud S, Phys. Rev. A67 (2003);
043811 Jaekel M T and Reynaud S, J. Phys. I 1, 1395 (1991). quant-ph/0101067.
[10] Bulgac A, Magierski P and Wirzba A, Phys. Rev. D73 (2006) 025007.
[11] Lippmann B and Schwinger J, Phys. Rev. 79, 469(1950).
[12] Kenneth O and Klich I, Phys. Rev. B. 78, 014103, 2008
[13] Candelas P and Deutsch D, Phys. Rev. D20 (1979),3063
[14] Feinberg J, Mann A and Revzen M, Annals of Physics (New York) 288(2001)103-136
[15] Lifshitz E M and Pitaevskii L P, Statistical Physics, Pt. 2, Pergamon, Oxford, 1984.
[16] van Kampen N G, Nijboer B R A and Schram K, Phys. Lett. A, Volume 26, Issue 7, 307 (1968).
[17] Mello P A and Stone D, Phys. Rev. B, 44, 8 (1991).
[18] Li H and Kardar M, Phys. Rev. Lett. 67, 3275 (1991);
Li H and Kardar M, Phys. Rev. A 46, 6490 (1992).
[19] Simon B, *Trace ideals and their applications*, LMS vol 35. Cambridge, New York, NY, 1979.
[20] Reed M and Simon B, Methods of mathematical physics 3: Scattering Theory, Academic Press 1979.
[21] Jaffe R L, Williamson L R, Annals Phys. 282 (2000) 432-448

Journal of Physics: Conference Series **161** (2009) 012021 doi:10.1088/1742-6596/161/1/012021

Numerical Calculation of the Force on some Generalized Casimir Pistons

Martin Schaden

Department of Physics, Rutgers University, Newark, NJ 07102

E-mail: mschaden@andromeda.rutgers.edu

Abstract. In this talk I presented numerical calculations of the Casimir force due to a scalar field on a piston in a cylinder of radius r with a spherical cap of radius $R > r$. Geometrical subtractions give a finite interaction energy. Due to reflection positivity, the vacuum force on the piston by a scalar field satisfying Dirichlet boundary conditions is attractive for these geometries, but the strength and short-distance behavior of the force depends strongly on the shape of the piston casing. For a cylindrical casing with a hemispherical head of large radius, the attractive force on the piston is inversely proportional to the square of the height of the piston.

1. Introduction

Below I define finite subtracted vacuum (interaction) energies that determine Casimir forces[1] in piston geometries. A similar geometrical approach was first used by Power[2] to obtain the well-known Casimir force between two parallel plates without regularizing potentially infinite zero-point energies. There has been renewed interest in such piston geometries[3] because the electromagnetic vacuum self-energy of a cube (as that of a sphere[4]) is positive[5]. The force on a partition in a parallelepiped nevertheless is attractive[6, 7, 8] at any position.

Semi-classical considerations suggest that the Casimir force on a piston depends strongly on the shape of the casing[9]. I here numerically calculate the vacuum force due to a massless scalar field satisfying Dirichlet boundary conditions for some generalized Casimir pistons by generalizing the world-line approach to Casimir interaction energies of Gies et al.[10] to include geometries with connected boundaries. In Casimir pistons all domains are bounded and the mathematical treatment in fact is much simpler and clear-cut than for the scattering situation with a continuous spectrum. A finite subtracted vacuum energy gives rise to the force on the piston. In the examples studied here, all domains are convex and numerical computations are vastly simplified by considering the convex hulls of Brownian bridges[11].

2. World-line approach for connected bounded domains

Consider the heat kernel operator $\mathfrak{K}_{\mathfrak{D}}(\beta) = e^{\beta\triangle/2}$ for the Laplacian $\triangle$ with Dirichlet boundary conditions on a bounded domain $\mathfrak{D} \subset \mathbb{R}^3$. The spectrum of eigenvalues $\{\lambda_n > 0, n \in \mathbb{N}\}$ of the negative Laplace operator in this case is discrete, real and positive. The corresponding spectral function (or trace of the heat kernel),

$$\phi_{\mathfrak{D}}(\beta) = \mathrm{Tr}\,\mathfrak{K}_{\mathfrak{D}}(\beta) = \sum_{n\in\mathbb{N}} e^{-\beta\lambda_n/2}, \tag{1}$$

Journal of Physics: Conference Series **161** (2009) 012021 doi:10.1088/1742-6596/161/1/012021

is finite for $\beta > 0$. It has the well-known[12, 13, 14, 15] high-temperature (short-time) expansion,

$$\phi_{\mathfrak{D}}(\beta \sim 0) \sim \frac{1}{(2\pi\beta)^{3/2}} \sum_{n=0}^{\infty} (2\pi\beta)^{n/2} a_n(\mathfrak{D}) + \mathcal{O}(e^{-l^2/\beta}). \tag{2}$$

For smoothly bounded domains, the heat kernel coefficients $a_n(\mathfrak{D})$ in this series are integrals over powers of the local curvature and reflect average geometric properties of the domain and its boundary[16, 17], such as the volume $\mathcal{V}_{\mathfrak{D}} = a_0(\mathfrak{D})$ and the surface area $\mathcal{S}_{\mathfrak{D}} = -4a_1(\mathfrak{D})$ of $\mathfrak{D}$. Since $a_4(\mathfrak{D}) \neq 0$ implies a logarithmic divergent vacuum energy, the vanishing of this coefficient is crucial for an unambiguously defined Casimir energy. Non-analytic and (for $\beta \sim 0$) exponentially suppressed contributions to the asymptotic expansion of $\phi_{\mathfrak{D}}(\beta)$ are associated with classical periodic- and diffractive- orbits[18] of minimal length l.

The world-line approach to Casimir energies[10] relies on the fact[16, 19] that the spectral function for a bounded flat Euclidean domain $\mathfrak{D}$ can be expressed in terms of its support of standard Brownian bridges. In three dimensions,

$$\phi_{\mathfrak{D}}(\beta) = \int_{\mathfrak{D}} \frac{d\mathbf{x}}{(2\pi\beta)^{3/2}} \mathcal{P}[\ell_\beta(\mathbf{x}) \subset \mathfrak{D}] \ , \tag{3}$$

where $\ell_\beta(\mathbf{x}) = \{\mathbf{B}_\tau(\mathbf{x},\beta), 0 \leq \tau \leq \beta; \mathbf{B}_0(\mathbf{x},\beta) = \mathbf{B}_\beta(\mathbf{x},\beta) = \mathbf{x}\}$ is a standard Brownian bridge from $\mathbf{x}$ to $\mathbf{x}$ in "proper time" β and $\mathcal{P}[\ell_\beta(\mathbf{x}) \subset \mathfrak{D}]$ denotes the probability that such a bridge is entirely within the bounded domain $\mathfrak{D}$. Note that $\mathcal{P}[\ell_\beta(\mathbf{x}) \subset \mathfrak{D}]/(2\pi\beta)^{3/2}$ is the Green function from $\mathbf{x}$ to $\mathbf{x}$ in time β of the associated diffusion problem with Dirichlet boundary conditions on $\partial\mathfrak{D}$.

Eq.(3) implies that the spectral function for a domain of finite volume is finite. Divergences arise only in the corresponding zero-point energy. For finite vacuum energies, the leading – five, in three dimensions,– heat kernel coefficients in Eq.(2) would have to vanish. Although impossible to achieve for a single domain, one can improve the asymptotic behavior by considering a (finite) linear combination of spectral functions for domains $\{\mathfrak{D}_k; k = 0, 1, \ldots, M\}$

$$\tilde{\phi}(\beta) = \sum_k c_k \phi_{\mathfrak{D}_k}(\beta) \ . \tag{4}$$

When the coefficients c_k are such that

$$\sum_k c_k a_i(\mathfrak{D}_k) = 0 \quad \text{for} \quad i = 0, \ldots, 4, \tag{5}$$

the "interaction" vacuum energy,

$$\mathcal{E}_{\text{int}} = -\pi \int_0^\infty \frac{d\beta}{(2\pi\beta)^{3/2}} \tilde{\phi}(\beta) = \sum_k c_k \mathcal{E}_{\text{vac}}(\mathfrak{D}_k) \ , \tag{6}$$

is finite because the integrand is $\mathcal{O}(\beta^{-1/2})$ for $\beta \sim 0$ when Eq.(5) holds. The $\mathcal{E}_{\text{vac}}(\mathfrak{D}_k) \sim \frac{1}{2}\sum_n \sqrt{\lambda_n}$ in Eq.(6) are the (divergent) formal zero-point energies of a massless scalar field satisfying Dirichlet boundary conditions on the individual bounded domains $\mathfrak{D}_k$. The linear combination $\mathcal{E}_{\text{int}}$ of these vacuum energies is the difference in the zero-point energy of domains with the same total volume, total surface area, average curvature, topology,etc...

The subtraction procedure outlined above is preferable to conventional regularization procedures for numerical computations, because it can be implemented as a restriction on the set of paths. One thus avoids the numerically difficult computation of small differences in large quantities. Note that this subtraction method is readily generalized to Neumann and metallic boundary conditions.

Some heat kernel coefficients may vanish due to (idealized) boundary conditions and do not have to be subtracted geometrically. Thus, for ideal metallic boundaries, the surface contribution vanishes[20]. For smooth ultrathin metallic boundaries in vacuum without creases or conical singularities in Euclidean space, a_2 and a_4 also vanish. Since a_3 does not depend on dimensions of the cavity, a finite Casimir energy in this case can be defined by subtracting the volume term only, that is by subtracting the vacuum energy in the absence of boundaries. However, the first five heat kernel coefficients in general will not vanish if any of these idealized conditions are relaxed and the volume subtraction then is not sufficient to obtain a finite Casimir energy[21, 22].

It is interesting (and useful numerically) to note that for any *convex* domain $\mathfrak{D}$ a Brownian bridge $\ell_\beta(\mathbf{x})$ is wholly within $\mathfrak{D}$ only if its convex hull $\mathfrak{H}[\ell_\beta(\mathbf{x})]$ is entirely contained in $\mathfrak{D}$, that is

$$\mathcal{P}[\ell_\beta(\mathbf{x}) \subset \mathfrak{D}] = \mathcal{P}[\mathfrak{H}[\ell_\beta(\mathbf{x})] \subset \mathfrak{D}] \text{ for } \mathfrak{D} \text{ convex} . \tag{7}$$

In the continuum limit, a standard Brownian bridge process satisfies the stochastic differential equation[19, 23],

$$d\mathbf{B}_\tau(\mathbf{x},\beta) = \frac{\mathbf{x} - \mathbf{B}_\tau(\mathbf{x},\beta)}{\beta - \tau} d\tau + d\mathbf{W}_\tau, \quad \text{with} \quad \mathbf{B}_0(\mathbf{x},\beta) = \mathbf{x} \tag{8}$$

where $\mathbf{W}_\tau$ is a (3-dimensional) standard Wiener process, i.e. each component of $\mathbf{W}_\tau$ is normally distributed about the origin $\mathbf{0}$ with variance τ. Translational invariance and the scaling property of random walks (and their convex hulls) imply that,

$$\mathcal{P}[\ell_\beta(\mathbf{x}) \subset \mathfrak{D}] = \mathcal{P}[\ell_1(\mathbf{0})]\Theta[\mathbf{x} + \sqrt{\beta}\ell_1(\mathbf{0}) \subset \mathfrak{D}] = \mathcal{P}[\ell_1(\mathbf{0})]\Theta[\mathbf{x} + \sqrt{\beta}\mathfrak{H}[\ell_1(\mathbf{0})] \subset \mathfrak{D}] . \tag{9}$$

Here multiplication and addition operations distribute over all members of a point-set, that is they scale and translate the whole loop or hull. Using Eq.(9) to make the dependence on $\mathbf{x}$ and β explicit in Eq.(3) and inserting the result in Eq.(6) one obtains,

$$\begin{aligned} \mathcal{E}_{\text{int}} &= -\frac{1}{8\pi^2}\int d\mathbf{x} \langle \int_0^\infty \frac{d\beta}{\beta^3} \sum_k c_k \Theta[\mathbf{x} + \sqrt{\beta}\mathfrak{H}[\ell_1(\mathbf{0})] \subset \mathfrak{D}_k] \rangle_{\ell_1(\mathbf{0})} \\ &= -\frac{1}{4\pi^2} \langle \int d\mathbf{q} \int_0^\infty d\lambda \sum_k c_k \Theta[\mathbf{q} + \mathfrak{H}[\ell_1(\mathbf{0})] \subset \lambda\mathfrak{D}_k] \rangle_{\ell_1(\mathbf{0})} , \end{aligned} \tag{10}$$

where the expectation $\langle \ldots \rangle_{\ell_1(\mathbf{0})}$ is with respect to unit loops (standard Brownian bridges over unit time). Eq.(10) is the basic formula[1] used for computing Casimir interaction energies in the world-line approach[10]. Although the straightforward interpretation of an integrated interaction energy density is lost by the change in integration variables $\mathbf{x} = \mathbf{q}/\lambda, \beta = \lambda^{-2}$, the last expression for $\mathcal{E}_{\text{int}}$ in Eq.(10) is slightly better adapted for numerical evaluation.

3. Convex Domains and Convex Hulls

Fig. 1 depicts a suitable combination of convex domains for computing the interaction vacuum energy $\mathcal{E}_{\text{int}}$ of a cylindrical Casimir piston of radius r with a cap of radius $R \geq r$. A finite overall length L ensures that the volumes of all convex domains is finite.

The main advantage of using the convex hull in numerical computations rather than the unit-loops themselves, is that the hull is a vastly reduced point-set that retains all the relevant

[1] The overall proportionality constant in Eq.(10) is 4 times that in[10] because the unit loops here are defined by a standard Brownian bridge process rather than one of twice the variance.

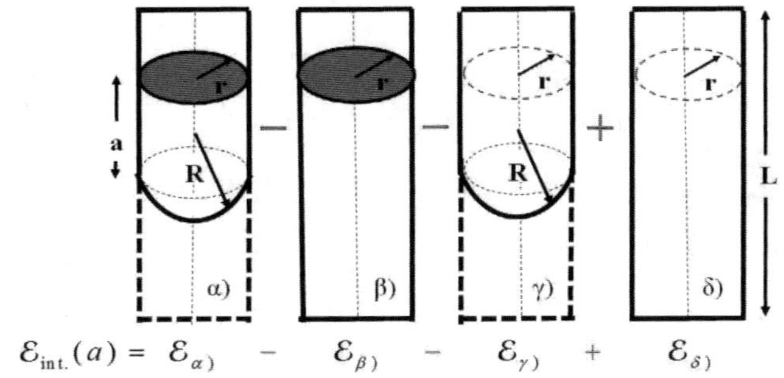

Figure 1. The interaction Casimir energy $\mathcal{E}_{\text{int}}(a)$ for a cylindrical cavity of radius r with a cap of radius $R \geq r$ and a moveable planar piston at height a. $\mathcal{E}_{\text{int}}(a)$ is the difference in vacuum energy of the cavity with and without piston compared to this difference for a cylinder of the same radius r. This difference of differences in vacuum energies is finite for all values of $a, r, L > 2r$ and $R \geq r$. $\mathcal{E}_{\text{int}}$ does not require regularization and for $L \gg a$ only $\mathcal{E}_{\alpha)}$ depends on the position of the piston. Solid lines denote surfaces on which Dirichlet boundary conditions are imposed. Thick dashed lines in Figs. 1α) and 1γ) indicate where the surface of the cylinder would be: Brownian bridges that pierce these surfaces give no contribution to $\mathcal{E}_{\text{int}}(a)$, leading to condition (11) for large $L \gg 2r$. Note that all five different bounded domains are *convex* and bounded for finite L.

information for the partition function on convex domains, but discards the irrelevant sequential ordering and the great number of interior points of a loop. The number of vertices of the convex hull increases only logarithmically with the number of points defining the loop. The convex hull of a loop given by a thousand points on average has ~ 55 vertices and that of one with a million points has about 190 vertices (see Fig. 2). The time required by efficient algorithms to compute the hull (with v vertices) of n points is proportional to $n\log(v)$ [24]. In evaluating the integrations in Eq.(10) for sufficiently complex convex domains $\mathfrak{D}$ like those in Fig. 1, one only requires the convex hulls.

To evaluate the spatial- and scale- integrals in Eq.(10) for a given set of vertices defining a hull, $\mathfrak{H}[\ell_1(\mathbf{0})] = \{\mathbf{v}_i = (x_i, y_i, z_i); i = 1, \ldots, v\}$, note that (for $L \gg r$) the only significant contribution to the interaction energy $\mathcal{E}_{\text{int}}$ of the combination of domains shown in Fig. 1 comes from

loops (hulls) that pierce the piston and the piston head, but not the cylinder. (11)

The cylindrical symmetry of the domains in Fig. 1 allows one to rotate the offset $\mathbf{q}$ of any loop (hull) to the positive x-axis and perform one angular integral trivially, resulting in

$$\mathcal{E}_{\text{int}}(\text{cyl.piston}) = -\frac{1}{2\pi}\langle \int_0^\infty \rho d\rho \int_{-\infty}^\infty dz \int_0^\infty d\lambda \sum_k c_k \Theta[(\rho, 0, z) + \mathfrak{H}[\ell_1(\mathbf{0})] \subset \lambda\mathfrak{D}_k]\rangle_{\ell_1(\mathbf{0})} \,. \quad (12)$$

Further details on the numerical algorithm used to compute Eq.(12) are given in [11]. To verify the accuracy of the algorithm and estimate systematic errors, I compared (see Fig. 3) the numerical results for $R \gg r \gg d$ with the known analytic Casimir energy due to a scalar satisfying Dirichlet boundary conditions on two parallel circular plates of area $\mathcal{S} = \pi r^2$ (half the electromagnetic Casimir energy[1] of this configuration). For $R \gtrsim r \gg a$ the numerical results were compared with the semiclassical estimate of the asymptotic behavior in[11],

$$\mathcal{E}(a \ll r \leq R \leq 1.05r) = -\frac{\hbar c}{96\pi a}\left(\frac{\sqrt{R^2 - r^2}}{a} + 1 + \mathcal{O}(a/r)\right) \,. \quad (13)$$

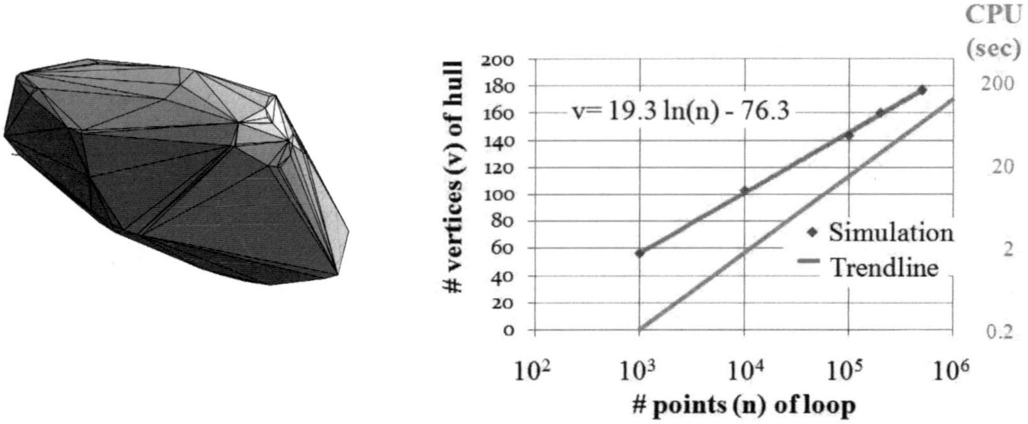

Figure 2. On the left is a typical triangulated hull of a Brownian bridge defined by 10^5 points. It has 286 faces and 145 vertices and the shape of an irregularly cut rhinestone with large triangular facettes and intricate corners. The average number of vertices (v) of the convex hull of a Brownian bridge given by n points can be read off the left axis of the graph using the upper (red) curve. Data points are from numerical simulations. The inset is the formula for the (red) trendline that best describes the data. The right axis and lower (green) line give the average CPU-time used to compute the hulls on a laptop with a 2GHz processor. Note that the left axis is linear while the right axis is logarithmic.

Including only statistical errors, the asymptotic numerical data for $r \leq R \leq 1.02r$ is best reproduced by,

$$\mathcal{E}_{\text{int}}(\tfrac{a}{r} \ll \tfrac{r}{R} \lesssim 1) \sim -\hbar c \left(\frac{0.00395(5)\sqrt{R^2 - r^2}}{a^2} + \frac{0.00326(4)}{a} + \mathcal{O}(a/r) \right) . \tag{14}$$

The coefficients of the $\sqrt{R^2 - r^2}/a^2$ and $1/a$ terms are comparable and the leading $0.00326(4)/a^2$-behavior of the attractive force on the piston with a hemispherical head differs by less than 2% from the semiclassical estimate in Eq.(13). This 2% discrepancy is as good an estimate of the systematic errors of the numerical calculation as any I can give at this level of accuracy (the statistical error of the calculations is about 1.2%). Fig. 6 shows the *difference* between the numerical results and the asymptotic semiclassical energy for the hemispherical Casimir piston in a linear plot. Note that the residual force is small and repulsive.

4. Discussion

The Casimir force on a piston by a massless scalar field satisfying Dirichlet boundary conditions depends qualitatively on the shape of the casing in the systems of Fig. 1. Whereas the attractive force for small piston height $a \ll r$ is proportional to r^2/a^4 for a flat cylinder head, it is proportional to $\sqrt{R^2 - r^2}/a^3$ for heads with a curvature radius $r < R \ll r^2/a$ and for a hemispherical piston with ($R = r$) is inversely proportional to a^2. Note that the asymptotic behavior of Eq.(13) implies that one percent deviation in radius from a hemispherical cylinder head doubles the Casimir force on the piston at a height $a \sim 0.1r$. Fig. 3 indicates that the Casimir force on the piston depends qualitatively on R/r and at small piston height a/r may range over several orders in magnitude.

The interaction Casimir energy defined in Eq.(10) is *a priori* finite (and does not require regularization) when the conditions of Eq.(5) are satisfied. Only a portion of the vacuum energy

Journal of Physics: Conference Series **161** (2009) 012021 doi:10.1088/1742-6596/161/1/012021

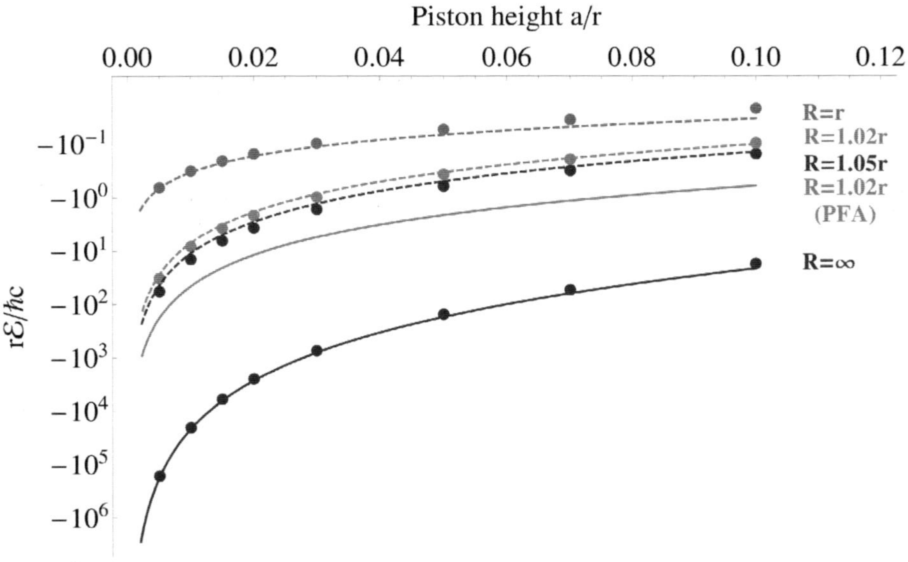

Figure 3. The dimensionless Casimir interaction energy $r\mathcal{E}_{\text{int}}(a)/(\hbar c)$ for the cylindrical cavities of radius r with caps of radius $R \geq r$ of Fig. 1 as a function of the rescaled height a/r of the moveable piston. Dots indicate numerical results (dot size does not represent errors, which are too small to show on this logarithmic plot). Solid lines are piston-based Proximity Force Approximation (PFA) estimates (note that the PFA overestimates the interaction energy by an order of magnitude for $R = 1.02r$). Dashed curves correspond to the asymptotic semiclassical estimate of Eq.(13). Note the many orders of magnitude between a piston with a flat (bottom, purple) and a hemispherical (top, red) cylinder head.

is computed that includes all its dependence on the piston height a and determines the force on the piston. To relate this interaction vacuum energy of a massless scalar to the spectral function of the Laplace operator, leading terms in the high-temperature expansion have to be canceled. This was here achieved by subtracting spectral functions of domains with the global characteristics represented by the first few (five in three dimensions) heat kernel coefficients. This subtraction procedure in particular implies that finite Casimir energies are differences in the vacuum energy of domains with the same topology, same volume, same average curvature etc...

The linear combination of vacuum energies for $\mathcal{E}_{\text{int}}$ in the present examples is shown in Fig. 1 – for $L \gg r$ only $\mathcal{E}_{\alpha)}$ depends on the height a of the piston. The Casimir energy due to a massless Dirichlet scalar in this case is given by the *positive* probability measure for Brownian paths that satisfy Eq.(11). This implies an attractive Casimir force on the piston for any separation a and cap radius $R > r$. A similar argument implies that the Casimir force by a Dirichlet scalar is attractive for a large class of piston geometries for which a condition on the Brownian bridges like Eq.(11) holds – including pistons and cylinders of arbitrary cross-section and form, with cylinder heads that are entirely contained within the cylinder. Cylinder heads that extend beyond the cylinder have also been examined[25]. Other classes of loops contribute in this case and the interaction energy can have either sign.

A semiclassical analysis of the geometries presented here reproduced[11] the numerical results for the force on the piston to better than 2% for $a/r < 0.1$, it in particular gives the correct asymptotic behavior of the force for $a/r \to 0$. By contrast, the PFA is inapplicable and overestimates the force on the piston by an order of magnitude for a hemispherical head. For a hemispherical cap the attractive Casimir force on the piston $-\hbar c/(96\pi a^2)$ does not depend on the radius $r = R \gg a$ of the cylinder and cap and mimics the electrostatic force on a metallic piston

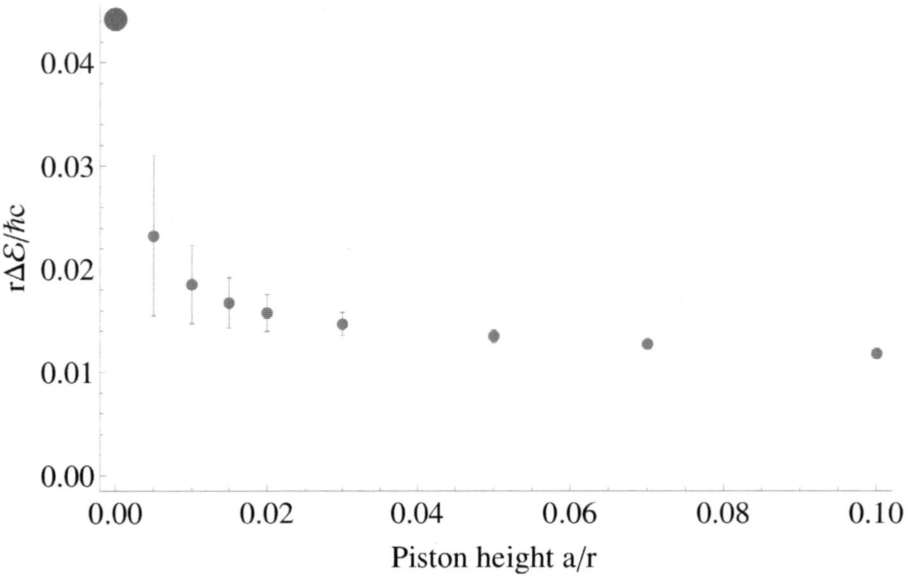

Figure 4. The difference $\Delta\mathcal{E}(a) = \mathcal{E}_{\rm int}(a) + \hbar c/(96\pi a)$ of the numerical- and asymptotic semiclassical-estimate for the scalar Casimir interaction energy of a hemispherical piston. The energy difference is plotted in dimensionless units against the piston height a in units of the cylinder radius r. Only statistical errors of the numerical calculation are shown. The numerical results are those of Fig. 3 for $R = r$ and were obtained from the convex hulls of 10^5 unit loops, each given by 10^5 points. The (red) dot at $a/r = 0$ is the semiclassical contribution to this difference due to periodic orbits[11].

exerted by a charge $q^2/(\hbar c) = 1/(48\pi) \sim 1/150.8 \lesssim \alpha_{EM}$ on the axis of the cylinder at $z = 0$. This paradoxical situation arises only for the Casimir energy of a massless scalar field satisfying Dirichlet (Neumann) boundary conditions. Semiclassically the asymptotic contribution to the interaction energy due to a scalar field satisfying Neumann boundary conditions is the same in magnitude as that in Eq.(13) but of the opposite (positive) sign. This for a hemispherical piston at small a/r dominant contribution to the interaction energy due to a Dirichlet scalar therefore is absent in the semiclassical approximation for metallic boundary conditions[9]. Contrary to the Dirichlet case, the reflection positivity argument fails for metallic and/or Neumann conditions, because these introduce correlations in the fluctuations on either side of the piston. The numerical (and semiclassical) attraction obtained here for all the piston configurations studied, therefore should not be taken to imply attraction for other, more realistic boundary conditions. In view of the fact that small variations in the geometry can change the Casimir force by many orders of magnitude, it remains a challenge to develop numerical methods that reliably estimate electromagnetic Casimir forces for complicated geometries.

Acknowledgments

I would like to thank the organizers of the International Workshop "60 Years of Casimir Effect" in Brasilia for a very engaging conference and I. Klich for some illuminating discussions on the implications of reflection positivity in the case of Dirichlet boundary conditions. This investigation was supported by the National Science Foundation with Grant No. PHY-0555580.

References

[1] Casimir H B G 1948 *Proc. Kon. Nederland. Akad. Wetensch.* B**51** 793
Lifshitz E M 1955 *Zh. Eksp. Teor. Fiz.* **29** 94 and *Sov. Phys. JETP* **2** 73 (1956)

Dzyaloshinskii I E, Lifshitz E M and Pitaevskii L 1961 *Adv. Phys.* **10** 165
[2] Power E A 1964 *Introductory Quantum Electrodynamics* (New York: Elsevier) Appendix I
[3] Cavalcanti R M 2004 *Phys. Rev.* D **69** 065015
[4] Boyer T H 1968 *Phys. Rev.* **174** 1764
Milton K A, DeRaad L L and Schwinger J S 1978 *Annals Phys.* **115** 388
[5] Lukosz W 1971 *Physica* **56** 109
[6] Hertzberg M P, Jaffe R L, Kardar M and Scardicchio A, *Phys. Rev. Lett.* **95**, 250402; 2007 *Phys. Rev.* D **76** 045016
[7] Kenneth O and Klich I 2006 Phys. Rev. Lett. **97** 160401
[8] Bachas C P 2007 J. Phys. A **40** 9089
[9] Schaden M and Mateescu L 2007 Weak and repulsive Casimir Forces in Piston Geometries *Preprint* arXiv:0705.3435
[10] Gies H and Langfeld K 2002 *Int. J. Mod. Phys.* A **17** 966 (*Preprint* hep-th/0112198)
Gies H, Langfeld K and Moyaerts L 2003 *J. High Energy Phys.* JHEP0306(2003)018; *ibid* 2003 Proceedings of QFEXT'03 (Norman, Sept. 2003) 203 (*Preprint* hep-th/0311168)
Gies H and Klingmuller K 2005 Proceedings of QFEXT'05 (Barcelona, Sept. 2005) (*Preprint* hep-th/0511092); *ibid* 2006 *Phys. Rev.* D**74** 045002 (*Preprint* quant-ph/0601094)
See Schubert C 2001 *Phys. Rept.* **355** 73 for a review of the worldline formalism
[11] Schaden M 2008 Numerical Analysis of Generalized Casimir Pistons *Preprint* arXiv:0810.1046
[12] Greiner P 1971 *Arch. Rat. Mech. Anal.* **41** 163
[13] Gilkey P B 1984 *Invariance Theory, the Heat Equation, and the Atiyah-Singer Index Theorem* (Wilmington: Publish or Perish), reprinted 1995 (Boca Raton: CRC Press)
[14] Kirsten K 2002 *Spectral functions in Mathematics and Physics* (Boca Raton: Chapman & Hall/CRC Press)
[15] Fulling S A 2007 *SIGMA* **3** 094
[16] Kac M 1966 *Amer. Math. Monthly* **73** Part II 1
[17] Vassilevich D V 2003 Physics Rep. **388** 279 and references therein.
[18] Brack M and Bhaduri R 1997 *Semiclassical Physics* (Reading: Addison-Wesley)
[19] Stroock D W 1993 *Probability Theory, an Analytic View* (Cambridge:Cambridge Univ. Press.)
[20] Balian R and Duplantier B 1977 Annals Phys. **104** 300
ibid 1978 Annals Phys. **112** 165
ibid 2002 Proceedings of the 15th SIGRAV Conference on General Relativity and Gravitational Physics held at Villa Mondragone, Monte Porzio Catone,Italy, September 9-12 (2002) (*Preprint* quant-ph/0408124)
[21] Deutsch D and Candelas P 1979 *Phys. Rev.* D**20** 3063
Candelas P 1982 *Ann. Phys. (NY)* **143** 241 ;*ibid* 1986 *Ann. Phys. (NY)* **167** 257
[22] Graham N, Jaffe R L, Khemani V, Quandt M, Scandurra M and Weigel H 2002 *Nucl.Phys.* B**645** 49
Graham N, Jaffe R L, Quandt M, Schröder O and Weigel H 2004 Nucl. Phys. B**677** 379
[23] Øksendahl B 2000 *Stochastic Differential Equations 5th Edition* (New York: Springer) p.75
Rogers L C G and Williams D 1994 *Diffusions, Markov Processes, and Martingales. 2nd Edition* Vol. 2 (Chichester: J. Wiley & Sons)
[24] Berg M de, Cheong O, Kreveld M van and Overmars M D 2008 *Computational Geometry Algorithms and Applications, 3rd Edition* (Berlin: Springer 2000)
[25] Schaden M 2008 On the Direction of Casimir Forces *Preprint* arXiv:0808.3966; to appear in *Phys. Rev. Lett.*

Journal of Physics: Conference Series **161** (2009) 012022 doi:10.1088/1742-6596/161/1/012022

Weak Coupling Casimir Energies for Finite Plate Configurations

Jef Wagner, Kimball A Milton, Prachi Parashar

Oklahoma Center for High Energy Physics and Homer L. Dodge Department of Physics and Astronomy, University of Oklahoma, Norman, OK, 73019, USA

E-mail: wagner@nhn.ou.edu, milton@nhn.ou.edu, prachi@ou.edu

Abstract. We derive and use an extremely simplified formula for the interaction Casimir energy for two separate bodies in the weak coupling regime for massless scalar fields. We derive closed form solutions for a general arrangement of two δ-function plates finite in one direction and infinite in another. We examine the situation of two parallel plates finite in both transverse directions.

1. Introduction

Recently, Emig et al. [1, 2] rederived the multiple scattering formalism and used it to calculate new results for Casimir energies between disjoint bodies. Similar techniques have been employed for many years, perhaps starting with Renne in 1971 [3], who rederived the Lifshitz formula [4]. More recently many papers have used multiple scattering techniques to examine the correction to the proximity force approximation for different situations [5, 6, 7, 8, 9, 10].

Casimir calculations have conventionally been carried out assuming that parallel plates had infinite extent. Gies and Klingmüller have found corrections to the infinite-size approximation for the cases of perfectly conducting boundary conditions [11]. We will here use the multiple scattering formalism to examine δ-function plate geometries with finite lengths for scalar fields subject to very weakly coupled boundaries.

The weak coupling regime is interesting to study because the formula for the Casimir energy simplifies greatly, and in many cases is amenable to closed form solutions. This introduction gives a very quick derivation of the weak-coupling form of the Casimir energy for massless scalar fields. For a more complete derivation see Kenneth and Klich [12, 13], or Milton and Wagner [10].

The vacuum expectation value for the action of a scalar field can be given by the standard formula

$$W = -\frac{\mathrm{i}}{2}\,\mathrm{Tr}\ln\frac{\mathcal{G}}{\mathcal{G}_0}, \tag{1}$$

where $\mathcal{G}$ is the Green's function that satisfies the wave equation of the scalar field, including any interactions with any objects or potentials. $\mathcal{G}_0$ is the free Green's function that satisfies the wave equation with the same boundary conditions at infinity as the full Green's function but without any interaction with background potentials. For a time-independent system we can use the condition that $W = -\int dtE$ to identify the energy as

$$E = \frac{i}{2}\int \frac{d\omega}{2\pi} \operatorname{Tr}\ln \frac{G}{G_0}, \tag{2}$$

where G is the Fourier-transformed Green's function given by

$$\mathcal{G}(x, x') = \int \frac{d\omega}{2\pi} e^{-i\omega(t-t')} G(\boldsymbol{x}, \boldsymbol{x}'), \tag{3}$$

and the trace if over spatial coordinates. Given two nonoverlapping potentials such that $V(x) = V_1(x) + V_2(x)$, we can define an interaction energy as the energy of the full system less the energy of each potential acting alone, $E_{\text{Int}} = E(V_1 + V_2) - E(V_1) - E(V_2)$. This simplified expression can be written as

$$E_{\text{Int}} = -\frac{i}{2}\int \frac{d\omega}{2\pi} \operatorname{Tr}\ln\left(1 - G_0 T_1 G_0 T_2\right), \tag{4}$$

where T_i is the scattering matrix for the ith potential defined as $T_i = V_i(1 - G_0 V_i)^{-1}$.

By formally expanding out the logarithm in (4), we can think about the formula as describing successively more scattering events between the two objects. The weak coupling expansion keeps only the first term of the expansion of the logarithm, essentially describing only a single scattering between the objects. Additionally, in the weak-coupling regime the scattering matrix can be approximated simply by the potential, $T \approx V$. This results in a very simplified weak-coupling single-scattering approximation to the energy ($\omega \to i\zeta$),

$$E_{\text{Int}} = -\frac{1}{4\pi}\int d\zeta \operatorname{Tr} G_0 V_1 G_0 V_2. \tag{5}$$

2. 2+1 Spatial Dimensions

If the potentials are independent of the z direction then we can further simplify the interaction energy. By dividing out the infinite length in the z direction we obtain an energy per unit length, which we will represent in this paper by the fraktur symbol $\mathfrak{E}$. By Fourier transforming in the z direction we get

$$\mathfrak{E} = -\frac{1}{2}\int \frac{d\zeta}{2\pi}\int \frac{dk_z}{2\pi} \operatorname{Tr} g_0 V_1 g_0 V_2, \tag{6}$$

where g_0 is given by

$$G_0(\boldsymbol{r} - \boldsymbol{r}'; \zeta) = \int \frac{dk_z}{2\pi} e^{ik_z(z-z')} g_0(\boldsymbol{r}_\perp - \boldsymbol{r}'_\perp; \kappa), \tag{7}$$

and κ is defined by $\kappa^2 = \zeta^2 + k_z^2$. The two dimensional Green's function is explicitly

$$g_0(\boldsymbol{r}_\perp - \boldsymbol{r}'_\perp; \kappa) = \frac{1}{2\pi} K_0(\kappa|\boldsymbol{r}_\perp - \boldsymbol{r}'_\perp|), \tag{8}$$

yielding a weak-coupling form for the energy per unit length of

$$\mathfrak{E} = -\frac{1}{32\pi^4}\int d\zeta dk_z \int d^2r \int d^2r' K_0^2(\kappa|\boldsymbol{r}_\perp - \boldsymbol{r}'_\perp|) V_1(\boldsymbol{r}_\perp; \zeta) V_2(\boldsymbol{r}'_\perp; \zeta). \tag{9}$$

In the case that the potentials are independent of the imaginary frequency ζ this simplifies even further to

$$\mathfrak{E} = -\frac{1}{32\pi^3}\int d^2r \int d^2r' \frac{V_1(\boldsymbol{r}_\perp) V_2(\boldsymbol{r}'_\perp)}{|\boldsymbol{r}_\perp - \boldsymbol{r}'_\perp|^2}. \tag{10}$$

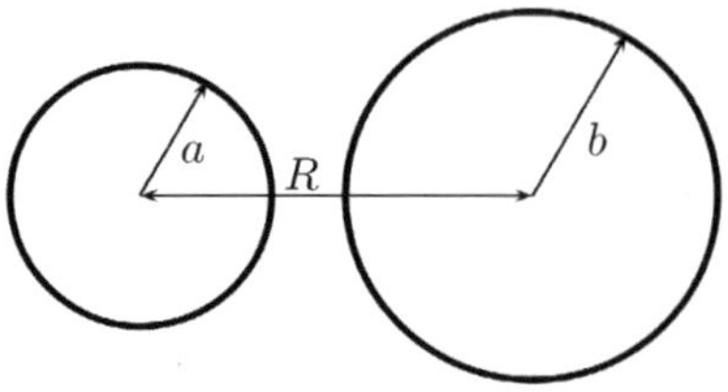

Figure 1. Two cylinders of radii a and b, their centers separated by a distance R.

To demonstrate the simplicity of this formula we will rederive the Casimir energy for two cylinders [10]. Assume that the two cylinders have radii a and b, and their centers are separated by a distance R as shown in figure 1. This situation can be represented by potentials $V_1 = \lambda_1\delta(r-a)$ and $V_2 = \lambda_2\delta(r'-b)$ where r and r' are radial coordinates in cylindrical polar coordinate systems centered on the respective cylinders. Using (10), these potentials yield an energy per unit length

$$\mathfrak{E} = -\frac{\lambda_1\lambda_2 ab}{32\pi^3}\int_0^{2\pi} \mathrm{d}\theta \int_0^{2\pi} \mathrm{d}\theta' \frac{1}{R^2+a^2+b^2-2aR\cos\theta+2bR\cos\theta'-2ab\cos(\theta-\theta')}. \tag{11}$$

With a simple change in angular coordinates to $u=\theta-\theta'$ and $v=\frac{\theta+\theta'}{2}$ this expression can be integrated to yield the exact closed form [10]

$$\mathfrak{E} = -\frac{\lambda_1\lambda_2 ab}{8\pi}\frac{1}{\sqrt{(R^2-(a-b)^2)\,(R^2-(a+b)^2)}}. \tag{12}$$

2.1. General Configuration of Plates

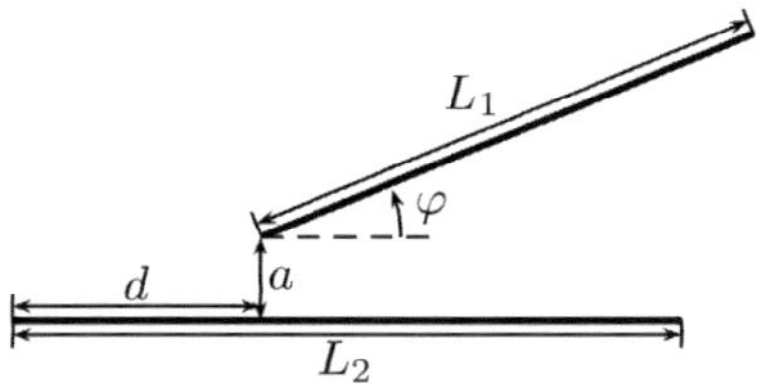

Figure 2. Two finite plates of length L_1 and L_2 in a general configuration. In a coordinate system centered at the edge of the top plate φ is the relative angle between the plates, a is the perpendicular distance the lower plate is shifted down, and d is the lateral distance the edge of the lower plate is shifted to the left.

Two finite plates of a general configuration as shown in figure 2 can be represented by the potentials in cylindrical coordinates, with origin at the left edge of plate 1,

$$V_1 = \lambda_1\delta(\theta-\varphi)\frac{\Theta(L_1-r)}{r}, \tag{13a}$$

$$V_2 = \lambda_2\delta(y+a)\Theta(x+d)\Theta(L_2-d-x). \tag{13b}$$

Here $\Theta(x)$ is the step function,

$$\Theta(x) = \begin{cases} 1, & x>0, \\ 0, & x<0. \end{cases} \tag{14}$$

Using these potentials with (10), we get the following expression for the energy per unit length,

$$\mathfrak{E} = -\frac{\lambda_1\lambda_2}{32\pi^3}\int_0^{L_1} dr \int_{-d}^{L_2-d} dx \frac{1}{(x - r\cos\varphi)^2 + (a + r\sin\varphi)^2}. \tag{15}$$

This integral can be done exactly, yielding a closed form for the general configuration,

$$\mathfrak{E} = -\frac{\lambda_1\lambda_2}{32\pi^3\sin\varphi}\left[\mathrm{Ti}_2\left(\frac{L_2-d}{a}, \cot\varphi\right) - \mathrm{Ti}_2\left(\frac{L_2-d-L_1\cos\varphi}{a+L_1\sin\varphi}, \cot\varphi\right)\right.$$
$$\left. -\mathrm{Ti}_2\left(\frac{-d}{a}, \cot\varphi\right) + \mathrm{Ti}_2\left(\frac{-d-L_1\cos\varphi}{a+L_1\sin\varphi}, \cot\varphi\right)\right], \tag{16}$$

where Ti_2 is the generalized inverse tangent integral[1] defined by

$$\mathrm{Ti}_2(x, a) = \int_0^x dy \frac{\mathrm{ArcTan}\, y}{y + a}. \tag{17}$$

2.2. Torque

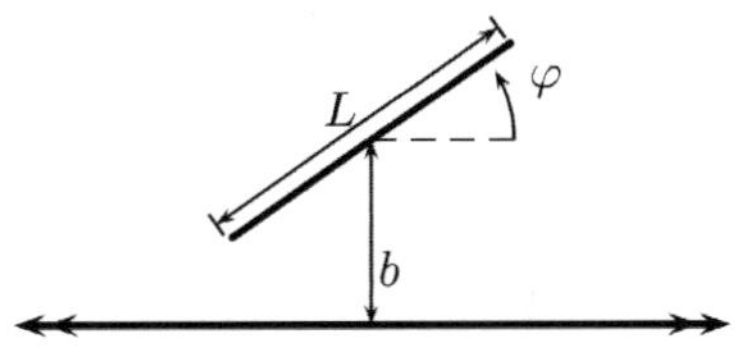

Figure 3. A finite plate of length L, above an infinite plate, The center of the finite plate is a perpendicular distance b above the infinite plate, and the finite plate makes an angle φ with respect to the infinite plate.

It would be desirable to examine the rotational stability of parallel plates. This can be done simply by looking at the sign of the torque for small angular displacements, while leaving the center of mass stationary. Although one can get a general form for the torque from (16), it is much simpler to restructure the problem to isolate the torque as shown in figure 3. This is the same as the "Casimir pendulum" problem studied by Scardicchio et al., who used the optical approximation [15].

Given a situation of a tilted plate of length L over an infinite plate ($L_1 \to L$, $L_2 \to \infty$, $d \to -\infty$ in figure 2), we can very easily isolate the torque that the finite plate experiences. The energy per unit length is given by

$$\mathfrak{E} = -\frac{\lambda_1\lambda_2}{32\pi^2\sin\varphi}\ln\left(\frac{b + \frac{L}{2}\sin\varphi}{b - \frac{L}{2}\sin\varphi}\right). \tag{18}$$

The torque per unit length $\mathfrak{T}$ found by taking the negative derivative of the energy with respect to the tilt angle. This gives an expression for the torque on the plates as

$$\mathfrak{T} = -\frac{\lambda_1\lambda_2\cos\varphi}{32\pi^2\sin\varphi}\left(\frac{1}{\sin\varphi}\ln\left(\frac{b + \frac{L}{2}\sin\varphi}{b - \frac{L}{2}\sin\varphi}\right) - \frac{Lb}{b^2 - \frac{L^2}{4}\sin^2\varphi}\right). \tag{19}$$

[1] The generalized inverse tangent integral is related to the dilogarithm function, and much information about it can be found in [14].

Journal of Physics: Conference Series **161** (2009) 012022 doi:10.1088/1742-6596/161/1/012022

From the expression for the torque we can see some clear qualitative features. The torque has a zero value at $\varphi = 0$, and a quick evaluation shows that the first derivative is positive, of value

$$\left.\frac{\partial \mathfrak{T}}{\partial \varphi}\right|_{\varphi=0} = \frac{\lambda_1 \lambda_2}{192\pi^2} \frac{L^3}{b^3}, \tag{20}$$

signifying an unstable equilibrium. For values of L such that $L > 2b$, the torque diverges as $\sin\varphi$ approaches $2b/L$. This is simply the result of the fact that the plates would touch in this situation. If $L < 2b$ the torque has another zero at $\varphi = \frac{\pi}{2}$, with a first derivative of

$$\left.\frac{\partial \mathfrak{T}}{\partial \varphi}\right|_{\varphi=\frac{\pi}{2}} = -\frac{\lambda_1 \lambda_2}{32\pi^2} \left(\frac{Lb}{b^2 - \frac{L^2}{4}} - \ln\left(\frac{b + \frac{L}{2}}{b - \frac{L}{2}} \right) \right), \tag{21}$$

which is negative for all $L < 2b$, meaning a stable equilibrium. Therefore a finite flat plate suspended above another infinite plate will tend to orient itself perpendicular to the plate if left to rotate about its center of mass. Physically, the reason for this is clear: because the magnitude of the Casimir force decreases with distance, to minimize the energy for a fixed center of mass, the smaller plate wants to rotate so as to place the shortest side closest to the infinite plate. (For the situation when the thickness of the plate is finite, see [16].)

2.3. Parallel Plates

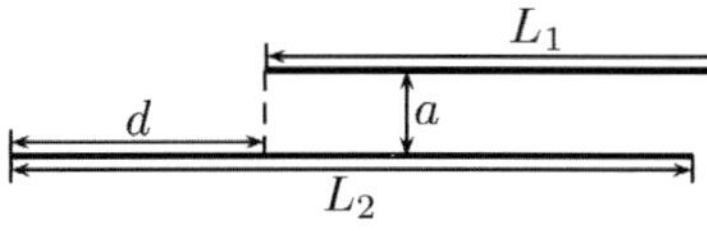

Figure 4. Parallel plates

Parallel plates are perhaps the most interesting special case. We can compare the exact expressions for energy and force to those for infinite parallel plates, getting corrections for finite size. In addition the parallel plates case, due to its simplicity, lends itself well to studying both normal and lateral forces.

Consider the same setup as in the general case shown in figure 2, simply letting φ go to zero, as shown in figure figure 4. The energy per unit length can be derived directly from (16) by using the identity

$$\lim_{a\to\infty} a\mathrm{Ti}_2(x, a) = \int dx \mathrm{ArcTan}\, x, \tag{22}$$

yielding an integral form for the energy per unit length

$$\mathfrak{E} = -\frac{\lambda_1 \lambda_2}{32\pi^3} \left[\int_{\frac{L_2-d-L_1}{a}}^{\frac{L_2-d}{a}} + \int_{\frac{d}{a}}^{\frac{d+L_1}{a}} \right] dx \mathrm{ArcTan}\, x. \tag{23}$$

Although an indefinite integral for the arctangent exists, this form is perhaps more illuminating because all the physical quantities are in the limits. The forces, which are given as derivatives of the energy, are all given in terms of arctangents.

Equation (23) yields closed forms for the normal force between the plates and the lateral force experienced by the plates by taking the negative derivative with respect to a or d, respectively.

The general form of the normal force, defined as $\mathfrak{F}_a = -\partial\mathfrak{E}/\partial a$, is

$$\mathfrak{F}_a = -\frac{\lambda_1\lambda_2}{32\pi^3 a^2}\left[(L_2-d)\mathrm{ArcTan}\left(\frac{L_2-d}{a}\right) - (L_2-d-L_1)\mathrm{ArcTan}\left(\frac{L_2-d-L_1}{a}\right) - d\mathrm{ArcTan}\left(\frac{d}{a}\right) + (d+L_1)\mathrm{ArcTan}\left(\frac{d+L_1}{a}\right)\right]. \quad (24)$$

In the limiting case of the plates getting very close together we expect to recover the result for the pressure for infinite parallel plates times the area exposed. By mathematically taking $a \to 0$, we use the large argument expansion of the inverse tangent,

$$\mathrm{ArcTan}\,(x) = \frac{\pi}{2} - \frac{1}{x} + \frac{1}{3}\frac{1}{x^3} + \cdots, \qquad \text{for } x \to \infty, \quad (25)$$

to recover the expected result plus corrections to that result. Because the limiting form of the arctangent depends on the sign of the argument, the single general equation can give several different answers depending on the size and position of the plates. For the situation shown in figure 4 the limiting form is

$$\mathfrak{F}_a = -\frac{\lambda_1\lambda_2}{32\pi^2 a^2}\left((L_2-d) + \mathcal{O}(a^3)\right), \quad (26)$$

and the first correction is zero. However, if the plates are the same size and aligned the limiting form of the force with the first correction is

$$\mathfrak{F}_a = -\frac{\lambda_1\lambda_2}{32\pi^2 a^2}\left(L - \frac{1}{\pi}2a + \mathcal{O}(a^3)\right). \quad (27)$$

If we let one end of both plates extend off into infinity then we can get the edge correction for two aligned plates. This correction is

$$\frac{\mathfrak{F}_a/\mathfrak{F}_0 - 1}{2a} = \frac{1}{\pi}. \quad (28)$$

The general form of the lateral force, similarly defined as $\mathfrak{F}_d = -\partial\mathfrak{E}/\partial d$, is

$$\mathfrak{F}_d = -\frac{\lambda_1\lambda_2}{32\pi^3 a}\left[\mathrm{ArcTan}\left(\frac{L_2-d-L_1}{a}\right) - \mathrm{ArcTan}\left(\frac{L_2-d}{a}\right) - \mathrm{ArcTan}\left(\frac{d}{a}\right) + \mathrm{ArcTan}\left(\frac{d+L_1}{a}\right)\right]. \quad (29)$$

From the exact expression for the lateral force, we find there is only one equilibrium position, occurring at $d = \frac{L_1-L_2}{2}$, where the derivative of the force is negative:

$$\left.\frac{\partial\mathfrak{F}_d}{\partial d}\right|_{d=\frac{L_1-L_2}{a}} = -\frac{\lambda_1\lambda_2}{16\pi^3}\frac{L_1L_2}{\left(a^2+\left(\frac{L_1+L_2}{2}\right)^2\right)\left(a^2+\left(\frac{L_1-L_2}{2}\right)^2\right)}, \quad (30)$$

signifying a stable equilibrium. The position and qualitative behavior is as expected, the plate have an stable equilibrium when they are symmetrically aligned.

We are also interested in how the lateral force behaves if the plates are very close together. To study that we simply take the limit as $a \to 0$. Assuming without loss of generality that $L_2 > L_1$, to lowest order the force is

$$\mathfrak{F}_d = \begin{cases} +\frac{\lambda_1\lambda_2}{16\pi^2 a} & \text{for } d > 0 \text{ and } d > L_2 - L_1 \\ 0 & \text{for } d > 0 \text{ and } 0 < d < L_2 - L_1 \\ -\frac{\lambda_1\lambda_2}{16\pi^2 a} & \text{for } d < 0 \end{cases} \tag{31}$$

This is what we would expect if we approximated the energy simply as the energy per area between the two infinite plates times the area exposed between the two plates, and took the derivative of this very simple approximation as the force.

3. Three spatial dimensions

Until now we have worked in 2+1 dimensions, meaning that the potentials had infinite length in one direction. So the finite plates considered in the previous section were more like two infinite ribbons. In this section we will work with plates of finite area.

We start by working out the $TGTG$ formula (5) in three dimensions. In three dimensions the form of the Green's function is even easier to work with than in two dimensions,

$$G_0(\boldsymbol{r}, \boldsymbol{r}') = \frac{1}{4\pi}\frac{e^{-|\zeta||\boldsymbol{r}-\boldsymbol{r}'|}}{|\boldsymbol{r}-\boldsymbol{r}'|}. \tag{32}$$

This gives a form of the energy

$$E = -\frac{1}{64\pi^3}\int_{-\infty}^{\infty} \mathrm{d}\zeta \int \mathrm{d}^3r \int \mathrm{d}^3r' \frac{e^{-2|\zeta||\boldsymbol{r}-\boldsymbol{r}'|}V_1(\boldsymbol{r};\zeta)V_2(\boldsymbol{r}';\zeta)}{|\boldsymbol{r}-\boldsymbol{r}'|^2}, \tag{33}$$

and if the potentials are independent of ζ then the erpression simplifies further to

$$E = -\frac{1}{64\pi^3}\int \mathrm{d}^3r \int \mathrm{d}^3r' \frac{V_1(\boldsymbol{r})V_2(\boldsymbol{r}')}{|\boldsymbol{r}-\boldsymbol{r}'|^3}. \tag{34}$$

If we restrict ourselves to parallel plate δ-function potentials of any general shape the energy can written in an even simpler form,

$$E = -\frac{\lambda_1\lambda_2}{64\pi^3}\int_{A_1} \mathrm{d}^2r_\perp \int_{A_2} \mathrm{d}^2r'_\perp \left[a^2 + (\boldsymbol{r}_\perp - \boldsymbol{r}'_\perp)^2\right]^{-3/2}, \tag{35}$$

where A_1 and A_2 are the areas of the two plates, and a, again, is the separation between the plates. If we let one of the areas, let it be A_2, tend to infinity, then the energy for a single finite plate above an infinite plate is

$$E = -\frac{\lambda_1\lambda_2}{32\pi^2 a}A_1, \tag{36}$$

exactly the energy per area from the Lifshitz formula times the area of the finite plate. For weak coupling plates, if one of the plates is infinite, and the other finite, then there is no correction to the Lifshitz formula. This is not unexpected, and is a result of the fact that the weak coupling approximation is the same as pairwise summation.

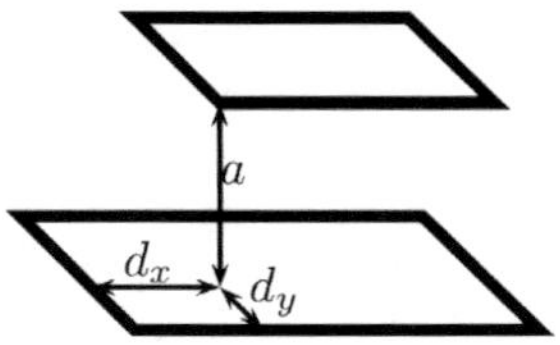

Figure 5. Two finite parallel rectangular plates. The sizes of the plates are $L_{1x} \times L_{1y}$ for the top plate, and $L_{2x} \times L_{2y}$ for the bottom plate. The lower left corner of the upper plate is displaced a distance d_x in the x direction and d_y in the y.

3.1. Rectangular Parallel Plates

For two rectangular parallel plates, as shown in figure 5, the interaction energy is given by the integral

$$E = -\frac{\lambda_1\lambda_2}{64\pi^3}\int_{d_x}^{L_{1x}+d_x} \mathrm{d}x \int_0^{L_{2x}} \mathrm{d}x' \int_{d_y}^{L_{1y}+d_y} \mathrm{d}y \int_0^{L_{2y}} \mathrm{d}y' \left[a^2 + (x-x')^2 + (y-y')^2\right]^{-3/2}. \tag{37}$$

This expression can be partially integrated and rewritten as

$$E = \frac{-\lambda_1\lambda_2 a}{64\pi^3}\left[\int_{\frac{L_{1x}+d_x-L_{2x}}{a}}^{\frac{L_{1x}+d_x}{a}} + \int_{\frac{d_x}{a}}^{\frac{d_x-L_{2x}}{a}}\right]\mathrm{d}x\left[\int_{\frac{L_{1y}+d_y-L_{2y}}{a}}^{\frac{L_{1y}+d_y}{a}} + \int_{\frac{d_y}{a}}^{\frac{d_y-L_{2y}}{a}}\right]\mathrm{d}y\mathrm{ArcTan}\left(\frac{xy}{\sqrt{1+x^2+y^y}}\right). \tag{38}$$

The two-dimensional indefinite integral in the equation is given by

$$\begin{aligned}\int \mathrm{d}x \int \mathrm{d}y\mathrm{ArcTan}\left(\frac{xy}{\sqrt{1+x^2+y^2}}\right) &= xy\mathrm{ArcTan}\left(\frac{xy}{\sqrt{1+x^2+y^2}}\right) \\ &+x\ln\left(x+\sqrt{1+x^2+y^2}\right) + y\ln\left(y+\sqrt{1+x^2+y^2}\right) \\ &-\frac{1}{2}x\ln(1+y^2) - \frac{1}{2}y\ln(1+x^2) - \sqrt{1+x^2+y^2}.\end{aligned} \tag{39}$$

The final closed-form expression for the energy of the two rectangular parallel plates is somewhat messy, consisting of the above indefinite integral evaluated at 16 different combinations of variables.

The normal and lateral forces can again be given by the derivatives of the energy with respect to the separation a or to the displacement (this time either d_x or d_y).

The lateral force from the plates has a stable equilibrium when the centers of the two plates are aligned. However, first derivatives of the force can be different for displacements from the equilibrium position in the x and y directions depending on the geometry.

Perhaps the most interesting property of this system as in section 2.3 to examine what happens to the attractive force between the plates as the plates get very close together. For very small separations we should get an expression for the force as a power series in a where the first term is the pressure given by the Lifshitz formula times the area between the plates,

$$F_a = -\frac{\lambda_1\lambda_2}{32\pi^2 a^2}A(1 + c_1 a + c_2 a^2 \cdots). \tag{40}$$

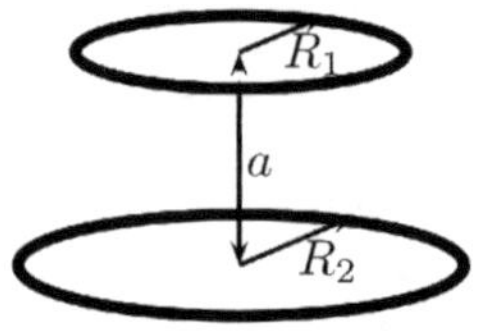

Figure 6. Two coaxial disks, separated by a distance a. The radii for the two disks are R_1 and R_2.

Using the large argument expansion for the arctangent (25), it is possible to get such an expression for the two plate arrangement, although the expressions for the area and the correction terms depend on the layout of the plates. For a situation in which the upper plate is completely above the lower plate, with none of the edges aligned, the area is given as $A = L_{1x}L_{1y}$ and the first correction term is $c_1 = 0$. For a situation where both plates are the same size, and they are exactly aligned ($d_x = d_y = 0$) then the area is $A = L_xL_y = L_{1x}L_{1y} = L_{2x}L_{2y}$ and

$$c_1 = -\frac{1}{\pi}\frac{2(L_x + L_y)}{L_xL_y}. \tag{41}$$

3.2. Parallel Disks

Instead of asking how two parallel rectangular plates attract, we could just have easily asked how two disks attract. The interaction Casimir energy for two coaxial disks separated by a distance a is given by

$$E = -\frac{\lambda_1\lambda_2}{32\pi^2}\int_0^{R_1} \mathrm{d}r \int_0^{R_2} \mathrm{d}r' \int_0^{2\pi} \mathrm{d}\theta \frac{rr'}{[a^2 + r^2 + r'^2 - 2rr'\cos\theta]^{3/2}}. \tag{42}$$

This expression can be integrated term by term in a series expansion in powers of r and r'. The energy can then be expressed as

$$E = -\frac{\lambda_1\lambda_2}{32\pi}\frac{R_1^2R_2^2}{a^3}\sum_{m=0}^{\infty}\sum_{n=0}^{m} A_{m,n}\left(\frac{R_1}{a}\right)^{2(m-n)}\left(\frac{R_2}{a}\right)^{2n}, \tag{43}$$

where

$$A_{m,n} = \frac{1}{2}\left(\frac{-1}{4}\right)^m \binom{2(m+1)}{m+1}\left[\binom{m}{n}^2 - \binom{m}{n+1}\binom{m}{n-1}\right]. \tag{44}$$

The power series is convergent, so we can simply take the derivative of each term to get the force between the plates. By using the asymptotics of the power series in the limit as $a \to 0$ we recover the expected result and get corrections to the infinite plate result. For two different sized disks, if $R_1 < R_2$ then $A = \pi R_1^2$ and $c_1 = 0$. For two equal sized disks where $R_1 = R_2 = R$, $A = \pi R^2$ and

$$c_1 = -\frac{1}{\pi}\frac{2\pi R}{\pi R^2}. \tag{45}$$

4. Conclusions

The weak-coupling regime greatly simplifies Casimir calculations, and even more so the interaction energy between two bodies. The simplicity allows us to obtain closed-form solutions to the energy and force between some nontrivial geometries. These closed-form solutions can help us in understanding the mechanics of these systems, such as the torque on two finite plates studied here. Also these solutions can help us in understanding the limits of certain approximations, such as the proximity force approximation or the correction to the attractive force between plates of finite size.

For the three cases of parallel plates studied here, the 2+1 D parallel plates, the rectangular parallel plates, and the parallel co-axial disks, we get very similar results in the limit as the plates get very close together. If the edges of the plates do not align then the first correction term $c_1 = 0$ and the area is the area of overlap of the plates. This fact can be reconciled with the fact that for a finite plate over an infinite plate the exact result for the attractive force is simply the pressure from the Lifshitz formula times the area of the plate. We would expect this to be a good approximation if one plate were much larger than the other, which corresponds to the first correction being zero. If one plate is even slightly larger, in the limit as the separation goes to zero the difference in the size of the plates is still large in comparison to the separation.

In the case of the edges of the two plates aligning exactly (that is, plates of the same size and shape directly above one another) then the area is simply the area of the plates and the first correction term takes the form

$$c_1 = -\frac{1}{\pi}\frac{\text{Perimeter}}{\text{Area}}. \tag{46}$$

This is a general property of this system, and can be visualized by realizing that in the limit as a goes to zero locally along any edge the system will appear to be two semi-infinite plates with their edges aligned. Therefore we might expect the correction to be a proportional to the perimeter of the plates, with the constant of proportionality given by (28).

In addition, these closed-form results will act as simple test cases for numerical studies.

Acknowledgments

We thank K.V. Shajesh and S.A. Fulling for helpful remarks. This work is supported in part by the US National Science Foundation and the US Department of Energy.

References

[1] Emig T and Jaffe R L 2008 *J. Phys. A: Math. Theor.* **41** 164001 (*Preprint* 0710.5104)
[2] Emig T, Graham N, Jaffe R L and Kardar M 2008 *Phys. Rev. D* **77** 025005 (*Preprint* 0710.3084)
[3] Renne M J 1971 *Physica* **56** 125
[4] Lifshitz E M 1955 *JETP* **29** 94 [*Sov. Phys. JETP* 1955 **2** 74]
[5] Bulgac A, Magierski P and Wirzba A 2006 *Phys. Rev. D* **73** 025007 (*Preprint* hep-th/0511056)
[6] Wirzba A, Bulgac A and Magierski P 2006 *J. Phys. A: Math. Gen.* **39** 6815 (*Preprint* quant-ph/0511057)
[7] Bordag M 2007 *Phys. Rev. D* **75** 065003 (*Preprint* quant-ph/0611243)
[8] Bordag M 2006 *Phys. Rev. D* **73** 125018 (*Preprint* hep-th/0602295)
[9] Bordag M, Geyer B, Klimchitskaya G L and Mostepanenko V M 2006 *Phys. Rev. B* **74** 205431
[10] Milton K A and Wagner J 2008 *J. Phys. A: Math. Theor.* **41** 155402 (*Preprint* 0712.3811)
[11] Gies H and Klingmuller K 2006 *Phys. Rev. Lett.* **97** 220405 (*Preprint* quant-ph/0606235)
[12] Kenneth O and Klich I 2006 *Phys. Rev. Lett.* **97** 160401 (*Preprint* quant-ph/0601011)
[13] Kenneth O and Klich I 2007 (*Preprint* 0707.4017)
[14] Lewin L 1981 *Polylogarithms and associated functions.* (New York, Oxford: North Holland)
[15] Scardicchio A and Jaffe R L 2005 *Nucl. Phys. B* **704** 552 (*Preprint* quant-ph/0406041)
[16] Milton K A, Parashar P and Wagner J 2008 (*Preprint* 0806.2880)

Journal of Physics: Conference Series **161** (2009) 012023 doi:10.1088/1742-6596/161/1/012023

Partition functions of three-dimensional quantum gravity and the black hole entropy

A A Bytsenko [(a)] **and M E X Guimarães** [(b)]

[(a)] Departamento de Física, Universidade Estadual de Londrina
Caixa Postal 6001, Londrina-Paraná, Brazil

E-mail: `abyts@uel.br`

[(b)] Instituto de Física, Universidade Federal Fluminense,
Av. Gal. Milton Tavares de Souza s/n, Niterói-RJ, Brazil

E-mail: `emilia@if.uff.br`

Abstract. We analyze aspects of the holographic principle relevant to the quantum gravity partition functions in Euclidean sector of AdS_3. The sum of known contributions to the partition functions can be presented exactly (including corrections) in the form where the Patterson-Selberg zeta function involves.

1. Introduction

Recently quantum gravity partition functions in the locally three-dimensional Anti-de Sitter (AdS_3) space-times have been analyzed in detail [1, 2]. In general such an analysis relies on the relation of the relevant product of determinants to Ray-Singer and Reidemeister torsion. For an irreducible flat connection on a compact manifold without boundary the torsion is simply a number, but for a manifold with boundary (such manifolds were considered in [1, 2]) the torsion must be understood as the measure on a certain moduli space associated with boundary. We suspect that computation of the one-loop correction using its interpretation via torsion is rather complicate. Nevertheless, the one-loop correction we may study by expressing the determinants in terms of the appropriate heat kernels, which can be evaluated by a images method (also one can use the Selberg trace formula applied to the tensor Laplacian). It has been shown that the symmetry group of AdS_3 gravity (with appropriate boundary conditions) is generated by the Virasoro algebra [3], and the one-loop partition function of gravity in AdS_3 does endeed the partition function of a conformal field theory (CFT_2) in two dimensions [2].

Now let us ask to what extent there may be $\mathrm{AdS}_3/\mathrm{CFT}_2$ correspondence, if it is completely correct. It is known that a simple geometrical structure of three-dimensional gravity (and black holes) allows exact computations since its Euclidean counterpart is locally isomorphic to the constant curvature hyperbolic space. Since the correspondence between quantum gravity partition function in AdS_3 and contribution to the partition function of CFT_2, we respect also correspondence between spectral functions related to Euclidean AdS_3 and modular-like functions (a Poincaré series) associated with a complex Riemann surface. We assume that this correspondence occur when complex variables of spectral functions take values on a Riemann surface (i.e. on the conformal boundary of AdS_3).

Thus, the main purpose of this work is to analyze a correspondence between the partition functions (and spectral functions) of three-dimensional quantum gravity and CFT in two-dimensions and provide the holographic principle relevant to this coprrespondence.

2. Expository remarks on partition functions

In this section there are no original results; we discuss partition functions for three-dimensional gravity and $\mathcal{N}=1$ supergravity mostly followed [1].

Quantum graity in three dimensions. It has been shown that the contribution to the partition function of pure gravity in a space-time asymptotic to AdS_3 comes from smooth geometries $M = \Gamma\backslash AdS_3$, where Γ is a discrete subgroup of $SO(3,1)$. To be more precise, it comes from geometries $M_{c,d}$ (see for detail [1]), where c and d are a pair of relatively prime integers, $c \geq 0$, and a pair (c,d) identified with $(-c,-d)$. The manifolds $M_{c,d}$ are all diffeomorphic to each other, and therefore the contribution $\mathbf{Z}_{c,d}(\tau)$ to the partition function can be expressed in terms of any one of them, say $\mathbf{Z}_{0,1}(\tau)$, by a modular transformation. It gives the following formula: $\mathbf{Z}_{c,d}(\tau) = \mathbf{Z}_{0,1}((a\tau+b)/(c\tau+d))$, where

$$\mathbf{Z}_{0,1}(\tau) = |q\bar{q}|^{-k}\prod_{m=2}^{\infty}|1-q^m|^{-2}\,. \tag{1}$$

The modulos of a Riemann surface Σ of genus one (the conformal boundary of AdS_3) is defined up to $\gamma\tau = (a\tau+b)/(c\tau+d)$ with $\gamma = \left(\begin{smallmatrix} a & b \\ c & d \end{smallmatrix}\right) \in SL(2,\mathbb{Z})$. The partition function as the sum of known contributions of states of left- and right-moving modes in the conformal field theory takes the form [1]

$$\mathbf{Z}(\tau) = \sum_{c,d}\mathbf{Z}_{c,d}(\tau) = \sum_{c,d}\mathbf{Z}_{0,1}((a\tau+b)/(c\tau+d))\,. \tag{2}$$

NS sector of $\mathcal{N}$=1 supergravity. We will consider only the basic case of $\mathcal{N}=1$ supergravity. Thus the symmetry group $SL(2,\mathbb{R})\times SL(2,\mathbb{R})$ of AdS_3 is replaced by $OSp(1|2)\times OSp(1|2)$, where $OSp(1|2)$ is a supergroup whose bosonic part is $Sp(2,\mathbb{R}) = SL(2,\mathbb{R})$. The boundary CFT has $(1,1)$ supersymmetry ($\mathcal{N}=1$ supersymmetry for both left- and right-movers). There are a few closely related choices of possible partition function, $\mathrm{Tr}\,\exp(-\beta H - i\theta J)$ or $\mathrm{Tr}\,(-1)^F\exp(-\beta H - i\theta J)$. This trace could be computed in either the Neveu-Schwarz (NS) or the Ramond (R) sector. One can compute these partition functions by summing over three-manifolds M that are locally AdS_3 and whose conformal boundary is a Riemann surface Σ of genus one. In addition, the four possible partition functions associated with NS or R sectors (with or without an insertion of $(-1)^F$) correspond to the four spin structures on Σ. An element g of $G = SL(2,\mathbb{R})$ acts on spin structure by

$$g\begin{bmatrix}\mu\\ \nu\end{bmatrix} \longrightarrow \begin{bmatrix}a & b\\ c & d\end{bmatrix}\begin{bmatrix}\mu\\ \nu\end{bmatrix},\quad \mu,\nu\in(1/2)\mathbb{Z}/\mathbb{Z}\,, \tag{3}$$

where the four spin structures on the two-torus Σ is represented by the column vector in (3), and μ,ν taking the values $(1/2)$ for antiperiodic (NS) boundary conditions and 0 for periodic (R) ones.

Taking into account the choice of the spin structure on Σ, one can sum over choices of M such that the given spin structure on Σ does extend over M. The NS spin structure on Σ is compatible with $M_{0,1}$, and therefore $M_{0,1}$ contributes to traces in the NS sector, not the R sector. The partition function of left- and right-moving exitations is $F(q,\overline{q}) = \mathrm{Tr}_{\mathrm{NS}}\exp(-\beta H - i\theta J)$, and the contribution to $F(q,\overline{q})$ associated with $M_{0,1}$ becomes [1]

$$F_{0,1} = F_{0,1}^{(\mathrm{ground})}\cdot\widehat{F}_{0,1}(\tau) \equiv \left|q^{-k^*/2}\right|^2\cdot\left|\prod_{n=2}^{\infty}\frac{1+q^{n-1/2}}{1-q^n}\right|^2\,, \tag{4}$$

where the contribution $F_{0,1}^{(\rm ground)} \equiv \left|q^{-k^*/2}\right|^2$ is related to the ground state energy. The complete function $F(\tau)$ can be computed by summing $F_{0,1}$ over modular images with $c+d$ odd. It corresponds to spin structure with $\mu=\nu=1/2$. Thus,

$$F\left[\begin{array}{c}\frac{1}{2}\\ \frac{1}{2}\end{array}\right](\tau) = \sum_{c,d|c+d \text{ odd}} \widehat{F}_{0,1}((a\tau+b)/(c\tau+d)). \qquad (5)$$

Let us also analyze partition functions with other spin structures. If we let $\mu=0$, $\nu=1/2$, then we get $G(q,\overline{q}) = {\rm Tr}_{\rm NS}\,(-1)^F\exp(-\beta H - i\theta J)$, and the contribution of $M_{0,1}$ to this partition function is obtained by reversing the sign of all fermionic contributions in (4):

$$G_{0,1}(\tau) = G_{0,1}^{(\rm ground)}\cdot\widehat{G}_{0,1}(\tau) \equiv \left|q^{-k^*/2}\right|^2\cdot\left|\prod_{n=2}^{\infty}\frac{1-q^{n-1/2}}{1-q^n}\right|^2. \qquad (6)$$

For the spin structure with $\mu=0$ and $\nu=1/2$ one has

$$G\left[\begin{array}{c}0\\ \frac{1}{2}\end{array}\right](\tau) = \sum_{c,d|d \text{ odd}} \widehat{G}_{0,1}((a\tau+b)/(c\tau+d))\,. \qquad (7)$$

Note that the summand in (5) and (21) does not depend on the choice a, b [1]. A modular transformation $\tau\rightarrow\tau+1$ exchanges pair $(\mu,\nu)=(0,1/2)$ with $(\mu,\nu)=(1/2,1/2)$; in particular, $F(\tau)=G(\tau+1)=F(\tau+2)$.

R sector of $\mathcal{N}$=1 supergravity. One can compute the Ramond partition function $K = {\rm Tr}_{\rm R}\exp(-\beta H - i\theta J)$ for $\mu=1/2$, $\nu=0$, so $K(\tau)=G(-1/\tau)$. This completes characterization list of three of the four partition functions. In supersymmetric theory with discrete spectrum, the fourth partition function $I={\rm Tr}_{\rm R}(-1)^F\exp(-\beta H-i\theta J)$ is an integer, independent of β and θ (it can be interpreted as the index of a supersymmetry generator). This function has to be computed using the odd spin structure, the one with $\mu=\nu=0$. Typically in three-dimensional gravity the partition function I vanishes, since the odd spin structure does not extend over any three-manifold with boundary Σ.

3. Holomorphic factorization

3.1. Spectral functions of hyperbolic geometry

The Euclidean sector of AdS_3 has an orbifold description $\mathbf{H}_\Gamma = \Gamma\backslash H^3$. The complex unimodular group $G=SL(2,\mathbb{C})$ act on real hyperbolic three-space H^3 in a standard way, namely for $(x,y,z)\in H^3$ and $g\in G$, $g\cdot(x,y,z)=(u,v,w)\in H^3$. Thus for $r=x+iy$, $g=\left[\begin{array}{cc}a & b\\ c & d\end{array}\right]$,

$$u+iv = \frac{(ar+b)\overline{(cr+d)}+a\bar{c}z^2}{|cr+d|^2+|c|^2z^2}, \qquad w = \frac{z}{|cr+d|^2+|c|^2z^2}\,. \qquad (8)$$

Here the bar denotes the complex conjugation. Let $\Gamma\in G$ be the discrete group of G. Define the discrete group Γ,

$$\begin{aligned}\Gamma &= \{{\rm diag}(e^{2n\pi({\rm Im}\,\tau+i{\rm Re}\,\tau)},\ e^{-2n\pi({\rm Im}\,\tau+i{\rm Re}\,\tau)}) : n\in\mathbb{Z}\} = \{\mathfrak{g}^n : n\in\mathbb{Z}\}\,,\\ \mathfrak{g} &= {\rm diag}(e^{2\pi({\rm Im}\,\tau+i{\rm Re}\,\tau)},\ e^{-2\pi({\rm Im}\,\tau+i{\rm Re}\,\tau)})\,.\end{aligned} \qquad (9)$$

One can define the Selberg zeta function for the group $\Gamma = \{\mathfrak{g}^n : n \in \mathbb{Z}\}$ generated by a single hyperbolic element of the form $\mathfrak{g} = \mathrm{diag}(e^z, e^{-z})$, where $z = a + ib$ for $a, b > 0$. In fact we will take $a = 2\pi \mathrm{Im}\,\tau$, $b = 2\pi \mathrm{Re}\,\tau$. For the standard action of $SL(2, \mathbb{C})$ on H^3 one has

$$\mathfrak{g}\begin{bmatrix} x \\ y \\ z \end{bmatrix} = \begin{bmatrix} e^a & 0 & 0 \\ 0 & e^a & 0 \\ 0 & 0 & e^a \end{bmatrix}\begin{bmatrix} \cos(b) & -\sin(b) & 0 \\ \sin(b) & \cos(b) & 0 \\ 0 & 0 & 1 \end{bmatrix}\begin{bmatrix} x \\ y \\ z \end{bmatrix}. \tag{10}$$

Therefore $\mathfrak{g}$ is the composition of a rotation in $\mathbb{R}^2$ with complex eigenvalues $\exp(\pm ib)$ and a dilation $\exp(a)$. The following zeta function can be attached to $\mathbf{H}_\Gamma$ (see for detail [4]):

$$Z_\Gamma(s) := \prod_{\substack{k_1,k_2 \geq 0 \\ k_1,k_2 \in \mathbb{Z}}}^{\infty} [1 - (e^{i\theta})^{k_1}(e^{-i\theta})^{k_2} e^{-(k_1+k_2+s)\ell}]\,. \tag{11}$$

We should emphasize that $\mathbf{H}_\Gamma$ is also the geometry of a Euclidean three-dimensional black hole. In (11) $\ell = 2\pi \mathrm{Im}\,\tau = 2\pi r_+$, $\theta = 2\pi \mathrm{Re}\,\tau = 2\pi |r_-|$, where $r_+ > 0$, and $r_- \in i\mathbb{R}\,(i^2 = -1)$ are the outer and inner horizons of a black hole. $Z_\Gamma(s)$ is an entire function of s, whose zeros are precisely the complex numbers $\zeta_{n,k_1,k_2} = -(k_1 + k_2) + i\,(k_1 - k_2)\,\theta/\ell + 2\pi i n/\ell$, $(n \in \mathbb{Z})$, and whose logarithm for $\mathrm{Re} s > 0$ is given by [4]

$$\log Z_\Gamma(s) = -\frac{1}{4}\sum_{n=1}^{\infty} \frac{e^{-n\ell(s-1)}}{n[\sinh^2\left(\frac{\ell n}{2}\right) + \sin^2\left(\frac{\theta n}{2}\right)]} = -\frac{1}{4}\sum_{n=1}^{\infty} \frac{e^{-2n\pi \mathrm{Im}\,\tau(s-1)}}{n|\sin(n\pi\tau)|^2}\,. \tag{12}$$

3.2. Quantum corrections

For three-dimensinal gravity in real hyperbolic space the one-loop partition function, as a product of holomorphic and antiholomorphic functions, has been calculated in [2], and the result is:

$$\widehat{\mathbf{Z}}(\tau) = \left[\prod_{m=2}^{\infty}[(1-q^m)_{\mathrm{hol}}\prod_{m=2}^{\infty}(1-\overline{q}^m)_{\mathrm{antihol}}\right]^{-1} = \prod_{m=2}^{\infty}|1-q^m|^{-2}. \tag{13}$$

We may evaluate $\log \widehat{\mathbf{Z}}(\tau)$ to get the formula

$$\begin{aligned} \log \widehat{\mathbf{Z}}(\tau) &= \sum_{n=1}^{\infty} \frac{e^{-2n\pi \mathrm{Im}\,\tau}\cos(4n\pi \mathrm{Re}\,\tau) - e^{-4n\pi \mathrm{Im}\,\tau}\cos(2n\pi \mathrm{Re}\,\tau)}{2n[\sinh^2(n\pi \mathrm{Im}\tau) + \sin^2(n\pi \mathrm{Re}\tau)]} \\ &= \log\left\{\left[\frac{Z_\Gamma(3 - i\frac{\mathrm{Re}\tau}{\mathrm{Im}\tau})}{Z_\Gamma(2 - 2i\frac{\mathrm{Re}\tau}{\mathrm{Im}\tau})}\right]_{\mathrm{hol}} \cdot \left[\frac{Z_\Gamma(3 + i\frac{\mathrm{Re}\tau}{\mathrm{Im}\tau})}{Z_\Gamma(2 + 2i\frac{\mathrm{Re}\tau}{\mathrm{Im}\tau})}\right]_{\mathrm{antihol}}\right\}. \end{aligned} \tag{14}$$

The expression (14) is invariant under the transformation $\tau \to \tau + 1$. We would like to comment on the sum over geometries. The partition function, including the contribution from the Brown-Henneaux excitations, have the form

$$\begin{aligned} \mathbf{Z}(\tau) = \sum_{c,d} \mathbf{Z}_{c,d}(\gamma\tau) &= \sum_{c,d}\left| q^{-k}\prod_{n=2}^{\infty}(1-q^n)^{-1}\right|_{\gamma}^{2} \\ &= \sum_{c,d}\left\{|q\overline{q}|^{-k}\cdot\left[\frac{Z_\Gamma(3-it)}{Z_\Gamma(2-2it)}\right]_{\mathrm{hol}} \cdot \left[\frac{Z_\Gamma(3+it)}{Z_\Gamma(2+2it)}\right]_{\mathrm{antihol}}\right\}_{\gamma}. \end{aligned} \tag{15}$$

Table 3.2. List of partition functions

$\log\prod_{m=2}^{\infty}(1-q^m) = -\sum_{n=1}^{\infty}\frac{q^{2n}}{(1-q^n)}$	$\prod_{m=2}^{\infty}(1-q^m) = \frac{Z_\Gamma(2-2it)}{Z_\Gamma(3-it)}$
$\log\prod_{m=2}^{\infty}(1-\overline{q}^m) = -\sum_{n=1}^{\infty}\frac{\overline{q}^{2n}}{(1-\overline{q}^n)}$	$\prod_{m=2}^{\infty}(1-\overline{q}^m) = \frac{Z_\Gamma(2+2it)}{Z_\Gamma(3+it)}$
$\log\prod_{m=2}^{\infty}(1+q^m) = -\sum_{n=1}^{\infty}\frac{(-1)^n q^{2n}}{n(1-q^n)}$	$\prod_{m=2}^{\infty}(1+q^m) = \frac{Z_\Gamma(2-2it-i\xi)}{Z_\Gamma(3-it-i\xi)}$
$\log\prod_{m=2}^{\infty}(1+\overline{q}^m) = -\sum_{n=1}^{\infty}\frac{(-1)^n \overline{q}^{2n}}{n(1-\overline{q}^n)}$	$\prod_{m=2}^{\infty}(1+\overline{q}^m) = \frac{Z_\Gamma(2+2it-i\xi)}{Z_\Gamma(3+it-i\xi)}$
$\log\prod_{m=2}^{\infty}(1-q^{m-\frac{1}{2}}) = \sum_{n=1}^{\infty}\frac{q^{\frac{3n}{2}}}{n(1-q^n)}$	$\prod_{m=2}^{\infty}(1-q^{m-\frac{1}{2}}) = \frac{Z_\Gamma(\frac{5}{2}-\frac{1}{2}it)}{Z_\Gamma(\frac{3}{2}-\frac{3}{2}it)}$
$\log\prod_{m=2}^{\infty}(1-\overline{q}^{m-\frac{1}{2}}) = \sum_{n=1}^{\infty}\frac{\overline{q}^{\frac{3n}{2}}}{n(1-\overline{q}^n)}$	$\prod_{m=2}^{\infty}(1-\overline{q}^{m-\frac{1}{2}}) = \frac{Z_\Gamma(\frac{5}{2}+\frac{1}{2}it)}{Z_\Gamma(\frac{3}{2}+\frac{3}{2}it)}$
$\log\prod_{m=2}^{\infty}(1+q^{m-\frac{1}{2}}) = \sum_{n=1}^{\infty}\frac{(-1)^n q^{\frac{3n}{2}}}{n(1-q^n)}$	$\prod_{m=2}^{\infty}(1+q^{m-\frac{1}{2}}) = \frac{Z_\Gamma(\frac{3}{2}-\frac{3}{2}it-i\xi)}{Z_\Gamma(\frac{5}{2}-\frac{1}{2}it-i\xi)}$
$\log\prod_{m=2}^{\infty}(1+\overline{q}^{m-\frac{1}{2}}) = \sum_{n=1}^{\infty}\frac{(-1)^n \overline{q}^{\frac{3n}{2}}}{n(1-\overline{q}^n)}$	$\prod_{m=2}^{\infty}(1+\overline{q}^{m-\frac{1}{2}}) = \frac{Z_\Gamma(\frac{3}{2}+\frac{3}{2}it-i\xi)}{Z_\Gamma(\frac{5}{2}+\frac{1}{2}it-i\xi)}$

Here $t = \mathrm{Re}\,\tau/\mathrm{Im}\,\tau$ and $|...|_\gamma$ denote the transform of an expression $|...|$ by γ. As we have seen (Eq. (14)) the quantum contribution to $\widehat{\mathbf{Z}}(\tau)$ is invariant under $\tau \to \tau+1$, and therefore the summand in (15) is independent of the choice of a and b in γ. Note that sum over c and d in (15) should be thought of as a sum over the coset $PSL(2,\mathbb{Z})/\mathbb{Z} \equiv (SL(2,\mathbb{Z})/\{\pm 1\})/\mathbb{Z}$. Since $ad-bc=1$, we get $(a\tau+b)/(c\tau+d) = a/c\ - 1/c(c\tau+d)$, one may show that

$$\mathrm{Im}\,(\gamma\tau) = \frac{\mathrm{Im}\,\tau}{|c\tau+d|^2}\,, \quad \mathrm{Re}\,(\gamma\tau) = \frac{a}{c} - \frac{c\mathrm{Re}\,\tau + d}{c|c\tau+d|^2}\,. \tag{16}$$

The final sums (15) are divergent. These kind of divergences also have been encountered in similar sums associated with three-dimensional gravity [5, 6, 7].

We will now analyze the partiton function in more details. Our goal here will be to repeat the analysis of Sect. 2 by using the spectral functions representation. Useful partition functions are presented in Table 3.2, where $\xi \equiv (2\mathrm{Im}\,\tau)^{-1}$. Since

$$\begin{aligned}\log\prod_{m=2}^{\infty}|1+q^{m-\frac{1}{2}}|^2 &= -\sum_{n=1}^{\infty}\frac{(-1)^n|q|^{-n}[q^{\frac{3n}{2}}+\overline{q}^{\frac{3n}{2}} - |q|^{2n}(q^{\frac{n}{2}}+\overline{q}^{\frac{n}{2}})]}{4n|\sin(n\pi\tau)|^2}\\ &= -\sum_{n=1}^{\infty}\frac{(-1)^n[e^{-n\pi\mathrm{Im}\,\tau}\cos(3n\pi\mathrm{Re}\,\tau) - e^{-3n\pi\mathrm{Im}\,\tau}\cos(n\pi\mathrm{Re}\,\tau)]}{2n|\sin(n\pi\tau)|^2}\,,\end{aligned} \tag{17}$$

then, using the result from Table 3.2, we get

$$\begin{aligned}\widehat{F}_{0,1}(\tau) &= \left[\frac{\prod_{n=2}^{\infty}(1+q^{n-1/2})}{\prod_{n=2}^{\infty}(1-q^{n})}\right]_{\rm hol}\cdot\left[\frac{\prod_{n=2}^{\infty}(1+\overline{q}^{n-1/2})}{\prod_{n=2}^{\infty}(1-\overline{q}^{n})}\right]_{\rm antihol}\\ &= \left[\frac{Z_\Gamma(\frac{3}{2}-\frac{3}{2}it-i\xi)Z_\Gamma(3-it)}{Z_\Gamma(\frac{5}{2}-\frac{1}{2}it-i\xi)Z_\Gamma(2-2it)}\right]_{\rm hol}\left[\frac{Z_\Gamma(\frac{3}{2}+\frac{3}{2}it-i\xi)Z_\Gamma(3+it)}{Z_\Gamma(\frac{5}{2}+\frac{1}{2}it-i\xi)Z_\Gamma(2+2it)}\right]_{\rm antihol}.\end{aligned} \tag{18}$$

The complete function $F(\tau)$ becomes

$$\begin{aligned}F\left[\begin{array}{c}\frac{1}{2}\\ \frac{1}{2}\end{array}\right](\tau) &= \sum_{c,d|c+d\ \rm odd} F_{0,1}^{(\rm ground)}(\gamma\cdot\tau)\widehat{F}_{0,1}(\gamma\cdot\tau)\\ &= \sum_{c,d|c+d\ \rm odd} F_{0,1}^{(\rm ground)}(\gamma\cdot\tau)\left\{\left[\frac{Z_\Gamma(\frac{3}{2}-\frac{3}{2}it-i\varphi)Z_\Gamma(3-it-i\eta)}{Z_\Gamma(\frac{5}{2}-\frac{1}{2}it-i\varphi)Z_\Gamma(2-2it-i\eta)}\right]_{\rm hol}\right\}_\gamma\\ &\times\left\{\left[\frac{Z_\Gamma(\frac{3}{2}+\frac{3}{2}it-i\varphi)Z_\Gamma(3+it-i\eta)}{Z_\Gamma(\frac{5}{2}+\frac{1}{2}it-i\varphi)Z_\Gamma(2+2it-i\eta)}\right]_{\rm antihol}\right\}_\gamma,\end{aligned} \tag{19}$$

where $\varphi\equiv\xi+\eta\equiv(\frac{1}{2}+\frac{d}{c})/{\rm Im}\,\tau$. Let us consider the partition function with other spin structures,

$$\begin{aligned}\log\prod_{m=2}^{\infty}|1-q^{m-\frac{1}{2}}|^2 &= \sum_{n=1}^{\infty}\frac{|q|^{-n}[q^{\frac{3n}{2}}+\overline{q}^{\frac{3n}{2}}-|q|^{2n}(q^{\frac{n}{2}}+\overline{q}^{\frac{n}{2}})]}{4n|\sin(n\pi\tau)|^2}\\ &= \sum_{n=1}^{\infty}\frac{e^{-n\pi{\rm Im}\,\tau}\cos(3n\pi{\rm Re}\,\tau)-e^{-3n\pi{\rm Im}\,\tau}\cos(n\pi{\rm Re}\,\tau)}{2n|\sin(n\pi\tau)|^2}.\end{aligned} \tag{20}$$

Then the contribution $\widehat{G}_{0,1}(\tau)$ becomes

$$\begin{aligned}\widehat{G}_{0,1}(\tau) &= \left[\frac{\prod_{n=2}^{\infty}(1-q^{n-1/2})}{\prod_{n=2}^{\infty}(1-q^{n})}\right]_{\rm hol}\cdot\left[\frac{\prod_{n=2}^{\infty}(1-\overline{q}^{n-1/2})}{\prod_{n=2}^{\infty}(1-\overline{q}^{n})}\right]_{\rm antihol}\\ &= \left[\frac{Z_\Gamma(\frac{5}{2}-\frac{1}{2}it)Z_\Gamma(3-it)}{Z_\Gamma(\frac{3}{2}-\frac{3}{2}it)Z_\Gamma(2-2it)}\right]_{\rm hol}\left[\frac{Z_\Gamma(\frac{5}{2}+\frac{1}{2}it)Z_\Gamma(3+it)}{Z_\Gamma(\frac{3}{2}+\frac{3}{2}it)Z_\Gamma(2+2it)}\right]_{\rm antihol}.\end{aligned} \tag{21}$$

To this end we have

$$\begin{aligned}G\left[\begin{array}{c}0\\ \frac{1}{2}\end{array}\right](\tau) &= \sum_{c,d|c+d\ \rm odd} G_{0,1}^{(\rm ground)}(\gamma\cdot\tau)\widehat{G}_{0,1}(\gamma\cdot\tau)\\ &= \sum_{c,d|c+d\ \rm odd} G_{0,1}^{(\rm ground)}(\gamma\cdot\tau)\left\{\left[\frac{Z_\Gamma(\frac{5}{2}-\frac{1}{2}it-i\varphi)Z_\Gamma(3-it-i\eta)}{Z_\Gamma(\frac{3}{2}-\frac{3}{2}it-i\varphi)Z_\Gamma(2-2it-i\eta)}\right]_{\rm hol}\right\}_\gamma\\ &\times\left\{\left[\frac{Z_\Gamma(\frac{5}{2}+\frac{1}{2}it-i\varphi)Z_\Gamma(3+it-i\eta)}{Z_\Gamma(\frac{3}{2}+\frac{3}{2}it-i\varphi)Z_\Gamma(2+2it-i\eta)}\right]_{\rm antihol}\right\}_\gamma.\end{aligned} \tag{22}$$

Quantum corrections to the black hole entropy. It has been claimed [1] that the formula for $\mathbf{Z}_{0,1}(\tau)$ is one-loop exact. Nevertheless it explicit calculation by evaluating the relevant one-loop determinants is not simple. One of the efforts to do so has been undertaken in [8]: The one-loop correction to the Bekenstein-Hawking entropy for the non-spinning black hole was calculated by expressing the determinants in terms of the appropriate heat kernels. However, the final result obtained in [8] is not completely correct (in fact the authors used the cutt-off regularization procedure of the divergent volume of the fundamental domain, and as a consequence the structure of the group actions on a real hyperbolic space has been changed). The purpose of this section is to get the correct contribution to the black hole entropy in terms of the Patterson-Selberg spectral function.

For our necessity remind that in (1) $24k = c_L = c_R = c$, and c is the central charge of a conformal field theory. Then $q = \exp[2\pi i\tau] = \exp[2\pi(-\mathrm{Im}\tau + i\mathrm{Re}\tau)]$ such that $\mathbf{Z}_{\rm cl}(\tau) = |q\overline{q}|^{-k} = \exp[4\pi k\mathrm{Im}\tau]$ corresponds to the classical prefactor of the partition function. The black hole partition function can be found by applying the modular transformation $\tau \rightarrow -1/\tau$ to the partition function $\mathbf{Z}_{0,1}(\tau)$ of thermal AdS_3. In fact the partition function (1) is a canonical ensemble partition function, so one may compute the black hole entropy using the following formula

$$\begin{aligned} S(\beta,\theta) &= \log \mathbf{Z}_{1,0} - \beta \mathbf{Z}_{1,0}^{-1}\frac{\partial\,\mathbf{Z}_{1,0}}{\partial\beta} = \left(1-\beta\frac{\partial}{\partial\beta}\right)\log\,\mathbf{Z}_{1,0} \\ &= \left(1-\beta\frac{\partial}{\partial\beta}\right)\log\left\{\mathbf{Z}_{\rm cl}(\tau)\cdot\left[\frac{Z_\Gamma(3-it)}{Z_\Gamma(2-2it)}\right]_{\rm hol}\cdot\left[\frac{Z_\Gamma(3+it)}{Z_\Gamma(2+2it)}\right]_{\rm antihol}\right\}_{(\tau\rightarrow-1/\tau)}, \end{aligned} \quad (23)$$

where $\beta \equiv \mathrm{Im}\,\tau = \ell/2\pi$ and $t = \mathrm{Re}\,\tau/\mathrm{Im}\,\tau = \theta/2\pi\beta$. In the limit $\theta \rightarrow 0$, i.e. for the non-spinning black hole, the entropy becomes

$$S(\beta,0) = \left(1-\beta\frac{\partial}{\partial\beta}\right)\log\left\{\mathbf{Z}_{\rm cl}(\tau)\cdot\left[\frac{Z_\Gamma(3)}{Z_\Gamma(2)}\right]^2_{(\theta=0)}\right\}_{(\tau\rightarrow-1/\tau)}. \quad (24)$$

3.3. Results on the holographic principle

According to the holographic principle, there exist strong ties between certain field theories on a manifold ("bulk space") and on its boundary (at infinity). A few mathematically exact results relevant to that program are the following. The class of Euclidean AdS_3 spaces (such as we have considered here) are quatients of the real hyperbolic space by a discrete group (a Schottky group). The boundary of these spaces can be compact oriented surfaces with conformal structure (compact complex algebraic curves). The results which can be regarded as manifestation of the holography principle are:

- There is a correspondence between spectral functions of hyperbolic three-geometry with it spectrum on a complex Riemann surface and Poincaré series associated with conformal structure in two dimensions. (The set of scatering poles coinsides with the zeros of a spectral function $Z_\Gamma(s)$; thus encoded in $Z_\Gamma(s)$ is the spectrum of a three-dimensional model).
- An explicit correspondence exists between a certain class of fields in the bulk space (gravity) and the class of fields of conformal theory on the boundary.

Acknowledgments

We are grateful to Professor L. Bonora for useful discussions. The authors would like to thank the Conselho Nacional de Desenvolvimento Científico e Tecnológico (CNPq) for support.

References

[1] Maloney A and Witten E, *Quantum Gravity Partition Function In Three Dimensions*, arXiv:0712.0155.
[2] Giombi S, Maloney A and Yin X, *One-loop Partition Functions of 3D Gravity*, arXiv:hep-th/0804.1773.
[3] Brown J D and Henneaux M, *Central Charges in the Canonical Realization of Asymptotic Symmetries: An Example from Three-Dimensional Gravity*, Commun. Math. Phys. **104** (1986) 207.
[4] Bytsenko A A, Guimarães M E X and Williams F L, *Spectral Functions for BTZ Black Hole Geometry*, Lett. Math. Phys. **79** (2007) 203; [arXiv:hep-th/0609102].
[5] Dijkgraaf R, Maldacena J, Moore G and Verlinde E, *A Black Hole Farey Tail*, arXiv:hep-th/0005003.
[6] Kleban M, Porrati M and Rabadan R, *Poincare Recurrences and Topological Diversity*, JHEP **0410**, (2004) 030; [arXiv:hep-th/0407192].
[7] Manschot J, AdS_3 *Partition Functions Reconstructed*, arXiv:hep-th/0707.1159.
[8] Bytsenko A A, Vanzo L, Zerbini S, *Quantum Correction to the Entropy of the (2+1)-Dimensional Black Hole*, Phys. Rev. D **57** (1998) 4917; [arXiv:gr-qc/9710106].

Bogoliubov transformation for quantum fields in $(\mathbb{S}^1)^d \times \mathbb{R}^{D-d}$ topology and applications to the Casimir effect

F C Khanna

Physics Department, Theoretical Physics Institute, University of Alberta, Edmonton, Alberta T6G 2J1 Canada
TRIUMF 4004, Westbrook mall, Vancouver, British Columbia V6T 2A3, Canada

E-mail: khanna@phys.ualberta.ca

J M C Malbouisson

Instituto de Física, Universidade Federal da Bahia, 40210-340, Salvador, Bahia, Brazil

E-mail: jmalboui@ufba.br

A E Santana

Instituto de Física, Universidade de Brasília, 70910-900, Brasília, DF, Brazil

E-mail: asantana@fis.unb.br

Abstract. A Bogoliubov transformation accounting simultaneously for spatial compactification and thermal effects is introduced. The fields are described in a $\Gamma_D^d = \mathbb{S}^{1_1} \times \cdots \times \mathbb{S}^{1_d} \times \mathbb{R}^{D-d}$ topology, and the Bogoliubov transformation is derived by a generalization of the thermofield dynamics formalism, a real-time finite-temperature quantum field theory. We consider the Casimir effect for Maxwell and Dirac fields and for a non-interacting massless QCD at finite temperature. For the fermion sector in a cubic box, we analyze the temperature at which the Casimir pressure changes its sign from attractive to repulsive. This critical temperature is approximately 200 MeV when the edge of the cube is of the order of the confining lengths ($\approx 1 : fm$) for quarks in baryons.

Keywords: Thermofield dynamics, Bogoliubov transformation, Compactification, Casimir effect

1. Introduction

In the literature there are several studies considering the quantum field theory formulated in flat spaces, with non-trivial topologies. This is the case of space-time considered as a simply or non-simply connected D-dimensional manifold with topology of type $\Gamma_D^d = \mathbb{S}^{1_1} \times \cdots \times \mathbb{S}^{1_d} \times \mathbb{R}^{D-d}$. The central characteristic of this kind of theory is that the topological structure of the space-time imposes modifications on the boundary conditions on fields and their Green functions, but it does not modify the local field equations. One known example of this kind of method is a quantum field at finite temperature, T, described by the Matsubara formalism. In this case the thermal effect is a path-integral calculated on $\mathbb{S}^1 \times \mathbb{R}^{D-1}$, where $\mathbb{S}^1$ is a circumference of length $\beta = 1/T$ [1, 2]. As a consequence, the thermal field theory can be thought, in a generalized way, as a mechanism to deal simultaneously with spatial constraints and thermal effects in a field theory model. In this context, the $\lambda\phi^4$ theory, including considerations on the mass generation due to topological effects, was studied in Ref. [1, 3], while the temperature and topological aspects were analyzed in Ref. [2]. More recently, these ideas have been developed in the context of the Matsubara formalism [4, 5, 6, 7] as well as in thermofield dynamics (TFD); and applied to the Casimir effect considering the electromagnetic and fermion fields within a box [8, 9, 10]), to the $\lambda\phi^4$ model as the Ginsburg-Landau theory for superconductors [4, 5], and to the Gross-Neveu model [6, 7]. In the present paper, we apply this methodology to calculate the Casimir effect for a massless fermion field in a 3-dimensional box (a cube), by using a generalized TFD-Bogoliubov transfomation, accounting for space compactification and the temperature effect.

An interest in the Casimir effect for fermions is the analysis of deconfinement in particle physics [11]–[24]. On the other hand, for this problem, the TFD-Bogoliubov transformation, as a real time formalism, gives rise to a useful theoretical ingredient, since the Green function is written naturally in two terms. One is the divergent free-space contribution; the other describes the compactification effects. This fact, not so direct in the imaginary time, allows one to treat the renormalization procedure in a convenient way [10].

TFD relies on two basic ingredients [25, 26, 27]. First, one defines a doubling of the original Fock space of the system leading to the expanded space $\mathcal{H}_T = \mathcal{H} \otimes \widetilde{\mathcal{H}}$. This doubling is carried out by associating to each operator a acting on $\mathcal{H}$ two operators in $\mathcal{H}_T$, A and $\widetilde{A}$, which are connected by the tilde (dual) conjugation rules

$$\begin{aligned}(A_i A_j)\tilde{} &= \widetilde{A}_i \widetilde{A}_j, \\ (cA_i + A_j)\tilde{} &= c^* \widetilde{A}_i + \widetilde{A}_j, \\ (A_i^\dagger)\tilde{} &= (\widetilde{A}_i)^\dagger, \\ (\widetilde{A}_i)\tilde{} &= -\xi A_i,\end{aligned}$$

with $\xi = -1$ for bosons and $\xi = +1$ for fermions. The physical variables are described by non-tilde operators. The other basic ingredient of TFD is a Bogoliubov transformation, $B(\alpha)$, introducing a rotation in the tilde and non-tilde variables, in such a way that thermal effects emerge from a condensate state. In the standard doublet notation [27], we write

$$(A^r(\alpha)) = \begin{pmatrix} A(\alpha) \\ \xi \widetilde{A}^\dagger(\alpha) \end{pmatrix} = B(\alpha) \begin{pmatrix} A \\ \xi \widetilde{A}^\dagger \end{pmatrix}, \tag{1}$$

$(A^r(\alpha))^\dagger = \left(A^\dagger(\alpha) \, , \, \widetilde{A}(\alpha) \right)$, with the Bogoliubov transformation given by

$$B(\alpha) = \begin{pmatrix} u(\alpha) & -v(\alpha) \\ \xi v(\alpha) & u(\alpha) \end{pmatrix}, \tag{2}$$

where $u^2(\alpha) + \xi v^2(\alpha) = 1$. The usual parametrization of the Bogoliubov transformation in TFD is obtained by setting $\alpha = \beta = T^{-1}$ and by requiring that $\langle 0, \tilde{0} | a^\dagger(\alpha) a(\alpha) | 0, \tilde{0} \rangle$ (with $a^\dagger$ and a being the creation and the annihilation operators) gives either the Bose or the Fermi distribution, i.e.

$$u(\beta) = \left(1 + \xi e^{-\beta\varepsilon} \right)^{-\frac{1}{2}} \quad , \quad v(\beta) = \left(e^{\beta\varepsilon} + \xi \right)^{-\frac{1}{2}} . \tag{3}$$

We consider the TFD approach for free fields aiming to extend the Bogoliubov transformation to account also for spatial compactification effects. The main application of our general discussion is the Casimir effect for cartesian confining geometries at finite temperature. For the sake of simplicity of notation, usually one identifies: $A \equiv a$ and $\widetilde{A} \equiv \tilde{a}$.

2. Maxwell and massless Dirac fields in TFD

For a quantum field described by a Lagrangian density $\mathcal{L}$, the TFD Lagrangian density is given by $\mathcal{L}_T = \mathcal{L} - \widetilde{\mathcal{L}}$, where $\widetilde{\mathcal{L}}$ is the tilde conjugate of $\mathcal{L}$ for the tilde fields. For the free-massless boson (Klein-Gordon) field, the two-point Green function in the doubled space is given by

$$\begin{aligned} G_0^{(ab)}(x-x') &= \langle 0,\tilde{0}|\mathrm{T}\left[\phi(x)^a\phi(x')^b\right]|0,\tilde{0}\rangle \\ &= \frac{-1}{(2\pi)^4}\int d^4k\; G_0^{(ab)}(k)\; e^{-ik\cdot(x-x')}, \end{aligned} \tag{4}$$

where
$$G_0^{(ab)}(k) = \begin{pmatrix} G_0(k) & 0 \\ 0 & \xi G_0^*(k) \end{pmatrix} = \begin{pmatrix} \frac{1}{k^2+i\epsilon} & 0 \\ 0 & \frac{-\xi}{k^2-i\epsilon} \end{pmatrix}$$

(here $\xi = -1$, for we have bosons). In the configuration space, we have explicitly
$$G_0^{(1,1)}(x-x') = G_0(x-x') = \frac{-i}{(2\pi)^2}\frac{1}{(x-x')^2 - i\varepsilon}.$$

The extension of this result for the electromagnetic field is simple. The free-photon propagator is written as

$$iD_{\mu\nu}^{(ab)}(x-x') = \langle 0,\tilde{0}|T[A_\mu^a(x)A_\nu^b(x')]|0,\tilde{0}\rangle = g_{\mu\nu}G_0^{(ab)}(x-x').$$

For the free-massless Dirac field, the doubled Green function is given by

$$S_0^{(ab)}(x-x') = \begin{pmatrix} S_0(x-x') & 0 \\ 0 & -S_0^*(x'-x) \end{pmatrix},$$

where $S_0(x-x') = -i\langle 0|T[\psi(x)\overline{\psi}(x')]|0\rangle = -i\gamma_\mu\partial^\mu G_0(x-x')$.

Now we use an α-dependent Bogoliubov transformation, performing a rotation among tilde and non-tilde variables. The α-dependent Green functions for the Klein-Gordon, Dirac and Maxwell fields are obtained through the Bogoliubov transformation given by Eq. (2) acting on $G_0^{(ab)}(k)$. (Notice that a Bogoliubov transformation has to be defined for each mode, k, of the field.) We have
$$G_0^{(ab)}(k;\alpha) = B_k^{-1(ac)}(\alpha)G_0^{(cd)}(k)B_k^{(db)}(\alpha); \tag{5}$$

explicitly, the components of $G_0^{(ab)}(k;\alpha)$ are given by

$$\begin{aligned} G_0^{11}(k;\alpha) &= G_0(k) + \xi v_k^2(\alpha)[G_0^*(k) - G_0(k)], && (6) \\ G_0^{12}(k;\alpha) &= G_0^{21}(k;\alpha) = \xi v_k(\alpha)u_k(\alpha)[G_0^*(k) - G_0(k)], && (7) \\ G_0^{22}(k;\alpha) &= \xi G_0^*(k) + v_k^2(\alpha)[G_0(k) - G_0^*(k)]. && (8) \end{aligned}$$

When $\xi = +1$, we obtain the auxiliary doubled two-point function which must be used for calculating the fermion propagator.

We are concerned with the α-dependent energy-momentum tensor, obtained from the usual expressions by replacing the fields by the Bogoliubov transformed counterparts. Actually, to treat thermal (and space confinement) effects, we will consider the renormalized vacuum expectation value of the α-dependent energy-momentum tensor defined, by subtracting the value corresponding to the free space at zero temperature, as

$$\mathcal{T}^{\mu\nu(ab)}(x;\alpha) = \langle 0,\tilde{0}|T^{\mu\nu(ab)}(x;\alpha)|0,\tilde{0}\rangle - \langle 0,\tilde{0}|T^{\mu\nu(ab)}(x)|0,\tilde{0}\rangle. \tag{9}$$

The physical results are obtained from the (11)-component. The doubled operators describing the energy-momentum tensor of free Maxwell and Dirac fields are given, respectively, by

$$\begin{aligned} T_{\rm M}^{\mu\nu(ab)}(x;\alpha) &= -F^{\mu\lambda(ab)}(x;\alpha)F^{\nu(ab)}_{\ \ \lambda}(x;\alpha) + \frac{1}{4}g^{\mu\nu}F^{(ab)}_{\lambda\rho}(x;\alpha)F^{\rho\lambda(ab)}(x;\alpha)\,, \quad (10) \\ T_{\rm D}^{\mu\nu(ab)}(x;\alpha) &= \frac{i}{2}(\overline{\psi}^a(x;\alpha)\gamma^\mu\partial^\nu\psi^b(x;\alpha) - \partial^\nu\overline{\psi}^a(x;\alpha)\gamma^\mu\psi^b(x;\alpha))\,, \quad (11) \end{aligned}$$

where $F^{(ab)}_{\mu\nu} = \partial_\mu A^a_\nu - \partial_\nu A^b_\mu$ is the electromagnetic field tensor.

For accounting thermal effects, one takes $\alpha = \beta = T^{-1}$. In this case $v_k^2(\alpha)$ (given in Eq. (3) for one mode, with $k_0 = \varepsilon$) is written as

$$v_k^2(\beta) = \sum_{l=1}^{\infty}(-\xi)^{l+1}e^{-\beta k_0 l}\,, \quad (12)$$

so that Eq. (6) becomes

$$G_0^{11}(k;\beta) = G_0(k) + \sum_{l=1}^{\infty}(-\xi)^{l+1}e^{-\beta k_0 l}[G_0^*(k) - G_0(k)]. \quad (13)$$

In the space coordinate, we have

$$\begin{aligned} \bar{G}_0^{11}(x-x';\beta) &= \sum_{l=1}^{\infty}(-\xi)^{l+1}\left[G_0^*(x'-x-i\beta l\widehat{n}_0)\right. \\ &\qquad \left. - G_0(x-x'-i\beta l\widehat{n}_0)\right]\,, \quad (14) \end{aligned}$$

where $\bar{G}_0^{11}(x-x';\beta) = G_0^{11}(x-x';\beta) - G_0(x-x')$ and $\widehat{n}_0 = (1,0,0,0)$ is a time-like vector. This expression is very useful for calculating $T^{\mu\nu(11)}(x;\beta)$.

For the electromagnetic field, we have

$$\begin{aligned} \langle 0,\tilde{0}|T_{\rm M}^{\mu\nu(ab)}(x;\beta)|0,\tilde{0}\rangle &= -i\left\{\Gamma^{\mu\nu}(x,x')G^{(ab)}(x-x';\beta)\right. \\ &\qquad \left.\left. +2\left(\widehat{n}_0^\mu\widehat{n}_0^\nu - \frac{1}{4}g^{\mu\nu}\right)\delta(x-x')\delta^{ab}\right\}\right|_{x\to x'}\,, \end{aligned}$$

where $\Gamma^{\mu\nu}(x,x') = 2(\partial^\mu\partial'^\nu - \frac{1}{4}g^{\mu\nu}\partial^\rho\partial'_\rho)$. This leads to

$$\begin{aligned} T_M^{\mu\nu(11)}(\beta) &= -i\left\{\Gamma^{\mu\nu}(x,x')\bar{G}_0^{11}(x-x';\beta)\right\}\big|_{x\to x'} \\ &= -\frac{2}{\pi^2}\sum_{l=1}^{\infty}\frac{g^{\mu\nu} - 4\widehat{n}_0^\mu\widehat{n}_0^\nu}{(\beta l)^4} = \frac{-\pi^2}{45\beta^4}(g^{\mu\nu} - 4\widehat{n}_0^\mu\widehat{n}_0^\nu)\,, \quad (15) \end{aligned}$$

where we have used the Riemann *zeta*-function $\zeta(4) = \sum_{l=1}^{\infty} l^{-4} = \pi^4/90$. As expected, $E(T) = T_M^{00(11)}(\beta) = \frac{1}{15}\pi^2T^4$ gives the correct energy density of the photon gas at temperature T, the blackbody radiation, formulae.

For the free-massless fermion field, we obtain

$$\begin{aligned} \langle 0,\tilde{0}|T_{\rm D}^{\mu\nu(ab)}(x;\beta)|0,\tilde{0}\rangle &= \gamma^\mu\partial^\nu S^{(ab)}(x-x')|_{x'\to x} \\ &= -4i\partial^\mu\partial^\nu G_0^{(ab)}(x-x')|_{x'\to x}\,, \end{aligned}$$

leading to

$$\begin{aligned}\mathcal{T}_D^{\mu\nu(11)}(\beta) &= -4i\partial^\mu\partial^\nu[\bar{G}_0^{11}(x-x';\beta)]_{x'\to x} \\ &= \frac{4}{\pi^2}\sum_{l=1}^{\infty}(-1)^l\left[\frac{g^{\mu\nu}-4\hat{n}_0^\mu\hat{n}_0^\nu}{(\beta l)^4}\right]. \end{aligned} \tag{16}$$

¿From this tensor, using $\sum_{l=1}^{\infty}(-1)^l l^{-4} = -7\pi^4/720$, we recover the well known result for the internal energy density of the Dirac field at temperature T, $E(T) = \mathcal{T}_D^{00(11)}(\beta) = \frac{7\pi^2}{60}T^4$. In the following section we introduce a more general Bogoliubov transformations taking into account not only temperature, but also the space compactification of a field.

3. Generalized Bogoliubov transformations

The preceding results show that the equilibrium TFD is equivalent to the Matsubara imaginary-time formalism, which has been used also to consider spatial compactification in field theoretical models [4, 5, 6]. Similarly, as pointed out in Introduction, confined fields can be treated with TFD by choosing appropriately the parameter α in the Bogoliubov transformation [8, 9, 10]. To see how this works, replace β and k_0 in Eq. (12) by $\alpha = i2L$ and k_3, corresponding to confinement along the z-axis, writing

$$v_k^2(L) = \sum_{l=1}^{\infty}(-\xi)^{l+1}e^{-i2Lk_3 l}. \tag{17}$$

Using this v_k^2 in Eq. (6) and performing the inverse Fourier transform, we get

$$\begin{aligned}\bar{G}_0^{11}(x-x';L) &= \sum_{l=1}^{\infty}(-\xi)^{l+1}\left[G_0^*(x'-x-2Ll\hat{n}_3)\right. \\ &\qquad \left.-\,G_0(x-x'-2Ll\hat{n}_3)\right]\end{aligned} \tag{18}$$

where $\bar{G}_0^{11}(x-x';L) = G_0^{11}(x-x';L) - G_0(x-x')$ and $\hat{n}_3 = (0,0,0,1)$. For this situation, similar steps as those leading to Eqs. (15) and (16) give

$$\mathcal{T}_M^{\mu\nu(11)}(L) = -\frac{2}{\pi^2}\sum_{l=1}^{\infty}\frac{g^{\mu\nu}+4\hat{n}_3^\mu\hat{n}_3^\nu}{(2Ll)^4} = -\frac{\pi^2}{720L^4}(g^{\mu\nu}+4\hat{n}_3^\mu\hat{n}_3^\nu), \tag{19}$$

$$\mathcal{T}_D^{\mu\nu(11)}(L) = \frac{4}{\pi^2}\sum_{l=1}^{\infty}(-1)^l\left[\frac{g^{\mu\nu}+4\hat{n}_3^\mu\hat{n}_3^\nu}{(2Ll)^4}\right] = -\frac{7\pi^2}{2880}[g^{\mu\nu}+4\hat{n}_3^\mu\hat{n}_3^\nu]. \tag{20}$$

The Casimir effect for the electromagnetic field between parallel metallic plates can be obtained from Eq. (19); the Casimir energy and pressure are

$$E(L) = \mathcal{T}_M^{00(11)}(L) = -\frac{\pi^2}{720L^4}, \quad P(L) = \mathcal{T}_M^{33(11)}(L) = -\frac{\pi^2}{240L^4}.$$

Similarly, from Eq. (20), we find the Casimir energy and pressure for the Dirac field confined between parallel plates, with anti-periodic boundary conditions, as:

$$E(L) = \mathcal{T}_D^{00(11)}(L) = -\frac{7\pi^2}{2880L^4}; \quad P(L) = \mathcal{T}_D^{33(11)}(L) = -\frac{7\pi^2}{960L^4}.$$

These results demonstrate explicitly the usefulness of the Bogoliubov transformation to treat confined fields as a generalization of TFD. From the above considerations, a question emerges naturally: what should be the appropriate generalization of the Bogoliubov transformation to account for simultaneously space compactification and thermal effects?

Such a generalization must reproduce, for example, the known results for the Casimir effect in the case of the parallel plates geometry at finite temperatures. Since energy is an additive quantity, we expect to have L- and T-dependent contributions plus a mixed (LT-dependent) contribution representing the interference of the two effects. In the next Section, we will show that the proper extension of expressions (12) and (17), for this case, is

$$\begin{aligned} v_k^2(\beta, L) &= \sum_{l_0=1}^{\infty}(-\xi)^{l_0+1}e^{-\beta k_0 l_0} + \sum_{l_3=1}^{\infty}(-\xi)^{l_3+1}e^{-i2Lk_3l_3} \\ &+2\sum_{l_0,l_3=1}^{\infty}(-\xi)^{l_0+l_3+2}e^{-\beta k_0 l_0 - i2Lk_3l_3}\,. \end{aligned} \tag{21}$$

To treat the general situation, compatible with cartesian geometries, we will consider the (1+N)-dimensional Minkowski space. Then, taking $\alpha = (\alpha_0, \alpha_1, \alpha_2, ..., \alpha_N)$, we write

$$\begin{aligned} v_k^2(\alpha) &= \sum_{s=1}^{N+1}\sum_{\{\sigma_s\}}\left(\prod_{n=1}^{s} f(\alpha_{\sigma_n})\right)2^{s-1} \\ &\times \sum_{l_{\sigma_1},...,l_{\sigma_s}=1}^{\infty}(-\xi)^{s+\sum_{r=1}^{s} l_{\sigma_r}}\exp\{-\sum_{j=1}^{s}\alpha_{\sigma_j}l_{\sigma_j}k_{\sigma_j}\}, \end{aligned} \tag{22}$$

where $f(\alpha_j) = 0$ for $\alpha_j = 0$, $f(\alpha_j) = 1$ otherwise and $\{\sigma_s\}$ denotes the set of all combinations with s elements, $\{\sigma_1, \sigma_2, ...\sigma_s\}$, of the first $N+1$ natural numbers $\{0, 1, 2, ..., N\}$, that is all subsets containing s elements, which we choose to write in an ordered form with $\sigma_1 < \sigma_2 < \cdots < \sigma_s$. Inserting this $v_k^2(\alpha)$ into Eq. (6) and taking the inverse Fourier transform, we obtain

$$\begin{aligned} \bar{G}_0^{11}(x-x';\alpha) &= \sum_{s=1}^{N+1}\sum_{\{\sigma_s\}}\left(\prod_{n=1}^{s} f(\alpha_{\sigma_n})\right)\sum_{l_{\sigma_1},...,l_{\sigma_s}=1}^{\infty}(-\xi)^{s+\sum_{r=1}^{s} l_{\sigma_r}} \\ &\times 2^{s-1}\Bigg[G_0^*(x'-x-i\sum_{j=1}^{s}\eta_{\sigma_j}\alpha_{\sigma_j}l_{\sigma_j}\widehat{n}_{\sigma_j}) \\ &- G_0(x-x'-i\sum_{j=1}^{s}\eta_{\sigma_j}\alpha_{\sigma_j}l_{\sigma_j}\widehat{n}_{\sigma_j})\Bigg]\Bigg|_{x'\to x}, \end{aligned} \tag{23}$$

where $\eta_{\sigma_j} = +1$, if $\sigma_j = 0$, and $\eta_{\sigma_j} = -1$ for $\sigma_j = 1, 2, ..., N$. To get the physical situation at finite temperature and spatial confinement, α_0 has to be taken as a positive real number while α_n, for $n = 1, 2, ..., N$, must be pure imaginary of the form $i2L_n$; in these cases, one finds that $\alpha_j^{*2} = \alpha_j^2$.

Considering such choices for parameters α_j and using the explicit form of $\bar{G}_0^{11}(x-x';\alpha)$ in the 4-dimensional space-time (corresponding to $N = 3$), we obtain the renormalized α-dependent energy-momentum tensor in the general case, for both Maxwell and Dirac fields:

$$\begin{aligned}
\mathcal{T}_M^{\mu\nu(11)}(\alpha) &= -i\left\{\Gamma^{\mu\nu}(x,x')\bar{G}_0^{11}(x-x';\alpha)\right\}\big|_{x\to x'} \\
&= -\frac{2}{\pi^2}\sum_{s=1}^{4}\sum_{\{\sigma_s\}}\left(\prod_{n=1}^{s} f(\alpha_{\sigma_n})\right)2^{s-1} \\
&\times \sum_{l_{\sigma_1},\ldots,l_{\sigma_s}=1}^{\infty}\left[\frac{g^{\mu\nu}}{[\sum_{j=1}^{s}\eta_{\sigma_j}(\alpha_{\sigma_j}l_{\sigma_j})^2]^2}\right. \\
&\left. -\frac{2\sum_{j,r=1}^{s}(1+\eta_{\sigma_j}\eta_{\sigma_r})(\alpha_{\sigma_j}l_{\sigma_j})(\alpha_{\sigma_r}l_{\sigma_r})\hat{n}_{\sigma_j}^{\mu}\hat{n}_{\sigma_r}^{\nu}}{[\sum_{j=1}^{s}\eta_{\sigma_j}(\alpha_{\sigma_j}l_{\sigma_j})^2]^3}\right]; \qquad (24)
\end{aligned}$$

$$\begin{aligned}
\mathcal{T}_D^{\mu\nu(11)}(\alpha) &= -4i\partial^\mu\partial^\nu[\bar{G}_0^{11}(x-x';\alpha)]_{x'\to x} \\
&= -\frac{4}{\pi^2}\sum_{s=1}^{4}\sum_{\{\sigma_s\}}\left(\prod_{n=1}^{s} f(\alpha_{\sigma_n})\right)\sum_{l_{\sigma_1},\ldots,l_{\sigma_s}=1}^{\infty}(-1)^{s+\sum_{r=1}^{s}l_{\sigma_r}} \\
&\times 2^{s-1}\left[\frac{g^{\mu\nu}}{[\sum_{j=1}^{s}\eta_{\sigma_j}(\alpha_{\sigma_j}l_{\sigma_j})^2]^2}\right. \\
&\left. -\frac{2\sum_{j,r=1}^{s}(1+\eta_{\sigma_j}\eta_{\sigma_r})(\alpha_{\sigma_j}l_{\sigma_j})(\alpha_{\sigma_r}l_{\sigma_r})\hat{n}_{\sigma_j}^{\mu}\hat{n}_{\sigma_r}^{\nu}}{[\sum_{j=1}^{s}\eta_{\sigma_j}(\alpha_{\sigma_j}l_{\sigma_j})^2]^3}\right]. \qquad (25)
\end{aligned}$$

Notice that the results obtained so far (Eqs. (15) and (19) for the Maxwell field and Eqs. (16) and (20) for the Dirac field) are particular cases of the above expressions, corresponding to $\alpha = (\beta, 0, 0, 0)$ and $\alpha = (0, 0, 0, i2L)$ respectively. Another important aspect is that $\mathcal{T}^{\mu\nu(11)}(\alpha)$ is traceless in both cases, as it should be. Now, we will apply these general results to some specific examples.

4. Casimir effect for parallel plates at finite temperature

As the first example of the development of the last Section, we now consider the electromagnetic field satisfying the Dirichlet boundary condition on parallel planes (metallic plates), normal to the z-direction, at finite temperature. In this case, $v_k^2(\alpha)$ is given by Eq. (21) with $\xi = -1$ (corresponding to the choice $\alpha = (\beta, 0, 0, i2L)$) and Eq. (24) reduces to

$$\begin{aligned}
\mathcal{T}_M^{\mu\nu(11)}(\beta, L) = -\frac{2}{\pi^2}\Bigg\{&\sum_{l_0=1}^{\infty}\frac{g^{\mu\nu}-4\hat{n}_0^\mu\hat{n}_0^\nu}{(\beta l_0)^4}+\sum_{l_3=1}^{\infty}\frac{g^{\mu\nu}+4\hat{n}_3^\mu\hat{n}_3^\nu}{(2Ll_3)^4} \\
&+2\sum_{l_0,l_3=1}^{\infty}\frac{(\beta l_0)^2[g^{\mu\nu}-4\hat{n}_0^\mu\hat{n}_0^\nu]+(2Ll_3)^2[g^{\mu\nu}+4\hat{n}_3^\mu\hat{n}_3^\nu]}{[(\beta l_0)^2+(2Ll_3)^2]^3}\Bigg\}. \qquad (26)
\end{aligned}$$

It follows then that the Casimir energy ($\mathcal{T}_M^{00(11)}$) and pressure ($\mathcal{T}_M^{33(11)}$) are given by [8]

$$E(\beta, L) = \frac{\pi^2}{15\beta^4} - \frac{\pi^2}{720L^4} + \frac{4}{\pi^2}\sum_{l_0,l_3=1}^{\infty}\frac{3(\beta l_0)^2-(2Ll_3)^2}{[(\beta l_0)^2+(2Ll_3)^2]^3}, \qquad (27)$$

$$P(\beta, L) = \frac{\pi^2}{45\beta^4} - \frac{\pi^2}{240L^4} + \frac{4}{\pi^2}\sum_{l_0,l_3=1}^{\infty}\frac{(\beta l_0)^2-3(2Ll_3)^2}{[(\beta l_0)^2+(2Ll_3)^2]^3}. \qquad (28)$$

The first two terms of these expressions reproduce Eqs. (15) and (19), giving the blackbody and the Casimir contributions for the energy and the pressure, separately. The last term represents the interplay between the two effects. These results have been obtained before with the use of mode-sum techniques and the image method [**?**, 18].

Notice that the positive black-body contributions for the energy E and pressure P dominate in the high-temperature limit, while the energy and the pressure are negative for low T. From Eq. (28), we determine the critical curve ($\beta_c = \chi_0 L$) for the transition from negative to positive values of P, by searching for the value of the ratio $\chi = \beta/L$ for which the pressure vanishes; this value, χ_0, is the solution of the transcendental equation

$$\frac{\pi^2}{45}\frac{1}{\chi^4} - \frac{\pi^2}{240} + \frac{4}{\pi^2}\sum_{l,n=1}^{\infty}\frac{(\chi l)^2 - 3(2n)^2}{[(\chi l)^2 + (2n)^2]^3} = 0, \tag{29}$$

given, numerically, by $\chi_0 \simeq 1.316$. In the next section we explore such an analysis where four spatial dimensions are compactified.

5. Casimir effect for the fermion sector of a non-interacting massless QCD

In this section we analyze the Casimir effect for a simplified quantum chromodynamics (QCD) model. The QCD Lagrangian is given by

$$\mathcal{L} = \overline{\psi}(x)[iD_\mu\gamma^\mu - m]\psi(x) - \frac{1}{4}F_{\mu\nu}F^{\mu\nu} - \frac{1}{2\alpha}(\partial^\mu A^r_\mu(x))^2 + A^r_\mu(x)t^r J^\mu(x)\,,$$

where

$$F^r_{\mu\nu} = \partial_\mu A^r_\nu(x) - \partial_\nu A^r_\mu(x) + gc^{rsl}A^s_\mu(x)A^l_\nu(x),$$

$F_{\mu\nu} = \sum\limits_r F^r_{\mu\nu}t^r$, is the field tensor describing the gluons; t^r and c^{rsl} are, respectively, the generators and the structure constants of $SU(3)$; $D_\mu = \partial_\mu + igA_\mu = I\partial_\mu + igA^r_\mu(x)t^r$ is the covariant derivative; $\psi(x)$ stands for the quark field, including the flavor and color components. The term $\frac{1}{2\alpha}(\partial^\mu A^r_\mu(x))^2$ is the gauge fixing term.

We consider an approximation for $\mathcal{L}$ describing some features of a massless baryon-free quark-gluon plasma, confined in space under the static bag model condition, corresponding to the adiabatic expansion of the plasma. In this case one discards, as a zero-order approximation, the interactions and the quark mass. Then the energy-momentum tensor for the quark field is given by

$$\begin{aligned} T^{\mu\nu}_q(x) &= \{in_c\sum_f \overline{\psi}(x)\gamma^\mu\partial'^\nu\psi(x')\}|_{x'\to x}, \\ &= \{in_c n_f\overline{\psi}(x)\gamma^\mu\partial'^\nu\psi(x')\}|_{x'\to x}, \end{aligned}$$

where n_c and n_f are the number of colors and flavors in the $SU(3)$ non-abelian gauge theory. With $T^{\mu\nu}_q(x)$ explicitly written, we introduce $\mathcal{T}^{\mu\nu(ab)}_q(x;\alpha)$ which is given by

$$\mathcal{T}^{\mu\nu(ab)}_q(x;\alpha) = \langle T^{\mu\lambda(ab)}_q(x;\alpha)\rangle - \langle T^{\mu\lambda(ab)}_q(x)\rangle,$$

where

$$\begin{aligned} \langle T^{\mu\lambda(ab)}_q(x)\rangle &= -i4n_c n_f\partial'^\nu\partial^\mu G^{(ab)}_0(x-x')|_{x'\to x}, \\ \langle T^{\mu\lambda(ab)}_q(x;\alpha)\rangle &= -i4n_c n_f\partial'^\nu\partial^\mu G^{(ab)}_0(x-x';\alpha)|_{x'\to x}, \end{aligned}$$

and so, for the quark field

$$
\begin{aligned}
\mathcal{T}_q^{\mu\nu(11)}(\alpha) &= -i4n_c n_f \partial^\mu \partial^\nu [\bar{G}_0^{11}(x-x';\alpha)]_{x'\to x} \\
&= -\frac{4n_c n_f}{\pi^2} \sum_{s=1}^{4} \sum_{\{\sigma_s\}} \left(\prod_{n=1}^{s} f(\alpha_{\sigma_n}) \right) \sum_{l_{\sigma_1},\ldots,l_{\sigma_s}=1}^{\infty} (-1)^{s+\sum_{r=1}^{s} l_{\sigma_r}} \\
&\times 2^{s-1} \left[\frac{g^{\mu\nu}}{[\sum_{j=1}^{s} \eta_{\sigma_j}(\alpha_{\sigma_j} l_{\sigma_j})^2]^2} \right. \\
&\left. - \frac{2\sum_{j,r=1}^{s}(1+\eta_{\sigma_j}\eta_{\sigma_r})(\alpha_{\sigma_j} l_{\sigma_j})(\alpha_{\sigma_r} l_{\sigma_r}) \hat{n}_{\sigma_j}^{\mu} \hat{n}_{\sigma_r}^{\nu}}{[\sum_{j=1}^{s} \eta_{\sigma_j}(\alpha_{\sigma_j} l_{\sigma_j})^2]^3} \right] . \qquad (30)
\end{aligned}
$$

For gluon field, in the approximation considered here, we have basically the same tensor for the electromagnetic field up to the color number, n_g. For this reason, we focus our analysis on the fermion sector only.

Taking the system in a Γ_4^4 topology, the quark field is physically analyzed in a cubic box of edge L at finite temperature, T ($\beta = T^{-1}$), satisfying anti-periodic boundary conditions in the four dimensions. In this case we have $\alpha = (\beta, iL, iL, iL)$ and

$$
T_q^{33(11)}(\beta, L) = n_c n_f \, f(\chi) \, \frac{1}{L^4}
$$

where

$$
\begin{aligned}
f(\chi) &= \frac{1}{\pi^2} \left\{ \mathcal{C}_f + \frac{7\pi^4}{180} \frac{1}{\chi^4} + 16 \sum_{l,n=1}^{\infty} \frac{(-1)^{l+n}}{[\chi^2 l^2 + n^2]^2} \right. \\
&+8 \sum_{l,n=1}^{\infty} (-1)^{l+n} \frac{\chi^2 l^2 - 3n^2}{[\chi^2 l^2 + n^2]^3} - 16 \sum_{l,n,r=1}^{\infty} \frac{(-1)^{l+n+r}}{[\chi^2 l^2 + n^2 + r^2]^2} \\
&-32 \sum_{l,n,r=1}^{\infty} (-1)^{l+n+r} \frac{\chi^2 l^2 + n^2 - 3r^2}{[\chi^2 l^2 + n^2 + r^2]^3} \\
&\left. + 32 \sum_{l,n,r,q=1}^{\infty} (-1)^{l+n+r+q} \frac{\chi^2 l^2 + n^2 + r^2 - 3q^2}{[\chi^2 l^2 + n^2 + r^2 + q^2]^3} \right\}
\end{aligned}
$$

with $\chi = \frac{\beta}{L}$ and

$$
\mathcal{C}_f = -8 \sum_{l,n=1}^{\infty} \frac{(-1)^{l+n}}{[l^2+n^2]^2} + \frac{16}{3} \sum_{l,n,r=1}^{\infty} \frac{(-1)^{l+n+r}}{[l^2+n^2+r^2]^2} - \frac{7\pi^4}{180} \approx -5.67
$$

The pressure changes from negative to positive as the temperature is increased. Indeed, for $T \to 0$, we have

$$
T_q^{33(11)}(L) = n_c n_f \frac{\mathcal{C}_f}{\pi^2} \frac{1}{L^4} < 0.
$$

For $T \to \infty$, the pressure is dominated by the term $\simeq T^4$,which is positive. The value of χ for which the pressure changes from negative to positive is

$$
\chi_c \approx 1.00 \Rightarrow T_c(L) \simeq \frac{1}{L}
$$

so that for $L = 1fm \longrightarrow T_c(L) \simeq 200$ MeV. This result is indicative that the Casimir effect plays a role in the deconfinement process of the hadronic matter. Although this role has been pointed out before [18], we find this by a different method. This fact reinforces the use of the generalized Bogoliubov transformation, that describes the Casimir effect as a condensate.

6. Concluding remarks

In this paper we have used generalizations of the TFD Bogoliubov transformation, describing fields in topology of the type $\Gamma_D^d = S^{1_1} \times ... \times S^{1_d} \times R^{D-d}$, to address the Casimir effect at finite temperature for both boson and fermion fields. It is worth emphasize that the procedure developed here is simpler and more direct than the standard techniques, such as the sum of modes and the image method, in particular to introduce the renormaliation procedure. This is a direct consequence of the structure of the Green function, which is written in two parts: one describing the divergent term due to the free space-time and the other due to the compactification effect. The result is that physical quantities are introduced as functions of the temperature and the compactification lengths. Notice that this procedure can describe systems satisfying other boundary conditions than those ones studied here. In this case, other types of Bogoliubov transformation describing fields in Γ_D^d can be defined, in particular, as the counterpart of the method studied in Ref. [1] for twisted boundary conditions.

As an application of this technique, we have analyzed the fermion sector of a massless baryon-free quark-gluon plasma, confined in a topology Γ_D^d, corresponding to an adiabatic expansion of the plasma. The main result is that by raising the temperature, there is a critical value for which the pressure changes sign. The transition from negative to positive Casimir pressure is at $T \simeq 200$Mev, for a length $L \simeq 1\,\mathrm{fm}^{-1}$. This result is another way to show the importance of the Casimir pressure for the deconfinement of the hadronic matter. A detailed analysis of these aspects and the nature of the generalized Bogoliubov transformation, including a discussion about scalar twisted fields, will be considered elsewhere.

Acknowledgments

Financial support from CNPq (Brazil) and NSERC (Canada) is acknowledged.

References

[1] Birrell N D and Ford L H, Phys. Rev. D **22**, 330 (1980).
[2] Ford L H and Svaiter N F, Phys. Rev D **51**, 6981 (1995).
[3] Ford L H and Yoshimura T, Phys. Lett. A **70**, 89 (1979).
[4] Malbouisson A P C and Malbouisson J M C, J. Phys. A: Math. Gen. **35**, 2263 (2002).
[5] Malbouisson A P C, Malbouisson J M C and Santana A E, Nucl. Phys. B **631**, 83 (2002).
[6] Malbouisson A P C, Malbouisson J M C, Santana A E and da Silva J C, Phys. Lett. B **583**, 373 (2004).
[7] Malbouisson A P C, Khanna F C, Malbouisson J M C and Santana A E, **36**, 1165 (2006).
[8] da Silva J C, Khanna F C, Matos Neto A and Santana A E, Phys. Rev. A **66**, 052101 (2002) .
[9] Queiroz H, da Silva J C, Khanna F C, Malbouisson J M C, Revzen M and Santana A E, Ann. Phys. (N.Y.) **317**, 220 (2005).
[10] Santana A E, Malbouisson J M C, Malbouisson A P C and Khanna F C, *Thermal Field Theory: Algebraic Aspects and Applications to Confined Systems*, and Khanna F C, Malbouisson J M C and Santana A E, *Thermofield Dynamics: Generalized Bogoliubov Transformation and Applications to the Casimir Effect*, in Non-Linear Dynamics and Fundamental Interactions, Khanna F and Matrasulov D Eds. (Kluver/Springer, N. York, 2005).
[11] Milonni P W, *The Quantum Vacuum* (Academic, Boston, 1993).
[12] Mostepanenko V M and Trunov N N, *The Casimir Effect and its Applications* (Clarendon, Oxford, 1997).
[13] Bordag M, Mohideen U and Mostepanenko V M, *New Developments in Casimir Effect*, quant-ph/0106045, Phys. Rep. **353**, 1 (2001).
[14] Levin F S and Micha D A (Eds.), *Long Range Casimir Forces* (Plenum, New York, 1993).
[15] Seife C, Science **275**, 158 (1997).
[16] Boyer T H, Am. J. Phys. **71**, 990 (2003).

[17] Milton K A, *The Casimir Effect: Physical Manifestations of Zero Point Energy*, hep-th/9901011 (W. Scientific, Singapore, 2001).
[18] Plunien G, Müeller B and Greiner W, Phys. Rep. **134** , 87(1986).
[19] Saito K, Z. Phys. C **50**, 69 (1991).
[20] Gundersen S A and Ravndal F, Ann. Phys. (N.Y.) **182**, 90 (1988).
[21] Lütken C A and Ravndal F, J. Phys. A: Math. Gen. **21**, L793 (1988).
[22] Lütken C A and Ravndal F, J. Phys. G: Nucl. Phys. **10**, 123 (1984).
[23] Ravndal F and Tollefsen D, Phys. Rev. D **40**, 4191 (1989).
[24] De Paola R D M, Rodrigues R B and Svaiter N F, Mod. Phys. Lett. **44**, 2353 (1999).
[25] Takahashi Y and Umezawa H, Coll. Phenomena **2** (1975) 55 (Reprinted in Int. J. Mod. Phys. 10, 1755 (1996)).
[26] Umezawa H, Matsumoto H and Tachiki M, *Thermofield Dynamics and Condensed States* (North-Holland, Amsterdan, 1982).
[27] Umezawa H, *Advanced Field Theory: Micro, Macro and Thermal Physics* (AIP, New York, 1993).

Journal of Physics: Conference Series **161** (2009) 012025 doi:10.1088/1742-6596/161/1/012025

Fermionic Casimir effect in de Sitter spacetime

A A Saharian

Department of Physics, Yerevan State University, 1 Alex Manoogian Street, 0025 Yerevan, Armenia

E-mail: saharian@ictp.it

Abstract. The Casimir densities are investigated for a massive spinor field in de Sitter spacetime with an arbitrary number of toroidally compactified spatial dimensions. The vacuum expectation value of the energy-momentum tensor is presented in the form of the sum of corresponding quantity in the uncompactified de Sitter spacetime and the part induced by the non-trivial topology. The latter is finite and the renormalization is needed for the first part only. The asymptotic behavior of the topological term is investigated in the early and late stages of the cosmological expansion. When the comoving lengths of the compactified dimensions are much smaller than the de Sitter curvature radius, to the leading order the topological part coincides with the corresponding quantity for a massless fermionic field and is conformally related to the corresponding flat spacetime result with the same topology. In this limit the topological term dominates the uncompactified de Sitter part and the back-reaction effects should be taken into account. In the opposite limit, for a massive field the asymptotic behavior of the topological part is damping oscillatory.

PACS numbers: 04.62.+v, 04.20.Gz, 04.50.-h, 11.10.Kk

1. Introduction

Many of high energy theories of fundamental physics are formulated in higher dimensional spacetimes and it is commonly assumed that the extra dimensions are compactified. In particular, the idea of compactified dimensions has been extensively used in supergravity and superstring theories. From an inflationary point of view universes with compact spatial dimensions, under certain conditions, should be considered a rule rather than an exception [1]. These models may play an important role by providing proper initial conditions for inflation (for physical motivations of considering compact universes see also [2]). As it was argued in [3], there are many reasons to expect that in string theory the most natural topology for the universe is that of a flat compact three-manifold. The quantum creation of the universe having toroidal spatial topology is discussed in [4] and in references [5] within the framework of various supergravity theories. In addition to the theoretical work, there has been a large activity to search for signatures of non-trivial topology by identifying ghost images of galaxies, clusters or quasars. Recent progress in observations of the cosmic microwave background provides an alternative way to observe the topology of the universe [6]. If the scale of periodicity is close to the particle horizon scale then the changed appearance of the microwave background sky pattern offers a sensitive probe of the topology.

The compactification of spatial dimensions leads to a number of interesting quantum effects which include instabilities in interacting field theories [7], topological mass generation [8, 9, 10]

and symmetry breaking [10, 11]. In the case of non-trivial topology the boundary conditions imposed on fields give rise to the modification of the spectrum for vacuum fluctuations and, as a result, to the Casimir-type contributions in the vacuum expectation values of physical observables (for the topological Casimir effect and its role in cosmology see [12, 13]). A characteristic feature of models with compactified dimensions is the presence of moduli fields which parametrize the size and the shape of the extra dimensions and the Casimir effect has been used for the generation of the effective potential for these moduli. The Casimir energy can also serve as a model for dark energy needed for the explanation of the present accelerated expansion of the universe (see [14, 15, 16] and references therein).

Quantum field theory in de Sitter (dS) spacetime has been extensively studied during the past two decades. Much of early interest was motivated by the questions related to the quantization of fields on curved backgrounds. dS spacetime has a high degree of symmetry and numerous physical problems are exactly solvable on this background. The importance of this theoretical work increased by the appearance of the inflationary cosmology scenario [17]. In most inflationary models an approximately dS spacetime is employed to solve a number of problems in standard cosmology. During an inflationary epoch quantum fluctuations in the inflaton field introduce inhomogeneities and may affect the transition toward the true vacuum. These fluctuations play a central role in the generation of cosmic structures from inflation. More recently astronomical observations of high redshift supernovae, galaxy clusters and cosmic microwave background [18] indicate that at the present epoch the Universe is accelerating and can be well approximated by a world with a positive cosmological constant. If the Universe would accelerate indefinitely, the standard cosmology would lead to an asymptotic dS universe. Hence, the investigation of physical effects in dS spacetime is important for understanding both the early Universe and its future.

One-loop quantum effects for various spin fields on the background of dS spacetime have been discussed by several authors (see, for instance, [19]-[34] and references therein). The effects of the toroidal compactification of spatial dimensions in dS spacetime on the properties of quantum vacuum for a scalar field with general curvature coupling parameter are investigated in [35, 36, 37] (the quantum effects in braneworld models with dS spaces and in higher-dimensional brane models with compact internal spaces were discussed in [38, 39]). In the present talk, based on [40, 41] we present the results on the Casimir effect for a fermionic field on background of dS spacetime with spatial topology $\mathrm{R}^p \times (\mathrm{S}^1)^q$.

The paper is organized as follows. In the next section the plane wave eigenspinors in $(D+1)$-dimensional dS spacetime with an arbitrary number of toroidally compactified dimensions are constructed. In section 3 these eigenspinors are used for the evaluation of the vacuum expectation value of the energy-momentum tensor in both cases of the fields with periodicity and antiperiodicity conditions along compactified dimensions. The behavior of these quantities is investigated in asymptotic regions of the parameters. The special case of topology $\mathrm{R}^{D-1}\times \mathrm{S}^1$ with the corresponding numerical results is discussed in section 4. The main results are summarized in section 5.

2. Plane wave eigenspinors in de Sitter spacetime with compactified dimensions

We consider a quantum fermionic field ψ on background of $(D+1)$-dimensional de Sitter spacetime, dS_{D+1}, described by the line element

$$ds^2 = dt^2 - e^{2t/\alpha} \sum_{l=1}^{D} (dz^l)^2, \tag{1}$$

where the parameter α in the expression for the scale factor is related to the corresponding cosmological constant Λ by the formula $\alpha^2 = D(D-1)/(2\Lambda)$. We assume that the spatial coordinates z^l, $l = p+1, \ldots, D$, are compactified to S^1 of the length L_l: $0 \leqslant z^l \leqslant L_l$, and for the other coordinates we have $-\infty < z^l < +\infty$, $l = 1, \ldots, p$. Hence, we consider the spatial topology $\mathrm{R}^p \times (\mathrm{S}^1)^q$, where $q = D - p$.

The dynamics of the field is governed by the covariant Dirac equation

$$i\gamma^\mu \nabla_\mu \psi - m\psi = 0 \ , \ \nabla_\mu = \partial_\mu + \Gamma_\mu, \tag{2}$$

where $\gamma^\mu = e^\mu_{(a)}\gamma^{(a)}$ are the generalized Dirac matrices and Γ_μ is the spin connection. The latter is given in terms of the flat-space Dirac matrices $\gamma^{(a)}$ by

$$\Gamma_\mu = \frac{1}{4}\gamma^{(a)}\gamma^{(b)}e^\nu_{(a)}e_{(b)\nu;\mu} \ . \tag{3}$$

In these relations $e^\mu_{(a)}$ are the tetrad components defined by $e^\mu_{(a)}e^\nu_{(b)}\eta^{ab} = g^{\mu\nu}$, with η^{ab} being the Minkowski spacetime metric tensor. In the $(D+1)$-dimensional flat spacetime the Dirac matrices are $N \times N$ matrices with $N = 2^{[(D+1)/2]}$, where the square brackets mean the integer part of the enclosed expression. We will assume that these matrices are given in the chiral representation:

$$\gamma^{(0)} = \begin{pmatrix} 1 & 0 \\ 0 & -1 \end{pmatrix}, \ \gamma^{(a)} = \begin{pmatrix} 0 & \sigma_a \\ -\sigma_a^+ & 0 \end{pmatrix}, \ a = 1, 2, \ldots, D. \tag{4}$$

From the anticommutation relations for the Dirac matrices one has $\sigma_a\sigma_b^+ + \sigma_b\sigma_a^+ = 2\delta_{ab}$.

In the discussion below we will denote the position vectors along the uncompactified and compactified dimensions by $\mathbf{z}_p = (z^1, \ldots, z^p)$ and $\mathbf{z}_q = (z^{p+1}, \ldots, z^D)$. One of the characteristic features of the field theory on backgrounds with non-trivial topology is the appearance of inequivalent types of fields with the same spin [42]. For fermion fields the boundary conditions along the compactified dimensions can be either periodic (untwisted field) or antiperiodic (twisted field). First we consider the field with periodicity conditions (no summation over l):

$$\psi(t, \mathbf{z}_p, \mathbf{z}_q + L_l\mathbf{e}_l) = \psi(t, \mathbf{z}_p, \mathbf{z}_q), \tag{5}$$

where $l = p+1, \ldots, D$ and $\mathbf{e}_l$ is the unit vector along the direction of the coordinate z^l. The case of a fermionic field with antiperiodicity conditions will be discussed below. We are interested in the effects of non-trivial topology on the vacuum expectation value (VEV) of the energy-momentum tensor of the fermionic field assuming that the field is prepared in the Bunch-Davies vacuum state (also called Euclidean vacuum). In order to evaluate this VEV, we will use the direct mode-summation procedure. In this approach the knowledge of the complete set of properly normalized eigenspinors, $\{\psi_\beta^{(+)}, \psi_\beta^{(-)}\}$, is needed. Here the collective index β presents a set of quantum numbers specifying the solutions. By virtue of spatial translation invariance the spatial part of the eigenfunctions $\psi_\beta^{(\pm)}$ can be taken in the standard plane wave form $e^{\pm i\mathbf{k}\mathbf{z}}$. We will decompose the wave vector into the components along the uncompactified and compactified dimensions, $\mathbf{k} = (\mathbf{k}_p, \mathbf{k}_q)$ with

$$k = \sqrt{\mathbf{k}_p^2 + \mathbf{k}_q^2}. \tag{6}$$

For a spinor field with periodicity conditions along the compactified dimensions the corresponding wave vector has the components

$$\mathbf{k}_q = (2\pi n_{p+1}/L_{p+1}, \ldots, 2\pi n_D/L_D), \tag{7}$$

where $n_{p+1}, \ldots, n_D = 0, \pm 1, \pm 2, \ldots$.

Choosing the basis tetrad in the form

$$e_{\mu}^{(0)} = \delta_{\mu}^{0},\ e_{\mu}^{(a)} = e^{t/\alpha}\delta_{\mu}^{a},\ a = 1, 2, \ldots, D, \tag{8}$$

for the positive frequency plane wave eigenspinors corresponding to the Bunch-Davies vacuum and satisfying the periodicity conditions along the compactified dimensions we have

$$\psi_{\beta}^{(+)} = A_{\beta}\eta^{(D+1)/2}e^{i\mathbf{k}\cdot\mathbf{r}}\begin{pmatrix} H_{1/2-i\alpha m}^{(1)}(k\eta)w_{\sigma}^{(+)} \\ -i(\mathbf{n}\cdot\boldsymbol{\sigma})H_{-1/2-i\alpha m}^{(1)}(k\eta)w_{\sigma}^{(+)} \end{pmatrix}, \tag{9}$$

where $\beta = (\mathbf{k}, \sigma)$, $\mathbf{n} = \mathbf{k}/k$, and $w_{\sigma}^{(+)}$, $\sigma = 1, \ldots, N/2$, are one-column matrices having $N/2$ rows with the elements $w_{l}^{(\sigma)} = \delta_{l\sigma}$. In (9), $H_{\nu}^{(1)}(x)$ is the Hankel function, $\boldsymbol{\sigma} = (\sigma_1, \sigma_2, \ldots, \sigma_D)$, and we are using the notation

$$\eta = \alpha e^{-t/\alpha},\ 0 \leqslant \eta < \infty. \tag{10}$$

Note that $\tau = -\eta$ is the conformal time coordinate, in terms of which the dS line element (1) takes the conformally flat form. Similarly, the negative frequency eigenspinors have the form

$$\psi_{\beta}^{(-)} = A_{\beta}\eta^{(D+1)/2}e^{-i\mathbf{k}\cdot\mathbf{r}}\begin{pmatrix} i(\mathbf{n}\cdot\boldsymbol{\sigma})H_{-1/2+i\alpha m}^{(2)}(k\eta)w_{\sigma}^{(-)} \\ H_{1/2+i\alpha m}^{(2)}(k\eta)w_{\sigma}^{(-)} \end{pmatrix}, \tag{11}$$

with $w_{\sigma}^{(-)} = iw_{\sigma}^{(+)}$.

The coefficients A_{β} in the expressions for the eigenspinors are determined from the orthonormalization condition

$$\int d^{D}x\,\sqrt{\gamma}\psi_{\beta}^{(\pm)+}\psi_{\beta'}^{(\pm)} = \delta_{\beta\beta'}, \tag{12}$$

with γ being the determinant of the spatial metric. On the right-hand side of this condition the symbol $\delta_{\beta\beta'}$ is understood as the Dirac delta function for the continuous indices and the Kronecker delta for the discrete ones. By making use of the Wronskian relation for the Hankel functions, we can see that

$$A_{\beta}^{2} = \frac{ke^{\pi\alpha m}}{2^{p+2}\pi^{p-1}V_{q}\alpha^{D}}, \tag{13}$$

where $V_q = L_{p+1}\cdots L_D$ is the volume of the compactified subspace. For a massless fermionic field the Hankel functions in (9) and (11) are expressed in terms of exponentials and we have the standard conformal relation $\psi_{\beta}^{(\pm)} = (\eta/\alpha)^{(D+1)/2}\psi_{\beta}^{(\mathrm{M})(\pm)}$ between eigenspinors defining the Bunch-Davies vacuum in dS spacetime and the corresponding eigenspinors $\psi_{\beta}^{(\mathrm{M})(\pm)}$ for the Minkowski spacetime with spatial topology $\mathrm{R}^{p} \times (\mathrm{S}^{1})^{q}$.

The plane wave eigenspinors for a twisted spinor field are constructed in a similar way. For this field we have antiperiodicity conditions along the compactified dimensions:

$$\psi(t, \mathbf{z}_p, \mathbf{z}_q + L_l\mathbf{e}_l) = -\psi(t, \mathbf{z}_p, \mathbf{z}_q). \tag{14}$$

The corresponding eigenspinors are given by expressions (9), (11), where now the components of the wave vector along the compactified dimensions are given by the formula

$$\mathbf{k}_q = (\pi(2n_{p+1}+1)/L_{p+1}, \ldots, \pi(2n_D+1)/L_D), \tag{15}$$

with $n_{p+1}, \ldots, n_D = 0, \pm1, \pm2, \ldots$, and

$$k^2 = \mathbf{k}_p^2 + \sum_{l=p+1}^{D}\left[\pi(2n_l+1)/L_l\right]^2. \tag{16}$$

Note that the physical wave vector is given by the combination $\mathbf{k}\eta/\alpha$.

3. Vacuum expectation value of the energy-momentum tensor

The compactification of the spatial dimensions leads to the modification of the spectrum for zero-point fluctuations of fields and as a result of this the VEVs of physical observables are changed. Among the most important quantities characterizing the properties of the vacuum is the expectation value of the energy-momentum tensor. In addition to describing the physical structure of the quantum field at a given point, the energy-momentum tensor acts as a source of gravity in the Einstein equations and plays an important role in modelling a self-consistent dynamics involving the gravitational field. In order to study the one-loop topological effects in the VEV of the energy-momentum tensor we will use the direct mode summation approach. The corresponding mode-sum has the form

$$\langle 0|T_{\mu\nu}|0\rangle = \frac{i}{2}\int d\mathbf{k}_p \sum_{\mathbf{k}_q,\sigma}[\bar{\psi}^{(-)}_{\mathbf{k},\sigma}\gamma_{(\mu}\nabla_{\nu)}\psi^{(-)}_{\mathbf{k},\sigma} - (\nabla_{(\mu}\bar{\psi}^{(-)}_{\mathbf{k},\sigma})\gamma_{\nu)}\psi^{(-)}_{\mathbf{k},\sigma}], \tag{17}$$

where the eigenspinors $\psi^{(-)}_{\beta} = \psi^{(-)}_{\mathbf{k},\sigma}$ are given by (11) and $\bar{\psi}^{(-)}_{\mathbf{k},\sigma} = \psi^{(-)+}_{\mathbf{k},\sigma}\gamma^0$ is the Dirac adjoint. For a fermionic field with periodic boundary conditions, substituting the eigenspinors (11) into this formula, for the energy density and the vacuum stresses we find (no summation over l)

$$\begin{aligned}\langle 0|T^0_0|0\rangle &= \frac{2^{-p}\eta^{D+1}\alpha^{-D-1}}{\pi^{p/2-1}\Gamma(p/2)V_q}\int_0^\infty dk_p\, k_p^{p-1}\sum_{n_{p+1}=-\infty}^{+\infty}\cdots\sum_{n_D=-\infty}^{+\infty}k^2 \\ &\times\mathrm{Re}\left[H^{(1)}_{1/2+i\alpha m}(k\eta)H^{(2)\prime}_{-1/2+i\alpha m}(k\eta) - H^{(1)}_{-1/2+i\alpha m}(k\eta)H^{(2)\prime}_{1/2+i\alpha m}(k\eta)\right], \qquad (18)\\ \langle 0|T^l_l|0\rangle &= \frac{2^{1-p}\eta^{D+1}\alpha^{-D-1}}{\pi^{p/2-1}\Gamma(p/2)V_q}\int_0^\infty dk_p\, k_p^{p-1}\sum_{n_{p+1}=-\infty}^{+\infty}\cdots\sum_{n_D=-\infty}^{+\infty}k_l^2 \\ &\times\mathrm{Re}\left[H^{(1)}_{1/2+i\alpha m}(k\eta)H^{(2)}_{1/2+i\alpha m}(k\eta)\right], \qquad (19)\end{aligned}$$

with $l = 1, 2, \ldots, D$, and the off-diagonal components vanish. The VEVs given by (18) and (19) are divergent and need some regularization procedure. To make them finite we can introduce a cut-off function $\varphi_\lambda(k)$ with the cut-off parameter λ, which decreases sufficiently fast with increasing k and satisfies the condition $\varphi_\lambda(k) \to 1$, for $\lambda \to 0$.

As the next step, we apply to the series over n_{p+1} in (18) and (19) the Abel-Plana summation formula [12, 43]

$$\sum_{n=0}^{\infty}{}' f(n) = \int_0^\infty dx\, f(x) + i\int_0^\infty dx\,\frac{f(ix)-f(-ix)}{e^{2\pi x}-1}, \tag{20}$$

where the prime means that the term $n = 0$ should be halved. The term in the VEV with the first integral in the right-hand side of this formula corresponds to the energy-momentum tensor in dS spacetime with topology $\mathrm{R}^{p+1}\times(\mathrm{S}^1)^{q-1}$. As a result, the VEV of the energy-momentum tensor is presented in the decomposed form

$$\langle T^l_k\rangle_{p,q} = \langle T^l_k\rangle_{p+1,q-1} + \Delta_{p+1}\langle T^l_k\rangle_{p,q}. \tag{21}$$

Here $\langle T^l_k\rangle_{p+1,q-1}$ is the VEV of the energy-momentum tensor for the topology $\mathrm{R}^{p+1}\times(\mathrm{S}^1)^{q-1}$ and the part (no summation over l)

$$\begin{aligned}\Delta_{p+1}\langle T^l_l\rangle_{p,q} &= \frac{N\eta^{D+2}L_{p+1}}{(2\pi)^{(p+1)/2}V_q\alpha^{D+1}}\sum_{n=1}^{\infty}\sum_{n_{p+2}=-\infty}^{+\infty}\cdots\sum_{n_D=-\infty}^{+\infty}\int_0^\infty dx\, x \\ &\times\frac{\mathrm{Re}[I^2_{-1/2-i\alpha m}(\eta x) - I^2_{1/2+i\alpha m}(\eta x)]}{\cosh(\alpha m\pi)(L_{p+1}n)^{p+1}}f^{(l)}_p(nL_{p+1}\sqrt{x^2+k^2_{\mathbf{n}_{q-1}}}), \qquad (22)\end{aligned}$$

is due to the compactness of the $(p+1)$th dimension. In (22), $I_\nu(x)$ and $K_\nu(x)$ are the modified Bessel functions and we have introduced the notations

$$f_\nu(x) \equiv x^\nu K_\nu(x),\ k_{\mathbf{n}_{q-1}}^2 = \sum_{l=p+2}^{D} (2\pi n_l/L_l)^2. \tag{23}$$

For the separate components the functions $f_p^{(l)}(y)$ in formula (22) have the form

$$\begin{aligned} f_p^{(l)}(y) &= f_{(p+1)/2}(y),\ l = 0, 1, \ldots, p, \\ f_p^{(p+1)}(y) &= -pf_{(p+1)/2}(y) - y^2 f_{(p-1)/2}(y), \\ f_p^{(l)}(y) &= k_l^2 (nL_{p+1})^2 f_{(p-1)/2}(y),\ l = p+2, \ldots, D, \end{aligned} \tag{24}$$

where $k_l = 2\pi n_l/L_l$. From (22) we see that the topological part depends on the variable η and the length scales L_l in the combinations L_l/η. Noting that $a(\eta)L_l$ is the comoving length with $a(\eta) = \alpha/\eta$ being the scale factor, we conclude that the topological part of the energy-momentum tensor is a function of comoving lengths of the compactified dimensions.

The topological term (22) is finite and by the recurrence relation (21) the renormalization procedure for the VEV of the energy-momentum tensor is reduced to the renormalization of the corresponding VEV in uncompactified dS spacetime. It can be seen that the topological part is traceless for a massless field and is covariantly conserved: $\left(\Delta_{p+1}\langle T_l^k\rangle_{p,q}\right)_{;k} = 0$. From the first relation in (24) we see that the vacuum stresses along the uncompactified dimensions are equal to the energy density,

$$\Delta_{p+1}\langle T_0^0\rangle_{p,q} = \Delta_{p+1}\langle T_1^1\rangle_{p,q} = \cdots = \Delta_{p+1}\langle T_p^p\rangle_{p,q}, \tag{25}$$

and, hence, in the uncompactified subspace the equation of state for the topological part of the energy-momentum tensor is of the cosmological constant type. Note that the topological parts are time-dependent and they break the dS symmetry.

For a massless fermionic field, the modified Bessel functions in (22) are expressed in terms of elementary functions and after the integration over x one finds (no summation over l)

$$\Delta_{p+1}\langle T_l^l\rangle_{p,q} = \frac{2N(\eta/\alpha)^{D+1}}{(2\pi)^{p/2+1} V_q L_{p+1}^{p+1}} \sum_{n=1}^{\infty} \sum_{n_{p+2}=-\infty}^{+\infty} \cdots \sum_{n_D=-\infty}^{+\infty} \frac{g_p^{(l)}(nL_{p+1}k_{\mathbf{n}_{q-1}})}{n^{p+2}}, \tag{26}$$

with the notations

$$\begin{aligned} g_p^{(l)}(y) &= f_{p/2+1}(y),\ l = 0, 1, \ldots, p, \\ g_p^{(p+1)}(y) &= -(p+1)f_{p/2+1}(y) - y^2 f_{p/2}(y), \\ g_p^{(l)}(y) &= (nL_{p+1}k_l)^2 f_{p/2}(y),\ l = p+2, \ldots, D. \end{aligned} \tag{27}$$

The massless fermionic field is conformally invariant in any dimension and in this case the problem under consideration is conformally related to the corresponding problem in the Minkowski spacetime with spatial topology $\mathrm{R}^p \times (\mathrm{S}^1)^q$. Formula (26) could also be obtained from the relation $\Delta_{p+1}\langle T_i^i\rangle_{p,q} = a^{-D-1}(\eta)\Delta_{p+1}\langle T_i^i\rangle_{p,q}^{(\mathrm{M})}$, with $a(\eta)$ being the scale factor. Comparing expression (26) with the corresponding formula from [36] for a conformally coupled massless scalar field, we see that the following relation takes place: $\Delta_{p+1}\langle T_k^l\rangle_{p,q} = -N\Delta_{p+1}\langle T_k^l\rangle_{p,q}^{(\mathrm{scalar})}$.

After the repetitive application of the recurrence formula (21), the VEV of the energy-momentum tensor for the topology $R^p \times (S^1)^q$ is presented in the form

$$\langle T_l^k \rangle_{p,q} = \langle T_l^k \rangle_{\mathrm{dS,ren}} + \langle T_l^k \rangle_{\mathrm{c}},\ \langle T_l^k \rangle_{\mathrm{c}} = \sum_{l=1}^{q} \Delta_{D+1-l} \langle T_l^k \rangle_{D-l,l}, \tag{28}$$

where $\langle T_l^k \rangle_{\mathrm{dS,ren}}$ is the renormalized VEV in the uncompactified dS spacetime and $\langle T_l^k \rangle_{\mathrm{c}}$ is the topological part. For $D = 3$ the renormalized VEV of the energy-momentum tensor for fermionic field in uncompactified dS spacetime is investigated in [24] (see also [12]). The corresponding expression has the form

$$\langle T_l^k \rangle_{\mathrm{dS,ren}}^{(D=3)} = \frac{\delta_l^k}{16\pi^2 \alpha^4} \left\{ 2m^2\alpha^2(m^2\alpha^2+1)[\ln(m\alpha) - \mathrm{Re}\,\psi(im\alpha)] + m^2\alpha^2/6 + 11/60 \right\}, \tag{29}$$

where $\psi(x)$ is the logarithmic derivative of the gamma-function. In the limit of zero mass, only the last term in curly brackets contributes and the corresponding energy density is positive. For large values of the mass we have:

$$\langle T_l^k \rangle_{\mathrm{dS,ren}}^{(D=3)} \approx -\frac{\delta_l^k}{960\pi^2\alpha^6 m^2},\ m\alpha \gg 1, \tag{30}$$

and the energy density is negative.

In higher dimensions a closed expression can be obtained for even values of D. In this case for the renormalized VEV of the energy-momentum tensor in uncompactified dS spacetime we have the formula [41]

$$\langle T_l^k \rangle_{\mathrm{dS,ren}} = \frac{2Nm^{D+1}\delta_l^k}{(4\pi)^{(D+1)/2}(D+1)} \frac{\Gamma((1-D)/2)}{e^{2\pi m\alpha}+1} \prod_{j=0}^{D/2-1} \left[1 + \frac{(j+1/2)^2}{m^2\alpha^2} \right]. \tag{31}$$

For a massless fermionic field this tensor vanishes. We could expect this result, since in odd-dimensional spacetimes the trace anomaly is absent. For large values of the mass, $m\alpha \gg 1$, the VEV for the energy-momentum tensor is exponentially suppressed. This is in contrast with the case of even dimensional spacetimes, where the suppression in the large mass limit is power-law. In particular, for 5-dimensional de Sitter spacetime formula (31) leads to the result

$$\langle T_l^k \rangle_{\mathrm{dS,ren}} = \frac{m^5 \delta_l^k}{15\pi^2} \frac{1}{e^{2\pi m\alpha}+1} \left(1 + \frac{5}{2m^2\alpha^2} + \frac{9}{16m^4\alpha^4} \right),\ D = 4. \tag{32}$$

The corresponding energy density is positive. In figure 1 we have plotted the vacuum energy density $\langle T_0^0 \rangle_{\mathrm{dS,ren}}$ as a function of the parameter $m\alpha$ for spatial dimensions $D = 3, 4, 6$.

In uncompactified dS spacetime the energy-momentum tensor $\langle T_l^k \rangle_{\mathrm{dS,ren}}$ is time-independent and corresponds to a gravitational source of the cosmological constant type. These properties directly follow from the dS invarinace of the Bunch-Davies vacuum state. Combining with the initial cosmological constant Λ, one-loop effects in uncompactified dS spacetime lead to the effective cosmological constant

$$\Lambda_{\mathrm{eff}} = \Lambda + 8\pi G_{D+1} \langle T_0^0 \rangle_{\mathrm{dS,ren}}, \tag{33}$$

where G_{D+1} is the gravitational constant in $(D+1)$-dimensional spacetime. In the discussion above we have assumed that the quantum state of a fermionic field is the Bunch-Davies vacuum state. In [44] it was shown that for a scalar field with a wide range of mass and

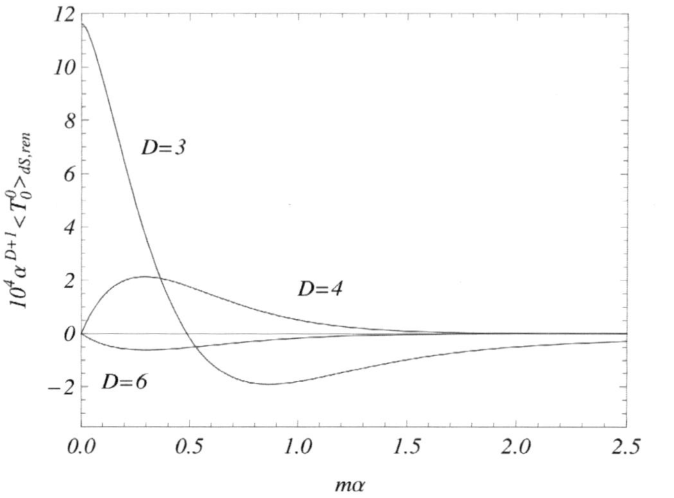

Figure 1. The vacuum energy density in uncompactified dS spacetime, $\langle T_0^0\rangle_{\text{dS,ren}}$, as a function of the parameter $m\alpha$ for spatial dimensions $D = 3, 4, 6$.

curvature coupling parameter the expectation values of the energy-momentum tensor in arbitrary physically admissable states approaches the expectation value in the Bunch-Davies vacuum at late times.

Now we return to the consideration of the topological part in the VEV of the energy-momentum tensor and discuss its behavior in the asymptotic regions of the parameters. For small values of the comoving length $a(\eta)L_{p+1}$ with respect to the dS curvature radius, $a(\eta)L_{p+1} \ll \alpha$, to the leading order the topological part in the VEV of the energy-momentum tensor coincides with that for a massless field given by formula (26). In particular, the topological part of the vacuum energy density is positive. This limit corresponds to the early stages of the cosmological evolution, $t \to -\infty$, and at these stages the total VEV is dominated by the topological part. In this limit the back-reaction effects of the topological terms are important and these effects can change the dynamics essentially (for a discussion of back-reaction effects from vacuum fluctuations on the dynamics of dS spacetime see, for instance, [45] and references therein).

For large values of the comoving length, $a(\eta)L_{p+1} \gg \alpha$, the leading term in the topological part is given by the expression (no summation over l):

$$\Delta_{p+1}\langle T_l^l\rangle_{p,q} \approx \frac{NB_l\cos[2mt - 2\alpha m\ln(\alpha/L_{p+1}) - \phi_l]}{2^{p/2}\pi^{(p+1)/2}V_q L_{p+1}^{p+1}\cosh(\alpha m\pi)e^{(D+1)t/\alpha}}. \tag{34}$$

Here the coefficients B_l and the phases ϕ_l are defined by the formula

$$B_l e^{i\phi_l} = \frac{2^{-i\alpha m}}{\Gamma(1/2+i\alpha m)}\sum_{n=1}^{\infty}\sum_{n_{p+2}=-\infty}^{+\infty}\cdots\sum_{n_D=-\infty}^{+\infty}\frac{h_p^{(l)}(nL_{p+1}k_{\mathbf{n}_{q-1}})}{n^{p+2+2i\alpha m}}, \tag{35}$$

with the notations

$$\begin{aligned} h_p^{(l)}(x) &= f_{p/2+1+i\alpha m}(x),\ l = 0, 1, \ldots, p, \\ h_p^{(p+1)}(x) &= -(p+1+2i\alpha m)f_{p/2+1+i\alpha m}(x) - x^2 f_{p/2+i\alpha m}(x), \\ h_p^{(l)}(x) &= (L_{p+1}k_l n)^2 f_{p/2+i\alpha m}(x),\ l = p+2, \ldots, D. \end{aligned} \tag{36}$$

Relation (34) describes the asymptotic behavior of the topological part in the late stages of cosmological evolution corresponding to the limit $t \rightarrow +\infty$. In this limit the behavior of the topological part for a massive spinor field is damping oscillatory. The damping factor in the amplitude and the oscillation frequency are the same for all terms in the sum of (28) and the total topological term behaves as

$$\langle T_l^k \rangle_{\rm c} \propto \delta_l^k e^{-(D+1)t/\alpha} \cos\left(2mt + \phi_{\rm c}'\right), \; t \rightarrow +\infty. \tag{37}$$

As the vacuum energy-momentum tensor for uncompactified dS spacetime is time-independent, we have similar damping oscillations in the total energy-momentum tensor $\langle T_l^k \rangle_{\rm dS,ren} + \langle T_l^k \rangle_{\rm c}$. This type of oscillations are absent in the case of a massless field when the topological parts decay monotonically as $e^{-(D+1)t/\alpha}$. Note that in the case of a scalar field the vanishing of the topological part is monotonic or oscillatory in dependence of the mass and the curvature coupling parameter of the field [35, 36].

For a fermionic field with antiperiodicity conditions (14), the VEV of the energy-momentum tensor is found in a way similar to that for the periodicity conditions. The corresponding formulae for the topological parts are obtained from those for the field with periodicity conditions inserting the factor $(-1)^n$ in the summation over n and replacing the definition for $k_{\mathbf{n}_{q-1}}^2$ by

$$k_{\mathbf{n}_{q-1}}^2 = \sum_{l=p+2}^{D} \left[\pi(2n_l+1)/L_l\right]^2. \tag{38}$$

In situations where the main contribution comes from the term with $n = 1$, the topological parts in the VEV of the energy-momentum tensor for fields with periodicity and antiperiodicity conditions have opposite signs.

4. Special case

In the special case of a single compactified dimension with the length L (topology $\mathrm{R}^{D-1} \times \mathrm{S}^1$), for the topological part in the VEV of the energy-momentum tensor for an untwisted field we have (no summation over l)

$$\begin{aligned} \langle T_l^l \rangle_{\rm c} &= \frac{N(\eta/L)^D}{(2\pi)^{D/2}\alpha^{D+1}\cosh(\alpha m\pi)} \sum_{n=1}^{\infty} \frac{1}{n^D} \int_0^{\infty} dx\, x \\ &\times f_{D-1}^{(l)}(nxL/\eta) \mathrm{Re}[I_{-1/2-i\alpha m}^2(x) - I_{1/2+i\alpha m}^2(x)]. \end{aligned} \tag{39}$$

For a massless fermionic field this formula reduces to (no summation over l)

$$\langle T_l^l \rangle_{\rm c} = \frac{N\zeta(D+1)}{\pi^{(D+1)/2}} \left(\frac{\eta}{\alpha L}\right)^{D+1} \Gamma\left(\frac{D+1}{2}\right), \; \langle T_D^D \rangle_{\rm c} = -D\langle T_0^0 \rangle_{\rm c}, \tag{40}$$

with $l = 0, 1, \ldots, D-1$ and $\zeta(x)$ being the Riemann zeta function. In accordance with the asymptotic analysis given before, the expressions (40) describe the asymptotic behavior of the topological part in the VEV of the energy-momentum tensor for a massive field at early stages of the cosmological expansion. The asymptotics at late stages are obtained from general formula (34) and have the form (no summation over l)

$$\langle T_l^l \rangle_{\rm c} \approx NC_l \frac{\cos[2mt - 2\alpha m \ln(\alpha/L) - \phi_l]}{\pi^{D/2}\cosh(\alpha m\pi)(Le^{t/\alpha})^{D+1}}, \tag{41}$$

where now C_l and ϕ_l are defined by the relations

$$\begin{aligned} C_l e^{i\phi_l} &= \frac{\Gamma((D+1)/2+i\alpha m)}{\Gamma(1/2+i\alpha m)}\zeta(D+1+2i\alpha m),\ l=0,1,\dots,D-1, \\ C_D e^{i\phi_D} &= -(D+2i\alpha m)C_0 e^{i\phi_0}. \end{aligned} \tag{42}$$

For a massless field, (41) is reduced to the exact result (40). The corresponding formulae for a twisted spinor field are obtained by adding the coefficient $(2^{-D}-1)$ in the first formula of (40) and the coefficient $(2^{-D-2i\alpha m}-1)$ in the first formula of (42).

For $D=3$ the functions $f_{D-1}^{(l)}(x)$ in (39) are expressed in terms of exponentials and after the summation over n we find (no summation over l)

$$\langle T_l^l \rangle_{\rm c} = \frac{(\eta/L)^3}{\pi\alpha^4\cosh(\alpha m\pi)}\int_0^\infty dx\, xG_l(Lx/\eta){\rm Re}[I^2_{-1/2-i\alpha m}(y) - I^2_{1/2+i\alpha m}(y)], \tag{43}$$

where the following notations are introduced

$$\begin{aligned} &G_0(y) = y^2\ln(1-e^{-y}),\ G_3(y) = G_0(y) - 2G_1(y), \\ &G_1(y) = G_2(y) = y{\rm Li}_2(e^{-y}) + {\rm Li}_3(e^{-y}), \end{aligned} \tag{44}$$

with ${\rm Li}_n(z) = \sum_{k=1}^\infty z^k/k^n$ being the polylogarithm function. In particular, for a massless fermionic field we have

$$\langle T_k^l \rangle_{\rm c} = \frac{2\pi^2}{45}\left(\frac{\eta}{\alpha L}\right)^4 {\rm diag}(1,1,1,-3). \tag{45}$$

In the case of a twisted field and for the topology $\mathrm{R}^2\times\mathrm{S}^1$ one has (no summation over l)

$$\langle T_l^l \rangle_{\rm c} = \frac{(\eta/L)^3}{\pi\alpha^4\cosh(\alpha m\pi)}\int_0^\infty dx\, xG_l^{(\rm tw)}(Lx/\eta){\rm Re}[I^2_{-1/2-i\alpha m}(y) - I^2_{1/2+i\alpha m}(y)], \tag{46}$$

with the notations

$$\begin{aligned} &G_0^{(\rm tw)}(y) = y^2\ln(1+e^{-y}),\ G_3^{(\rm tw)}(y) = G_0^{(\rm tw)}(y) - 2G_l^{(\rm tw)}(y), \\ &G_l^{(\rm tw)}(y) = y{\rm Li}_2(-e^{-y}) + {\rm Li}_3(-e^{-y}),\ l=1,2. \end{aligned} \tag{47}$$

For a massless twisted field one has the result

$$\langle T_k^l \rangle_{\rm c} = -\frac{7\pi^2}{180}\left(\frac{\eta}{\alpha L}\right)^4 {\rm diag}(1,1,1,-3). \tag{48}$$

Note that for a massless field in dS_5 having spatial topology $\mathrm{R}^3\times\mathrm{S}^1$ the topological part in the VEV of the energy-momentum tensor has the form

$$\langle T_k^l \rangle_{{\rm c},m=0} = \frac{b\zeta(5)}{\pi^2}\left(\frac{\eta}{\alpha L}\right)^5 {\rm diag}(1,1,1,1,-4), \tag{49}$$

where $b=3$ for the field with periodicity condition and $b=-45/16$ for the field with antiperiodicity condition. This topology corresponds to the original Kaluza-Klein model.

In figure 2 we have plotted the topological parts in the VEVs of the energy density and the vacuum stress along the compactified dimension for untwisted (full curves) and twisted (dashed curves) spinor fields as functions of the ratio L/η for the value of the parameter $\alpha m=1$. The numbers near the curves indicate the component of the energy-momentum tensor ($l=0$ for the energy density and $l=D$ for the stress). The left panel is for dS_4 with topology $\mathrm{R}^2\times\mathrm{S}^1$ and

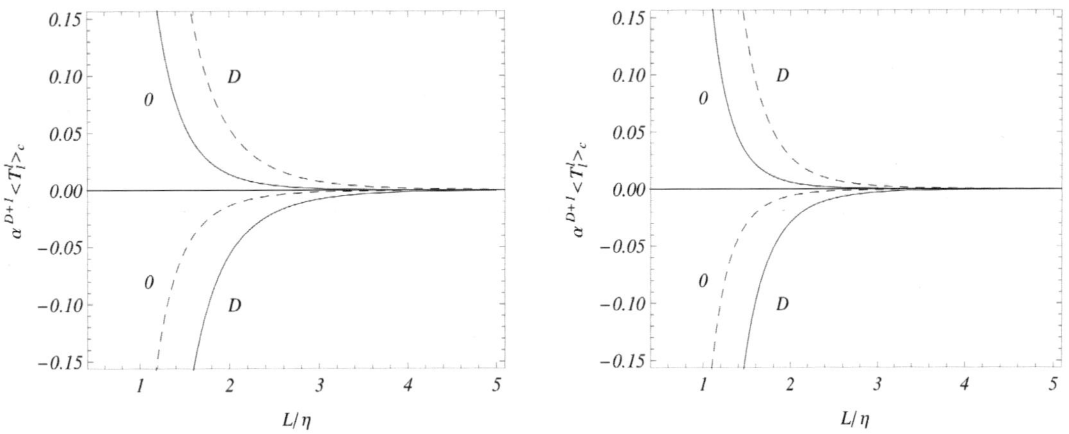

Figure 2. The topological parts in the VEVs of the energy density ($l = 0$) and the vacuum stress along the compactified dimension ($l = D$) in units of the dS curvature scale, $\alpha^{D+1}\langle T_l^l\rangle_c$, as functions of the ratio L/η for the value of the parameter $\alpha m = 1$ and in the special case of spatial topology $R^{D-1} \times S^1$. Full/dashed curves correspond to spinor fields with periodicity/antiperiodicity conditions along the compactified dimension. The left (right) panel corresponds to $D = 3$ ($D = 4$).

the right panel is for dS_5 having spatial topology $R^3 \times S^1$. Note that the ratio $L/\eta = a(\eta)L/\alpha$ is the comoving length of the compactified dimension in units of the dS curvature radius. For a massless fermionic field in dS spacetime the corresponding VEVs are given by formulae (45), (48) and (49). As we have explained before, in the limit $L/\eta \ll 1$, these expressions are the leading terms in the corresponding asymptotic expansion for the VEV of the energy-momentum tensor of the massive field.

From the asymptotic formula (34) it follows that for large values of the ratio L/η the topological part oscillates. As in the scale of figure 2 the oscillations are not well seen, we illustrate this oscillatory behavior in figure 3, where the topological parts in the energy density are plotted for untwisted (full curves) and twisted (dashed curves) fermionic fields L/η for $\alpha m = 4$. As in figure 2, the left panel is for dS_4 with topology $R^2 \times S^1$ and the right panel is for dS_5 with spatial topology $R^3 \times S^1$. The first zero with respect to L/η and the distance between neighbor zeros decrease with increasing values of the parameter αm.

5. Conclusion

We have investigated the Casimir densities for a massive spinor field in $(D + 1)$-dimensional dS spacetime with an arbitrary number of toroidally compactified spatial dimensions assuming that the field is prepared in the Bunch-Davies vacuum state. For the evaluation of the vacuum expectation value of the energy-momentum tensor the mode-summation procedure is employed. In this approach we need the corresponding eigenspinors satisfying appropriate boundary conditions along the compactified dimensions. These eigenspinors are constructed in section 2 for fields with both periodicity and antiperiodicity conditions. By using the eigenspinors and applying to the mode-sum the Abel-Plana formula, the VEV for the spatial topology $R^p \times (S^1)^q$ is presented in the form of the sum of the corresponding quantity in the topology $R^{p+1} \times (S^1)^{q-1}$ and of the part which is induced by the compactness of $(p + 1)$th dimension. For the field with periodicity conditions the topological part is given by formula (22). The corresponding formula for the field with antiperiodicity conditions is obtained from that for the field with periodicity

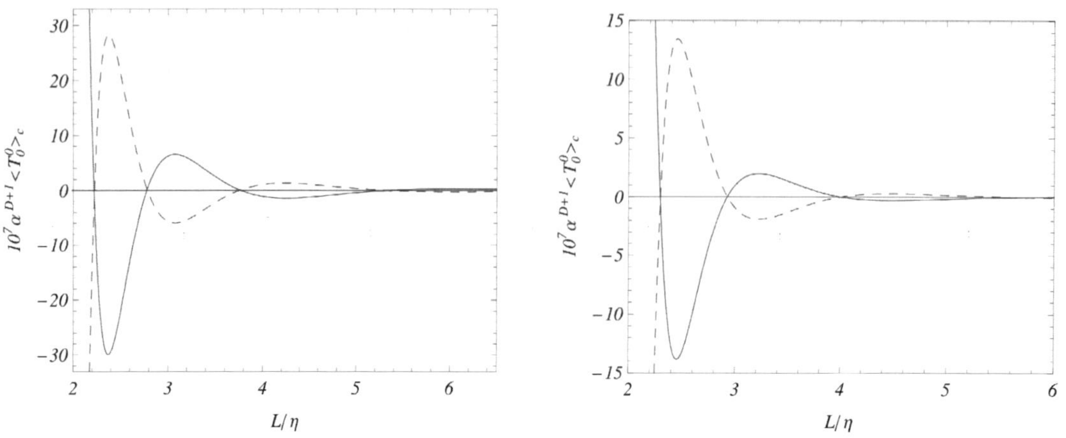

Figure 3. The topological part in the VEV of the energy density as a function of the ratio L/η for the value of the parameter $\alpha m = 4$ and in the special case of spatial topology $\mathrm{R}^{D-1} \times \mathrm{S}^1$. Full/dashed curves correspond to spinor fields with periodicity/antiperiodicity conditions along the compactified dimension. The left (right) panel corresponds to $D = 3$ ($D = 4$).

conditions inserting the factor $(-1)^n$ in the summation over n and replacing the definition for $k^2_{\mathbf{n}_{q-1}}$ by (38).

The topological part is finite and the renormalization procedure for the VEV of the energy-momentum tensor is reduced to that for the uncompactified dS spacetime. This part is time-dependent and breaks the dS symmetry. The corresponding vacuum stresses along the uncompactified dimensions coincide with the energy density and, hence, in the uncompactified subspace the equation of state for the topological part of the energy-momentum tensor is of the cosmological constant type. For a massless fermionic field the problem under consideration is conformally related to the corresponding problem in the Minkowski spacetime with spatial topology $\mathrm{R}^p \times (\mathrm{S}^1)^q$ and we have the standard relation $\langle T^l_k \rangle_{\mathrm{c}} = a^{-(D+1)}(\eta) \langle T^l_k \rangle^{(\mathrm{M})}_{\mathrm{c}}$ between the topological terms.

For a massive fermionic field, in the limit when the comoving length of a compactified dimension is much smaller than the dS curvature radius, the topological part in the VEV of the energy-momentum tensor coincides with the corresponding quantity for a massless field and is conformally related to the VEV in toroidally compactified Minkowski spacetime. In particular, the topological part in the vacuum energy density is positive for an untwisted fermionic field. This limit corresponds to the early stages of the cosmological evolution and the topological part dominates over the uncompactified dS part. At these stages the back-reaction effects of the topological term are important and these effects can essentially change the dynamics of the model. In the opposite limit, when the comoving lengths of the compactified dimensions are large with respect to the dS curvature radius, in the case of a massive field the asymptotic behavior of the topological part is damping oscillatory and to the leading order is given by formula (34). These formula describes the behavior of the topological part in the late stages of the cosmological expansion. As the uncompactified dS part is time-independent, we have similar oscillations in the total VEV as well. Note that this type of oscillatory behavior is absent for a massless fermionic field.

Acknowledgments
The work was supported by the Armenian Ministry of Education and Science Grant No. 119 and by Conselho Nacional de Desenvolvimento Científico e Tecnológico (CNPq).

References

[1] Linde A 2004 *J. Cosmol. Astropart. Phys.* JCAP10(2004)004
[2] Starkman G D 1998 *Class. Quantum Grav.* **15** 2529 (1998)
Cornish N J, Spergel D N and Starkman G D 1998 *Class. Quantum Grav.* **15** 2657
[3] McInnes B 2005 *Nucl. Phys.* B **709** 213
McInnes B 2006 *Nucl. Phys.* B **748** 309
[4] Zeldovich Y B and Starobinsky A A 1984 *Sov. Astron. Lett.* **10**, 135
[5] Goncharov Yu P and Bytsenko A A 1985 *Phys. Lett.* B **160** 385
Goncharov Yu P and Bytsenko A A 1986 *Phys. Lett.* B **169** 171
Goncharov Yu P and Bytsenko A A 1986 *Nucl. Phys.* B **271** 726
Goncharov Yu P and Bytsenko A A 1987 *Class. Quant. Grav.* **4**, 555
[6] Levin J 2002 *Phys. Rep.* **365**, 251
Cornish N J, D.N. Spergel D N, Starkman G D and Komatsu E 2004 *Phys. Rev. Lett.* **92** 201302
[7] Ford L H 1980 *Phys. Rev.* D **22** 3003
[8] Ford L H and Yoshimura T 1979 *Phys. Lett.* A **70** 89
[9] Toms D J 1980 *Phys. Rev.* D **21** 928
[10] Toms D J 1980 *Phys. Rev.* D **21** 2805
[11] Odintsov S D 1988 *Sov. J. Nucl. Phys.* **48** 1148
Buchbinder I L and Odintsov S D 1989 *Int. J. Mod. Phys.* A **4** 4337
Buchbinder I L and Odintsov S D 1989 *Fortschr. Phys.* **37** 225
Buchbinder I L, Odintsov S D and Dergaleo V P 1989 *Theor. Math. Phys.* **80** 150
[12] Grib A A, Mamayev S G and Mostepanenko V M 1994 *Vacuum Quantum Effects in Strong Fields* (St. Petersburg: Friedmann Laboratory Publishing)
Mostepanenko V M and Trunov N N 1997 *The Casimir Effect and Its Applications* (Oxford: Clarendon)
[13] Bordag M, Mohidden U and Mostepanenko V M 2001 *Phys. Rep.* **353** 1
Milton K A 2002 *The Casimir Effect: Physical Manifestation of Zero-Point Energy* (Singapore: World Scientific)
Elizalde E, Odintsov S D, Romeo A, Bytsenko A A and Zerbini S 1994 *Zeta regularization techniques with applications* (Singapore: World Scientific)
Duff M J, Nilsson B E W and Pope C N 1986 *Phys. Rep.* **130** 1
Bytsenko A A, Cognola G, Vanzo L and Zerbini S 1996 *Phys. Rep.* **266** 1
[14] Milton K A 2003 *Grav. Cosmol.* **9** 66
[15] Elizalde E 2006 *J. Phys. A: Math. Gen.* **39** 6299
[16] Green B and Levin J 2007 *J. High Energy Phys.* JHEP11(2007)096
Burikham P, Chatrabhuti A, Patcharamaneepakorn P and Pimsamarn K 2008 *J. High Energy Phys.* JHEP07(2008)013
[17] Linde A D 1990 *Particle Physics and Inflationary Cosmology* (Switzerland: Harwood Academic Publishers)
[18] Riess A G et al. 2007 *Astrophys. J.* **659** 98
Spergel D N et al. 2007 *Astrophys. J. Suppl. Ser.* **170** 377
Seljak U, Slosar A and McDonald P 2006 *J. Cosmol. Astropart. Phys.* JCAP10(2006)014
Komatsu E et al. 2008 Five-Year Wilkinson Microwave Anisotropy Probe (WMAP) Observations: Cosmological Interpretation *Preprint* arXiv:0803.0547
[19] Chernikov N A and Tagirov E A 1968 *Ann. Inst. Henri Poincaré* **9** 109
Tagirov E A 1973 Ann. Phys. **76** 561
[20] Candelas P and Raine D J 1975 *Phys. Rev.* D **12** 965
[21] Dowker J S and Critchley R 1976 *Phys. Rev.* D **13** 224
Dowker J S and Critchley R 1976 *Phys. Rev.* D **13** 3224
[22] Bunch T S and Davies P C W 1978 *Proc. R. Soc. London* A **360** 117
[23] Birrell N D 1979 *J. Phys. A: Math. Gen.* **12** 337
[24] Mamayev S G 1981 *Sov. Phys. J.* **24** 63
[25] Vilenkin A and Ford L H 1982 *Phys. Rev.* D **26** 1231
[26] Allen B 1983 *Nucl. Phys.* B **226** 228
Allen B 1985 *Ann. Phys.* **161** 152
[27] Ford L H 1985 *Phys. Rev.* D **31** 710

[28] Kirsten K and Garriga J 1993 Phys. Rev. D **48** 567
[29] Esposito G, Miele G and Rosa L 1994 *Class. Quantum Grav.* **11** 2031
[30] Prokopec T and Woodart R P 2003 *J. High Energy Phys.* JHEP10(2003)059
Prokopec T, Tornkvist O and Woodart R P 2003 *Ann. Phys.* **303** 251
Prokopec T and Woodart R P 2004 *Ann. Phys.* **312** 1
Prokopec T and Puchwein E, *J. Cosmol. Astropart. Phys.* JCAP04(2004)007
[31] Cognola G, Elizalde E, Nojiri S, Odintsov S D and Zerbini S 2005 *J. Cosmol. Astropart. Phys.* JCAP02(2005)010
[32] Finelli F, Marozzi G, Vacca G P and Venturi G 2005 *Phys. Rev.* D **71** 023522
[33] Dolgov A and Pelliccia D N 2006 *Nucl. Phys.* B **734** 208
[34] Birrell N D and Davies P C W 1982 *Quantum Fields in Curved Space* (Cambridge: Cambridge University Press)
[35] Saharian A A and Setare M R 2008 *Phys. Lett.* B **659** 367
[36] Bellucci S and Saharian A A 2008 *Phys. Rev.* D **77** 124010
[37] Saharian A A 2008 Casimir effect in toroidally compactified de Sitter spacetime *Preprint* arXiv:0808.3803
[38] Nojiri S, Odintsov S D and Zerbini S 2000 *Class. Quantum. Grav.* **17** 4855
Naylor W and Sasaki M 2002 *Phys. Lett.* B **542** 289
Elizalde E, Nojiri S, Odintsov S D and Ogushi S 2003 *Phys. Rev.* D **67** 063515
Moss I G, Naylor W, Santiago-German W and Sasaki M 2003 *Phys. Rev.* D **67** 125010
Uzawa K 2003 *Prog. Theor. Phys.* **110** 457
Saharian A A and Setare M R 2004 *Phys. Lett.* B **584** 306
Setare M R 2005 *Phys. Lett.* B **620** 111
Minamitsuji M, Naylor W and Sasaki M 2006 *Nucl. Phys.* B **737** 121
Setare M R 2006 *Phys. Lett.* B **637** 1
[39] Flachi A, Garriga J, Pujolàs O and Tanaka T 2003 *J. High Energy Phys.* JHEP08(2003)053
Flachi A and Pujolàs O 2003 *Phys. Rev.* D **68** 025023
Saharian A A 2006 *Phys. Rev.* D **73** 044012
Saharian A A 2006 *Phys. Rev.* D **73** 064019
Saharian A A 2006 *Phys. Rev.* D **74** 124009
[40] Saharian A A 2008 *Class. Quantum Grav.* **25** 165012
[41] Bezerra de Mello E R and Saharian A A 2008 Fermionic vacuum densities in higher-dimensional de Sitter spacetime *Preprint* arXiv:0808.0614
[42] Isham C J 1978 *Proc. R. Soc. London* A **362** 383
Isham C J 1978 *Proc. R. Soc. London* A **364**, 591
[43] Saharian A A 2007 The Generalized Abel-Plana Formula with Applications to Bessel Functions and Casimir Effect *Preprint* arXiv:0708.1187
[44] Anderson P R, Eaker W, Habib S, Molina-París C and Mottola E 2000 *Phys. Rev.* D **62** 124019
[45] Pérez-Nadal G, Roura A and Verdaguer E 2008 *Class. Quantum Grav.* **25** 154013

Journal of Physics: Conference Series **161** (2009) 012026 doi:10.1088/1742-6596/161/1/012026

Gravitational vacuum energy in our recently accelerating universe

Sidney Bludman
Departamento de Astronomia, Universidad de Chile, Santiago, Chile
E-mail: sbludman@das.uchile.cl

Abstract. We review current observations of the homogeneous cosmological expansion which, because they measure only kinematic variables, cannot determine the dynamics driving the recent accelerated expansion. The minimal fit to the data, the flat ΛCDM model, consisting of cold dark matter and a cosmological constant, interprets 4Λ geometrically as a classical spacetime curvature constant of nature, avoiding any reference to quantum vacuum energy. (The observed Uehling and Casimir effects measure forces due to QED vacuum polarization, but not any quantum material vacuum energies.) An Extended Anthropic Principle, that Dark Energy and Dark Gravity be indistinguishable, selects out flat ΛCDM. Prospective cosmic shear and galaxy clustering observations of the growth of fluctuations are intended to test whether the 'dark energy' driving the recent cosmological acceleration is static or moderately dynamic. Even if dynamic, observational differences between an additional negative-pressure material component within general relativity (Dark Energy) and low-curvature modifications of general relativity (Dark Gravity) will be extremely small.

1. Homogeneous expansion of our universe has recently accelerated

CMB, baryon oscillation, supernova and weak lensing cosmographic measurements of the homogeneous expansion history of our universe are summarized in consistent with either a static finely-tuned cosmological constant or a dynamic mechanism, which may be material Dark Energy or low-curvature modifications of Einstein gravity (Dark Gravity). The homogeneous expansion measures only kinematic variables, in particular the evolution of the cosmological deceleration $q := -a\ddot{a}/\dot{H}^2 = dH^{-1}/dt - 1$, which can be given a cosmological fluid description: If the expansion rate $H^2 := (8\pi G_N/3)\rho := \rho/3M_P^2$, is used to define a cosmological 'mass density', 'enthalpy density', 'pressure' and 'equation of state' are also defined:

$$\rho + P := -2M_P^2\dot{H}, \qquad P := -M_P^2(3H^2 + 2\dot{H}), \qquad w := P/\rho = -(3H^2 + 2\dot{H})/3H^2. \quad (1)$$

In these definitions, $\rho := 3M_P^2H^2$ can be separated into the ordinary cold dark matter component ρ_m and a 'dark energy' $\rho_{DE} := \rho - \rho_m$, which may be a new negative-pressure material Dark Energy or a Dark Gravity modification to Einstein general relativity.

The observed expansion history then fixes either a fine-tuned scalar field Dark Energy potential or a fine-tuned Dark Gravity modification to the GR Friedmann expansion rate [1]. The observations of the homogeneous expansion history $H^2(t)$ do not distinguish between Dark Energy and Dark Gravity and cannot fix the underlying dynamics driving the recent acceleration since redshift $z \lesssim 0.46$.

2. Concordance model: cosmological constant as a classical constant of nature

The minimal model fit to the observed 5-year WMAP CBR, Type Ia supernova, and Baryon Acoustic Oscillation galaxy distribution data is the flat ΛCDM model consisting of cold dark matter and a cosmological constant [2]:

$$\Omega_b, \quad \Omega_c, \quad \Omega_\Lambda = \qquad 0.0462 \pm 0.00115, \quad 0.233 \pm 0.013, \qquad 0.721 \pm 0.015, \tag{2}$$

in which the Ω_i are the present baryon, cold dark matter, and cosmological constant or 'Dark Energy' densities, each measured in units $\rho_{cr} := 3H_0^2 M_P^2$. This one-parameter fit is fully consistent with constant Λ or Dark Energy enthalpy $(\rho + P)/\rho = 1 + w = 0 \pm 10\%$, or 'equation of state' $w = -1$.

The simplest interpretation of constant Λ, due to Einstein, is as a classical spacetime curvature 4Λ, giving *geometric structure* to empty spacetime. The alternative interpretation, due to Zeldovich, interprets constant Λ as a uniform gravitational vacuum energy density, and brings forth the Cosmological Constant Problem: why quantum matter vacuum energy fluctuations apparently do not gravitate. (The Uehling and Casimir *forces* on electric charges or conducting, semi-conducting or dielectric materials show only the polarization of the QED vacuum, but not the QED vacuum energy itself [3]). The Cosmological Constant Problem is a *geometric* problem, only for ambitious theorists who want to calculate Λ [4]. So far, this attempt to calculate Λ along with other constants of nature has been fruitless. Pragmatically, there is no problem if Λ is accepted as a constant of nature like the observed electron mass or fine-structure constant.

3. Observers present existence selects among values of Λ

There still remains the Cosmological Coincidence Problem: Why do we live at so late a time that the material energy density has diluted down to a comparable value $\sim \Lambda/2$? This is a *material* problem, calling for an understanding of the observers' role in cosmology. An Enhanced Weak Anthropic Principle, that observers exist now but cannot distinguish between Dark Energy and Dark Gravity, would select out the Concordance Model ΛCDM, consistent with current observations and predicting that there will be no dynamical effects on the growth of fluctuations.

4. Only growth of fluctuations can resolve cosmodynamic degeneracy

The challenge therefore is to show whether the Cosmological 'Constant' is truly static or dynamic, either by showing $w \neq -1$ or by observing the growth of cosmological fluctuations. We already know [2] that, for constant $w(z)$

$$-0.11 \leq w \leq 0.14(95\%CL), \quad \text{present spatial curvature} - 0.0175 \leq \Omega_k \leq 0.0085(95\%CL), \tag{3}$$

with weaker constraints on time-varying Dark Energy. Improved measurements of the homogeneous expansion can never prove $w = -1$ exactly. Because the 'dark energy' evolution at low red-shifts is at most quasi-static, any dynamical effects on the growth of fluctuations will be minimal, unless $w(z)$ evolves strongly at high red-shifts $z > 5$. Difficult as they will be, projected observations of galaxy clustering and of gravitational weak gravitational lensing cosmic shear are needed to test ΛCDM by potentially distinguishing static from dynamic 'dark energy', and dynamic Dark Energy from Dark Gravity.

Acknowledgement

SAB acknowledges support from the Millennium Center for Supernova Science through grant P06-045-F funded by Programa Bicentenario de Ciencia y Tecnología de CONICYT and Programa Iniciativa Científica Milenio de MIDEPLAN.

Journal of Physics: Conference Series **161** (2009) 012026 doi:10.1088/1742-6596/161/1/012026

References

[1] Bludman S 2004 *Phys. Rev. D* **69** 122002
[2] Spergel D *et al.* 2006 ArXiv:astro-ph/0603449, Tables 2, 9, [WMAP3]
[3] Jaffe R 2005 *Phys. Rev. D* **72** 021301 arXiv:hep-th/0503158v1
[4] Bludman S and Ruderman M A 1987 *Phys. Rev. Lett.* **38** 255

Dynamical Casimir effect: some theoretical aspects

V V Dodonov

Instituto de Física, Universidade de Brasília, Caixa Postal 04455, 70910-900 Brasília, DF, Brazil

E-mail: vdodonov@fis.unb.br

Abstract.
This is a brief description of some recent achievements in the theory of dynamical Casimir effect, mainly in connection with the experiment which is under preparation in the University of Padua. The first part of this paper is devoted to the theory of quantum damped oscillator with arbitrary time dependence of the frequency and damping coefficient. New results for the mean number of created photons, its variance and photon distribution function are given. The second part is devoted to calculations of the time-dependent shift of resonance frequency of an electromagnetic cavity due to strong variations of dielectric properties in a thin layer near an ideally conducting wall. A simple analytical formula for this shift is derived. It generalizes the known Schwinger–Bethe–Casimir result. The influence of different parameters on the photon generation rate is discussed. A brief review of recent publications on the subject is also included.

1. Introduction

The so-called *Dynamical Casimir Effect* (DCE), i.e., a generation of photons from vacuum due to the motion of uncharged boundaries, was a subject of numerous theoretical studies for almost 40 years. An extensive list of publications until 2000 can be found in the review [1], whereas the papers published in the period from 2001 to 2004 were briefly discussed in [2]. Also, the status of the research in this area by 2005 was demonstrated in several papers published in the special issue of Journal of Optics B [3]. Therefore here I shall concentrate mainly on the publications from 2005 to 2008 and main results obtained for this period.

Among the most important publications of the past years I would like to mark out the papers [4–8] containing different concrete experimental proposals. The scheme of [6, 7] is based on the suggestion (formulated for the first time in [9, 10]) to excite true surface vibrations of a cavity wall in the GHz band and to use a beam of atoms passing through the cavity as a detector. Another idea, proposed by the MIR group of the university of Padua [4, 5], is to simulate a motion of a boundary, using an effective electron-hole 'plasma mirror', created periodically on the surface of a semiconductor slab by illuminating it with a sequence of short laser pulses. If the interval between pulses exceeds the recombination time of carriers in the semiconductor, a highly conducting layer will periodically appear and disappear on the surface of the semiconductor film, thus simulating periodical displacements of the boundary. A different scheme of simulating the dynamical Casimir effect, where periodical changes of the cavity eigenfrequency can be achieved by changing the surface impedance of a *superconducting film* illuminated by laser pulses, was proposed recently in [8]. However, it seems that the scheme of [4, 5] is the most promising one from the point of view of reaching the result in the nearest future. Therefore I concentrate only on this scheme.

Journal of Physics: Conference Series **161** (2009) 012027 doi:10.1088/1742-6596/161/1/012027

It is worth remembering that the history of the dynamical Casimir effect began almost 40 years ago, when Moore showed that the motion of ideal boundaries can result in creation of quanta of the electromagnetic field from the initial vacuum state [11]. But he concluded that the effect should be extremely small, if the velocities of boundaries are much less than the velocity of light. However, even earlier, a possibility of a significant amplification of *classical* electromagnetic fields inside cavities with *oscillating* boundaries under the conditions of parametric resonance was pointed out for the first time by Askar'yan in 1962 [12]. Later, a possibility of enhancement of vacuum (zero point) fluctuations under the conditions of resonance between field modes and oscillations of boundaries was discussed in [13, 14]. However, the first evaluations of the effect gave unrealistic numbers for two reasons: (i) approximate perturbation approaches, used in that papers, are invalid, as a matter of fact, under resonance conditions; (ii) the chosen amplitudes of oscillations of the cavity length were many orders of magnitude bigger than those which could be actually achieved in practice.

More precise and realistic calculations were performed only in 1990s in the frameworks of different approaches [9, 10, 15–18]. It was shown that a significant amount of photons could be created from vacuum, if boundaries of a high-Q cavity perform small oscillations at a frequency which is multiple of some cavity eigenfrequency. In particular, if a plane boundary of a *three-dimensional* cavity performs *harmonical* oscillations with an amplitude a at the frequency $\omega_w = 2\omega_0$, where ω_0 is the eigenfrequency of the lowest electromagnetic mode in the cavity with fixed geometry, then the mean number of photons created from vacuum in this mode is given by the formula [9, 10]

$$\langle n\rangle(t) = \sinh^2\left(\varepsilon\omega_0 t\eta^3\right), \tag{1}$$

where $\varepsilon = a/\lambda$ is the maximal relative displacement of the boundary (with respect to the wavelength $\lambda = 2\pi c/\omega_0$) and $\eta = \lambda/(2L_0) < 1$ is a numerical coefficient, which depends on the cavity geometry (L_0 is the average distance between vibrating walls).

Formula (1) can be derived from a general solution for a quantum harmonic oscillator with an arbitrary time-dependent frequency obtained for the first time in the seminal paper by Husimi [19] in 1953, if one remembers that field modes behave as a set of harmonic oscillators. Husimi showed that all dynamical properties of the *quantum* oscillator are determined by the fundamental system of solutions of the *classical* equation of motion

$$\ddot{\varepsilon} + \omega^2(t)\varepsilon = 0. \tag{2}$$

In particular, if $\omega(t) = \omega_i$ for $t \to -\infty$ and initially (at $t \to -\infty$) the oscillator was in the vacuum state, then the mean energy at the moment t equals

$$\mathcal{E}(t) = \frac{1}{4}\left[|\dot{\varepsilon}(t)|^2 + \omega^2(t)|\varepsilon(t)|^2\right], \tag{3}$$

where the function $\varepsilon(t)$ satisfies equation (2) and the initial condition

$$\varepsilon_{t\to-\infty} = \omega_i^{-1/2} e^{-i\omega_i t}. \tag{4}$$

Formula (3) holds for an arbitrary function $\omega(t)$, provided this function is *real*, i.e., the quantum evolution is *unitary*.

According to (1), one of the most important parameters which determines a possible number of created photons is an achievable value of the wall displacement amplitude. For the cavity dimensions of the order of $1 \div 100$ cm, the field resonance frequencies ($\omega_0/2\pi$) belong to the band from 30 GHz to 300 MHz. An idea of [9, 10] was not to force the wall to oscillate as a whole at such a high frequency, but to excite oscillations of the *surface* of the cavity wall. In such a case, the amplitude a of a standing acoustic wave at frequency $\omega_w = 2\omega_0$ (coinciding with the amplitude of oscillations of the free surface) is related to the relative deformation amplitude δ inside the wall as $\delta = \omega_w a/v_s$, where v_s is the sound velocity. Since usual materials cannot bear the deformations exceeding the value $\delta_{max} \sim 10^{-2}$, the maximal possible velocity of the boundary is $v_{max} \sim \delta_{max} v_s \sim 50$ m/s (independent of the frequency). The maximal relative displacement $\varepsilon = a/L_0$ is $\varepsilon_{max} \sim (v_s/2\pi c)\delta_{max} \sim 3\cdot 10^{-8}$ for the lowest mode with the frequency $\omega_0 \sim c\pi/L_0$. Then, taking $\varepsilon = 10^{-9}$, $\omega_0/(2\pi) = 10\,\mathrm{GHz}$ and $\eta^3 = 1/3$, in $t = 1$ s one could get a big number $\sinh^2(10) \sim 10^8$ photons in an empty cavity. However, in such a case one needs a cavity with the Q-factor of the order of 10^{10}. Moreover, it is necessary to maintain the resonance condition, which means that the frequency of the wall oscillations must not deviate from $2\omega_0/(2\pi)$ by more than $\delta/(2\pi) < \varepsilon\,\omega_0\eta^3/(2\pi) \sim 3\,\mathrm{Hz}$ during the time 1 s.

On the other hand, what we really need to create photons from vacuum, it is a possibility to change the resonance frequency in a periodical way. But this can be achieved not only by changing the geometry, but by changing the electric properties of the walls or some medium inside the cavity. Hence the idea of simulating DCE and other quantum effects arose about two decades ago in the article by Yablonovitch [20], who proposed to use a medium with a rapidly decreasing in time refractive index ('plasma window') to simulate the so-called Unruh effect. Also, he pointed out that fast changes of electric properties can be achieved in semiconductors illuminated by laser pulses. This idea was propagandized by Man'ko [21], who proposed to use semiconductors with time-dependent properties to produce an analogue of the *nonstationary Casimir effect* (see also [22,23]). A more developed scheme, based on the creation of an electron-hole 'plasma mirror' inside a semiconductor slab, illuminated by a femtosecond laser pulse, was proposed in [24] (in the single-pulse case). But only recently a possibility of creating an effective 'plasma mirror' in a semiconductor slab was confirmed experimentally [25].

Quantum effects caused by a time dependence of properties of thin slabs inside resonance cavities were studied by several authors [26–29]. However, only very simple models of the media were considered in that papers: lossless homogeneous dielectrics with time-dependent permeability [29], ideal dielectrics or ideal conductors suddenly removed from the cavity [26,27] or infinitely thin conducting slabs modeled by δ-potentials with time-dependent strength [28] (this model of 'plasma sheet' was introduced in [30]). Moreover, all that models, as well as the estimations of the photon generations rate based on the simple formula (1), did not take into account inevitable losses inside the semiconductor slab during the excitation-recombination process. This is the immediate consequence of the fact that the dielectric permeability $\epsilon(x)$ of the semiconductor medium is a *complex function*: $\epsilon = \epsilon_1 + i\epsilon_2$, where $\epsilon_2 = 2\sigma/f_0$, σ and f_0 being the conductivity (in the CGS units) and frequency in Hz, respectively. Good conductors have $\epsilon_2 \sim 10^8$ at microwave frequencies, which is essentially bigger than $\epsilon_1 \sim 1 \div 10$. Although ϵ_2 is negligibly small in the non-excited semiconductor at low temperatures, it rapidly and continuously grows up to the values of the order of $10^5 \div 10^6$ during the laser pulse, returning to zero after the recombination time. In order to predict possible results of experiments and to suggest optimal choices of different parameters, one has to take into account the internal dissipation in the slab, which means that one has to generalize the Husimi solution to the case of *damped* quantum nonstationary oscillator. Such a generalization is considered in sections 2 and 3. In addition to the formulas for the mean number of quanta created in the process of parametric excitation, obtained in previous papers [2,31–36], I give several important new results. Namely, I calculate the distribution function of created quanta and the variance of their number, which characterizes fluctuations in the quantum state of the field mode. The distribution function turns out to be *larger* than in the thermal state. Such states are called sometimes *super-chaotic states*.

Section 4 is devoted to a new approximate formula for the resonance frequency shift in the cavity, which holds (in contradistinction to all previous studies) even for *big* changes of the complex dielectric permeability. The knowledge of this frequency shift is absolutely necessary for calculating the number of created quanta. Corresponding estimations are given in section 5.

But before going to the main goal and concrete calculations, it seems useful to provide a list of other publications related to the dynamical Casimir effect, which appeared during the period from 2005 to 2008. The role of different boundary conditions was studied in [37] for a scalar field in 1+3 dimensions with the Neumann boundary condition on a single moving deformed mirror. A functional approach was applied to the problem of a single moving dispersive mirror in 1+1 dimensions in [38]. A comparison of the Dirichlet and Neumann conditions for a single relativistic mirror in 1+1 dimensions was made in [39]. The spectrum of radiation from a single dynamically deforming mirror was calculated in [40]. A scalar field between two mirrors in 1+1 dimensions with different (Dirichlet or Neumann) conditions on the mirrors was considered in [41–43]. The case of a 3D cavity with conducting and permeable oscillating plates (described by means of the Dirichlet and Neumann boundary conditions) was studied in [44, 45]. The Robin boundary conditions in 1+1 dimensions were considered in [46,47]. The one-dimensional cavity with one and two oscillating mirrors was considered within the framework of the 'optical' approach in [48]. Generalizations of Moore's approach to the one-dimensional vibrating cavities were considered in [49]. A one-dimensional *uniformly contracting* cavity was studied in [50,51]. The authors of [41,52] pointed out on a possible physical realization of one-dimensional models in the case of TEM modes in cylindrical waveguides. Oscillating *spherical* cavities were considered in [41, 53, 54]. A multiple scale analysis was used in studies on oscillating ideal cavities in three dimensions in [55, 56], giving the same results that were obtained during the preceding decade. The Hamiltonian approach to the dynamical Casimir effect was revised in [57–61] and applied to the calculation of the spectrum of created particles in [62, 63] and to cosmological problems in [64]. Possible cosmological manifestations of the Dynamical Casimir Effect were studied also in [65,66]. A comparison of analytical and numerical results was made in [67–69]. The relationship between DCE, Unruh-Hawking and other effects was analyzed in [70–72]. An account of nonlinear effects in the DCE was made in [73]. Photon creation in nonstationary media was considered in [74,75]. Different parametric processes which can be thought as analogs of the DCE were studied in [76–85]. Methods of detection of Casimir photons were considered in [86]. The dynamical *Casimir–Polder* effect was the subject of studies [87–90]. The phenomena of decoherence and entanglement in connection with DCE were studied in [91–95].

2. Quantum damped oscillator with time-dependent parameters in the Heisenberg–Langevin approach

An immediate consequence of the time variation of electromagnetic properties of the cavity walls is the time dependence of the eigenmode frequencies. Hence it follows a simple idea that one could understand the main features of the behavior of the quantum field in the cavity by considering a single selected mode and describing it as a quantum oscillator with 'instantaneous' time-dependent frequency [96, 97]. Later on, it was justified (see, e.g., [9, 10, 98, 99]) for three-dimensional cavities without accidental degeneracy of the spectrum of eigenmode frequencies and for harmonic variations of the effective frequency. Thus I *assume* that even in the presence of dissipation and non-monochromatic periodical variations, the field problem still can be reduced approximately to the dynamics of a *single selected mode* described in the classical limit as a harmonic oscillator with time-dependent *complex* frequency $\omega_c(t) = \omega(t) - i\gamma(t)$, which can be found from the solution of the classical electrodynamical problem by taking the *instantaneous* geometry and material properties (as was done in the non-dissipative case in [28, 29]).

The scheme presented below was developed in [2, 31, 32]. It is a generalization of the *quantum noise operator* approach, first proposed in [100–103] for systems with time-independent parameters, to the case of arbitrary time dependence of the frequency and damping coefficient. In this approach, dissipative quantum systems are described within the framework of the

Heisenberg–Langevin operator equations. In the case concerned these equations can be written as

$$d\hat{x}/dt = \hat{p} - \gamma_x(t)\hat{x} + \hat{F}_x(t), \qquad d\hat{p}/dt = -\gamma_p(t)\hat{p} - \omega^2(t)\hat{x} + \hat{F}_p(t). \tag{5}$$

Here $\hat{x}$ and $\hat{p}$ are the dimensionless quadrature operators of the selected mode, normalized in such a way that the mean number of photons equals

$$\mathcal{N} = \frac{1}{2}\langle \hat{p}^2 + \hat{x}^2 - 1\rangle. \tag{6}$$

In other words, in the subsequent formulas ω and γ are the frequency and damping coefficient normalized by the initial frequency ω_i. The *two noncommuting* noise operators $\hat{F}_x(t)$ and $\hat{F}_p(t)$ are necessary to preserve the canonical commutator between the Heisenberg operators $\hat{x}(t)$ and $\hat{p}(t)$ (however, it is supposed that $\hat{F}_x(t)$ and $\hat{F}_p(t)$ commute with $\hat{x}$ and $\hat{p}$). At first glance, the presence of two extra terms $-\gamma_x(t)\hat{x} + \hat{F}_x(t)$ in the first equation in (5) seems unusual (from the point of view of the classical theory of Brownian motion). However, these terms arise quite naturally in quantum optics, for example, if one rewrites the standard Heisenberg–Langevin equation of motion for the annihilation operator $d\hat{a}/dt = (-i\omega - \gamma)\hat{a} + \hat{F}_a$ [103] in terms of quadrature components.

The system of linear equations (5) can be solved explicitly for arbitrary time-dependent functions $\gamma_{x,p}(t)$, $\omega(t)$ and $\hat{F}_{x,p}(t)$. It is convenient to represent the solutions in the form

$$\hat{x}(t) = \hat{x}_s(t) + \hat{X}(t), \qquad \hat{p}(t) = \hat{p}_s(t) + \hat{P}(t), \tag{7}$$

where the first terms represent the solutions of the homogeneous parts of equations (5) (the subscript 's' stands for 'source'):

$$\hat{x}_s(t) = e^{-\Gamma(t)}\left\{\hat{x}_0 \mathrm{Re}\left[\xi(t)\right] - \hat{p}_0 Im\left[\xi(t)\right]\right\}, \tag{8}$$

$$\hat{p}_s(t) = e^{-\Gamma(t)}\left\{\hat{x}_0 \mathrm{Re}\left[\eta(t)\right] - \hat{p}_0 Im\left[\eta(t)\right]\right\}. \tag{9}$$

Here $\hat{x}_0$ and $\hat{p}_0$ are the initial values of operators at $t = 0$ (taken as the initial instant) and

$$\Gamma(t) = \int_0^t \gamma(\tau)d\tau, \qquad \gamma(t) = \frac{1}{2}\left[\gamma_x(t) + \gamma_p(t)\right]. \tag{10}$$

The operators $\hat{X}(t)$ and $\hat{P}(t)$ represent the influence of the stochastic forces (note that the 'source' and 'noise' parts of the solutions commute with each other):

$$\begin{pmatrix} \hat{X}(t) \\ \hat{P}(t) \end{pmatrix} = e^{-\Gamma(t)} \int_0^t d\tau e^{\Gamma(\tau)} \begin{pmatrix} a_x^x(t;\tau) & a_x^p(t;\tau) \\ a_p^x(t;\tau) & a_p^p(t;\tau) \end{pmatrix} \begin{pmatrix} \hat{F}_x(\tau) \\ \hat{F}_p(\tau) \end{pmatrix}, \tag{11}$$

where

$$a_x^x(t;\tau) = \mathrm{Im}\left[\xi(t)\eta^*(\tau)\right], \qquad a_x^p(t;\tau) = \mathrm{Im}\left[\xi^*(t)\xi(\tau)\right], \tag{12}$$

$$a_p^x(t;\tau) = \mathrm{Im}\left[\eta(t)\eta^*(\tau)\right], \qquad a_p^p(t;\tau) = \mathrm{Im}\left[\eta^*(t)\xi(\tau)\right]. \tag{13}$$

Function $\xi(t)$ is the special solution to equation (2) with $\omega^2(t)$ replaced by the effective frequency

$$\omega_{ef}^2(t) = \omega^2(t) + \dot{\delta}(t) - \delta^2(t), \qquad \delta(t) = \frac{1}{2}\left[\gamma_x(t) - \gamma_p(t)\right]. \tag{14}$$

This special solution is selected by the initial condition $\xi(t) = \exp(-it)$ for $t \to -\infty$, which is equivalent to fixing the value of the Wronskian:

$$\xi\dot{\xi}^* - \dot{\xi}\xi^* = 2i. \tag{15}$$

Journal of Physics: Conference Series **161** (2009) 012027 doi:10.1088/1742-6596/161/1/012027

The function $\eta(t)$ is defined as

$$\eta(t) = \dot{\xi}(t) + \delta(t)\xi(t). \tag{16}$$

It satisfies the identity following from (15)

$$\mathrm{Im}\,[\xi(t)\eta^*(t)] \equiv 1. \tag{17}$$

It seems natural to identify the functions $\omega(t)$ and $\gamma(t)$ in equations (5), (10) and (14) with the real and imaginary parts of the instantaneous complex cavity eigenfrequency $\omega_c(t) = \omega(t) - i\gamma(t)$. An immediate consequence of equations (7)-(9), (15) and (17) is the formula

$$[\hat{x}(t), \hat{p}(t)] = i\hbar e^{-2\Gamma(t)} + \left[\hat{X}(t), \hat{P}(t)\right]. \tag{18}$$

Using equations (11) and (17), one can verify that the commutator $[\hat{x}(t), \hat{p}(t)] = i\hbar$ is preserved exactly for arbitrary functions $\omega(t)$ and $\gamma(t)$, if one assumes that the noise operators are *delta-correlated* (the Markov approximation) with the following commutation relations:

$$\left[\hat{F}_x(t), \hat{F}_p(t')\right] = 2i\hbar\gamma(t)\delta(t-t'), \qquad \left[\hat{F}_x(t), \hat{F}_x(t')\right] = \left[\hat{F}_p(t), \hat{F}_p(t')\right] = 0. \tag{19}$$

Indeed, under these conditions one obtains

$$\left[\hat{X}(t), \hat{P}(t)\right] = e^{-2\Gamma(t)} \int_0^t 2i\hbar\gamma(t) e^{2\Gamma(\tau)} d\tau = i\hbar \left[1 - e^{-2\Gamma(t)}\right].$$

In contrast to the classical Langevin equations, which contain a single stochastic force, in the quantum case one must use *two* noise operators, otherwise the canonical commutation relations cannot be saved. The Markov approximation implies the relations

$$\langle \hat{F}_x(t)\hat{F}_x(t')\rangle = \delta(t-t')\chi_{xx}(t), \qquad \langle \hat{F}_p(t)\hat{F}_p(t')\rangle = \delta(t-t')\chi_{pp}(t), \tag{20}$$

$$\langle \hat{F}_x(t)\hat{F}_p(t') + \hat{F}_p(t')\hat{F}_x(t)\rangle = 2\delta(t-t')\chi_s(t). \tag{21}$$

Strictly speaking, the 'noise coefficients' $\chi_{xx}(t)$, $\chi_{pp}(t)$ and $\chi_s(t)$ must be derived from some 'microscopical' model, which takes into account explicitly (1) the coupling of the field mode with electron–hole pairs inside the semiconductor slab and (2) the coupling of electrons and holes with phonons or other quasiparticles, responsible for the damping mechanisms. Unfortunately, it seems that no model of this kind was considered until now. Nonetheless, some conclusions on the relations between the noise coefficients can be made, if one calculates the second-order moments of the quadrature operators. Namely, using equations (11), (17), (20) and (21) one can obtain the following *exact* expressions for the mean values of the operators $\hat{x}^2(t)$ and $\hat{p}^2(t)$ (from now on we shall use dimensionless variables, putting $\hbar = 1$; besides, we replace the symbol of function $f(t)$ by a short form f_t):

$$\begin{aligned}\langle \hat{x}^2(t)\rangle &= e^{-2\Gamma(t)} \left\{\langle \hat{x}^2\rangle_0 \left[\mathrm{Re}\,(\xi_t)\right]^2 + \langle \hat{p}^2\rangle_0 \left[\mathrm{Im}\,(\xi_t)\right]^2 - \langle \hat{x}\hat{p} + \hat{p}\hat{x}\rangle_0 \mathrm{Re}\,(\xi_t)\,\mathrm{Im}\,(\xi_t)\right\} \\ &\quad + e^{-2\Gamma(t)} \int_0^t d\tau e^{2\Gamma(\tau)} \left\{\chi_{xx}(\tau)\left[\mathrm{Im}\,(\xi_t\eta_\tau^*)\right]^2 + \chi_{pp}(\tau)\left[\mathrm{Im}\,(\xi_t\xi_\tau^*)\right]^2\right. \\ &\quad \left. - 2\chi_s(\tau)\mathrm{Im}\,(\xi_t\eta_\tau^*)\,\mathrm{Im}\,(\xi_t\xi_\tau^*)\right\},\end{aligned} \tag{22}$$

$$\begin{aligned}\langle \hat{p}^2(t)\rangle &= e^{-2\Gamma(t)} \left\{\langle \hat{x}^2\rangle_0 \left[\mathrm{Re}\,(\eta_t)\right]^2 + \langle \hat{p}^2\rangle_0 \left[\mathrm{Im}\,(\eta_t)\right]^2 - \langle \hat{x}\hat{p} + \hat{p}\hat{x}\rangle_0 \mathrm{Re}\,(\eta_t)\,\mathrm{Im}\,(\eta_t)\right\} \\ &\quad + e^{-2\Gamma(t)} \int_0^t d\tau e^{2\Gamma(\tau)} \left\{\chi_{xx}(\tau)\left[\mathrm{Im}\,(\eta_t\eta_\tau^*)\right]^2 + \chi_{pp}(\tau)\left[\mathrm{Im}\,(\eta_t\xi_\tau^*)\right]^2\right. \\ &\quad \left. - 2\chi_s(\tau)\mathrm{Im}\,(\eta_t\eta_\tau^*)\,\mathrm{Im}\,(\eta_t\xi_\tau^*)\right\}.\end{aligned} \tag{23}$$

Let us consider the case of time-independent frequency $\omega = \omega_i = 1$ and time-independent damping and noise coefficients. Assuming that $\gamma \ll 1$ (small damping) one can neglect the correction $\delta^2 \sim \gamma^2$ in function $\omega_{ef}(t)$ (14) and use the solution $\xi(t) = \exp(-it)$. Then integrals in equations (22) and (23) can be calculated exactly. Supposing that $\gamma_{x,p} \sim \gamma$ and $\chi_{jk} \sim \gamma$ (in accordance with the fluctuation–dissipation theorem) and neglecting terms proportional to γ^2, one can obtain the following mean values at $t \to \infty$:

$$\langle \hat{x}^2 \rangle_\infty = \frac{1}{4\gamma}\left[\chi_{xx} + \chi_{pp} + 2\gamma_p \chi_s\right], \qquad \langle \hat{p}^2 \rangle_\infty = \frac{1}{4\gamma}\left[\chi_{xx} + \chi_{pp} - 2\gamma_x \chi_s\right], \tag{24}$$

$$\langle \hat{x}\hat{p} + \hat{p}\hat{x} \rangle_\infty = \frac{1}{2\gamma}\left[\gamma_x \chi_{pp} - \gamma_p \chi_{xx}\right]. \tag{25}$$

We see that the steady-state moments of the second order coincide with the thermodynamical equilibrium values

$$\langle \hat{x}^2 \rangle_{eq} = \langle \hat{p}^2 \rangle_{eq} = 1/2 + \langle n \rangle_{th}, \qquad \langle \hat{x}\hat{p} + \hat{p}\hat{x} \rangle_{eq} = 0 \tag{26}$$

(where $\langle n \rangle_{th}$ is the mean number of quanta in the thermal state) with the accuracy of the order of γ^2 (i.e., without *linear* corrections with respect to the damping coefficients), provided the noise coefficients are chosen as follows

$$\chi_s = 0, \qquad \chi_{xx} = \gamma_x G, \qquad \chi_{pp} = \gamma_p G \tag{27}$$

where

$$G = 1 + 2\langle n \rangle_{th} = \coth\left(\frac{\hbar\omega_i}{2k_B\Theta}\right) \tag{28}$$

and Θ is the temperature of the reservoir.

It is important that the noise and damping coefficients cannot be quite arbitrary and independent. Namely, the coefficients χ_{pp} and χ_{xx} must obey the restriction $\chi_{pp}\chi_{xx} \geq \gamma^2$, which prevents from possible violations of the positivity of the density matrix during the evolution [104, 105]. Therefore we must demand the fulfillment of the inequality

$$G^2 \gamma_x \gamma_p \geq (\gamma_x + \gamma_p)^2 / 4. \tag{29}$$

At zero temperature of the reservoir ($G = 1$) inequality (29) can be satisfied only for

$$\gamma_x = \gamma_p = \gamma. \tag{30}$$

In the case of nonzero temperature (when $G > 1$), the positivity of the density matrix can be preserved for unequal damping coefficients γ_x and γ_p. Their exact values could be calculated in the frameworks of the 'microscopical' theory, which does not exist yet. Therefore I *assume* in this paper that equation (30) holds for nonzero temperatures, as well, so that the diffusion and damping coefficients are given by the relations

$$\chi_s = 0, \qquad \chi_{xx} = \chi_{pp} = \gamma G. \tag{31}$$

In such a case, $\delta \equiv 0$, $\eta(t) \equiv \dot{\xi}(t)$ and the effective frequency in equation (14) coincides exactly with $\omega(t)$. For this special set of coefficients, the stationary asymptotical values of the second-order statistical moments coincide *exactly* with the equilibrium values (26) for an *arbitrary* (not necessarily small) function $\gamma(t)$ (if $\omega = const$). This will be shown in subsections 2.1 and 2.2.

The main arguments in favor of the assumption (31) are based on the analysis of different 'microscopical' models describing the interaction of a selected harmonic oscillator (field mode) with a *fixed* frequency with an 'environment' described by means of multidimensional *quadratic* Hamiltonians of the most general form. Namely, it was shown in [104, 106, 107] that time-independent damping and noise coefficients, satisfying all the requirements (i.e., not allowing violations of the positivity of the density matrix), arise in these models in the only case: when the coupling between the selected and the 'bath' oscillators has the so-called 'rotating wave approximation' (RWA) form $\sum_j \hat{a}\hat{b}_j^\dagger + H.c.$ (this special kind of coupling is considered in all textbooks on quantum optics, see, e.g., [108]). Under this restriction, the models of this kind result in the set of coefficients given by equations (30) and (31). It is worth remembering in this connection that the usual justification for the exclusion of 'antirotating' terms $\hat{a}\hat{b}_j + H.c.$ is that such terms give rapidly oscillating corrections to the equations of motion, whose influence becomes small after averaging over many periods of the field mode oscillations. The problem is that the duration of laser pulses in the MIR experiment is much smaller than the period of the field oscillations. Therefore, strictly speaking, one cannot exclude a possibility that the assumption (31) is not correct. These observations show the necessity of a rigorous microscopical theory of the dynamical Casimir effect in dissipative media with time-dependent parameters.

2.1. The mean number of created quanta

The formula for the time dependence of the mean number of quanta (6) can be split in two parts

$$\mathcal{N}(t) = \mathcal{N}_s(t) + \mathcal{N}_r(t) \tag{32}$$

where the first term depends on the initial state ('signal') while the second term is determined completely by the interaction with the reservoir (it is represented by the integrals containing the coefficients χ_{jk}). The contribution of noise to the mean number of quanta under the condition (31) is given by the formula

$$\mathcal{N}_r(t) = E(t)J(t) - \mathrm{Re}\left[\tilde{E}^*(t)\tilde{J}(t)\right], \tag{33}$$

where

$$J(t) = Ge^{-2\Gamma(t)}\int_0^t d\tau e^{2\Gamma(\tau)}\gamma(\tau)E(\tau), \qquad \tilde{J}(t) = Ge^{-2\Gamma(t)}\int_0^t d\tau e^{2\Gamma(\tau)}\gamma(\tau)\tilde{E}(\tau) \tag{34}$$

and

$$E(t) = \frac{1}{2}\left[|\xi(t)|^2 + |\dot{\xi}(t)|^2\right], \qquad \tilde{E}(t) = \frac{1}{2}\left[\xi^2(t) + \dot{\xi}^2(t)\right]. \tag{35}$$

In the special case of the initial *coherent* state $|\alpha\rangle$ (which corresponds to the initial 'classical' signal in the cavity), the 'signal' contribution is given by the formula

$$\mathcal{N}_s^{(coh)}(t) = e^{-2\Gamma(t)}\left\{(\mathrm{Re}[\alpha\xi(t)])^2 + \left(\mathrm{Re}[\alpha\dot{\xi}(t)]\right)^2 + \frac{1}{2}E(t)\right\} - \frac{1}{2}. \tag{36}$$

For the initial thermal state we have

$$\mathcal{N}_s^{(th)}(t) = \frac{1}{2}\left\{G_0 e^{-2\Gamma(t)}E(t) - 1\right\}. \tag{37}$$

One should remember that formulas (33), (36) and (37) hold for sufficiently big values of time t, when the time dependent normalized frequency $\omega(t)$ returns to its initial value $\omega(-\infty) = 1$.

2.2. Fluctuations of the number of quanta

Fluctuations of the number of created quanta are characterized by the variance

$$\sigma_N = \langle \hat{\mathcal{N}}^2 \rangle - \langle \hat{\mathcal{N}} \rangle^2, \qquad \hat{\mathcal{N}} = \frac{1}{2}\left(\hat{p}^2 + \hat{x}^2 - 1\right). \tag{38}$$

Using the decomposition (7) we can write

$$\begin{aligned} 4\sigma_N \;=\; & \langle \hat{P}^4 \rangle - \langle \hat{P}^2 \rangle^2 + \langle \hat{X}^4 \rangle - \langle \hat{X}^2 \rangle^2 + \langle \hat{P}^2\hat{X}^2 + \hat{X}^2\hat{P}^2 \rangle - 2\langle \hat{P}^2 \rangle \langle \hat{X}^2 \rangle \\ & + \langle \hat{p}_s^4 \rangle - \langle \hat{p}_s^2 \rangle^2 + \langle \hat{x}_s^4 \rangle - \langle \hat{x}_s^2 \rangle^2 + \langle \hat{p}_s^2\hat{x}_s^2 + \hat{x}_s^2\hat{p}_s^2 \rangle - 2\langle \hat{p}_s^2 \rangle \langle \hat{x}_s^2 \rangle \\ & + 4\langle \hat{p}_s^2 \rangle \langle \hat{P}^2 \rangle + 4\langle \hat{x}_s^2 \rangle \langle \hat{X}^2 \rangle + 4\langle \hat{p}_s\hat{x}_s \rangle \langle \hat{P}\hat{X} \rangle + 4\langle \hat{x}_s\hat{p}_s \rangle \langle \hat{X}\hat{P} \rangle. \end{aligned} \tag{39}$$

Let us consider, for example, the quantity $\langle \hat{P}^4 \rangle$. In view of equation (11), it is given by the four-fold integral containing different average values of the stochastic force operators of the form $\langle \hat{F}_i(\tau_1)\hat{F}_j(\tau_2)\hat{F}_k(\tau_3)\hat{F}_l(\tau_4) \rangle$. We assume that stochastic forces are *Gaussian*. Then the fourth-order average values are factorized in the sum over all different products of the second-order moments, so that we can use the following formula:

$$\begin{aligned} \langle \hat{F}_i(\tau_1)\hat{F}_j(\tau_2)\hat{F}_k(\tau_3)\hat{F}_l(\tau_4) \rangle \;=\; & \chi_{ij}(\tau_1)\chi_{kl}(\tau_3)\delta(\tau_1-\tau_2)\delta(\tau_3-\tau_4) \\ & + \chi_{ik}(\tau_1)\chi_{jl}(\tau_2)\delta(\tau_1-\tau_3)\delta(\tau_2-\tau_4) \\ & + \chi_{il}(\tau_1)\chi_{jk}(\tau_3)\delta(\tau_1-\tau_4)\delta(\tau_2-\tau_3). \end{aligned} \tag{40}$$

As a consequence, all four-fold integrals are also factorized in the sums of the products of two ordinary integrals of the following structure:

$$I_{\mu\lambda}^{\nu\rho} = e^{-2\Gamma(t)} \int_0^t d\tau e^{2\Gamma(\tau)} a_\mu^\nu(\tau) a_\lambda^\rho(\tau) \chi_{\nu\rho}(\tau), \qquad \mu, \nu, \lambda, \rho = x, p. \tag{41}$$

The functions $a_\mu^\nu(t)$ were defined in (12) and (13). Making the only assumption $\chi_s = 0$, one obtains the following noise contribution to the second-order moments of $\hat{X}$ and $\hat{P}$:

$$\langle \hat{P}^2 \rangle = I_{pp}^{xx} + I_{pp}^{pp}, \qquad \langle \hat{X}^2 \rangle = I_{xx}^{xx} + I_{xx}^{pp}. \tag{42}$$

Besides, the following relations hold due to identities $\chi_{xp} = -\chi_{px} = i\gamma$ and (17):

$$I_{ab}^{px} + I_{ba}^{xp} \equiv 0, \qquad a, b = x, p, \tag{43}$$

$$I_{xp}^{px} + I_{xp}^{xp} = -I_{px}^{xp} - I_{px}^{px} = \frac{i}{2}\theta(t), \qquad \theta(t) \equiv 1 - e^{-2\Gamma(t)}. \tag{44}$$

Taking into account (42)-(44) and the relation $I_{ab}^{\nu\nu} = I_{ba}^{\nu\nu}$, one can write the part of σ_N given by the first line in equation (39) (which depends on the noise operators only) as follows:

$$\sigma_N^r(t) = \frac{1}{2}\left[\left(I_{pp}^{xx} + I_{pp}^{pp}\right)^2 + \left(I_{xx}^{xx} + I_{xx}^{pp}\right)^2 + \left(I_{px}^{xx} + I_{px}^{pp}\right)^2 + \left(I_{xp}^{xx} + I_{xp}^{pp}\right)^2\right] - \frac{1}{4}\theta^2(t). \tag{45}$$

If $\chi_{pp} = \chi_{xx}$, then (45) can be simplified as follows,

$$\sigma_N^r(t) = (2E^2 - 1)J^2 + E^2|\tilde{J}|^2 + \mathrm{Re}\left(\tilde{E}^2\tilde{J}^{*2} - 4E\tilde{E}J\tilde{J}^*\right) - \frac{1}{4}\theta^2(t), \tag{46}$$

where the functions $J(t)$, $\tilde{J}(t)$, $E(t)$ and $\tilde{E}(t)$ were defined in (34) and (35).

For the initial *thermal* state characterized by parameter G_0 (which can be different from the reservoir factor G) we obtain the following expressions for the sums of terms in the second ('signal') and third ('signal–noise') lines of equation (39):

$$\sigma_N^s(t) = \frac{1}{4}e^{-4\Gamma(t)}\left(G_0^2\left[2E^2(t)-1\right)\right]-1\right), \tag{47}$$

$$\sigma_N^{s-r}(t) = e^{-2\Gamma(t)}\left\{G_0 J(t)\left[2E^2(t)-1\right)\right]-2G_0E(t)\mathrm{Re}\left[\tilde{E}^*(t)\tilde{J}(t)\right]-\frac{1}{2}\theta(t)\right\}. \tag{48}$$

The correctness of formulas (46)-(47) can be verified in the special case of relaxation from one temperature (characterized by parameter G_0) to another (with parameter G) without changing the frequency $\omega = const = 1$. In this case $E \equiv 1$, $\tilde{E} = \tilde{J} \equiv 0$, $J(t) = G\theta(t)/2$, so that we obtain for an arbitrary function $\gamma(t)$

$$\mathcal{N}(t) = \frac{1}{2}\left[G_{ef}(t)-1\right], \qquad G_{ef}(t) = G\theta(t) + G_0\left[1-\theta(t)\right] \tag{49}$$

and

$$\sigma_N(t) = \frac{1}{4}\left[G_{ef}^2(t)-1\right] \equiv \mathcal{N}(t)[\mathcal{N}(t)+1], \tag{50}$$

as it must be for thermal states.

3. Periodical variations of parameters

Having in mind applications to the dynamical Casimir effect, we are interested in the special case when the functions $\omega(t)$ and $\gamma(t)$ have the form of *periodical* pulses with the periodicity T, separated by intervals of time with $\omega = 1$ and $\gamma = 0$ (we neglect the damping of the field between pulses, supposing that the quality factor of the cavity is big enough). Then the integrals in equation (34) are reduced to the sums of n (the total number of pulses) integrals taken between the initial and final time moments of each pulse t_i and t_f. In the interval between the kth and $(k+1)$th pulses the solution to equation (2) can be written as

$$\varepsilon_k(t) = a_k e^{-it} + b_k e^{it}, \qquad a_0 = 1, \quad b_0 = 0, \tag{51}$$

where a_k and b_k are constant coefficients. Consequently, during these intervals functions $E(t)$ and $\tilde{E}(t)$ assume constant values

$$E_k = |a_k|^2 + |b_k|^2, \qquad \tilde{E}_k = 2a_kb_k. \tag{52}$$

Evidently, every two sets of the nearest constant coefficients, (a_{k-1}, b_{k-1}) and (a_k, b_k), are related by means of a linear transformation

$$\begin{pmatrix} a_k \\ b_k \end{pmatrix} = M_k \begin{pmatrix} a_{k-1} \\ b_{k-1} \end{pmatrix}. \tag{53}$$

Four complex elements of the 2×2 matrix M_k for the single pulse (which can be treated also as an effective 'potential barrier', if one wishes to interprete (2) as an analog of the stationary Schrödinger equation) can be expressed through two complex amplitude reflection coefficients and two complex amplitude transmission coefficients, which connect the 'plane waves' coming from the 'left' and from the 'right'. Namely, if the pulse begins at $t = 0$ at terminates at $t = t_*$, then one can write two independent solutions of equation (2) for $t < 0$ and $t > t_*$ as (remember that we assume the constant initial and final value of the frequency to be equal to $\omega_i = 1$)

$$\varepsilon^{(-)}(t) = \begin{cases} e^{it} + r_-e^{-it}, & t<0 \\ s_-e^{it}, & t>t_* \end{cases} \qquad \varepsilon^{(+)}(t) = \begin{cases} s_+e^{-it}, & t<0 \\ e^{-it}+r_+e^{it}, & t>t_* \end{cases} \tag{54}$$

The reflection coefficients $r_\pm$ and transmission coefficients $s_\pm$ are not independent, because equation (2) is invariant with respect to complex conjugation (the frequency $\omega(t)$ is real). The following relations hold [109] (the simplest way to obtain them is to calculate Wronskians for suitable pairs of independent solutions):

$$s_- = s_+ \equiv s, \qquad r_- s^* + r_+^* s = 0, \qquad |r_-|^2 + |s_-|^2 = |r_+|^2 + |s_+|^2 = 1. \tag{55}$$

Comparing equations (51) and (53) with (54) and taking into account identities (55), one can express the elements of matrix M_k as follows:

$$M_k^{(0)} = \left\| \begin{array}{cc} f_k & g_k^* \\ g_k & f_k^* \end{array} \right\|, \qquad f_k \equiv s^{-1}, \quad g_k \equiv r_+/s, \tag{56}$$

where the superscript '(0)' means that the phases of reflection and transmission coefficients correspond to the pulse starting at the moment $t = 0$ (the subscript k is supressed in coefficients s and $r_\pm$). If the kth pulse begins at the moment t_{k-1} (so that $t_0 = 0$), then one should make the time shift $t \to t - t_{k-1}$ in equation (54). This means that matrix M_k in (53) can be expressed through the matrix $M_k^{(0)}$ as

$$M_k = \Phi_{k-1}^\dagger M_k^{(0)} \Phi_{k-1}, \qquad \Phi_k \equiv \left\| \begin{array}{cc} \exp(it_k) & 0 \\ 0 & \exp(-it_k) \end{array} \right\|. \tag{57}$$

An important consequence of identities (55) is the condition of *unimodularity* of matrix M_k

$$\det M_k = |f_k|^2 - |g_k|^2 \equiv 1 \tag{58}$$

which is equivalent to the Wronskian identity (15).

For n pulses shifted in time with respect to the initial instant $t = 0$ by t_k, $k = 1, \ldots, n-1$, we have the relation (here $\mathcal{M}_n$ is the total *transfer matrix*, whereas $M_n^{(0)}$, $M_{n-1}^{(0)}$, ..., $M_1^{(0)} \equiv M_1$ are the matrices describing the action of individual pulses)

$$\begin{pmatrix} a_n \\ b_n \end{pmatrix} = \mathcal{M}_n \begin{pmatrix} a_0 \\ b_0 \end{pmatrix}, \qquad \mathcal{M}_n = \Phi_{n-1}^\dagger M_n^{(0)} \Phi_{n-1} \Phi_{n-2}^\dagger M_{n-1}^{(0)} \cdots \Phi_1^\dagger M_2^{(0)} \Phi_1 M_1. \tag{59}$$

For strictly periodic pulses all matrices $M_k^{(0)}$ coincide with M_1 and $\Phi_k \equiv \Phi_1^k$, so that

$$\mathcal{M}_n = \Phi^{\dagger n} (\Phi M_1)^n, \qquad \Phi \equiv \left\| \begin{array}{cc} \exp(iT) & 0 \\ 0 & \exp(-iT) \end{array} \right\|, \tag{60}$$

where T is the periodicity of pulses. Since $\det(\Phi M_1) = 1$, one can use the well-known formula for the powers of any two-dimensional unimodular matrix S (see, e.g., [110])

$$S^n = U_{n-1}(z) S - U_{n-2}(z) E, \qquad z \equiv \frac{1}{2}\mathrm{Tr} S, \tag{61}$$

where E means the unit matrix and $U_n(z)$ is the Tchebyshev polynomial of the second kind. In the case involved one has $z = \frac{1}{2}\mathrm{Tr}(\Phi M_1) = \mathrm{Re}[f \exp(iT)]$ (with $f \equiv f_1$). An amplification can happen if $|z| > 1$. Then it is convenient to use the parametrization

$$\frac{1}{2}\mathrm{Tr}(\Phi M_1) = \mathrm{Re}[f \exp(iT)] = \cosh(\nu). \tag{62}$$

Using (59)–(62), one can arrive at the following expressions [2]:

$$a_n = f\,\frac{\sinh(n\nu)}{\sinh(\nu)}e^{-iT(n-1)} - \frac{\sinh[(n-1)\nu]}{\sinh(\nu)}e^{-iTn}, \qquad b_n = g\,\frac{\sinh(n\nu)}{\sinh(\nu)}e^{iT(n-1)}. \tag{63}$$

If $\mathrm{Re}[f\exp(iT)] > 1$, then ν is real parameter. If $\mathrm{Re}[f\exp(iT)] < -1$, then one can write $\nu = \tilde{\nu} + i\pi$, where $\tilde{\nu}$ is real. The maximal values of $|\mathrm{Re}[f\exp(iT)]|$ correspond to the cases of strict resonance with

$$T = T_{res} = \frac{1}{2}T_0\,(m - \varphi/\pi)\,, \tag{64}$$

where $f = |f|\exp(i\varphi)$, T_0 is the period of oscillations in the selected field mode and $m = 1, 2, \ldots$ (even values of m correspond to $\mathrm{Re}[f\exp(iT)] > 1$, whereas odd values of m correspond to $\mathrm{Re}[f\exp(iT)] < -1$). Introducing the parameter

$$\delta = \omega_0\,(T - T_{res}) \tag{65}$$

characterizing a detuning from the strict resonance, one can write

$$\cosh(\nu) = |f|(-1)^m\cos(\delta). \tag{66}$$

One can check the fulfillment of the identity $|a_n|^2 - |b_n|^2 \equiv 1$ as a consequence of the initial identity $|f|^2 - |g|^2 \equiv 1$.

Hereafter we confine ourselves to the simplest case of the *strict resonance* (a more general treatment can be found in [32]). Then $|f|^2 = \cosh^2(\nu)$ and $|g|^2 = \sinh^2(\nu)$, so that

$$a_k = \cosh(k\nu)e^{-ikT}, \qquad b_k = \sinh(k\nu)e^{iT(k-1)+i\phi}, \tag{67}$$

$$E_k = \cosh(2k\nu), \qquad \tilde{E}_k = \sinh(2k\nu)e^{i(\phi-T)}, \tag{68}$$

where ϕ is the phase of complex number g.

3.1. *Approximate formulas for effective single-pulse reflection and transmission coefficients*

For small variations of the frequency $\omega(t)$ one can write a solution to equation (2) in the form (generalizing the standard WKB solution and following the approach of [111])

$$\varepsilon_W(t) = \frac{1+\zeta(t)}{\sqrt{\omega(t)}}\exp[-i\Omega(t)] + \rho_1(t)\exp[i\Omega(t)], \qquad \Omega(t) = \int_0^t \omega(\tau)d\tau. \tag{69}$$

Putting function (69) with $\zeta(t) = 0$ into equation (2) one arrives at the inhomogeneous equation

$$\dot{\rho}_1 + \frac{\dot{\omega}}{2\omega}\rho_1 = \frac{i\ddot{\sigma}}{2\omega}\exp[-2i\Omega(t)] + \frac{i\ddot{\rho}_1}{2\omega}, \qquad \sigma(t) \equiv [\omega(t)]^{-1/2}. \tag{70}$$

It is assumed that $|\rho_1| \ll \sigma$ and $\rho_1 = 0$ if $\omega = const$. This means that the second term in (69) describes a weak reflection of the 'wave' going in the negative t-direction (caused by deviations of function $\omega(t)$ from a constant value), but it is not a part of another fundamental solution of the second order differential equation (2). Since function $\rho_1(t)$ is determined by the inhomogeneous term in equation (69), it is reasonable to suppose that the rates of time variations of $\rho_1(t)$ are more or less the same as for the function $\sigma(t)$. Then one can suppose also that the second term in the right-hand side of equation (70) (which contains the second derivative of ρ_1) is much smaller than the first one. Neglecting this term (as was done in [111]), one arrives at the first order differential equation, whose solution, satisfying the condition $\rho_1 \equiv 0$ for $t < 0$ (when $\dot{\omega}(t) \equiv 0$), is as follows,

$$\rho_1(t) = \frac{\rho(t)}{\sqrt{\omega(t)}}, \qquad \rho(t) = \frac{i}{2}\int_0^t \sigma(\tau)\frac{d^2\sigma}{d\tau^2}\exp[-2i\Omega(\tau)]d\tau. \tag{71}$$

If $\omega(t)$ returns to its initial value for $t \geq t_f$, then $\rho(t)$ given by (71) also does not depend on time for $t \geq t_f$. Comparing equations (54), (56) and (71), one finds that for a single pulse

$$g = \rho(t_f)\exp\left[i\Omega(t_f) - i\omega_0 t_f\right], \qquad f = [1 + \zeta(t_f)]\exp\left[i\omega_0 t_f - i\Omega(t_f)\right].$$

Writing

$$\omega(t) = \omega_0[1 + \chi(t)], \qquad |\chi| \leq \chi_0 \ll 1 \tag{72}$$

(the function $\chi(t)$ should not be confused with the noise coefficients χ_{xx} and χ_{pp}), one obtains the following expression for the phase of the inverse single-pulse transmission coefficient f:

$$\varphi = -\omega_0 \int_{t_i}^{t_f} \chi(t)dt \tag{73}$$

(the sign of the phase is corrected here, compared with previous papers [2, 32, 33, 35, 36, 112]). The function $\zeta(t)$ can be found if one calculates the Wronskian for two independent solutions: $\varepsilon(t)$ (69) and its complex conjugate partner. Neglecting higher-order terms, containing squares of function $\zeta(t)$ and products of $\rho(t)$ and $\zeta(t)$ (or their derivatives), and demanding the fulfilment of identity (15), one can obtain the following result:

$$\zeta(t) = \frac{1}{2}|\rho(t)|^2 + i\zeta_I(t), \qquad \zeta_I(t) = \int_0^t \mathrm{Im}(\rho\dot{\rho}^*)d\tau - \int_0^t \frac{\ddot{\sigma}}{2\sigma}d\tau. \tag{74}$$

For $t = t_f$ the second integral in function $\zeta_I(t)$ can be written as (using the integration by parts and taking into account that $\dot{\sigma}(0) = \dot{\sigma}(t_f) = 0$)

$$\int_0^{t_f} \frac{\ddot{\sigma}}{2\sigma}d\tau = \int_0^{t_f} \frac{\dot{\sigma}^2}{2\sigma^2}d\tau.$$

Consequently, the imaginary part of $\zeta(t_f)$ is of the order of χ^2, so that it can be considered as a small (of an order of χ_0^2) correction to the phase φ (73) of the coefficient f. Formula (71) can be simplified for $t = t_f$, if one performs an integration by parts, taking into account that function $\dot{\sigma}$ equals zero for $t = 0$ and $t = t_f$. Neglecting the integral containing the term $\dot{\sigma}^2$ (since it has an order of χ_0^2), one obtains

$$\rho(t_f) \approx -\int_0^{t_f} \frac{\dot{\sigma}}{\sigma}\exp[-2i\Omega(\tau)]d\tau.$$

Making the replacement $\dot{\sigma}/\sigma \approx -\dot{\chi}/2$ (neglecting again the terms of the order of χ_0^2) and making one more integration by parts, one can arrive at the final simple formula (since $\chi(0) = \chi(t_f) = 0$)

$$\rho(t_f) \approx i\omega_0 \int_0^{t_f} \chi(t)e^{-2i\omega_0 t}dt. \tag{75}$$

The phase $\Omega(t)$ is replaced here by the product $\omega_0 t$, because an account of the additional term $\int \chi(\tau)d\tau$ in the phase is equivalent to corrections of the order of χ_0^2 for $\rho(t_f)$, which were neglected in the derivation of (75). Thus we obtain the following elements of matrix $M_1^{(0)}$ (56) satisfying the unimodularity condition with an accuracy of the order of χ_0^2

$$g \approx i\omega_0 \int_0^{t_f} \chi(t)e^{-2i\omega_0 t}dt. \qquad f = \left(1 + \frac{1}{2}|g|^2\right)e^{i\varphi} \tag{76}$$

The phase φ is given by (73). The validity of the approximation used can be checked if one puts function $\rho_1(t)$ given by formula (71) in the right-hand side of equation (70) and integrates this equation again with the additional inhomogeneous part. It can be shown that corrections to the solution arising due to the new term are, indeed, of an order of χ_0^2 at $t = t_f$. ¿From (66) and (76) one obtains the formula

$$\nu^2 = \tilde{\nu}^2 = |g|^2 - \delta^2. \qquad (77)$$

3.2. The contribution of noise

To find the contribution of the kth pulse to the functions $J(T)$ and $\tilde{J}(t)$ one should use the solution $\varepsilon(t)$ in the form $a_k\varepsilon_W(t - t_{k-1}) + b_k\varepsilon_W^*(t - t_{k-1})$ and calculate the integrals defined in (34) from t_{k-1} to t_k. But it is clear from the explicit form of function $\varepsilon_W(t)$ (69) obtained in the preceding section that functions $E(t)$ and $\tilde{E}(t)$ are close to constant values (52) for $t_{k-1} \le t \le t_k$, with small corrections of the order of χ_0. Neglecting these corrections, one can calculate the integrals over the duration of each pulse exactly (since $\gamma(t) = d\Gamma/dt$). Consequently, after n pulses one has

$$J_n = \frac{G}{2}e^{-2\Lambda n}\left(1 - e^{-2\Lambda}\right)\sum_{k=1}^{n} e^{2\Lambda k}E_k, \qquad \tilde{J}_n = \frac{G}{2}e^{-2\Lambda n}\left(1 - e^{-2\Lambda}\right)\sum_{k=1}^{n} e^{2\Lambda k}\tilde{E}_k \qquad (78)$$

where $J_n \equiv J(nT)$, $\tilde{J}_n \equiv \tilde{J}(nT)$ and

$$\Lambda = \int_{t_i}^{t_f} \gamma(\tau)d\tau. \qquad (79)$$

Numerical evaluations confirmed a high accuracy of this approximation.

Using (68) one can obtain the following explicit formulas for the sums in equation (78):

$$J_n = A_n + B_n, \qquad \tilde{J}_n = e^{i(\phi - T)}\left(A_n - B_n\right), \qquad (80)$$

$$A_n = \frac{G\Lambda}{4(\nu + \Lambda)}\left(e^{2n\nu} - e^{-2n\Lambda}\right), \qquad B_n = \frac{G\Lambda}{4(\nu - \Lambda)}\left(e^{-2n\Lambda} - e^{-2n\nu}\right). \qquad (81)$$

I take into account that $\Lambda, \nu \ll 1$. In particular, the difference $1 - \exp(-2\Lambda)$ is replaced by 2Λ.

3.3. Mean number of created quanta and its variance

In view of equations (33), (68) and (80), the mean number of 'noise' quanta created after n pulses in the resonance case is given by a simple formula

$$\mathcal{N}_r(n) = A_n e^{-2n\nu} + B_n e^{2n\nu}, \qquad (82)$$

where $\mathcal{N}_r(n)$ stands for the value $\mathcal{N}_r(t)$ of the function (33) at the moment $t = nT$ (I hope that such a replacement will not lead to a confusion). Adding to the quantity (82) the expression (37), we obtain the total number of quanta created from the initial thermal state:

$$\mathcal{N}^{(th)}(n) = e^{2n\nu}\left(B_n + \frac{G_0}{4}e^{-2\Lambda n}\right) + e^{-2n\nu}\left(A_n + \frac{G_0}{4}e^{-2\Lambda n}\right) - \frac{1}{2}. \qquad (83)$$

Only the first term in the right-hand side of this expression grows with time (the number of pulses n), and for $2n\nu \gg 1$ one arrives at the following asymptotical formula:

$$\mathcal{N}^{(th)}(n) = \frac{1}{4}e^{2n(\nu - \Lambda)}\left(G_0 + \frac{G\Lambda}{\nu - \Lambda}\right) + \mathcal{O}(1). \qquad (84)$$

We see that photons can be generated provided $\nu > \Lambda$.

Combining formulas (46)-(48) with (68) and (80) we obtain the variance of the number of photons in the case of resonance excitation of the initial thermal state:

$$\sigma_N^{(th)}(n) = 2e^{4n\nu}\left(B_n + \frac{G_0}{4}e^{-2\Lambda n}\right)^2 + 2e^{-4n\nu}\left(A_n + \frac{G_0}{4}e^{-2\Lambda n}\right)^2 - \frac{1}{4}. \tag{85}$$

The leading asymptotical term of this expression for $n(\nu-\Lambda) \gg 1$ is

$$\sigma_N^{(th)}(n) \approx \frac{1}{8}e^{4n(\nu-\Lambda)}\left(G_0 + \frac{G\Lambda}{\nu-\Lambda}\right)^2 \approx 2\left[\mathcal{N}^{(th)}(n)\right]^2 \tag{86}$$

This means that the fiels mode goes asymptotically to the so called 'superchaotic' [113, 114] quantum state, whose statistics is essentially different from the statistics of the initial thermal state, characterized by formula (50).

3.4. Photon statistics

Due to the linearity of the Heisenberg–Langevin equations of motion (5), any initial *Gaussian* state remains Gaussian in the process of evolution [105]. In particular, this is true for initial *thermal* states. The photon statistics in the Gaussian states was studied in [105, 115, 116]. If the mean values of the quadrature operators (or electric and magnetic fields) are equal to zero, then this statistics is determined completely by two parameters (we assume here $\hbar = 1$),

$$\tau = \sigma_{xx} + \sigma_{pp} \equiv 1 + 2\mathcal{N}, \qquad \Delta = \sigma_{xx}\sigma_{pp} - \sigma_{px}^2 \equiv 1/(4\mu^2), \tag{87}$$

where $\mu \equiv \mathrm{Tr}(\hat{\rho}^2)$ is the *purity* of the (Gaussian) quantum state of the field mode described by the statistical operator $\hat{\rho}$. The photon distribution function $f(m) \equiv \langle m|\hat{\rho}|m\rangle$ (i.e., the probability to detect m quanta in the state $\hat{\rho}$) can be expressed in terms of the Legendre polynomials [105]:

$$f(m) = \frac{2}{\sqrt{1+2\tau+4\Delta}}\left(\frac{1+4\Delta-2\tau}{1+4\Delta+2\tau}\right)^{m/2} P_m\left(\frac{4\Delta-1}{\sqrt{(4\Delta+1)^2-4\tau^2}}\right). \tag{88}$$

The variance of the photon number distribution for an arbitrary Gaussian state (with zero quadrature mean values) is given by the formula [105]

$$\sigma_N = \frac{1}{2}\tau^2 - \Delta - \frac{1}{4}. \tag{89}$$

Comparing this formula with (83) and (85) we find

$$\Delta(n) = \left(2B_n + \frac{G_0}{2}e^{-2\Lambda n}\right)\left(2A_n + \frac{G_0}{2}e^{-2\Lambda n}\right). \tag{90}$$

In the asymptotical regime $n\nu \gg 1$ we have

$$\Delta(n) \approx \frac{G\Lambda}{4(\nu+\Lambda)}\left(\frac{G\Lambda}{\nu-\Lambda} + G_0\right)e^{2n(\nu-\Lambda)}, \qquad \frac{2\Delta}{\tau} \approx \frac{\Delta}{\mathcal{N}} \approx \frac{G\Lambda}{\nu+\Lambda}. \tag{91}$$

Formula (88) is exact. However, since we are interested in the cases of large numbers of photons created due to the parametric resonance (when $m > 1000$), it is convenient to use its asymptotical forms for $m \gg 1$. Note that the argument of the Legendre polynomial in (88) is always outside the interval $(-1, 1)$, being equal to 1 only for thermal states with $\tau = 2\sqrt{\Delta}$. Therefore it is convenient to use the following asymptotical formula [117]:

$$P_m(\cosh\xi) \approx \left(\frac{\xi}{\sinh\xi}\right)^{1/2} I_0\left([m+1/2]\,\xi\right) \tag{92}$$

(where $I_0(z)$ is the modified Bessel function), because it holds even for complex values of variable ξ, provided $\mathrm{Re}\xi \geq 0$ and $|\mathrm{Im}\xi| < \pi$. The parameter ξ is given by the formula

$$\xi = \ln\left(\frac{4\Delta - 1 + 2\sqrt{\tau^2 - 4\Delta}}{\sqrt{(4\Delta+1)^2 - 4\tau^2}}\right). \tag{93}$$

Under realistic conditions the ratio $2\Delta/\tau$, given by equation (91), is of the order of unity for $\tau \gg 1$ (see section 5). Consequently, parameter ξ is also of the order of unity. In such a case, the function $I_0(x)$ in (92) can be replaced by its asymptotical form $I_0(x) \approx (2\pi x)^{-1/2}\exp(x)$. Then, using (93) to calculate $\sinh(\xi)$, we obtain an approximate formula

$$f(m) \approx \left[\pi(m+1/2)\sqrt{\tau^2 - 4\Delta}\right]^{-1/2}\left(\frac{4\Delta - 1 + 2\sqrt{\tau^2 - 4\Delta}}{4\Delta + 1 + 2\tau}\right)^{m+1/2}. \tag{94}$$

Obviously, one can neglect the term $4\Delta \ll \tau^2$ in the first factor. The fraction in the second factor can be simplified as follows under the same conditions:

$$\frac{4\Delta - 1 + 2\sqrt{\tau^2 - 4\Delta}}{4\Delta + 1 + 2\tau} = \frac{4\Delta - 1 + 2\tau - 4\Delta/\tau + \mathcal{O}(1/\tau)}{4\Delta + 1 + 2\tau} = 1 - \frac{1}{\tau} + \mathcal{O}(1/\tau^2).$$

Replacing $(1-x)^m \approx \exp(-mx)$ for $x \ll 1$, we arrive finally at the following simple formula:

$$f(m) \approx \frac{\exp[-(m+1/2)/\tau]}{\sqrt{\pi\tau(m+1/2)}} \approx \frac{\exp[-(m+1/2)/(2\mathcal{N})]}{\sqrt{2\pi\mathcal{N}(m+1/2)}}. \tag{95}$$

It holds under the conditions $\tau \approx 2\mathcal{N} \gg 1$ and $m \gg 1$. Using the Euler–MacLaurin summation formula, one can verify that the distribution function (95) has the correct normalization with an accuracy $\mathcal{O}(\tau^{-1/2})$:

$$\sum_{m=0}^{\infty} f(m) \approx \int_0^{\infty} f(m)dm + \mathcal{O}[f(0)] \approx \int_0^{\infty}\frac{\exp(-x/\tau)}{\sqrt{\pi\tau x}}dx + \mathcal{O}(\tau^{-1/2}) = 1 + \mathcal{O}(\tau^{-1/2}).$$

With the same accuracy, the moments of the distribution function are given by the formula

$$\langle m^k\rangle \equiv \sum_{m=0}^{\infty} m^k f(m) \approx \int_0^{\infty} x^k \frac{\exp(-x/\tau)}{\sqrt{\pi\tau x}}dx = \tau^k\frac{(2k-1)!!}{2^k} \approx \mathcal{N}^k(2k-1)!!\,. \tag{96}$$

For $k = 2$ equation (96) reproduces the result (86): $\sigma_N = \langle m^2\rangle - \langle m\rangle^2 \approx 2\mathcal{N}^2$.

4. Frequency shift of the cavity mode

We see that the time-dependent relative frequency shift $\chi(t)$ is the main ingredient of the theory of DCE. The problem of shift of resonance frequencies of electromagnetic cavities due to perturbations of geometry and material properties of the walls or internal parts of the cavities was considered for the first time by Müller [118]. It is remarkable that it was also studied (a few years later) by two persons whose names are inseparable forever from the Casimir effect:

Schwinger [119] and Casimir himself [120], although that papers are less known than the others, and the results waited for their use in this field for more than half a century (for different other applications see, e.g., [121–127]). In particular, the cavity perturbation techniques is one of the main methods used in studies on photoexcitation of semiconductors [128, 129].

The main formula for the resonance frequency shift in electromagnetic cavities with ideal boundaries was derived in [118–120]. We consider its special form in the case of nonmagnetic materials, assuming that the magnetic permeability $\mu \equiv 1$, so that the magnetic vectors $\mathbf{B}$ and $\mathbf{H}$ coincide everywhere (we use the Gauss system of units). Considering monochromatic electric and magnetic fields of the form $\mathbf{E}(\mathbf{r},t) = \mathbf{E}_0(\mathbf{r})\exp(-i\omega_0 t)$ and $\mathbf{H}(\mathbf{r},t) = \mathbf{H}_0(\mathbf{r})\exp(-i\omega_0 t)$, we have the following set of equations for complex vector fields $\mathbf{E}_0(\mathbf{r})$ and $\mathbf{H}_0(\mathbf{r})$ in an empty cavity:

$$\mathrm{rot}\,\mathbf{E}_0 = \frac{i\omega_0}{c}\mathbf{H}_0 \tag{97}$$

$$\mathrm{rot}\,\mathbf{H}_0 = -\frac{i\omega_0}{c}\mathbf{E}_0 \tag{98}$$

For a cavity filled in with a medium described by means of a complex dielectric function $\varepsilon(\mathbf{r}) = \varepsilon_1(\mathbf{r}) + i\varepsilon_2(\mathbf{r})$ we have equations without the subscript '0', where ω is the shifted resonance frequency:

$$\mathrm{rot}\,\mathbf{E} = \frac{i\omega}{c}\mathbf{H} \tag{99}$$

$$\mathrm{rot}\,\mathbf{H} = -\frac{i\omega}{c}\,\varepsilon\,\mathbf{E} \tag{100}$$

Now we multiply equation (99) by the complex conjugate function $\mathbf{H}_0^*$, equation (100) by $-\mathbf{E}_0^*$, the complex conjugate equation (97) by $\mathbf{H}$ and the complex conjugate equation (98) by $-\mathbf{E}$. Taking the sum of these four new equations we have

$$\mathbf{H}\,\mathrm{rot}\,\mathbf{E}_0^* - \mathbf{E}_0^*\,\mathrm{rot}\,\mathbf{H} + \mathbf{H}_0^*\,\mathrm{rot}\,\mathbf{E} - \mathbf{E}\,\mathrm{rot}\,\mathbf{H}_0^* = \frac{i}{c}\left[\delta\omega\left(\mathbf{E}\mathbf{E}_0^* + \mathbf{H}\mathbf{H}_0^*\right) + \omega\,\delta\varepsilon\,\mathbf{E}\mathbf{E}_0^*\right], \tag{101}$$

where

$$\delta\omega = \omega - \omega_0, \qquad \delta\varepsilon(\mathbf{r}) = \varepsilon(\mathbf{r}) - 1. \tag{102}$$

Now we integrate both sides of equation (101) over the total volume of the cavity, taking into account the identity

$$\mathrm{div}[\mathbf{a}\times\mathbf{b}] \equiv \mathbf{b}\,\mathrm{rot}\,\mathbf{a} - \mathbf{a}\,\mathrm{rot}\,\mathbf{b}. \tag{103}$$

Due to the Gauss theorem, the volume integral in the left-hand side can be transformed into the surface integral over the total surface of the cavity

$$\int_{walls}\left([\mathbf{H}\times\mathbf{E}_0^*] - [\mathbf{E}\times\mathbf{H}_0^*]\right)d\mathbf{s}. \tag{104}$$

But this integral is equal to zero, because the scalar product of the vector integrand and the vector surface differential $d\,\mathbf{s}$ only depends on tangential components of vectors $\mathbf{E}$ and $\mathbf{E}_0$, which turn into zero on the surface of an ideal cavity. Thus we arrive at the *exact* formula

$$\frac{\delta\omega}{\omega} = -\frac{\int\delta\varepsilon(\mathbf{r})\,\mathbf{E}\mathbf{E}_0^*\,dV}{\int\left(\mathbf{E}\mathbf{E}_0^* + \mathbf{H}\mathbf{H}_0^*\right)dV}. \tag{105}$$

If the magnetic permeability μ is also different from zero, one should add to the numerator of the fraction in the r.h.s. of (105) the integral $\int\delta\mu(\mathbf{r})\,\mathbf{H}\mathbf{H}_0^*\,dV$ (this general formula is called sometimes 'Bethe–Schwinger formula'). A good discussion of formula (105) and its applications can be found in [122].

It is assumed usually that the unperturbed fields $\mathbf{E}_0(\mathbf{r})$ and $\mathbf{H}_0(\mathbf{r})$ are known at all points of the empty cavity. But the perturbed fields $\mathbf{E}(\mathbf{r})$ and $\mathbf{H}(\mathbf{r})$ *are not known*, as a rule (except for the simplest cases when they can be obtained by scaling the fields $\mathbf{E}_0(\mathbf{r})$ and $\mathbf{H}_0(\mathbf{r})$ in *all* points [122]). If the variation $\delta\varepsilon(\mathbf{r})$ is small everywhere, then one can believe that the perturbed field $\mathbf{E}(\mathbf{r})$ is close to $\mathbf{E}_0(\mathbf{r})$ and replace $\mathbf{E}(\mathbf{r})$ by $\mathbf{E}_0(\mathbf{r})$ in the r.h.s. of (105). Besides, if $\delta\omega \ll \omega_0$, then one can replace ω by ω_0 in the denominator of the fraction in the l.h.s., arriving at a simple approximate formula given in many textbooks [130, 131],

$$\frac{\delta\omega}{\omega_0} \approx -\frac{\int \delta\varepsilon(\mathbf{r})\,\mathbf{E}_0\mathbf{E}_0^*\,dV}{\int (\mathbf{E}_0\mathbf{E}_0^* + \mathbf{H}_0\mathbf{H}_0^*)\,dV} = -\frac{\int \delta\varepsilon(\mathbf{r})\,\mathbf{E}_0\mathbf{E}_0^*\,dV}{2\int \mathbf{E}_0\mathbf{E}_0^*\,dV} \tag{106}$$

(the second equality holds due to the well known equality of energies of electric and magnetic fields in ideal resonance cavities, $\int \mathbf{E}_0\mathbf{E}_0^*\,dV = \int \mathbf{H}_0\mathbf{H}_0^*\,dV$).

If the function $\delta\varepsilon(\mathbf{r})$ is different from zero only inside some small volume $\delta V \ll V$ (where V is the total cavity volume), then it is clear that the ratio $\delta\omega/\omega_0$ is small even for big values of this function. Moreover, it seems reasonable to believe that the difference between functions $\mathbf{E}(\mathbf{r})$ and $\mathbf{E}_0(\mathbf{r})$ remains small outside the volume δV [120, 122, 125]. Then assuming that the energy of the field concentrated in the volume δV is still much less than the total energy of the field in the cavity, one can replace in the first approximation $\mathbf{E}(\mathbf{r})$ by $\mathbf{E}_0(\mathbf{r})$ in the denominator of the r.h.s. of (105). But this certainly cannot be done in the numerator if $\delta\varepsilon(\mathbf{r})$ is big (otherwise the frequency shift can become arbitrarily big for $\delta\varepsilon \to \infty$, which is obviously not true).

4.1. An approximation for small field variations: TE mode in a cylindrical cavity

Note that the fields depend not only on the coordinate vector $\mathbf{r}$, but also on the frequency ω which enters the Maxwell equations (99) and (100), so that we should write in fact the electric field as $\mathbf{E}(\mathbf{r};\omega)$. Let us suppose that for some geometrical configurations small variations of frequency are accompanied by *small* variations of the electric field. Then using the approximation

$$\mathbf{E}(\mathbf{r};\omega) \approx \mathbf{E}_0(\mathbf{r};\omega_0) + \left.\frac{\partial \mathbf{E}_0(\mathbf{r};\omega)}{\partial\omega}\right|_{\omega=\omega_0} \delta\omega \tag{107}$$

and putting this expression in the numerator of the fraction in the r.h.s. of equation (105) (but neglecting the term with a derivative over ω in the denominator, where it is not multiplied by the possibly big quantity $\delta\varepsilon$), we arrive at the following generalization of (106):

$$\frac{\delta\omega}{\omega_0} \approx -\frac{\int \delta\varepsilon(\mathbf{r})\,\mathbf{E}_0\mathbf{E}_0^*\,dV}{\int \left[2\mathbf{E}_0\mathbf{E}_0^* + \omega_0\delta\varepsilon(\mathbf{r})\mathbf{E}_0^*\,\partial\mathbf{E}_0(\mathbf{r};\omega)/\partial\omega|_{\omega=\omega_0}\right]dV}. \tag{108}$$

Now we may expect that the frequency shift will remain finite even if $\delta\varepsilon \to \infty$.

Let us show that formula (108) gives a correct result in the case of a thin semiconductor layer on a plane surface of a cylindrical cavity in the case of TE polarization of the electromagnetic field (this case was studied in detail, using different approaches, in [2, 31, 32]). Having in mind that the laser radiation is absorbed in a very thin layer of the width $l = \alpha^{-1} \sim 10^{-4}$–$10^{-2}$ cm (where α is the absorption coefficient), which is much smaller than the characteristic scale of spatial variations of the field $\mathbf{E}_0$ in an empty cavity (the resonance frequency $f = 2.5$ GHz corresponds to the free space wavelength $\lambda = 12$ cm) and that the creation of electron-hole pairs results in a high conductivity without a noticeable change of the real part of dielectric susceptibility, we can assume that the function $\delta\varepsilon(\mathbf{r})$ inside the slab can be approximated as

$$\delta\varepsilon(\mathbf{r}) = \tilde{\varepsilon}_1 + iBg(\mathbf{r}_\perp)\delta(z), \qquad \tilde{\varepsilon}_1 = \varepsilon_1 - 1, \tag{109}$$

where ε_1 is the lattice dielectric constant of a non-excited semiconductor ($\varepsilon_1 \approx 13$ for GaAs), z is the coordinate in the direction perpendicular to the slab surface (which is assumed to be $z = 0$) and $\mathbf{r}_\perp$ is the two-dimensional vector in the surface plane. The function $g(\mathbf{r}_\perp) \leq 1$ takes into account a possible non-uniformity of the laser pulse in the transverse plane, whereas the coefficient B is equal to the maximal total concentration of electron-hole pairs in the slab per unit surface area (we ascribe the value $\mathbf{r}_\perp = \mathbf{0}$ to this point on the surface, so that $g(\mathbf{0}) = 1$):

$$B = \int_0^D \mathrm{Im}\delta\varepsilon(\mathbf{0}, z)dz = \frac{4\pi b_{tot}|e|}{\omega_0}\int_0^D n(\mathbf{0}, z)dz\,, \qquad g(\mathbf{r}_\perp) = \frac{\int_0^D n(\mathbf{r}_\perp, z)dz}{\int_0^D n(\mathbf{0}, z)dz}. \tag{110}$$

Here D is the slab thickness, n is the concentration of electron-hole pairs, $b_{tot} = b_e + b_h$ is the total mobility of carriers related to the pair and e is the electron charge. Then (108) leads to the following formula for the relative frequency shift with respect to the empty cavity:

$$\left.\frac{\delta\omega}{\omega_0}\right|_{empty} \approx -\frac{iB\mu + \varphi_1}{1 + iB\nu}, \tag{111}$$

where

$$\mu = \frac{1}{\mathcal{W}}\int_\Sigma g(\mathbf{r}_\perp)\left|\mathbf{E}_0(\mathbf{r}_\perp, 0)\right|^2 dS, \tag{112}$$

$$\nu = \frac{\omega_0}{\mathcal{W}}\int_\Sigma g(\mathbf{r}_\perp)\mathbf{E}_0^*(\mathbf{r}_\perp, 0;\omega)\left.\frac{\partial\mathbf{E}_0(\mathbf{r}_\perp, 0;\omega)}{\partial\omega}\right|_{\omega=\omega_0} dS, \tag{113}$$

$$\mathcal{W} = 2\int_{cav}\mathbf{E}_0\mathbf{E}_0^*\,dV, \qquad \varphi_1 = \frac{1}{\mathcal{W}}\int_{slab}\tilde{\varepsilon}_1\mathbf{E}_0\mathbf{E}_0^*\,dV. \tag{114}$$

The surface integrals in (112) and (113) are taken over the area Σ of the slab (the plane $z = 0$). The term with φ_1 is neglected in the denominator in equation (111), because it is obviously much smaller than unity for limited values of ε_1.

For a non-illuminated semiconductor slab we have $B = 0$, since we consider the case of a low temperature, when the proper conductivity of the semiconductor can be neglected. In this case the resonance frequency is shifted from the empty cavity value by the quantity $-\omega_0\varphi_1$. Consequently, the relative resonance frequency shift between the cavities with illuminated and non-illuminated semiconductors is

$$\frac{\delta\omega}{\omega_0} \approx \frac{iB\nu\left(\varphi_1 - \mu/\nu\right)}{1 + iB\nu}. \tag{115}$$

The maximal relative frequency shift χ_{max} corresponds to the case $B \to \infty$:

$$\chi_{max} = \varphi_1 - \mu/\nu. \tag{116}$$

On the other hand, taking the same limit in equation (111) we arrive at the frequency shift corresponding to the case when an ideal boundary of an empty cavity is shifted by the distance D inside the cavity: $\chi_{id} = -\mu/\nu$. But the formula for χ_{id} is well known since the papers [118–120]:

$$\chi_{id} = \frac{1}{\mathcal{W}}\int_{slab}\left(|\mathbf{H}_0|^2 - |\mathbf{E}_0|^2\right)dV\,. \tag{117}$$

Consequently,

$$\chi_{max} = \frac{1}{\mathcal{W}}\int_{slab}\left(|\mathbf{H}_0|^2 + (\varepsilon_1 - 2)|\mathbf{E}_0|^2\right)dV \tag{118}$$

and we can exclude the coefficient ν, expressing it through μ and χ_{id}, which can be calculated more easily. Then the actual frequency shift can be written as

$$\frac{\delta\omega}{\omega_0} \approx \frac{A\chi_{max}}{A+i} = \frac{A(A-i)}{A^2+1}\chi_{max}, \tag{119}$$

where

$$A = B\mu/\chi_{id} = B\frac{\int_\Sigma g(\mathbf{r}_\perp)\left|\mathbf{E}_0(\mathbf{r}_\perp,0)\right|^2 dS}{\int_{slab}\left(|\mathbf{H}_0|^2 - |\mathbf{E}_0|^2\right) dV}. \tag{120}$$

The structure of equation (119) is exactly the same as was obtained in [2,31,32]) through finding solutions of the Helmholtz equation in different parts of the cavity and their adjustment on the surface of the slab. Moreover, it can be shown that formulas (118) and (120) give the same values for the coefficients χ_{max} and A as in the cited papers in the case of TE-mode (when the electric field is parallel to the slab surface, so that $|\mathbf{H}_0| \gg |\mathbf{E}_0|$ inside the slab). Formula (119) was verified in an experiment [132].

5. Evaluations of the photon generation rate

The time dependence of parameters $A(t)$ (120) and $B(t)$ (110) is determined by the time dependence of the integral concentration of carriers created by the laser pulse $\int n(z,t)dz$ over the semiconductor slab. It can be found from equations which take into account, besides the photo-absorption, the effect of diffusion and different recombination processes. In the simplest case we have [33]

$$\partial n/\partial t = \nabla\cdot(Y\nabla n) + (\alpha\zeta/E_s)I(t)e^{-\alpha z} - \beta_1 n. \tag{121}$$

Here Y is the coefficient of ambipolar diffusion, α is the absorption coefficient of the laser radiation inside the layer, E_s is the energy gap of the semiconductor (which is close to the energy of laser photons), $I(t)$ is time-dependent intensity of the laser pulse which enters the slab (it can be less than the intensity of the pulse outside the slab, because the reflection coefficient from the semiconductor surface can be rather big, due to the big value of the dielectric constant $\varepsilon_1 \sim 10$; however, the reflection can be diminished if some quarter-wavelength film is put on the surface), $\zeta \le 1$ is the efficiency of the photo-electron conversion and β_1 is the trap-assisted recombination coefficient. We have disregarded nonlinear terms $-\beta_3 n^3 - \beta_2 n^2$ in the right-hand side of (121), because for modeling the DCE one needs very small recombination times, of the order of $T_r \equiv \beta_1^{-1} \sim 20 \div 30\,\text{ps}$. Under these conditions, the contribution of neglected terms is several orders of magnitude smaller than that of the term $\beta_1 n$ [33]. In the most general case, one should use the function $I(t - z/v)$ instead of $I(t)$, where v is the group velocity. But for materials with high absorption coefficients the coordinate dependence can be neglected, if the duration of each pulse is of the order of a few picoseconds. Since equation (121) is linear, it can be solved exactly. The details of calculations can be found in [33, 36].

Here we give the most important results of our previous studies. The maximal photon generation rate is expected under the following conditions.

1) An ideal surface of the semiconductor slab, when the effect of surface recombination can be neglected, so that the boundary condition to equation (121) is $\partial n/\partial z|_{z=0} = 0$.

2) A high absorption coefficient $\alpha \gg D^{-1}$, where D is the slab thickness, so that the presence of the second boundary does not affect the carrier distribution near the irradiated surface.

3) A very short duration of the laser pulse: much less than the recombination time T_r.

4) A uniform illumination of the semiconductor slab.

Under these conditions, the time-dependent function $A(t)$ in equation (119) has a simple form

$$A(\tau) = A_0 e^{-\tau/Z}, \tag{122}$$

where the following dimensionless variables and parameters are introduced:

$$\tau = \omega_0 t \qquad Z = \frac{\omega_0}{\beta_1} = \frac{2\pi T_r}{T_0} \qquad A_0 = \frac{8\pi^2|eb|\zeta WD}{c^2 T_0 E_s S}. \tag{123}$$

Here W is the total energy of the single laser pulse, S is the surface area of the semiconductor slab, T_0 is the period of oscillations in the chosen field mode, b is the effective mobility of a pair of carriers (electrons and holes) and e is the electron charge (we use the Gauss system of units).

According to equation (84), the rate of the photon generation is determined by the difference $\nu - \Lambda$. In the case of the fundamental TE-mode of a rectangular cavity with the longitudinal length L and the transverse length B, this difference can be expressed as follows [2, 33],

$$\nu - \Lambda = \eta^3 \Delta\, F, \qquad \Delta = 2D/\lambda_0, \qquad \eta = \lambda_0/(2L), \qquad F = \tilde{\nu} - \tilde{\Lambda}, \tag{124}$$

where $\lambda_0 = cT_0$ is the wave length of the excited field mode.

The coefficients $\tilde{\nu}$ and $\tilde{\Lambda}$ are given by the expressions following from (76), (79) and (119):

$$\tilde{\nu} = Z\left|\int_0^\infty \frac{dx\, e^{-2iZx}\,[A_0\exp(-x)]^2}{1+[A_0\exp(-x)]^2}\right|, \tag{125}$$

$$\tilde{\Lambda} = Z\int_0^\infty \frac{dx\, A_0\exp(-x)}{1+[A_0\exp(-x)]^2} = Z\tan^{-1}(A_0). \tag{126}$$

The integral in (125) can be calculated analytically in terms of the Gauss hypergeometric function [36]. The behavior of function $F(A_0, Z)$ was studied in detail in [2]. It was shown that the generation of quanta is impossible ($F < 0$) if $Z > 0.54$ or $A_0 < 4$. For moderate values of parameter A_0, the maximal values of F are achieved for $Z \approx 0.3$, which is an optimal value. The corresponding recombination time is close to $T_r^{opt} = T_0/(2\pi^2)$. For $T_0 = 400$ ps (or $f_0 = 2.5$ GHz) this means that $T_r^{opt} \approx 20$ ps, and in no case T_r should exceed 35 ps. The value $T_r^{opt} \approx 20$ ps is quite realistic from the point of view of the available technology (and it was confirmed in recent preliminary measurements [5]).

The optimal value of the parameter A_0 can be found in the following way. According to (123), this parameter is proportional to the energy of the laser pulse. On the other hand, if we fix the number of photons which can be created after n pulses, then formula (84) shows that $n \approx const/F$. Consequently, the function $B_0(A_0, Z) = A_0/F(Z, A_0)$ is proportional to the *total energy* of all necessary laser pulses. Choosing for each value of A_0 the optimal value of parameter Z, we obtained [2] the following best choice of parameters (marked with an asterisk), corresponding to the minimal total energy of all pulses:

$$A_{0*} = 11.3, \qquad Z_* = 0.29, \qquad F_* = 0.18, \qquad \tilde{\nu}_* = 0.61, \qquad \tilde{\Lambda}_* = 0.43. \tag{127}$$

Then formula (84) with $G = G_0 = 1$ (i.e., for zero absolute temperature of the initial field state and the cavity walls) and $n\Lambda \gg 1$ can be written as

$$\mathcal{N}_n \approx \frac{\tilde{\nu}}{4F}\exp\left(2\eta^3 F\Delta\, n\right), \qquad \mathcal{N}_{*n} \approx 0.85\,\exp\left(0.36\,\eta^3\Delta\, n\right). \tag{128}$$

A specific property of highly doped semiconductors (which are necessary to achieve a small recombination time) is a low mobility of carriers at zero absolute temperature (because the scattering on charged impurities results in the contribution to the mobility $b \sim \Theta^{3/2}$ for $\Theta \to 0$). Therefore it seems that an optimal strategy could be to prepare somehow only the resonance field mode in the vacuum state, while maintaining the walls at some finite temperature Θ chosen according to the following requirements: a high mobility of carriers, an optimal recombination time and a high quality factor of the cavity. Note that for the frequency $\omega_0/(2\pi) = 2.5\,\text{GHz}$ the temperature gain factor G defined in (28) has the numerical value $G \approx 17\Theta$ if $\Theta > 1$ is expressed in Kelvins. Then formula (84) with $G \gg G_0 = 1$ becomes

$$\mathcal{N}_n \approx \frac{G\tilde{\Lambda}}{4F} \exp\left(2\eta^3 F \Delta\, n\right), \qquad \mathcal{N}_{*n} \approx 10\,\Theta \exp\left(0.36\,\eta^3 \Delta\, n\right) \tag{129}$$

and the ratio of the number of created photons to the mean number of thermal photons at temperature Θ is

$$\frac{\mathcal{N}_{*n}}{\langle n \rangle_{th}(\Theta)} \approx 1.2 \exp\left(0.36\,\eta^3 \Delta\, n\right). \tag{130}$$

An idea to use the initial thermal state of the *field* to facilitate the observation of the dynamical Casimir effect was put forward in [18]. We emphasize that here we assume that the initial field state is *vacuum*, and the amplification of the photon generation rate is due to the finite temperature of the *walls*.

An optimal value of the geometrical factor η can be determined as follows [32, 36]. The resonance wavelength λ, corresponding to the mode TE_{101} with the lowest eigenfrequency of the rectangular cavity, is related to the cavity length L and the biggest transverse dimension B as $\lambda = 2LB/\sqrt{L^2 + B^2}$ (the third dimension should be taken much smaller than L and B in order to diminish the illuminated surface of the slab; this choice excludes automatically the TM mode for the lowest eigenfrequency). Consequently, $B = \lambda/(2\sqrt{1-\eta^2})$. For the fixed values of parameter A_0 and the smallest transverse dimension, the energy of the pulse is proportional to the surface area, i.e., B, whereas the necessary number of pulses depends on L as $\eta^{-3} \sim L^3$. Therefore the total energy is proportional to the product BL^3. Minimizing this product for the fixed value of λ, which is equivalent to maximization of the function $f(\eta) = \eta^3\sqrt{1-\eta^2}$, we find the optimal value $\eta_{opt} = \sqrt{3}/2 = 0.866$, which corresponds to $L = \lambda/\sqrt{3} \approx 7\,\text{cm}$ and $B = \lambda = 12\,\text{cm}$ (for $\omega_0/(2\pi) = 2.5\,\text{GHz}$). However, since it can be difficult to illuminate such a long plate, we should consider other reasonable values of η. Fortunately, the profile of function $f(\eta)$ is rather flat in the vicinity of point η_{opt}. For this reason, the main requirement is to avoid resonances with other cavity eigenfrequencies for the given modulation depth Δ (we considered $D = 2\,\text{mm}$ and $\Delta = 1/30$). These resonances can happen due to the highly anharmonical profile of the frequency shift function (119), and their presence can diminish significantly the photon generation rate in the fundamental mode under consideration [99, 133]. The analysis shows that the best choice, permitting to avoid at least 4 accidental resonances, corresponds to the values of η in the interval between 2/3 and 3/4. Taking $\eta = 3/4$ (i.e., $L = 8\,\text{cm}$ and $B = 9\,\text{cm}$) we have $\eta^3 = 0.42$ and $f(\eta_1) = 0.28$, which is not very far from the maximal value $f(\eta_{opt}) = 0.325$.

Taking $\Delta = 1/30$, $\eta = 3/4$ and using (128), we evaluate that 10^4 photons can be created from vacuum at zero temperature of the cavity walls after about 2000 laser pulses. This number can be reduced to 500 if the walls are maintained at the liquid nitrogen temperature $\Theta = 77\,\text{K}$, according to formula (129) (in this case $\langle n \rangle_{th}(\Theta) \approx 650$).

If the third dimension can be reduced to 2.5 mm [134], then taking $T_r = 20\,\text{ps}$, $b = 0.7\,\text{m}^2\text{V}^{-1}\text{s}^{-1} \approx 2.1 \times 10^6$ CGS units and $E_g = 1.4\,\text{eV}$ (GaAs) we find that the optimal value of parameter $A_0 \approx 10$ can be achieved for the energy of a single laser pulse about 0.1 mJ. In such a case, the total energy of 500 pulses must be about 50 mJ. It can be diminished if materials with a higher mobility can be found.

What can happen if conditions 1)–4) are not satisfied completely? This can be taken into account by using the function $A(t)$ in the form

$$A(\tau) = \exp(-\tau/Z)\mathcal{J}(\tau) \tag{131}$$

with some 'formfactor' $\mathcal{J}(\tau)$. In this general case integrals generalizing (125) and (126) can be calculated only numerically. For example, the general boundary condition to equation (121) in the case of non-ideal surface is $Y\partial n/\partial z|_{z=0} = Rn(0)$, where R is the surface recombination velocity. This case was analyzed in detail in [33]. The corresponding formfactor has the form

$$\mathcal{J}_g(\tau,h) = \frac{1}{g-1}\left[g e^{h\tau}\mathrm{Erfc}\left(\sqrt{h\tau}\right) - e^{h\tau g^2}\mathrm{Erfc}\left(g\sqrt{h\tau}\right)\right], \qquad g = R/(\alpha Y), \quad h = \alpha^2 Y/\omega_0,$$

where $\mathrm{Erfc}(x)$ is the complementary error function. It was shown that for nonzero surface recombination velocity an increase of parameter h (or the absorption coefficient) diminishes the value of F. Also, an increase of h diminishes the maximal possible values of the recombination time (parameter Z) for which $F > 0$ (if $h \gg 1$). Usual mechanically polished semiconductor surfaces have the surface recombination velocity $R \sim 10^6 \div 10^7\,\mathrm{cm/s}$ [135]. Using the Einstein relation $Y = k_B\Theta|b/e|$ we estimated the diffusion coefficient at $\Theta = 4\,\mathrm{K}$ as $Y \sim 3\,\mathrm{cm}^2/\mathrm{s}$ for $b \sim 1\,\mathrm{m}^2\mathrm{V}^{-1}\mathrm{s}^{-1}$. Thus for usual surfaces the parameter g is greater than unity, moreover, it can be much greater than unity for the values of absorption coefficient smaller than $10^6\,\mathrm{cm}^{-1}$. Therefore in order to have sufficiently big values of the amplification factor F, parameter h must be smaller than unity (by an order of magnitude). For the frequency 2.5 GHz this means that the absorption coefficient α should not exceed the value of an order of $10^5\,\mathrm{cm}^{-1}$. Using some special procedures (e.g., etching the surface) one can reduce the surface recombination velocity to the values of the order of $10^2 \div 10^3\,\mathrm{cm/s}$ [135]. In such a case one has $g < 1$ for $\alpha > 10^2\,\mathrm{cm}^{-1}$, so that the influence of parameters g and h becomes insignificant. One can also diminish parameter g by means of increasing the diffusion coefficient (or the mobility). But in this case parameter h *increases* in the same proportion. It was shown in [33] that one should avoid big values of α and R, trying to maintain the parameter $M = gh = R\alpha/\omega_0$ in the region $M < 1$. For instance, preferable values of α are less than $10^3\,\mathrm{cm}^{-1}$ for 'bad' surfaces with $R \sim 10^7\,\mathrm{cm/s}$. Or one should keep $R < 10^4\,\mathrm{cm/s}$ if $\alpha \sim 10^6\,\mathrm{cm}^{-1}$.

Another reason for introducing a formfactor in equation (131) is a finite duration of laser pulses or non-uniformity of illumination. This subject was studied in [36], where different functions $\mathcal{J}(t)$ were considered. It was concluded that the non-uniformity of illumination or temporal spread of laser pulses will not deteriorate significantly the rate of photon generation if the recombination time is in the interval 10-20 ps.

In the derivation of the formulas for the frequency shift in section 4 we assumed that only the imaginary part of the complex dielectric function is changed due to the generation of electron-hole pairs in the semiconductor slab. A change of the real part can be taken into account, if one replaces the factor B in equations (109)–(120) by $B(1-i\mu)$, where $\mu = \mathrm{Re}(\Delta\varepsilon)/\mathrm{Im}(\Delta\varepsilon)$ and $\Delta\varepsilon$ is the total change of the dielectric function [35, 36]. In the case of the Drude model we have

$$b = e\tau_c/m_{ef}, \qquad \mu = \omega_0\tau_c = \omega_0 b m_{ef}/|e|,$$

where τ_c is the mean time between collisions and m_{ef} an effective mass of carriers. For realistic values $b \sim 1\,\mathrm{m}^2\mathrm{V}^{-1}\mathrm{s}^{-1}$, $m_{ef} \sim m_0$ (the mass of free electron) and $\omega_0/(2\pi) = 2.5\,\mathrm{GHz}$ we obtain $\mu \sim 0.1 \ll 1$. It was shown in [35] that the amplification coefficient $F = \tilde{\nu} - \tilde{\Lambda}$ *decreases* as function of parameter μ in the interval $0 < \mu < 1$, although it can become significantly bigger than $F(0)$ for $\mu \gg 1$. The latter case corresponds to the plasma model considered in [28–30] (see also a recent paper [136]), where the role of dissipation (the imaginary part of dielectric function) was neglected. However, at low frequencies (of an order of a few GHz) this can happen only in materials with extremely high mobilities of carriers, $b \gg |e|/(m_{ef}\omega_0) \sim 10\,\mathrm{m}^2\mathrm{V}^{-1}\mathrm{s}^{-1}$, which probably do not exist.

6. Summary

I have described several theoretical achievements of the past few years, which are connected with planned experiments on the observation of the dynamical Casimir effect. One of them is the construction of a consistent model of a damped nonstationary quantum oscillator with arbitrary time-dependent frequency and damping coefficients, based on the generalization of the Senitzky–Schwinger–Haus–Lax noise operator approach. New results given in this paper include formulas for the distribution function of photons created in a resonantly excited field mode and the fluctuations of the number of photons.

Another achievement consists in a generalization of the Müller–Schwinger–Bethe–Casimir formula for the shift of the complex resonance frequency of a cavity to the case of strong variations of the complex dielectric permeability inside a thin inhomogeneous slab. The new formula can be applied to thin semiconductor slabs irradiated by laser pulses of arbitrary shapes in space and time. The theoretical analysis and concrete numerical evaluations performed in this and previous papers show that an experimental demonstration of the dynamical Casimir effect is a quite feasible task which can be achieved in the nearest future.

Acknowledgments

A partial support of the Brazilian agency CNPq is acknowledged.

References

[1] Dodonov V V 2001 Nonstationary Casimir effect and analytical solutions for quantum fields in cavities with moving boundaries *Modern Nonlinear Optics* (*Advances in Chemical Physics Series* vol 119, part 1) ed M W Evans (New York: Wiley) pp 309–94 (*Preprint* arXiv: quant-ph/0106081)

[2] Dodonov V V and Dodonov A V 2005 Quantum harmonic oscillator and nonstationary Casimir effect *J. Russ. Laser Res.* **26** 445–83

[3] Barton G, Dodonov V V and Man'ko V I (guest editors) 2005 Special issue on Nonstationary Casimir Effect and Quantum Systems with Moving Boundaries *J. Opt. B: Quantum Semiclass. Opt.* **7** S1–126

[4] Braggio C, Bressi G, Carugno G, Del Noce C, Galeazzi G, Lombardi A, Palmieri A, Ruoso G and Zanello D 2005 A novel experimental approach for the detection of the dynamic Casimir effect *Europhys. Lett.* **70** 754–60

[5] Agnesi A, Braggio C, Bressi G, Carugno G, G Galeazzi G, Pirzio F, Reali G, Ruoso G and Zanello D 2008 MIR status report: an experiment for the measurement of the dynamical Casimir effect *J. Phys. A: Math. Theor.* **41** 164024

[6] Kim W-J, Brownell J H and Onofrio R 2006 Detectability of dissipative motion in quantum vacuum via superradiance *Phys. Rev. Lett.* **96** 200402

[7] Brownell J H, Kim W J and Onofrio R 2008 Modelling superradiant amplification of Casimir photons in very low dissipation cavities *J. Phys. A: Math. Theor.* **41** 164026

[8] Segev E, Abdo B, Shtempluck O, Buks E and Yurke B 2007 Prospects of employing superconducting stripline resonators for studying the dynamical Casimir effect experimentally *Phys. Lett.* A **370** 202–6

[9] Dodonov V V 1995 Photon creation and excitation of a detector in a cavity with a resonantly vibrating wall *Phys. Lett.* A **207** 126–32

[10] Dodonov V V and Klimov A B 1996 Generation and detection of photons in a cavity with a resonantly oscillating boundary *Phys. Rev.* A **53** 2664–82

[11] Moore G T 1970 Quantum theory of the electromagnetic field in a variable-length one-dimensional cavity *J. Math. Phys.* **11** 2679–91

[12] Askar'yan G A 1962 Interaction between laser radiation and oscillating surfaces *Sov. Phys. – JETP* **15** 1161–2

[13] Rivlin L A 1979 Note on parametric birth of photons in model problems of cosmology *Sov. J. Quant. Electron.* **9** 1320–1

[14] Sarkar S 1988 Moving mirrors and nonclassical light *Photons and Quantum Fluctuations* ed E R Pike and H Walther (Bristol: Hilger) pp 151–72

[15] Lambrecht A, Jaekel M T and Reynaud S 1996 Motion induced radiation from a vibrating cavity *Phys. Rev. Lett.* **77** 615–18

[16] Dodonov V V 1998 Resonance photon generation in a vibrating cavity *J. Phys. A: Math. Gen.* **31** 9835–54

[17] Dalvit D A R and Mazzitelli F D 1999 Creation of photons in an oscillating cavity with two moving mirrors *Phys. Rev.* A **59** 3049–59
[18] Plunien G, Schützhold R and Soff G 2000 Dynamical Casimir effect at finite temperature *Phys. Rev. Lett.* **84** 1882–5
[19] Husimi K 1953 Miscellanea in elementary quantum mechanics. II *Prog. Theor. Phys.* **9** 381–402
[20] Yablonovitch E 1989 Accelerating reference frame for electromagnetic waves in a rapidly growing plasma: Unruh–Davies–Fulling–De Witt radiation and the nonadiabatic Casimir effect *Phys. Rev. Lett.* **62** 1742–5
[21] Man'ko V I 1992 Time-dependent invariants and nonclassical light *Quantum Measurements in Optics* (*NATO ASI Series B: Physics* vol 282) ed P Tombesi and D F Walls (New York: Plenum) pp 239–46
[22] Grishchuk L P, Haus H A and Bergman K 1992 Generation of squeezed radiation from vacuum in the cosmos and the laboratory *Phys. Rev.* D **46** 1440–9
[23] Okushima T and Shimizu A 1995 Photon emission from a false vacuum of semiconductors *Japan. J. Appl. Phys.* **34** 4508–10
[24] Lozovik Y E, Tsvetus V G and Vinogradov E A 1995 Parametric excitation of vacuum by use of femtosecond laser pulses *Phys. Scripta* **52** 184–90
[25] Braggio C, Bressi G, Carugno G, Lombardi A, Palmieri A, Ruoso G and Zanello D 2004 Semiconductor microwave mirror for a measurement of the dynamical Casimir effect *Rev. Sci. Instrum.* **75** 4967–70
[26] Cirone M, Rzążewski K and Mostowski J 1997 Photon generation by time-dependent dielectric: A soluble model *Phys. Rev.* A **55** 62–6
[27] Cirone M and Rzążewski K 1999 Electromagnetic radiation in a cavity with a time-dependent mirror *Phys. Rev.* A **60** 886–92
[28] Crocce M, Dalvit D A R, Lombardo F C and Mazzitelli F D 2004 Model for resonant photon creation in a cavity with time-dependent conductivity *Phys. Rev.* A **70** 033811
[29] Uhlmann M, Plunien G, Schützhold R and Soff G 2004 Resonant cavity photon creation via the dynamical Casimir effect *Phys. Rev. Lett.* **93** 193601
[30] Barton G and Calogeracos A 1995 On the quantum electrodynamics of a dispersive mirror. I. Mass shifts, radiation, and radiative reaction *Ann. Phys.* (NY) **238** 227–67
[31] Dodonov V V 2005 Time-dependent quantum damped oscillator with 'minimal noise': application to the nonstationary Casimir effect in non-ideal cavities *J. Opt. B: Quantum Semiclassical Opt.* **7** S445–51
[32] Dodonov V V and Dodonov A V 2006 Nonstationary Casimir effect in a cavity with periodical time-dependent conductivity of a semiconductor mirror *J. Phys. A: Math. Gen.* **39** 6271–81
[33] Dodonov V V and Dodonov A V 2006 QED effects in a cavity with time-dependent thin semiconductor slab excited by laser pulses, *J. Phys. B: At. Mol. Opt. Phys.* **39** S749–66
[34] Dodonov V V and Dodonov A V 2006 The Heisenberg–Langevin model of a quantum damped harmonic oscillator with time-dependent frequency and damping coefficients *J. Russ. Laser Res.* **27** 379–88
[35] Dodonov V V and Dodonov A V 2006 Generation of photons from vacuum in a cavity with time-dependent eigenfrequency and damping coefficient *Acta Phys. Hung. B – Quantum Electronics* **26** 29–36
[36] Dodonov V V and Dodonov A V 2008 Theory of the dynamical Casimir effect in nonideal cavities with time-dependent parameters *J. Phys.: Conference Series* **99** 012006
[37] Montazeri M and Miri M 2005 Motion-induced radiation from a dynamically deforming mirror: Neumann boundary condition *Phys. Rev.* A **71** 063814
[38] Fosco C D, Lombardo F C and Mazzitelli F D 2007 Quantum dissipative effects in moving mirrors: A functional approach *Phys. Rev.* D **76** 085007
[39] Alves D T, Granhen E R and Lima M G 2008 Quantum radiation force on a moving mirror with Dirichlet and Neumann boundary conditions for a vacuum, finite temperature, and a coherent state *Phys. Rev.* D **77** 125001
[40] Montazeri M and Miri M 2008 Radiation from a dynamically deforming mirror immersed in the electromagnetic vacuum *Phys. Rev.* A **77** 053815
[41] Dalvit D A R, Mazzitelli F D and Millan X O 2006 The dynamical Casimir effect for different geometries *J. Phys. A: Math. Gen.* **39** 6261–70
[42] Alves D T, Farina C and Granhen E R 2006 Dynamical Casimir effect in a resonant cavity with mixed boundary conditions *Phys. Rev.* A **73** 063818
[43] Alves D T and Granhen E R 2008 Energy density and particle creation inside an oscillating cavity with mixed boundary conditions *Phys. Rev.* A **77** 015808
[44] Sarabadani J and Miri M 2006 Mechanical response of the quantum vacuum to dynamic deformations of a cavity *Phys. Rev.* A **74** 023801
[45] Sarabadani J and Miri M 2007 Motion-induced radiation in a cavity with conducting and permeable plates *Phys. Rev.* A **75** 055802
[46] Mintz B, Farina C, Neto P A M and Rodrigues R B 2006 Casimir forces for moving boundaries with Robin

conditions *J. Phys. A: Math. Gen.* **39** 6559–65
[47] Farina C 2006 The Casimir effect: Some aspects *Braz. J. Phys.* **36** 1137–49
[48] Wegrzyn P 2006 An optical approach to the dynamical Casimir effect *J. Phys. B: At. Mol. Opt. Phys.* **39** 4895–4903
[49] Wegrzyn P 2007 Exact closed-form analytical solutions for vibrating cavities *J. Phys. B: At. Mol. Opt. Phys.* **40** 2621–40
[50] Fedotov A M, Lozovik Y E, Narozhny N B and Petrosyan A N 2006 Dynamical Casimir effect in a one-dimensional uniformly contracting cavity *Phys. Rev.* A **74** 013806
[51] Fedotov A M, Narozhny N B, Petrosyan A N and Lozovik Y E 2007 On the dynamical Casimir effect in a one-dimensional contracting cavity *Laser Phys.* **17** 310–5
[52] Crocce M, Dalvit D A R, Lombardo F C and Mazzitelli F D 2005 Hertz potentials approach to the dynamical Casimir effect in cylindrical cavities of arbitrary section *J. Opt. B: Quantum Semiclassical Opt.* **7** S32–9
[53] Mazzitelli F D and Millan X O 2006 Photon creation in a spherical oscillating cavity *Phys. Rev.* A **73** 063829
[54] Pascoal F, Céleri L C, Mizrahi S S and Moussa M H Y 2008 Dynamical Casimir effect for a massless scalar field between two concentric spherical shells *Phys. Rev.* A **78** 032521
[55] Setare M R and Dinani H T 2008 Particle creation in oscillating cavities with cubic and cylindrical geometry *Acta Phys. Polon.* B **39** 771–80
[56] Yuce C and Ozcakmakli Z 2008 The dynamical Casimir effect for two oscillating mirrors in 3D *J. Phys. A: Math. Theor.* **41** 265401
[57] Haro J 2006 Pair production in a cavity with moving boundaries *J. Phys. A: Math. Gen.* **39** 6433–40
[58] Haro J and Elizalde E 2006 Hamiltonian approach to the dynamical Casimir effect *Phys. Rev. Lett.* **97** 130401
[59] Haro J 2007 Dynamical Casimir effect for scalar fields I (particle creation) *Int. J. Theor. Phys.* **46** 1003–19
[60] Haro J 2007 Dynamical casimir effect for scalar fields II (energy calculation) *Int. J. Theor. Phys.* **46** 951–71
[61] Haro J and Elizalde E 2007 Physically sound Hamiltonian formulation of the dynamical Casimir effect *Phys. Rev.* D **76** 065001
[62] Haro J and Elizalde E 2008 Dynamical Casimir effect and the black body spectrum *J. Phys. A: Math. Theor.* **41** 032002
[63] Haro J and Elizalde E 2008 Black hole collapse simulated by vacuum fluctuations with a moving semitransparent mirror *Phys. Rev.* D **77** 045011
[64] Elizalde E 2008 Dynamical Casimir effect with semi-transparent mirrors, and cosmology *J. Phys. A: Math. Theor.* **41** 164061
[65] Durrer R and Ruser M 2007 Dynamical Casimir effect in braneworlds *Phys. Rev. Lett.* **99** 071601
[66] Ruser M and Durrer R 2007 Dynamical Casimir effect for gravitons in bouncing braneworlds *Phys. Rev.* D **76** 104014
[67] Ruser M 2005 Vibrating cavities: a numerical approach *J. Opt. B: Quantum Semiclassical Opt.* **7** S100–15
[68] Ruser M 2006 Numerical approach to the dynamical Casimir effect *J. Phys. A: Math. Gen.* **39** 6711–23
[69] Ruser M 2006 Numerical investigation of photon creation in a three-dimensional resonantly vibrating cavity: Transverse electric modes *Phys. Rev.* A **73** 043811
[70] Rosu H C 2005 Quantum vacuum radiation and detection proposals *Int. J. Theor. Phys.* **44** 493–528
[71] Sorge F 2006 Dynamical Casimir Effect in a kicked box *Int. J. Mod. Phys.* A **21** 6173–82
[72] Mendonça J T, Brodin G and Marklund M 2008 Vacuum effects in a vibrating cavity: time refraction, dynamical Casimir effect, and effective Unruh acceleration *Phys. Lett.* A **372** 5621–4
[73] Srivastava Y N, Widom A, Sivasubramanian S and Pradeep Ganesh M 2006 Dynamical Casimir effect instabilities *Phys. Rev.* A **74** 032101
[74] Shvartsburg A B 2005 Optics of nonstationary media *Physics–Uspekhi* **48** 797–823
[75] Mendonça J T and Guerreiro A 2005 Time refraction and the quantum properties of vacuum *Phys. Rev.* A **72** 063805
[76] Tozzo C, Krämer M and Dalfovo F 2005 Stability diagram and growth rate of parametric resonances in Bose-Einstein condensates in one-dimensional optical lattices *Phys. Rev.* A **72** 023613
[77] Ciuti C, Bastard G and Carusotto I 2005 Quantum vacuum properties of the intersubband cavity polariton field *Phys. Rev.* B **72** 115303
[78] Lawandy N M 2006 Scattering of vacuum states by dynamic plasmon singularities: generating photons from vacuum *Optics Lett.* **31** 3650–2
[79] De Liberato S, Ciuti C and Carusotto I 2007 Quantum vacuum radiation spectra from a semiconductor microcavity with a time-modulated vacuum Rabi frequency *Phys. Rev. Lett.* **98** 103602
[80] Takashima K, Hatakenaka N, Kurihara S and Zeilinger A 2008 Nonstationary boundary effect for a quantum flux in superconducting nanocircuits *J. Phys. A: Math. Theor.* **41** 164036
[81] Bialynicki-Birula I and Bialynicka-Birula Z 2008 Electromagnetic radiation by gravitating bodies *Phys. Rev.* A **77** 052103

[82] Carusotto I, Antezza M, Bariani F, De Liberato S and Ciuti C 2008 Optical properties of atomic Mott insulators: From slow light to dynamical Casimir effects *Phys. Rev.* A **77** 063621
[83] Saito H and Hyuga H 2008 Dynamical Casimir effect for magnons in a spinor Bose-Einstein condensate *Phys. Rev.* A **78** 033605
[84] Bialynicki-Birula I and Bialynicka-Birula Z 2008 Dynamical Casimir effect in oscillating media *Phys. Rev.* A **78** 042109
[85] Zagoskin A M, Il'ichev E, McCutcheon M W, Young J F and Nori F 2008 Controlled generation of squeezed states of microwave radiation in a superconducting resonant circuit *Phys. Rev. Lett.* **101** 253602
[86] Sarkisyan E M, Petrosyan K G, Oganesyan K B, Hakobyan A A, Saakyan V A, Gevorkian S G, Izmailyan N S and Hu C K 2008 Detection of Casimir photons with electrons *Laser Phys.* **18** 621–4
[87] Westlund P O and Wennerström H 2005 Photon emission from translational energy in atomic collisions: A dynamic Casimir-Polder effect *Phys. Rev.* A **71** 062106
[88] Passante R, Persico F and Rizzuto L 2007 Nonlocal field correlations and dynamical Casimir–Polder forces between one excited- and two ground-state atoms *J. Phys. B: At. Mol. Opt. Phys.* **40** 1863–74
[89] Rizzuto L, Passante R and Persico F 2007 Nonlocal properties of dynamical three-body Casimir–Polder forces *Phys. Rev. Lett.* **98** 240404
[90] Vasile R and Passante R 2008 Dynamical Casimir–Polder force between an atom and a conducting wall *Phys. Rev.* A **78** 032108
[91] Dodonov A V, Dodonov V V and Mizrahi S S 2005 Separability dynamics of two-mode Gaussian states in parametric conversion and amplification *J. Phys. A: Math. Gen.* **38** 683–96
[92] Andreata M A and Dodonov V V 2005 Dynamics of entanglement between field modes in a one-dimensional cavity with a vibrating boundary *J. Opt. B: Quantum Semiclassical Opt.* **7** S11–20
[93] Schützhold R and Tiersch M 2005 Decoherence versus dynamical Casimir effect *J. Opt. B: Quantum Semiclassical Opt.* **7** S120–5
[94] Dodonov V V, Andreata M A and Mizrahi S S 2005 Decoherence and transfer of quantum states of field modes in a one-dimensional cavity with an oscillating boundary *J. Opt. B: Quantum Semiclassical Opt.* **7** S468–79
[95] Celeri L C, Pascoal F, de Ponte M A and Moussa M H Y 2008 Number of particle creation and decoherence in the nonideal dynamical Casimir effect at finite temperature *Preprint* arXiv: 0805.0786
[96] Man'ko V I 1991 The Casimir effect and quantum vacuum generator *J. Sov. Laser Res.* **12** 383–5
[97] Sassaroli E, Srivastava Y N and Widom A 1994 Photon production by the dynamical Casimir effect *Phys. Rev.* A **50** 1027–34
[98] Schützhold R, Plunien G and Soff G 1998 Trembling cavities in the canonical approach *Phys. Rev.* A **57** 2311–8
[99] Crocce M, Dalvit D A R and Mazzitelli F D 2001 Resonant photon creation in a three-dimensional oscillating cavity *Phys. Rev.* A **64** 013808
[100] Senitzky I R 1960 Dissipation in quantum mechanics. The harmonic oscillator *Phys. Rev.* **119** 670–9
[101] Schwinger J 1961 Brownian motion of a quantum oscillator *J. Math. Phys.* **2** 407–32
[102] Haus H A and Mullen J A 1962 Quantum noise in linear amplifiers *Phys. Rev.* **128** 2407–13
[103] Lax M 1966 Quantum noise. IV. Quantum theory of noise sources *Phys. Rev.* **145** 110–45
[104] Dodonov V V and Man'ko V I 1987 Density matrices and Wigner functions of quasiclassical quantum systems *Group Theory, Gravitation and Elementary Particle Physics (Proceedings of Lebedev Physics Institute, vol 167)* ed A A Komar (Commack: Nova Science) p 7–101
[105] Dodonov V V 2003 Parametric excitation and generation of nonclassical states in linear media *Theory of Nonclassical States of Light* ed V V Dodonov and V I Man'ko (London: Taylor & Francis) pp 153–218
[106] Dodonov V V, Man'ko O V and Man'ko V I 1995 Quantum nonstationary oscillator: models and applications *J. Russ. Laser Res.* **16** 1–56
[107] Dodonov V V, Man'ko O V and Man'ko V I 2002 Reduced density matrices of oscillator systems *Preprint* arXiv: quant-ph/0210030
[108] Louisell W H 1973 *Quantum Statistical Properties of Radiation* (New York: Wiley)
[109] Faddeev L D 1958 On the relation between the S-matrix and potential for the one-dimensional Schroedinger equation *Sov. Phys. – Doklady* **3** 747–751
[110] Born M and Wolf E 1975 *Principles of Optics* (Cambridge: Cambridge University Press) section 1.6.5
[111] Ginzburg V L 1967 *Propagation of Electromagnetic Waves in Plasmas* (Moscow: Nauka) pp 292-4
[112] Dodonov A V and Dodonov V V 2005 Resonance generation of photons from vacuum in cavities due to strong periodical changes of conductivity in a thin semiconductor boundary layer *J. Opt. B: Quantum Semiclassical Opt.* **7** S47–58
[113] McNeil K J and Walls D F 1975 Possibility of observing enhanced photon bunching from two photon emission *Phys. Lett.* A **51** 233–4

[114] Sotskii B A and Glazachev B I 1981 Entropy properties of superrandom fields *Opt. Spectrosc.* **50** 582–4
[115] Dodonov V V, Man'ko O V and Man'ko V I 1994 Photon distribution for one-mode mixed light with a generic Gaussian Wigner function. *Phys. Rev.* A **49** 2993–3001
[116] Dodonov V V and Man'ko V I 1994 New relations for two-dimensional Hermite polynomials *J. Math. Phys.* **35** 4277–94
[117] Olver F W J 1974 *Asymptotics and Special Functions* (New York: Academic Press) pp 463–74
[118] Müller J 1939 Untersuchungen über Electromagnetische Hohlraume *Hochfrequenztechnik und Elektroakustik* **54** 157–61
[119] Bethe H A and Schwinger J 1943 Perturbation theory for cavities *National Defense Research Committee report D1-117* (Cornell University)
[120] Casimir H B G 1951 On the theory of electromagnetic waves in resonant cavities *Philips Res. Rep.* **6** 162–82
[121] Hsieh H-T, Goldey J M and Brown S C 1954 A resonant cavity study of semiconductors *J. Appl. Phys.* **25** 302–7
[122] Spencer E G, LeCraw R C and Ault L A 1957 Note on cavity perturbation theory *J. Appl. Phys.* **28** 130–2
[123] Karpova O V 1959 On an absolute method of measurement of dielectric properties of a solid using a Π-shaped resonator *Sov. Phys. – Solid State* **1** 220–8
[124] Eldumiati I I and Haddad G I 1972 Cavity perturbation techniques for measurement of the microwave conductivity and dielectric constant of a bulk semiconductor material *IEEE Trans. Microwave Theory Tech.* **MTT-20** 126–32
[125] Chao S-H 1985 Measurements of microwave conductivity and dielectric constant by the cavity perturbation method and their errors *IEEE Trans. Microwave Theory Tech.* **MTT-33** 519–26
[126] Peligrad D-N, Nebendahl B, Kessler C, Mehring M, Dulčić A, Požek M and Paar D 1998 Cavity perturbation by superconducting films in microwave magnetic and electric fields *Phys. Rev.* B **58** 11652
[127] Peligrad D-N, Nebendahl B, Mehring M, Dulčić A, Požek M and Paar D 2001 General solution for the complex frequency shift in microwave measurements of thin films *Phys. Rev.* B **64** 224504
[128] Arndt G D, Hartwig W H and Stone J L 1968 Photodielectric detector using a superconducting cavity *J. Appl. Phys.* **39** 2653–6
[129] Grabtchak S and Cocivera M 1998 Microwave response due to light-induced changes in the complex dielectric constant of semiconductors *Phys. Rev.* B **58** 4701–7
[130] Van Bladel J 1964 *Electromagnetic Fields* (New York: McGraw-Hill) chap 10
[131] Landau L D and Lifshitz E M 1982 *Electrodynamics of Continuous Media* (Moscow: Nauka) section 90
[132] Braggio C, Bressi G, Carugno G, Dodonov A V, Dodonov V V, Ruoso G and Zanello D 2007 Resonance frequency shift in a cavity with a thin conducting film near a conducting wall *Phys. Lett.* A **363** 33–7
[133] Dodonov A V and Dodonov V V 2001 Nonstationary Casimir effect in cavities with two resonantly coupled modes *Phys. Lett.* A **289** 291–300
[134] Braggio C and Carugno G 2008 private communication
[135] Aspnes D E 1983 Recombination at semiconductor surfaces and interfaces *Surface Science* **132** 406–21
[136] Kawakubo T and Yamamoto K 2008 A standard Hamiltonian formulation for the dynamical Casimir effect *Preprint* arXiv: 0808.3618

MIR: an experiment for the measurement of the dynamical Casimir effect

A Agnesi[1,2], C Braggio[3], G Bressi[2], G Carugno[4], F Della Valle[5], G Galeazzi[3,6], G Messineo[5], F Pirzio[1,2], G Reali[1,2], G Ruoso[6], D Scarpa[1,2] and D Zanello[7]

[1] Dipartimento di Elettronica, Università di Pavia, Via Ferrata 1, 27100 Pavia, Italy
[2] INFN sez. di Pavia, via Bassi 6, 27100 Pavia, Italy
[3] Dipartimento di Fisica, Università di Padova, via Marzolo 8, 35131 Padova, Italy
[4] INFN sez. di Padova, via Marzolo 8, 35131 Padova, Italy
[5] INFN sez. di Trieste and Dipartimento di Fisica, Università di Trieste, Via Valerio 2, 34127 Trieste, Italy
[6] INFN lab. naz. Legnaro, viale dell'Università 2, 35020 Legnaro (Pd), Italy
[7] INFN sez. di Roma, P.le A. Moro 2, 00185 Roma, Italy

E-mail: Caterina.Braggio@pd.infn.it

Abstract. In this paper we report the status of the experiment MIR (Motion Induced Radiation), aimed at the experimental verification of the dynamical Casimir effect. The stringent theoretical requirements to observe the effect are satisfied in a scheme in which the conductivity of a semiconductor inside a superconducting microwave resonant cavity is varied in time. Free carriers in the semiconductor are periodically excited and recombine at 5 GHz. In this process vacuum and thermal photons are parametrically amplified in a time interval of 200-500 ns.

1. Introduction

The study of the quantum vacuum has seen, in the recent years, an increase in the number of dedicated experiments, following the pioneering results of Lamoreaux on the Casimir effect [1]. The interest on this subject is motivated by the possibility of attaining a more cogent verification of Quantum Electrodynamics (QED), giving at the same time results in other fields of fundamental physics research. Casimir effect experiments can, for example, shed light onto the field of non Newtonian forces at small separation [2], while experiments which sense optical properties of vacuum under external fields [3] are able to search for new small mass particles. We are interested in the Casimir effect in the presence of accelerated boundaries, which is best known as the dynamic Casimir effect [4]. While in the static Casimir effect [5] the action of the vacuum results in a force acting between two bodies, in this experiment we try to reveal directly the presence of a non-empty vacuum by using a specifically designed device to amplify the virtual vacuum photons and to produce real electromagnetic radiation. The 'amplifier' is nothing else than a boundary undergoing an oscillation, and hence radiating energy due to the dissipative action against the vacuum photons. The most obvious realization of this amplifier would be a single mirror moving with periodic motion at a frequency Ω. Unfortunately this system produces an undetectable number of photons N, approximately given by [6]: $N \sim \Omega T(v/c)^2$ where T is

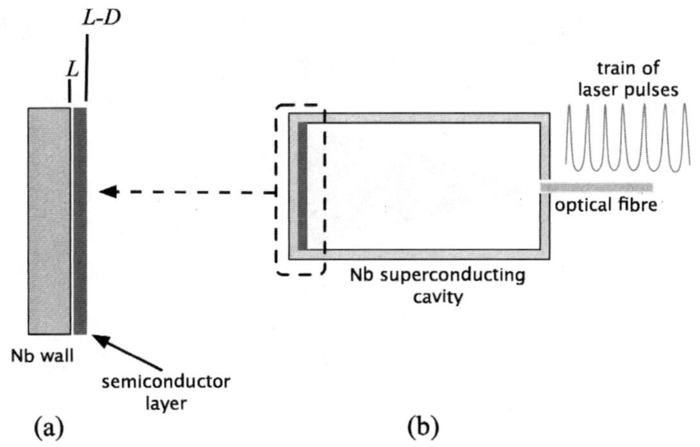

Figure 1. Principle of the experimental apparatus. On the left a larger view of the cavity end wall covered with a semiconductor layer is shown. Light renders the semiconductor surface reflecting thus generating a variation D of the cavity length L. Repeated pulses mimic an harmonic oscillation of the wall.

the observation time, v the maximum velocity and c the speed of light. Using for the parameters the utmost values $\Omega \sim$ GHz and $(v/c) \sim 10^{-8}$, for an observation time $T = 1$ s the expected number of photons of frequency Ω would be much less than 1. Such a small number of photons is undetectable, but it can be increased by realizing the moving boundary as part of a resonant cavity with quality factor Q. If the resonant frequency ν of the cavity fulfills the parametric resonance condition, i.e. $2\nu = \Omega$, the number of photons would then be QN [4, 7], which might be in the range of detectability. In our approach the cavity is a high-Q superconducting niobium cavity inside which the moving boundary is realized with a time variable mirror. Lozovik [8] and Yablonovitch [9] suggested to change the reflectivity of a semiconductor by using short intense laser pulses: following this idea we developed a scheme where a semiconductor switches from complete transparency to total reflection at a well defined frequency by means of an amplitude modulated laser beam [10]. This is depicted in Figure 1.

The semiconductor is illuminated by a train of laser pulses, with a repetition frequency twice the resonance frequency of the cavity. Each pulse has enough energy to produce in the semiconductor a plasma reflecting the electromagnetic radiation at frequency ν [11]. If the recombination time of the semiconductor is short enough, at the end of each pulse the plasma disappears. In this way we will have achieved an effective motion of the boundary, which satisfies the parametric resonance condition. If the number of consecutive pulses is sufficient, a measurable signal will appear in the resonant cavity.

A complete parametrization of our set-up is currently under study by V.V. Dodonov [12]. In his calculation all the parameters describing the apparatus are taken into account. The most critical ones are those concerning the behaviour of the semiconductor and the duration of the process of parametric amplification. The reference values used in the following come from Dodonov's calculations.

2. Experimental set-up

A detailed description of the apparatus and of the detection technique has been given in [10], here only a brief summary is given.

Figure 2. *Left*: picture of a rectangular resonant cavity. The cavity is open, the piece on the right is the top cover with the groove to hold the semiconductor. *Right*: the cavity inside the cryostat, equipped with motors to control the antenna position. The box on the lower left part of the picture contains a cryogenic amplifier and a directional coupler. The box is closed and filled with exchange helium gas to provide better cooling of the electronics.

2.1. Resonant cavity and the cryogenic system

The core of the apparatus is a superconducting niobium cavity, shown in figure 2. We tested several different cavity shapes; in figure 2 we show a rectangular one with dimensions $9\times1\times8$ cm^3, optimised for a large dynamic Casimir effect signal [13]. One of the walls of the cavity can be removed and contains the groove for positioning the semiconductor slab. The cavity resonance frequency is approximately $\nu = 2.35$ GHz and the Q value is in excess of 10^6 with the semiconductor inserted. The resonance frequency can be varied in order to match the laser repetition rate by controlled insertion of a sapphire slab into the cavity.

The energy stored in the cavity mode is detected by a loop antenna connected to an heterodyne receiver; the antenna position is adjustable and is controlled by two computer driven cryogenic motors. The cavity and the first stage of amplification of the receiver are kept in contact with the bottom of a 50 l liquid helium cryostat. Inside the chamber (shown in figures 2 and 3) He gas at a pressure of 1 mbar allows a better thermal stabilisation of the components in the presence of heat sources (amplifier bias, laser pulses over the semiconductor surface). The cryostat can be operated in the $1-8$ K temperature range.

2.2. Semiconductor

There are two theoretical requirements which optimise the photon production in our apparatus: the recombination time τ, which should be of the order of 20 ps, and the mobility μ, whose required value is in the range $3000-5000$ cm^2/(V s). As the combination of these two characteristics is not at all common in semiconductor materials, we had to test a lot of GaAs samples grown with different methods before finding a valuable procedure. Starting from semi-insulating GaAs with $\tau\sim1$ ns and the required mobility, it is possible to reduce τ in a well defined way while keeping μ rather constant. This is done by irradiating the semi-insulating GaAs sample with fast neutrons (MeV energy range). The dose determines the recombination time, and a dose of $\sim10^{15}$ neutrons/cm^2 results in a $\tau\sim20$ ps. Irradiation is performed at the laboratorio ENEA in Rome, and high precision measurement (about 1 ps resolution) of the recombination time is performed with the THz pump and probe technique, at cryogenic temperature, by the group of Prof. Krotkus at the University of Vilnius. This measurement

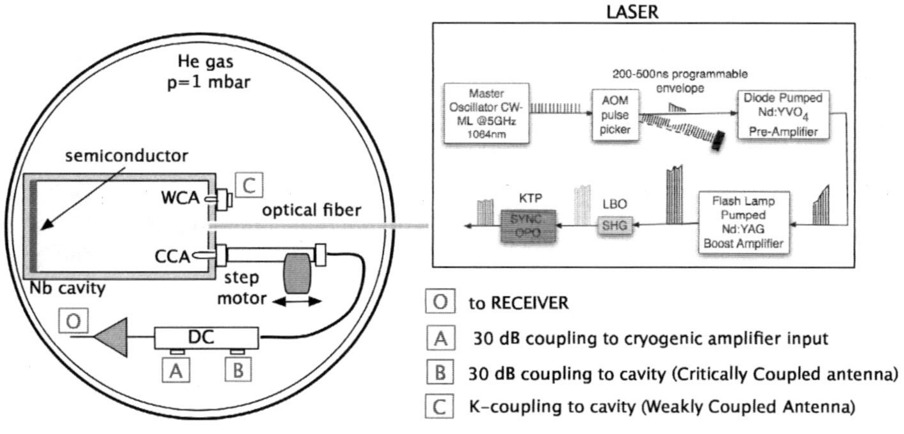

Figure 3. *Left*: schematic view of the experimental apparatus. *Right*: scheme of the laser system (see text).

provides also an estimate of the mobility, which degrades only slightly with respect to the value measured before irradiation, as obtained also in [14].

2.3. Laser

The laser must deliver a train of a few thousands pulses (macro-pulse of 200 – 500 ns duration) with single pulse energy of the order of 100 μJ energy and a few picoseconds duration. Laser wavelength must be around 800 nm wavelength to excite carriers across the GaAs bandgap. The time interval between consecutive pulses must be about 200 ps, corresponding to a frequency of 5 GHz. These parameters are not available in commercial laser systems, and for this reason the light source was realised in-house.

In figure 3 (right side) we show the Nd:YVO_4 laser generating a stable cw (λ=1064 nm) mode-locking train of 6 ps pulses at approximately 5 GHz. This part, called the main oscillator, is pumped by a 1-W laser diode emitting at 808 nm [15, 16, 17]. The seed is injected into a pulse picker system, which selects a number of pulses thus generating the macro-pulse to be fed into a two stage optical amplifier [17]. The pre-amplifier stage comprises a couple of Nd:YVO_4 slabs, each side-pumped by a pulsed 150 W peak power laser diode array in a grazing incidence, total internal reflection configuration. Nd:YAG flash-lamp pumped rods are used for the final power amplification stage, for a total single pass gain of 90 dB. Second-harmonic conversion allows synchronous pumping of an optical parametric oscillator, obtaining up to 40 mJ. A further amplification stage to reach the final required value of energy/pulse is currently under development.

3. Expected sensitivity and background effects

The energy stored in the cavity after excitation decays exponentially in time, with a time constant $\tau_U = 1/(\pi\Delta\nu)$, where $\Delta\nu$ is the linewidth of the cavity. The decay signal travels to an heterodyne receiver, that is shown in figure 4.

We use a cryogenic amplifier kept at liquid helium temperature to increase the sensitivity of the measurement. In fact, the sensitivity of the receiver is primarily determined by the noise temperature T_e of the first stage of amplification, which has a nominal value $T_e = 5$ K. With such a low noise temperature, the system sensitivity is limited by the temperature and the length of the cable which connects the input of the amplifier to the cavity. This is the reason why the

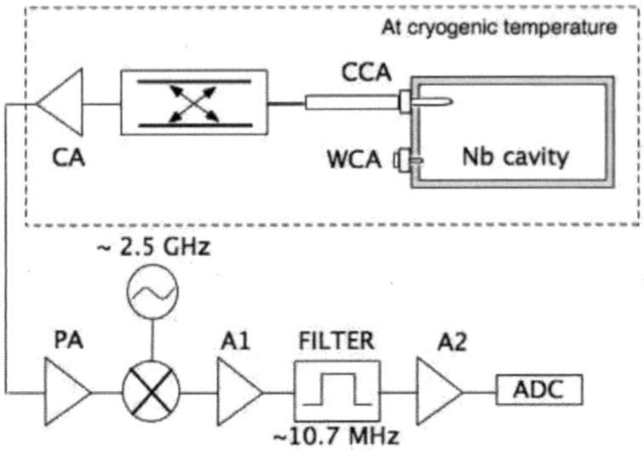

Figure 4. Detailed scheme of the electronics of the MIR experiment. A microwave cavity, equipped with a weakly coupled input antenna (WCA) and a critically coupled antenna (CCA), is placed inside a liquid helium cryostat. A directional coupler connects the CCA to the receiver. The cryogenic amplifier CA is followed by a post amplifier PA; the mixer provides frequency down-conversion of the signal while a filter between the two amplifiers A1 and A2 rejects noise outside the cavity bandwidth. All the components following the cryogenic amplifier are at room temperature.

cable is anchored to the bottom of the liquid helium vessel and is kept as short as possible. The results obtained with different cavities indicate that the actual noise temperature of the system varies between 7 and 10 K.

We measure the noise temperature of the receiver with the variable temperature load method, which has been described in detail elsewhere [18]. The calibration procedure with a 50 Ω load allows us to obtain the precise value of the gain G of the receiver and the overall noise temperature T_e, as shown in figure 5. The inserted table shows the gain G and the noise temperature T_e obtained by fitting the data with equation $P/B = k_B G(T + T_e)$, where P is the noise power within the measurement bandwidth B of the spectrum analyser, k_B is the Boltzmann constant and T the temperature of the calibration load. When we connect a cavity in place of the resistor we measure the value represented by the encircled dot in Fig. 5. This noise level can be converted, dividing it by the measured gain, in a noise density in input to the amplifier of $(2.0 \pm 0.2) \cdot 10^{-22}$ W/Hz, which corresponds to approximately 100 microwave photons at equilibrium.

The parametric amplification process described in this paper can be characterised by a gain G_V which can be estimated from the parameters of the apparatus [12, 13]: the total number of photons produced is $N = G_V(1+2N_{th})$, where N_{th} is the number of thermal photons in the mode considered and $(1 + 2N_{th})$ is the thermal factor [19]. Since our receiver is capable to detect a hundred photons, a gain G_V of about 100 must be reached to see the amplified vacuum photons. The contribution to N of the vacuum photons G_V for our apparatus has not been measured yet, but it should not be critical [12, 13]. It will be measured by charging the cavity before the parametric excitation, but in principle it can be measured also using the thermal photons present at equilibrium in the cavity at different temperatures in the range $1 - 8$ K. To have an estimate of the number of photons distributed in the cavity mode TE_{101}, we have to consider that the photon wavelength in a microwave cavity is comparable to the cavity dimensions. In this case the number of thermal photons N_{th} at equilibrium at temperature T is

$$N_{th} = \frac{1}{e^{\frac{h\nu}{k_B T}} - 1} \simeq \frac{k_B T}{h\nu}$$

where h is the Planck constant and ν the cavity resonance frequency. For example, at 4 K a

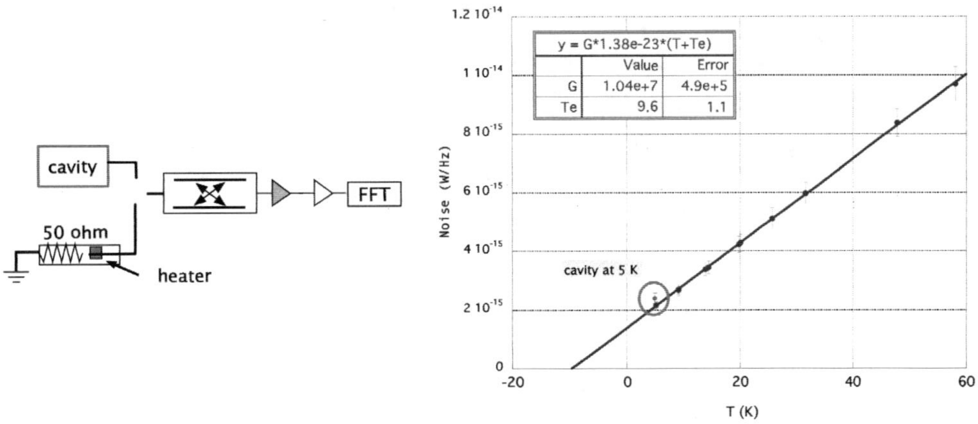

Figure 5. *Left*: scheme of the apparatus for the noise and calibration measurement at LHe temperature. Measurements are performed with either the cold cavity or the variable temperature resistor connected to the receiver electronics. *Right*: noise power density of the receiver measured with a load resistor in place of the cavity plotted as a function of the temperature. The continuous line is a linear fit of the experimental data. The encircled value corresponds to the measured noise power density peak of a reentrant cylindrical cavity.

thermal contribution of 33 photons is expected. A number of other procedures are also under study to disentangle a real signal from a systematic effect. These include detuning of laser frequency out of the parametric resonance, variation of the thickness and the recombination time of the semiconductor and possible new shapes of the resonant microwave cavity.

Acknowledgments
The authors would like to thank Prof. A. Krotkus from SELITEC (Semiconductor LIght TEChnologies) in Vilnius and R. Adomavicius for characterisation of the irradiated SI GaAs samples. We are also indebted to E. Berto (Università di Padova) and D. Corti (INFN Padova) for technical support.

References
[1] Decca *et al.*, Phys. Rev. D **75**, 077101 (2007);
Chen *et al.*, Phys. Rev. B **76**, 035338 (2007);
for a review see Bordag M, Mohideen U and Mostepanenko V M, Phys. Rep. **353**, 1 (2001);
Lamoreaux S K, Rep. Progr. Phys. **68**, 201 (2005)
[2] Onofrio R, New J. Phys. **8**, 237 (2006)
[3] Bregant M *et al.*, Phys. Rev. D **78**, 032006 (2008)
[4] Moore G T, J. Math. Phys. **11**, 2679 (1970);
Fulling S A and Davies P C W, Proc. R. Soc. London A **348**, 393 (1976);
Schwinger J, Proc. Natl. Acad. Sci. U.S.A. **89**, 4091 (1992);
Kardar M and Golestanian R, Rev. Mod. Phys. **71**, 1233 (1999)
Dodonov V V, in *Modern Nonlinear Optics*, edited by Evans M W, Adv. Chem. Phys. Ser. **119**, 309, Wiley, New York (2001)
[5] Bressi G *et al.* , Phys. Rev. Lett. **88**, 041804 (2002)
[6] Jaekel M T, Lambrecht A and Reynaud S, in Proceedings of the Ninth Marcel Grossmann Meeting, Ed. by Gurzadyan V G, Jantzen R T and Ruffini R, pag. 1447 (2002) World Scientific
[7] Lambrecht A, Jaekel M T and Reynaud S, Phys. Rev. Lett. **77**, 615 (1996);
Crocce M, Dalvit D A R and Mazzitelli F D, Phys. Rev. A **64**, 013808 (2001);
Schaller G, Schützhold R, Plunien G and Soff G, Phys. Rev. A **66**, 023812 (2002)

[8] Lozovik Yu E, Tsvetus V G and Vinograd E A, JETP Lett. **61**, 723 (1995);
Lozovik Yu E, Tsvetus V G and Vinograd E A, Phys. Scripta **52**, 184 (1995)
[9] Yablonovitch E, Phys. Rev. Lett. **62**, 1742 (1989)
[10] Braggio C *et al.*, Europhys. Lett. **70**, 754 (2005)
[11] Braggio C, Bressi G, Carugno G, Lombardi A, Palmieri A, Ruoso G and Zanello D, Rev. Sci. Instrum. **75**, 4967 (2004);
Braggio C, Bressi G, Carugno G, Dodonov A V, Dodonov V V, Galeazzi G, Ruoso G and Zanello D, Phys. Lett. A **363**, 33 (2007)
[12] Dodonov V V and Dodonov A V, J. Phys. B **39**, S749 (2006)
[13] Dodonov V V and Dodonov A V, J. Phys. A **39**, 6271 (2006)
[14] Foulon F *et al.*, J. Appl. Phys. **88**, 3634 (2000)
[15] Agnesi A *et al.*, Opt. Express **13**, 5302 (2005)
[16] Agnesi A *et al.*, Opt. Express **14**, 9244 (2006)
[17] Agnesi A *et al.*, Opt. Express. **16**, 15811 (2008).
[18] Braggio C *et al.*, submitted to IEEE Trans. Micr. Th. Tech.
[19] Plunien G *et al.*, Phys. Rev. Lett. **84**, 1882 (2000).

Journal of Physics: Conference Series **161** (2009) 012029 doi:10.1088/1742-6596/161/1/012029

Photon creation from vacuum and interactions engineering in nonstationary circuit QED

A V Dodonov

Departamento de Física, Universidade Federal de São Carlos, 13565-905, São Carlos, SP, Brazil

Abstract. We study theoretically the nonstationary circuit QED system in which the artificial atom transition frequency, or the atom-cavity coupling, have a small periodic time modulation, prescribed externally. The system formed by the atom coupled to a single cavity mode is described by the Rabi Hamiltonian. We show that, in the dispersive regime, when the modulation periodicity is tuned to the 'resonances', the system dynamics presents the dynamical Casimir effect, resonant Jaynes-Cummings or resonant Anti-Jaynes-Cummings behaviors, and it can be described by the corresponding effective Hamiltonians. In the resonant atom-cavity regime and under the resonant modulation, the dynamics is similar to the one occurring for a stationary two-level atom in a vibrating cavity, and an entangled state with two photons can be created from vacuum. Moreover, we consider the situation in which the atom-cavity coupling, the atomic frequency, or both have a small non-periodic time modulation, and show that photons can be created from vacuum in the dispersive regime. Therefore, an analog of the dynamical Casimir effect can be simulated in circuit QED, and several photons, as well as entangled states, can be generated from vacuum due to the anti-rotating term in the Rabi Hamiltonian.

E-mail: adodonov@df.ufscar.br

1. Introduction

Over the last few decades nonstationary processes in cavities have received considerable attention. One such process is the nonstationary or dynamical Casimir Effect (DCE) – in particular, the creation of photons from vacuum, or another initial field state, in cavity whose geometry [1–5] or material properties [4–9] have a periodic time dependence, with the modulation frequency close to twice the unperturbed field eigenfrequency. Nowadays, DCE in cavities is a well studied problem, with a variety of theoretical predictions concerning the number and the statistics of created photons, as well as the influence of detuning, dissipation, boundary conditions, geometry and non-periodicity of the modulation (see [10–12] for an extensive list of references). To date, DCE has not been observed in laboratory, however several concrete proposals have appeared over the last years [13–18], with some of them being currently implemented experimentally [19–21].

The interest in nonstationary processes in cavities reappeared over the last 5 years due to the progress in the field of Cavity Quantum Electrodynamics (cavity QED [22, 23]) in the condensed matter systems, e.g. semiconductor quantum dots [24–29], polar molecules [30, 31] and superconducting circuits [32–34] coupled to high-Q resonators, the latest architecture known as *circuit* QED [33, 35]. In cavity QED, the effective 2-level atom is coupled to the field inside the resonator via the dipole interaction, allowing for observation of the light-matter interaction at the level of single photons and single atoms. The new ingredient in the solid state cavity QED, in particular circuit QED, is the possibility of engineering and manipulating the properties of the artificial 2-level atom and the resonator [33, 36, 37], as well as the interaction strength between them [38], either during the fabrication, or *in situ*.

Recently, the strong resonant and the strong dispersive coupling limits between the artificial atom and a single cavity mode were observed experimentally in circuit QED [33, 39] and other solid state architectures [24–27, 29]. Moreover, the single photon source [40], single artificial-atom maser [41], multiphoton Fock states [42] and interaction between two artificial atoms (qubits) [37, 43] were implemented experimentally, among many other important achievements (see [35] for more references). Besides, the circuit QED architecture benefits from robust readout schemes of the atomic and the resonator states [38,42,44–48], relatively low dissipative losses [39], state preparation techniques [49] and real time manipulation of the atomic transition frequency via electric and magnetic fields [33, 37, 42] or non-resonant microwave fields [43, 48].

Harnessing the tunability of the atomic transition frequency, the authors in [50, 51] proposed an experimental implementation of the Landau-Zener sweeps in circuit QED, when the atomic frequency increases linearly in time. This allows for generation of single photons, as well as cavity-atom entangled states. In a similar direction, the generation of the quantum vacuum radiation by modulating periodically the vacuum Rabi frequency of an intersubband transition in a doped quantum well system embedded in a planar microcavity was considered in [17,18,52,53], and emission of photon pairs was predicted for 'resonant' modulation frequencies. A preliminary theoretical study of the feasibility of realizing the dynamical Casimir effect with a quantum flux qubit in superconducting quantum nanocircuits, as well as the detection of the generated photons, was reported in [54]. On the other hand, the freedom of controlling in real-time the atomic frequency is currently being explored to couple/decouple one or several qubits to/from the cavity mode in order to implement quantum logic operations [37, 43, 48].

In this paper, following the original proposal [55], we study nonstationary processes in the solid state cavity QED, where a single cavity mode is coupled to a single artificial atom whose transition frequency, or atom-cavity coupling, have a small periodic time modulation, prescribed externally. Such a control over the transition frequency, with compatible modulation frequency, can be achieved in circuit QED with present or near-future technology [37,42,48]. We show that in the dispersive atom-cavity regime and under the resonance conditions one obtains completely different effective regimes of the atom-field interaction, which can be approximately described by the resonant Anti-Jaynes-Cummings (AJC), resonant Jaynes-Cummings (JC) or the dynamical Casimir effect (DCE) Hamiltonians. Moreover, in the resonant atom-cavity regime, the system dynamics resembles the behavior of the DCE in the vibrating cavity containing a resonant two-level atom, and entangled states with at most two photons can be generated from vacuum.

We also consider the case, in which the atomic frequency, the atom-cavity coupling parameter, or both have a small non-periodic time modulation, prescribed externally. Namely, we suppose that the modulation is given by the sum of two harmonic functions with different amplitudes and frequencies. We deduce an effective Hamiltonian in the dispersive regime, and show that photon generation from vacuum is possible for fine tuned modulation frequencies.

Thus, we demonstrate the possibility of simulating the DCE in circuit QED using a single nonstationary atom, instead of a macroscopic dielectric medium as in Refs. [6–9]. As applications, it could be possible to create excitations, either photonic or atomic, from the initial vacuum state $|g, 0\rangle$, generate nonclassical states of light and realize transitions between the states $\{|g, n\rangle, |e, n \pm 1\rangle\}$ in the dispersive regime, where $|g\rangle$ and $|e\rangle$ are the atomic ground and excited states, respectively, and $|n\rangle$ is the Fock state of the cavity field. A related problem was recently studied in [56], where it was suggested that lasing behavior and the creation of a highly non-thermal population of the oscillator, as well as the cooling, could be implemented using an analogous scheme in the near future.

2. Nonstationary circuit QED with periodic modulation

We assume that the atomic transition frequency $\Omega(t)$ is given by

$$\Omega(t) = \Omega_0 + \varepsilon f_t \,. \tag{1}$$

Here Ω_0 denotes the bare atomic frequency, ε is a small modulation amplitude, $\varepsilon \ll \Omega_0$, and f_t is an arbitrary periodic function of time, prescribed externally

$$f_t = \sum_{k=0}^{\infty} \left[s_k \sin k\eta t + c_k \cos k\eta t \right], \tag{2}$$

where η is the modulation frequency and $\{s_k, c_k\}$ form a set of coefficients describing the time modulation. The cavity frequency ω and the atom-cavity coupling parameter g_0 are constant, so at the charge degeneracy point [38] the system is described by the Rabi Hamiltonian (RH)

$$H = H_0(t) + g_0(a + a^\dagger)(\sigma_+ + \sigma_-), \tag{3}$$

with a ($a^\dagger$) being the cavity annihilation (creation) operator, $\sigma_+ = |e\rangle\langle g|$ and $\sigma_- = |g\rangle\langle e|$. The free Hamiltonian is

$$H_0(t) = \omega \hat{n} + \frac{\Omega(t)}{2}\sigma_z, \tag{4}$$

where $\hat{n} = a^\dagger a$ is the photon number operator, $\sigma_z = |e\rangle\langle e| - |g\rangle\langle g|$ and we assume $\hbar = 1$. In the stationary case, $\varepsilon = 0$, one can perform the Rotating Wave Approximation (RWA) and obtain the standard Jaynes-Cummings (JC) Hamiltonian [57], which has been verified in several experiments over the last few years [33, 39, 58–60]. However, in the non-stationary case, as well as under strong dephasing noise [61], the anti-rotating term $(a\sigma_- + a^\dagger\sigma_+)$ cannot always be eliminated. Moreover, it is responsible for producing an analog of the DCE and creating photonic and atomic excitations from vacuum under modulation 'resonance' conditions, as shown below.

In the interaction picture with respect to $H_0(t)$ the interaction Hamiltonian reads

$$H_I = g_0\left(e^{i\Xi_-} a\sigma_+ + e^{i\Xi_+} a^\dagger \sigma_+ + h.c.\right), \tag{5}$$

where h.c. stands for the Hermitian conjugate and

$$\Xi_\pm \equiv \int_0^t d\tau \left[\Omega(\tau) \pm \omega\right]. \tag{6}$$

All the information about the influence of the external modulation on the system dynamics is contained in the time-dependent coefficients $\exp(i\Xi_\pm)$, which can be significantly simplified by tuning the modulation frequency η to the 'resonances'. We have explicitly

$$g_0 e^{i\Xi_\pm} = g e^{i\Delta_\pm t} \sum_{l=0}^{\infty} \frac{1}{l!}\left[\frac{\varepsilon}{\eta}\sum_{k=1}^{\infty}\left(\Lambda_k e^{-ik\eta t} - \Lambda_k^* e^{ik\eta t}\right)\right]^l, \tag{7}$$

where we defined the complex coupling constant

$$g \equiv g_0 \exp\left[i\frac{\varepsilon}{\eta}\sum_{k=1}^{\infty} k^{-1} s_k\right] \tag{8}$$

and parameters

$$\Lambda_k \equiv -\frac{c_k + i s_k}{2k} \tag{9}$$

$$\Delta_\pm \equiv (\Omega_0 + \varepsilon c_0) \pm \omega. \tag{10}$$

Journal of Physics: Conference Series **161** (2009) 012029 doi:10.1088/1742-6596/161/1/012029

3. Anti-Jaynes-Cummings (AJC) resonance

The 'Anti-Jaynes-Cummings' (AJC) resonance occurs for

$$\eta = \eta_{AJC} \equiv \Delta_+ - \xi, \tag{11}$$

where $|\xi| \ll \Delta_+$ is a small 'resonance shift' to be adjusted afterwards. Assuming a reasonable experimental condition $\varepsilon/\eta \ll 1$, we expand $g_0 \exp(i\Xi_\pm)$ in (5) to the first order in ε/η and make RWA, obtaining

$$H_I \simeq g\left(\theta e^{i\xi t} a^\dagger \sigma_+ + e^{i\Delta_- t} a\sigma_+\right) + h.c., \tag{12}$$

where the modulation induced dimensionless coupling is

$$\theta \equiv \Lambda_1 \frac{\varepsilon}{\eta}. \tag{13}$$

3.1. Dispersive regime

In the dispersive regime, $g_0\sqrt{\langle\hat{n}\rangle}/|\Delta_-| \ll 1$, where $\langle\hat{n}\rangle$ is the mean photon number, the Hamiltonian (12) can be approximated by [57]

$$H_I^{(1)} \simeq \left(g\theta e^{i\xi t} a^\dagger \sigma_+ + h.c.\right) + \delta\left(\hat{n} + 1/2\right)\sigma_z\ , \tag{14}$$

where

$$\delta = \delta_0 + \mathcal{O}\left(g_0^2/\Delta_+\right) \tag{15}$$

and

$$\delta_0 = \frac{g_0^2}{\Delta_-} \tag{16}$$

is the standard dispersive shift [57]. After the unitary time-dependent transformation defined by the operator

$$U = \exp\left[i\xi t\sigma_z/2\right] \tag{17}$$

we get the effective AJC Hamiltonian

$$H_{AJC} = U^{-1} H_I^{(1)} U - iU^{-1}\frac{\partial U}{\partial t} \simeq \frac{\xi + \delta\left(1 + 2\hat{n}\right)}{2}\sigma_z + \left(g\theta a^\dagger \sigma_+ + h.c.\right). \tag{18}$$

For the field initially in the Fock state, by adjusting the resonance shift ξ in order to make $|\xi + \delta(1 + 2\hat{n})|$ small compared to $|g\theta|$, one obtains approximately the resonant AJC Hamiltonian

$$H_{AJC}^{(r)} \simeq g\theta a^\dagger \sigma_+ + g^*\theta^* a\sigma_-. \tag{19}$$

From the physical point of view, the external modulation supplies the energy $\omega + \Omega_0$ necessary to create one photon and one atomic excitation simultaneously. Thus, one can create the superposition of states $|e, 1\rangle$ and $|g, 0\rangle$ starting from the initial vacuum state $|g, 0\rangle$ and fine tuning the resonance shift.

In figure 1 we show the exact dynamics for the AJC resonance, obtained through the numerical integration of the Schrödinger equation with the initial Hamiltonian (3), setting the experimental circuit QED parameters $\Omega_0/\omega = 1.4$, $g_0/\omega = 2 \cdot 10^{-2}$ and assuming $\varepsilon/\omega = 0.2$. In all the simulations throughout this paper we consider the harmonic modulation, $f_t = \sin\eta t$. In the figure 1a we plot the number of created photons

$$N \equiv \langle\hat{n}\rangle - n_0, \tag{20}$$

for the initial Fock state $|g, n_0\rangle$ as function of the dimensionless time $|\theta|\, g_0 t$, assuming the resonance shift $\xi = -2\delta_0\,(1+n_0)$. In figure 1b we plot the atom excitation probability P_e. From figures 1a and 1b one can observe the simultaneous generation of one photon and one atomic excitation, in agreement with the effective AJC Hamiltonian (18). However, the maximal mean number of created photons is slightly less than 1 and depends on the initial photon number n_0, meaning that the resonance shift was not adjusted precisely. In figure 1c we plot N for the initial number state $|g, n_0 = 5\rangle$ for different resonance shifts $\xi = -2\delta_0\,(1+n_0+x)$, $x = 0, 1, 2, 3, 4$, demonstrating that it is possible to find an optimum ξ which reproduces the resonant AJC Hamiltonian and allows for the creation of one photon and the atomic excitation simultaneously.

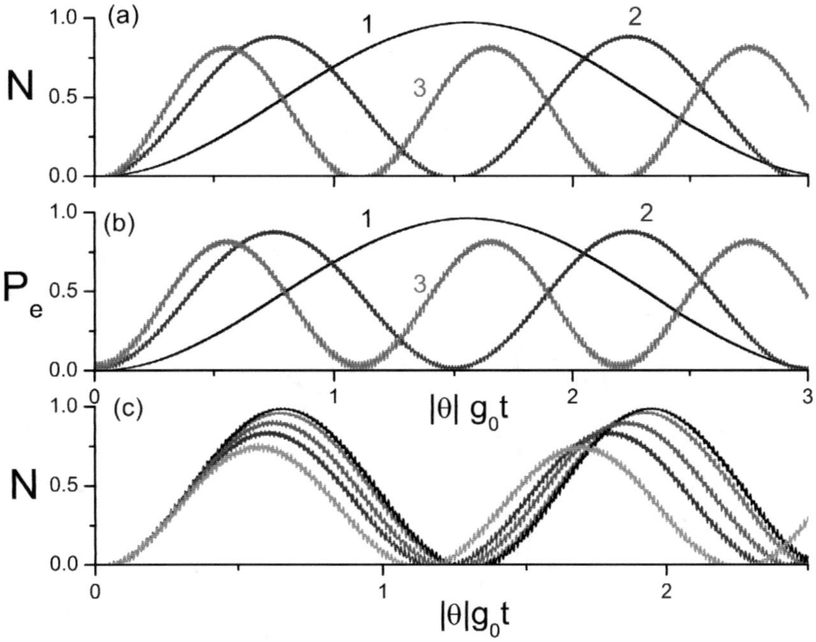

Figure 1. The results of the numerical integration of the Schrödinger equation with the Rabi Hamiltonian (3) for the modulation frequency $\eta = \Delta_+ + 2\delta_0\,(1+n_0)$. We see an AJC behavior, as given by the effective Hamiltonian (18). **a)** The number of created photons N and **b)** the atom excitation probability P_e versus the dimensionless time for the initial Fock states $|g, n_0\rangle$, with $n_0 = 0$ (line 1), $n_0 = 3$ (line 2) and $n_0 = 6$ (line 3). **c)** The dependence of N on the small change x in the resonance shift $\xi = -\delta\,(1+n_0+x)$ for the initial number state $|g, n_0 = 5\rangle$. For $|\theta|\, g_0 t$ from 2 to 2.3 the curves correspond to $x = 4, 0, 3, 1, 2$, from below. By adjusting x one can optimize the photon generation process.

3.2. Resonant regime

In the resonant regime, $|\Delta_-|/g_0 \ll 1$, the effective Hamiltonian method is not applicable. So we apply the method of slowly varying amplitudes on the Hamiltonian (12), repeating the procedure [62] employed originally for studying the photon generation from vacuum due to the DCE in a vibrating cavity containing a resonant (stationary) two-level atom [63–65]. We find that for the initial state $|g, 0\rangle$ the photon generation occurs for two possible resonance shifts

$$\xi_{\pm} = \Delta_-/2 \pm \sqrt{2}g_0 \tag{21}$$

and one gets the following non-zero probabilities P_{an}, with $a = (e, g)$ denoting the atomic state and n the photon number

$$\begin{aligned} P_{g0} &\approx \cos^2(\chi t) \\ P_{e1} &\approx \sin^2(y+q)\sin^2(\chi t) \\ P_{g2} &\approx \cos^2(y-q)\sin^2(\chi t)\,. \end{aligned} \tag{22}$$

Here

$$\chi \approx |g\theta|\,\sin(y+q)\,, \quad \tan y \approx \left(\frac{2\sqrt{2}g_0 + \Delta_-}{2\sqrt{2}g_0 - \Delta_-}\right)^{1/2} \tag{23}$$

and $q = 0$ $(\pi/2)$ for ξ_- (ξ_+).

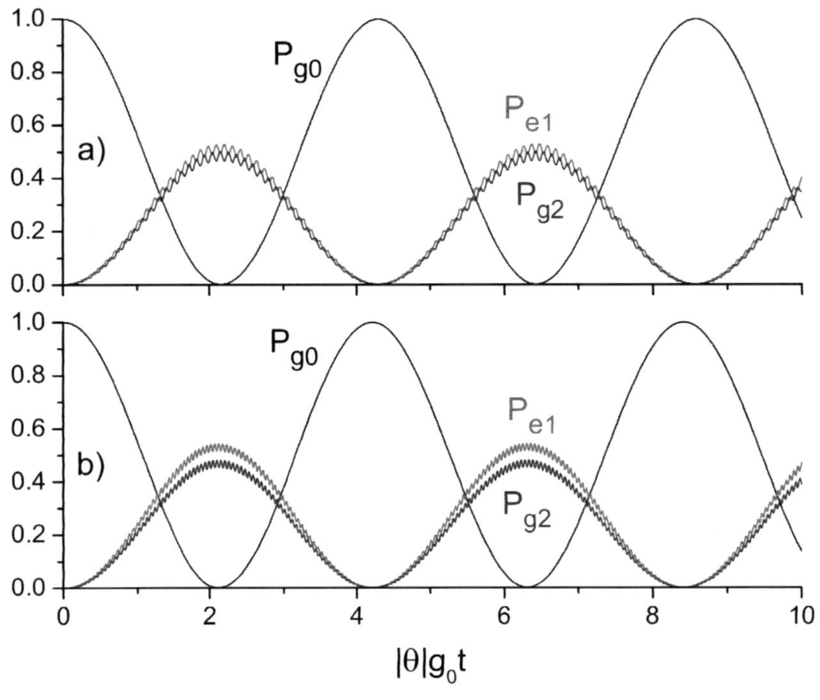

Figure 2. The time dependence of the probabilities P_{g0}, P_{e1} and P_{g2}, obtained by the numerical integration of the Schrödinger equation with the Rabi Hamiltonian (3) in the resonant atom-cavity regime for the AJC resonance with properly adjusted frequency shift. **a)** $\Delta_- = 0$, $\xi = \xi_-$ and $\varepsilon/\omega_r = 2 \cdot 10^{-1}$. **b)** $\Delta_- = 10^{-1}g_0$, $\xi = \xi_-$ and $\varepsilon/\omega_r = 10^{-1}$. Not more than 2 photons can be generated from vacuum, and the atom excitation probability is limited by the value $\sim 1/2$, in agreement with equation (22).

We illustrate this behavior in figure 2, where we show P_{g0}, P_{e1} and P_{g2} in the resonant atom-cavity regime under the AJC resonance. In figure 2a we consider $\Delta_- = 0$ and $\xi = \xi_-$, using the parameters $g_0/\omega = 4 \cdot 10^{-2}$ and $\varepsilon/\omega = 2 \cdot 10^{-1}$. In figure 2b we set $\Delta_- = 10^{-1} g_0$, $\xi = \xi_-$, $g_0/\omega = 4 \cdot 10^{-2}$ and $\varepsilon/\omega = 10^{-1}$. The results are in excellent agreement with the theoretical predictions, equation (22), demonstrating that a superposition of states $|g,0\rangle$, $|e,1\rangle$ and $|g,2\rangle$ is created from the initial vacuum state $|g,0\rangle$. This dynamics is similar to the one studied in the context of DCE, where a resonant (stationary) two-level atom is fixed inside a cavity whose boundary is oscillating with the frequency close to 2ω [62]. In both cases not more than two photons can be created from the vacuum state $|g,0\rangle$ and the probability of exciting the atom is limited by the value 1/2. This similarity is not surprising, because for the AJC resonance the modulation frequency is $\eta \approx \Delta_+ \approx 2\omega$, and the artificial atom plays the role of the stationary two-level atom and the cavity modulating mechanism at the same time.

4. Jaynes-Cummings (JC) resonance

In the dispersive regime the 'Jaynes-Cummings' (JC) resonance occurs for

$$\eta = \eta_{JC} \equiv |\Delta_-| - \xi. \tag{24}$$

For $\Delta_- > 0$, repeating the steps that led us to (18), we get the effective JC Hamiltonian

$$H_{JC} \simeq \frac{\xi + \delta\,(1 + 2\hat{n})}{2}\sigma_z + (g\theta\; a\sigma_+ + h.c.)\,. \tag{25}$$

If $\Delta_- < 0$, we obtain (25) upon the replacements $\theta \to -\theta^*$ and $\xi \to -\xi$. Thus, by employing the JC resonance and adjusting the resonance shift ξ, one can couple the subspaces $\{|g,n\rangle, |e,n-1\rangle\}$ when the atom and the field are far off resonant, since the external modulation supplies the energy difference $|\Omega_0 - \omega|$ necessary to couple the atom and the cavity field.

The JC behavior is illustrated in figure 3 for the initial Fock state $|g,n_0\rangle$. In figures 3a and 3b we plot N and $P_g = 1 - P_e$, respectively, using the same parameters as in figure 1 and setting $\eta = \Delta_- + 2\delta_0\,(1 + n_0)$. As expected from the effective JC Hamiltonian (25) one has transitions between the states $\{|g,n\rangle, |e,n-1\rangle\}$ for large detuning, $|\Delta_-| \gg g$. Thus, in the dispersive regime one could engineer entangled states with several photons from the initial vacuum state $|g,0\rangle$ by alternating between the AJC and JC resonances and controlling the time interval of each resonance.

5. Dynamical Casimir effect (DCE) resonance

In the dispersive regime, the 'dynamical Casimir effect' (DCE) resonance occurs for

$$\eta = \eta_{DCE} \equiv \Delta_+ - \Delta_- - 2\xi = 2\omega - 2\xi. \tag{26}$$

Performing RWA in the interaction Hamiltonian (5) we obtain to the first order in ε/η

$$H_I \simeq g e^{i\Delta_- t} a\sigma_+ + g\theta e^{i(\Delta_- + 2\xi)t} a^\dagger \sigma_+ + h.c. \tag{27}$$

Applying the unitary time-dependent transformation defined by the operator

$$U_1 = \exp\left\{ i\left[\xi\hat{n} + (\Delta_- + \xi)\,\sigma_z/2\right] t\right\}, \tag{28}$$

we obtain the time-independent Hamiltonian consisting of the JC Hamiltonian plus the anti-rotating term multiplied by the adjustable coupling $g\theta$

$$H_I^{(1)} \simeq \xi\hat{n} + \frac{\Delta_- + \xi}{2}\sigma_z + \left(g a\sigma_+ + g\theta a^\dagger \sigma_+ + h.c.\right)\,. \tag{29}$$

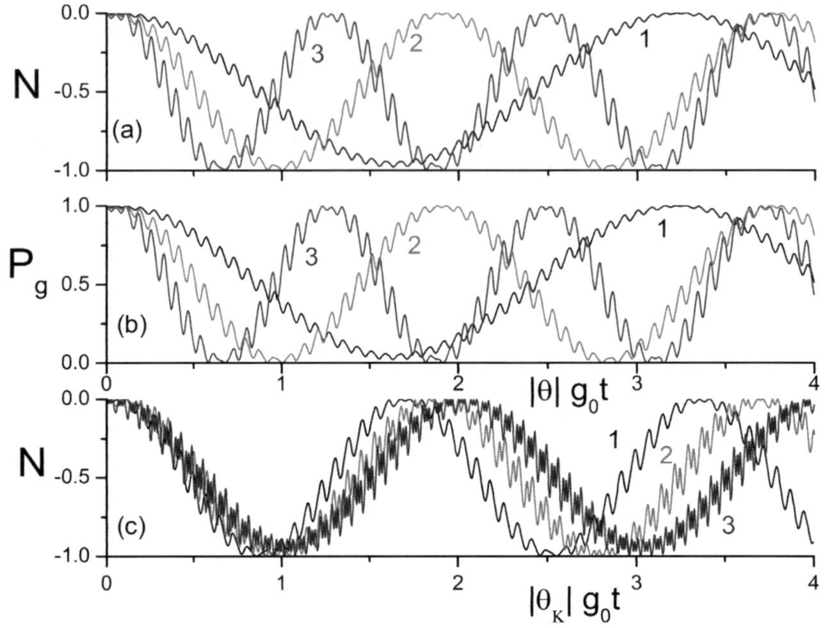

Figure 3. The results of the numerical integration of the Schrödinger equation with the Rabi Hamiltonian (3) for the JC resonance, $\eta = \Delta_- + 2\delta_0\,(1+n_0)$, considering initial number states $|g, n_0\rangle$ with $n_0 = 1$ (line 1), $n_0 = 3$ (line 2) and $n_0 = 7$ (line 3). **a)** The number of created photons N and **b)** the probability of finding the atom in the ground state P_g as functions of the dimensionless time demonstrate a transfer between the states $|g, n_0\rangle$ and $|e, n_0 - 1\rangle$ in the dispersive regime. **c)** N for the initial number state $|g, 4\rangle$ and three lower JC resonances ($\eta = \eta_-^{(K)}$, $K = 1, 2, 3$ as shown, see section 6) for the resonance shift $\xi = -10\delta_0$.

Since $|g/\Delta_-| \ll 1$, we can obtain an effective Hamiltonian by applying a sequence of small unitary transformations on $H_I^{(1)}$ [66] and performing the Hausdorff expansion in some small parameter at each step. Here our expansion parameter is $\epsilon = |g/\Delta_-|$ and, assuming that $\theta \sim \mathcal{O}(\epsilon)$, we consider terms up to the second order in ϵ. Applying first the 'rotating' unitary transformation defined by the time-independent operator

$$U_r = \exp\left[\left(ga\sigma_+ - h.c.\right)/\Delta_-\right] \tag{30}$$

followed by the 'antirotating' transformation defined by the operator

$$U_a = \exp\left[\left(g\theta a^\dagger \sigma_+ - h.c.\right)/\left(\Delta_- + 2\xi\right)\right], \tag{31}$$

we obtain the effective Hamiltonian

$$H_{eff} = U_a U_r H_I^{(1)} U_r^\dagger U_a^\dagger \tag{32}$$

of the following form (considering terms up to the second order in ϵ)

$$H_{eff} \simeq \left(\xi + \delta\sigma_z\right)\hat{n} + \delta\sigma_z\left(\theta^* a^2 + h.c.\right) + \frac{\Delta_- + \delta + \xi}{2}\sigma_z - \frac{2\delta}{\Delta_-}\left(ga\hat{n}\sigma_+ + h.c.\right) + \mathcal{O}\left(\epsilon^3\right). \tag{33}$$

Finally, moving to a rotating frame by means of the unitary time-dependent transformation defined by the operator

$$U_2 = \exp\left[-i\left(\Delta_- + \delta + \xi\right)\sigma_z t/2\right], \tag{34}$$

we obtain

$$H_{DCE} \simeq \left(\xi + \delta\sigma_z\right)\hat{n} + \delta\sigma_z(\theta a^{\dagger 2} + \theta^* a^2) - \frac{2\delta}{\Delta_-}\left(ge^{i\Delta_- t}a\hat{n}\sigma_+ + h.c.\right) + \mathcal{O}\left(\epsilon^3\right), \tag{35}$$

where we neglected $(\delta + \xi)$ compared to Δ_- in the exponential of the third term.

This is the main result of this section. The first two terms of the effective Hamiltonian (35) form the DCE part and the remaining terms represent corrections, oscillating with a high frequency $\sim \Delta_-$. For small $\langle\hat{n}\rangle \ll n_c \sim (\Delta_-/g)^2$, the contribution of these corrections is relatively small and σ_z becomes approximately a constant. Assuming that initially the atom is in the ground state, then for $\langle\hat{n}\rangle \ll n_c$ one has $\sigma_z \approx -1$ and the right-hand side of equation (35) transforms into the DCE Hamiltonian [4,5,11]

$$H_{DCE} \simeq \left(\xi - \delta\right)\hat{n} - \delta(\theta a^{\dagger 2} + \theta^* a^2). \tag{36}$$

If the atom is initially in the excited state, a similar effective Hamiltonian is obtained through the substitution $\delta \to -\delta$. Therefore, by adjusting the frequency shift to $\xi = \pm\delta$, depending on the initial atomic state [1], we have photon pairs creation from vacuum, as well as field amplification, due to an analog of the DCE.

In this case the DCE is simulated by the atomic transition frequency modulation through the strong atom-cavity coupling. However, the photon generation process is not steady because after several photons have been created the third and higher order terms become important and the photon generation is interrupted. Indeed, the third term describes the non-resonant atomic excitation by means of the absorption of photons from the cavity. Nevertheless, the Hamiltonian (35) shows that it is possible to simulate DCE and generate several photons from vacuum using a single artificial atom.

5.1. Qualitative explanation and numerical results

This phenomenon can be qualitatively understood as follows. In the dispersive regime the atom acts as an effective non-linear capacitance [38], pulling the cavity frequency to approximately

$$\omega\left(t\right) \approx \omega + \frac{\sigma_z g_0^2}{\Delta_- + \varepsilon f_t} \approx \left(\omega + \sigma_z\delta\right) - \sigma_z\delta\frac{\varepsilon}{\Delta_-}f_t\,. \tag{37}$$

Consequently, it is expected that the periodic modulation of f_t with the modulation frequency close to $\eta \approx 2\left(\omega \pm \delta\right)$ will lead to DCE [11,62], for which the photons are generated as long as the modulation is present. The energy 2ω necessary to create pairs of photons is provided through the atomic frequency modulation and the resonance shift ξ must be adjusted [2] in order to get the constructive interference on the cavity field [11,12]. However, in our case the atom gets entangled with the field due to the third term in (35), by which the atom acquires some probability of being excited through photon absorption and the photon generation process cannot continue asymptotically due to the loss of constructive interference [the first term in (35) becomes non-zero]. This is different from the usual DCE situation, in which the properties of the macroscopic linear, lossless and nondispersive dielectric medium inside the cavity are modulated [6–9] and the field mode does not get entangled with individual atoms.

[1] If initially the atom is in the superpositions of states $|g\rangle$ and $|e\rangle$, we can choose any of the signs to obtain photon generation. However, the photon generation is optimized if initially the atom is exactly in $|g\rangle$ or $|e\rangle$.

[2] One could change ξ adiabatically in time to enhance the resonance.

In figure 4 we present the results of numerical integration of the Schrödinger equation with the Rabi Hamiltonian (3) for the DCE resonance with properly adjusted frequency shift, $\eta = 2(\omega - \delta_0)$, using the parameters $\Omega_0/\omega = 1.4, g_0/\omega = 2 \cdot 10^{-2}, \varepsilon/\omega = 4 \cdot 10^{-1}$. We show the number of created photons N (figure 4a) and the atom excitation probability P_e (figure 4b, where the fast oscillations have been averaged out) versus the dimensionless time $|\delta\theta|\, t$ for the initial ground atomic state $|g\rangle$ and the field number states $|n_0\rangle$ ($n_0 = 0, 1$) and coherent states $|g, \alpha\rangle$ ($|\alpha|^2 = 1/2, 1$). For initial times, while $\langle\hat{n}\rangle \ll n_c \sim 400$, there is photon generation and amplification, as expected. However, the atom gets entangled with the field and P_e grows as time goes on, suffering fast oscillations (not shown for clarity), as expected from equation (35).

In figures 4c and 4d we show N and P_e, respectively, for larger times and different resonance shifts $\xi = x\delta_0$, $x = \{0.95, 1, 1.05, 1.15, 1.25\}$, considering the initial state $|g, 0\rangle$. In the first place, we see that the photon generation is very sensitive to the resonance shift in the vicinity of $\xi = \delta$, and near the 'resonance', $x \approx 1$, there is a substantial photon creation from vacuum. However, the photon generation process ceases after several photons have been created and the photon number decreases, although the photon generation process restarts afterwards (data not shown). P_e resembles the behavior of N, but remains small for all the times. We also verified numerically that similar results are obtained for different atomic initial states, in agreement with equation (35). Therefore, for $\eta = 2(\omega \pm \delta)$ one obtains photon generation from vacuum, as well as photon number amplification, analogously to the dynamical Casimir effect [11].

6. Discussion of results for periodic modulation

In previous sections we have considered only the first order resonances. In general, the K-th order resonances occurs for an integer K when

$$\eta = \eta_i^{(K)} \equiv K^{-1}\eta_i, \tag{38}$$

where η_i stands for the AJC, JC and DCE resonances. In this case one recovers the previous results upon substitutions

$$\delta \to \delta_K \quad \text{and} \quad \theta \to \theta_K, \tag{39}$$

as can be deduced from equations (5) and (7). For instance, to obtain θ_K, one expands both sums on right-hand side of equation (7) and make RWA, keeping the terms which oscillate with the lowest frequency ξ. So θ_K contains contributions due to the nonharmonic shape of the modulation (e.g. $\Lambda_K\, \varepsilon/\eta$), as well as due to the powers of (ε/η) [e.g. $(\Lambda_1\, \varepsilon/\eta)^K/K!$]. The effective dispersive shift δ_K is obtained analogously, by first performing RWA in (7) and keeping terms which oscillate with frequencies higher than ξ, and then calculating the effective Hamiltonian, as in equations (14) and (15). So δ_K contains the contribution of many terms besides δ_0, among them the Bloch-Siegert shift g^2/Δ_+ [66] and powers of (ε/η). As an example, in figure 3c we show N versus $|\theta_K|\, g_0 t$ for the initial number state $|g, 4\rangle$ and the first three JC resonances, considering the resonance shift $\xi = -10\delta_0$. This example illustrates that higher order resonances are readily accessible. However, for the higher order resonances the effective dispersive shift δ_K should be carefully evaluated either numerically or experimentally, otherwise there is a risk of missing the exact resonance, since $|\theta_K|$ becomes smaller and there is less freedom in committing small errors in the resonance shift ξ.

Our results can be directly transposed to the situation where $\Omega(t) = \Omega$ is constant and $g_0 = g_0(t)$ has a periodic time modulation,

$$g_0(t) = g_0 + \varepsilon f_t. \tag{40}$$

In this case, in the interaction picture the interaction Hamiltonian is

$$H_I = \left[(g_0 + \varepsilon c_0) - \varepsilon \sum_{k=1}^{\infty} k\left(\Lambda_k^* e^{ik\eta t} + \Lambda_k e^{-ik\eta t}\right)\right]\left(e^{i(\Omega-\omega)t} a\sigma_+ + e^{i(\Omega+\omega)t} a^\dagger \sigma_+ + h.c.\right). \tag{41}$$

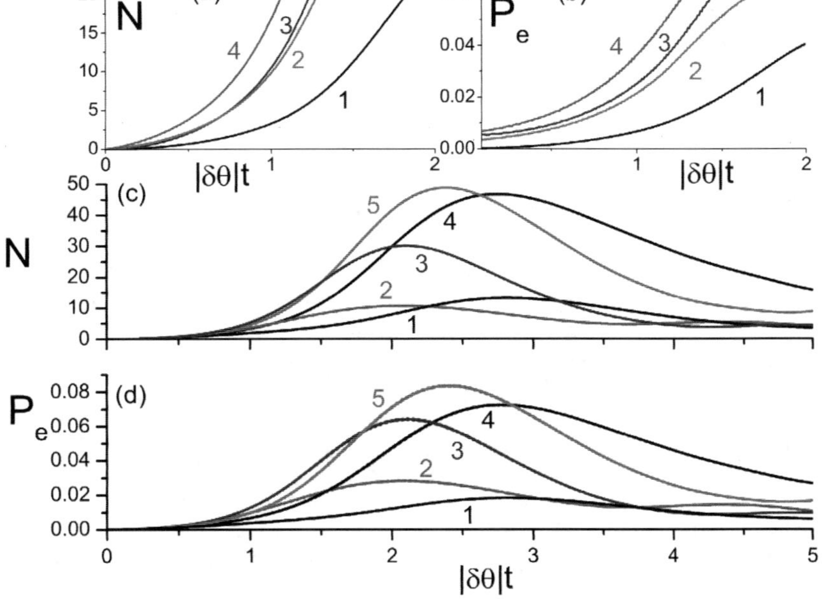

Figure 4. The initial stage of the time evolution of **a)** the number of created photons N and **b)** the atom excitation probability P_e, obtained by the numerical integration of the Schrödinger equation with the Rabi Hamiltonian (3) for $\eta = 2(\omega_r - \delta_0)$, for the initial Fock states $|g, n_0\rangle$ with $n_0 = 0$ (line 1), $n_0 = 1$ (line 2) and coherent states $|g, \alpha\rangle$ with $|a|^2 = 1/2$ (line 3) and $|a|^2 = 1$ (line 4). There is photon creation due to an analog of the dynamical Casimir effect, as described by the first two terms in the effective Hamiltonian (35). **c)** N and **d)** P_e for larger times and initial state $|g, 0\rangle$. Here $\eta = 2(\omega_r - x\delta_0)$ with $x = 0.95$ (line 1), $x = 1.25$ (line 2), $x = 1.15$ (line 3), $x = 1$ (line 4) and $x = 1.05$ (line 5). The photon creation is very sensitive to the resonance shift, and the generation process is interrupted due to the atom-field entanglement, as expected from the third term in equation (35).

If one expands $\exp(i\Xi_\pm)$ in equation (7) up to the first order in ε/η, Hamiltonian (5) becomes equivalent to Hamiltonian (41), so the results obtained above for $\Omega(t)$ also hold for $g_0(t)$ after making appropriate substitutions. The main difference between these two cases is that in the $\Omega(t)$ case the higher order K-th resonances occur both due to the powers of ε/η and non-zero coefficients Λ_K, while in the $g_0(t)$ case the higher order resonances are due only to non-zero coefficients Λ_K. Therefore, for a harmonic modulation, $f_t = \sin \eta t$, in the $\Omega(t)$ case one has higher order resonances, but in the $g_0(t)$ case there are only the first order resonances. Finally, if the cavity frequency ω is modulated periodically, with constant Ω and g_0, the AJC and JC resonances also occur, besides the well known DCE resonance [62–65].

An experimental verification of this scheme seems possible in circuit QED architecture with superconducting qubits and coplanar waveguide resonators [38], where one can adjust the system parameters *in situ* via electric and magnetic fields, as demonstrated in several experiments [33, 37, 43]. Moreover, several schemes to read the cavity and the atomic states are currently available [38, 39, 47] (or under investigation [48]). The main task would be finding a way to modulate periodically the atomic transition frequency with a stable modulation frequency $\eta \sim 10\,\mathrm{GHz}$, what is also within an experimental reach [37]. One could also use this scheme to couple M identical qubits (e.g. superconducting 2-level atoms [37, 43] or a cloud of polar molecules [30, 31]) to the same cavity mode and modulate the frequency of M atoms simultaneously, since in this case the effective coupling is increased to $\sqrt{M} g_0$.

One important point we did not analyze here is the dissipation and decoherence of both the artificial atom and the cavity due to the noisy solid state environment [48, 67]. Recent experiments achieved experimental values $\{\kappa/\omega < 10^{-4}, \gamma/\omega < 10^{-3}, \gamma_{ph}/\omega < 10^{-3}\}$ [39], where κ is the cavity decay rate, γ is the atomic decay rate and γ_{ph} is the atomic pure dephasing rate. To deal with dissipation in a qualitative manner, we compare the rates of the photon generation from vacuum for each resonance to the dissipation rates. We take the current experimental value of the coupling constant $g_0/\omega \approx 2 \cdot 10^{-2}$ [39] and assume $\varepsilon/\omega \sim \Delta_-/\omega \sim 10^{-1}$ to make the estimative. The photon creation rate for the first order DCE resonance is roughly $|\delta\theta|/\omega \sim 10^{-4}$ [equation (35)], and for the first order AJC resonance is roughly $g_0 |\theta|/\omega \sim 10^{-3}$ [equation (18)]. Both these values are larger than the dissipation and decoherence rates (or of the same order of magnitude). Therefore, the photon creation due to modulation of $\Omega(t)$ or $g_0(t)$ seems possible in the future.

7. A glance at non-periodic modulation

Let us now consider that the coupling parameter $g(t)$ has a bichromatic external modulation

$$g(t) = g_0 + 2\varepsilon \sin[(\Delta_+ - 2\xi)t] + 2\varepsilon_- \sin[(\Delta_- - 2\xi_-)t], \tag{42}$$

where g_0 is the bare coupling parameter, $|\varepsilon|, |\varepsilon_-| \ll g_0$ are small modulation amplitudes and ξ, ξ_- are the 'resonance shifts'. Here $\Delta_\pm$ is given by equation (10) with $c_0 = 0$ (i.e., $\Delta_\pm = \Omega \pm \omega$) and we assume $\Delta_- > 0$. The system Hamiltonian is given by equation (3) with time-dependent $g(t)$ and constant ω and Ω. In the dispersive regime, we obtain the effective Hamiltonian

$$H_{eff} \simeq \delta(\hat{n} + 1/2)\sigma_z + (i\varepsilon e^{2i\xi t} a^\dagger \sigma_+ + i\varepsilon_- e^{2i\xi_- t} a\sigma_+ + h.c.). \tag{43}$$

Performing the unitary time-dependent transformation

$$U_3 = \exp\{i[(\xi + \xi_-)\sigma_z/2 + (\xi - \xi_-)\hat{n}]t\} \tag{44}$$

we get the time-independent Hamiltonian combining the 'rotating' and 'anti-rotating' terms with adjustable couplings

$$\begin{aligned} H_{np} &\simeq U_3^{-1} H_{eff} U_3 - iU_3^{-1}\frac{\partial U_3}{\partial t} \\ &= \frac{\xi + \xi_- + \delta(1 + 2\hat{n})}{2}\sigma_z + (\xi - \xi_-)\hat{n} + \left(i\varepsilon a^\dagger \sigma_+ + i\varepsilon_- a\sigma_+ + h.c.\right). \end{aligned} \tag{45}$$

We take equal resonance shifts in order to cancel the second term in (45), $\xi = \xi_-$, so for $\varepsilon_- = 0$ ($\varepsilon = 0$) we obtain the effective Anti-Jaynes-Cummings (Jaynes-Cummings) Hamiltonian. A more interesting regime occurs when both ε and ε_- are different from zero. For $\varepsilon = \varepsilon_-$ we obtain

$$H_+ \simeq \frac{2\xi + \delta(1 + 2\hat{n})}{2}\sigma_z + i\varepsilon(a + a^\dagger)(\sigma_+ - \sigma_-), \tag{46}$$

while for $\varepsilon = -\varepsilon_-$ we get (46) with the last term replaced by $i\varepsilon\left(a - a^\dagger\right)(\sigma_+ + \sigma_-)$. The Hamiltonian (46) cannot be integrated exactly, so below we solve it numerically and compare the results to the ones obtained using the initial Rabi Hamiltonian (3).

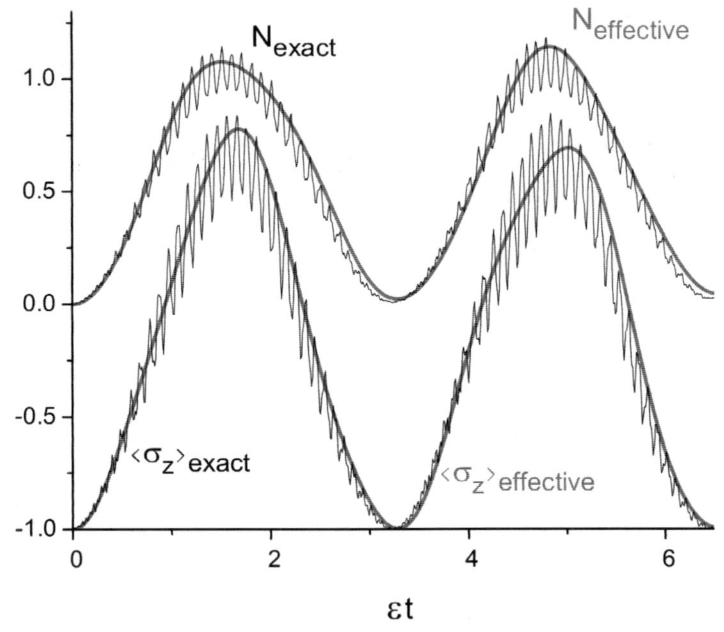

Figure 5. The mean number of created photons N and atomic population inversion $\langle s_z \rangle$ obtained numerically for the exact [equation (3)] and effective [equation (46)] Hamiltonians for the resonance shift $\xi = -\delta_0$. The exact N and $\langle \sigma_z \rangle$ show rapid oscillations, while the effective ones are smooth.

To obtain photon creation from the initial vacuum state $|g, 0\rangle$, we have to make the exponent multiplying σ_z in (46) as small as possible by adjusting the resonance shift ξ. In figure 5 we show the mean number of created photons N and the population inversion $\langle s_z \rangle$ as function of dimensionless time εt for $\varepsilon = \varepsilon_-$ and $\xi = -\delta_0$, calculated numerically for the initial Hamiltonian (3) and the effective Hamiltonian (46) using the parameters $\Omega/\omega = 1.3$, $g_0/\omega = 5 \cdot 10^{-2}$ and $\varepsilon/\omega = 5 \cdot 10^{-3}$. One can see that the effective Hamiltonian H_+ describes well the exact dynamics, and there is a quasi-periodic generation of a few photons, together with the atomic excitation. Qualitatively, this occurs because for initial times the Hamiltonian (46) becomes

$$H_+ \simeq -\sqrt{2}\,\varepsilon x \sigma_y, \tag{47}$$

where x is the cavity field quadrature operator and σ_y is the Pauli pseudo-spin operator

$$x \equiv \frac{a + a^\dagger}{\sqrt{2}}, \quad \sigma_y \equiv \frac{\sigma_+ - \sigma_-}{i}, \tag{48}$$

and the photons are generated from vacuum quadratically in time. However, after ~ 1 photons have been created, the exponent multiplying σ_z in (46) becomes large, so the photon creation process goes out of resonance and it is interrupted.

To create a large number of photons from vacuum, we can adopt an 'active' approach, by which the resonance shift ξ is adjusted *adiabatically* as function of time, so the effective

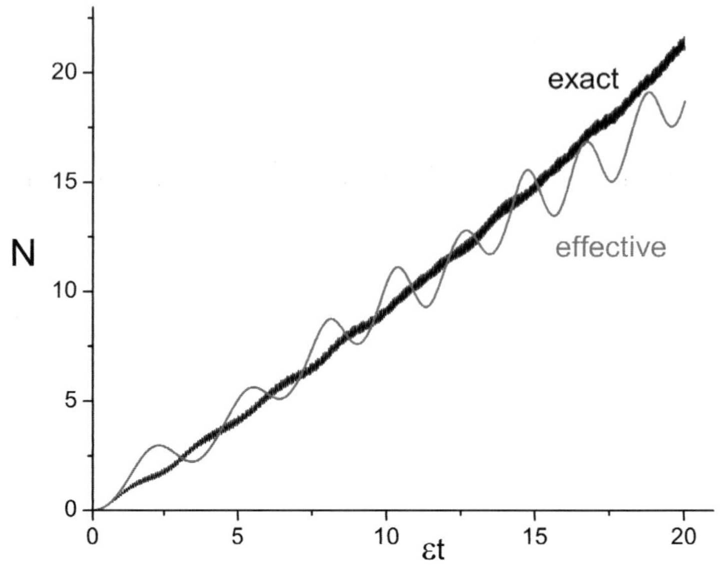

Figure 6. The mean number of created photons N obtained numerically for the exact [equation (3)] and effective [equation (46)] Hamiltonians for the resonance shift $\xi = \tilde{\xi}(t)$ [equation (49)]. The effective N shows slow oscillations, while the exact one does not.

Hamiltonian (45) is valid. In this way, we continuously adjust the resonance shift to the number of photons in the cavity, provided we choose an appropriate functional form of ξ. In figure 6 we show the mean number of created photons N from the initial state $|g,0\rangle$ for the Rabi Hamiltonian (3) and the effective Hamiltonian (46), using the parameters $\Omega/\omega = 1.4$, $g_0/\omega = 5\cdot 10^{-2}$, $\varepsilon/\omega = 5\cdot 10^{-3}$ and setting

$$\xi = \tilde{\xi}(t) = -\left(\delta_0 - 3g_0^2/\Delta_+\right)(1/2 + \varepsilon t)\,. \tag{49}$$

One can see that a significant amount of photons can be generated from vacuum, provided the resonance shift is adjusted adiabatically, and the effective Hamiltonian H_+ describes well the exact dynamics.

Finally, notice that these results also hold if, instead of the coupling parameter g, one modulates the atomic transition frequency Ω, or both g and Ω simultaneously, one with frequency $\sim \Delta_+$, and another with $\sim \Delta_-$.

8. Conclusions

In conclusion, we studied the nonstationary solid state cavity QED system in which the atomic transition frequency or the atom-cavity coupling have a small periodic time modulation, prescribed externally. We showed that in the dispersive regime and under the resonant modulations, the Rabi Hamiltonian (which describes the system dynamics) can be significantly simplified, resulting in three different regimes that can be described by the Anti-Jaynes-Cummings, Jaynes-Cummings or the dynamical Casimir effect Hamiltonians. Moreover, in the resonant atom-cavity regime, entangled states with two photons can be created from the vacuum state $|g,0\rangle$ under the corresponding resonance, analogously to the dynamical Casimir effect in a vibrating cavity containing a resonant two-level atom.

We also showed that the photon generation from vacuum occurs for a small non-periodic time modulation of the atom-cavity coupling, the atomic transition frequency, or both in circuit QED. We deduced an effective Hamiltonian for the bichromatic modulation in the dispersive regime, and demonstrated that it describes well the exact dynamics. Numerical simulations confirmed that several photons can be generated from vacuum provided the modulation frequencies are fine tuned adiabatically.

This study illustrates the importance of the anti-rotating terms in the Rabi Hamiltonian, ignored in the Jaynes-Cummings model – here this term is responsible for photon generation from vacuum and field amplification. Moreover, one could engineer effective interactions in nonstationary circuit QED by means of modulating the system parameters. As applications, this scheme can be used to verify photon creation from vacuum in nonstationary circuit QED due to an analog of the dynamical Casimir effect using a single atom, as well as off-resonant population transfer between the states $\{|g,n\rangle, |e,n\pm 1\rangle\}$ and generation of entangled states with several photons.

Acknowledgments

Work supported by FAPESP (SP, Brazil, contract No. 04/13705-3), ITAMP (at Harvard University and Smithsonian Astrophysical Observatory) and (partially) by National Science Foundation.

References

[1] Sassaroli E, Srivastava Y N and Widom A 1994 Photon production by the dynamical Casimir effect *Phys. Rev.* A **50** 1027
[2] Dalvit D A R and Mazzitelli F D 1998 Renormalization-group approach to the dynamical Casimir effect *Phys. Rev.* A **57** 2113
[3] Dodonov V V 1998 Dynamical Casimir effect in a nondegenerate cavity with losses and detuning *Phys. Rev.* A **58** 4147
[4] Law C K 1994 Resonance response of the quantum vacuum to an oscillating boundary *Phys. Rev. Lett.* **73** 1931
[5] Law C K 1995 Interaction between a moving mirror and radiation pressure: A Hamiltonian formulation *Phys. Rev.* A **51** 2537
[6] Yablonovitch E 1989 Accelerating reference frame for electromagnetic waves in a rapidly growing plasma: Unruh–Davies–Fulling–De Witt radiation and the nonadiabatic Casimir effect *Phys. Rev. Lett.* **62** 1742
[7] Schwinger J 1993 Casimir light: a glimpse *Proc. Nat. Acad. Sci. USA* **90** 958
[8] Dodonov V V, Klimov A B and Nikonov D E 1993 Quantum phenomena in nonstationary media *Phys. Rev.* A **47** 4422
[9] Johnston H and Sarkar S 1995 Moving mirrors and time-varying dielectrics *Phys. Rev.* A **51** 4109
[10] Barton G Dodonov V V and Man'ko V I 2005 The nonstationary Casimir effect and quantum systems with moving boundaries *J. Opt. B: Quant. Semiclass. Opt.* **7** S1
[11] Dodonov V V 2001 Nonstationary Casimir effect and analytical solutions for quantum fields in cavities with moving boundaries *Modern Nonlinear Optics* (*Advances in Chemical Physics Series* vol 119, part 1) ed M W Evans (New York: Wiley) pp 309–94 (*Preprint* arXiv: quant-ph/0106081)
[12] Dodonov V V and Dodonov A V 2005 Quantum harmonic oscillator and nonstationary Casimir effect *J. Rus. Las. Res.* **26** 445
[13] Dodonov A V and Dodonov V V 2005 Resonance generation of photons from vacuum in cavities due to strong periodical changes of conductivity in a thin semiconductor layer *J. Opt. B: Quant. Semiclass. Opt.* **7** S47
[14] Dodonov V V and Dodonov A V 2006 The nonstationary Casimir effect in a cavity with periodical time-dependent conductivity of a semiconductor mirror *J. Phys. A: Math. and General* **39** 6271
[15] Kim W-J, Brownell J H Onofrio R 2006 Detectability of dissipative motion in quantum vacuum via superradiance *Phys. Rev. Lett.* **96** 200402
[16] Segev E, Abdo B, Shtempluck O, Buks E and Yurke B 2007 Prospects of employing superconducting stripline resonators for studying the dynamical Casimir effect experimentally *Phys. Lett. A* **370** 202

[17] De Liberato S, Ciuti C and Carusotto I 2007 Quantum vacuum radiation spectra from a semiconductor microcavity with a time-modulated vacuum Rabi frequency *Phys. Rev. Lett.* **98** 103602
[18] Carusotto I, Antezza M, Bariani F, De Liberato S and Ciuti C 2008 Optical properties of atomic Mott insulators: From slow light to dynamical Casimir effects *Phys. Rev.* A **77** 063621
[19] Braggio C, Bressi G, Carugno G, Lombardi A, Palmieri A, Ruoso G and Zanello D 2004 Semiconductor microwave mirror for a measurement of the dynamical Casimir effect *Rev. Sci. Instr.* **75** 4967
[20] Braggio C, Bressi G, Carugno G, Del Noce C, Galeazzi G, Lombardi A, Palmieri A, Ruoso G and Zanello D 2005 A novel experimental approach for the detection of the dynamical Casimir effect *Europhys. Lett.* **70** 754
[21] Agnesi A, Braggio C, Bressi G, Carugno G, Galeazzi G, Pirzio F, Reali G, Ruoso G Zanello D 2008 MIR status report: an experiment for the measurement of the dynamical Casimir effect *J. Phys. A: Math. Theor.* **41** 164024
[22] Raimond J , Brune M and Haroche S 2001 Manipulating quantum entanglement with atoms and photons in a cavity *Rev. Mod. Phys.* **73** 565
[23] Mabuchi H and Doherty A C 2002 Cavity quantum electrodynamics: Coherence in context *Science* **298** 1372
[24] Reithmaier J P, Sek G, Loffler A, Hofmann C, Kuhn S, Reitzenstein S, Keldysh L V, Kulakovskii V D, Reinecke T L and Forchel A 2004 Strong coupling in a single quantum dot-semiconductor microcavity system *Nature* **432** 197
[25] Yoshie T, Scherer A, Hendrickson J, Khitrova G, Gibbs H M, Rupper G, Ell C, Shchekin O B and Deppe D G 2004 Vacuum Rabi splitting with a single quantum dot in a photonic crystal nanocavity *Nature* **432** 200
[26] Englund D, Faraon A, Fushman I, Stoltz N, Petroff P and Vuckovic J 2007 Controlling cavity reflectivity with a single quantum dot *Nature* **450** 857
[27] Fushman I, Englund D, Faraon A, Stoltz N, Petroff P and Vuckovic J 2008 Controlled phase shifts with a single quantum dot *Science* **320** 769
[28] Badolato A, Hennessy K, Atature M, Dreiser J, Hu E, Petroff P M and Imamoglu A 2005 Deterministic coupling of single quantum dots to single nanocavity modes *Science* **308** 1158
[29] Hennessy K, Badolato A, Winger M, Gerace D, Atature M, Gulde S, Falt S, Hu E L and Imamoglu A 2007 Quantum nature of a strongly coupled single quantum dot-cavity system *Nature* **445** 896
[30] Andre A, Demille D, Doyle J M, Lukin M D, Maxwell S E, Rabl P, Schoelkopf R J and Zoller P 2006 A coherent all-electrical interface between polar molecules and mesoscopic superconducting resonators *Nature Physics* **2** 636
[31] Rabl P, DeMille D, Doyle J M, Lukin M D, Schoelkopf R J and Zoller P 2006 Hybrid quantum processors: Molecular ensembles as Quantum memory for solid state circuits *Phys. Rev. Lett.* **97** 033003
[32] Chiorescu I, Bertet P, Semba K, Nakamura Y, Harmans C J P M and Mooij J E 2004 Coherent dynamics of a flux qubit coupled to a harmonic oscillator *Nature* **431** 159
[33] Wallraff A, Schuster D I, Blais A, Frunzio L, Huang R S, Majer J, Kumar S, Girvin S M and Schoelkopf R J 2004 Strong coupling of a single photon to a superconducting qubit using circuit quantum electrodynamics *Nature* **431** 162
[34] Johansson J, Saito S, Meno T, Nakano H, Ueda M, Semba K and Takayanagi H 2006 Vacuum Rabi oscillations in a macroscopic superconducting qubit LC oscillator system *Phys. Rev. Lett.* **96** 127006
[35] Zagoskin A and Blais A 2007 Superconducting qubits *Phys. Can.* **63** 215 (*Preprint* arXiv: 0805.0164)
[36] Sandberg M, Wilson C M, Persson F, Johansson G, Shumeiko V, Duty T and Delsing P 2008 In-situ frequency tuning of photons stored in a high Q microwave cavity *Preprint* arXiv: 0801.2479.
[37] Sillanpää M A, Park J I and Simmonds R W 2007 Coherent quantum state storage and transfer between two phase qubits via a resonant cavity *Nature* **449** 438
[38] Blais A, Huang R-S, Wallraff A, Girvin S M and Schoelkopf R J 2004 Cavity quantum electrodynamics for superconducting electrical circuits: An architecture for quantum computation *Phys. Rev.* A **69** 062320
[39] Schuster D I, Houck A A, Schreier J A, Wallraff A, Gambetta J M, Blais A, Frunzio L, Majer J, Johnson B, Devoret M H, Girvin S M and Schoelkopf R J 2007 Resolving photon number states in a superconducting circuit *Nature* **445** 515
[40] Houck A A, Schuster D I, Gambetta J M, Schreier J A, Johnson B R, Chow J M, Frunzio L, Majer J, Devoret M H, Girvin S M and Schoelkopf R J 2007 Generating single microwave photons in a circuit *Nature* **449** 328
[41] Astafiev O, Inomata K, Niskanen A O, Yamamoto T, Pashkin Yu A, Nakamura Y and Tsai J S 2007 Single artificial-atom lasing *Nature* **449** 588
[42] Hofheinz M, Weig E M, Ansmann M, Bialczak R C, Lucero E, Neeley M, O'Connell A D, Wang H, Martinis J M and Cleland A N 2008 Generation of Fock states in a superconducting quantum circuit *Nature* **454** 310

[43] Majer J, Chow J M, Gambetta J M, Koch J, Johnson B R, Schreier J A, Frunzio L, Schuster D I, Houck A A, Wallraff A, Blais A, Devoret M H, Girvin S M and Schoelkopf R J 2007 Coupling superconducting qubits via a cavity bus *Nature* **449** 443

[44] Schuster D I, Wallraff A, Blais A, Frunzio L, Huang R-S, Majer J, Girvin S M and Schoelkopf R J 2005 ac Stark shift and dephasing of a superconducting qubit strongly coupled to a cavity field *Phys. Rev. Lett.* **94** 123602

[45] Gambetta J, Blais A, Schuster D I, Wallraff A, Frunzio L, Majer J, Devoret M H, Girvin S M and Schoelkopf R J 2006 Qubit-photon interactions in a cavity: Measurement-induced dephasing and number splitting *Phys. Rev.* A **74** 042318

[46] Gambetta J, Braff W A, Wallraff A, Girvin S M and Schoelkopf R J 2007 Protocols for optimal readout of qubits using a continuous quantum nondemolition measurement *Phys. Rev.* A **76** 012325

[47] Wallraff A, Schuster D I, Blais A, Frunzio L, Majer J, Devoret M H, Girvin S M and Schoelkopf R J 2005 Approaching unit visibility for control of a Superconducting qubit with dispersive readout *Phys. Rev. Lett.* **95** 060501

[48] Blais A, Gambetta J, Wallraff A, Schuster D I, Girvin S M, Devoret M H and Schoelkopf R J 2007 Quantum-information processing with circuit quantum electrodynamics *Phys. Rev.* A **75** 032329

[49] Leek P J, Fink J M, Blais A, Bianchetti R, Göppl M, Gambetta J M, Schuster D I, Frunzio L, Schoelkopf R J and Wallraff A 2007 Observation of Berry's phase in a solid-state qubit *Science* **318** 1889

[50] Saito K, Wubs M, Kohler S, Hänggi P and Kayanuma Y 2006 Quantum state preparation in circuit QED via Landau-Zener tunneling *Europhys. Lett.* **76** 22

[51] Wubs M, Kohler S and Hanggi P 2007 Entanglement creation in circuit QED via Landau–Zener sweeps *Physica E - Low-Dim. Syst. Nanostr.* **40** 187

[52] Ciuti C and Carusotto I 2007 On the ultrastrong vacuum Rabi coupling of an intersubband transition in a semiconductor microcavity *J. Appl. Phys.* **101** 081709

[53] Ciuti C, Bastard G and Carusotto I 2005 Quantum vacuum properties of the intersubband cavity polariton field *Phys. Rev.* B **72** 115303

[54] Takashima K, Hatakenaka N, Kurihara S and Zeilinger A 2008 Nonstationary boundary effect for a quantum flux in superconducting nanocircuits *J. Phys. A: Math. Theor.* **41** 164036

[55] Dodonov A V, Céleri L C, Pascoal F, Lukin M D and Yelin S F 2008 Photon generation from vacuum in non-stationary circuit QED *Preprint* arXiv: 0806.4035

[56] Hauss J, Fedorov A, André S, Brosco V, Hutter C, Kothari R, Yeshwanth S, Shnirman A and Schön G 2008 Dissipation in circuit quantum electrodynamics: lasing and cooling of a low-frequency oscillator *New J. Phys.* **10** 095018

[57] Schleich W P 2001 *Quantum Optics in Phase Space* (Wiley-VCH Verlag, Berlin)

[58] Fink J M, Göppl M, Baur M, Bianchetti R, Leek P J, Blais A and Wallraff A 2008 Climbing the Jaynes-Cummings ladder and observing its $\sqrt{n}$ nonlinearity in a cavity QED system *Nature* **454** 315

[59] Schuster I, Kubanek A, Fuhrmanek A, Puppe T, Pinkse P W H, Murr K and Rempe G 2008 Nonlinear spectroscopy of photons bound to one atom *Nature Physics* **4** 382

[60] Bishop L S, Chow J M, Koch J, Houck A A, Devoret M H, Thuneberg E, Girvin S M and Schoelkopf R J 2008 Nonlinear response of the vacuum Rabi resonance *Preprint* arXiv: 0807.2882

[61] Werlang T, Dodonov A V, Duzzioni E I and Villas-Bôas C J 2008 Rabi model beyond the rotating-wave approximation: Generation of photons from vacuum through decoherence *Phys. Rev.* A **78** 053805

[62] Dodonov V V 1995 Photon creation and excitation of a detector in a cavity with a resonantly vibrating wall *Phys. Lett.* A **207** 126

[63] Fedotov A M, Narozhny N B and Lozovik Y E 2000 'Shaking' of an atom in a non-stationary cavity *Phys. Lett.* A **274** 213

[64] Narozhny N B, Fedotov A M and Lozovik Y E 2001 Dynamical Lamb effect versus dynamical Casimir effect *Phys. Rev.* A **64** 053807

[65] Narozhny N B, Fedotov A M and Lozovik Y E 2003 Dynamical Casimir and Lamb effects and entangled photon states *Las. Phys.* **13** 298

[66] Klimov A B and Chumakov S M 2008 *A Group Theoretical Approach to Quantum Optics* (Wiley-VCH, Berlin)

[67] Houck A A, Schreier J A, Johnson B R, Chow J M, Koch J, Gambetta J M, Schuster D I, Frunzio L, Devoret M H, Girvin S M and Schoelkopf R J 2008 Controlling the spontaneous emission of a superconducting transmon qubit *Phys. Rev. Lett.* **101** 080502

Journal of Physics: Conference Series **161** (2009) 012030 doi:10.1088/1742-6596/161/1/012030

Radiation from neutral atoms and mirrors following prescribed trajectories

Alex Calogeracos

Department of Physics & Astronomy, University of Sussex, Brighton BN1 9QH, England
Physics Division, Air Force Academy, TG1010 Dekelia Air Force Base, Greece

E-mail: a.calogeracos@yahoo.co.uk

Abstract. Discussions of the Unruh effect are often confused by the unrealistic nature of the trajectory (acceleration ad infinitum) and by the invocation of accelerated observers. We present a formalism that allows the calculation of radiation emission from accelerated atoms and mirrors in time-dependent perturbation theory moving along asymptotically inertial trajectories. The description is unambiguously set in terms pertaining to the laboratory frame. In the limit of constant acceleration of infinite duration one recovers Unruh's result.

1. Introduction

I shall review some relatively recent work (Barton and Calogeracos 1995, 2005, 2008) on radiation emission by neutral systems such as mirrors and atoms. Consideration of such effects first took place in the seventies in a series of papers seeking deep connections between fundamental quantum physics, thermodynamics, and gravity. The accelerating trajectory of a mirror can provide a useful simulation of a black hole if one wishes to calculate for instance the spectrum of photons created during gravitational collapse. One often requires the (scalar) electromagnetic field to strictly vanish at the position of the mirror thus introducing perfect reflection at all frequencies. The ensuing underlying conformal invariance of the theory allows an elegant mathematical treatment. The absence however of a Hamiltonian formulation renders difficult the comparison with any realistic model which would incorporate partial transmission and dispersion.

A similar distance from experimentally attainable scenarios sometimes marks the work that followed Unruh's (1976) seminal observation that "an accelerated observer perceives the vacuum as a thermal bath of photons". If an atom moves through the vacuum (i.e. through the ground state of the quantized Maxwell field) at constant proper acceleration of magnitude α for τ ranging from $-\infty$ to $+\infty$ (τ denotes the proper time) then in its instantaneous rest frame the atom will experience the vacuum as if the atom were at rest, and interacting with a heat bath at temperature T_U given by

$$k_B T_U = \alpha\hbar/2\pi c, \qquad (1)$$

the subscript U standing for "Unruh". One consequence of the statement is that an atom accelerated in vacuum should undergo spontaneous excitation, as if by absorbing a heat bath photon; such photons will also affect the rates of downwards transitions. The statement preceding equation (1) tells us very little about what one ought to expect in the case of a realistic

Journal of Physics: Conference Series **161** (2009) 012030 doi:10.1088/1742-6596/161/1/012030

trajectory i.e. one starting and ending with vanishing acceleration. In this context however we must take note of the work by Marzlin and Audretsch (1998) who calculate the excitation rate of an atom performing uniform *circular* motion. For a thorough review of previous work on the Unruh effect and for a discussion of various interpretational issues see Crispino, Higuchi and Matsas (2008).

In our work we look at semi-realistic models of mirrors and atoms interacting with the electromagnetic field and following assigned trajectories. In particular our programme is based on the following requirements: (a) the theory should possess an underlying relativistically covariant Lagrangean formulation, (b) the resulting Hamiltonian is used to compute transition amplitudes using nothing more exotic than conventional time-dependent perturbation theory, (iii) asymptotically inertial trajectories (i.e. uniform velocity at $t = \pm\infty$) only need be considered.

Although atoms and mirrors seem not to have very much in common and although the velocity regimes will be very different (relativistic speeds for the former and non-relativistic for the latter) it will transpire that the technicalities are similar. For either mirror or atom our programme is divided to three stages. First we construct a Hamiltonian describing the system interacting with the radiation field. Second we construct an *effective* Hamiltonian describing the system while subject to the constraint that forces it along a prescribed trajectory. It turns out that this step is technically somewhat delicate and it eventually leads to an effective Hamiltonian coinciding with the Routhian. Third we define the process and calculate the spectrum of emitted radiation in the case of a mirror and the rate for an upwards transition in the case of a ground state atom. These steps are presented in sections 2, 3, and 4 for the mirror and in sections 5, 6, and 7 for an atom.

2. A Hamiltonian formulation for a dispersive mirror.

As a preliminary step consider a scalar electromagnetic field interacting with a fixed mirror. A convenient Lagrangean is provided by

$$L_0 = \int dx \left\{ \frac{1}{2}\left(\frac{\partial\phi}{\partial t}\right)^2 - \frac{1}{2}\left(\frac{\partial\phi}{\partial x}\right)^2 - \gamma\phi^2\delta(x-\xi) \right\}$$

where the subscript in L_0 reminds us that the mirror is fixed and ξ refers to the mirror's position. The parameter γ controls the dispersivity of the mirror in a way to be discussed presently.

Variation with respect to the field yields the wave equation

$$-\frac{\partial^2\phi}{\partial t^2} + \frac{\partial^2\phi}{\partial x^2} - 2\gamma\phi\delta(x-\xi) = 0$$

Integrating across the singularity at the mirror's position we obtain

$$\frac{\partial\phi}{\partial x}(\xi+) - \frac{\partial\phi}{\partial x}(\xi-) = 2\gamma\phi(\xi) \tag{2}$$

For left-incident waves (right-incident are treated similarly) of frequency ω we write

$$\psi_{L\omega}(x) = \theta(-x)\left[e^{i\omega x} + R(\omega)e^{-i\omega x}\right] + \theta(x)T(\omega)e^{i\omega x} \tag{3}$$

where θ is the Heaviside step-function and R, T are reflection and transmission coefficients that are determined by substituting (3) in condition (2):

$$R(\omega) = -\frac{i\gamma}{\omega + i\gamma}, \qquad T(\omega) = \frac{\omega}{\omega + i\gamma}. \tag{4}$$

Thus γ indeed controls the mirror's transparency in an obvious way.

A covariant Lagrangean describing a dispersive moving mirror coupled to the radiation field is given by (Barton and Calogeracos (1995))

$$L = -\left(M + \gamma\phi^2(\xi,t)\right)\left(1-\dot{\xi}^2\right)^{1/2} + \int dx \left\{ \frac{1}{2}\left(\frac{\partial\phi}{\partial t}\right)^2 - \frac{1}{2}\left(\frac{\partial\phi}{\partial x}\right)^2 \right\} \tag{5}$$

where $\xi(t), \dot{\xi}(t)$ are the mirror's position and momentum respectively. The first term in the Lagrangean (5) coincides with that describing a scalar particle in relativistic mechanics with the particle mass augmented by the extra field contribution $\gamma\phi^2(\xi,t)$ and the second term is the familiar electromagnetic Lagrangean. The jump condition appropriate to a moving mirror (the generalization of (2)) is obtained by variation of ϕ in both terms in (5) and taking into account contributions arising from integration by parts of the second term. We quote the Hamiltonian derived from (5)

$$H = \sqrt{p^2 + \left(M + \gamma\phi^2(\xi,t)\right)^2} + H_{rad} \tag{6}$$

and the version appropriate for small velocities

$$H_{NR} = M + \gamma\phi^2(\xi,t) + H_{rad} + \frac{p^2}{2M}\left(1 - \gamma\phi^2(\xi,t)/M\right) \tag{7}$$

where p is the momentum conjugate to the mirror's position ξ and H_{rad} is the Hamiltonian of a massless scalar field

$$H_{rad} = \int dx \frac{1}{2}\left\{ \Pi^2 + \left(\frac{\partial\phi}{\partial x}\right)^2 \right\} \tag{8}$$

(Π is the momentum conjugate to ϕ).

The Hamiltonian (7) has been used to compute mass and energy shifts of the mirror arising from its interaction with the radiation field. However in this paper we shall discuss only questions that arise when the motion of the mirror is prescribed.

3. The effective Hamiltonian for a mirror following a prescribed trajectory.

In this section we take the mirror to follow a prescribed trajectory thus in the sequel $\xi(t)$ ceases to be an operator and becomes a c-number. The technical complications that we face at this stage must be appreciated. We may trivially rewrite wavefunctions (3) so that they refer to a mirror at a fixed position ξ; however the fact that the wavefunctions do depend on ξ in an explicit manner makes the passage to a perturbative scheme highly non-obvious. The first step is to restart with Lagrangean (5) but now construct the *Routhian* (see for example Landau and Lifshitz (1976), para 41 or Sommerfeld (1952), para 42. We Legendre-transform with respect to the field variables only without touching ξ and $\dot{\xi}$:

$$H^{(1)} = -\left(M + \gamma\phi^2(\xi,t)\right)\left(1-\dot{\xi}^2\right)^{1/2} + H_{rad.} \tag{9}$$

The reader may recall that the Routhian $H^{(1)}$ is to be used as a Lagrangean to yield the equations of motion for ξ and $\dot{\xi}$ and as a Hamiltonian to yield the equations of motion for the field variables. The second step is to realize that the present problem may be profitably examined b y going over to a frame co-moving with the mirror. The most convenient way to do this in a quantum-mechanical language is via the use of an *explicitly time-dependent* canonical transformation. It is reminded that if

$$U(t) = \exp\left(iS(t)\right) \tag{10}$$

is such a transformation the Hamiltonian (call it H_{old}) transforms to a new Hamiltonian H_{new} given by

$$H_{new} = U^{-1} H_{old} U + \Delta, \qquad \Delta = i\frac{\partial U^{-1}}{\partial t} U. \tag{11}$$

If $\partial S/\partial t$ commutes with $S(t)$ then $i\partial U^{-1}/\partial t = (\partial S/\partial t)\, U$ and (11) assumes the simpler form

$$H_{new} = U^{-1} H_{old} U + \Delta, \qquad \Delta = \frac{\partial S}{\partial t}. \tag{12}$$

New states $|\overline{\psi}\rangle$ are connected to old states $|\psi\rangle$ via

$$|\overline{\psi}\rangle = U^{-1}(t)\, |\psi\rangle \tag{13}$$

and their time-evolution is governed by

$$i\frac{\partial}{\partial t} |\overline{\psi}\rangle = H_{new} |\overline{\psi}\rangle . \tag{14}$$

We now define a unitary transformation

$$U_{tr} = \exp\left(-i\xi(t) P_{rad}\right) \tag{15}$$

where P_{rad} is the momentum operator for the radiation field. We can readily check that

$$U_{tr}^{-1} \phi\left(\xi\right) U_{tr} = \phi\left(0\right).$$

In other words the mirror is now fixed at the origin of the coordinate system (this amounts to the transition to a co-moving frame). The transform $\overline{H}$ of the Routhian (9) is (using the abbreviation $B \equiv d\xi/dt$)

$$\overline{H} = H_{rad} - \left(M + \gamma\phi^2\left(\xi,t\right)\right)\left(1 - B^2\right)^{1/2} - B(t) P_{rad} \tag{16}$$

and the non-relativistic version is

$$\overline{H}_{NR} = H_{rad} + \gamma\phi^2(0) - B(t) P_{rad} \equiv \overline{H}_0 - B(t) P_{rad}. \tag{17}$$

Observe that the last term in (16) is due to the Δ term in (12) (compare exponents in (10) and (15)). The $\overline{H}_0$ part of the effective Hamiltonian is diagonalized by the modes (3). The last term is to be used perturbatively in the velocity B; the presence of P_{rad} entails the creation of pairs of photons under non-trivially varying B.

4. The spectrum of the radiation emitted by the mirror.

Hamiltonian (17) was used in Barton and Calogeracos (1995) to calculate the energy spectrum of the emitted radiation and the radiative reaction force. Here we shall discuss the former. We expand the field ϕ in terms of left-incident and right-incident modes $\psi^{(\lambda)}$ (see (3)) multiplied by creation and annihilation operators $a_\lambda^\dagger$, a_λ in the standard way and denote by $|\lambda\rangle$ the eigenstates of $\overline{H}_0$. We observe a $|\lambda\rangle$ is not a photon state; rather it is connected to a photon via $U(t)$ of equation (13). The phase carried by the eigenstates of $\overline{H}_0$ will prove crucial in the discussion of the Unruh effect. We assume that assume that at time $t = 0$ we start with the vacuum and write the solution of (14) (with $H_{new} = \overline{H}_{NR}$) in the form

$$|\overline{\psi}(t)\rangle = |0\rangle + \frac{1}{2}\sum_{\lambda\lambda'} C(\lambda\lambda'; t) e^{-i(\omega+\omega')t} |\lambda\lambda'\rangle \tag{18}$$

Standard time-dependent perturbation theory then yields

$$C(\lambda\lambda';t) = i\left\langle\lambda\lambda'\right| P_{rad}\left|0\right\rangle J_{\omega\omega'}(t)$$

where

$$J_{\omega\omega'}(t) = \int_0^t dt' e^{i(\omega+\omega')t'} B(t').$$

In the case of an asymptotically inertial trajectory and *after* the acceleration has stopped the time-dependent part of the phase of $\left|\overline{\psi}(t)\right\rangle$ equals just the energy ω of the corresponding photon state. Thus the total radiated energy is

$$\mathcal{E}_{rad}(t) = \frac{1}{2}\sum_{\lambda\lambda'}\left|\left\langle\lambda\lambda'\right| P_{rad}\left|0\right\rangle\right|^2 \left|J_{\omega\omega'}(t)\right|^2 \left(\omega+\omega'\right) e^{-\alpha(\omega+\omega')}$$

where α is a high-frequency cut-off. The calculation of the matrix element is straightforward and the result is given in terms of the reflection coefficient (4):

$$\mathcal{E}_{rad}(t) = \frac{1}{\pi^2}\int_0^\infty\int_0^\infty d\omega d\omega' \frac{\omega\omega'}{(\omega+\omega')}\left|R(\omega)\right|^2\left|J_{\omega\omega'}(t)\right|^2 e^{-\alpha(\omega+\omega')}$$

5. The Hamiltonian formulation for an atom whose nucleus follows a prescribed trajectory.

5.1. Generalities.

Our model includes a scalar radiation field interacting with a scalar nucleon and with a spin 1/2 electron and resides in three spatial dimensions. It is presented in some detail in Barton and Calogeracos (2005), (2008). I shall not duplicate the papers just cited; rather I will try to emphasize the principles underlying the calculation. The question is stated in very unambiguous terms. We consider an atom that starts at proper time $-\infty$ is in its ground state i with no photons present and moves along the z axis at velocity $-B_0$, then goes through an intermediate acceleration from $\tau = -\tau_0$ to time $\tau = \tau_0$ when the velocity is given by

$$B = c\tanh(\alpha\tau/c) \qquad (19)$$

and ends up at uniform velocity $+B_0$ at times $\tau > \tau_0$ (τ_0, α and B_0 are connected by the condition that $B(\tau)$ be continuous at $\tau = \tau_0$). We wish to calculate the rate for the electron to make a transition to an excited state f with the emission of a photon having wave-vector $\mathbf{k}$. In the limit of $\tau \to \infty$ the trajectory becomes Unruh's hyperbolic trajectory and we can compare our answer with his. Observe that the question is phrased wholly in terms of quantities pertaining to inertial observers and that there is no mention of accelerated observers.

The Routhian $H^{(1)}$ (the analog of (9)) describing the radiation field coupled to the electron and the nucleon with the nucleon's trajectory $\mathbf{R}(t)$ to be prescribed is readily written down:

$$H^{(1)} = H_{rad} + (M + Zg\phi(\mathbf{R}))\sqrt{1-B^2} + H_{electron} \qquad (20)$$

The electromagnetic couplings of the electron and nucleon are coupling g and Zg respectively, M is the nuclear mass, and the electron Hamiltonian $H_{electron}$ reads

$$H_{electron} = \alpha\cdot\mathbf{p} + \beta m + \beta g\phi(\mathbf{r}) \qquad (21)$$

Observe the mass-like coupling of the electron to the radiation field.

5.2. The sequence of transformations

We now perform a sequence of unitary transformations, the first of them U_{tr} being the exact analog of (15):

$$U_{tr} = \exp\left(-i\mathbf{R}(t)\cdot(\mathbf{P}_{rad}+\mathbf{p})\right) \tag{22}$$

where the exponent features the sum of the electron momentum and the radiation field momentum operators. Then $H^{(1)}$ is transformed via (12) to

$$H^{(2)}(\mathbf{B}) = (M + Zg\phi(\mathbf{0}))\sqrt{1-B^2} + [H_{rad} - \mathbf{B}\cdot\mathbf{P}_{rad}] + [\alpha\cdot\mathbf{p} - \mathbf{B}\cdot\mathbf{p} + \beta m + \beta g\phi(\mathbf{r})] \tag{23}$$

The $\mathbf{B}$-proportional terms in (23) are the exact analogs of the last term in (16).

Our Hamiltonian has to provide the binding potential for the electron. This is supplied in the form of a Lienard-Wiechert potential by the so-called *passive source transformation* U_{ps} that eliminates the $\phi(\mathbf{0})$ term in (23), thus decoupling the nucleon from the field, and also provides the electron-photon interaction that induces the electronic transitions (such transformations are well-known from old fashioned nucleon-meson interactions; see e. g. Wentzel (2005)). We write

$$U_{ps} = \exp\left(iS_{ps}\right), \qquad S_{ps} = -i\sum\left[a_{\mathbf{k}}\alpha^*_{\mathbf{k}} - a^\dagger_{\mathbf{k}}\alpha_{\mathbf{k}}\right] \tag{24}$$

where $a_{\mathbf{k}}, a^\dagger_{\mathbf{k}}$ are the creation and annihilation operators defined by the expansion

$$\phi(\mathbf{x}) = \sum_{\mathbf{k}}\left[a_{\mathbf{k}}\phi_{\mathbf{k}} - a^\dagger_{\mathbf{k}}\phi^*_{\mathbf{k}}(\mathbf{x})\right] \tag{25}$$

where $\phi_{\mathbf{k}}$ are plane waves. The c-numbers appearing in (24) are designed so as to eliminate the $\phi(\mathbf{0})$ and they obviously depend on the velocity.[1] We shall only quote the Hamiltonian $H^{(3)}(\mathbf{B})$ to which U_{ps} leads

$$H^{(3)}(\mathbf{B}) = \underbrace{\left[\Delta M\sqrt{1-B^2} + \Delta_{ps}\right]}_{drop} + [H_{rad} - \mathbf{B}\cdot\mathbf{P}_{rad}] + H^{(3)}_{atom}(\mathbf{B}) + H^{(3)}_{int} \tag{26}$$

where

$$H^{(3)}_{atom}(\mathbf{B}) \equiv \alpha\cdot\mathbf{p} - \mathbf{B}\cdot\mathbf{p} + \beta m + \beta V_{\mathbf{B}}(\mathbf{r}) \tag{27}$$

$$H^{(3)}_{int} \equiv \beta g\phi(\mathbf{r}) \tag{28}$$

The binding potential appearing in $H^{(3)}_{atom}(\mathbf{B})$ as a result of U_{ps} is

$$V_{\mathbf{B}}(\mathbf{r}) = -\frac{Zg^2/4\pi}{\left[r^2_{\parallel}/\left(1-B^2(t)\right) + r^2_{\perp}\right]^{1/2}} \tag{29}$$

where the $\parallel, \perp$ subscripts refer to direction parallel and perpendicular to $\mathbf{B}$. The ΔM-proportional term in (26) is a mass-shift arising from the nucleon-photon interaction and is uninteresting for current purposes. The Δ_{ps} term in (26) is of course the second term in (11) as provided by U_{ps}. Recall the remark following (25) that U_{ps} depends on the velocity $\mathbf{B}$ so Δ_{ps} gives rise to *acceleration-dependent* terms in the Hamiltonian. Because of the presence of creation and annihilation operators in S_{ps} (24) the term Δ_{ps} in fact gives rise to radiation emission *from the accelerating nucleon* (as opposed to radiation from electronic transitions).

[1] The c-numbers $\alpha_{\mathbf{k}}$ are given in BCI. Transformation (24) would have to be implemented in the case of vanishing velocity $\mathbf{B} = \mathbf{0}$ in order to get rid of the $\phi(\mathbf{0})$ term and provide the binding of the electron to the *static* course.

Journal of Physics: Conference Series **161** (2009) 012030 doi:10.1088/1742-6596/161/1/012030

The third transformation $\hat{U}_n(\mathbf{B})$results as follows. Let $|n;\mathbf{B}\rangle$ be the nth eigenstate of $H^{(3)}_{atom}(\mathbf{B})$ and $\varepsilon_n(\mathbf{B})$ the corresponding eigenvalue. According to (27) they satisfy

$$\{-\gamma_0\varepsilon_n(\mathbf{B}) - \gamma_0\mathbf{B}\cdot\mathbf{p} + \gamma\cdot\mathbf{p} + m + V_{\mathbf{B}}(\mathbf{r})\}\,|n;\mathbf{B}\rangle = 0 \tag{30}$$

Thus the atomic Hamiltonian (27) reads

$$H^{(3)}_{atom}(\mathbf{B}) = \sum_n |n;\mathbf{B}\rangle\,\varepsilon_n(\mathbf{B})\,\langle n;\mathbf{B}| \tag{31}$$

At zero velocity $|n;\mathbf{0}\rangle$ and $\varepsilon_n \equiv \varepsilon_n(\mathbf{0})$ satisfy

$$\{-\gamma_0\varepsilon_n + \gamma\cdot\mathbf{p} + m + V_{\mathbf{0}}(\mathbf{r})\}\,|n;\mathbf{0}\rangle = 0 \tag{32}$$

The transformation $\hat{U}_n(\mathbf{B})$ by definition satisfies

$$|n;\mathbf{B}\rangle = \hat{U}_n(\mathbf{B})\,|n;\mathbf{0}\rangle$$

and is explicitly constructed in BCI where it is also shown that

$$\varepsilon_n(\mathbf{B}) = \varepsilon_n\sqrt{1-B^2} \tag{33}$$

I restrict myself to a few comments. Transformation $\hat{U}_n(\mathbf{B})$ involves (a) a rescaling of $r_{\parallel}$ so that the denominator of (29) becomes just r, (b) a product of a spinor boost and a gauge transformation so that we get rid of the $\gamma_0\mathbf{B}\cdot\mathbf{p}$ in (30), (c) an overall factor that ensures the correct normalization of the wavefunctions

$$\int d^3r\psi^{(n)\dagger}(\mathbf{r},\mathbf{B})\,\psi^{(q)}(\mathbf{r},\mathbf{B}) = \delta_{nq}$$

where

$$\psi^{(q)}(\mathbf{r},\mathbf{B}) \equiv \langle\mathbf{r}|n;\mathbf{B}\rangle$$

These requirements are met by

$$\hat{U}_n(\mathbf{B}) = \left(1-B^2\right)^{1/4} U_{sc}\hat{U}_{spin}W_n \tag{34}$$

where the operators appearing above satisfy

$$U_{sc}^{-1}\left\{r_{\parallel}, r_{\perp}\right\}U_{sc} = \left\{r_{\parallel}\sqrt{1-B^2}, r_{\perp}\right\} \tag{35}$$

$$\hat{U}_{spin}^{-1}\left\{\gamma_0,\gamma_{\parallel},\gamma_{\perp}\right\}\hat{U}_{spin} = \left\{\frac{\gamma_0 + B\gamma_{\parallel}}{\sqrt{1-B^2}}, \frac{\gamma_{\parallel} + B\gamma_0}{\sqrt{1-B^2}}, \gamma_{\perp}\right\} \tag{36}$$

$$W_n = \exp\left\{i\varepsilon_n(\mathbf{B})\frac{\mathbf{B}\cdot\mathbf{r}}{\sqrt{1-B^2}}\right\} \tag{37}$$

Observe that these transformations carry their own Δ terms (see (11)) which are *acceleration-dependent* and in fact describe acceleration stresses within the atom; they will be interesting when if it ever comes to contemplating an actual experiment. Note also that transformation $\hat{U}_n(\mathbf{B})$ is *not* unitary because the spinor boost (see (b) above) is not.

The final unitary transformation connects the sets of states $\{|n;\mathbf{B}\rangle\}$ and $\{|n;\mathbf{0}\rangle\}$ and reads

$$\mathcal{U}(\mathbf{B}) = \sum_n |n;\mathbf{B}\rangle\langle n;\mathbf{0}| = \sum_n \hat{U}_n(\mathbf{B})|n;\mathbf{0}\rangle\langle n;\mathbf{0}| \tag{38}$$

and is unitary by construction:

$$\mathcal{U}^{-1}(\mathbf{B}) = \sum_n |n;\mathbf{0}\rangle\langle n;\mathbf{B}| = \sum_n |n;\mathbf{0}\rangle\langle n;\mathbf{0}|\hat{U}_n^\dagger(\mathbf{B}) = \mathcal{U}^\dagger(\mathbf{B})$$

(recall the non-unitarity of $\hat{U}_n$ and observe the presence of $\hat{U}_n^\dagger(\mathbf{B})$ in the last equality above.

We can now at last construct the final Hamiltonian by transforming $H^{(3)}(\mathbf{B})$ of (26):

$$\begin{aligned} H^{(4)}(\mathbf{B}) &= \mathcal{U}^{-1}(\mathbf{B})H^{(3)}(\mathbf{B})\mathcal{U}(\mathbf{B}) = \\ &[H_{rad} - \mathbf{B}\cdot\mathbf{P}_{rad}] + \sqrt{1-B^2(t)}\sum_n |n;\mathbf{0}\rangle\varepsilon_n\langle n;\mathbf{0}| \\ &+\sqrt{1-B^2(t)}\sum_{n,q}|n;\mathbf{0}\rangle\langle n;\mathbf{0}|\exp(-i(\varepsilon_n-\varepsilon_q)\mathbf{B}\cdot\mathbf{r}) \\ &\times g\beta\phi\left(r_\parallel\sqrt{1-B^2},\mathbf{r}_\perp\right)|q;\mathbf{0}\rangle\langle q;\mathbf{0}| \end{aligned} \tag{39}$$

We should here pause to record what the end product of this rather long subsection is. We started with the rather forbidding Hamiltonian (23) and ended with (39) where (i) the square bracket is already familiar from the mirror problem, (ii) a diagonal term describing *bound* electrons, (iii) the third term is the descendant of the interaction (28). The factor $\sqrt{1-B^2(t)}$ describes time-dilation. Recall that the Hamiltonians we have been quoting are descendants of the original Routhian (20) hence the eigenvalues are *frequencies* rather than energies. In the next subsection we shall look closer to the interplay between the frequencies that govern photon and electron states resulting from the first two terms in (39).

5.3. Evolution of photon and electron states under the free Hamiltonian and Doppler shifts. Let $|\mathbf{k},t\rangle$ be an eigenstate to the first term $H_{rad} - \mathbf{B}\cdot\mathbf{P}_{rad}$ of (39) corresponding to momentum $\mathbf{k}$. The Schroedinger equation gives the time dependence

$$|\mathbf{k},t\rangle = \exp\{i[\mathbf{R}(t)-\mathbf{R}(0)]\cdot\mathbf{k}\}\exp(-ikt)|\mathbf{k},t=0\rangle \tag{40}$$

and in the case of uniform motion

$$|\mathbf{k},t\rangle = \exp(-it(k-\mathbf{B}\cdot\mathbf{k}))|\mathbf{k},t=0\rangle \tag{41}$$

Similarly the time-evolution of the nth eigenstate $\left|\psi^{(n)(4)}\right\rangle$ of the second term in (39)

$$H_{atom}^{(4)}(\mathbf{B}) = \sqrt{1-B^2(t)}\sum_n |n;\mathbf{0}\rangle\varepsilon_n\langle n;\mathbf{0}| \tag{42}$$

evolves according to

$$\left|\psi^{(n)(4)}(t)\right\rangle = \exp(-i\varepsilon_n\tau)|n;\mathbf{0}\rangle, \qquad \tau \equiv \int^t dt'\sqrt{1-B^2(t')} \tag{43}$$

Given the remark at the end of the previous subsection the occurrence of the proper time τ in (43) is not unexpected. It is of interest to see how the auxiliary state $\left|\psi^{(n)(4)}(t)\right\rangle$ is expressed in terms of electron eigenstates of the original Hamiltonian (in the language of (13) we wish to construct $|\psi\rangle$ given $|\overline{\psi}\rangle$):

$$\left|\psi^{(n)}(t)\right\rangle \equiv U_{tr}(\mathbf{R})U_{ps}\mathcal{U}(\mathbf{B})\left|\psi^{(n)(4)}(t)\right\rangle$$

We recall that U_{ps} does not act on electronic degrees of freedom and use (38), (34) and (43):

$$\left|\psi^{(n)}(t)\right\rangle = \exp\left(-i\varepsilon_n\tau\right)U_{tr}(\mathbf{R})\left(1-B^2\right)^{1/4}U_{sc}\hat{U}_{spin}W_n\left|n;\mathbf{0}\right\rangle \tag{44}$$

The meaning of the above becomes much more transparent if we write the Schroedinger wavefunction (the inner product of the above with $\langle\mathbf{r}|$) and let the above operators act on $\langle\mathbf{r}|$):

$$\begin{aligned}\left\langle\mathbf{r}|\psi^{(n)}(t)\right\rangle &= \left(1-B^2\right)^{1/4}\exp\left(-i\varepsilon_n\tau\right)\exp\left(\frac{i\varepsilon_n B\left(r_\parallel-R_\parallel\right)}{\sqrt{1-B^2}}\right)\\ &\times\hat{U}_{spin}\left(\mathbf{B}\right)\psi_n\left(\frac{r_\parallel-R_\parallel}{\sqrt{1-B^2}},\mathbf{r}_\perp-\mathbf{R}_\perp;\mathbf{0}\right)\end{aligned} \tag{45}$$

where

$$\psi_n\left(\mathbf{r};\mathbf{0}\right)\equiv\left\langle\mathbf{r}|n;\mathbf{0}\right\rangle \tag{46}$$

is the standard electron wavefunction for an atom at rest.

An insight as to the nature of the photon and electron eigenstates of the auxiliary Hamiltonian may be gained as follows. Consider a downwards transition $q \rightarrow n+\gamma$ (q,n are electron states) taking place in flight while the nucleon has velocity $\mathbf{B}$. Lowest order time-dependent perturbation provides a δ-function that connects frequencies and so it leads to

$$\left(\varepsilon_q-\varepsilon_n\right)\sqrt{1-B^2}=k-\mathbf{B}\cdot\mathbf{k} \tag{47}$$

or

$$k=\frac{k_0\sqrt{1-B^2}}{1-\mathbf{B}\cdot\hat{\mathbf{k}}} \tag{48}$$

which is indeed the standard expression for the Doppler shift for a source of velocity $\mathbf{B}$.

6. Acceleration-induced transitions.

6.1. The amplitude

The treatment in the previous section should have made clear how the phase factors in electron and photon states for a time-dependent $\mathbf{B}$ may well result in the spontaneous (i.e. in the absence of incident photons) excitation of an atom. I shall give an account of how the calculation proceeds (for more details see BCII). At this stage we do not yet commit ourselves to the hyperbolic (Unruh) trajectory. The specific process we are interested in is an atomic up transition from $i=1s$ to $f=2pm$ (in spectroscopic notation), accompanied by the emission of a single photon $\mathbf{k}$ into the initial vacuum. The transition is induced by the last term in (39) (the rest of the terms form the free Hamiltonian whose eigenstates were discussed in the previous section). Then, in an obvious notation, the standard perturbative Ansatz, starting bare at $t=-\infty$, i.e. with $c_{q\mathbf{k}}(-\infty)=0$, reads

$$\mid\psi;t>=\mid i>\mid 0>\exp\left\{-i\varepsilon_i\int_0^t dt'\sqrt{1-B^2(t')}\right\}$$

$$+\sum_q \int d^3k \mid q >\mid \mathbf{k} > \exp\left\{-i\int_0^t dt'\left[k - \mathbf{B}(t')\cdot\mathbf{k} + \varepsilon_q\sqrt{1-B^2(t')}\right]\right\} c_{q\mathbf{k}}(t),$$

where $\mid 0 >$ is the vacuum (no-photon) state and $\mid \mathbf{k} >$ the state with one photon having wave-vector $\mathbf{k}$. This yields the first-order solution

$$c_{f\mathbf{k}}(\infty) = -i\int_{-\infty}^{\infty} dtF < f \mid H_{int}(t) \mid i > \exp\left\{i\int_0^t dt'\left[\varepsilon_{fi}\sqrt{1-B^2(t')} + k - \mathbf{B}(t')\cdot\mathbf{k}\right]\right\}, \quad (49)$$

where the auxiliary function $F(\tau)$, of the form $F = \exp(-\lambda\,|\tau|)$, switches off the interaction adiabatically and covariantly as $|\tau| \to \pm\infty$. The switching factor is needed in principle to make sense of the outer integrals in (49) over the initial and final uniform velocity stages, but disappears from the end-results upon taking the limit $\lambda \to 0$ afterwards.

The matrix element in (49) is

$$\frac{g}{4\pi^{3/2}k^{1/2}}\sqrt{1-B^2} < f \mid \beta\exp\left\{-i\left[\varepsilon_{fi}\mathbf{B}\cdot\mathbf{r} + k_{\|}r_{\|}\sqrt{1-B^2} + \mathbf{k}\cdot\mathbf{r}_{\perp}\right]\right\} \mid i > . \quad (50)$$

and so in terms of proper time τ

$$c_{f\mathbf{k}}(\infty) = -\frac{ig}{4\pi^{3/2}k^{1/2}}\int_{-\infty}^{\infty} d\tau F(\tau)\exp\left\{i\int_0^\tau d\tau'\left[\varepsilon_{fi} + \frac{k - \mathbf{B}(\tau')\cdot\mathbf{k}}{\sqrt{1-B^2(\tau')}}\right]\right\}$$

$$\times < f \mid \beta\exp\left\{-i\left[\varepsilon_{fi}B(\tau)r_{\|} + k_{\|}r_{\|}\sqrt{1-B^2(\tau)} + \mathbf{k}\cdot\mathbf{r}_{\perp}\right]\right\} \mid i > . \quad (51)$$

¿From here on we consider only the rectilinear trajectory of the type described just before (19). We introduce the scaled variables

$$h \equiv c\varepsilon_{fi}/\hbar\alpha, \qquad \kappa \equiv \mathbf{k}c^2/\alpha. \quad (52)$$

(h ought not to be confused with Planck's constant). The electron acceleration in the Bohr model is of the order of 0.90×10^{23} m/s^2 whereas laboratory accelerations are unlikely to exceed those of order 10^6 m/s^2 achievable by the ultracentrifuge. Thus we confine our attention to

$$h \gg 1. \quad (53)$$

The result (76) and a close look at the properties of the modified Bessel function featuring in the latter also entail

$$\kappa/h \sim \mathcal{O}(1)\ . \quad (54)$$

The atomic matrix element in (17) is given by

$$< 2p, \pm1 \mid ... \mid 1s > \quad \to \quad Mak_{\pm}, \qquad M = \pm i128/243. \quad (55)$$

Then

$$c_{f\mathbf{k}} = \frac{gaM}{4\pi^{3/2}}\cdot\frac{k_{\pm}}{k^{1/2}}L, \quad (56)$$

where

$$L = \left\{\int_{-\infty}^{-\tau_0} + \int_{-\tau_0}^{\tau_0} + \int_{\tau_0}^{\infty}\right\} d\tau\exp(-iI(\tau)) = L_i + L_a + L_f \quad (57)$$

requires the phases

$$I(\tau) \equiv \int_0^\tau d\tau'\left\{\varepsilon_{fi} + \frac{k - B(\tau')k_3}{\sqrt{1-B^2(\tau')}}\right\}. \quad (58)$$

It is straightforward to evaluate $I(\tau)$ both in the initial and final stages $|\tau| > \tau_0$, and in the acceleration stage $|\tau| < \tau_0$ defined in (19).

The amplitudes $L_{i,f}$ are easy to evaluate, by virtue of the adiabatic switching factors F, and because (58) shows that the phases $I_{i,f}$ are linear in x. We define the polar angle ϑ of κ with respect to $\mathbf{B}$:

$$\kappa_3 = \kappa\cos\vartheta, \qquad \kappa_\perp = \kappa\sin\vartheta. \tag{59}$$

For nonzero x_0 the two amplitudes combine into

$$L_{if} \equiv L_i + L_f = \frac{(-2i/\alpha)\exp\{i\kappa_3(\cosh x_0 - 1)\}}{[(h+\kappa\cosh x_0)^2 - \kappa_3^2\sinh^2 x_0]}$$

$$\times\{-i(h+\kappa\cosh x_0)\sin(\kappa\sinh x_0 + hx_0) + \kappa_3\sinh x_0\cos(\kappa\sinh x_0 + hx_0)\}\,. \tag{60}$$

We turn to the amplitude L_a and write

$$L_a = \exp\{i\,[hx_3 - \kappa_3]\}\,J/\alpha. \tag{61}$$

where

$$J \equiv \int_{u_1}^{u_2} du\exp\{-i\,[hu + \kappa_\perp\sinh(u)]\}\,, \quad u_2 = x_0 + x_3, \quad u_1 = -x_0 + x_3. \tag{62}$$

The quantity x_3 is defined by

$$x_3 \equiv \log\left[\frac{\kappa - \kappa_3}{\kappa_\perp}\right] = \log\left[\tan(\vartheta/2)\right] \tag{63}$$

As $x_0 \to \infty$ clearly L_i, L_f and therefore L_{if} vanish like $\exp(-x_0)$

$$|L_{if}|^2 \to \frac{16\exp(-2x_0)}{\alpha^2\kappa_\perp^4}\left\{\kappa^2\sin^2[\kappa\sinh x_0 + hx_0] + \kappa_3^2\cos^2[\kappa\sinh x_0 + hx_0]\right\}, \tag{64}$$

and so we need consider only L_a:

$$\lim_{x_0\to\infty} L = \lim_{x_0\to\infty} L_a \equiv L_\infty. \tag{65}$$

The limit $x_0 \to \infty$ entails $u_2 \to \infty$ and $u_1 \to -\infty$. In virtue of the increasingly fast oscillation of the integrand, the integral J remains convergent, so that

$$J_\infty \equiv \lim_{x_0\to\infty} J = \int_{-\infty}^{\infty} du\exp\{-i\,[hu + \kappa_\perp\sinh(u)]\} =$$

$$= 2\int_0^\infty du\,\{\cos[hu]\cos[\kappa_\perp\sinh(u)] - \sin[hu]\sin[\kappa_\perp\sinh(u)]\}\,. \tag{66}$$

The relations (Abramowitz & Stegun 1965, eqs 9.6.22 and 9.6.24)

$$K_\nu(x) = \sec(\nu\pi/2)\int_0^\infty ds\cos(x\sinh s)\cosh(\nu s)$$

$$= \csc(\nu\pi/2)\int_0^\infty ds\sin(x\sinh s)\sinh(\nu s), \qquad (|\mathrm{Re}\nu| < 1,\ x > 0), \tag{67}$$

then lead to

$$J_\infty = 2\{\cos(ih\pi/2) + i\sin(ih\pi/2)\}\,K_{ih}(\kappa_\perp) = 2\exp(-h\pi/2])K_{ih}(\kappa_\perp), \tag{68}$$

$$L_\infty = \frac{2}{\alpha}\exp\{i\,[hx_3 - \kappa_3]\}\exp(-\pi h/2)K_{ih}(\kappa_\perp). \tag{69}$$

For transversely emitted photons (69) reduces to

$$L_\infty(\vartheta = \pi/2) = (2/\alpha)\exp(-\pi h/2)K_{ih}(\kappa). \tag{70}$$

6.2. The probability rate.
The transition probability is defined by

$$P_f(\tau) = \int d^3k P_{f\mathbf{k}}(\tau), \qquad P_{f\mathbf{k}}(\tau) = |c_{f\mathbf{k}}(\tau)|^2 \tag{71}$$

Equations (56), (70) and (71) lead to

$$P_f = \frac{g^2 a^2 |M|^2}{16\pi^3} \int d^3k \frac{k_\perp^2}{k} |L|^2, \tag{72}$$

We restrict ourselves to large durations of the acceleration and accordingly use the limiting form of L (65) which we substitute in expression (72) for the transition probability:

$$P_f(\tau) = \frac{g^2 a^2 |M|^2 \alpha^4}{16\pi^3} \int d^3\kappa \frac{\kappa_\perp^2}{\kappa} |L_\infty(\tau)|^2 \tag{73}$$

We define the transition rate

$$W = \lim_{\tau\to\infty} \frac{1}{2}\frac{dP_f(\tau)}{d\tau} = \lim_{x_0\to\infty} \frac{\alpha}{2}\frac{dP_f}{dx_0} \tag{74}$$

(where the factor $1/2$ stems from the fact that the duration of the acceleration is 2τ). Relation (73) leads to

$$W = \frac{g^2 a^2 |M|^2 \alpha^5}{16\pi^3} \lim_{x_0\to\infty} \int d^3\kappa \frac{\kappa_\perp^2}{\kappa} \mathrm{Re}\left\{ L_\infty^* \frac{dL_\infty}{dx_0} \right\} \tag{75}$$

Eventually using (69)

$$W = \frac{g^2 a^2 |M|^2 \alpha^3}{2\pi^2} e^{-\pi h} \int d\kappa_\perp \kappa_\perp^3 K_{ih}^2(\kappa_\perp) \tag{76}$$

We make use of the remarkable integral (Prudnikov et al 1990)

$$\int_0^\infty dx x^3 K_{ih}^2(x) = \frac{1}{3} |\Gamma(2+ih)|^2 = \frac{1}{3}(1+h^2)\frac{\pi h}{\sinh(\pi h)}\,. \tag{77}$$

to obtain the final expression for the transition rate

$$W = \frac{g^2}{3\pi} a^2 |M|^2 \varepsilon_{fi}^3 \frac{(1+1/h^2)}{e^{2\pi h}-1}. \tag{78}$$

Notice the Planckian factor, with $2\pi h = 2\pi\varepsilon_{fi}/\alpha = \varepsilon_{fi}/k_B T_U$ featuring the Unruh temperature. The $1/h^2$ in the numerator of (78) is a correction to the Unruh result (see also Marzlin and Audretsch (1998)).

7. Conclusion

We reviewed a relativistic formalism that allowed us to prescribe the trajectory $\mathbf{R}(t)$ of a mirror or of the nucleus of a bound state interacting with the radiation field. We were then able to calculate the emission the pattern of emitted radiation in the case of a mirror and the transition rate for atomic transitions using time-dependent perturbation theory. The processes were unambiguously described in terms of quantities pertaining to the laboratory frame. In the case of the atom we examined an asymptotically inertial trajectory and we computed the transition rate for an acceleration-induced upwards transition; in other words we started with an atom in its ground state and computed the rate for emission of a photon plus transition of the electron to an excited state. The energy is of course put in by the external agent that accelerates the nucleon. In the limit of constant acceleration of infinite duration the atom behaves as if it were stationary in a heat-bath of photons at the Unruh temperature.

Acknowledgments
The author wishes to thank the organizers of the Brasilia June 2008 meeting on the Casimir effect for a wonderful hospitality and for financial support.

References

[1] Abramowitz M and Stegun I A 1965 *Handbook of Mathematical Functions* (New York: Dover)
[2] Barton G and Calogeracos A 1995 *Annals of Physics* **238** 227
[3] Barton G and Calogeracos A 2005 *J. Opt. B* **7** S21 (referred to as BCI)
[4] Barton G and Calogeracos A 2008 *J. Phys A: Math. Theor.* **41** 164030 (referred to as BCII)
[5] Calogeracos A and Barton G 1995 *Annals of Physics* **238** 268
[6] Crispino L C B, Higuchi A and Matsas G E A 2008 *Rev. Mod. Phys.* **80** 787
[7] Landau L D and Lifshitz E M 1976 *Mechanics* (Oxford: Pergamon)
[8] Marzlin K-P and Audretsch J 1998 *Phys. Rev.* D **57** 1045
[9] Prudnikov A P, Brychkov Yu A and Marichev O I 1990 *Integrals and Series* (Amsterdam: Gordon & Breach) vol 2 p 385
[10] Sommerfeld A 1952 *Mechanics*; (New York: Academic Press)
[11] Unruh W G 1976 *Phys. Rev.* D **14** 870
[12] Wentzel G 1949 *Quantum Theory of Fields* (New York: Interscience)

Journal of Physics: Conference Series **161** (2009) 012031 doi:10.1088/1742-6596/161/1/012031

Energy level shifts of a uniformly accelerated atom in the presence of boundary conditions

L Rizzuto, S Spagnolo

Dipartimento di Scienze Fisiche ed Astronomiche, Università di Palermo, Via Archirafi,36 Palermo, Italy

E-mail: `lucia.rizzuto@fisica.unipa.it`

E-mail: `spagnolo@fisica.unipa.it`

Abstract.
We discuss the radiative level shifts of an atom moving with uniform acceleration near an infinite reflecting plate. We first consider the case of a two-level system interacting with a massless scalar field in the vacuum state. The acceleration of the two-level atom is supposed in a direction parallel to the conducting plate. We evaluate the contribution of vacuum fluctuations and of the radiation reaction field to the energy shift of the atomic levels, and discuss their behaviour as a function of the atomic acceleration and of the atom-plate distance. Then, we investigate the more general case of an hydrogen atom accelerating near a perfectly reflecting plate and interacting with the electromagnetic field in the vacuum state.

1. Introduction

A remarkable consequence of Quantum Field Theory (QFT) is that the vacuum state, which is the state of lowest energy of the field, possesses an infinite energy and exhibits fluctuations of the fields, even at zero temperature. These fluctuations are at the origin of many observable effects, such as atomic energy-level shifts, Casimir effect and Casimir-Polder forces. Casimir effect and Casimir-Polder forces are long-range interactions between neutral polarizable bodies (see [1, 2] for a review). These effects have been extensively investigated under very general conditions, for example in the presence of boundary conditions and/or at temperature different from zero [3]. Fluctuations of the atom-wall Casimir-Polder force have been also investigated [4,5]. Also, both the Casimir effect and the atom-wall Casimir-Polder forces have been recently measured with remarkable accuracy [6–8].

Recently, the properties of quantum vacuum in accelerated frames have been investigated also. In particular, the problem of the interaction of an accelerated neutral atom (or of a charge) with a quantum field has received much attention in recent years. This problem has been investigated in connection with the Unruh effect, according to which accelerated atoms perceive vacuum fluctuations as a thermal bath at temperature $T_U = \hbar a/(2\pi c k_B)$, a being the atomic acceleration [9, 10]. Actually, the problem of the appearance of the vacuum in an accelerated frame is a widely controversial problem [11,12]. A closely related phenomenon is the dynamical Casimir effect, which concerns with the emission of electromagnetic radiation from a single accelerated mirror in the vacuum. These phenomena clearly show that the quantum vacuum possesses dynamic properties that are at variance with those of vacuum in classical physics: quantum vacuum can be perturbed and these perturbations leads to observable effects.

Radiative properties of accelerated atoms in vacuum have been extensively investigated from a theoretical point of view until very recently [13–20]. Unfortunately, both the Unruh effect and the dynamical Casimir effect are extremely weak and their detection is very difficult. For example, in order to obtain Unruh radiation corresponding to a temperature of 1 K, an acceleration of the order of 10^{22} cm/s^2 would be necessary. Experimental schemes have been recently proposed for detecting phenomena related to the Unruh effect, aimed at enhancing the Unruh radiation under very specific circumstances [21].

This paper is devoted to discuss the Casimir-Polder interaction between a uniformly accelerated atom and an infinite conducting plate. We are interested in investigating if thermal effects due to the acceleration of the atom can modify the Casimir-Polder interaction between the atom and the infinite plate. This is indeed expected because, as it is well-known, Casimir-Polder interactions are directly related to vacuum field fluctuations [22]. The interest on this subject is also related to the fact that the static Casimir-Polder interaction between an atom at rest and a wall has been recently measured with good precision [7,8,23,24]. This suggests the possibility of an indirect detection of the Unruh effect by a measurement of the Casimir-Polder interaction between the accelerated atom and the reflecting plate. Also dynamical Casimir-Polder forces have recently attracted much interest in literature [25–29].

In this paper we first consider a neutral two-level atom interacting with a massless scalar field and uniformly accelerated in a direction parallel to an infinite plate. We calculate the energy level shift of the accelerated atom in the presence of the mirror [30]. As it is well-known, the presence of the reflecting plate changes vacuum field fluctuations. The correction to the Lamb-shift of the atom contains terms depending from the atom-mirror distance, yielding the atom-wall Casimir-Polder potential. We identify the contributions of vacuum fluctuations and of radiation reaction to the Casimir-Polder interaction, and discuss their dependence on the acceleration of the atom in the limits of small and large accelerations. The relation with the Unruh effect is then considered, as well as observability of the results obtained.

Then we extend our investigations to the case of an hydrogen atom interacting with the electromagnetic field in the vacuum state and in the presence of a conducting wall. We calculate the energy level shifts of the atom and we compare the results with those obtained in the case of a scalar field.

2. The energy level shift of a uniformly accelerated two-level system interacting with a massless scalar field in the presence of an infinite plate

Consider a two-level system interacting with a massless scalar field in the vacuum state and in the presence of a perfectly reflecting plate located at $z = 0$. The Hamiltonian describing this system in the instantaneous inertial frame of the atom is [14,15,30] ($\hbar = c = 1$)

$$H(\tau) = H_A(\tau) + H_F(\tau) + H_{AF}(\tau) \tag{1}$$

where τ is the proper time and

$$H_A(\tau) = \omega_0 S_z(\tau) \tag{2}$$

$$H_F(\tau) = \sum_{\mathbf{k}} \omega_k a^\dagger_{\mathbf{k}} a_{\mathbf{k}} \frac{dt}{d\tau} \tag{3}$$

$$H_{AF}(\tau) = \mu S_2(\tau)\phi(x(\tau)) \tag{4}$$

$a_{\mathbf{k}}$ and $a^\dagger_{\mathbf{k}}$ are the bosonic operators of the scalar field, and μ is the atom-field coupling constant. We have also introduced the pseudospin operators $S_z = \frac{1}{2}(\mid e\rangle\langle e \mid - \mid g\rangle\langle g \mid)$, and $S_2 = \frac{i}{2}(S_- - S_+)$, where $S_- =| g\rangle\langle e |$ and $S_+ =| e\rangle\langle g |$ are the atomic lowering and raising operators. Finally, $\phi(\mathbf{x}, t)$ is the scalar field operator,

$$\phi(\mathbf{x}, t) = \sum_{\mathbf{k}} \sqrt{\frac{1}{2V\omega_k}} f(\mathbf{k}, \mathbf{x}) \left[a_{\mathbf{k}}(t) + a^\dagger_{\mathbf{k}}(t)\right] \tag{5}$$

where $f(\mathbf{k}, \mathbf{x})$ are the appropriate mode functions taking into account Dirichlet boundary conditions for the field operator. The mode functions satisfy the normalization condition

$$\frac{1}{V} \int d^3\mathbf{x} f(\mathbf{k}, \mathbf{x}) f(\mathbf{k}', \mathbf{x}) = \delta_{\mathbf{k},\mathbf{k}'} \tag{6}$$

The Hamiltonian $H_F(\tau)$ in (3) governs the evolution of the field in terms of the proper time τ in the instantaneous inertial frame of the atom. It reduces to the usual free-field Hamiltonian in the simple case of an atom at rest, where $dt/d\tau = 1$.

We are interested in evaluating the vacuum field fluctuations and radiation reaction field contributions to the energy level shift of the accelerated atom. As discussed in [14, 15, 30], these quantities can be obtained from an effective Hamiltonian H^{eff} consisting of the sum of two terms

$$H^{eff}_{vf} = \frac{i\mu^2}{2} \int_{\tau_0}^{\tau} d\tau' C^F(x(\tau), x(\tau'))[S^f_2(\tau'), S^f_2(\tau)], \tag{7}$$

$$H^{eff}_{rr} = -\frac{i\mu^2}{2} \int_{\tau_0}^{\tau} d\tau' \chi^F(x(\tau), x(\tau'))\{S^f_2(\tau'), S^f_2(\tau)\}, \tag{8}$$

where

$$C^F(x(\tau), x(\tau')) = \frac{1}{2}\langle 0 \mid \{\phi^f(x(\tau)), \phi^f(x(\tau'))\} \mid 0\rangle \tag{9}$$

$$\chi^F(x(\tau), x(\tau')) = \frac{1}{2}\langle 0 \mid [\phi^f(x(\tau)), \phi^f(x(\tau'))] \mid 0\rangle \tag{10}$$

are the correlation function and the linear susceptibility of the field, respectively, and $[,]$ and $\{,\}$ denote commutator and anticommutator.

The expectation values of H^{eff}_{vf} and H^{eff}_{rr} on a generic atomic state $\mid a\rangle$ give the vacuum fluctuations and the radiation reaction contributions to the radiative shift of the atomic level a

$$(\delta E_a)_{vf} = -i\mu^2 \int_{\tau_0}^{\tau} d\tau' C^F(x(\tau), x(\tau'))(\chi^A)_a(\tau, \tau'), \tag{11}$$

$$(\delta E_a)_{rr} = -i\mu^2 \int_{\tau_0}^{\tau} d\tau' \chi^F(x(\tau), x(\tau'))(C^A)_a(\tau, \tau') \tag{12}$$

where $(C^A)_a(\tau, \tau')$ and $(\chi^A)_a(\tau, \tau')$ are, respectively, the symmetric correlation function and the linear susceptibility of the atom in the state $|a\rangle$

$$(C^A)_a(\tau, \tau') = \frac{1}{2}\langle a \mid \{S^f_2(\tau), S^f_2(\tau')\} \mid a\rangle$$
$$= \frac{1}{2}\sum_b \mid \langle a \mid S_2(0) \mid b\rangle \mid^2 (e^{i\omega_{ab}(\tau-\tau')} + e^{-i\omega_{ab}(\tau-\tau')}) \tag{13}$$

$$(\chi^A)_a(\tau, \tau') = \frac{1}{2}\langle a \mid [S^f_2(\tau), S^f_2(\tau')] \mid a\rangle$$
$$= \frac{1}{2}\sum_b \mid \langle a \mid S_2(0) \mid b\rangle \mid^2 (e^{i\omega_{ab}(\tau-\tau')} - e^{-i\omega_{ab}(\tau-\tau')}) \tag{14}$$

Thus the calculation of the energy level shifts of accelerated atom reduces to the calculation of the statistical functions for the atom and the field.

Let us suppose that the atom is at a distance z_0 from the mirror and that it accelerates along the x direction. In the laboratory frame, its trajectory is described, as a function of the proper time τ, by the equations

$$t(\tau) = \frac{1}{a}\sinh(a\tau),\ \ x(\tau) = \frac{1}{a}\cosh(a\tau),$$
$$y(\tau) = 0\ \ z(\tau) = z_0 \tag{15}$$

where a is the proper acceleration. The statistical functions $C^F(x(\tau), x(\tau'))$ and $\chi^F(x(\tau), x(\tau'))$, in the presence of the reflecting plane boundary, can be calculated from the Wightman function, $G(x(\tau), x(\tau'))$ satisfying the Dirichlet boundary conditions on the mirror, $\phi(x)\mid_{z=0}= 0$,

$$G(x(\tau), x(\tau')) = \langle 0 \mid \phi(x(\tau))\phi(x(\tau')) \mid 0\rangle \tag{16}$$

This function describes the field correlations at two different points of space-time, $x(\tau)$ and $x(\tau')$. In the presence of a boundary, it consists of the sum of two terms: the empty-space contribution $(G^{(0)}(x(\tau), x(\tau')))$ and a part which depends on the presence of the boundary $(G^{(b)}(x(\tau), x(\tau')))$ [31]

$$\begin{aligned} G(x(\tau), x(\tau')) &= G^{(0)}(x(\tau), x(\tau')) + G^{(b)}(x(\tau), x(\tau')) \\ &= \frac{1}{2\pi^2}\left[\frac{1}{(\Delta(\mathbf{x}))^2 - (\Delta t - i\eta)^2} - \frac{1}{(\overline{\Delta(\mathbf{x})})^2 - (\Delta t - i\eta)^2}\right], \end{aligned} \tag{17}$$

$(\eta \to 0^+)$ where we have introduced the variables $\Delta(\mathbf{x}) =\mid \mathbf{x}(\tau) - \mathbf{x}(\tau')\mid$ (the difference between atomic coordinates $\mathbf{x}(\tau)$ taken at two different proper times) and $\overline{\Delta(\mathbf{x})} =\mid \mathbf{x}(\tau) - \sigma\mathbf{x}(\tau')\mid$ where $\sigma\mathbf{x}(\tau')$ is the point corresponding to the reflection of point $\mathbf{x}(\tau')$ on the mirror, with

$$\sigma = \begin{pmatrix} 1 & 0 & 0 \\ 0 & 1 & 0 \\ 0 & 0 & -1 \end{pmatrix} \tag{18}$$

Finally $\Delta t = t(\tau) - t(\tau')$. Substituting (15) into (17), we obtain

$$\begin{aligned} G(x(\tau), x(\tau')) &= -\frac{1}{16\pi^2}\frac{a^2}{\sinh^2\frac{a(\tau-\tau'-i\eta)}{2}} \\ &+\frac{1}{16\pi^2}\frac{a^2}{\sinh^2\frac{a(\tau-\tau'-i\eta)}{2} - z_0^2a^2}, \end{aligned} \tag{19}$$

The symmetrical correlation function and the linear susceptibility of the field can be obtained from (19). After some algebra (detailed calculations are given in [30]) we get

$$\begin{aligned} C^F(x(\tau), x(\tau')) &= -\frac{1}{8\pi^2}\left\{\int_0^\infty d\omega\omega\coth(\frac{\pi\omega}{a})\left(e^{i\omega(\tau-\tau')} + e^{-i\omega(\tau-\tau')}\right)\right. \\ &-\frac{1}{2z_0N^{1/2}}\int_0^\infty d\omega\left(\coth(\frac{\pi\omega}{a})\right. \\ &\left.\left.\times\sin\left(\frac{2\omega}{a}\sinh^{-1}(az_0^2)\right)\left(e^{i\omega(\tau-\tau')} + e^{-i\omega(\tau-\tau')}\right)\right)\right\} \end{aligned} \tag{20}$$

and

$$\chi^F(x(\tau), x(\tau')) = -\frac{1}{8\pi^2}\left\{\int_0^\infty d\omega\omega\left(e^{i\omega(\tau-\tau')} - e^{-i\omega(\tau-\tau')}\right)\right.$$
$$-\frac{1}{2z_0N^{1/2}}\int_0^\infty d\omega \sin\left(\frac{2\omega}{a}\sinh^{-1}(az_0)\right)$$
$$\left.\left(e^{i\omega(\tau-\tau')} - e^{-i\omega(\tau-\tau')}\right)\right\} \tag{21}$$

where $N = 1 + z_0^2a^2$. Substituting Eqs. (13), (14), (20) and (21) into (11) and (12) and taking the limits $\tau_0 \to -\infty$, $\tau \to \infty$, we obtain

$$(\delta E_a)_{vf} = \frac{\mu^2}{8\pi^2}\sum_b |\langle a \mid S_2(0) \mid b\rangle|^2$$
$$\times\int_0^\infty d\omega\omega\left(1 - \frac{1}{2z_0\omega N^{1/2}}\sin\left(\frac{2\omega\sinh^{-1}(z_0a)}{a}\right)\right)$$
$$\times P\left[\frac{1}{\omega+\omega_{ab}} - \frac{1}{\omega-\omega_{ab}}\right]\left(1 + \frac{2}{e^{2\pi\omega/a}-1}\right) \tag{22}$$

and

$$(\delta E_a)_{rr} = -\frac{\mu^2}{8\pi^2}\sum_b |\langle a \mid S_2(0) \mid b\rangle|^2$$
$$\times\int_0^\infty d\omega\omega\left(1 - \frac{1}{2z_0\omega N^{1/2}}\sin\left(\frac{2\omega\sinh^{-1}(z_0a)}{a}\right)\right)$$
$$\times P\left[\frac{1}{\omega-\omega_0} + \frac{1}{\omega-\omega_{ab}}\right], \tag{23}$$

where P denotes the principal part. These expressions give the contributions of vacuum fluctuations and of the radiation reaction field to the energy-level shift of the accelerated atom near the mirror. In contrast to the case of the unbounded space (where the radiation reaction term is not affected by the acceleration of the atom), in the present case, both contributions explicitly depend on the acceleration of the atom, due to the presence of the plate,. In the limit $z_0 \to \infty$, the function $f(z_0) = 1 - \frac{1}{2z_0\omega N^{1/2}}\sin\left(\frac{2\omega\sinh^{-1}(z_0a)}{a}\right)$ tends to 1 and the results obtained for an accelerated atom in unbounded space are recovered [15]. As discussed in [30] Eq.(22) shows that the effect of the uniform acceleration is a thermal-like correction with the Unruh temperature $T_U = \hbar a/2\pi ck_B$, due to the term containing the $\coth(\pi\omega/a)$ function in the symmetric correlation function.

We now focus our attention on the Casimir-Polder interaction between the two-level atom and the wall. In analogy with the case of an atom at rest near a plate, this quantity can be obtained considering only the z_0-dependent terms in Eqs. (22) and (23). For a ground-state atom we get

$$E_{CP} = E_{CP}^{(vf)} + E_{CP}^{(rr)} \tag{24}$$

where

$$E_{CP}^{(vf)} = -\frac{\mu^2}{8\pi^2}\frac{1}{8z_0N^{1/2}}$$
$$\int_0^\infty d\omega\sin\left(\frac{2\omega\sinh^{-1}(z_0a)}{a}\right)$$
$$\times P\left[\frac{1}{\omega-\omega_0} - \frac{1}{\omega+\omega_0}\right]\left(1 + \frac{2}{e^{2\pi\omega/a}-1}\right) \tag{25}$$

and

$$E_{CP}^{(rr)} = \frac{\mu^2}{8\pi^2} \frac{1}{8z_0 N^{1/2}} \times \int_0^\infty d\omega \sin\left(\frac{2\omega \sinh^{-1}(z_0 a)}{a}\right) \times P\left[\frac{1}{\omega - \omega_0} + \frac{1}{\omega + \omega_0}\right] \tag{26}$$

Using the relation

$$f(\frac{2\omega_0}{a}\sinh^{-1}(z_0 a)) = \int_0^\infty d\omega \frac{\sin\left(\frac{2\omega \sinh^{-1}(z_0 a)}{a}\right)}{\omega + \omega_0} \tag{27}$$

the expressions above can be written in the following form

$$E_{CP}^{(vf)} = -\frac{\mu^2}{8\pi^2} \frac{1}{8z_0 N^{1/2}} \times \left(-2f(\frac{2\omega_0}{a}\sinh^{-1}(z_0 a)) + \pi\cos(\frac{2\omega_0}{a}\sinh^{-1}(z_0 a)) - \frac{a}{\omega_0}\left(\cos\left(\frac{2\omega_0 \sinh^{-1}(z_0 a)}{a}\right) - 1\right)\right) \tag{28}$$

and

$$E_{CP}^{(rr)} = \frac{\mu^2}{8\pi^2} \frac{1}{8z_0 N^{1/2}} \cos\left(\frac{2\omega_0}{a}\sinh^{-1}(z_0 a)\right) \tag{29}$$

As expected, the Casimir-Polder interaction depends explicitly on the acceleration of the atom. It is now interesting to investigate the behaviour of the Casimir-Polder interaction as a function of a and z_0, in the limits $(2\omega_0/a)\sinh^{-1}(z_0 a) \ll 1$ and $(2\omega_0/a)\sinh^{-1}(z_0 a) \gg 1$. These two limits define two different *regions* of the space, $z_0 \ll a^{-1}\sinh(a/2\omega_0)$ and $z_0 \gg a^{-1}\sinh(a/2\omega_0)$, respectively, in analogy with the near- and the far-zone limits of the inertial atom-wall Casimir-Polder interaction. In other words, in the case of accelerated atoms, we can define a new *near-zone* and a new *far-zone* limit for the Casimir-Polder interaction, which depend on the acceleration of the atom too. We now investigate the behavior of the Casimir-Polder interaction in these two regions and in the limits $a < \omega_0$ and $a > \omega_0$. This is equivalent to considering the two cases, $k_B T < \omega_0$ and $k_B T > \omega_0$.

In the limit of acceleration small compared with ω_0, the two regions defined above coincide with the usual near-zone and far-zone of the stationary Casimir-Polder interaction for an inertial atom. ¿From Eqs. (28) and (29) we obtain

$$E_{CP}^{(vf)} = -\frac{\mu^2}{8\pi^2} \frac{1}{8z_0} \times (-2f(2\omega_0 z_0) + \pi\cos(2\omega_0 z_0)) \tag{30}$$

and

$$E_{CP}^{(rr)} = \frac{\mu^2}{8\pi^2} \frac{1}{8z_0} \pi\cos(2\omega_0 z_0) \tag{31}$$

where we have approximated $2\omega_0/a \sinh^{-1}(z_0 a) \sim 2\omega_0 z_0$. Thus, in the limit of small acceleration we recover, as expected, the well-known stationary atom-wall Casimir-Polder potential,

$$E_{CP} = \frac{\mu^2}{8\pi^2}\frac{\pi}{8z_0} \qquad \textit{near-zone} \tag{32}$$

$$E_{CP} = \frac{\mu^2}{8\pi^2}\frac{1}{8\omega_0 z_0^2} \qquad \textit{far-zone.} \tag{33}$$

This shows that, in the case of a scalar field, the Casimir-Polder force between an atom at rest and a plate is repulsive.

We now consider the case $a \geq \omega_0$. For typical values of the atomic transition frequency ($\omega_0 \sim 10^{15}$ s^{-1}), this corresponds to an acceleration larger than $a \sim 10^{25}$ cm/s^2. In this limit, $2\omega_0/a \sinh^{-1}(z_0 a) \ll 1$ and from Eqs. (27) and (28) we obtain

$$E_{CP} = E_{CP}^{(vf)} + E_{CP}^{(rr)} \sim \frac{\mu^2}{8\pi^2}\frac{1}{8z_0\sqrt{1+(z_0 a)^2}} \times \left(\pi + \frac{a}{\omega_0}\left(\cos\left(\frac{2\omega_0}{a}\sinh^{-1}(z_0 a)\right) - 1\right)\right) \tag{34}$$

which explicitly depends on the acceleration a. For distances such that $z_0 a \ll 1$, we have

$$E_{CP} = E_{CP}^{(vf)} + E_{CP}^{(rr)} \sim \frac{\mu^2}{8\pi^2}\frac{1}{8z_0} \times \left(\pi + \frac{a}{\omega_0}\left(\cos\left(\frac{2\omega_0}{a}\sinh^{-1}(z_0 a)\right) - 1\right)\right) \tag{35}$$

In the limit $z_0 a \gg 1$, we obtain

$$E_{CP} \sim \frac{\mu^2}{8\pi^2}\frac{1}{8z_0^2 a} \times \left(\pi + \frac{a}{\omega_0}\left(\cos\left(\frac{2\omega_0}{a}\sinh^{-1}(z_0 a)\right) - 1\right)\right) \tag{36}$$

which gives the Casimir-Polder interaction between the accelerated atom and the wall for high accelerations. The most striking effect of the acceleration of the atom is the presence of an oscillatory term in the interaction energy, which modulates the interaction as a function of z_0 and a. This oscillatory behaviour is reminiscent of the stationary Casimir-Polder interaction between an excited atom and a mirror, where a spatially oscillating term is also present [32]. This can be explained observing that the limit $a \geq \omega_0$ corresponds to a temperature $T \geq \omega_0/k_B$. In such limit the excitation probability of the atom is nonvanishing and this is reflected in the oscillatory behaviour of the Casimir-Polder interaction. We emphasize that in the *far-zone* limit the Casimir-Polder interaction is essentially due to the vacuum-fluctuation contribution, where a "thermal" term is present, due to the acceleration of the atom. Thus our results show that thermal effects of acceleration may induce observable effects in the far-zone Casimir-Polder interaction between an accelerated atom and a wall, at least in the case of a scalar field. In fig.1, the energy shift E_{CP} is plotted as a function of the atom-plate distance z_0, for different values of atomic acceleration. A comparison between the different curves shows that the oscillatory behaviour of potential E_{CP} drastically decreases for acceleration smaller than 10^{24} cm/s^2. Also, depending on the distance z_0, the Casimir-Polder interaction can be attractive or repulsive, in contrast with the stationary case of a ground-state atom, where the force is always repulsive. Similar features have been recently discussed in the case of dynamical Casimir-Polder interaction between an atom and a plate [29]. In Fig. 2, we have plotted the quantity $E_{CP}^{a\neq 0}/E_{CP}^{a=0}$, as a function of z_0. We observe that for given values of z_0 the correction to the potential due to the atomic acceleration is of the same order of the static Casimir-Polder interaction between an atom at rest and a plate.

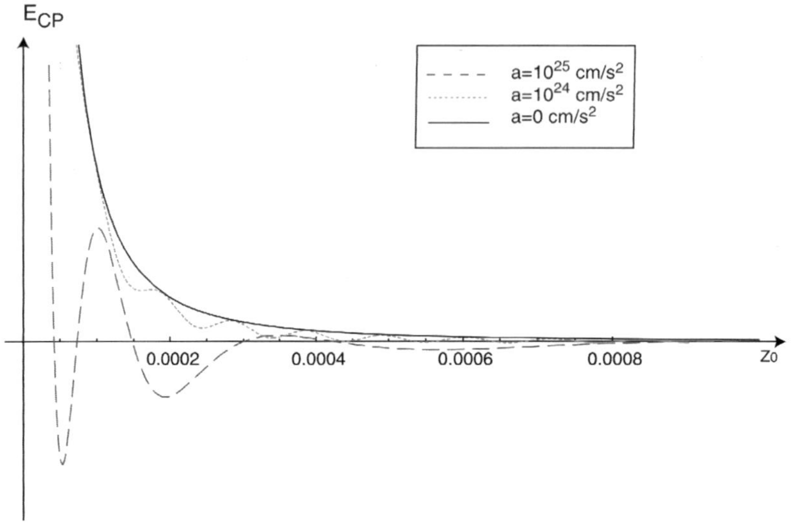

Figure 1. Casimir-Polder interaction as a function of the atom-plate distance z_0 and for different values of the atomic acceleration. Red, cyan and black (long-dashed-dot, dashed and continuous lines, respectively) represent the interaction for $a = 10^{25}\text{cm/s}^2$, $a = 10^{24}\text{cm/s}^2$, $a = 0$ cm/s^2, respectively.

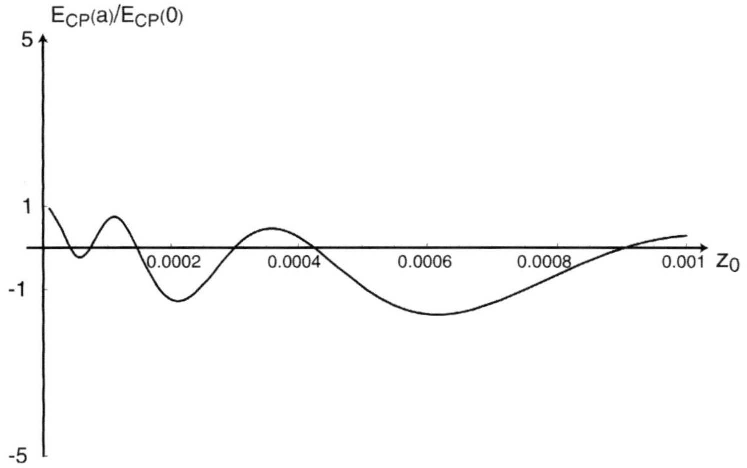

Figure 2. Ratio between the atom-plate Casimir-Polder interaction with accelerated atom ($a = 10^{25}\text{cm/s}^2$) and the atom-plate Casimir-Polder interaction with the atom at rest.

We conclude comparing our results with the Casimir-Polder interaction between an atom at rest and a plate immersed in a thermal bath. As it is well-known [33], the atom-wall Casimir-Polder interaction at temperature T is proportional to the temperature of the bath. A comparison of the results in [33] with Eqs. (34)-(36), immediately shows that the Casimir-Polder interaction between the accelerated atom and the plate is qualitatively different from the static counterpart at the Unruh temperature T_U, because of the nontrivial dependence from the acceleration a in (35) and (36). This consideration indicates that, in general, uniformly accelerated atoms behave differently from static ones in a thermal bath at the Unruh temperature. In a different context, this aspect has been discussed in [19, 20]

3. Energy shifts of an accelerated hydrogen atom interacting with the electromagnetic field in the presence of a conducting plate

In this section we extend our investigation to the case of the electromagnetic field. Recently, the radiative level shifts of an accelerated atom interacting with the electromagnetic field in the vacuum state have been considered [16] and significant differences compared to the case of scalar field have been found. This motivates us to consider the case of a uniformly accelerated hydrogen atom interacting with the electromagnetic field in the presence of a perfectly reflecting plate.

Let us consider a uniformly accelerated hydrogen atom interacting with the electromagnetic field in the vacuum state near a perfectly conducting plate. The Hamiltonian that describes the time evolution of this system with respect to the proper time τ, in the multipolar coupling scheme, is [16]

$$H(\tau) = H_A(\tau) + H_F(\tau) + H_{AF}(\tau) \tag{37}$$

where

$$H_A(\tau) = \sum_m \omega_m \sigma_{mm}(\tau) \tag{38}$$

$$H_F(\tau) = \sum_{\mathbf{k}j} \omega_k a^\dagger_{\mathbf{k}j} a_{\mathbf{k}j} \frac{dt}{d\tau} \tag{39}$$

$$H_{AF} = -e\mathbf{r}(\tau) \cdot \mathbf{E}(x(\tau)) = -e \sum_{\ell m} \mathbf{r}_{\ell m} \cdot \mathbf{E}(x(\tau)) \sigma_{\ell m}(\tau) \tag{40}$$

where $\sigma_{\ell m} = \mid \ell\rangle\langle m \mid$ and $e\mathbf{r} = \sum_{\ell m} \mu_{\ell m} \sigma_{\ell m}$ is the atomic electric dipole moment.

We wish to evaluate the energy level shift of the atom accelerating along the x direction parallel to the plate. Our approach generalizes the method used by Takagi [34] to the case in which boundary conditions are present. The trajectory of the atom is described by equations (15) of the previous section.

In order to calculate the statistical functions of the field, we consider the Wightman's function for the photon field in the presence of a boundary condition, which can be expressed as a sum of two terms

$$\begin{aligned} \langle_M 0 \mid A_\alpha(x) A_\beta(x') \mid 0_M\rangle &= \langle_M 0 \mid A_\alpha(x) A_\beta(x') \mid 0_M\rangle_0 + \langle_M 0 \mid A_\alpha(x) A_\beta(x') \mid 0_M\rangle_b \\ &= \eta_{\alpha\beta} G^{(0)}(x, x') + (\eta_{\alpha\beta} - 2n_\alpha n_\beta) G^{(b)}(x, x') \end{aligned} \tag{41}$$

where $A_\alpha(x)$ is the electromagnetic 4-vector potential, $G^{(0)}(x, x')$ and $G^{(b)}(x, x')$ are the Wightman's functions of the scalar field discussed in the previous section. Also, $x = (t, \mathbf{x})$, $\mid 0_M\rangle$ is the Minkowski vacuum, $\eta_{\alpha\beta} = diag(-1, 1, 1, 1)$ and $n_\alpha = (0, 0, 0, 1)$. We now focus our attention only on the boundary-dependent term, which is relevant for the calculation of the Lamb shift of the accelerated atom in the presence of a reflecting plate. The Wightman function for the fields (in the laboratory frame) follows from differentiating Eq.(41),

$$\begin{aligned} G^{(b)}_{\alpha\beta\gamma\delta}(x,x') &= \langle_M 0 \mid F_{\alpha\beta}(x)F_{\gamma\delta}(x) \mid 0_M\rangle_b \\ &= (\eta_{\beta\delta}\partial_\alpha\partial'_\gamma - \eta_{\beta\gamma}\partial_\alpha\partial'_\delta - \eta_{\alpha\delta}\partial_\beta\partial'_\gamma + \eta_{\alpha\gamma}\partial_\beta\partial'_\delta) \\ &-2(n_\beta n_\delta\partial_\alpha\partial'_\gamma - n_\beta n_\gamma\partial_\alpha\partial'_\delta - n_\alpha n_\delta\partial_\beta\partial'_\gamma + n_\alpha n_\gamma\partial_\beta\partial'_\delta) \\ &\times G^{(b)}(x,x') \end{aligned} \tag{42}$$

where $F_{\alpha\beta}(x) = \partial_\alpha A_\beta - \partial_\beta A_\alpha$.

We are now interested in calculating the two-point correlation function of the electric field in the reference frame co-moving with the atom. The electric field $\hat{\mathbf{E}}(\tau)$ observed at proper time τ by the atom accelerating along the x direction (at a fixed distance z_0 from the plate) is related by a Lorentz transformation to the electromagnetic field in the laboratory frame

$$\hat{E}_i = \hat{F}_{i0} = (\Lambda_\tau)_i^\alpha(\Lambda_\tau)_0^\beta F_{\alpha\beta} \tag{43}$$

where $\hat{E}$ indicates the electric field in the instantaneously inertial frame, and

$$\Lambda = \begin{pmatrix} \cosh(a\tau) & -\sinh(a\tau) & 0 & 0 \\ -\sinh(a\tau) & \cosh(a\tau) & 0 & 0 \\ 0 & 0 & 1 & 0 \\ 0 & 0 & 0 & 1 \end{pmatrix} \tag{44}$$

After some algebra we obtain

$$\begin{aligned} g^{(b)}(\tau-\tau') &= \langle_M 0 \mid \hat{\mathbf{E}}_\ell(\tau)\hat{\mathbf{E}}_m(\tau') \mid 0_M\rangle_b \\ &= \frac{a^4}{16\pi^2}\left[\delta_{\ell m}\left(\sinh^2(a(\tau-\tau')/2) + a^2 z_0^2\right) - 2n_\ell n_m \sinh^2(a(\tau-\tau')/2)\right] \\ &\times \frac{1}{\left[\sinh^2(a(\tau-\tau')/2) + a^2 z_0^2\right]^3} - \frac{a^4}{16\pi^2}\left[2a^2 z_0^2\left(\delta_{\ell m} - k_\ell k_m\right) + 2a z_0\left(n_\ell k_m + k_\ell n_m\right)\right] \\ &\times \frac{\sinh^2(a(\tau-\tau')/2)}{\left[\sinh^2(a(\tau-\tau')/2) + a^2 z_0^2\right]^3} \end{aligned} \tag{45}$$

where $k_i = (1,0,0)$. This expression reduces to the two-point field correlation function obtained for an atom at rest near a reflecting plate in the limit of $a \to 0$.

¿From relation (45) we can obtain the symmetrical and anti-symmetrical correlation function. In analogy with the case of the scalar field, we may express the field correlation function and the field susceptibility as integrals over frequencies. After some algebra, we obtain

$$\begin{aligned} C^F_{\ell m}(x(\tau),x(\tau')) &= \frac{1}{2}\langle_M 0 \mid \{E_\ell(x(\tau)), E_m(x(\tau'))\} \mid 0_M\rangle \\ &= -\frac{1}{8\pi^2}\frac{1}{(2z_0)^3 N}\int_0^\infty \left[\sigma_{\ell n} f_{nm}(\omega;z_0,a) + 2a^2 z_0^2 \sigma_{\ell n} g_{nm}(\omega;z_0,a) + a z_0 h_{\ell m}(\omega;z_0,a)\right] \\ &\times \coth\left(\frac{\pi\omega}{a}\right)\left(\mathrm{e}^{i\omega(\tau-\tau')} + \mathrm{e}^{-i\omega(\tau-\tau')}\right) \end{aligned} \tag{46}$$

and

$$\begin{aligned} \chi^F_{\ell m}(x(\tau),x(\tau')) &= i\langle_M 0 \mid \{E_\ell(x(\tau)), E_m(x(\tau'))\} \mid 0_M\rangle \\ &= -i\frac{1}{8\pi^2}\frac{1}{(2z_0)^3 N}\int_0^\infty d\omega\left\{-2\sigma_{\ell n} f_{nm}(\omega;z_0,a) - 2a^2 z_0^2 \sigma_{\ell n} g_{nm}(\omega;z_0,a) - 2a z_0 h_{\ell m}(\omega;z_0,a)\right\} \\ &\times \left(\mathrm{e}^{i\omega(\tau-\tau')} - \mathrm{e}^{-i\omega(\tau-\tau')}\right) \end{aligned} \tag{47}$$

where

$$f_{nm}(\omega; z_0, a) = \frac{1}{\sqrt{N}}\left\{(\delta_{nm} - n_n n_m)(2\omega z_0)^2 \sin(2\omega/a \sinh^{-1}(z_0 a))\right.$$
$$\left.+ (\delta_{nm} - 3n_n n_m)\left[\frac{2z_0\omega}{\sqrt{N}}\cos(2\omega/a \sinh^{-1}(z_0 a)) - \frac{1}{N}\sin(2\omega/a \sinh^{-1}(z_0 a))\right]\right\}; \quad (48)$$
$$g_{nm}(\omega; z_0, a) = \frac{1}{N}\left\{(2\delta_{nm} - 3n_n n_m)(2\omega z_0)\cos(2\omega/a \sinh^{-1}(z_0 a)) - \frac{1}{\sqrt{N^3}}\right.$$
$$\left.\times\left[(\delta_{nm} - 4n_n n_m) - 2a^2 z_0^2(\delta_{nm} - 2n_n n_m)\right]\sin(2\omega/a \sinh^{-1}(z_0 a))\right\}; \quad (49)$$
$$h_{\ell m}(\omega; z_0, a) = a z_0 (a z_0 (\delta_{\ell m} - k_\ell k_m) + n_\ell k_m + k_\ell n_m)$$
$$\times\left[(2\omega z_0)^2 \sin(2\omega/a \sinh^{-1}(z_0 a)) - \frac{2\omega z_0}{N}(1 - 2a^2 z_0^2)\cos(2\omega/a \sinh^{-1}(z_0 a))\right.$$
$$\left.+\frac{1 + 4a^2 z_0^2}{\sqrt{N^3}}\sin(2\omega/a \sinh^{-1}(z_0 a))\right] \quad (50)$$

We are now ready to calculate the energy level shifts of the accelerated atom in the presence of an infinite reflecting plate. The symmetric and antisymmetric correlation functions for the accelerated hydrogen atom are

$$C^A_{\ell m}(\tau, \tau') = \tfrac{1}{2}\langle a \mid \{\mathbf{r}_\ell(\tau), \mathbf{r}_m(\tau')\} \mid a\rangle$$
$$= \tfrac{1}{2}\textstyle\sum_b\left[\langle a \mid r_\ell(0) \mid b\rangle\langle b \mid r_m(0) \mid a\rangle e^{i\omega_{ab}(\tau-\tau')} + \langle a \mid r_m(0) \mid b\rangle\langle b \mid r_\ell(0) \mid a\rangle e^{-i\omega_{ab}(\tau-\tau')}\right] \quad (51)$$
$$\chi^A_{\ell m}(\tau, \tau') = i\langle a \mid [\mathbf{r}_\ell(\tau), \mathbf{r}_m(\tau')] \mid a\rangle$$
$$= i\textstyle\sum_b\left[\langle a \mid r_\ell(0) \mid b\rangle\langle b \mid r_m(0) \mid a\rangle e^{i\omega_{ab}(\tau-\tau')} - \langle a \mid r_m(0) \mid b\rangle\langle b \mid r_\ell(0) \mid a\rangle e^{-i\omega_{ab}(\tau-\tau')}\right] \quad (52)$$

Using Eqs.(46),(47),(51) and (52) in (11) and (12), we obtain

$$(\delta E_a)^{VF}_{(b)} = -\frac{1}{4\pi^2}\sum_b\left(\mu^{ab}_\ell \mu^{ba}_m\right)\frac{1}{2N(2z_0)^3}\int_0^\infty d\omega$$
$$\times\left(\sigma_{\ell n} f_{nm}(\omega; z_0, a) + 2a^2 z_0^2 \sigma_{\ell n} g_{nm}(\omega; z_0, a) + z_0 a h_{\ell m}(\omega; z_0, a)\right)$$
$$\times \coth(\pi\omega/a) P\left(\frac{1}{\omega + \omega_{ab}} - \frac{1}{\omega - \omega_{ab}}\right) \quad (53)$$

and

$$(\delta E_a)^{RR}_{(b)} = \tfrac{1}{4\pi^2}\textstyle\sum_b\left(\mu^{ab}_\ell \mu^{ba}_m\right)\tfrac{1}{2N(2z_0)^3}\int_0^\infty d\omega$$
$$\times\left(\sigma_{\ell n} f_{nm}(\omega; z_0, a) + 2a^2 z_0^2 \sigma_{\ell n} g_{nm}(\omega; z_0, a) + z_0 a h_{\ell m}(\omega; z_0, a)\right) P\left(\tfrac{1}{\omega+\omega_{ab}} + \tfrac{1}{\omega-\omega_{ab}}\right) \quad (54)$$

Equations (53) and (54) give the correction to the energy level shifts of the accelerated atom due to the presence of boundary.

The total energy level shift is obtained by summing the contributions of vacuum fluctuations and of the radiation reaction field

$$\delta E_a = (\delta E_a)^{VF} + (\delta E_a)^{RR} \quad (55)$$

Equations (53) and (54) have several interesting properties. First, while in the case of the unbounded space only the contribution of vacuum fluctuations depends on the acceleration, in the presence of a boundary both vacuum fluctuations and radiation reaction depend on the atomic acceleration, in analogy with the case of the scalar field discussed in the previous section. In particular, in the limit of small acceleration, they reduce to the usual Lamb-shift for an atom at rest near a reflecting plate. Secondly, when z_0 goes to infinity the boundary-dependent terms vanish and we obtain the results for the Lamb-shift of an accelerated atom in the unbounded space [16]

$$(\delta E_a)^{VF} = \frac{1}{12\pi^2}\sum_b \mid \langle a \mid \mu \mid b\rangle \mid^2 \int_0^\infty d\omega\omega^3(1+\frac{a^2}{\omega^2})\coth(\pi\omega/a)P(\frac{1}{\omega+\omega_{ab}}-\frac{1}{\omega-\omega_{ab}}) \qquad (56)$$

and

$$(\delta E_a)^{RR} = -\frac{1}{12\pi^2}\sum_b \mid \langle a \mid \mu \mid b\rangle \mid^2 \int_0^\infty d\omega\omega^3 P(\frac{1}{\omega+\omega_{ab}}+\frac{1}{\omega-\omega_{ab}}) \qquad (57)$$

Finally, the effect of the acceleration is not simply a *thermal* correction given by a term $\coth(\pi\omega/a)$. In fact, the functions $f_{\ell n}(\omega; z_0, a)$, $g_{\ell n}(\omega; z_0, a)$ and $h_{\ell m}(\omega; z_0, a)$ depend explicitly on the atomic acceleration. Thus, in analogy with the scalar field case, also in the case of electromagnetic field we find a non-thermal correction to the vacuum fluctuation contribution, due to the presence of the boundary. This point will be discussed in detail in a future publication.

4. Conclusions

In this paper we have discussed the energy level shifts of a uniformly accelerated two-level system interacting with a massless scalar field in the presence of an infinite plate with Dirichlet boundary conditions. In particular, we have considered the contributions of vacuum fluctuations and of radiation reaction to the Casimir-Polder interaction energy and we have discussed their dependence on the acceleration of the two-level atom, in the limits of *near-* and *far-zone*. We have shown that the atom-wall Casimir-Polder interaction in the limit of small accelerations coincides with the stationary atom-wall Casimir-Polder potential. In contrast, for high atomic accelerations the Casimir-Polder interaction depends explicitly on the acceleration of the atom and exhibits an oscillatory behaviour in space. Therefore it appears that effects due to the acceleration of the atom may become evident in the atom-wall Casimir-Polder interaction. Finally, we have discussed the case of an hydrogen atom moving with constant acceleration near a perfectly conducting plate and interacting with the electromagnetic field in the vacuum state. The results obtained have been compared with those of the scalar field case.

Acknowledgements

The authors wish to thank Prof. F. Persico for commenting the manuscript. The authors wish to thank also Prof. R. Passante, Prof. A. Calogeracos and Prof. G. Carugno for interesting discussions on the subject. Partial support by Ministero dell'Università e della Ricerca Scientifica e Tecnologica and by Comitato Regionale di Ricerche Nucleari e di Struttura della Materia is acknowledged.

References

[1] Compagno G, Passante R, Persico F, *Atom-Field Interactions and Dressed Atoms*, Cambridge University Press, Cambridge (1995)
[2] Salam A, Int. Rev. Phys. Chem. **27**, 405 (2008)
[3] Passante R, Spagnolo S, Phys. Rev. A **76**, 042112 (2007)
[4] Wu C-H, Chung-I Kuo, Ford L-H, Phys. Rev. A **65**, 062102 (2002)
[5] Messina R, Passante R, Phys. Rev. A **76**, 032107 (2007)
[6] Bressi G, Carugno G, Onofrio R, Ruoso G, Phys. Rev. Lett. **88**, 041804 (2002)
[7] Sukenik C I, Boshier M G, Cho D, Sandoghdar V, Hinds E A, Phys. Rev. Lett. **70**, 560 (1993)
[8] Druzhinina V and DeKieviet M, Phys. Rev. Lett. **91**, 193202 (2003)

[9] Davies P C W, J. Phys. A: Math. Gen. **8**, 609 (1975)
[10] Unruh W G, Phys. Rev. D **14**, 870 (1976)
[11] Jaekel M T and Reynaud S, Quant. Semicl. Opt. **7**, 499 (1995)
[12] Hu B L, Roura A, Phys. Rev. Lett. **93**, 129301 (2004)
[13] Boyer T H, Phys. Rev. D **21**, 2137 (1980)
[14] Audretsch J, Muller R, Phys. Rev. A **50**, 1755 (1994)
[15] Audretsch J, Muller R, Phys. Rev. A **52**, 629 (1995)
[16] Passante R, Phys. Rev. A **57**, 1590 (1998)
[17] Yu H and Lu S, Phys. Rev. D **72**, 064022 (2005)
[18] Yu H and Zhu Z, Phys. Rev. D **74**, 044032 (2006)
[19] Zhang J and Yu H, Phys. Rev. D **75**, 104014 (2007)
[20] Obadia N, Phys. Rev. D **76**, 045013 (2007)
[21] Belyanin A, Kocharovsky V V, Capasso F, Fry E, Zubairy M S and Scully M O, Phys. Rev. A **74**, 023807 (2006)
[22] Power E A and Thirunamachandran T, Phys. Rev. A **4**8, 4761 (1993)
[23] Obrecht J M, Wild R J, Antezza M, Pitaevskii L P, Stringari S and Cornell E A, Phys. Rev. Lett. **98**, 063201 (2007)
[24] Bordag M, Mohideen U, Mostepanenko V M, Phys. Rep. **353**, 1 (2001)
[25] Shresta S, Hu B L, Phillips N G, Phys. Rev. A **68**, 062101 (2003)
[26] Rizzuto R, Passante R and Persico F, Phys. Rev. A **70**, 012107 (2004)
[27] Per-Olof Westlund and Wennerstrom H, Phys. Rev. A **71**, 062106 (2005)
[28] Rizzuto L, Passante R and Persico F, Phys. Rev. Lett. **98**, 240404 (2007)
[29] Vasile R and Passante R, Phys. Rev. A **78**, 032108 (2008)
[30] Rizzuto L, Phys. Rev. A **76**, 062114 (2007)
[31] Birrel N D, Davies P C W, *Quantum Field in curved space*, Cambridge Monographs on Mathematical Physics, Cambridge 1982
[32] Messina R, Passante R, Rizzuto L, Spagnolo S and Vasile R, J. Phys. A: Math. Theor., **41**, 164031 (2008)
[33] Boyer T H, Phys. Rev. A **11**, 1650 (1975)
[34] Takagi S, Progr. Theor. Phys. Suppl. **88**, 1 (1986)

Journal of Physics: Conference Series **161** (2009) 012032 doi:10.1088/1742-6596/161/1/012032

Time evolution of the energy density inside a one-dimensional non-static cavity with a vacuum, thermal and a coherent state

D T Alves[1], E R Granhen[2], M G Lima[1], H O Silva[1] and A L C Rego[1]

[1] Faculdade de Física, Universidade Federal do Pará, 66075-110, Belém, PA, Brazil

[2] Centro Brasileiro de Pesquisas Fsicas, Rua Dr. Xavier Sigaud, 150, 22290-180, Rio de Janeiro, RJ, Brazil

Abstract. We study the time evolution of the energy density for a real massless scalar field in a two-dimensional spacetime, inside an expanding (not oscillating) cavity, and also the force exerted by the field on the moving boundary. Our calculations are based on the exact numerical approach proposed by Cole and Schieve. Considering Neumann and Dirichlet boundary conditions, we investigate the following initial states of the field: vacuum, thermal and the coherent state.

1. Introduction

Moore, in the context of a real massless scalar field inside a non-static cavity, obtained an exact formula for the expected value of the energy-momentum tensor (in a two-dimensional spacetime), given in terms of a functional equation, usually called Moore's equation [1]. For this equation, there is no general technique of analytical solution, but exact analytical solutions for particular movements of the boundary [2–4], and also approximate analytical solutions (see, for instance [5, 6]) have been obtained. The dynamical Casimir effect has been investigated for a thermal bath as the initial field state in a non-static cavity [5, 7–10], and also for a coherent state [9, 11–13]. Squeezed states have also been considered [9].

In the context of the Dynamical Casimir effect, the problem of boundaries oscillating with small amplitudes is commonly related to the problem of photon creation in laboratory, and it is usually investigated via approximate analytical solutions of the Moore equation [5, 6], and also via other perturbative methods [14]. On the other hand, the investigation of non-oscillating movements, with large amplitudes, are frequently motivated by problems of particle creation in cosmological models and radiation emitted by collapsing black holes [2, 15, 16], and the related energy density has been investigated just for particular laws of motion for which the Moore equation can be solved exactly [2, 16]. For general non-oscillating trajectories, an approach based on the numerical method to solve exactly the Moore equation for a general law of motion, proposed by Cole and Schieve [17], can be useful. This method gives results in agreement with those obtained by perturbative approaches [17, 18], and also can give results for models that are not solvable via perturbative techniques, as the problem of expanding cavities.

In the present paper, we apply a technique of calculation based on the Cole-Schieve approach [17, 19], and examine the energy density in an expanding one-dimensional cavity, with a law of motion for which there is no exact analytical solution for the correspondent Moore equation. Beyond the vacuum, we also consider thermal and coherent states as the initial field states.

The article is organized as follows. In section 2 we discuss the general field solution and the law of motion of the boundary. In section 3 we investigate the non-static cavity for a vacuum, thermal and a coherent state. In section 4 we summarize our results.

2. General field solution and law of motion

Let us start considering the field satisfying the Klein-Gordon equation (we assume throughout this paper $\hbar = c = k_B = 1$) and obeying conditions imposed at the static boundary located at $x = 0$, and also at the moving boundary's position at $x = L(t)$, where $x = L(t)$ is a prescribed law for the moving boundary and $L(t < 0) = L_0$, with L_0 being the length of the cavity in the static situation. As done in Ref. [20], we consider four types of boundary conditions. The Dirichlet-Neumann (DN) boundary condition imposes Dirichlet condition at the static boundary, whereas the space derivative of the field taken in the instantaneously co-moving Lorentz frame vanishes (Neumann condition) at the moving boundary's position. We also consider: Dirichlet-Dirichlet (DD), Neumann-Neumann (NN) and Neumann-Dirichlet (ND) boundary conditions. A general solution of the wave equation can be written as [20]:

$$\hat{\psi}(t,x) = \lambda(\hat{A} + \hat{B}\psi^{(0)}) + \sum_{n=1-2\beta}^{\infty} \left[\hat{a}_n \psi_n(t,x) + H.c.\right], \tag{1}$$

where the field modes $\psi_n(t,x)$ are given by

$$\psi_n(t,x) = \frac{1}{\sqrt{4(n+\beta)\pi}} \left[\gamma e^{-i(n+\beta)\pi R(v)} + \gamma^* e^{-i(n+\beta)\pi R(u)}\right], \tag{2}$$

with $\psi^{(0)} = [R(v) + R(u)]/2$ (see Ref. [21]), $u = t - x$, $v = t + x$, and R satisfying the functional equation $R[t + L(t)] - R[t - L(t)] = 2$, which is the Moore equation. The operators $\hat{A}$ and $\hat{B}$ satisfy the commutation rules $\left[\hat{A}, \hat{B}\right] = i$, $\left[\hat{A}, \hat{a}_n\right] = \left[\hat{B}, \hat{a}_n\right] = 0$. The NN solution is recovered for $\lambda = \gamma = 1$ and $\beta = 0$. The other three cases are recovered if $\lambda = 0$ and: $\beta = 0$ and $\gamma = i$ for the DD case; $\beta = 1/2$ and $\gamma = i$ for the DN case; $\beta = 1/2$ and $\gamma = 1$ for the ND case.

In the present paper we investigate the following particular trajectory $x = L(t)$, based on the one proposed by Walker and Davies [22]:

$$t = L_0 - L + A\left(e^{-2(L_0 - L)/B} - 1\right)^{1/2}, \tag{3}$$

valid for $t \geq 0$,where A and B are positive constants, with $A > B$ (so that $|\dot{L}| < 1$). For $A = 2$ and $B = 1$ (values also chosen in Ref. [22]) the trajectory is showed in Fig. 1. This trajectory has some interesting properties: it has a discontinuity of the boundary acceleration when $t = 0$ ($L = 1$) (as showed in Fig. 1(b)), but it is smooth and asymptotically static for $t \to \infty$ (see Fig. 1 (a)). With this feature, the "in" and "out" states can be well defined and the average number of produced particles can be calculated. In addition, we have a discontinuity of the boundary acceleration at one spacetime point only, in contrast with other laws of motion investigated in the literature, which have discontinuities of the acceleration at the two points: in the beginning and also at the end of the time interval in which the boundary is accelerated. According to Walker and Davies, asymptotically static trajectories brought the advantage of avoiding certain pathologies related to the radiation emitted by a moving mirror with abrupt acceleration [22].

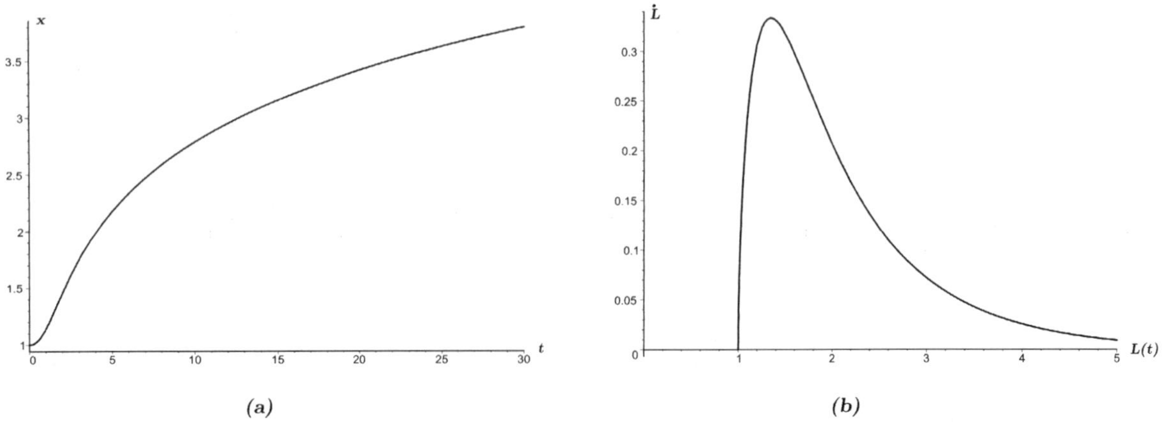

Figure 1. (a) Moving mirror trajectory defined by Eq. (3); **(b)** Moving mirror velocity (vertical axis) as function of the boundary position.

3. Vacuum, thermal and coherent states

As examples of initial field states such that the density matrix is diagonal in the Fock basis, let us consider the vacuum and the thermal state. It can be shown that the expected value of the energy density operator $\mathcal{T} = \langle \hat{T}_{00}(t,x)\rangle$ can be split in $\mathcal{T} = \mathcal{T}_{\text{vac}} + \mathcal{T}_{\text{non-vac}}$, where $\mathcal{T}_{\text{vac}}$ is the contribution to the energy density due the vacuum part and $\mathcal{T}_{\text{non-vac}}$ is the non-vacuum contribution due to the real particles in the initial state of the field. Hereafter we consider the averages $\langle ...\rangle$ taken over initial field states annihilated by $\hat{B}$. Let us start considering the vacuum as the initial field state ($\mathcal{T}_{\text{non-vac}} = 0$). The vacuum contribution to the energy density inside the oscillating cavity can be written as [1, 2, 20] $\mathcal{T}_{\text{vac}} = -f(v) - f(u)$, where:

$$f = \frac{|\gamma|^2}{24\pi}\left\{\frac{R'''}{R'} - \frac{3}{2}\left(\frac{R''}{R'}\right)^2 + \pi^2\left[\frac{1}{2} - 3(\beta - \beta^2)\right]R'^2\right\}. \tag{4}$$

For the static situation the function R is given by $R(z) = z/L_0$ and its first derivative is a constant $R'(z) = 1/L_0$. ¿From this equation we get (see [1,2,23] the known static Casimir forces $\mathcal{F}^{(s)}_{\text{vac}}$ acting on the right boundary:

$$\mathcal{F}^{(s)\,\text{DD}}_{\text{vac}} = \mathcal{F}^{(s)\,\text{NN}}_{\text{vac}} = -\pi/(24L_0^2),\ \ \mathcal{F}^{(s)\,\text{DN}}_{\text{vac}} = \mathcal{F}^{(s)\,\text{ND}}_{\text{vac}} = \pi/(48L_0^2), \tag{5}$$

where the superscripts DD, NN, DN and ND mean the types of boundary conditions considered in the calculations. In Fig. 2(a) we plot, for both DD and NN cases and the vacuum as the initial field state, the time evolution of the actual force $\mathcal{F}_{\text{vac}} = \mathcal{T}_{\text{vac}}[t, L(t)]$ acting on the moving boundary (solid line) for each position L, whereas the dotted line shows the value of the static Casimir force $-\pi/[24L(t)^2]$ which would act on the boundary if it was static at the position $x = L$. In analogous manner, in Fig. 2(b) we plot, for the DN and ND cases, the time evolution of the force acting on the moving boundary (solid line), whereas the dotted line shows the repulsive static Casimir force $\pi/[48L(t)^2]$. We see in Fig. 2 that, for DD and NN, and also for DN and ND, the radiation force acting on the moving boundaries are the same:

$$\mathcal{F}^{\text{DD}}_{\text{vac}} = \mathcal{F}^{\text{NN}}_{\text{vac}},\ \ \mathcal{F}^{\text{DN}}_{\text{vac}} = \mathcal{F}^{\text{ND}}_{\text{vac}}. \tag{6}$$

Fig. 2 also shows the dynamical force approaching the static Casimir one in the limit $t \to \infty$.

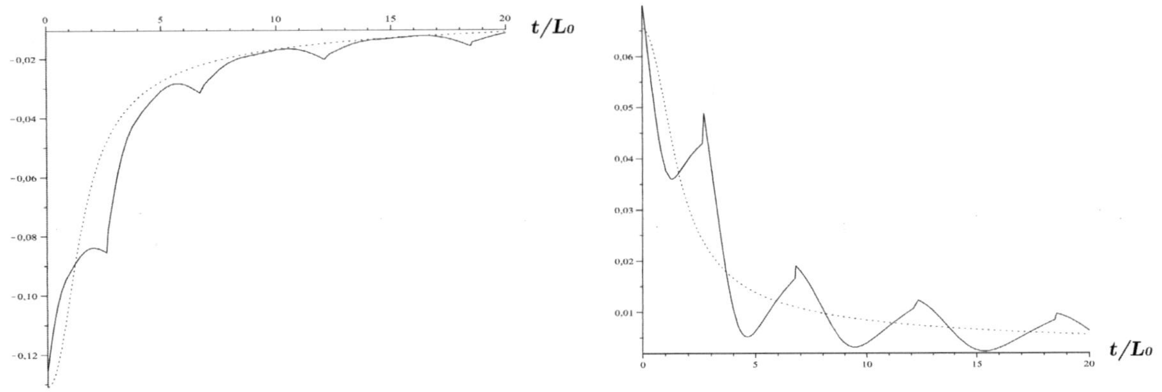

(a) *(b)*

Figure 2. **(a)** The quantum force $\mathcal{F}_{\text{vac}}$ acting on the moving boundary (solid line) and the attractive static Casimir force $\mathcal{F}^{(s)}_{\text{vac}}$ (dotted line) for the DD or NN cases; **(b)** The quantum force $\mathcal{F}_{\text{vac}}$ acting on the moving boundary (solid line) for the DN or ND cases, and the repulsive static Casimir force $\mathcal{F}^{(s)}_{\text{vac}}$ (dotted line). In both cases the vacuum was considered as the initial field state. The discontinuities in the derivative occur at $t_1 \approx 2.67, t_2 \approx 6.78, t_3 \approx 12.20, t_4 \approx 18.56$.

In Fig. 2 we can see discontinuities of the derivatives. These discontinuities always occur when the front of the wave in the energy density meets the right boundary. When $t = 0$ the right boundary starts to move, interacting with the vacuum field, and generating a wave in the energy density, propagating leftward in the cavity. The wave will be reflected back by the left (static) boundary and propagates rightward until meeting the right boundary at time $t = t_1 \approx 2.67$. This value can be obtained by solving the equation $t_1 - L_0 = L(t_1)$. From this instant on, the right boundary will interact also with this reflected wave, what generates the mentioned discontinuity. In general, the wave in the energy density, after reflections, will meet the right boundary at the instants $t = t_i$, where $i = 1, 2, ...$, which can be obtained by solving the equation

$$t_i - t_{i-1} - L(t_{i-1}) = L(t_i) \tag{7}$$

with $t_0 = 0$.

In Fig. 3(a) we show the energy density at the point $x = L_0/2$, as function of time, for DD or NN cases (solid line), and also for DN or ND cases (dotted line). ¿From $t = 0$ to $t = 1/2$ we observe a constant value, correspondent to the static Casimir energy density. At $t = 1/2$ the wave in the energy density arrives at the point $x = L_0/2$. The jump observed at this point is related to abrupt motion started by the right boundary, in the sense that there is a discontinuity of the boundary acceleration as showed in Fig. 1(b). After reflected by the left boundary, this discontinuity point propagates and can be observed again at $t = 1.5$, an so on, after successive reflections. The energy density at the point $x = L_0/2$ goes to zero, since, for our law of motion, the length of the cavity goes to infinity and the motion is asymptotically static as $t \to \infty$. In Fig. 3(b) we show the energy for all points in the cavity for $t = 30L_0$.

Now, let us consider the thermal state as the initial field state. For this case we have $\langle \hat{a}^\dagger_n \hat{a}_{n'} \rangle = \delta_{nn'} \overline{n}(n, \beta)$ and $\langle \hat{a}_n \hat{a}_{n'} \rangle = \langle \hat{a}^\dagger_n \hat{a}^\dagger_{n'} \rangle = 0$, where $\overline{n}(n, \beta) = [\exp(\kappa(n + \beta)) - 1]^{-1}$ and $\kappa = 1/T$. In Fig. 4 we compare the behavior of the force $\mathcal{F}_T = \mathcal{T}_{\text{non-vac}}[t, L(t)]$ for $T = 1$ and $\mathcal{F}_{\text{vac}}$, obtaining that the difference between $\mathcal{F}^{\text{DD}}_T$ and $\mathcal{F}^{\text{DN}}_T$ is a scale factor. We also see that

$$\mathcal{F}^{\text{DD}}_T = \mathcal{F}^{\text{NN}}_T, \quad \mathcal{F}^{\text{DN}}_T = \mathcal{F}^{\text{ND}}_T. \tag{8}$$

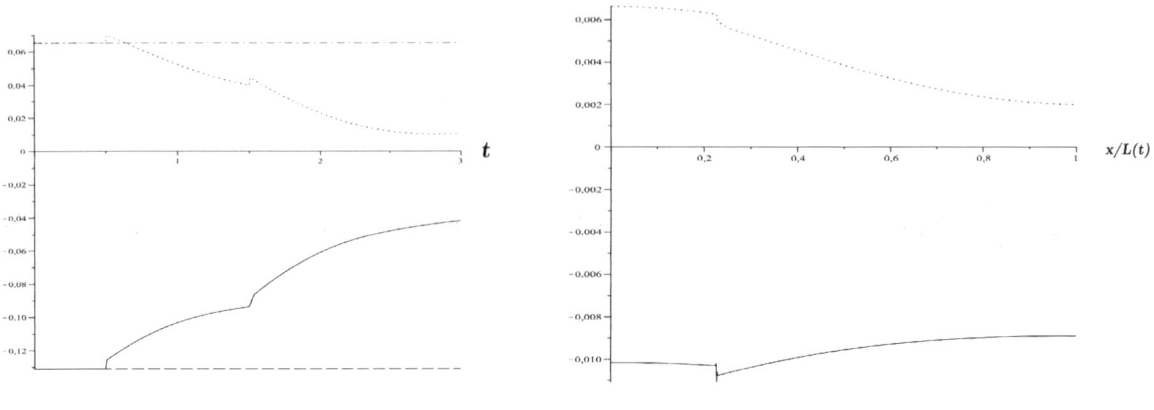

(a) *(b)*

Figure 3. (a) The energy density for DD or NN cases (solid line) and also for DN or ND cases (dotted line), at the point $x = L_0/2$, as function of time. The dashed and dashed-point lines represents, respectively, the static Casimir energy densities for DD-NN and ND-DN cases. **(b)** The energy density for for DD or NN cases (solid line) and also for DN or ND cases (dotted line), as function of the normalized position $x/L(t)$ in the cavity, for a time $t = 30L_0$. In both cases the vacuum was considered as the initial field state.

Note that the discontinuities of the derivative visualized in Fig. 4 occur for the same values t_i given in Eq. 7.

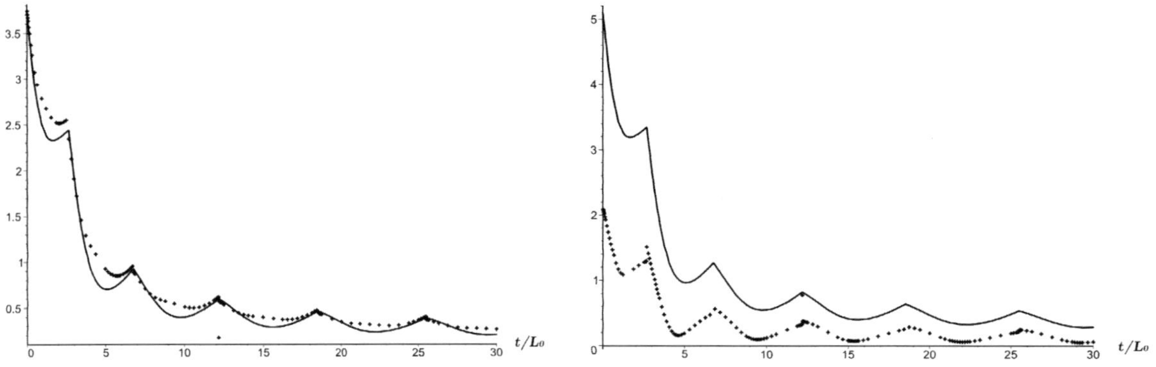

(a) *(b)*

Figure 4. (a) The quantum forces $\mathcal{F}_T$ (solid line) and -30$\times\mathcal{F}_{\text{vac}}$ (dotted line) for DD or NN cases, as function of time; **(b)** The quantum forces $\mathcal{F}_T$ (solid line) and $30 \times \mathcal{F}_{\text{vac}}$ (dotted line) for DN or ND cases, as function of time. In both cases the thermal bath with temperature $T = 1$ was considered as the initial field state.

The coherent state, an example of a non-diagonal state ($\langle \hat{a}_n \hat{a}_{n'} \rangle \neq 0$), can be defined as an eigenstate of the annihilation operator: $\hat{a}_n \left|\alpha\right\rangle = \alpha\delta_{nn_0} \left|\alpha\right\rangle$, where $\alpha = |\alpha|e^{i\theta}$ and n_0 is related to the frequency of the excited mode [24]. In Fig. 5 we visualize the behavior of the force $\mathcal{F}_\alpha = \mathcal{T}_{\text{non-vac}}[t, L(t)]$ for $\alpha = 1$ and $\theta = 0$. Note that, for the coherent state, the symmetry between the DD and NN cases, and also between DN and ND cases, is broken in the sense that

$$\mathcal{F}_\alpha^{\rm DD} \neq \mathcal{F}_\alpha^{\rm NN}, \ \ \mathcal{F}_\alpha^{\rm DN} \neq \mathcal{F}_\alpha^{\rm ND}. \tag{9}$$

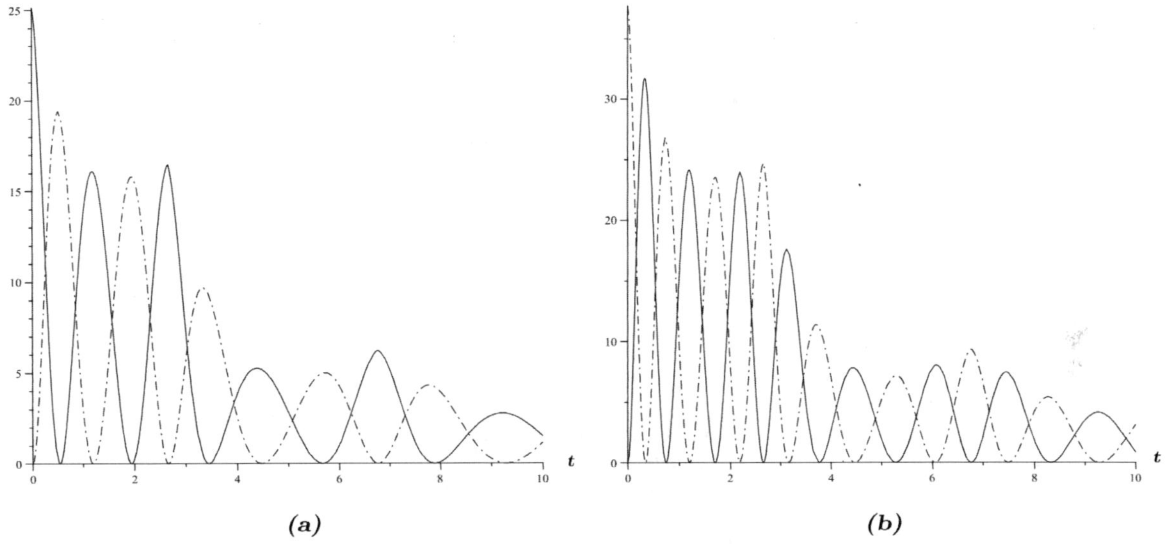

Figure 5. The coherent force $\mathcal{F}_\alpha$ acting on the moving boundary (vertical axis), as function of time, for a coherent state with $|\alpha| = 1$ and $\theta = 0$. **(a)** The DD case (solid line) and the NN case (dashed-point line); **(b)** The DN case (solid line) and the ND case (dashed-point line).

Even in the static case ($t < 0$), for a certain value of α and θ, the energy density for the coherent state is spacetime dependent, and also different if we consider NN, DD, ND or DN cases [20]. In contrast, for vacuum and thermal initial states, the energy density is a constant, having a same value for NN and DD cases, and also for ND and DN cases [20]. These differences between coherent (non-diagonal) and thermal-vacuum (diagonal) states in the static situation propagate along the dynamical situation ($t > 0$), as viewed in Eqs. (6), (8) and (9).

4. Final comments

In summary, using a technique of calculation based on the Cole-Schieve approach, we obtained the behavior of the energy density inside an expanding cavity, and also the behavior of the dynamical force acting on the moving boundary. We also considered Dirichlet and Neumann boundary conditions We observed that the energy density and the force on the moving mirror are not affected if we change NN by DD (or DN by ND) boundary conditions, for vacuum or thermal initial field states (diagonal states). On the other hand, the same invariance is not observed for the coherent (non-diagonal) state.

The number of produced particles in this model will be calculated further.

Acknowledgments

The authors acknowledge V. V. Dodonov for useful suggestions. This work was supported by FAPESPA, CNPq and CAPES - Brazil.

References

[1] Moore G T 1970 *J. Math. Phys.* **11** 2679
[2] Fulling S A and Davies P C W 1976 *Proc. R. Soc. London* A **348** 393
[3] Law C K 1994 *Phys. Rev. Lett.* **73** 1931
[4] Wu Y, Chan K W, Chu M C and Leung P T 1999 *Phys. Rev.* A **59** 662; Wegrzyn P 2007 *J. Phys.* B **40** 2621
[5] Dodonov V V, Klimov A B and Nikonov D E 1993 *J. Math. Phys.* **34** 2742
[6] Dalvit D A R and Mazzitelli F D 1998 *Phys. Rev.* A **57** 2113
[7] Jaekel M T and Reynaud S 1993 *J. Phys. I (France)* **3** 339; Jaekel M T and Reynaud S 1993 *Phys. Lett.* A **172** 319; Machado L A S, Maia Neto P A and Farina C 2002 *Phys. Rev.* D **66** 105016.
[8] Plunien G, Schutzhold R and Soff G 2000 *Phys. Rev. Lett.* **84** 1882
[9] Alves D T, Granhen E R and Lima M G 2008 *Phys. Rev.* D **77** 125001
[10] Hui J, Qing-Yun S and Jian-Sheng W 2000 *Phys. Lett.* A **268** 174; Schutzhold R, Plunien G and Soff G 2002 *Phys. Rev.* A **65** 043820; Schaller G, Schtzhold R, Plunien G and Soff G 2002 *Phys. Rev.* A **66** 023812
[11] Alves D T, Farina C and Maia Neto P A 2003 *J. Phys.* A **36** 11333
[12] Dodonov V V, Klimov A B and Man'ko V I 1990 *Phys. Lett.* A **149** 225
[13] Andreata M A and Dodonov V V 2000 *J. Phys.* A **33** 3209
[14] Razavy M and Terning J 1985 *Phys. Rev.* D **31** 307; Calucci G 1992 *J. Phys.* A **25** 3873; Law C K 1994 *Phys. Rev.* A **49** 433; Law C K 1995 *Phys. Rev.* A **51** 2537; Dodonov V V and Klimov A B 1996 *Phys. Rev. A* **53** 2664; Mundarain D F and Maia Neto P A 1998 *Phys. Rev.* A **57** 1379
[15] Haro J 2005 *J. Phys.* A **38** L307
[16] Castagnino M and Ferraro R 1984 *Ann. Phys.* **154** 1-23
[17] Cole C K and Schieve W C 1995 *Phys. Rev.* A **52** 4405
[18] Alves D T and Granhen E R 2008 *Phys. Rev.* A **77** 015808
[19] Cole C K and Schieve W C 2001 *Phys. Rev.* A **64** 023813-1
[20] Alves D T, Granhen E R, Lima M G and Silva H O *Exact solution for the energy density inside a non-static cavity with an arbitrary initial field state* (to be published)
[21] Dalvit D A R, Mazzitelli F D and Milln O 2006 *J. Phys.* A **39** 6261
[22] Walker W R and Davies P C W 1982 *J. Phys.* A **15** L477
[23] Boyer T H 2003 *Am. J. Phys.* **71** 990
[24] Glauber R J 1963 *Phys. Rev.* **131** 2766; Glauber R J 1963 *Phys. Rev. Lett.* **10** 84

Journal of Physics: Conference Series **161** (2009) 012033 doi:10.1088/1742-6596/161/1/012033

Quantum radiation force on a moving mirror for a thermal and a coherent field

D T Alves[1], **E R Granhen**[2], **M G Lima**[1] **and A L C Rego**[1]
[1] Faculdade de Física, Universidade Federal do Pará, 66075-110, Belém, PA, Brazil
[2] Centro Brasileiro de Pesquisas Fsicas, Rua Dr. Xavier Sigaud, 150, 22290-180, Rio de Janeiro, RJ, Brazil

Abstract.
We consider a real massless scalar field in a two-dimensional spacetime, satisfying Dirichlet or Neumann boundary condition at the instantaneous position of a moving boundary. For a relativistic particular motion proposed by Walker-Davies (J. Phys. A, **15**, L477 (1982)), we obtain the exact formulas for the radiation force on the boundary for thermal and coherent states.

1. Introduction

Thirty years ago, several authors started the investigation about the moving mirror radiation problem (see Refs. [1–7]). In the context of a real scalar field in a two dimensional spacetime, with the Dirichlet boundary condition imposed to the field at the moving boundary position $x = z(t)$, Fullling and Davies [3] obtained an exact formula for the finite physical part of the expected value of the energy-momentum tensor, assuming the initial state as the vacuum. Several mirror trajectories considered in the literature, including those investigated in Ref. [3], assumed the boundary accleration starting or dropping abruptly, generating an infinite energy density, or a delta pulse of energy emitted from the discontinuity point. Desiring to study an exactly soluble model without these discontinuities, Walker and Davies [8] proposed a law of motion for which the mirror accelerates along a smooth and asymptotically static trajectory. Then, they obtained an exact solution for the radiation moving mirror problem. Their asymptotically static motion brought the advantage of avoid certain pathologies related to the radiation emitted by a moving mirror with abrupt acceleration.

In the present paper, we investigated the model proposed by Walker and Davies [8], but considering the initial field state as thermal or a coherent one, and also Dirichlet and Neumann boundary conditions imposed to the field. The present work is based on the exact results obtained in Ref. [9].

2. Thermal and Coherent Forces for a Particular Moving Mirror Law

In Ref. [8], Walker and Davies proposed the following law of motion of the mirror:

$$t = -z \pm A\left(e^{-2z/B} - 1\right)^{1/2}, \tag{1}$$

where A and B are constants, with $A > B$ so that $|\dot{z}| < 1$. This is a smooth and asymptotically static motion (see Fig. 1) at $t \to \pm\infty$. The mirror velocity can be relativistic (see Fig. 2) for the parameters $A = 2$ and $B = 1$. This kind of motion enables us to write the expression for the mirror velocity as:

$$\dot{z} = -\frac{Bp(u)}{Bp(u) + p(u)^2 + A^2}, \tag{2}$$

where the function $p(u)$ is obtained from $p(u) = 2\tau(u) - u$ and $\tau(u)$ from $u = \tau(u) - z[\tau(u)]$. From Eq. (1) we get $u = B\ln(p(u)^2/A^2 + 1) + p(u)$ [8], and also:

$$z = -\frac{B}{2}\ln(p(u)^2/A^2 + 1). \tag{3}$$

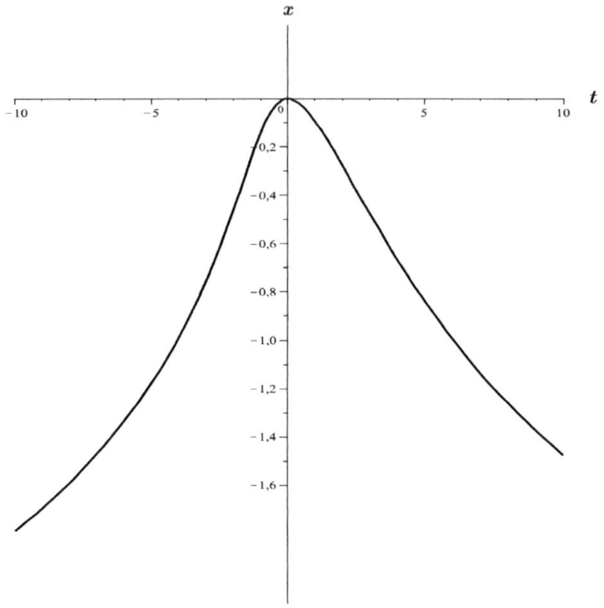

Figure 1. The smooth and asymptocally motion proposed by Walker and Davies, with A=2 and B=1.

Let us now examine the radiation force on the moving mirror when a thermal bath with temperature T is considered as the initial field state. For this case, we have: $\left\langle \hat{a}^\dagger_{\omega'}\hat{a}_\omega \right\rangle = \bar{n}(\omega)\delta(\omega - \omega')$ where $\bar{n}(\omega) = 1/(e^{\hbar\omega/T} - 1)$, with the Boltzmann constant equal to 1. Hereafter we consider the averages $\langle ... \rangle$ taken over an initial field state, assumed here, for simplicity, as being the same one for both sides of the mirror. Starting from the expected value of the energy density operator $\mathcal{T} = \langle \hat{T}_{00}(t,x) \rangle$, we can write the net force $F(t)$ acting on the moving boundary defined by (since $T_{00} = T_{11}$ in this model) :

$$F(t) = \mathcal{T}[t, z(t)]^{(-)} - \mathcal{T}[t, z(t)]^{(+)}, \tag{4}$$

where the superscript "+" indicates that the force acts on the right side of the boundary, whereas "-" indicates the force acting on the left side. We can split the net force in the following manner:

$$F = F_{vac} + F^{(T)}, \tag{5}$$

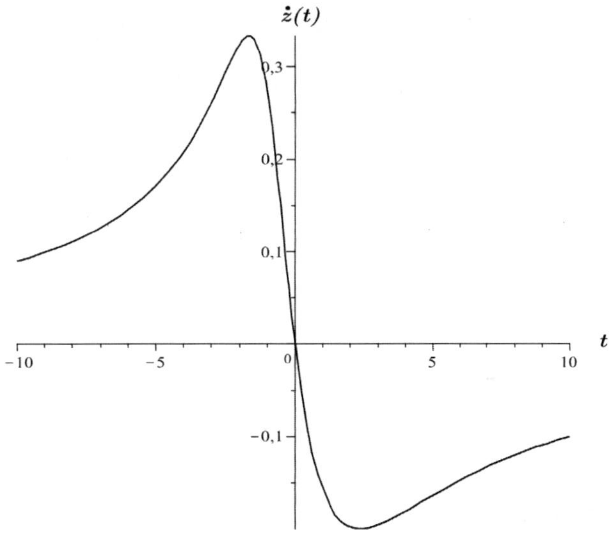

Figure 2. The relativistic mirror velocity for the parameters $A = 2$ and $B = 1$.

where F_{vac} stands for the vacuum contribution to the total force. For the thermal contribution, $F^{(T)}$, we get the exact formula [9]:

$$F^{(T)} = -\sigma_T \left[\dot{z} \frac{(1+\dot{z}^2)}{(1-\dot{z}^2)^2} \right] = \sum_{n=0}^{\infty} F^{(T)}_{(n)}, \tag{6}$$

where

$$F^{(T)}_{(n)} = -\sigma_T \sum_{n=0}^{\infty} (2n+1)\dot{z}^{2n+1} \tag{7}$$

and $\sigma^{(T)} = 2\pi T^2/3$ is the viscosity coefficient. ¿From Eq. (2) and (6) we get:

$$\begin{aligned} F^{(T)} &= \sigma^{(T)} \left\{ \frac{Bp(u)\left(Bp(u) + p(u)^2 + A^2\right)}{(2Bp(u)^3 + 2Bp(u)A^2 + p(u)^4 + 2p(u)^2 A^2 + A^4)^2} \right. \\ &\quad \left. \times \left(2B^2 p(u)^2 + 2Bp(u)^3 + 2Bp(u)A^2 + p(u)^4 + 2p(u)^2 A^2 + A^4\right) \right\}, \end{aligned} \tag{8}$$

In Eq. (6), truncating the series in $n = N$ $(N = 0, 1, 2, ...)$, we get the approximate force $\tilde{F}^{(T)}$:

$$\tilde{F}^{(T)} = \sum_{n=0}^{N} F^{(T)}_{(n)}. \tag{9}$$

For $n = 0$ we get:

$$F^{(T)} \approx F^{(T)}_{(0)} = -\sigma^{(T)}\dot{z} = \sigma^{(T)} \frac{Bp(u)}{Bp(u) + p(u)^2 + A^2}. \tag{10}$$

This formula shows that the force is proportional to the boundary velocity (the approximate thermal force depending on velocity was obtained by Jaekel and Reynaud [10]). In Fig. 3 we show the exact force (solid line) and the approximate one (dashed line). We see that for higher velocities (see also Fig. 2) a bigger discrepancy between exact and approximate values occurs. But for $t = 10$, the boundary velocity is around 0.1 of the light velocity and, besides this relativistic velocity, both exact and approximate formulas are in good agreement in this region. It is also remarkable that the thermal force is the same for Dirichlet or Neumann boundary condition [9].

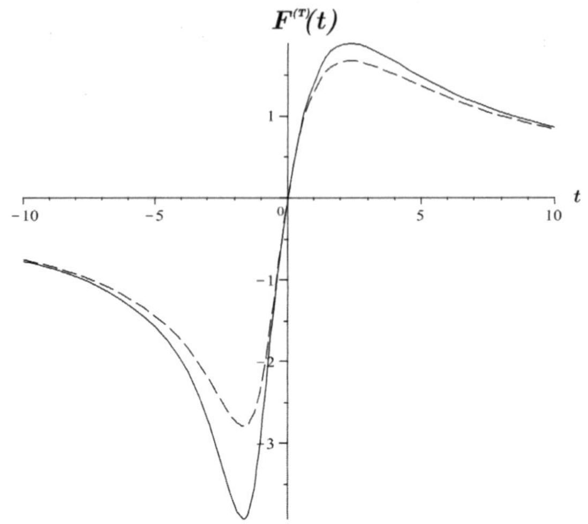

Figure 3. The exact thermal force $F^{(T)}$ (solid line) and the approximate thermal force $F^{(T)} \approx F^{(T)}_{(0)}$ (dashed line), for $T = 1$.

Now, let us investigate the coherent state as the initial field state. The coherent state of amplitude α is defined as an eigenstate of the annihilation operator: $\hat{a}_\omega |\alpha\rangle = \alpha\delta(\omega - \omega_0)|\alpha\rangle$, where $\alpha = |\alpha|\exp(i\theta)$ and ω_0 is the frequency of the excited mode. For this case, we can split the net force in the following manner:

$$F = F_{vac} + F^{(\alpha)}, \tag{11}$$

where:

$$F^{(\alpha)} = F^{(\alpha)}_{\langle\hat{a}^\dagger\hat{a}\rangle} + F^{(\alpha)}_{\langle\hat{a}\hat{a}\rangle}, \tag{12}$$

$$F^{(\alpha)}_{\langle\hat{a}^\dagger\hat{a}\rangle} = -\frac{4}{\pi}\omega_0 |\alpha|^2 \dot{z}(1+\dot{z}^2)/(1-\dot{z}^2)^2, \tag{13}$$

$$\begin{aligned} F^{(\alpha)}_{\langle\hat{a}\hat{a}\rangle} &= \pm\frac{\omega_0}{4\pi}|\alpha|^2 e^{-2i(\omega_0 t-\theta)}\left\{\left[e^{2i\omega_0 z(t)}\left(\frac{1-\dot{z}}{1+\dot{z}}\right)^2 - e^{-2i\omega_0 z(t)}\right]\right. \\ &\left. -\left[\left(\frac{1+\dot{z}}{1-\dot{z}}\right)^2 e^{-2i\omega_0 z(t)} - e^{2i\omega_0 z(t)}\right]\right\} + \text{c.c.}, \end{aligned} \tag{14}$$

and the sign "+" refers to Dirichlet and "-" to Neumann boundary conditions. Using Eq. (2) and (3) in Eqs.(13) and (14), we get the specific formulas:

$$\begin{aligned} F^{(\alpha)}_{\langle\hat{a}^\dagger\hat{a}\rangle} &= \sigma^{(\alpha)}\left\{\frac{Bp(u)(Bp(u)+p(u)^2+A^2)}{(2Bp(u)^3+2Bp(u)A^2+p(u)^4+2p(u)^2A^2+A^4)^2}\right. \\ &\left.\times(2B^2p(u)^2+2Bp(u)^3+2Bp(u)A^2+p(u)^4+2p(u)^2A^2+A^4)\right\}, \end{aligned} \tag{15}$$

$$F^{(\alpha)}_{\langle\hat{a}\hat{a}\rangle} = \pm\frac{\sigma^{(\alpha)}e^{-2i(\omega_0 t-\theta)}}{16}\left\{\frac{(p(u)^2+A^2)^2+(2Bp(u)+p(u)^2+A^2)^2}{(p(u)^2+A^2)^2}\left(\frac{p(u)^2+A^2}{A^2}\right)^{-i\omega_0 B}\right.$$
$$\left.-\frac{(p(u)^2+A^2)^2+(2Bp(u)+p(u)^2+A^2)^2}{(2Bp(u)+p(u)^2+A^2)^2}\left(\frac{p(u)^2+A^2}{A^2}\right)^{i\omega_0 B}\right\}+\text{c.c.}. \qquad (16)$$

Where $\sigma^{(\alpha)} = 4|\alpha|^2\omega_0/\pi$. From Eq. (15) we see that the force $F^{(\alpha)}_{\langle\hat{a}^\dagger\hat{a}\rangle}$ is the same for Dirichlet or Neumann conditions. On the other hand, in Eq. (16) (and also in Eq. (17)), the sign "+" refers to Dirichlet and "-" to Neumann boundary conditions.

If we consider simultaneously nonrelativistic velocities and small displacements (in the sense considered in Ref. [11]), the force $F^{(\alpha)}$ can be approximated as [9]:

$$F^{(\alpha)} \approx \tilde{F}^{(\alpha)} = -\frac{4\omega_0}{\pi}|\alpha|^2\left\{\dot{z}(t) \pm \left[\cos(2\omega_0 t-2\theta)\,\dot{z}(t) - \sin(2\omega_0 t-2\theta)\,\omega_0 z(t)\right]\right\}. \qquad (17)$$

Using Eq. (2) and (3) in Eqs.(17), we get:

$$\tilde{F}^{(\alpha)} = \sigma^{(\alpha)}\left\{\frac{Bp(u)}{Bp(u)+p(u)^2+A^2}\left[1\pm\cos(2(\omega_0 t-\theta))\right]\right.$$
$$\left.\mp\frac{\omega_0 B}{2}\ln\left(p^2/A^2+1\right)\sin(2(\omega_0 t-\theta))\right\}. \qquad (18)$$

In Fig. 4 we show the exact force (solid line) and the approximate one (dashed line) for the coherent case with Dirichlet boundary condition. Again we see that for higher velocities (see also Fig. 2) a bigger discrepancy between exact and approximate values occurs. But, here, the discrepant part presents the approximate coherent forces bigger than the exact one in absolute values. The inverse behavior was observed in the thermal case. It is also remarkable that for Neumann boundary condition, the coherent force oscillates in a different manner (see Fig. 5).

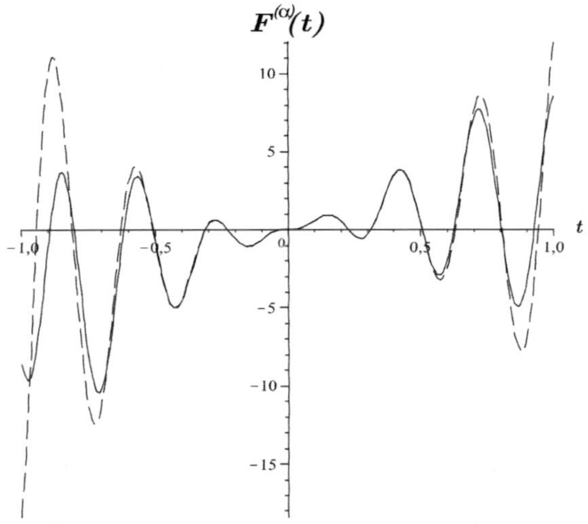

Figure 4. The exact coherent force $F^{(\alpha)}$ (solid line) and approximate coherent force $F^{(\alpha)} \approx \tilde{F}^{(\alpha)}$ (dashed line), for the Dirichlet case, with $\alpha = 1$, $\theta = \pi/2$ and $\omega_0 = 10$.

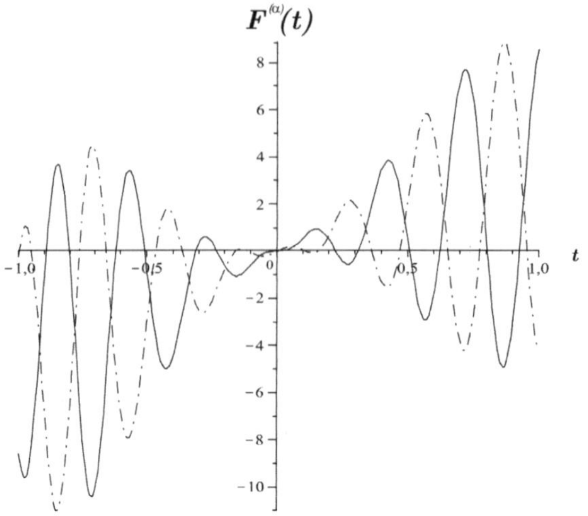

Figure 5. The exact coherent force for Dirichlet (solid line) and Neumann (dot-dashed line) boundary conditions, with $\alpha = 1$, $\theta = \pi/2$ and $\omega_0 = 10$.

3. Final comments

In Fig. 3 we see that $|F^{(T)}| \geq |F^{(T)}_{(0)}|$. In Eq. (6) we can see that $F^{(T)}$ is given in terms of a series in odd powers of $\dot{z}$, with each term in this series having the same sign. This enable us to write:

$$|F^{(T)}| = \sum_{n=0}^{\infty} |F^{(T)}_{(n)}|. \tag{19}$$

¿From the above equation, we can conclude that:

$$|F^{(T)}| \geq |\tilde{F}^{(T)}|. \tag{20}$$

The result in Eq. (20) is valid for any law of motion (not only for the one considered in this paper). In this context, the inequality in Eq. (20) explains the graphical behavior of the curves in Fig. 3. Note that in Figs. 2 and 3 $F^{(T)}$ is always opposite to the boundary motion, what can also be concluded via Eq. (6).

In Fig. 4 we can observe an inverse behavior: $|F^{(\alpha)}| \leq |\tilde{F}^{(\alpha)}|$. In fact, the formula (17) is obtained after considering small velocities and also small displacements. The latter consideration originates a linear dependence on $z(t)$ in Eq. (17). Then, when $t \to \pm\infty$, $|z| \to \infty$ and the amplitude of $\tilde{F}^{(\alpha)}$ grows up.

In summary, we studied the force exerted by the field on one moving mirror which imposes on a real massless scalar field in a two-dimensional spacetime a Dirichlet or Neumann boundary condition. We extended to the thermal and coherent initial field states the analytical results obtained by Walker and Davies for the force on the moving boundary. Numerical calculations were done, resulting in the graphical behavior of these forces.

Acknowledgments

We acknowledge the Referee for valuable suggestions. This work was supported by FAPESPA,CNPq and CAPES-Brazil.

References

[1] Moore G T 1970 *J Math. Phys.* **11** 2679

[2] DeWitt B S 1975 *Phys. Rep.* **19** 295
[3] Fulling S A and Davies P C W 1976 *Proc. R. Soc. London* **A** **348** 393
[4] Davies P C W and Fulling S A 1977 *Proc. R. Soc. London* **A** **354** 59
[5] Davies P C W and Fulling S A 1977 *Proc. R. Soc. London* **A** **356** 237
[6] Candelas P and Raine D J 1976 *J. Math. Phys.* **17** 2101
[7] Candelas P and Deutsch D J 1977 *Proc. R. Soc. London* **A** **354** 79
[8] Walker W R and Davies P C W 1982 *J. Phys.* A **15** L477
[9] Alves D T, Granhen E R and Lima M G 2008 *Phys. Rev.* D **77** 125001
[10] Jaekel M T and Reynaud S 1993 *J. Phys.* I (France) **3** 339
[11] Alves D T, Farina C and Maia Neto P A 2003 *J. Phys. A: Math. Gen.* **36** 11333

Journal of Physics: Conference Series **161** (2009) 012034 doi:10.1088/1742-6596/161/1/012034

The Phononic Casimir Effect: An Analog Model

L H Ford

Institute of Cosmology, Department of Physics and Astronomy
Tufts University, Medford, MA 02155

E-mail: ford@cosmos.phy.tufts.edu

N F Svaiter

Centro Brasiliero de Pesquisas Fisicas CBPF
Rua Dr. Xavier Sigaud 150, Rio de Janeiro, RJ, 22290 180, Brazil

E-mail: nfuxsvai@cbpf.br

Abstract. We discuss the quantization of sound waves in a fluid with a linear dispersion relation and calculate the quantum density fluctuations of the fluid in several cases. These include a fluid in its ground state. In this case, we discuss the scattering cross section of light by the density fluctuations, and find that in many situations it is small compared to the thermal fluctuations, but not negligibly small and might be observable at room temperature. We also consider a fluid in a squeezed state of phonons and fluids containing boundaries. We suggest that the latter may be a useful analog model for better understanding boundary effects in quantum field theory. In all cases involving boundaries which we consider, the mean squared density fluctuations are reduced by the presence of the boundary. This implies a reduction in the light scattering cross section, which is potentially an observable effect.

1. Introduction

It is well known that quantized sound waves, whose excitations are phonons, share several properties with relativistic quantum fields, such as the electromagnetic field. Here we will be primarily concerned with fluctuations in the phonon ground state. These fluctuations can share certain features with electromagnetic vacuum fluctuations, including the creation of Casimir-type forces [1]. The forces due to quantum or stochastic sound fluctuations have been discussed recently by several authors [2, 3, 4, 6, 7]. In this paper, we will study quantum density fluctuations in a fluid. The fluid in question could be either a classical or a quantum fluid, but we will consider only the regime where the dispersion relation is approximately linear. Thus the analog models which we consider are rather different than those which model the Hawking process [8, 9, 10]. The topic of hydrodynamic fluctuations has been reviewed, for example, in Refs. [11, 12]. In Sec. 2, we will discuss density fluctuations in a fluid in the phonon ground state without boundaries, and summarize work recently reported in Ref. [13]. It was argued there that light scattering by quantum density fluctuations might be observable, even in fluids at room temperature. The effects of a squeezed state of phonons will be briefly discussed in Sec. 3. In Sec. 4, we turn to the issue of the effects of boundaries in the fluid, and argue that this can be a useful analog model for the effects of boundaries on relativistic quantum fields. We also suggest that the boundary effects, at least in principle, create observable effects in the

Journal of Physics: Conference Series **161** (2009) 012034 doi:10.1088/1742-6596/161/1/012034

fluid case. We report on the result of several specific calculations for different geometries. The results will be summarized and discussed in Sec. 5.

2. Phonons and Density Fluctuations in a Fluid

2.1. Quantization and the Density Correlation Function

We consider the quantization of sound waves in a fluid with a linear dispersion relation, $\Omega_q = c_S\, q$, where Ω_q is the phonon angular frequency, q is the magnitude of the wave vector, and c_S is the speed of sound in the fluid. This should be a good approximation for wavelengths much longer than the interatomic separation. Let ρ_0 be the mean mass density of the fluid. Then the variation in density around this mean value is represented by a quantum operator, $\hat{\rho}(\mathbf{x}, t)$, which may be expanded in terms of phonon annihilation and creation operators as [14]

$$\hat{\rho}(\mathbf{x}, t) = \sum_{\mathbf{q}} (b_{\mathbf{q}} f_{\mathbf{q}} + b^\dagger_{\mathbf{q}} f^*_{\mathbf{q}}) \,, \tag{1}$$

where

$$f_{\mathbf{q}} = \sqrt{\frac{\hbar \omega \rho_0}{2 V c_S^2}}\; \mathrm{e}^{i(\mathbf{q}\cdot\mathbf{x} - \Omega_q\, t)} \,. \tag{2}$$

Here V is a quantization volume. The normalization factor in Eq. (2) can be fixed by requiring that the zero point energy of each mode be $\frac{1}{2}\hbar\Omega_q$ and using the expression for the energy density in a sound wave,

$$U = \frac{c_S^2}{\rho_0}\, \hat{\rho}^2 \,. \tag{3}$$

In the limit in which $V \to \infty$, we may write the density correlation function as

$$\langle \hat{\rho}(\mathbf{x}, t)\, \hat{\rho}(\mathbf{x}', t') \rangle = \frac{\hbar \rho_0}{16\pi^3 c_S^2} \int d^3 q\, \Omega_q\, \mathrm{e}^{i(\mathbf{q}\cdot\Delta\mathbf{x} - \Omega_q\, \Delta t)} \,, \tag{4}$$

where $\Delta\mathbf{x} = \mathbf{x} - \mathbf{x}'$ and $\Delta t = t - t'$. The integral may be evaluated to write the coordinate space correlation function as

$$\langle \hat{\rho}(\mathbf{x}, t)\, \hat{\rho}(\mathbf{x}', t') \rangle = -\frac{\hbar \rho_0}{2\pi^2 c_S} \frac{\Delta\mathbf{x}^2 + 3 c_S^2 \Delta t^2}{(\Delta\mathbf{x}^2 - 3 c_S^2 \Delta t^2)^3} \,. \tag{5}$$

This is of the same form as the correlation function for the time derivative of a massless scalar field in relativistic quantum field theory, $\langle \dot{\varphi}(\mathbf{x}, t)\, \dot{\varphi}(\mathbf{x}', t') \rangle$. (This analogy has been noted previously in the literature. See, for example, Ref. [15].) Apart from a factor of ρ_0, these two quantities may be obtained from one another by interchanging the speed of light c and the speed of sound c_S. If $c \to c_S$, then

$$\langle \dot{\varphi}(\mathbf{x}, t)\, \dot{\varphi}(\mathbf{x}', t') \rangle \to \rho_0 \, \langle \hat{\rho}(\mathbf{x}, t)\, \hat{\rho}(\mathbf{x}', t') \rangle \,. \tag{6}$$

In the limit of equal times, the density correlation function becomes

$$\langle \hat{\rho}(\mathbf{x}, t)\, \hat{\rho}(\mathbf{x}', t) \rangle = -\frac{\hbar \rho_0}{2\pi^2 c_S\, (\Delta\mathbf{x})^4} \,. \tag{7}$$

Thus the density fluctuations increase as $|\Delta\mathbf{x}|$ decreases. Of course, the continuum description of the fluid and the linear dispersion relation both fail as $|\Delta\mathbf{x}|$ approaches the interatomic separation. Also note the minus sign in Eq. (7). This implies that density fluctuations at different locations at equal times are anticorrelated. By contrast, when $c_S|\Delta t| > |\Delta\mathbf{x}|$, then

Journal of Physics: Conference Series **161** (2009) 012034 doi:10.1088/1742-6596/161/1/012034

$\langle \hat{\rho}(\mathbf{x},t)\,\hat{\rho}(\mathbf{x}',t)\rangle > 0$ and the fluctuations are positively correlated. This is complete analogy with the situation in the relativistic theory. Fluctuations inside the lightcone can propagate causally and tend to be positively correlated. Fluctuations in a fluid for which $c_S|\Delta t| < |\Delta \mathbf{x}|$ cannot have propagated from one point to the other, and are anti-correlated. This can be understood physically because an over density of fluid at one point in space requires an under density at a nearby point.

2.2. Light Scattering by Density Fluctuations

In Ref. [13], the cross section for the scattering of light by the zero point density fluctuations is computed for the case that the incident light angular frequency is large compared to the typical phonon frequency. The result is

$$\left(\frac{d\sigma}{d\Omega}\right)_{ZP} = \sqrt{2(1-\cos\theta)}\,\frac{\hbar\omega^5\,\mathcal{V}\,\eta^4}{32\pi^2 c^5 c_S \rho_0}\,(\hat{\mathbf{e}}_{\mathbf{k},\lambda}\cdot\hat{\mathbf{e}}_{\mathbf{k}',\lambda'})^2\,, \tag{8}$$

where θ is the scattering angle, $\mathcal{V}$ is the scattering volume, and η is the mean index of refraction of the fluid. The ω^5 dependence of the scattering cross section can be viewed as the product of the ω^4 dependence of Rayleigh-Brillouin scattering and one power of Ω_q, and hence of ω, coming from the spectrum of zero point fluctuations in the fluid. The factor of η^4 represents the influence of the fluid on light propagation before and after the scattering process, and arises as a product of a factor of η in the incident flux and a factor of η^3 in the density of final states [13]. Because light travels through the fluid at speeds much greater than the sound speed, light scattering reveals a nearly static distribution of density fluctuations. Thus we can regard Eq. (8) as a probe of the fluctuations described by Eq. (7). The scattering by zero point fluctuations is inelastic, with the creation of a phonon. Thus, the scattering described by Eq. (8) is strictly Brillouin rather than Rayleigh scattering.

This scattering by zero point density fluctuations should be compared to the effects of thermal density fluctuations. The ratio of the zero point to the thermal scattering can be expressed as

$$R \equiv \frac{(d\sigma/d\Omega)_{ZP}}{(d\sigma/d\Omega)_{TB}} = \sqrt{2(1-\cos\theta)}\left(\frac{\hbar\omega}{2k_BT}\right)\left(\frac{c_S}{c}\right)\eta^4\left[\rho_0\left(\frac{\partial\epsilon}{\partial\rho_0}\right)_S\right]^{-2}. \tag{9}$$

The index of refraction, η, and the quantity $\rho_0\,(\partial\epsilon/\partial\rho_0)_S$, which involves a derivative of the fluid dielectric function with respect to density at constant entropy are both of order unity. Hence R is primarily determined by the ratio of the photon energy to the thermal energy, and the ratio of the speed of sound to the speed of light.

As an example, consider the case of water at room temperature and violet light with a wavelength of $\lambda = 350nm$. In this case, we have $c_S = 1480m/s$ and $\eta = 1.4$ [16]. In addition, $\rho_0\,(\partial\epsilon/\partial\rho_0)_S = 0.79$ [17]. For back scattering, $\cos\theta = -1$, this leads to $R \approx 0.005$. Consequently, about 0.5% of the Stokes line is due to zero point motion effects. Although this is a small fraction, it may be detectable, and will increase at lower temperatures and shorter wavelengths.

3. Squeezed States of Phonons

Now we consider the case where the phonon field is not in the vacuum state, but rather a squeezed state. The squeezed states are a two complex parameter family of states, but we will focus attention on the case of the squeezed vacuum states $|\zeta\rangle$, labeled by a single complex squeeze parameter

$$\zeta = r\,\mathrm{e}^{i\delta}\,. \tag{10}$$

This set of states is of special interest because they are the states generated by quantum particle creation processes, and they can exhibit local negative energy densities. Consider the shift in the mean squared density fluctuations between the given state and the vacuum

$$\langle \hat{\rho}^2 \rangle_R = \langle \zeta | \hat{\rho}^2 | \zeta \rangle - \langle 0 | \hat{\rho}^2 | 0 \rangle \,, \tag{11}$$

the "renormalized" mean squared density fluctuation. The result for this quantity in a single mode squeezed vacuum state for a plane wave in the z-direction is

$$\langle \hat{\rho}^2 \rangle_R = \frac{\hbar\, \omega\, \rho_0}{c_S^2 V} \,\sinh r\, \left\{ \sinh r - \cosh r\, \cos[2(kz - \omega t) + \delta] \right\} . \tag{12}$$

Note that this quantity can be either positive or negative, but its time or space average is positive. The suppression of the local density fluctuations in a squeezed state is analogous to the creation of negative energy densities for a massless, relativistic field.

4. Boundaries

If we introduce an impenetrable boundary into the fluid, the phonon field will satisfy Neumann boundary conditions

$$\hat{\mathbf{n}} \cdot \nabla \delta\rho = 0 \tag{13}$$

as a consequence of the impenetrability. Thus there will be a Casimir force on the boundaries which is analogous to the Casimir force produced by electromagnetic vacuum effects. For example, consider two parallel plates, which will experience an attractive force per unit area of

$$\frac{F}{A} = \frac{\hbar\, c_S\, \pi^2}{480\, a^4} \,, \tag{14}$$

which is smaller than the electromagnetic case for perfect plates by a factor of $c_S/(2c)$, and is thus quite small in any realistic situation.

Henceforth, we consider the local effect of boundaries on mean squared density fluctuations, and now define $\langle \hat{\rho}^2 \rangle_R$ to be the change due to the presence of the boundary. This quantity is of interest both as an analog model for the effects of boundaries in quantum field theory, and in its own right. The shifts in density fluctuations are at least in principle observable in light scattering experiments.

Our interest in the phononic analog model is inspired by the fact that the study of boundary effects in quantum field theory is an active area of research, and has given rise to some recent controversies in the literature [18, 19]. One question is the nature of the physical cutoff which prevents singularities at the boundary. An example of the subtleties is afforded by the mean squared electric and magnetic fields near a dielectric interface. When the material is a perfect conductor, these quantities are proportional to z^{-4}, where z is the distance to the interface. Specifically, in Lorentz-Heaviside units their asymptotic forms are

$$\left\langle E^2 \right\rangle \sim \frac{3}{16\pi^2} \frac{1}{z^4} \tag{15}$$

and

$$\left\langle B^2 \right\rangle \sim -\frac{3}{16\pi^2} \frac{1}{z^4} \tag{16}$$

One might expect that a realistic frequency dependent dielectric function would remove this singularity, but this is not the case. Instead one finds [20] that

$$\left\langle E^2 \right\rangle \sim \frac{\sqrt{2}\omega_p}{32\pi} \frac{1}{z^3} \tag{17}$$

and

$$\left\langle B^2 \right\rangle \sim -\frac{5\omega_p^2}{96\pi}\frac{1}{z^2}, \tag{18}$$

where ω_p is the plasma frequency of the material. Thus some physical cutoff other than dispersion is required. For realistic materials, it is likely to be surface roughness, but fluctuations in the position of the boundary can also serve as a cutoff, as we showed several years ago [21]. In a fluid, there is always a physical cutoff at the interatomic separation, but in a given situation other physical cutoffs may be dominant. For the present, we will report the results of calculations of $\langle \hat{\rho}^2 \rangle_R$ for several different geometries, and leave a more detailed treatment of physical cutoffs for later work.

In the remainder of this section, we will quote several results for $\langle \hat{\rho}^2 \rangle_R$ in different geometries. The details of the calculations involved will be presented in a later paper [22].

4.1. One or Two Parallel Plane Boundaries

In both of these case, the renormalized density two-point function may be constructed by the method of images. For the case of a single plane, one finds

$$\langle \hat{\rho}^2 \rangle_R = -\frac{\hbar\,\rho_0\,c_S}{32\pi^2\,z^4} < 0 \tag{19}$$

where z is the distance to the boundary. For the case of two parallel planes, the result is

$$\langle \hat{\rho}^2 \rangle_R = -\frac{\hbar\,\rho_0\,c_S}{96a^4}\left[\frac{1}{15} + \frac{3 - 2\sin^2(\pi z/a)}{\sin^4(\pi z/a)}\right], \tag{20}$$

where a is the separation of the two planes, and z is the distance to one boundary. Note that both of these expression are negative everywhere. In the absence of a physical cutoff, both of these expressions diverge as z^{-4} near the boundaries, just as do the squared electric and magnetic fields near a perfectly reflecting plane.

4.2. A Three-Dimensional Torus

Here we consider a rectangular box with periodic boundary conditions in all three spatial directions, with periodicity lengths L_1, L_2 and L_3. Thus the three-dimensional space has the topology of $S^1 \times S^1 \times S^1$. This is closely related to the geometry of a waveguide, where the fluctuations of a relativistic scalar field were discussed by Rodrigues and Svaiter [23]. As in the parallel plane case, an image sum method may be employed, with the result

$$\langle \hat{\rho}^2 \rangle_R = -\frac{\hbar\,\rho_0\,c_S}{2\pi^2}\sum_{\ell,m,n}{}' \frac{1}{(\ell^2 L_1^2 + m^2 L_2^2 + n^2 L_3^2)^2}. \tag{21}$$

Here the prime on the summation indices denotes that the $\ell = m = n = 0$ term is omitted. Here $\langle \hat{\rho}^2 \rangle_R$ is a negative constant.

4.3. A Wedge

Consider two intersecting plane which are at an angle of α with respect to each other. Now consider a point inside of this wedge which is located at polar coordinates (r, θ), where r is the distance to the intersection line and $\theta < \alpha$. This geometry was treated for the relativistic case by Candelas and Deutsch [24], and we may transcript one of their results [their Eq. (5.39)] to find

$$\begin{aligned}\langle \hat{\rho}^2 \rangle_R &= -\frac{\hbar\,\rho_0\,c_S}{1440\pi^2\,r^4\,\sin^4(\pi\theta/\alpha)} \\ &\times \left\{(\pi-\alpha)(\pi+\alpha)\sin^2(\pi\theta/\alpha)[(\pi^2+11\alpha^2)\sin^2(\pi\theta/\alpha) - 30\pi^2] + 45\pi^4\right\}. \end{aligned} \tag{22}$$

Again, this quantity is negative everywhere.

4.4. A Cosmic String

As is well known, the space surrounding a cosmic string is a conical space with a deficit angle $\alpha < 2\pi$. Quantum field theory in this conical space has been discussed by many authors, beginning with Helliwell and Konkowski [25], and is similar to the wedge problem discussed above. We may follow the procedure in Ref. [25]. At a distance r from the apex, we find

$$\langle \hat{\rho}^2 \rangle_R = -\frac{\hbar\, \rho_0\, c_S}{1440\pi^2\, \alpha^4\, r^4}\,(2\pi - \alpha)(2\pi + \alpha)(11\alpha^2 + 4\pi^2)\,, \tag{23}$$

which is also negative everywhere.

4.5. Near the Focus of a Parabolic Mirror

The quantization of the electromagnetic field in the presence of a parabolic mirror was discussed by us in Refs. [26, 27], where a geometric optics approximation was employed to find the mean squared fields near the focus. This treatment lead to the result that these quantities are singular at the focus, diverging as an inverse power of the distance a to the focus. This result holds both for parabolic cylinders and for parabolas of revolution, and basically arises from the interference term of multiply reflected rays with nearly the same optical path length. The geometry is illustrated in Fig. 1. An incoming ray at an angle of θ reflects at an angle of θ' to reach the point P, which is a distance a from the focus F, as illustrated. The distance from the focus to the mirror itself is $b/2 \gg a$. There can be two values of θ', denoted α and β, corresponding to two reflected rays. The difference in the optical paths of these two rays is denoted by $\Delta\ell$.

The detailed expression for this distance $\Delta\ell$ used in Refs. [26, 27] is not quite correct, as was pointed out to us by Vuletic [28]. The corrected expression is

$$\Delta\ell = a\,\Big[\cos\gamma(\cos\alpha - cos\beta + \sin^2\alpha - \sin^2\beta) + \sin\gamma(\sin\alpha - \sin\beta + \sin\beta\cos\beta - \sin\alpha\cos\alpha)\Big]\,. \tag{24}$$

This is to be used in the expression obtained from geometric optics,

$$\langle E^2 \rangle = 45\pi^3 \int \frac{d\theta}{(\Delta\ell)^4}\,. \tag{25}$$

In this expression, $\alpha = \alpha(\theta)$, and $\beta = \beta(\theta)$, as will be discussed in Ref. [22].

A detailed discussion of the electromagnetic case will be given elsewhere. Here we are concerned with $\langle \hat{\rho}^2 \rangle_R$, which is obtained from Eq. (25) by letting $c \to c_S$ and dividing by 2, leading to a result of the form

$$\langle \hat{\rho}^2 \rangle_R = -\frac{\hbar\, \rho_0\, c_S\, C}{b\, a^3} < 0\,. \tag{26}$$

This, and the analogous expressions for $\langle E^2 \rangle$ and $\langle B^2 \rangle$, which also are proportional to $1/(b\,a^3)$, are striking in that they can be large when the focus is far from the mirror itself, $b \gg a$. This result is controversial, and seems to be in conflict with a general result by Fewster and Pfenning [29], which implies that quantities such as $\langle E^2 \rangle$ or $\langle \hat{\rho}^2 \rangle_R$ should be proportional to the inverse fourth power of the distance to the mirror, which is to say $\propto b^{-4}$ in this case. On the other hand, there is a simple physical argument to the contrary in this case, which we find compelling: the interference term between multiply reflected rays is slowly oscillating when $\Delta\ell \propto a$ is small, and should give a contribution proportional to an inverse power of a, as in Eq. (26). In any case, the study of the phononic case provides an additional theoretical, and potentially experimental, probe to better understand this issue.

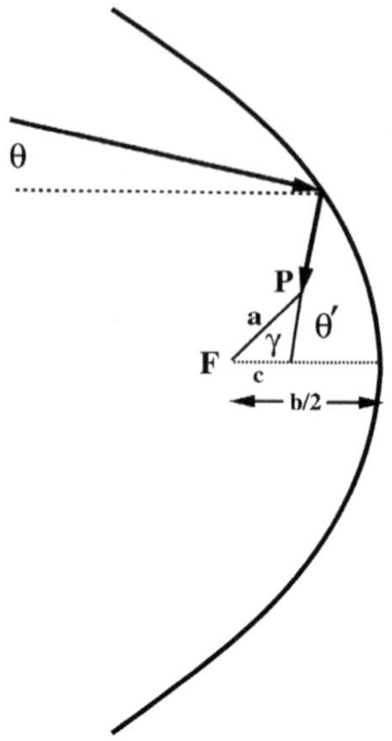

Figure 1. The geometry of rays reflecting from a parabolic mirror is illustrated. An incoming ray at an angle of θ reflects at an angle of θ' to reach the point P, which is a distance a from the focus F, and at an angle of γ.

5. Summary

In this paper, we have considered a classical or quantum fluid with a linear dispersion relation as a analog model for quantum fluctuations, and the effects of boundaries on these fluctuations. We have argued that the local density fluctuations in a fluid in the phonon ground state are potentially observable in light scattering experiments. This is of interest in its own right. We have reported the calculation of the change in these fluctuations due to squeezed states of phonons and due to the presence of boundaries. Squeezed states can produce either positive or negative values for $\langle \hat{\rho}^2 \rangle_R$, in a way which is analogous to the effects of squeezed states on the energy density of a relativistic quantum field. However, all of the boundary examples which we consider result in a negative $\langle \hat{\rho}^2 \rangle_R$. If we compare to the already negative result in Eq. (7), this seems to represent an increase in magnitude of an already negative quantity. If one were to scatter light from the fluid with a boundary, this implies a change in the scattering cross section, compared to the result for a boundary-free fluid, Eq. (8), at least for short wavelengths. In general, the calculation of the modified cross section in the presence of a boundary is a complex task. However, if the incident wavelength is short compared to the scale over which $\langle \hat{\rho}^2 \rangle_R$ varies, then we might the scattering cross section to increase in the presence of a boundary. A more detailed analysis of this effect is a subject for future study. It will also be of interest to consider the effects on non-linearity in the phonon dispersion relation on the questions considered here.

Acknowledgments

This work was supported in part by the National Science Foundation under Grant PHY-0555754 and by Conselho Nacional de Desenvolvimento Cientifico e Tecnologico do Brasil (CNPq).

[1] Dzyaloshinskii I E, Lifshitz E M and Pitaevski L P 1961 Adv. Phys. **10**, 165
[2] Larraza A 1998 Phys. Lett. A **248**, 151
[3] Bschorr O 1999 J. Acoust. Soc. Am. **106**, 3730
[4] Schäffer E and Steiner U 2002 Eur. Phys. J. E **8**, 347

[5] Roberts D C and Pomeau Y 2005 Phys. Rev. Lett. **95**, 145303
[6] Recati A, Fuchs J N, Peça C S and Zwerger W 2005 Phys. Rev. A **72**, 023616
[7] Lamoreaux S K 2008 (Preprint arxiv:0808.4000)
[8] Unruh W G 1981 Phys. Rev. Lett. **46** 1351; Unruh 1995 Phys. Rev. D **51** (Preprint gr-qc/9409008)
[9] Jacobson T 1991 Phys. Rev. D **44** 1731
[10] Barcelo C, Liberati S and Visser M 2005 Living Rev. Rel. **8** 12 (Preprint gr-qc/0505065)
[11] Foster D 1975 *Hydrodynamics Fluctuations, Broken Symmetry, and Correlation Functions*, (W. A. Benjamin, Massachusetts)
[12] Ortiz de Zarate M and Sengers J V 2006 *Hydrodynamic Fluctuations*, (Elsevier, Amsterdam)
[13] Ford L H and Svaiter N F 2008 (Preprint arXiv:0809.1851)
[14] See, for example, Lifshitz E M and Pitaevski L P 1969 *Statistical Physics, Part 2*, 2nd ed. (Pergamon, Oxford), Eq. (24.10).
[15] Fedichev P O and Fischer U 2004 Phys. Rev. A **69**, 033602
[16] *CRC Handbook of Chemistry and Physics* 2008 Lide D R, ed., 89th edition (CRC Press, Boca Raton), pp 14-44 and 10-251.
[17] Cummins H Z and Gammon R W 1966 J. Chem. Phys. **44**, 2785, Table III.
[18] Jaffe R L 2003 (Preprint arXiv:hep-th/0307014)
[19] Milton K A, Cavero-Pelaez I and Wagner J 2006 J. Phys. A **39**, 6543 (Preprint hep-th/0510236)
[20] Sopova V and Ford L H 2005 Phys. Rev. D **72**, 105010 (Preprint quant-ph/0504143)
[21] Ford L H and Svaiter N F 1998 Phys. Rev. D **58**, 065007 (Preprint quant-ph/9804056)
[22] Ford L H and Svaiter N F 2008 manuscript in preparation.
[23] Rodrigues R B and Svaiter N F 2003 Physica A **328**, 466
[24] Candelas P and Deutsch D 1979 Phys. Rev. D **20**, 3063
[25] Helliwell T M and Konkowski D A 1986 Phys. Rev. D **34**, 1918
[26] Ford L H and Svaiter N F 2000 Phys. Rev. A **62**, 062105 (Preprint quant-ph/0003129)
[27] Ford L H and Svaiter N F 2002 Phys. Rev. A **66**, 062106 (Preprint quant-ph/0204126)
[28] Vuletic V, private communication.
[29] Fewster C J and Pfenning M J 2006 J. Math. Phys. **47**, 082303

Journal of Physics: Conference Series **161** (2009) 012035 doi:10.1088/1742-6596/161/1/012035

The Casimir Effect in Biology: The Role of Molecular Quantum Electrodynamics in Linear Aggregations of Red Blood Cells

K Bradonjić
Physics Department, Boston University, Boston MA USA

J D Swain
Physics Department, Northeastern University, Boston MA USA
International Institute for Biophysics, Neuss, Germany

A Widom
Physics Department, Northeastern University, Boston MA USA

Y N Srivastava
Physics Department, Northeastern University, Boston MA USA
Physics Department & INFN, University of Perugia, Perugia Italy

E-mail: john.swain@cern.ch

Abstract. Despite the fact that red blood cells carry negative charges, under certain conditions they form cylindrical stacks, or "rouleaux". It is shown here that a form of the Casimir effect, generalizing the more well-known van der Waals forces, can provide the necessary attractive force to balance the electrostatic repulsion. Erythrocytes in plasma are modelled as negatively charged dielectric disks in an ionic solution, allowing predictions to be made about the conditions under which rouleaux will form. The results show qualitative agreement with observations which suggest that the basic idea is worth further pursuit. In addition to revealing a mechanism which may be widespread in biology at the cellular level, it also suggest new experiments and further applications to other biological systems, colloid chemistry and nanotechnology.

1. Introduction

Casimir forces between objects typically rise faster with decreasing separation than electrostatic forces raising the possibility that at sufficiently small distances they become dominant. We show by an explicit example that the aggregation of red blood cells is a biological situation which can be understood as an interplay between Casimir attraction of plate-like objects (red blood cells) and their mutual electrostatic repulsion. Once the idea is accepted that Casimir forces may play a role in biology, a whole new window on biological phenomena at the cellular level opens up, as well as the possiblity of novel new experiments on the Casimir effect in liquid media.

Journal of Physics: Conference Series **161** (2009) 012035 doi:10.1088/1742-6596/161/1/012035

2. Rouleaux Formation

It has been observed for many years that, under certain conditions, erythrocytes (red blood cells) form cylindrical stacks known as "linear aggregations", or "rouleaux"[1], sketched in figure 1. Each cell is a biconvex discoid (a disc whose opposite faces have been pushed together, forming indentations on each side), with a radius of some 6-8 μm. The maximum thickness near the edges is about 1 μm and the minimum at the centre is about a third that. One may view blood as an ionic plasma in which red blood cells, and some other cellular colloidal particles, are suspended. Employing direct measurements of colloidal particle mobility, the red blood cells have been shown to carry a net negative charge[2].

A long cylindrical stack of negatively charged red blood cells is shown schematically in figure 1. Attractive forces must exist in order to stabilize the rouleaux formation against the explosion which would take place if only the repulsive Coulomb interaction were existent[1],[2]. Our purpose is to argue that quantum radiation field Casimir forces provide the required rouleaux stability mechanism. The Casimir effect generalizes the well-known van der Waals forces by including electromagnetic retardation effects [3]. Casimir forces are simpler in nature than the selective long range dispersion forces suggested by Fröhlich[4, 5].

The essential physical principles underlying the Casimir forces are quite simple. When electromagnetic radiation interacts with condensed matter, the frequencies of the normal modes are Lamb shifted (though the use of the term "Lamb shift" is not so often used in the context of effects due to boundary conditions). Frequency shifts imply a change in the electromagnetic field free energy. In particular, at virtually zero temperature, the ground state zero point energy changes. A special case of such an energy shift may be found from the electromagnetic modes located between two parallel highly conducting plates. The energy shift then describes an attractive force between two plates and can be understood as being due to the change of zero-point modes when the distance between the plates is changed. The closer the plates, the more reduced is the vacuum energy and the greater the attractive force between the plates[3].

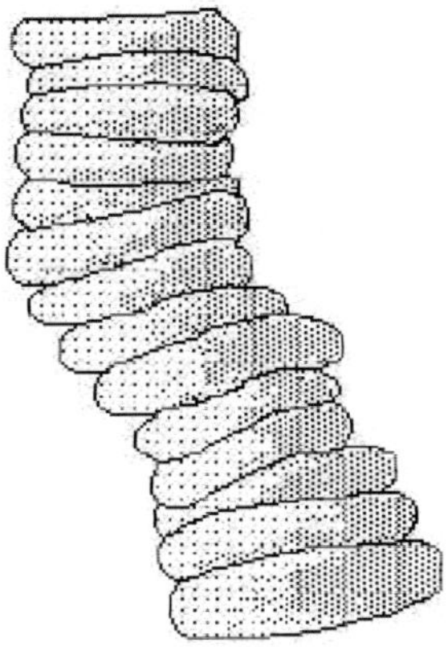

Figure 1. Although each red blood cell or slab has a net negative charge, in certain ionic surroundings the blood cells nevertheless tend to come in stacks or rouleaux structures. Such a stack is schematically drawn above. Despite their apparent closeness, each cell is strongly biconvex and the centres are significantly more separated than this view suggests.

3. Charged Dielectric Plates in an Ionic Solution

For the problem at hand, we consider two dielectric plates each having cross sectional area A and a dielectric constant ε_1. Between the two dielectric plates is a plasma medium of thickness d and dielectric constant ε_2[6]. The zero point electromagnetic field energy per unit plate area is calculable. If the dielectric plates both have a surface charge per unit area σ, then the electrostatic potential energy density may also be computed taking into account the Debye screening effects of a background plasma solution of ionization strength[1] I. It is assumed in both calculations[7] that $d^2 << A$. The final result for the total free energy per unit area u as a function of plate separation d may be written[8]

$$u(d) = \frac{\sigma^2 \Lambda}{2\varepsilon_2} \left\{ e^{-d/\Lambda} - \left(\frac{\pi^2 \hbar v \sqrt{\varepsilon_0 \varepsilon_2}}{360 \sigma^2 \Lambda} \right) \frac{1}{d^3} \right\}. \quad (1)$$

The length Λ describes electrolytic Debye screening and v is a sort of reflection coefficient $v = c\left[(\varepsilon_1 - \varepsilon_2)/(\varepsilon_1 + \varepsilon_2)\right]^2$. The first term on the right hand side of Eq.(1) describes the Coulomb repulsion between red blood cells while the second term on the right hand side of Eq.(1) describes the Casimir attraction between red blood cells. The relative strength of the two effects is quantitatively described by the dimensionless parameter

$$a = \left(\frac{\pi^2 \hbar v \sqrt{\varepsilon_0 \varepsilon_2}}{360 \sigma^2 \Lambda^4} \right) = \left(\frac{\pi^2 \hbar c \sqrt{\varepsilon_0 \varepsilon_2}}{360 \sigma^2 \Lambda^4} \right) \left[\frac{(\varepsilon_1 - \varepsilon_2)}{(\varepsilon_1 + \varepsilon_2)} \right]^2 \quad (2)$$

where the Debye screening length Λ is related to the ionization strength I via $\Lambda^2 = \{\varepsilon_2 k_B T / e^2 \tilde{I}\}$. The ionization strength in physical units is $\tilde{I} = \sum_a z_a^2 n_a$ where n_a is the number of ions per cubic meter having an ionic charge $z_a|e|$. In chemical units of moles per liter, one employs the ionization strength $I = [10^{-3}\text{meter}^3/\text{liter}](\tilde{I}/N_A)$ where Avogadro's N_A is the number of ions per mole.

The above theory assumes that the parallel plates are maximally overlapping. If the plates were to slide over one another yet remaining parallel but *not* maximally overlapping, then the energy would be increased. Thus, there is a force parallel to the plates tending to pull them back into a maximally overlapping state. This force represents a *lateral* Casimir effect (See, for example, references in [9] which deal with Casimir forces between corrugated planes. The case here is somewhat different, as it deals with finite plates, but certainly one expects an increasingly negative Casimir energy as the plates become more and more close to maximally overlapping[10]). It is centrally important in the context of the rouleaux formation problem since it supplies a force which will tend to align two plates so that they directly face each other with the largest possible area overlap. When many plates are stacked in a cylinder, the lateral force tends to give the cylinder transverse structural stability. With regard to compression stability, one must find a local minimum in the free energy per unit area $u(d)$ with respect to the distance between the plates. A meta-stable equilibrium distance between plates arises from the local minimum at d_0 found from equating the force per unit area to zero; i.e. $u'(d_0) = 0$. There exists a critical parameter $a_c \approx 1.57$ such that a local rouleaux minimum exists if $a < a_c$ and a local rouleaux minimum does not exist if $a > a_c$. The resulting free energy per unit area $u(d)$ is plotted in figure 2 for several values of a.

4. Comparison with Data

We fix the parameters in the model so that the plates correspond to erythrocytes (relative dielectric constant of 1.1857 and a charge density of 3500 esu/cm^2[1, 2] and nominal physical spacings observed of about 1 μm measuring between the centres of the cells) and work at room temperature. This leaves the ionic strength (I) and the permeability of the blood plasma relative

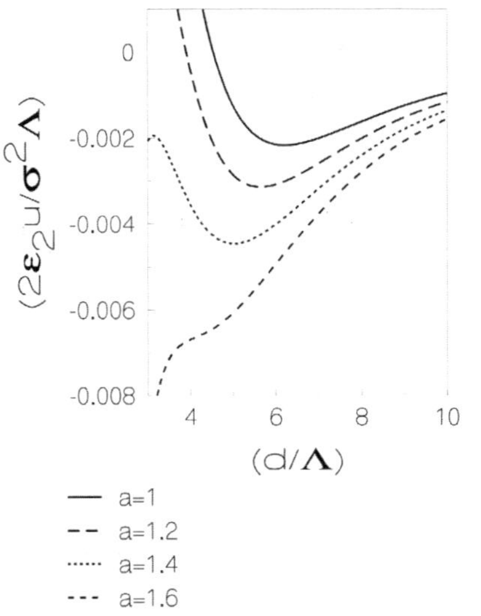

Figure 2. The variation of the free energy per unit area with the separation distance. The rouleaux formation can exist at the local minimum only if the parameter a in Eq.(2) is sufficiently small.

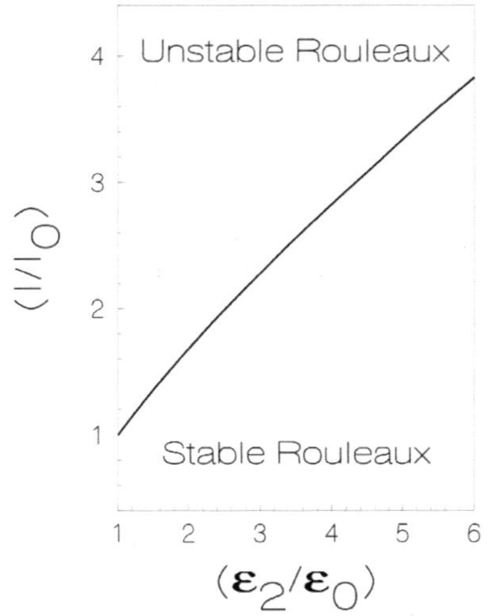

Figure 3. The phase diagram showing the regions in which room temperature rouleaux formations are stable and in which regions such rouleaux formations are unstable. The reference ionization strength $I_0 = 0.05$ *moles/liter* has been employed.

to the vacuum $(\varepsilon_2/\varepsilon_0)$ as free parameters – parameters which are controllable by changing composition of the fluid in which the erythrocytes are suspended. Assuming that rouleaux do form (the condition $a < a_c$ given above is satisfied) one can plot a phase diagram (see figure 3) in the $\varepsilon_2 - I$ plane, showing in which regions the rouleaux are stable. For stable rouleaux it is theoretically necessary to have sufficiently small ionic strength and/or sufficiently large values of $(\varepsilon_2/\varepsilon_0)$. The temperature dependence of the Casimir energy is very weak for the problem at hand, but may be of importance in other systems.

It is important to note that rouleaux formations are usually observed in significantly artificially diluted blood. It is also known that such formations can be unstable in solutions of sufficiently high ionic strength. It would be of great interest for this model to investigate quantitatively the complete phase diagram, especially as quite a number of approximations and idealizations go into figure 3.

We are keenly aware that there are many approximations that have to be made in order to compare any sort of calculation with data. Red blood cells are obviously complex heterogeneous dielectrics with complicated frequency responses, they are certainly not perfect disks, *etc.* That said, we do feel that the fact that one does not find wild disagreement with simple models and physiological data suggests that further research is of interest. Future work could include:

- the construction and study of model systems with more controllable geometries and dielectric properties (*i.e.* uniform disks of known dielectric properties and surfaces treated so they carry known charges);
- a more detailed study of the electromagnetic properties of real cells to enable more detailed calculations for real blood to be performed

- more careful treatment of deviations from flatness of the red blood cells (see, for example, [11] for work in that direction), including interactions between points of very near approach where the (non-retarded) van der Waals forces might provide a better description of the physics there, and, again, provide a more accurate description of real blood.

5. Implications

In addition to providing an possible explanation for the otherwise mysterious attractive force which compensates the electrostatic repulsion of erythrocytes in rouleaux, this work opens up the possibility of numerous experimental tests, both in blood and in novel colloids with nonspherical particles in suspension. The fact that the Casimir energies involved are significant suggests that studies of colloidal systems may provide novel approaches to studies of the Casimir effect which avoid the extremely difficult techniques employed in past studies[12, 13, 14]. It also suggests that the Casimir effect will be of importance in the burgeoning fields of nanotechnology and nanomedicine as devices approach the size of single cells[15, 16].

We also suggest that since living cells are able to change the local ion concentration nearby via various pumping mechanisms in or out of cell bodies, this offers a means of controlling adhesion or separation of cells from each other or from various boundaries at the cellular level in living organisms. Whether or not this is actually realized in nature is an open question, but an interesting one which might find broad applicability.

6. Acknowledgements

J.D.S. would like to thank the National Science Foundation for support, and the conference organizers for a wonderful meeting in Brasilia. He also would like to thank V. Voeikov for useful discussions on rouleaux formation, and G. Hyland for discussions on Fröhlich interactions. Y.S. would like to thank INFN, Sezione di Perugia, Italy, for financial support. K.B. would like to thank Eyad Katrangi for useful discussions. She was at Northeastern University when this work was started. We would also like to gratefully acknowledge the suggestions of anonymous referees for this paper.

References

[1] Chien S 1975 in *The Red Blood Cell (2nd Ed.)*, edited by Surgenor Douglas MacN. (Academic Press, New York), Vol.2, Chap.26, p. 1032;
Seaman G V F ibid., Vol.2, Chap. 27, 1136

[2] Eylar E H, Madoff M A, Brody O V and Oncley J L 1962 *The Journal of Biological Chemistry* **237** (6) 1992

[3] Mostepanenko V M and Trunov N N 1997 *The Casimir Effect and its Applications*, translated by R L Znajek, (Clarendon Press, Oxford)

[4] Fröhlich H 1972 *Phys. Lett.* **39A** 153

[5] Rowlands S, Sewchand L S, Lovlin R E, Beck J S and Enns E G 1981 *Phys. Lett.* **82A**, 436

[6] He J, Karlsson A, Swartling J and Andersson-Engels S 2003 *Light Scattering By Multiple Red Blood Cells*, Tech. Rep. LUTEDX/(TEAT-7117, Lund Inst. of Tech., Dept. of Electroscience http://www.tde.lth.se/teorel/Publications/TEAT-7000-series/TEAT-7117.pdf

[7] Verwey E J W and Overbeek J TH G 1999 *Theory of The Stability of Lyophobic Colloids*, (Dover Publications, Inc., Mineola, NY)

[8] Srivastava Y, Widom A and Friedman M H 1985 Microchips As Precision Quantum Electrodynamic Probes *Phys. Rev. Lett.* **55** 2246

[9] Golstanian R and Kardar M 1997 *Phys. Rev. Lett* **78** 3421;
Golestanian R and Kardar M 1998 *Phys. Rev.* **A58** 1713;
Chen F, Mohideen U, Klimchitskaya G L and Mostepanenko V 2002 *Phys. Rev. Lett.* **87** 101801;
Chen F, Mohideen U, Klimchitskaya G L and Mostepanenko V 2002 *Phys. Rev.* **A66** 032113.

[10] Fulling S A, Kaplan L, Kirsten K, Liu Z H and Milton K A 2008 Vacuum Stress and Closed Paths in Rectangles, Pistons, and Pistols, preprint arXiv:0806.2468v2 and references therein.

[11] Zaheer S, Rodriguez A, Johnson S G and Jaffe L 2007 *Phys. Rev.* **A76** 063816

[12] Sparnaay M J 1958 Measurement of attractive forces between flat plates *Physica* **24** 751

[13] Lamoreaux S K 1997 Demonstration of the Casimir Force in the 0.6 to 6 μm Range *Phys. Rev. Lett.* **78** 5
[14] Bressi G, Carugno G, Onofrio R and Ruoso G 2002 Measurement of the Casimir Force between Parallel Metallic Surfaces *Phys. Rev. Lett.* **88** 041804
[15] Freitas Jr. R A 1999 in *Nanomedicine Vol 1*, Landes Bioscience, Austin, TX
[16] Drexler K E 1982 *Nanosystems: Molecular Machinery, Manufacturing, and Computation* John Wiley and Sons, Inc., New York NY

Journal of Physics: Conference Series **161** (2009) 012036 doi:10.1088/1742-6596/161/1/012036

Fluctuation-induced self-force and violation of action–reaction in a nonequilibrium steady state fluid

Pascal R Buenzli

Departamento de Física, FCFM, Universidad de Chile, Santiago, Chile

E-mail: pbuenzli@dfi.uchile.cl

Abstract. We show that the fluctuations of a fluid driven out of equilibrium can induce a net force on a single asymmetric object immersed in it. The force originates in the restriction of the fluid's fluctuations at the object's boundaries, as in the Casimir effect. In contrast to the equilibrium situation, its emergence on a single obstacle is not ruled out by the second law of thermodynamics since the fluid is in a nonequilibrium state. We explicitly calculate this self-force on a deformed circle embedded in a fluid whose density fluctuations obey a stochastic reaction–diffusion equation. When two objects are considered, the presence of self-forces can violate the action–reaction principle. We illustrate this by calculating the internal Casimir-type forces between a circle and a plate. Their sum, instead of vanishing, provides the self-force exerting on the circle–plate assembly.

1. Introduction

Since the pioneering work of Casimir [1], it has been known that the microscopic fluctuations of a media can induce net forces between macroscopic objects immersed in it. The paradigmatic example is that of the zero-temperature Casimir force between two metallic plates, induced by the quantum fluctuations of the vacuum electromagnetic field [2, 3]. Many examples of similar forces have since been exhibited. Lifshitz [4] showed that additional contributions to the Casimir force arise when thermal fluctuations of the quantum field are considered. Thermal fluctuations are known to induce forces in critical fluids close to criticality, liquid crystals in the nematic phase, systems with a broken continuous symmetry and other systems [5, 6, 7, 8]. A necessary ingredient for such forces to arise is that the range of the correlations in the media be large enough to explore the macroscopic disturbances provoked by the embedded objects. Fluctuation-induced forces thereby strongly depend on these objects' shape [9, 10].

Most of the systems considered in this field are usually assumed to be in equilibrium states. Situations out of equilibrium, however, are also known to develop fluctuations that can have long ranges [11]. Examples are shear Couette flows, Rayleigh–Bénard instabilities, but also granular fluids [12], shaken fluids [13], reaction–diffusion fluids [14], or simply systems of the kind mentioned above, that are driven out of equilibrium [15, 16]. One thus expects that similar fluctuation-induced forces can be generated between inclusions in nonequilibrium systems [17, 18]. The characterization of these forces, however, does not rely on general first principles. Indeed, (equilibrium) thermodynamic potentials are generally not available. It is only recently that such forces have been obtained between two planar objects immersed into nonequilibrium driven systems [19, 16], granular fluids [20], or reaction–diffusion systems that break the detailed balance [21].

Journal of Physics: Conference Series **161** (2009) 012036 doi:10.1088/1742-6596/161/1/012036

Various properties of nonequilibrium fluctuation-induced forces can make them very interesting in applications. A recurrent theme in Casimir-like forces is the control of its sign (attraction or repulsion). Thermodynamic equilibrium reduces the number of parameters controlling the system to only a few (such as temperature, the density, etc.). These external parameters, on which the system's fluctuations depend, are then usually able to control only the *magnitude*, but not the *sign*, of the induced forces. (Thermal fluctuation-induced forces, e.g., are proportional to $k_B T$.) Nonequilibrium fluctuations, on the other hand, can be tuned by as many control parameters as one may wish and by modifying their relative strength (as we will see) or their properties one may achieve sign reversal of the forces (see also [22, 19])

In addition, we will see in this work that the extension of Casimir-like forces to fluctuating fluids driven out of equilibrium allows for two interrelated effects that arise in nontrivial geometries and that are not possible in equilibrium systems. Namely, nonvanishing forces and torques can be induced on single asymmetric obstacles and the action–reaction principle between two intruders can be violated. These phenomena can have significant consequences in experiments, as they lead to directed motion and unevenness in the measurements of the forces between two objects.

Whether at equilibrium or not, accurate experimental measurements of fluctuation-induced forces need to go beyond the idealized geometry of infinitely long plates, predominant in theory for its simplicity. While a long-studied topic, the objects' shape dependence of Casimir forces is notoriously difficult for nontrivial geometries. The most widely used technique by experimentalists to date relies on the so-called Derjaguin construction (proximity force theorem), which in essence integrates the two-plate expression of the force along the curved surfaces [23, 24]. An unbalance of action and reaction in a nonequilibrium situation would impeed the use of the Derjaguin approximation where it would normally be valid at equilibrium. Note that this unbalance does not seem to be systematic: in [16], e.g., the Casimir forces exerting between two plates of different material constitutions seem to satisfy the action-reaction principle. A violation of Newton's third law, however, has also been noted in depletion forces between identical spherical objects immersed into a flowing fluid [25]. It furthermore prevents the two-body forces from being derived from an effective smooth potential, in striking contrast to equilibrium situations. We will see that this unbalance directly results from the presence of self-forces.

Since nonequilibrium systems are thermodynamically open, it is not entirely unexpected that self-forces can appear. In fact, provided that both the microscopic time-reversibility and space rotation-invariance symmetries are broken, such forces have been implicitly suggested by the occurrence of sustained motions in other nonequilibrium contexts, such as in ratchets [26], Brownian motors [26, 27], molecular motors [28], or the adiabatic piston [29]. However, in these systems the space asymmetry usually lies in an external temperature gradient or another anisotropic field exerting on the object. More recently, the directed motion of an asymmetric object immersed into vibrated granular matter has been exhibited [30, 31]. Here, a direct calculation of the force exerting on single objects is presented from a "Casimir effect" point of view. Knowing these forces allows for a better understanding of the different effects at play in such sustained motions. It also makes possible the evaluation of additional stresses exerting on asymmetrical static structures in microdevices. Jointly with the sign reversal control of the force, it could be used as a powerful tailoring mechanism for the self-assembly of ordered structures.

2. A nonequilibrium fluid model

To illustrate these effects, we introduce here a rather simplified nonequilibrium fluid. We consider a reaction–diffusion system in which two chemical components, A and B, diffuse and react according to

$$A + B \xrightarrow{k_1} 2B, \qquad B \xrightarrow{k_2} A \tag{1}$$

with rates k_1 and k_2, respectively. The average densities ρ_A, ρ_B of these species thus satisfy the rate equations

$$\frac{\partial}{\partial t}\rho_A = D\nabla^2\rho_A - \frac{k_1}{\rho_{\rm tot}}\rho_A\rho_B + k_2\rho_B, \qquad \frac{\partial}{\partial t}\rho_B = D\nabla^2\rho_B + \frac{k_1}{\rho_{\rm tot}}\rho_A\rho_B - k_2\rho_B, \tag{2}$$

where D is the diffusion constant and $\rho_{\rm tot}$ is the total density. When $k_1 > k_2$, these equations admit the homogeneous stationary solution

$$\bar{\rho_A} = \frac{k_2}{k_1}\rho_{\rm tot}, \qquad \bar{\rho_B} = \rho_{\rm tot} - \bar{\rho_A}. \tag{3}$$

The A and B density departures from this solution, that are generically denoted in the following by Φ, both satisfy after linearization of (2) and addition of Langevin-type noises, the equation

$$\frac{\partial \Phi}{\partial t} = -\boldsymbol{\nabla}\cdot(-D\boldsymbol{\nabla}\Phi + \boldsymbol{\xi}_{\rm c}) - \gamma\Phi + \xi_{\rm nc}. \tag{4}$$

In this equation, $\gamma = k_1 - k_2 > 0$ is the rate at which the system relaxes to local equilibrium. The added conservative (c) and nonconservative (nc) random noises $\boldsymbol{\xi}_{\rm c}$ and $\xi_{\rm nc}$ are relevant to the fluid's fluctuations once it has relaxed to the steady state. The conservative noise corresponds to fluctuations in the diffusive flux whilst the nonconservative noise, to fluctuations in the local density production. These fluctuations are assumed to have zero average and to be delta-correlated in space and time:

$$\begin{aligned}\langle \xi_{\rm nc}(\mathbf{r},t)\xi_{\rm nc}(\mathbf{r}',t')\rangle &= \Gamma_{\rm nc}\,\delta(\mathbf{r}-\mathbf{r}')\delta(t-t'),\\ \langle \xi^\mu_{\rm c}(\mathbf{r},t)\xi^\nu_{\rm c}(\mathbf{r}',t')\rangle &= \Gamma_{\rm c}\,\delta_{\mu\nu}\,\delta(\mathbf{r}-\mathbf{r}')\delta(t-t'),\\ \langle \xi^\mu_{\rm c}(\mathbf{r},t)\xi_{\rm nc}(\mathbf{r}',t')\rangle &= 0,\end{aligned} \tag{5}$$

where $\mu,\nu = 1,2,3$ denote the cartesian coordinates of $\boldsymbol{\xi}_{\rm c}$. This model for the fluid is slightly improved compared to [32], where only nonconservative noise is considered. Let us add that other nonequilibrium systems in their steady state can be described by Eqs. (4)–(5) [19, 29].

In the presence of static objects in the fluid, the density fluctuations get constrained as no flow of matter can cross the rigid surfaces. Eq. (4) is thus supplemented by the non-flux condition

$$\mathbf{n}\cdot(-D\boldsymbol{\nabla}\Phi + \boldsymbol{\xi}_{\rm c}) = 0 \tag{6}$$

at the objects' surface, where $\mathbf{n}(\mathbf{r})$ is a unit normal vector chosen to point outward from the fluid's domain. The total force experienced by an immersed object results from integrating the fluid's average pressure along its surface. In a steady state, one may calculate this pressure p from a local equation of state that relates it to the density $p = p(\rho(\mathbf{r},t))$ [33, 34]. (Such a relation is experimentally measured in a number of cases of interest, e.g., in driven granular media [35].)[1] Assuming that this relation is expandable around the homogeneous reference density and that density fluctuations stay small, one has $\langle p\rangle = p_0 + \frac{p_0''}{2}\langle\Phi^2\rangle$, where p_0 and p_0'' are constant parameters depending on the equation of state. Since the homogeneous pressure p_0 cannot generate a force, the total force $\mathbf{F}_S$ induced on the object S is thus given by

$$\mathbf{F}_S = \frac{p_0''}{2}\int_S {\rm d}\sigma\;\mathbf{n}\;\langle\Phi^2\rangle. \tag{7}$$

[1] In the reaction–diffusion model, one may consider that $\rho_{\rm tot}$ in (3) is maintained exactly constant so that only the fluctuations $\Phi = \rho_A - \bar{\rho}_A$ of the density of A (for example) need to be considered.

To evaluate (7), one needs to calculate Φ as the solution of (4), (6). After a characteristic time of the order of γ^{-1} the system relaxes to a stationary state only affected by the random noises. Given a realization of $\boldsymbol{\xi}_{\mathrm{c}}$ and ξ_{nc}, the local density deviation then reads

$$\Phi(\mathbf{r},t) = \int \mathrm{d}t' \int_{\Omega} \mathrm{d}\mathbf{r}'\, G(\mathbf{r},\mathbf{r}',Dt-Dt')(-\boldsymbol{\nabla}\cdot\boldsymbol{\xi}_{\mathrm{c}} + \xi_{\mathrm{nc}})(\mathbf{r}',t') + \int \mathrm{d}t' \int_{\partial\Omega} d\sigma'\, G(\mathbf{r},\mathbf{r}',Dt-Dt')\mathbf{n}(\mathbf{r}')\cdot\boldsymbol{\xi}_{\mathrm{c}}(\mathbf{r}',t'), \tag{8}$$

where G is the Green function propagating the effect of a unit source localized at $\mathbf{r}'$ at time t' to the point $\mathbf{r}$ at time t. It is used here to forward both the conservative and nonconservative noises acting in the whole volume Ω of the fluid and the conservative noise acting on the boundaries $\partial\Omega$ by (6). In Fourier representation with respect to the scaled time $\tau = Dt$, G is solution of

$$(-\nabla^2 + \kappa^2 - i\omega)G(\mathbf{r},\mathbf{r}',\omega) = \delta(\mathbf{r}-\mathbf{r}') \tag{9}$$

$$\mathbf{n}(\mathbf{r})\cdot\boldsymbol{\nabla}G(\mathbf{r},\mathbf{r}',\omega)|_{\mathbf{r}\in\partial\Omega} = 0 \quad \forall\mathbf{r}'\in\Omega, \forall\omega \tag{10}$$

where $\kappa^{-1} \equiv (\gamma/D)^{-1/2}$ is the *correlation length* of the fluid. Using the properties of G and the correlations (5) of the noises, one can show from (8) (see details in [36]) that the static structure factor is given by

$$\big\langle\Phi(\mathbf{r},t)\Phi(\mathbf{r}',t)\big\rangle = \frac{\Gamma_{\mathrm{c}}}{2D}\delta(\mathbf{r}-\mathbf{r}') + \frac{\Gamma}{2D}G(\mathbf{r},\mathbf{r}'), \tag{11}$$

with $\Gamma \equiv \Gamma_{\mathrm{nc}} - \gamma\Gamma_{\mathrm{c}}/D$ and $G(\mathbf{r},\mathbf{r}') \equiv G(\mathbf{r},\mathbf{r}',\omega=0)$. The fluctuation-induced modification of the local pressure of the fluid at $\mathbf{r}$, given by taking the limit $\mathbf{r}'\to\mathbf{r}$ in the above formula, obviously diverges so that the expression (7) needs to be regularized due to the inaccuracy of the continuous model at microscopic distances.

At thermal equilibrium, the fluctuation intensities Γ_{c} and Γ_{nc} are determined by the temperature of the fluid via the fluctuation–dissipation theorem: $\Gamma_{\mathrm{c}} = 2k_{\mathrm{B}}TD$, $\Gamma_{\mathrm{nc}} = 2k_{\mathrm{B}}T\gamma$, so that $\Gamma = 0$. Thus, only microscopic (delta) correlations are left. Being independent of the immersed bodies, these correlations cannot induce any force on them and we will simply omit their contribution in (7). In contrast, when the reaction–diffusion fluid reaches a nonequilibrium steady state where the detailed balance condition is not fulfilled [14], Γ can be different from zero and correlations of mesoscopic range (of the order of κ^{-1}) can occur through G in (11). However, two other divergencies emerge: first in taking the limit $\mathbf{r}'\to\mathbf{r}$ on $G(\mathbf{r},\mathbf{r}')$ and then when evaluating $\mathbf{r}$ on the surface S in (7). The first divergency turns out to be independent of the immersed objects and can easily be removed by subtracting from G the Green function G_0 of the unconstrained fluid. The second divergency, on the other hand, is more difficult to handle as it inherently depends on the object's shape and is only compensated between different sides through the surface integral (7) [32]. A hard-core cutoff ϵ is introduced for that purpose and the regularized force reads

$$\mathbf{F}_S = \lim_{\epsilon\to 0} F_0\,\kappa \int_S \mathrm{d}\sigma\,\mathbf{n}\,[G - G_0]\,(\mathbf{r}-\epsilon\mathbf{n},\mathbf{r}-\epsilon\mathbf{n}), \tag{12}$$

where $F_0 \equiv \frac{p_0''\Gamma}{4D\kappa}$ has the dimension of a force. In two dimensions $G_0(\mathbf{r},\mathbf{r}') = \mathrm{K}_0(\kappa|\mathbf{r}-\mathbf{r}'|)/2\pi$ (K_0 is the modified Bessel function of order zero) and in three dimensions, $G_0(\mathbf{r},\mathbf{r}') = \exp\{-\kappa|\mathbf{r}-\mathbf{r}'|\}/4\pi|\mathbf{r}-\mathbf{r}'|$.

Since F_0 is proportional to Γ, one is not only able to turn the fluctuation-induced force (12) *on* or *off* by driving the system *out of* equilibrium or *at* equilibrium; one can also modify its *sign* by controlling the balance between Γ_{nc} and Γ_{c} through the external parameters characterizing the nonequilibrium state. This important feature can have very appreciable consequences in experiments and applications.

3. Two nonplanar systems

Formula (12) is illustrated in the following on two classes of objects immersed into the fluid. For simplicity, the examples are all two-dimensional, but similar conclusions can be drawn in three-dimensional cases. Here, we only state the main results; explicit calculations will be developed in [36].

3.1. Self-force and self-torque

We first consider the immersion of a single asymmetric object, a deformed circle, into the fluid. In an equilibrium system, no net force would be induced on such an object. Here, however, one is induced due to the nonequilibrium character of the fluid's fluctuations, that breaks the microscopic time-reversibility. This force is called "self-force" as there are no other rigid bodies in the fluctuating media.

The generic Green function in the presence of a single bounded object at the origin can be written from the differential equation (9) (at $\omega = 0$) in polar coordinates $\mathbf{r} = (\rho, \theta)$, $\mathbf{r}' = (\rho', \theta')$ as

$$G(\mathbf{r},\mathbf{r}') = G_0(\mathbf{r}-\mathbf{r}') + \sum_{m,n\in\mathbb{Z}} \frac{\mathrm{e}^{im\theta+in\theta'}}{2\pi}\, a_{mn}\, \mathrm{K}_m(\kappa\rho)\, \mathrm{K}_n(\kappa\rho'), \tag{13}$$

where K_m is the modified Bessel function of order m. (The two-dimensional bulk Green function $G_0(\mathbf{r}-\mathbf{r}')$ is expressed in polar coordinates by the addition theorem for K_0 [37].) The reality of $G(\mathbf{r},\mathbf{r}')$ and the reciprocity relation $G(\mathbf{r},\mathbf{r}') = G(\mathbf{r}',\mathbf{r})$ have been used and further imply that the coefficients a_{mn} satisfy $a_{mn} = a_{nm} = a^*_{-m,-n}$. These coefficients depend on the particular shape of the object. Considering a slightly deformed circle represented by a polar curve $R(\theta) = R + \eta s(\theta)$ (with $\eta \ll R, \kappa^{-1}$), they can be obtained perturbatively in η by the enforcement of the boundary condition (10) up to a certain order. The whole force (12) is then expanded in η (note that its dependence upon $R(\theta)$ occurs both through G and through the line integral). The first nonvanishing contribution is of second order in η. The absence of first order can be understood as follows: the only perturbation of the circle exhibiting a preferred direction comes from the dipolar Fourier mode; however, a small dipolar perturbation amounts to a circle that is merely shifted—it stays symmetric. At second order, on the other hand, the dipolar mode couples to another mode, and leads to a finite result [32]. For instance, superposing a dipolar and quadrupolar deformation as in

$$s(\theta) = 2s_1\cos(\theta) + 2s_2\cos(2\theta) \tag{14}$$

results in the shapes displayed in Figure 1, and an average force

$$\mathbf{F} = -F_0\, s_1\, s_2\, (\kappa\eta)^2 H_F(\kappa R)\, \hat{\mathbf{x}}, \tag{15}$$

is induced on them (up to $\mathrm{O}(\eta^3)$ terms). In (15), H_F is a dimensionless function obtained as the limit as $\epsilon \to 0$ of a nontrivial series of Bessel functions evaluated at either κR or $\kappa(R+\epsilon)$. It can be shown that the series is absolutely convergent provided $\epsilon > 0$ but the limit cannot be taken under the summation sign. Nevertheless, the numeric computation of this function is accurately fitted by $2/\kappa R$ in the whole range $0.1 \le \kappa R \le 100$ (see Figure 2). Note that the shapes displayed in Figure 1 experience opposite forces.

Following the same lines, one can calculate the torque induced by the nonequilibrium fluctuations of the fluid on a single asymmetric object. Choosing a deformation

$$s(\theta) = 2s_2\cos(2\theta) + 2s_4\sin(4\theta) \tag{16}$$

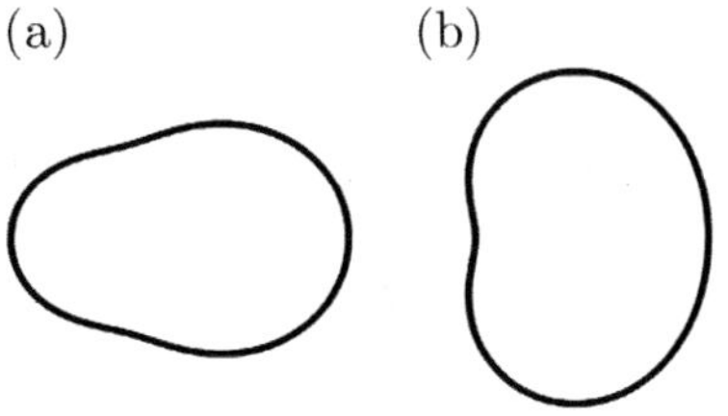

Figure 1. Typical shapes of a circle deformed by (14).
(a) $s_1, s_2 > 0$; (b) $s_1 > 0, s_2 < 0$.

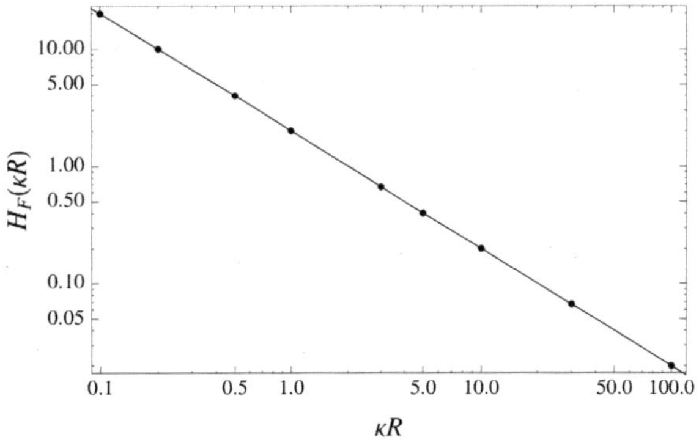

Figure 2. Numeric evaluations (dots) of the function $H_F(\kappa R)$, plotted along with $2/\kappa R$ (solid line).

of the circle (see Figure 3), the average torque up to $O(\eta^4)$ contributions is

$$\boldsymbol{T} = -\frac{F_0}{\kappa}\, s_2^2\, s_4\, (\kappa\eta)^3\, H_\tau(\kappa R)\, \hat{\boldsymbol{z}}. \tag{17}$$

The absence of second order $O(\eta^2)$ in (17) is associated to the antisymmetry of the torque under a reflection $s(\theta) \mapsto s(-\theta)$ of the deformation [36]. The dimensionless function H_τ is a nontrivial series of Bessel functions similar to H_F. However, it exhibits a different asymptotic behaviour for $\kappa R \ll 1$, where it approaches $\approx 80/\kappa R$, than for $\kappa R \gg 1$, where it approaches $\approx 50/\kappa R$, as can be seen in Figure 4. The transition between these regimes occurs at $\kappa R \approx 1$, i.e., when the correlation length κ^{-1} matches the size R of the deformed circle.

Figure 3. Typical shape of a circle deformed by (16) ($s_4 > 0$).

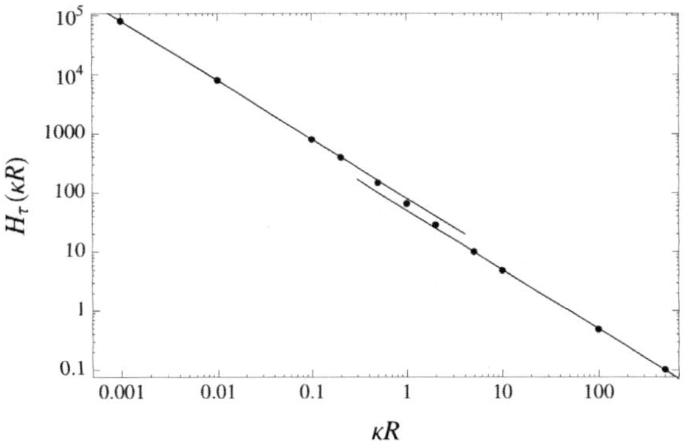

Figure 4. Numeric evaluations (dots) of the function $H_\tau(\kappa R)$, plotted along with $80/\kappa R$, $\kappa R \lesssim 1$ and $50/\kappa R$, $\kappa R \gtrsim 1$ (solid lines).

3.2. Two obstacles

When a second object is immersed into the fluid, the fluid's fluctuations are modified and the force already exerting on the first object will likewise be affected. If S and S' denote these two objects, the total force $\mathbf{F}_S$ on S can be decomposed as the contribution of the self-force $\mathbf{F}_S^0$, already present in the absence of S', plus a contribution $\mathbf{F}_{S\leftarrow S'} \equiv \mathbf{F}_S - \mathbf{F}_S^0$ due to the additional asymmetry provoked by the insertion of S'. From (12), one has

$$\mathbf{F}_{S\leftarrow S'} = F_0\,\kappa \int_S \mathrm{d}\sigma\,\mathbf{n}\left[G_{SS'} - G_S^0\right](\mathbf{r},\mathbf{r}), \tag{18}$$

where $G_{SS'}$ is the Green function in the presence of both S and S', and G_S^0 the Green function when only S is immersed into the fluid. In the expression (18), the hard-core cutoff ϵ has been removed: the divergencies present in both $G_{SS'}$ and G_S^0 cancel in the subtraction. The force (18) represents the inherent "two-body" force analogous to the Casimir interaction with zero set at infinitely separated objects. Considering the similar force $\mathbf{F}_{S'\leftarrow S}$ acting on S', one has

$$\mathbf{F}_{S\leftarrow S'} + \mathbf{F}_{S'\leftarrow S} = \mathbf{F}_{SS'} - \mathbf{F}_S^0 - \mathbf{F}_{S'}^0, \tag{19}$$

where $\mathbf{F}_{SS'}$ is the self-force exerting on the assembly $S \cup S'$ considered as a whole. Clearly, in the presence of self-forces, the right hand side of (19) does not necessarily vanish (as fluctuation-induced forces are not additive) and the action-reaction principle between the internal two-body forces can be violated.

To calculate $\mathbf{F}_{S\leftarrow S'}$ in a regime of large separation d between the objects, a multiple scattering approach can be devised [9]. The differential problem (9)–(10) can equivalently be written in integral form as

$$G(\mathbf{r},\mathbf{r}') = G_0(\mathbf{r}-\mathbf{r}') - \int_S \mathrm{d}\sigma_1\; G(\mathbf{r},\mathbf{r}_1)\,\mathbf{n}_1\cdot\boldsymbol{\nabla}_1 G_0(\mathbf{r}_1-\mathbf{r}'). \tag{20}$$

Recursively iterating this integral equation expands G as a series of multiple scatterings of the bulk green function G_0 on the surface S: using the abbreviation $G = G_0 + G * G_0$ for (20), one has $G = G_0 + G_0 * G_0 + G_0 * G_0 * G_0 + \ldots$. In the presence of the two objects S and S', G_0 is scattered on both surfaces. When the separation d is much larger than the correlation length κ^{-1}, the dominant terms in the expansion of $G_{SS'}$ are the ones with the least number of propagations between S and S'. Indeed, for $\mathbf{r} \in S$ and $\mathbf{r}' \in S'$, $G_0(\mathbf{r}-\mathbf{r}') = \mathrm{O}(\mathrm{e}^{-\kappa d})$. However, any number of scatterings from S to S or from S' to S' can be done without affecting the order as $\kappa d \to \infty$. The sum of these scatterings can be recast as G_S^0 and $G_{S'}^0$, respectively, and one can then show that

$$(G_{SS'} - G_S^0)(\mathbf{r},\mathbf{r})|_{\mathbf{r}\in S} \overset{\kappa d\to\infty}{\sim} G_{S'}^0 * G_S^0 + G_S^0 * G_{S'}^0 * G_S^0. \tag{21}$$

The Green function of the composite system can thus be calculated in a large-separation regime from the knowledge of the Green functions associated to the single objects, which are usually much easier to obtain.

As a concrete example, we consider a circle C of radius R in front of a thin, infinitely-long plate P. The distance d is taken as their separation at the closest point. An explicit expression for G_C^0 is straightforward to obtain from (13), with the result $a_{mn} = -[\mathrm{I}_m'(\kappa R)/\mathrm{K}_m'(\kappa R)]\delta_{n,-m}$, where I_m' and K_m' are the derivatives of the modified Bessel functions of order m. It can be checked that $G_P^0(\mathbf{r},\mathbf{r}') = G_0(\mathbf{r}-\mathbf{r}') + G_0(\mathbf{r}-\mathbf{r}'^*)$, where $\mathbf{r}'^*$ is the point symmetric to $\mathbf{r}'$ with respect to the plate. Using (18) and (21), one can then evaluate $\mathbf{F}_{C\leftarrow P}$ and $\mathbf{F}_{P\leftarrow C}$ in a regime where $\kappa^{-1} \ll d$. Note that since both the circle and the plate are symmetric, $\mathbf{F}_C^0 = \mathbf{F}_P^0 = 0$.

We performed such an evaluation with the additional assumption that $R \ll \kappa^{-1}$. From an asymptotic analysis based on small-κR expansions and steepest-descent values of integrals as $\kappa d \to \infty$, we find that these forces are given by

$$F_{C\leftarrow P} \sim -F_0 \frac{\sqrt{\pi}(\kappa R)^2 \mathrm{e}^{-2\kappa d}}{\sqrt{\kappa d}}, \qquad F_{P\leftarrow C} \sim -\frac{3}{2}F_{C\leftarrow P}, \qquad R \ll \kappa^{-1} \ll d, \tag{22}$$

on the axis perpendicular to the plate. The action–reaction principle between these forces is clearly violated and the circle-plate assembly experiences the nonzero global force

$$F_{CP} \;=\; F_{C\leftarrow P} + F_{P\leftarrow C} \;\sim\; \frac{1}{2}F_0\frac{\sqrt{\pi}(\kappa R)^2\mathrm{e}^{-2\kappa d}}{\sqrt{\kappa d}} \tag{23}$$

in this regime. Note that this violation is not a negligible effect: the force F_{CP} has the same order of magnitude as $F_{C\leftarrow P}$ and $F_{P\leftarrow C}$.

4. Concluding remarks

The extension of Casimir-like forces to fluctuating media driven out of equilibrium opens up the possibility for new effects that might have important consequences in experiments and applications. The contributions to the forces due to the nonequilibrium character of the fluctuations can be tunable by more parameters than in equilibrium and both their strength and sign may be controlled. Furthermore, in nontrivial geometries, nonequilibrium fluctuations can induce forces on single asymmetric rigid bodies and can break the action-reaction principle between two objects noticeably. These two consequences are clearly ruled out at equilibrium by the second law of thermodynamics.

The occurrence of a violation of the action-reaction principle impeeds that an effective interaction potential holds in nonequilibrium and could prevent the use of the Derjaguin approximation. The magnitude of this violation can be of the same order as the internal forces, so that special care should be exercised in measurements performed in asymmetric setups.

If the immersed objects are let free to move, the presence of self-forces would put them into directed motion as in the case of ratchets. Jointly with the control of the sign of the force, this could be used as a powerful tool to tailor micro-devices by self-assembling or to construct motor axles with external energy source. The dynamical properties of such devices, however, need a more in-depth analysis as their motion will affect the fluid's fluctuations and a self-dynamical interaction could take place.

Refining the initial model to describe more realistic fluids would clearly be needed to draw more quantitative predictions for systems such as colloidal solutions, granular fluids, dusty plasmas etc. Nevertheless, the simple model for the nonequilibrium fluid that we presented here allows to deal efficiently with the complexity of nonplanar geometries via a straightforward Green function formalism and multiple-scattering scheme.

Acknowledgments

This work has been made in collaboration with R. Soto and is supported by the grants *Fondecyt* 1061112, 3070037, *Fondap* 11980002 and *Snsf* PBEL2-116909.

References

[1] Casimir H G B 1948 On the attraction between two perfectly conducting plates *Proc. Kon. Ned. Akad. Wet.* **51**, 793–5
[2] Milonni P W 1994 *The Quantum Vacuum: An Introduction to Quantum Electrodynamics* (San Diego: Academic Press)
[3] Milton K A 2001 *Physical Manifestations of Zero-Point Energy* (Singapore: World Scientific)
[4] Lifshitz E M 1955 The theory of molecular attractive forces between solids *J. Exp. Th. Phys. USSR* **29**, 94–110. (English translation: 1956 *Sov. Phys. JETP* **2**, 73–83)
[5] Fisher M E and de Gennes P G 1978 Phenomena at the walls in a critical binary mixture *C. R. Acad. Sci. Paris* B **287**, 207–9
[6] Krech M 1999 Fluctuation-induced forces in critical fluids *J. Phys.: Condens. Matter* **11**, R391–412
[7] Ajdari A, Duplantier B, Hone D, Peliti L and Prost J 1992 "Pseudo-Casimir" effect in liquid crystals *J. Phys. II France* **2**, 487–501

[8] Kardar M and Golestanian R 1999 The "friction" of vacuum, and other fluctuation-induced forces *Rev. Mod. Phys.* **71** 1233–45
[9] Balian R and Duplantier B 2004 Geometry of the Casimir Effect, in *Recent developments in gravitational physics* vol 176, ed I Ciufiolini *et al*
[10] Plunien G, Müller B and Greiner W 1986 The Casimir effect *Phys. Reports* **134** 87–193
[11] Hohenberg P C and Halperin B I 1977 Theory of dynamic critical phenomena *Rev. Mod. Phys.* **49**, 435–79
[12] van Noije T P C and Ernst M H 1999 Randomly driven granular fluids: large-scale structure *Phys. Rev.* E **59**, 4326–41
[13] Voth G A, Bigger B, Buckley M R, Losert W, Brenner M P, Stone H A and Gollub J P 2002 Ordered clusters and dynamical states of particles in a vibrated fluid *Phys. Rev. Lett.* **88**, 234301
[14] Gardiner C W 2004, *Handbook of stochastic methods*, 3rd ed (Berlin: Springer)
[15] Emig T 2007 Casimir force driven ratchets *Phys. Rev. Lett.* **98**, 160801
[16] Antezza M, Pitaevskii L P, Stringari S and Svetovoy V B 2008 Casimir-Lifshitz force out of thermal equilibrium *Phys. Rev.* A **77**, 022901
[17] Golestanian R 2005 Fluctuation-induced forces in and out of equilibrium *Pramana – J. Phys.* **64**, 1029–38
[18] Whitesides G M and Grzybowski B 2002 Self-assembly at all scales *Science* **295**, 2418–21
[19] Bartolo D, Ajdari A and Fournier J-B 2003 Effective interactions between inclusions in complex fluids driven out of equilibrium *Phys. Rev. E* **67**, 061112
[20] Cattuto C, Brito R, Marini Bettolo Marconi U, Nori F and Soto R 2006 Fluctuation-induced Casimir forces in granular fluids *Phys. Rev. Lett.* **96**, 178001
[21] Brito R, Marini Bettolo Marconi U and Soto R 2007 Generalized Casimir forces in nonequilibrium systems *Phys. Rev.* E **76**, 011113
[22] Antezza M, Pitaevskii L P and Stringari S 2005 New asymptotic behavior of the surface–atom force out of thermal equilibrium *Phys. Rev. Lett.* **95**, 113202
[23] Derjaguin B 1934 Untersuchung über die Reibung und Adhäsion, IV *Kolloid-Z.* **69**, 155–64
[24] Hertlein C, Helden L, Gambassi A, Dietrich S and Bechinger C 2008 Direct measurement of critical Casimir forces *Nature* **451**, 172–5
[25] Dzubiella J, Löwen H and Likos C N 2003 Depletion forces in nonequilibrium *Phys. Rev. Lett.* **91**, 248301
[26] Van den Broeck C, Meurs P and Kawai R 2005 From Maxwell demon to Brownian motor *New J. Phys.* **7**, 10
[27] Dean Astumian R 1997 Thermodynamics and kinetics of a Brownian motor *Science* **276**, 917–22
[28] Jülicher F, Ajdari A and Prost J 1997 Modeling molecular motors *Rev. Mod. Phys.* **69**, 1269–81
[29] Najafi A and Golestanian R 2004 Forces induced by nonequilibrium fluctuations: the Soret-Casimir effect *Europhys. Lett.* **68**, 776–82
[30] Costantini G, Marini Bettolo Marconi U and Puglisi A 2007 Granular Brownian Ratchet Model *Phys. Rev.* E **75**, 061124
[31] Cleuren B and Van den Broeck C 2007 Granular Brownian Motor *Europhys. Lett.* **77**, 50003
[32] Buenzli P R and Soto R 2008 Violation of the action-reaction principle and self-forces induced by nonequilibrium fluctuations *Phys. Rev.* E **78**, 020102(R)
[33] Landau L D and Lifshitz E M 1984 *Fluid Mechanics* (Oxford: Pergamon Press)
[34] Ortiz de Zarate J M and Sengers J V 2006 *Hydrodynamic Fluctuations in Fluids and Fluid Mixtures* (Amsterdam: Elsevier)
[35] Clerc M G, Cordero P, Dunstan J, Huff K, Mujica N, Risso D and Varas G 2008 Liquid-solid-like transition in quasi-one-dimensional driven granular media *Nature Physics* **4**, 249–54
[36] Buenzli P R and Soto R 2008 In preparation
[37] Schwinger J, DeRaad L L Jr, Milton K A and Tsai W-Y 1998 *Classical Electrodynamics* (Boulder: Westview Press)

Journal of Physics: Conference Series **161** (2009) 012037 doi:10.1088/1742-6596/161/1/012037

The Casimir effect: from quantum to critical fluctuations

Andrea Gambassi

Max-Planck-Institut für Metallforschung, Heisenbergstr. 3, D-70569 Stuttgart, Germany and Institut für Theoretische und Angewandte Physik, Universität Stuttgart, Pfaffenwaldring 57, D-70569 Stuttgart, Germany.

E-mail: gambassi@mf.mpg.de

Abstract. The Casimir effect in quantum electrodynamics (QED) is perhaps the best-known example of fluctuation-induced long-ranged force acting on objects (conducting plates) immersed in a fluctuating medium (quantum electromagnetic field in vacuum). A similar effect emerges in statistical physics, where the force acting, e.g., on colloidal particles immersed in a binary liquid mixture is affected by the classical thermal fluctuations occurring in the surrounding medium. The resulting Casimir-like force acquires universal features upon approaching a critical point of the medium and becomes long-ranged at criticality. In turn, this universality allows one to investigate theoretically the temperature dependence of the force via representative models and to stringently test the corresponding predictions in experiments. In contrast to QED, the Casimir force resulting from critical fluctuations can be easily tuned with respect to strength and sign by surface treatments and temperature control. We present some recent advances in the theoretical study of the universal properties of the critical Casimir force arising in thin films. The corresponding predictions compare very well with the experimental results obtained for wetting layers of various fluids. We discuss how the Casimir force between a colloidal particle and a planar wall immersed in a binary liquid mixture has been measured with femto-Newton accuracy, comparing these experimental results with the corresponding theoretical predictions.

1. Introduction

This contribution focuses on the Casimir effect which arises when one confines fluctuations of a different nature compared to those originally considered by Casimir in his pioneering work [1]. Indeed we shall be concerned here with the fluctuations of thermal origin which occur close to a second-order phase transition point (*critical point*). The resulting *critical* Casimir effect is apparently less widely known than the corresponding one within quantum electrodynamics (QED). Nonetheless, the associated femto-Newton forces at the sub-micrometer scale can be put to work in soft matter systems and their high degree of tunability might even be exploited for concrete applications in colloidal suspensions. However, before focusing on the Casimir effect due to critical fluctuations we first discuss in subsec. 2.1 the effect within QED, following in spirit the original derivation by Casimir [1]. In subsec. 2.2 we then explain how this effect arises in statistical physics upon approaching a critical point, highlighting the analogies and differences between the two. In sec. 3 we present an overview of the available theoretical predictions which are relevant for the qualitative and quantitative interpretation of the experimental results. A short summary, with perspectives and applications is presented in sec. 4.

2. The Casimir effect

2.1. The effect in Quantum Electrodynamics

The Casimir effect is named after Hendrik Casimir who discovered, 60 years ago, that – quite surprisingly – two parallel, perfectly conducting and uncharged metallic plates in vacuum *attract* each other due to the *quantum fluctuations* of the electromagnetic fields, even at zero temperature $T = 0$ [1]. Indeed the plates (assumed to be perfectly conducting) effectively impose boundary conditions (BCs) on the electromagnetic fields so that, more specifically, $\mathbf{E}_{\parallel}$, $\mathbf{B}_{\perp} = 0$ where $\mathbf{E}_{\parallel}$ and $\mathbf{B}_{\perp}$ are the components of the electric $\mathbf{E}$ and magnetic $\mathbf{B}$ fields which are parallel and transverse to the surfaces of the plates. As a result of these boundary conditions, the *fluctuation modes* of the fields in the space within the two plates are allowed to have only a specific set of L-dependent wave-vectors, where L is the separation between the parallel plates. For example, in the case of Dirichlet BCs, the component $k_{\perp}$ of the wave-vector $\mathbf{k}$ perpendicular to the plates assumes only the quantized values $k_{\perp} = \pi n/L$ with $n = 1, 2, \ldots$. Roughly speaking, the "unbalance" between the pressure exerted by the allowed modes within the plates and the one exerted by the modes outside them is at the origin of the Casimir effect. This statement can be made quantitative by calculating the size-dependent energy $\mathcal{E}(L)$ of the fluctuation modes of the electromagnetic fields which are allowed in the portion of the vacuum within the plates. According to QED, this is given by

$$\mathcal{E}(L) = \sum_{\text{modes}} \frac{1}{2}\hbar c|\mathbf{k}_{\text{modes}}|\,, \tag{1}$$

where $\mathbf{k}_{\text{modes}}$ are the corresponding wave-vectors and c is the speed of light. For perfectly conducting plates of large transverse area S, the expansion of $\mathcal{E}(L)$ in decreasing powers of L takes the form (in three spatial dimensions)

$$\mathcal{E}(L) = \mathcal{E}_{\text{bulk}} + \mathcal{E}^{(1+2)}_{\text{surf}} + S\frac{\hbar c}{L^3}\left[-\frac{\pi^2}{1440} + O((\kappa L)^{-2})\right], \tag{2}$$

where $\mathcal{E}_{\text{bulk}} \propto SL^1$ is the energy associated to the electromagnetic field of the vacuum in the absence of the plates and within a volume $S \times L$; $\mathcal{E}^{(1+2)}_{\text{surf}} = \mathcal{E}^{(1)}_{\text{surf}} + \mathcal{E}^{(2)}_{\text{surf}} \propto SL^0$ is the sum of the energies $\mathcal{E}^{(i)}_{\text{surf}}$ associated to the introduction of each single plate, *separately*, in the vacuum; The next term in the expansion (2), i.e.,

$$\mathcal{E}_{\text{Cas}}(L) \equiv -\frac{\pi^2}{1440}S\frac{\hbar c}{L^3}\,, \tag{3}$$

is proportional to SL^{-3} and represents the *interaction energy* between the two plates which is due to their *simultaneous* presence in space. This is the Casimir term we are interested in and which is responsible for the celebrated Casimir force. Further terms in the expansion are of order $O((\kappa L)^{-2})$ compared to $\mathcal{E}_{\text{Cas}}$, where κ is a material-dependent parameter which describes the deviations of the plates from the perfectly conducting behavior due to the *finite* conductivity of the metal. In the expansion (2) the first term $\mathcal{E}_{\text{bulk}}$ is a property of the vacuum, the second one ($\mathcal{E}^{(1+2)}_{\text{surf}}$) depends on the material properties of the plates, whereas the third one, $\mathcal{E}_{\text{Cas}}$ in eq. (3), is *universal* in that it does not depend on the specific material which the plates are made of, but only on geometrical properties (S and L) and on fundamental constants ($\hbar c$). The relevance of higher-order terms in the expansion (2) depends on the material parameter κ.

The Casimir term, the third of the the expansion of the vacuum energy $\mathcal{E}(L)$ in decreasing powers of L, actually gives rise to a measurable effect: a small displacement δL of one of the two plates results in a change $\delta\mathcal{E}(L)$ in the energy of the fields within the plates and therefore in a force $F_{\rm in}(L) = -\delta\mathcal{E}/\delta L$ (pushing on the inner surface of each single plate). Focusing on one of the two plates, the contribution to the force $F_{\rm in}(L)$ due to the change of $\mathcal{E}_{\rm bulk}$ in $\mathcal{E}(L)$, i.e., $F_{\rm bulk} \equiv -\delta\mathcal{E}_{\rm bulk}/\delta L$, is actually canceled out by the force $F_{\rm out} = F_{\rm in}(L=\infty) = F_{\rm bulk}$ acting from the other side of the plate and due to the fluctuation modes in the corresponding half-space outside the plate. $\mathcal{E}_{\rm surf}$ does not change upon changing L so that the total force $F(L) = F_{\rm in}(L) - F_{\rm out}$ acting on each single plate is due only to the change in the Casimir energy:

$$\frac{F(L)}{S} = -\frac{\pi^2}{480}\frac{\hbar c}{L^4} \quad \text{for} \quad L \gg \kappa^{-1}\,. \tag{4}$$

The universal behavior encoded in this equation can only be detected at separations L between the plates which are much larger than the length scale κ^{-1}, being material-specific properties relevant at smaller distances. In particular, κ is related to the typical wave-vector at which the plates are no longer effective in imposing the boundary conditions on the fields. As a rough estimate, this scale is set by the plasma frequency ω_p of the metal (more detailed calculations support this argument [2]), i.e., $c\kappa \simeq \omega_p$, where $\omega_p \simeq 3\cdot 10^{15}$Hz for copper, yielding $\kappa^{-1} \simeq 0.3\mu$m. Attempts to verify experimentally eq. (4) started already in 1958 [3] but the first sound experimental confirmation came only 40 years later within the range of separations $0.5\mu\text{m} \lesssim L \lesssim 6\mu$m [2, 4]. Apart from the relevance of this effect for possible technological applications in micro- and nano-electromechanical systems (MEMS and NEMS, respectively), in which the associated force is responsible for *stiction*, its precise measurement could provide a test of the validity of some fundamental laws down to the nano-meter scale, see, e.g., ref. [5].

2.2. The effect in Statistical Physics

Thirty years after the seminal paper by Casimir, M. E. Fisher and P. -G. de Gennes published a note "On the phenomena at the walls in a critical binary mixture" [6] in which it was shown that Casimir-like effects (i.e., *fluctuation-induced forces*) arise also in statistical physics if a medium in which fluctuations of a certain nature take place is spatially confined [7]. In order to understand how these effects arise and what their relation is with the Casimir effect described in subsec. 2.1, we briefly recall here some basic facts about binary mixtures, their critical point and the effect of confinement. The schematic phase diagram of a liquid binary mixture in the bulk is depicted in fig. 1. At constant pressure, the relevant thermodynamic variables are the temperature T and the mass fraction c_A of one of the two components in the mixture. For a certain concentration c_A, the two components of the mixture are mixed at low temperatures and the resulting liquid solution looks homogeneous in a test tube (schematically represented on the left of the phase diagram). Upon increasing the temperature, however, the liquid demixes into an A- and a B-rich solution and becomes inhomogeneous in the test tube (represented on the right), where the two solutions are typically separated by an interface. The transition between the mixed and the demixed phase occurs at the solid first-order transition line. At the lowest point (CP) of this line the transition becomes critical, i.e., second-order. This point is referred to as the (lower) *critical point*. The mixed and demixed phases can be distinguished by monitoring the so-called *order parameter* of the transition, which can be identified with the deviation $\delta c_A(\mathbf{x}) \equiv c_A(\mathbf{x}) - c_A$ of the *local* concentration $c_A(\mathbf{x})$ of, say, component A from its average value in the mixture. Indeed, the thermal average $\langle \delta c_A(\mathbf{x})\rangle$ of the order parameter is homogeneous in the mixed phase, whereas it assumes two different values in the A-rich and B-rich solution which coexist in the demixed phase. The order parameter is characterized by a spatial correlation length ξ, so that, within each single phase, $\langle \delta c_A(\mathbf{x})\delta c_A(\mathbf{x}')\rangle - \langle \delta c_A(\mathbf{x})\rangle\langle \delta c_A(\mathbf{x}')\rangle \propto \exp\{-|\mathbf{x}-\mathbf{x}'|/\xi\}$, where ξ depends on T and c_A. Upon crossing the first-order transition line, the correlation length stays finite, whereas it diverges upon approaching the critical point. The physical behavior of the system at the scale set by ξ is actually determined by the fluctuations of the order parameter, which becomes the relevant physical quantity for the description of the system. In terms of it, one can

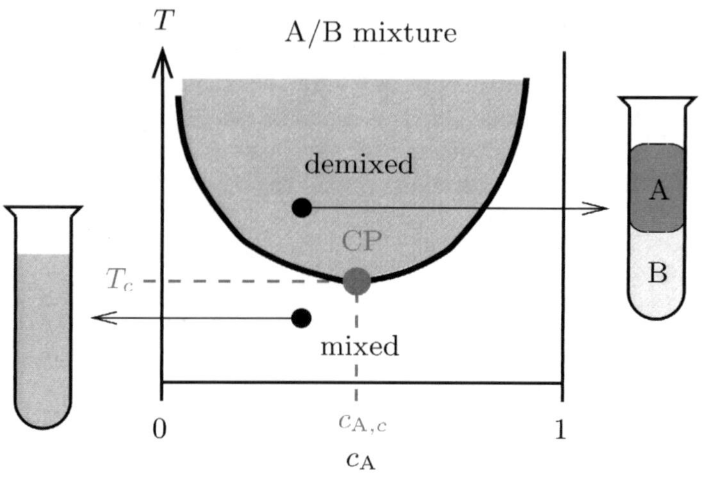

Figure 1. Schematic phase diagram of a binary liquid mixture with a lower critical point (CP). The schematic side views of a test tube filled with the binary liquid mixture in the mixed and in the demixed phase are shown on the left and on the right of the diagram, respectively.

express the effective free energy $\mathcal{F}$ of the system and determine its thermodynamic properties upon approaching the transition.

In order to understand the consequences of confining such a binary liquid mixture, we have first to consider the result of inserting a single plate (made up of, say, glass) into the mixture. In general one expects the plate to show preferential adsorption for one of the two components of the mixture, say A, due to quantitative differences in the interactions between each of them and the molecules of the plate. Accordingly, a plate in the mixture induces a local increase of the average order parameter $\langle \delta c_A(\mathbf{x}) \rangle$ close to it. Upon increasing the correlation length ξ (i.e., getting closer to the critical point), the plate effectively imposes boundary conditions on the order parameter [8, 9].

Table 1. Analogies and differences between the Casimir effect in (zero-temperature) QED and in Statistical Physics (see also footnote 1).

	QED	Stat. Phys
fluctuating quantity:	**E**, **B**	order parameter ϕ
excitation:	Quantum $\hbar c$ $(T = 0)$	Thermal(classical) $k_B T$ $(\hbar = 0)$
range of fluct.:	∞	finite: ξ $\xi \nearrow \infty$ close to CP
	$\Downarrow$	$\Downarrow$
	Confinement	
	$\Downarrow$	$\Downarrow$
	long-ranged force	range: ξ long-ranged at CP

Now that we have recalled how to describe the binary mixture in terms of a fluctuating order parameter (generically refereed to as ϕ in what follows) and the effects of inserted plates in terms of boundary conditions, it is easy to understand in which sense a Casimir-like interaction arises in these instances. Indeed (see table 1), the Casimir effect in QED emerges because there are some quantities (the fields $\mathbf{E}$ and $\mathbf{B}$) which "fluctuate" in space and time due to the quantum nature of the "medium" (actually the vacuum) in which their fluctuations take place. Accordingly, the relevant scale of the phenomenon is set by $\hbar c$ at zero temperature. The two-point correlation function of the fluctuations of these quantities are characterized by an algebraic decay in space, i.e., the associated range is infinite (due to the vanishing mass $m_\gamma = 0$ of the photon). When the spectrum of these fluctuations gets modified by external bodies (metallic plates) which impose boundary conditions on the fluctuating fields, the associated energy $\mathcal{E}$ becomes a function of the position of these boundaries and as a result an effective long-ranged Casimir force arises on them. In statistical physics, close to a critical point, the relevant fluctuating quantity is the order parameter ϕ of the phase transition ($\phi(\mathbf{x}) = \delta c_A(\mathbf{x})$ in the case of the binary mixture) and its fluctuations are of thermal nature and due to the coupling between the system and the thermal bath at temperature T. Accordingly, the relevant scale of the phenomenon is $k_{\mathrm{B}}T$ and the fluctuations are of classical nature (i.e., $\hbar$ plays no role). The range of these fluctuations is given by ξ and therefore is *finite* but tunable upon approaching the critical point. As in the case of QED, external bodies impose boundary conditions on the order parameter ϕ and therefore they affect the spectrum of its allowed fluctuations and the associated effective free energy $\mathcal{F}$ depends on the positions of these bodies. As a result, a force is expected to act on them, with a range set by the correlation length ξ. In table 1 we summarize analogies and differences between the Casimir effect occurring in QED and in statistical physics[1]. Even though we have been referring to the case of a binary liquid mixture, the line of argument presented above extends to critical phenomena in general. Indeed, consider a medium close to its critical point and confined between two parallel plates at a distance L (film geometry). The medium might be, e.g., pure ^{4}He or a ^{3}He-^{4}He mixture close to the superfluid transition, a binary liquid mixture, a Bose gas close to the Bose-Einstein condensation, liquid crystals etc. If the correlation length ξ of the fluctuations of the order parameter and the thickness L of the film are much larger than the microscopic length scale ℓ_{micr} (set, say, by the molecular scale of the system), the *free energy* $\mathcal{F}$ of the confined medium can be decomposed in decreasing powers of L for a fixed value of L/ξ and large transverse area S, in analogy to eq. (2) [10, 11]

$$\mathcal{F}(T,L) = \mathcal{F}_{\mathrm{bulk}} + \mathcal{F}^{(1+2)}_{\mathrm{surf}} + S\frac{k_{\mathrm{B}}T}{L^2}\Theta_{\parallel}(L/\xi) + \mathrm{corr.} \tag{5}$$

We have assumed here that the only relevant thermodynamic variable is the temperature T, which determines the correlation length $\xi \sim \xi_0|(T - T_c)/T_c|^{-\nu}$ where ξ_0 is a system-specific quantity and ν a *universal* exponent, in the sense specified below. Possible additional thermodynamic variables, such as the concentration c_A of the binary mixture, are assumed to be tuned to their critical values, i.e., $c_A = c_{A,c}$ with reference to fig. 1. In the expansion (5) the first term $\mathcal{F}_{\mathrm{bulk}}$ is the free energy of the *bulk* medium in a volume $S \times L$, i.e., $\mathcal{F}_{\mathrm{bulk}} \propto SL$, the second one ($\mathcal{F}^{(1+2)}_{\mathrm{surf}}$) represents the sums of the free energy costs $\mathcal{F}^{(i)}_{\mathrm{surf}} \propto S$ for the separate introduction of the two walls in the system, i.e., $\mathcal{F}^{(1+2)}_{\mathrm{surf}} = \mathcal{F}^{(1)}_{\mathrm{surf}} + \mathcal{F}^{(2)}_{\mathrm{surf}}$. The third term represents, as in eq. (2), the *interaction (free)energy* between the two walls. Compared to the analogous term in the case of QED we note, as expected, that $k_{\mathrm{B}}T$ replaces $\hbar c$ in providing the scale of the phenomenon [2] and that the interaction between the two walls is no longer long-ranged but its

[1] In some cases the correspondence illustrated in table 1 goes beyond the mere analogy. Indeed, for the simplest geometrical settings, such as film geometry, there is a mapping between the Casimir effect in QED in d spatial dimensions and the effect at the critical point of the so-called Gaussian model in $d+1$ dimensions.

[2] Due to the different engineering dimensions of $[\hbar c]$ = energy $\times$ length and $[k_{\mathrm{B}}T]$ = energy, and taking into

range is set by ξ via the dependence on L/ξ of the scaling function $\Theta_{\parallel}$ ([3]). This dependence can be understood as follows: the perturbation that the wall induces on the order parameter extends within a distance $\sim \xi$ far from it, so that the two walls can interact with each other only at separations $L \lesssim \xi$ whereas the interaction energy vanishes for $L \gg \xi$, so that $\Theta_{\parallel}(x \to \infty) = 0$.

The scaling function $\Theta_{\parallel}$ is *universal* (as the analogous term $\mathcal{E}_{\rm Cas}$ (3) in eq. (2)) in the sense that it depends only on some gross features of:

(a) the system in the bulk, such as the range and symmetries of the interaction and the kind of order parameter which describes the phase transitions,

(b) the surfaces which provide the confinement, such as the possible symmetries of the bulk system which they break, e.g., by favoring certain values of the order parameter at the boundaries. These preferences eventually translate into effective boundary conditions for the order parameter [8, 9]. In addition, $\Theta_{\parallel}$ depends on the shape of the boundaries, as we shall see below.

The general features at points (a) and (b) define the so-called *bulk* and *surface universality classes* of the confined system, respectively. This universality is typical of critical phenomena and allows one to investigate universal properties such as ν and $\Theta_{\parallel}$ by means of suitable representative models belonging to the same bulk and surface universality classes of the actual system one is interested in and which lend themselves for a simpler theoretical analysis. Analogously to what happens in QED, a small change δL of the distance between the confining surfaces leads to a change $\delta\mathcal{F}$ in the free energy and therefore to a force $-\delta\mathcal{F}/\delta L$ acting on the displaced wall. The L-independent contribution to this force due to $\mathcal{F}_{\rm bulk}$ in eq. (5) is counterbalanced by the same contribution acting from outside the walls when they are immersed in the critical medium, so that the net force $F(L,T)$ is given by (see eq. (5))

$$\frac{F(T,L)}{S} = \frac{k_{\rm B}T}{L^3}\vartheta_{\parallel}(L/\xi) \quad \text{for} \quad L,\, \xi \gg \ell_{\rm micr}, \tag{6}$$

where $\vartheta_{\parallel}(u) = -2\Theta_{\parallel}(u) + u\Theta'_{\parallel}(u)$ and it shares with $\Theta(u)$ the properties of universality. The universal scaling function $\vartheta_{\parallel}$ can be conveniently determined from the analysis of suitable representative models, either via field-theoretical methods or Monte Carlo simulations. In turn, the resulting predictions can be very stringently compared with the experimental determination of $\vartheta_{\parallel}$.

2.3. The force of quantum and critical fluctuations

In the preceding two sections we have briefly reviewed the origin of the Casimir effects due to quantum and critical fluctuations, highlighting the analogies and differences between them and the corresponding scaling and universal properties. In view of possible experimental investigations, however, it is important to estimate the expected magnitude of these effects. Assuming (quite unrealistically, indeed) that one is able to realize a system of two parallel plates of area $S = 1\,{\rm cm}^2$ at a distance $L = 1\,\mu{\rm m}$, the zero-temperature Casimir force resulting from quantum fluctuations is roughly estimated as $\sim S \times \hbar c/L^4 \simeq 6 \cdot 10^{-7}\,{\rm N}$, whereas the force due to critical fluctuations possibly occurring at room temperature ($T \simeq 300\,{\rm K}$) can be estimated as $\sim S \times k_{\rm B}T/L^3 \simeq 4 \cdot 10^{-7}\,{\rm N}$. Even though these appear to be quite small forces, their magnitudes are actually comparable with the weight $\simeq 2 \cdot 10^{-7}\,{\rm N}$ of a water droplet of

account the expected proportionality to S, the two effects are characterized by different powers in their dependence on L.

[3] The scaling function $\Theta_{\parallel}$ actually depends also on whether a certain correlation length ξ is realized above or below the bulk critical temperature T_c. This is made clear later on by expressing the scaling function as a function of an appropriate scaling variable x, c.f., eq. (7), instead of $u = L/\xi$, as done here.

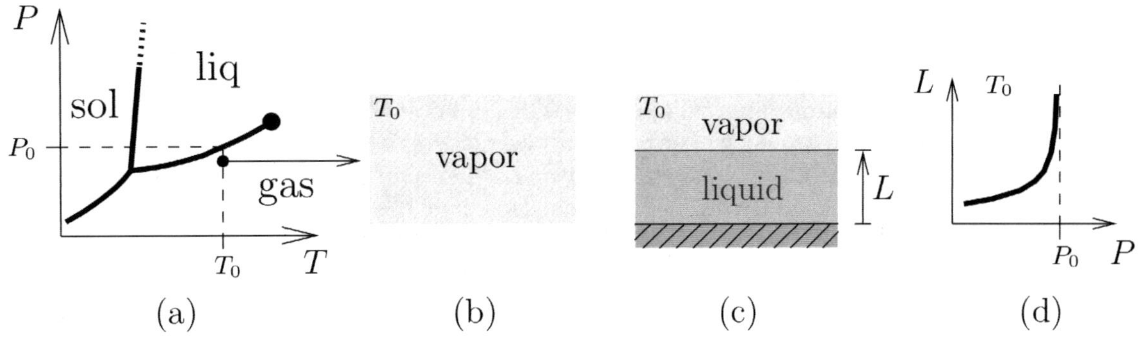

Figure 2. Formation of a *complete wetting* film of a fluid. Close to the liquid-gas phase transition (a), a vapor (b) condenses in the presence of a suitably chosen (solid) substrate, forming a liquid film (c), the thickness L of which diverges upon approaching the bulk condensation pressure P_0 (d).

half a millimeter in diameter. However, the realization of such a geometrical setting is actually extremely difficult due to the problem of maintaining the alignment between the two plates within the required accuracy. This problem is particularly difficult when trying to detect the Casimir effect in quantum electrodynamics and indeed this has been achieved in the parallel-plate geometry only quite recently [4]. For the detection of the critical Casimir effect the problem of alignment can be solved by taking advantage of physical phenomena, such as *wetting*, which naturally lead to the formation of confined films of fluids of a certain controllable and well-defined thickness L, determined by the thermodynamic parameters of the system, such as pressure and temperature [12, 13].

2.4. Wetting films and critical endpoints

In fig. 2(a) we depict schematically the phase diagram of a substance, as a function of the pressure P and temperature T, showing the generic solid (sol), liquid (liq), and gas phases. In order to obtain a liquid wetting film of this substance, the temperature T_0 and pressure P have to be chosen in such a way that the bulk system is in the vapor phase (fig. 2(b)) but close to the condensation transition, occurring at $P = P_0(T_0)$. If the vapor is taken in contact with a suitably chosen (solid) substrate (hatched grey area in fig. 2(c)), also at temperature T_0, a fluid film will condense on it as a consequence of the interaction of the substance with the substrate which makes the formation of a liquid layer thermodynamically favorable. The thickness L of the resulting film can be controlled acting on the undersaturation $\delta P = P - P_0(T_0)$ and is determined by van-der-Waals and dispersion forces. Depending on the choice of the substrate, L diverges smoothly upon approaching the bulk condensation pressure, i.e., for $\delta P \to 0$. This case is referred to as *complete wetting* and is characterized by the formation of a liquid film which can be made macroscopically thick, so that $L \gg \ell_{\rm micr}$. The phenomenon of wetting leads naturally to the formation of a liquid film of constant thickness L, in which the liquid is confined between the surface of the solid substrate and the liquid-vapor interface. The latter is macroscopically well defined sufficiently far from the liquid-vapor critical point. Even though the problem of the alignment between the confining surfaces is seemingly solved by wetting films, one has to be aware of the fact that the liquid-vapor interface actually fluctuates (*capillary fluctuations*) around its average position.

A possible *indirect* evidence of critical Casimir forces can be obtained by monitoring the thickness L of the wetting layer upon approaching a critical point which, in the bulk, occurs within the liquid phase, is close to the liquid-vapor first-order transition but is far enough from the liquid-vapor critical point. This is the case for *critical endpoints* (cep), which are located where a line of critical points (within the liquid phase, see, c.f., fig. 3(a)) meets the liquid-vapor transition line. Indeed, upon approaching a critical endpoint, the critical fluctuations of the associated order parameter, confined within the liquid wetting film of thickness L, give rise to the critical Casimir effect. The associated force adds up to the previously acting (van-der-Waals) forces in determining the equilibrium distance L of the liquid-vapor interface from the substrate and consequently it affects the thickness of the wetting film. The dependence of L on the thermodynamic control parameters such as pressure and temperature is the basis for the determination of the Casimir force within this approach, which was proposed and discussed in detail in ref. [13] (see also ref. [14]) and provided the theoretical motivation for the experimental investigations summarized in subsec. 3.1.

3. The critical Casimir effect at work

In this section we summarize the theoretical predictions for the scaling function of the critical Casimir force in thin films and we compare them with the available experimental results based on wetting films (subsec. 3.1). Then we discuss a more direct measurement of this force (subsec. 3.2).

3.1. Wetting films

In subsec. 2.4 we argued that it is possible to exploit critical endpoints in order to infer the critical Casimir force from the thickness of a complete wetting film. One of the most studied fluids with such a critical endpoint is pure ^{4}He, the phase diagram of which is sketched in fig. 3(a). Within the fluid phase ^{4}He undergoes a second-order phase transition between a *super*fluid (also refereed to as HeII) and a *norma*l (HeI) behavior upon increasing the temperature across the so-called λ-line. The order parameter ϕ of this phase transition is physically provided by the wave-function of the superfluid. The λ-line terminates at the critical endpoint (cep) on the fluid-vapor first-order transition line (which ends on the right at the liquid-vapor critical point) located at low pressure, as indicated by the closer view of fig. 3(a). As previously described, in order to detect critical Casimir forces one prepares the system in the vapor phase corresponding to the point P in the closer view of fig. 3(a) and exposes this vapor to a suitable substrate (e.g., Cu) in order to obtain a complete wetting film of thickness L. Then one changes the thermodynamic parameters in such a way to follow the thermodynamic path indicated as γ in fig. 3(a), along which the film thickness L would not change in the absence of the critical endpoint. The actual changes are therefore due to the action of the critical Casimir force on the liquid-vapor interface and from these changes it is possible to determine experimentally the scaling function $\vartheta_{\|}$ of the force (see eq. (6)) [13]. In order to predict theoretically the scaling function $\vartheta_{\|}$ one has to single out a theoretical model which belongs to the same *bulk* and *surface* universality class as the confined ^{4}He. It is well known that the universal aspects of the bulk critical behavior of ^{4}He are properly captured, upon approaching the λ-line, by the lattice XY model [15], which lends itself, e.g., for Monte Carlo simulations. Accordingly, the behavior of a film of ^{4}He close to the λ-line is captured by the XY model with film geometry, where the boundary conditions on the lattice are chosen to be in the same surface universality class (i.e., to lead to the same effective boundary conditions for the order parameter) as the actual confining boundaries, represented by the substrate-liquid and liquid-vapor interfaces. As anticipated, the order parameter ϕ of the superfluid transition is related to the wave-function of the superfluid component and therefore, as a wave-function, it has to be continuous in space and to vanish both inside the substrate and within the vapor, where no superfluid is present. As a result, ϕ vanishes at the boundaries, i.e., ϕ satisfies Dirichlet BCs at both surfaces (DD). On the lattice model these BCs are realized via free boundary conditions for the degrees of freedom. Having identified a model which belongs to the same bulk and surface universality class as the film of

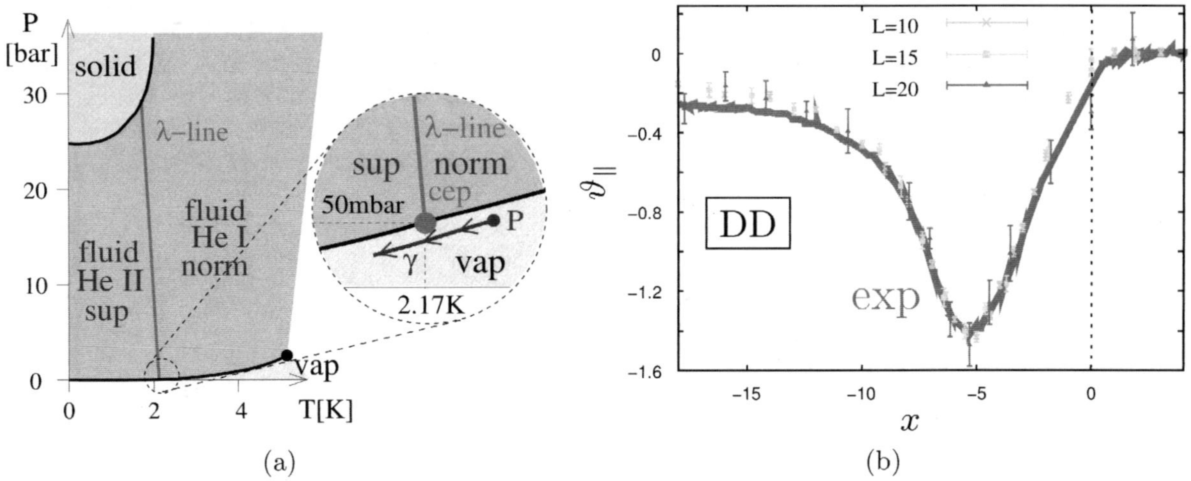

Figure 3. (a) Phase diagram of pure ^{4}He. The closer view highlights the presence of the critical endpoint (cep). (b) Universal scaling function $\vartheta_\parallel$ of the critical Casimir force within the three-dimensional XY universality class and Dirichlet-Dirichlet (DD) BCs, as a function of the scaling variable $x = (T/T_\lambda - 1)(L/\xi_0)^{1/\nu}$. The data points were obtained by Monte Carlo simulation of the XY model on the lattice in ref. [16], whereas the solid line represents the experimental results of ref. [20].

^{4}He, one can proceed to the determination of the scaling function $\vartheta_\parallel$. In fig. 3(b) the data points refer to the scaling function $\vartheta_\parallel$ as inferred from Monte Carlo simulations of the XY model in film geometry of various thicknesses and DD boundary conditions [16] (see also ref. [17]). The scaling function is plotted as a function of the proper scaling variable

$$x = t(L/\xi_0)^{1/\nu} \tag{7}$$

where $t \equiv (T - T_\lambda)/T_\lambda$, T is the temperature at which the lattice model is simulated and T_λ is the associated bulk critical value (corresponding to the λ-line in the actual system). L is the thickness of the film, ξ_0 the non-universal amplitude which controls the divergence of the bulk correlation length ξ and which can be determined independently via Monte Carlo simulation, and $\nu \simeq 0.66$ [15] is the universal critical exponent of the correlation length. The scaling variable x in eq. (7) is related to $u \equiv L/\xi$ in eq. (6) by $x = u^{1/\nu}$ for $t > 0$. As a confirmation of the scaling behavior in eq. (6), the scaling function $\vartheta_\parallel$ in fig. 3(b) does not depend on the thickness L of the film in which it has been numerically determined (as long as L is large enough compared to the lattice spacing) and the data sets corresponding to different sizes actually fall onto the same master curve. The predicted critical Casimir force turns out to be attractive in the whole range of temperatures ($\vartheta_\parallel < 0$), as expected it decays to zero for $L \gg \xi$, i.e., for $x \gg 1$, whereas it saturates to a non-vanishing value for $x \to -\infty$ due to the low-temperature long-range correlations maintained by Goldstone modes. Some of the qualitative features of the present force, such as the occurrence of a deep minimum below T_λ, can be understood within a simple mean-field approximation (see, e.g., refs. [18, 19]).

The very delicate experimental determination of the critical Casimir force acting on wetting films of ^{4}He on a Cu-substrate has been done in ref. [20], where the thickness $L \simeq 200 \ldots 300$ Å of the film as a function of the distance from the critical point has been determined by capacitance measurements. The corresponding experimental data are indicated by a solid line in fig. 3(b) and show a remarkably good agreement with the corresponding theoretical predictions based on Monte Carlo simulations of the lattice model. In determining the value of the abscissa x corresponding to a certain experimental point one has to use for ξ_0 the system-specific value which has been determined independently in the experiment on the basis of the behavior of the bulk correlation length. If this is properly done, then there are no free parameters which can be adjusted in the comparison. In this sense the test of the theoretical predictions is remarkably stringent.

Based on the values of the experimental parameters of ref. [20] one can estimate the critical Casimir pressure to be of the order of 2 Pa which, however, is still of quite difficult detection in those experimental conditions. In order to increase the magnitude of the critical Casimir force, which is proportional to the temperature T, it is convenient to look for critical endpoints occurring at higher temperatures compared to the one of ^{4}He, for example by considering wetting films of *classical* binary liquid mixture. It turns out that in these cases the boundary conditions can be tuned such that the resulting forces are *repulsive*. These forces could find application for compensating the attractive electrodynamic Casimir force (the sign of which cannot be easily controlled) and therefore avoid stiction in micro- and nano-machines (MEMS and NEMS).

The phase diagram of a binary liquid mixture is depicted in fig. 1 for a fixed value of the pressure P (i.e., fig. 1 is a cut of the phase diagram in the (c_A, T, P)-space). Upon decreasing the pressure the location of the critical point CP changes in the (c_A, T)-plane and eventually it meets at the critical endpoint the sheet corresponding to the transition between the liquid and the vapor of the mixture. This critical endpoint can be exploited in order to create a wetting film of the binary mixture [13], similarly to the case of ^{4}He. The scaling function $\vartheta_\parallel$ can therefore be inferred by monitoring the equilibrium thickness of the wetting layer as a function of the thermodynamic parameters. Also in this case, theoretical predictions for the scaling function $\vartheta_\parallel$ can be obtained by studying a suitable model which belongs to the same *bulk* and *surface* universality class as the confined classical binary mixture. It is well-known that the bulk critical behavior close to the demixing critical point in a binary mixture is properly captured by the lattice Ising model [15], which lends itself, e.g., for numerical studies via Monte Carlo simulations. The universal behavior of the binary mixture confined in the film is therefore captured by the Ising model in a film geometry, in which the boundaries are chosen to be in the same surface universality class as the actual confining surfaces constituted by the substrate-liquid and liquid-vapor interfaces. In subsec. 2.2 we have already mentioned the fact that the boundaries generically show preferential adsorption for one of the two components of the mixture. As a result, the order parameter $\delta c_A(\mathbf{x})$ of the binary mixture (see the discussion in subsec. 2.2) either increases (+) or decreases (−) upon approaching the boundaries and this tendency turns into effective boundary conditions close enough to the critical point [8, 9]. We shall refer to the case in which both boundaries preferentially adsorb the *same* component of the mixture as (++) and to the case of *opposite* preferences as (+−). Additional details of the strength of this preferences etc. turn out to be irrelevant sufficiently close to the critical point. In the lattice Ising model, these boundary conditions are realized by fixing the lattice degrees of freedom (± 1 spins) on the two opposing boundaries either to the same [(++)] or to opposite [(+−)] values. The resulting numerical prediction of the scaling function $\vartheta_\parallel$ is reported in fig. 4 as a function of the scaling variable x in eq. (7), where L is the thickness of the film, $\nu \simeq 0.63$ [15] is the universal critical exponent of the correlation length, $t = (T - T_c)/T_c$ where T is the temperature at which the lattice model is simulated and T_c its critical value. As it was the case for ^{4}He, the value of the non-universal amplitude ξ_0 for this lattice model is measured independently via Monte Carlo simulations. The collapse of data points referring to lattices of different thicknesses onto the same master curve is again a confirmation of the scaling behavior expected on the basis of eq. (6)[4]. The sign of the resulting Casimir force depends on the boundary conditions and

[4] In order to extract in this case the asymptotic scaling behavior from Monte Carlo data it is necessary to account

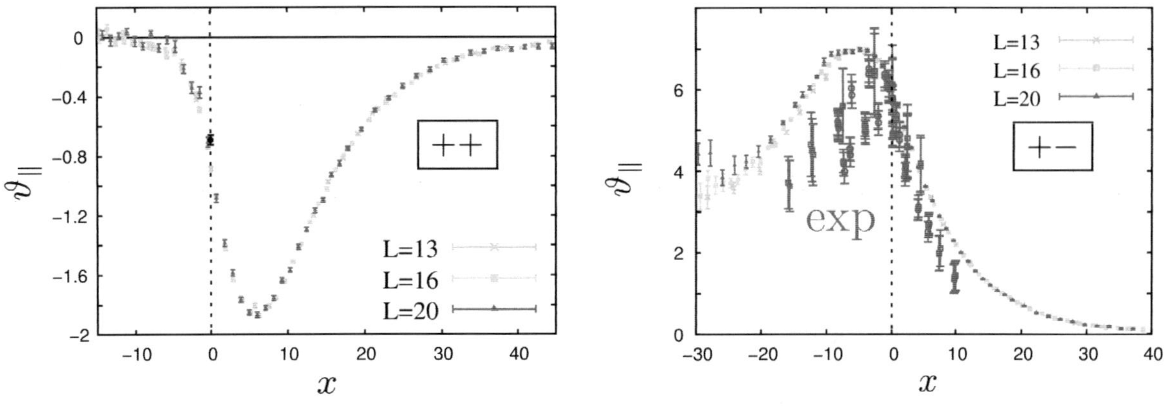

Figure 4. Universal scaling function $\vartheta_{\|}$ of the critical Casimir force within the three-dimensional Ising universality class and (++)/(+−) BCs, as a function of the scaling variable $x = t(L/\xi_0)^{1/\nu}$. The data points were obtained in ref. [16] by Monte Carlo simulation of the Ising model on the lattice, whereas experimental data for the case of (+−) BCs were obtained in ref. [22].

is *attractive* for (++) and *repulsive* for (+−), corresponding, respectively, to $\vartheta_{\|}(x) < 0$ and $\vartheta_{\|}(x) > 0$ in the whole range of temperatures. As expected, these scaling functions decay to zero for $x \to \pm\infty$ due to the fact that away from the critical point the correlation length becomes much smaller than the film thickness L both in the mixed (disordered in terms of the Ising model) and demixed (ordered) phase. In addition, the typical magnitude of the repulsive force for (+−) BCs is larger than the one of the attractive force for (++) boundary conditions. This is due to the fact that in the former case also the fluctuations of the position of the interface between the region with positive and negative values of $\langle \delta c_A(\mathbf{x}) \rangle$ contribute to the force acting on the confining surfaces. As in the case of ^{4}He some of the qualitative features of the force are already properly captured by a mean-field approximation [21]

The critical Casimir effect in wetting films of classical binary mixtures has been investigated in refs. [22, 23] where either ellipsometry [23] or X-ray scattering [22] techniques have been employed in order to determine the equilibrium thickness L of the wetting film. In particular in ref. [22] a binary mixture of methylcyclohexane (MC) and perfluoromethylcyclohexane (PFMC) has been used as the critical medium, with an upper critical point at temperature $T_c = 42.6°$C and molar fraction of PFMC $c_{\mathrm{PFMC},c} = 0.36$. The investigated wetting films were formed on a SiO_2/Si substrate and they had a typical thickness $L \simeq 100$ Å with the PFMC and MC preferring to be close to the liquid-vapor (+) and the liquid-substrate (−) interface, respectively. The corresponding experimental data for the inferred scaling function of the critical Casimir force are reported on the right panel of fig. 4. As in the comparison between theoretical and experimental predictions for ^{4}He, the value of the abscissa x corresponding to a certain experimental point has to be calculated according to eq. (7) taking into account the experimentally determined value of ξ_0 and the fact that $t \equiv (T - T_c)/T_c$ for an upper critical point. Note that, again, there are no adjustable parameters for the comparison. The agreement with the corresponding theoretical predictions is rather good (for a more detailed discussion, see ref. [16]).

All the evidence previously discussed for critical Casimir forces relies on the use of wetting films of (classical and quantum) fluids, a rather indirect way of detecting the effects of these

for corrections to scaling, see ref. [16] for details.

forces. In addition, the reliable quantitative interpretation of the observed variations of the thickness of these films in terms of the action of Casimir forces requires an independent and quite detailed knowledge of some material-dependent parameters which determine the van-der-Waals forces acting within the system. In view of these potential limitations, it would be desirable to measure *directly* the critical Casimir forces, e.g., by determining the associated potentials.

3.2. Brownian motion of colloids

The experimental evidence summarized in the previous subsection suggests that critical Casimir forces become relevant at the submicrometer scale, so that, on dimensional ground and for critical points at room temperature $T_c \simeq 300\,\mathrm{K}$ one estimates the typical scale of force as $\simeq k_B T_c/(0.1\,\mu\mathrm{m}) \simeq 40\,\mathrm{fN}$. These tiny forces are hardly detectable even with very sensitive methods such as atomic force microscopy but nonetheless they are still strong enough to affect the Brownian motion of particles of micrometer size immersed in a fluid (i.e., *colloids*) which can be used as detectors. Indeed, under the effect of external forces (such as critical Casimir forces, gravity, buoyancy, etc., see below) a colloidal particle tends to stay at the equilibrium position $\mathbf{x}_{eq}$ which corresponds to the minimum of the potential $\Phi(\mathbf{x})$ associated with the acting forces. However, due to the random collisions with the molecules of the surrounding fluid (in thermal equilibrium at temperature T), the particle actually fluctuates around $\mathbf{x}_{eq}$, with a probability density $\rho(\mathbf{x}) \propto \exp\{-\Phi(\mathbf{x})/(k_B T)\}$. Accordingly, the forces acting on a colloid can be inferred from the measurement of $\rho(\mathbf{x})$. This possibility suggests the study of a geometrical arrangement in which the critical fluctuations of the medium are restricted in the space between a spherical colloid and a planar surface rather than between two parallel surfaces, as considered so far. On the basis of the discussion of subsec. 2.2, one actually expects a Casimir force F to act on the sphere. If the minimal distance z between the surface of the sphere and the planar substrate is much smaller than the radius R of the sphere itself, the spherical surface can be approximated by a set of circular rings parallel to the opposing planar substrate (Derjaguin approximation) and therefore the force in this geometrical setting can be related to the one for parallel plates. Within this approximation, the z-dependent critical Casimir force F at temperature T acting on the sphere is given by (see ref. [25] for details)

$$F(z) = k_B T \frac{R}{z^2} \vartheta_{|\circ}(z/\xi) \quad \text{for} \quad z \ll R, \tag{8}$$

where the universal scaling function $\vartheta_{|\circ}$ can be calculated in terms of a suitable integral of the scaling function $\vartheta_{\|}$ in a film [26] and it retains the qualitative features of $\vartheta_{\|}$. The theoretical prediction for the scaling form of the Casimir potential $\Phi_C(z) = \int_z^\infty \mathrm{d}z' F(z') = k_B T (R/z) \Xi(z/\xi)$ immediately follows from the integration of eq. (8).

In order to measure tiny forces acting on a single colloid, one can take advantage of the Total Internal Reflection Microscopy (TIRM) [24], a very sensitive technique that is widely employed in the study of soft-matter systems [24]. With TIRM one monitors the Brownian motion of the colloid floating in a fluid and, as anticipated, from the associated statistics one infers the potential of the forces acting on the particle. The scheme of the experimental setup is depicted in fig. 5(a): an incoming visible laser beam is totally reflected at the substrate (glass)-fluid interface, so that an evanescent field of intensity $I_{ev}(x_\perp)$ penetrates the liquid with a typical exponential decay as a function of the distance $x_\perp$ from the interface, see fig. 5(b). The colloidal particle, of suitable refractive index and at a surface-to-surface distance z from the substrate, scatters light out of this evanescent field. The intensity I_{sc} of the scattered light is measured by a photomultiplier (PM) and turns out to be proportional to the intensity $I_{ev}(z)$ of the evanescent field. ¿From the time dependence of the scattered intensity $I_{sc}(t) \propto I_{ev}(z(t))$, represented in fig. 5(c), one can therefore calculate the time dependence of the particle-wall separation, i.e., $z(t)$. This dependence, depicted in fig. 5(d), reflects the Brownian motion of the particle under

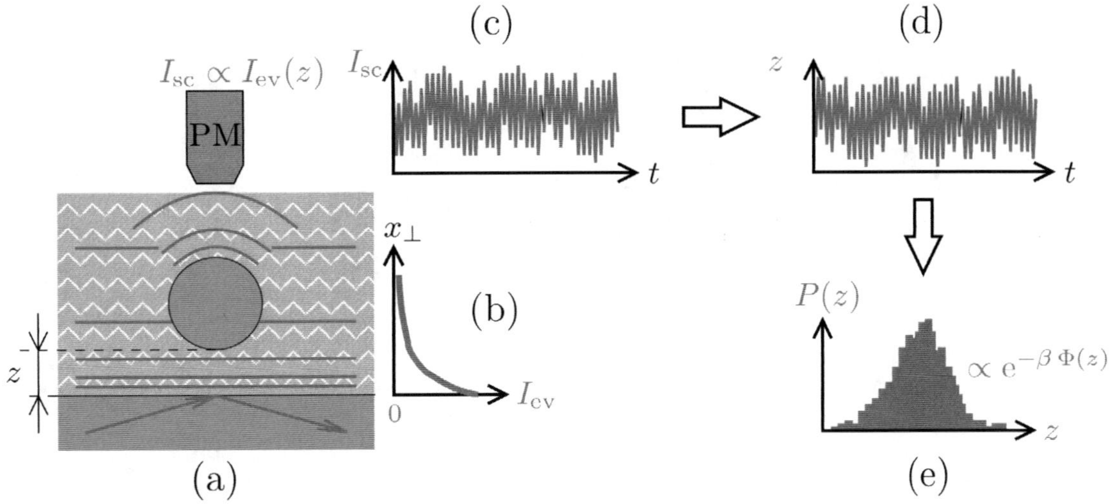

Figure 5. Determination of the potential $\Phi(z)$ of the forces acting on a single colloidal particle by means of the Total Internal Reflection Microscopy.

the effect of thermal fluctuations and possible additional forces. The probability distribution $P(z)$, as extracted from the histogram of $\{z(t)\}_{0\leq t\leq t_{\rm samp}}$ for sufficiently large sampling time $t_{\rm samp}$ (fig. 5(e)), is proportional to the Boltzmann factor $\exp\{-\beta\Phi(z)\}$ (with $\beta = (k_{\rm B}T)^{-1}$), where $\Phi(z)$ is the potential of the total force acting on the particle. Accordingly, $\Phi(z)$ can be calculated up to an irrelevant constant on the basis of the $P(z)$ inferred from the experimental data.

In order to provide a *direct* evidence of critical Casimir forces one can study a single colloidal particle immersed in a binary liquid mixture of water (W) and lutidine (L), close to a glass substrate [26]. In ref. [26] TIRM has been employed for the determination of the potential $\Phi(z)$ of the forces acting on the particle and for studying how Φ changes when the binary mixture is driven towards the critical point. The W-L mixture has a bulk phase diagram such as the one presented in fig. 1, with a lower critical point at $T_c = 34$°C and lutidine mass fraction $c_{{\rm L},c} \simeq 0.29$. The local excess concentration of lutidine $\delta c_{\rm L}(\mathbf{x})$ is a suitable order parameter for this demixing transition. In what follows we illustrate the experimental results in the case of a mixture at the critical concentration $c_{{\rm L},c}$ (the case $c_{\rm L} \neq c_{{\rm L},c}$ is discussed in refs. [25, 26]).

At a temperature T far below the critical value T_c, fluctuations of the order parameter are irrelevant and the potential $\Phi(z)$ of the forces acting on the particle is basically the sum of two contributions: (a) a screened electrostatic repulsion ($\Phi_{\rm el}(z) \propto e^{-\kappa z}$) between the colloid and the glass substrate, relevant in our experimental conditions for $z \lesssim 0.1\,\mu$m and (b) a linearly increasing potential due to the combined effect of buoyancy, gravity and optical pressures generated by the optical tweezer employed in the actual experimental setup[5]. This latter contribution has been subtracted from the curves presented in fig. 6, which show the dependence of the measured potential $\Phi(z)$ on the distance from the critical point. Panel (a) and (b) in fig. 6 refer to (colloid, substrate) = ($--$) and ($+-$) BCs, respectively. As expected, few hundreds mK far from the experimentally estimated critical temperature, the potential $\Phi(z)$ consists only of the electrostatic repulsion, which is negligible on this scale for $z \gtrsim 0.12\,\mu$m in (a) and $z \gtrsim 0.1\,\mu$m in (b). Upon increasing the temperature by few tens mK significant changes

[5] A third possible contribution, due to van-der-Waals forces turned out to be negligible in the experiment reported in refs. [26, 25].

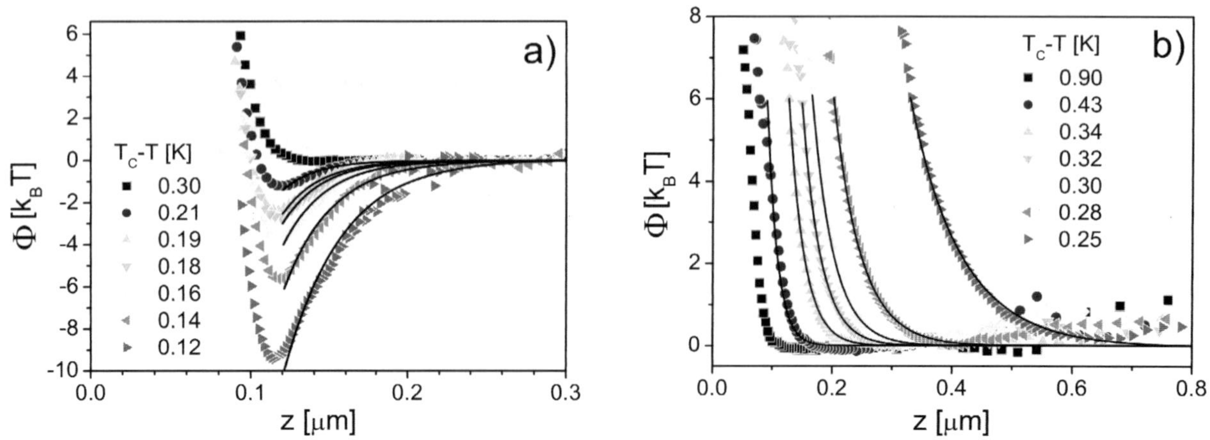

Figure 6. Temperature dependence of the potential $\Phi(z)$ of the forces acting on a single polystyrene colloidal particle at distance z of closest approach from a glass substrate. The particle and the substrate are immersed in a water-lutidine liquid mixture with critical composition and temperature T smaller than the critical one T_c. Data points refer to the experimental results (where gravity and buoyancy have been subtracted, see the main text) obtained for a substrate treated with NaOH in order to make it hydrophilic [$(-)$ BC for the order parameter] and (a) a colloid of diameter $2R = 2.4\,\mu$m absorbing preferentially water $(-)$ (b) a colloid of diameter $2R = 3.7\,\mu$m with preferential adsorption for lutidine $(+)$. The solid lines in (a) and (b) are the corresponding theoretical predictions for the critical Casimir potential with (colloid, substrate) $= (--)$ and $(+-)$ BCs, respectively. (From ref. [26].)

occur in the measured potentials: a strong attractive and repulsive force develops, respectively, in (a) and (b). In particular, in both cases one observes a change in the total potential of the order of 10 $k_{\rm B}T$, which is really significant on the scale of $k_{\rm B}T$ which characterizes the physics of colloids. The maximum value of the attractive force acting in case (a) is estimated at 600 fN. The strong temperature dependence of the measured potentials is a clear indication of the involvement of critical Casimir forces. This evidence is supported by the comparison with the corresponding theoretical predictions for $\Phi_C(z)$, shown as solid lines in fig. 6. In order to focus only on the contribution to $\Phi(z)$ due to the critical Casimir force, this comparison is presented only within the range of distances where the electrostatic contribution is negligible. The limited accuracy with which the critical temperature T_c has been experimentally determined in ref. [26] does not allow a reliable calculation of the correlation length ξ on the basis of the theoretical relation $\xi = \xi_0(1 - T/T_c)^{-0.63}$ and of the value $\xi_0^{(\rm exp)}$ of ξ_0 measured for the W-L mixture in the bulk by independent experiments [27]. Instead, for each single temperature T, the correlation length $\xi(T)$ can be determined in order to optimize the agreement between theory and experimental data. Then, separately for the two experiments, the value $\xi_0^{(\rm fit)}$ of ξ_0 and $T_c^{(\rm fit)}$ of T_c are determined in such a way to yield the best fit of the determined $\xi(T)$ with the expected algebraic behavior. In both cases the fact that $\xi_0^{(\rm fit)} = \xi_0^{(\rm exp)}$ within errorbars is a check of the significant agreement between theory and experiments. (See refs. [25, 26] for a more detailed discussion.) Interestingly enough, via a suitable chemical surface treatment, it is possible to change quite easily the substrate from hydrophilic $(-)$ to hydrophobic $(+)$ and therefore switch the Casimir force acting on the colloids in fig. 6(a) and (b), respectively, from attractive to repulsive and vice-versa (see refs. [25, 26]).

Journal of Physics: Conference Series **161** (2009) 012037 doi:10.1088/1742-6596/161/1/012037

Summing up, by measuring their effect on the Brownian motion of a colloidal particle, direct evidence of both attractive and repulsive critical Casimir forces has been provided. Based on the general arguments presented in subsec. 2.2 we expect such forces to act also between two or more colloids immersed in a near-critical mixture. Due to the strong non-additivity of these fluctuation-induced forces we expect interesting many-body effects.

4. Conclusions, perspectives and applications

4.1. Conclusions

In the previous sections we presented an overview of the Casimir effect due to critical fluctuations, highlighting its relation with the analogous effect in QED. We focused on the theoretical predictions which are relevant for the interpretation of the available and possibly forthcoming experiments, especially involving wetting films and colloidal particles. The comparison between the available theoretical and experimental results turns out to be very good. We did not attempt, however, to review of all the currently available relevant theoretical results and approaches, a task which is well beyond the scope of the present contribution. In what follows we mention only some of them.

The discussion in subsec. 2.2 clearly shows that Casimir-like effects are expected whenever the spectrum of fluctuations of a certain nature is changed by the presence of confining bodies, which consequently experience fluctuation-induced forces. The range of such forces is typically set by the correlation length ξ of the relevant fluctuations and a *universal* behavior has generally to be expected whenever $\xi, L \gg \ell_{\mathrm{micr}}$ where L is the typical distance between the boundaries and ℓ_{micr} a microscopic, system-dependent scale.

4.2. Perspectives

In view of the quite general conditions under which such Casimir forces arise, several examples can be found in statistical physics. They occur not only at the critical points previously discussed, or in systems belonging to other bulk and surface universality classes (e.g., liquid crystals or complex fluids [7, 28]), but also close to *tri*critical points. This instance has been studied both experimentally and theoretically [29, 18]. Due to the fact that the upper critical spatial dimensionality for a tricritical point is three, a mean-field approximation of suitable models captures rather well the features observed in experiments. Casimir-like forces result also from (long-ranged) fluctuations occurring in non-equilibrium steady states of different nature, from chemical reactions to granular matter, see, e.g., ref. [30], or from quantum fluctuations of the order parameter of a quantum phase transition [11]. However, at present, it is not clear how these predictions can be experimentally tested. On the other hand, there are experimentally relevant systems, such as Bose gases close to the Bose-Einstein condensation, which have not yet been studied with the aim of detecting critical Casimir forces, even though some quantitative theoretical predictions (see fig. 3(b)) are available beyond the case of ideal gases [31]. Thermal capillary fluctuations of the interface between two different liquids give rise to a Casimir force acting on two or more colloidal particles trapped at that interface [32]. In this case, the presence of the colloids alters the spectrum of these surface fluctuations, with the additional feature that the boundaries themselves at which the boundary conditions are imposed, represented by the contact line of the interface at the surface of the colloids, fluctuate (see ref. [32] for details). The resulting forces might be of relevance for understanding the topical problem of the interaction among colloids trapped at interfaces.

So far we have only considered homogeneous substrates and particles. However, in the presence of chemical or topographical modulations, such as the one depicted in fig. 7, novel phenomena are expected, including the emergence of lateral critical Casimir forces $F_{\|}$ and torques τ_{Cas}, still characterized by a universal scaling behavior. Some of the features of these forces have been studied in ref. [33] and might form the basis for the quantitative understanding of recent

 doi:10.1088/1742-6596/161/1/012037

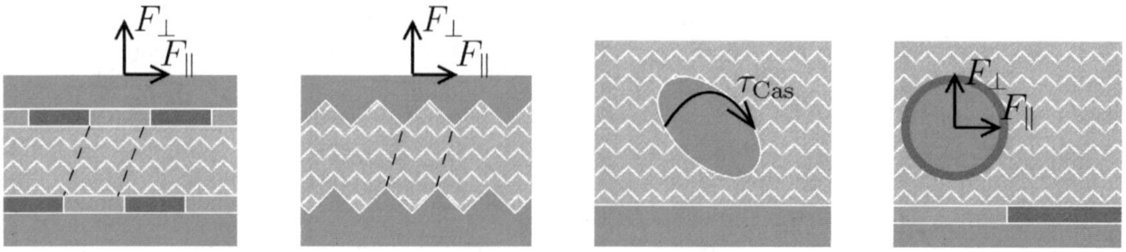

Figure 7. Possible chemically [(a), (d)] and topographically [(b)] patterned substrates in which a lateral critical Casimir force $F_{\|}$ arises in addition to the normal force $F_{\perp}$. A critical Casimir torque τ_{Cas} is expected to act on non-spherical particles [(c)].

experimental investigation of the behavior of a single colloidal particle immersed in a binary mixture and exposed to a patterned substrate [34].

In addition to the equilibrium static behavior of the critical Casimir effect, its equilibrium and non-equilibrium dynamics pose both theoretical and experimental challenges. Indeed, it is not even obvious how to define theoretically such forces when their definition based on the decomposition of the effective free energy $\mathcal{F}$ (see eq. (5)) is no longer viable due to the lack of a proper notion of free energy for dynamical phenomena. A rich dynamical behavior is actually emerging already in quite simple theoretical relaxational models [35]. In order to make contact with the dynamical information which might come from TIRM experiments, it remains to be seen how the time-dependence of the critical Casimir force affects the Brownian motion of the colloidal particle in conjunction with hydrodynamic and adsorption phenomena.

4.3. Applications?

In contrast with the interactions typically acting among colloids (e.g., electrostatic), critical Casimir forces show a striking temperature dependence, as we have reported in subsec. 3.2. This fact can possibly be exploited in order to control via minute temperature changes the phase behavior and aggregation phenomena in systems with dispersed colloids. In addition, not only the range of interaction can be easily controlled but also the sign of the resulting force. This can be achieved by surface treatments and does not require (as it does for the quantum mechanical Casimir effect) substantial changes or tuning of the properties of the bulk materials which constitute the immersed objects. This property might be exploited in order to neutralize the attractive quantum mechanical Casimir force responsible for the stiction which brings micro-electromechanical systems to a standstill. If these machines would work not in a vacuum but in a liquid mixture close to the critical point, the stiction could be prevented by tuning the critical Casimir force to be repulsive via a suitable coating of the various machine parts. In principle, by using optically removable or controllable coatings, one could very conveniently control the functioning of the microdevice without acting directly on it.

Acknowledgments

I am grateful to C. Bechinger, S. Dietrich, L. Helden, C. Hertlein, A. Maciołek and O. Vasilyev for the stimulating collaborations which lead to some of the results summarized here and to M. Oettel and D. Dantchev for a careful reading of the manuscript.

References

[1] Casimir H B 1948 *Proc. K. Ned. Akad. Wet.* **51** 793

[2] Lamoreaux S K 1997 *Phys. Rev. Lett.* **78** 5
[3] Sparnaay M J 1958 *Physica* **24** 751
[4] Bressi G, Carugno G, Onofrio R and Ruoso G 2002 *Phys. Rev. Lett.* **88** 041804
[5] Lamoreaux S K 2007 *Physics Today* **60** 40
Ball P 2007 *Nature* **447** 772
Lambrecht A 2002 *Physics World* **15** 29
[6] Fisher M E and de Gennes P G 1978 *C. R. Acad. Sci. Paris Ser. B* **287** 207
[7] Kardar M and Golestanian R 1999 *Rev. Mod. Phys.* **71** 1233
[8] Binder K 1983 *Phase Transitions and Critical Phenomena* vol 8 eds C Domb and J L Lebowitz (London: Academic) p 1
[9] Diehl H W 1986 *Phase Transitions and Critical Phenomena* vol 10 eds C Domb and J L Lebowitz (London: Academic) p 76
[10] Krech M 1994 *Casimir Effect in Critical Systems* (Singapore: World Scientific)
1999 *J. Phys.: Condens. Matter* **11** R391
[11] Brankov G, Tonchev N S and Danchev D M 2000 *Theory of Critical Phenomena in Finite-Size Systems* (Singapore: World Scientific)
[12] Dietrich S 1988 *Phase Transitions and Critical Phenomena* vol 12 eds C Domb and J L Lebowitz (New York: Academic) p 1
[13] Krech M and Dietrich S 1992 *Phys. Rev. A* **46** 1922
[14] Nightingale M P and Indekeu J O 1985 *Phys. Rev. Lett.* **54** 1824
Indekeu J O 1986 *J. Chem. Soc. Faraday Trans. II* **82** 1835
[15] Pelissetto A and Vicari E 2002 *Phys. Rep.* **368** 549
[16] Vasilyev O, Gambassi A, Maciołek A and Dietrich S 2007 *EPL* **80** 60009
Preprint arXiv:0812.0750 (2008)
[17] Hucht A 2007 *Phys. Rev. Lett.* **99** 185301
[18] Maciołek A, Gambassi A and Dietrich S 2007 *Phys. Rev. E* **76** 031124
[19] Zandi R, Shackell A, Rudnick J, Kardar M and Chayes L P 2007 *Phys. Rev. E* **76** 030601(R)
Zandi R, Rudnick J and Kardar M 2004 *Phys. Rev. Lett.* **93** 155302
[20] Garcia R and Chan M H W 1999 *Phys. Rev. Lett.* **83** 1187
Ganshin A, Scheidemantel S, Garcia R and Chan M H W 2006 *Phys. Rev. Lett.* **97** 075301
[21] Krech M 1997 *Phys. Rev. E* **56** 1642
[22] Fukuto M, Yano Y F and Pershan P S 2005 *Phys. Rev. Lett.* **94** 135702
[23] Rafaï S, Bonn D and Meunier J 2007 *Physica A* **386** 31
[24] Prieve D C 1999 *Adv. Colloid Interf. Sci.* **82** 93
[25] Gambassi A, Maciołek A, Hertlein C, Nellen U, Helden L, Bechinger C and Dietrich S in preparation (2008)
[26] Hertlein C, Helden L, Gambassi A, Dietrich S and Bechinger C 2008 *Nature* **451** 172
[27] Gülari E, Collings A F, Schmidt R L and Pings C J 1972 *J. Chem. Phys.* **56** 6169
[28] Ziherl P, Podgornik R and Žumer S 1998 *Chem. Phys. Lett.* **295** 99
Bartolo D, Long D and Fournier J-B 2000 *Europhys. Lett.* **49** 729
Uchida N 2001 *Phys. Rev. Lett.* **87** 216101
[29] Krech M and Dietrich s 1992 *Phys. Rev. A* **46** 1886
Garcia R and Chan M H W 2002 *Phys. Rev. Lett.* **88** 086101
Maciołek A and Dietrich S 2006 *Europhys. Lett.* **74** 22
[30] Buenzli P R and Soto R 2008 *Phys. Rev. E* **78** 020102(R)
Buenzli P R contribution to this volume
Brito R, Soto R and Marini Bettolo Marconi U 2007 *Granular Matter* **10** 29
[31] Gambassi A and Dietrich S 2006 *Europhys. Lett.* **74** 754
Martin P A and Zagrebnov V A 2006 *Europhys. Lett.* **73** 15
[32] Lehle H, Oettel M and Dietrich S 2006 *Europhys. Lett.* **75** 174
Lehle H and Oettel M 2007 *Phys. Rev. E* **75** 011602
[33] Golestanian R, Ajdari A and Fournier J-B 2001 *Phys. Rev. E* **64** 022701
Karimi Pour Haddadan F, Schlesener F and Dietrich S 2004 *Phys. Rev. E* **70** 041701
Karimi Pour Haddadan F and Dietrich S 2006 *Phys. Rev. E* **73** 051708
Sprenger M, Schlesener F and Dietrich S 2006 *J. Chem. Phys.* **124** 134703
Tröndle M, Harnau L and Dietrich S 2008 *J. Chem. Phys.* **129** 124761
[34] Soyka F, Zvyagolskaya O, Hertlein C, Helden L and Bechinger C 2008 *Phys. Rev. Lett.* **101** 208301
[35] Gambassi A and Dietrich S 2006 *J. Stat. Phys.* **123** 929
Gambassi A 2008 *Eur. Phys. J. B* **64** 379

Journal of Physics: Conference Series **161** (2009) 012038 doi:10.1088/1742-6596/161/1/012038

A frustrated non-contact rack-pinion-rack device

MirFaez Miri
Institute for Advanced Studies in Basic Sciences, Zanjan, 45195-1159, Iran
E-mail: miri@iasbs.ac.ir

Ramin Golestanian
Department of Physics and Astronomy, University of Sheffield, Sheffield S3 7RH, UK
E-mail: r.golestanian@sheffield.ac.uk

Abstract. A design is proposed for a mechanical device made of a nanoscale pinion sandwiched without contact between two racks that exert opposing forces rooted in the quantum fluctuations of the electromagnetic field via the lateral Casimir force. The built-in frustration in the design of the system helps it to react dramatically to minute changes in the geometrical features, which suggests that it could act as a good sensor. The noncontact nature of this device could help solve the infamous wear problem in nanoscale mechanical devices.

1. Introduction

The advances in the technology of making small mechanical devices is rapidly taking us to the point where we need to worry about the tiny components in the machines wearing out too quickly [1] or sticking to each other and jamming the machines [2, 3, 4]. While a number of different strategies are being pursued to tackle these tribological issues [5], one appealing possibility is to try and take advantage of physical mechanisms of force transduction that do not involve direct contact between the components of the system.

Recently, proposals have been put forward to use the Casimir force [6, 7] for such non-contact mechanical interactions. These include both the standard normal Casimir force [8, 9] and the lateral Casimir force that appears in systems with geometrical features such as surface corrugations [10, 11]. Since the normal Casimir force is susceptible to instabilities that might be difficult to avoid in machines with small parts [2], one might argue that the lateral Casimir force between corrugated surfaces offers a more promising prospect for such applications.

Motivated by this promise, a number of designs have been recently introduced for model mechanical systems that would work at the nano-scale using the lateral Casimir force as the main force transduction mechanism. These systems include a rack-and-pinion device with the rack moving at constant speed in one direction [12], a pinion in front of a laterally vibrating rack [13], a ratchet system with a modulating gap separation [14], and a more complicated system with two racks and one pinion [15]. These designs exhibit a plethora of novel and often nonintuitive behaviors, which could help inspire the development of alternative routes to mechanical engineering at the nano-scale. Alongside with the general development of ideas for mechanical devices using the lateral Casimir force, there has also been recent significant improvements in the field theoretical calculation of the magnitude of these forces [16, 17, 18].

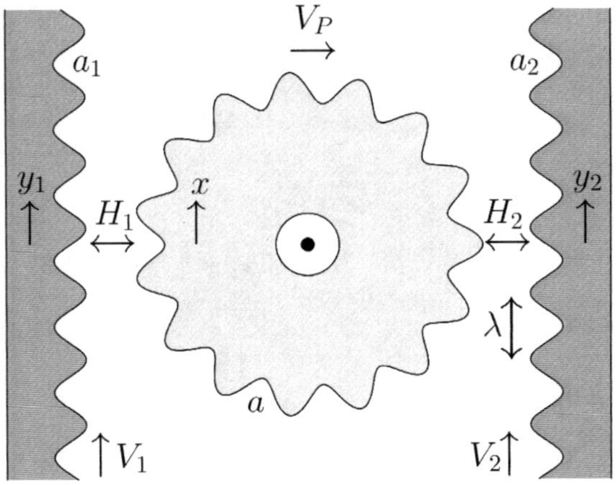

Figure 1. The schematics of the rack-pinion-rack device. The first (second) rack and the pinion have sinusoidal corrugations of wavelength λ and amplitude a_1 (a_2), and a, respectively. The choice of parallel rack velocities V_1 and V_2 (rather than opposite) frustrates the system, which is only possible because of the noncontact design. The pinion velocity V_P is taken as positive if it is in the direction shown, as a convention.

The calculation of the Casimir force is generally going to be a challenge for non-ideal geometries that might occur in the components of the small devices of the future, and we are still in need of developing theoretical formulations that can tackle this issue effectively [19].

Here, we report an extensive analysis of the behavior of a frustrated device made with two racks that are moving in the same direction, and a pinion that is coupled to them via the lateral Casimir force, as shown schematically in Fig. 1 [15]. The device has a number of geometrical features that can be controlled as parameters. Since the parameter space of the general asymmetric case is very large, we first focus on the fully symmetric device and analyze it in the dissipative regime, which is practically more relevant than the inertial regime both because it can be realized more easily and also because it will not exhibit chaotic behavior. We then focus on the heavily damped regime of the general asymmetric device, and find that the system could have five distinct behaviors: (i) the pinion could be locked with either rack-1 or rack-2, (ii) the pinion could move along with either rack-1 or rack-2 but with a lower average velocity, and (iii) the pinion could oscillate back and forth without choosing to go with either of the racks. The oscillatory regime could be used to generate a clock signal of tunable frequency.

The rest of the paper is organized as follows. Section 2 describes the general dynamical formulation of the motion of the pinion, and Sec. 3 analyzes the motion in the fully symmetric case where the problem is analytically tractable. This is followed by a report on numerical analysis of the system in it general geometrical form in Sec. 4, and some discussions in Sec. 5.

2. The Equation of Motion

The system (see Fig. 1) consists of two corrugated plates (racks) with corrugation amplitudes a_1 and a_2, separated by gaps of H_1 and H_2 from a corrugated cylinder (pinion) with corrugation amplitude a. The pinion experiences a lateral Casimir force of the form $F_{\rm lateral} = F \sin\left[2\pi(x-y)/\lambda\right]$ from each of the racks [10], where x and y represent the lateral positioning of the surfaces and λ is the wavelength of the corrugations. Note that the wavelength of the corrugations must be the same on all the three surfaces so that coherent coupling is possible. The amplitude of the lateral Casimir force F, which we can call the *Casimir grip*, depends on the geometric characteristics of the device and in particular the gap size and the amplitudes of corrugations [11, 12, 13, 20, 21, 22]. The Casimir forces from the two surface contacts add up to exert a net torque of $-RF_1 \sin\left[2\pi(x-y_1)/\lambda\right] - RF_2 \sin\left[2\pi(x+y_2)/\lambda\right]$ on the pinion, which plays a key role in the equation of motion for the principal coordinate of $x = R\theta$ that probes the dynamics of the pinion. Here θ is the angle of rotation and R is the radius of the pinion.

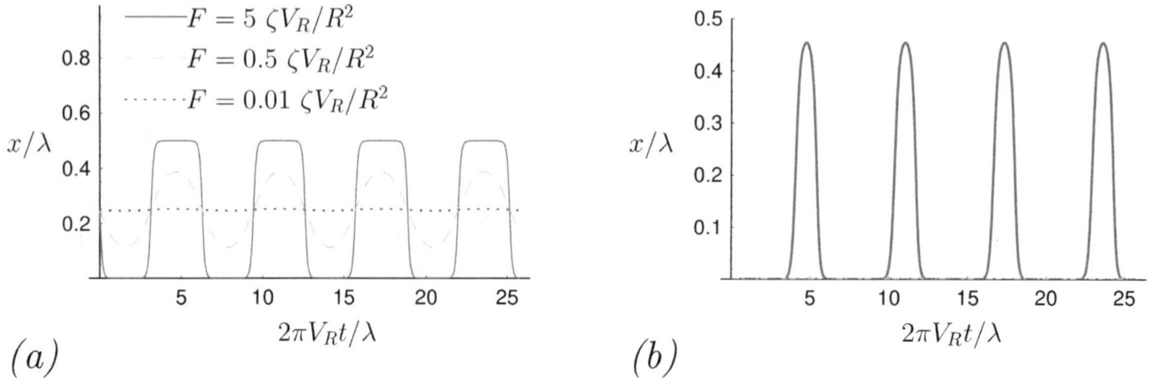

Figure 2. *(a)* Displacement of the pinion cogs as a function of time [from Eq. (4)] for the fully symmetric device, and $x_0 = 0.25\lambda$. For high values of the grip, the motion resembles a square-wave pattern. *(b)* The same (three) plots for $x_0 = 0.0001\lambda$.

Putting $y_1 = V_1 t$ and $y_2 = V_2 t$, the equation of motion reads

$$\frac{I}{R}\frac{d^2x}{dt^2} + \frac{\zeta}{R}\frac{dx}{dt} = -RF_1 \sin\left[\frac{2\pi(x - V_1 t)}{\lambda}\right] - RF_2 \sin\left[\frac{2\pi(x + V_2 t)}{\lambda}\right], \tag{1}$$

where I is the moment of inertia of the pinion about its major axis, and ζ is the rotational friction coefficient. Note that a constant offset in the argument of the sine functions in Eq. (1) due to the positioning of the pinion can be eliminated by a suitable choice of the origin for x and t. The moment of inertia of the pinion can be estimated as $I = \frac{1}{2}MR^2 = \frac{\pi}{2}\rho L R^4$, assuming it is a cylinder of mass M and density ρ, with L being the thickness (or height) of the pinion. We can also estimate the rotational friction coefficient ζ assuming that the main source of friction in the system comes from the lubrication at the axle (or pivot) on which the pinion is mounted. For an axle of radius r that is lubricated with a fluid layer of thickness h and viscosity η, we find $\zeta \simeq 2\pi\eta L r^3/h$.

We can gauge the relative importance of the inertial term and the friction term on the left hand side of Eq. (1), by scaling out the relevant dimensional parameters. We find that the ratio between the inertial term and the friction term is controlled by the dimensionless parameter $I^* \equiv 2\pi I V/(\zeta\lambda)$ where V is a typical rack velocity. We thus find that for $V \ll V_\times$, with the *crossover velocity* being defined as

$$V_\times = \frac{\zeta\lambda}{2\pi I} = \frac{2\eta\lambda r^3}{\pi\rho h R^4}, \tag{2}$$

the dynamics of the system is primarily controlled by the friction term, and the inertial term can be safely ignored. Using $\rho = 1.17$ gr/cm^3 (for silicone), $\eta = 10^{-3}$ Pa.s (for a lubricant as thick as water), and the geometrical parameters as $R = 1$ μm, $\lambda = 500$ nm, $r = 500$ nm, and $h = 100$ nm, we find $V_\times = 0.34$ m/s. In this paper we are going to assume that this condition is fulfilled, and study the dynamics of the system in the heavily damped limit.

The system has four tunable parameters: two rack velocities of V_1 and V_2 that can be directly controlled, and two Casimir grips of F_1 and F_2 that can be tuned by the gap sizes H_1 and H_2. We first consider the fully symmetric case where $V_1 = V_2$ and $F_1 = F_2$ due to its simplicity.

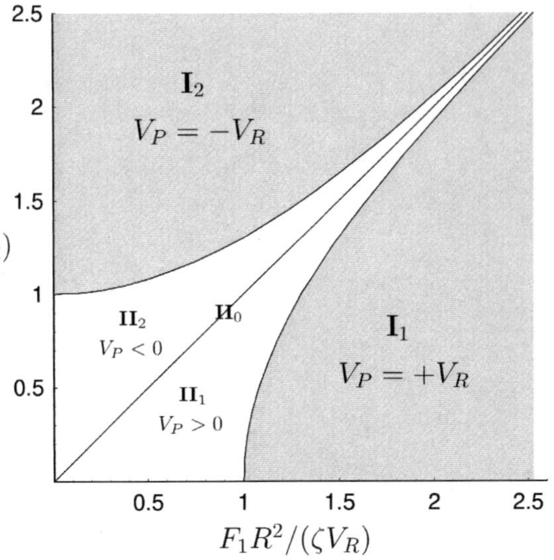

Figure 3. Phase diagram for the motion of the pinion, corresponding to $V_R = V_1 = V_2$. There are five different behaviors: $V_P = +V_R$ locked to the first rack, $V_P > 0$ moving with the first rack with skipping, $V_P = -V_R$ locked to the second rack, and $V_P < 0$ moving with the second rack with skipping. On the line $F_1 = F_2$ the pinion has a neutral oscillatory motion.

3. The Fully Symmetric Case

In the fully symmetric case of $V_R \equiv V_1 = V_2$ and $F \equiv F_1 = F_2$, the equation of motion reads

$$\frac{dx}{dt} = -2\left(\frac{R^2F}{\zeta}\right)\sin\left(\frac{2\pi x}{\lambda}\right)\cos\left(\frac{2\pi V_R t}{\lambda}\right), \tag{3}$$

and is symmetric under the transformation $x \leftrightarrow -x$. It follows that the pinion cannot choose a sense of rotation over the other, and thus performs purely oscillatory motion. Due to its simplicity, in this case the equation of motion can be solved analytically to yield

$$x = \frac{\lambda}{\pi}\tan^{-1}\left\{\tan\left(\frac{\pi x_0}{\lambda}\right)\exp\left[-\frac{2R^2F}{\zeta V_R}\sin\left(\frac{2\pi V_R t}{\lambda}\right)\right]\right\}, \tag{4}$$

where x_0 is the initial position of the pinion. Note that the sign of x_0 determines the sign of the whole solution $x(t)$. Equation (4) is plotted in Fig. 2 for different values of the Casimir grip F and the initial pinion position. It shows that the device can generate interesting periodic patterns such as a nearly perfect square wave or a train of spikes, with a frequency V_R/λ that can be easily controlled by the rack velocity. The solution [Eq. (4)] involves an exponential amplification factor that is controlled by the Casimir grip, which could help detect incredibly small displacements. This can be seen in the example plotted in the inset of Fig. 2, which shows a 10^4-fold amplification for $F = 5\ \zeta V_R/R^2$.

4. The General Case

The equation of motion for a general heavily damped rack-pinion-rack device reads

$$\frac{\zeta}{R^2}\frac{dx}{dt} = -F_1\sin\left[\frac{2\pi(x - V_1 t)}{\lambda}\right] - F_2\sin\left[\frac{2\pi(x + V_2 t)}{\lambda}\right]. \tag{5}$$

Equation (5) is symmetric under the combined transformation of $V_1 \leftrightarrow V_2$, $F_1 \leftrightarrow F_2$, and $x \leftrightarrow -x$. This symmetry greatly simplifies our numerical study of the behavior of the device in its large parameter space.

First we consider the case of equal rack velocities $V_R \equiv V_1 = V_2$. We expect our system to inherit the behavior of the previously studied design of a rack and pinion using the lateral

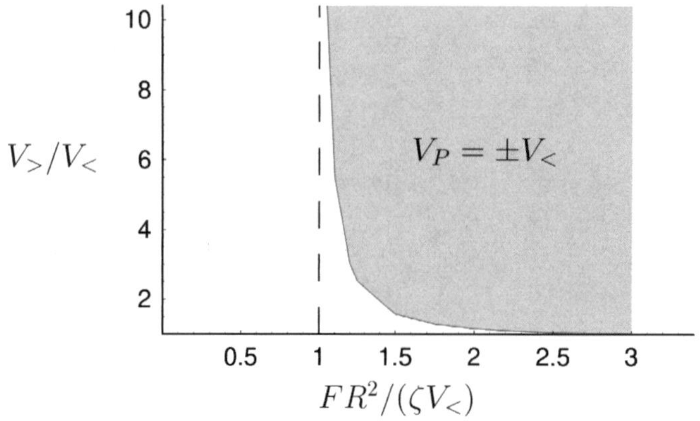

Figure 4. Phase diagram for the motion of the pinion, corresponding to $F \equiv F_1 = F_2$. In the region $V_P = \pm V_<$ the pinion is locked to the rack whose velocity is lower. Note that the phase boundary is about to asymptote to the horizontal line $V_>/V_< = 1$ at high values of $FR^2/(\zeta V_<)$.

Casimir force [12, 22]: The pinion can be locked onto a rack moving at a given velocity, if the friction force for such velocities can be balanced by the available Casimir grip. This means that the Casimir grips F_1 and F_2 introduce two *skipping* velocities $V_{S1} \sim F_1R^2/\zeta$ and $V_{S2} \sim F_2R^2/\zeta$. In the case of $F_2 < F_1$ and $V_R < V_{S1}$ ($F_1 < F_2$ and $V_R < V_{S2}$) it is possible for the pinion to lock onto the first (second) rack. Numerical studies summarized in Fig. 3 confirm our expectation. Figure 3 delineates the domains where the pinion is locked, as well as the phase boundary for the skipping transition. The system exhibits five different behaviors ranging from the pinion being locked to rack-1 (I_1) or rack-2 (I_2) to motion with skipping along rack-1 (II_1) or rack-2 (II_2). The phase boundary between the two II-phases (denoted in Fig. 3 as II_0) corresponds to the vanishing of the net pinion velocity and thus an oscillatory behavior similar to the symmetric device as described by the solution in Eq. (4) and Fig. 2. As expected, this phase diagram is symmetric with respect to the line $F_1 = F_2$.

In the case of equal forces $F \equiv F_1 = F_2$, the pinion follows one of the racks — the one whose velocity is lower — provided that the friction force for such velocities can be balanced by the available Casimir grip. It is natural to introduce $V_> = \text{maximum}(V_1, V_2)$ and $V_< = \text{minimum}(V_1, V_2)$ and present the phase diagram of the system in terms of $V_>/V_<$ and $FR^2/(\zeta V_<)$, although the symmetry of Eq. (5) under the combined transformation of $V_1 \leftrightarrow V_2$ and $x \leftrightarrow -x$ is not manifest. Figure 4 demonstrates the domains where the pinion is locked, as well as the phase boundary for the skipping transition. Note that the phase boundary of Fig. 4 is about to asymptote to a vertical line at small values of F or large values of $V_<$ (i.e. when the skipping velocity is quite small). The phase boundary also asymptotes to the horizontal line $V_>/V_< = 1$ at high values of $FR^2/(\zeta V_<)$. This implies that at large values of the Casimir grip F or small gap size H, the pinion motion is extremely sensitive and can reveal a slight difference in the rack velocities.

We have also studied the behavior of the device in the case of unequal Casimir grips and unequal rack velocities. Figure 5 shows a representative phase diagram for a specific section of the space of parameters. It can be seen that the system again exhibits the same five different behaviors: locked to rack-1 (I_1) or rack-2 (I_2), motion with skipping along rack-1 (II_1) or rack-2 (II_2), or oscillatory motion. The phase boundary between the two II-phases (shown in Fig. 5 as a dashed line and denoted as II_0) corresponds to the vanishing of the net pinion velocity and thus an oscillatory behavior similar to the symmetric device as described by the solution in Eq. (4) and Fig. 2. Note that the phase boundary between I_1 and II_1 in Fig. 5 is about to asymptote to a vertical line at lower values of F_1 below which the phase I_1 cannot exist (corresponding to V_{S1} discussed above). The inset of Fig. 5 shows the average pinion velocity for a typical section of the phase diagram (shown as dash-dotted line in Fig. 5), going from $V_P = -V_2$ to $V_P = V_1$ passing through two sharp transition points and zero. The singular nature of the

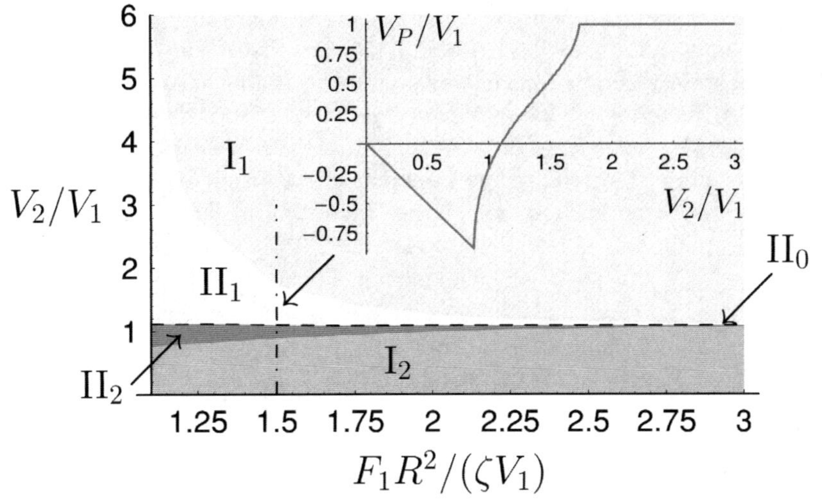

Figure 5. Phase diagram for the motion of the pinion, corresponding to $F_2 = F_1 + 0.05\ \zeta V_1/R^2$. There are five different behaviors: I_1 locked with rack-1, II_1 moving with rack-1 with skipping, II_0 neutral oscillatory motion (dashed line), II_2 moving with rack-2 with skipping, and I_2 locked with rack-2. Inset: pinion velocity versus rack velocity for $F_1 = 1.50\ \zeta V_1/R^2$ and $F_2 = 1.55\ \zeta V_1/R^2$ (corresponding to a vertical cut of the phase diagram at the dash-dotted line) showing the five different regimes.

skipping transitions is manifest in the cusps that appear in the velocity profile at the location of the two transitions.

5. Discussion

The rack-pinion-rack device demonstrates a rich variety of possible behaviors that are all possible thanks to the noncontact nature of the design and the built-in frustration. A key parameter in the system seems to be the skipping transition velocity which determines the limit of velocity that can be tolerated by the system in the locked phase.

We can estimate the skipping velocities using typical values for the Casimir grip, which in its turn is determined by the gap size and other geometrical and material features of the device [11, 12, 13, 20, 21]. For typical (and experimentally realized [23]) values of $a = 50$ nm (assumed for all surfaces) and $\lambda = 500$ nm, we find $F = 0.3$ pN for $H = 200$ nm assuming the boundaries are perfect metals. This yields a skipping velocity of $V_S \sim FR^2/\zeta = 3.8\ \mu$m/s. However, reducing the gap size (by only a factor of **two**) to $H = 100$ nm yields $F = 12$ pN and consequently $V_S \sim FR^2/\zeta = 150\ \mu$m/s (that is enhanced by a factor of **forty**). In practical situations with real metallic boundaries of smaller reflectivity, the Casimir grip tends to be slightly weaker, but the general behavior of the system will still be the same.

The clock signal is a particularly appealing feature of the device, which is the result of perfect balance between the forces in the system. Interestingly, such a signal does not necessarily need to come from the fully symmetric device, and as the phase diagram of Fig. 5 shows it can be obtained by tuning the system into the dashed line (II_0) transition boundary for any geometrical design. To get a strong coupling signal like the square wave or the train of spikes shown in Fig. 2, we need to have a rack velocity of say $V_R = FR^2/(5\zeta)$, which is equal to 30 μm/s for $H = 100$ nm. This yields $f = V_R/\lambda = 60$ Hz using the above value for the corrugation wavelength. Higher rack velocities can lead to higher frequencies, although the shape of the signal will change to a simple sinusoidal one as the coupling gets weaker progressively with increasing rack velocity.

In conclusion,, we have employed the coupling between corrugated surfaces due to the lateral Casimir force to propose a nanoscale mechanical device composed of two racks and a pinion. The noncontact nature of the interaction allows for the system to be made frustrated, which leads to a rich and sensitive phase behavior. Most notably, we have found that the device could be potentially useful as a mechanical sensor or amplifier, and it could also be used to make a mechanical clock signal of tunable frequency. The noncontact nature of this device, and other variants based on similar principles, could help in solving the wear problem in nano-scale mechanical devices.

Acknowledgments

One of us (RG) would like to thank Victor Dodonov and other organizers of the international workshop *60 Years of Casimir Effect* in Brasilia, for creating such a memorable event. It is a pleasure to acknowledge fruitful discussions with R.A.L. Jones. This work was supported by EPSRC under Grant EP/F036167/1.

References

[1] de Boer M P and Mayer T M 2001 *MRS Bull.* **26** 302
[2] Buks E and Roukes M L 2001 *Phys. Rev.* B **63** 033402
[3] Socoliuc A, Gnecco E, Maier S, Pfeiffer O, Baratoff A, Bennewitz R and Meyer E 2006 *Science* **313** 207
[4] Park J Y, Ogletree D F, Thiel P A and Salmeron M 2006 *Science* **313** 186
[5] Carpick R W 2006 *Science* **313** 184
[6] Casimir H B G 1948 *Proc. K. Ned. Akad. Wet.* **51** 793
[7] Bordag M, Mohideen U and Mostepanenko V M 2001 *Phys. Rep.* **353** 1
[8] Chan H B, Aksyuk V A, Kleiman R N, Bishop D J and Capasso F 2001 *Science* **291** 1941
[9] Chan H B, Aksyuk V A, Kleiman R N, Bishop D J and Capasso F 2001 *Phys. Rev. Lett.* **87** 211801
[10] Golestanian R and Kardar M 1997 *Phys. Rev. Lett.* **78** 3421
[11] Chen F, Mohideen U, Klimchitskaya G L and Mostepanenko V M 2002 *Phys. Rev. Lett.* **88** 101801
[12] Ashourvan A, Miri M F and Golestanian R 2007 *Phys. Rev. Lett.* **98** 140801
[13] Ashourvan A, Miri M F and Golestanian R 2007 *Phys. Rev.* E **75** 040103 (R)
[14] Emig T 2007 *Phys. Rev. Lett.* **98** 160801
[15] Miri M F and Golestanian R 2008 *Appl. Phys. Lett.* **92** 113103
[16] Lombardo F C, Mazzitelli F D and Villar P I 2008 *J. Phys. A* **41** 164009
[17] Cavero-Peláez I, Milton K A, Parashar P and Shajesh K V 2008 *Phys. Rev.* D **78** 065018
[18] Cavero-Peláez I, Milton K A, Parashar P and Shajesh K V 2008 *Phys. Rev.* D **78** 065019
[19] For an example, see: Golestanian R 2005 *Phys. Rev. Lett.* **95** 230601
[20] Emig T, Hanke A, Golestanian R and Kardar M 2003 *Phys. Rev. A* **67** 022114
[21] Rodrigues R B, Maia Neto P A, Lambrecht A and Reynaud S 2006 *Phys. Rev. Lett.* **96** 100402
[22] Ashourvan A, Miri M F and Golestanian R 2007 *J. Phys.: Conf. Ser.* **89** 012017
[23] Mohideen U 2008 *private communication*

Journal of Physics: Conference Series **161** (2009) 012039 doi:10.1088/1742-6596/161/1/012039

On the possibility of Casimir repulsion using Metamaterials

F S S Rosa

Theoretical Division, Los Alamos National Laboratory, Los Alamos, NM 87545, USA

E-mail: darosa@lanl.gov

Abstract. It is well known that the Casimir force between two half-spaces is dictated by their electromagnetic properties. In particular, when one of the half-spaces is mainly metallic or dielectric and the other is mainly magnetic, it is possible to show that the force is repulsive. This has attracted lots of interest towards the study of metamterials (MMs) in the context of Casimir effect, as their magnetic activity might help bring the idea of Casimir repulsion from the theoretical realm to experimental verification. In this paper we investigate the possibility of repulsion when the MM magnetic permeability is given not by a Drude-Lorentz behavior, but by a model put forward by Pendry et al. [16]. After introducing the model and deriving the necessary formulas, we show that it is impossible to achieve repulsion with such a model and present a qualitative discussion of why this is so.

1. Introduction

The last decade has witnessed an increased interest in Casimir physics [1, 2] thanks to improved precision measurements [3] of the force between material objects separated by micron and sub-micron gaps. In a few words, the Casimir force may be thought of as a consequence of changing the vacuum fluctuations spectrum through the insertion of material boundaries. While the Casimir force offers new possibilities for nanotechnology, such as actuation mediated by the quantum vacuum, it also presents some challenges, as micro and nanoelectromechanical systems (MEMS and NEMS) may stick together and cease to work due to the attractive nature of van der Waals and Casimir forces. Recent years have also witnessed a huge activity in the development of metamaterials (MMs) [4], boosted by the possibility that such engineered media may give rise to novel optical properties at selected frequency ranges, including negative refraction [5], perfect lensing [6], and cloaking [7], among others. Generally speaking, metamaterials are made of micro and nanostructures carefully designed to collectively present a particular electromagnetic feature. They are able to produce the aforementioned striking phenomena, inaccessible with natural materials, due to the significant magnetic activity built into them, starting at microwave frequencies and going all the way up to the optical range.

Substantial magnetic activity at high frequencies is exactly what brings Casimir physics and metamaterials close together. It is known for a long time that the attractive character of Casimir forces is not universal, that it actually depends crucially on the geometry and composition of a given setup. This naturally leads to speculations about the feasibility of customized systems displaying repusive Casimir forces, or maybe strongly reduced attraction, thus providing a so desired anti-"stiction" effect. An interesting possibility of flipping attraction into repulsion was first demonstrated by Boyer [8], where he showed that a perfect conducting plate repels a perfect magnetic one with vacuum in between. The requirement of having a perfect magnet was relaxed in subsequent works [9], where it was shown that non-ideal materials could give rise to repulsion (provided one was mainly electric and the other mainly magnetic), but even this more realistic scenario is constrained by the absence of natural strong magnets at high frequencies[1]. However, recent developments in nanofabrication have resulted in metamaterials with magnetic response in the visible range of the electromagnetic spectrum [10], fueling the hope for Casimir repulsion [11]. The expectation is that, by tuning this magnetic response to the right frequency range and making it strong enough, one could produce an experimentally measurable Casimir repulsion between, say, a MM slab and a thin metallic plate, or at least a significantly reduced attraction.

The major issue about Casimir repulsion and, more generally, about any degree of Casimir force control is that the Casimir interaction between real dispersive materials is a broadband frequency phenomenon, as shown by the Lifshitz formula expressing the force between two semispaces as an integral over all (imaginary) frequencies with an exponential cut-off c/d [12]. In previous works [13, 14] we have shown that this is a formidable obstacle on the route to Casimir repulsion, as typically the magnetic response of a MM is restricted to a narrow frequency band. In this work we would like to discuss an additional issue, briefly touched in [13] (see also [15]), concerning the modeling of magnetic activity in a given metamaterial. As we shall see below, this is a very important discussion since the Casimir force depends crucially on the choice of the proper model.

2. Metamaterials: a simple model

As our first step, we want to present a simple yet effective procedure of how to characterize the magnetic activity of a metamaterial. This is better illustrated by tackling a specific example, so let us consider the metamaterial depicted in Fig. 1 [16]. It consists on a periodic array of double layered conducting cylindrical sheets, each one being characterized by its radius r, separation s between the sheets and length L. The layers themselves have some structure - they are tinily split apart on opposite sides, in such a way to prevent currents of completing an entire loop (see fig. 2). The array itself has period b, in addition we assume that the cylinders are long ($r/L \ll 1$) and that there is little space between the two layers ($s/r \ll 1$) of a given cylinder. By turning on an external magnetic field $\mathbf{H}(t) = H(t)\hat{z} = H_0 e^{-i\omega t}\hat{z}$ parallel to the cylinders we induce circulating currents on them, that therefore induce a new magnetic field $\mathbf{H}_1$[2]. The total field is therefore

$$\mathbf{H}_T(\mathbf{x}) = \mathbf{H} + \mathbf{H}_I(\mathbf{x}) = (H + H_I(\mathbf{x}))\hat{z}\,, \tag{1}$$

where $\mathbf{x}$ is the position where the field is being evaluated and $H_I(\mathbf{x})$ is still to be determined. In this approximate model, our first assumption is that H_1, despite having different expressions inside and outside a given cylinder, is space independent in each of these two regions. Inside of a given cylinder, the field may be naturally divided into two contributions: the one that originates from that cylinder itself and another that comes from all the other cylinders combined, recalling here that the cylinders are finite in length and as such they "leak" some magnetic field to their surroundings. The first contribution is readily expressed in terms of the induced current per unit of length J with the help of Ampère's law, resulting in $H_I^{(1)} = J/c$. This of course holds exactly only for infinite cylinders, but is an excellent approximation for cylinders sufficiently long. For the second contribution we shall present an heuristic argument, based on the conservation of the number of field lines. We know that the field strength in a given location is roughly proportional

[1] There are plenty of natural strong magnets at low frequencies, but they become magnetically inert at frequencies well below c/d, where d is the distance between the half-spaces.

[2] In principle, this new field induces another current, which in turn induces another field and so on and so forth. These subsequent fields and currents, however, are very small and therefore neglected in this approximate model.

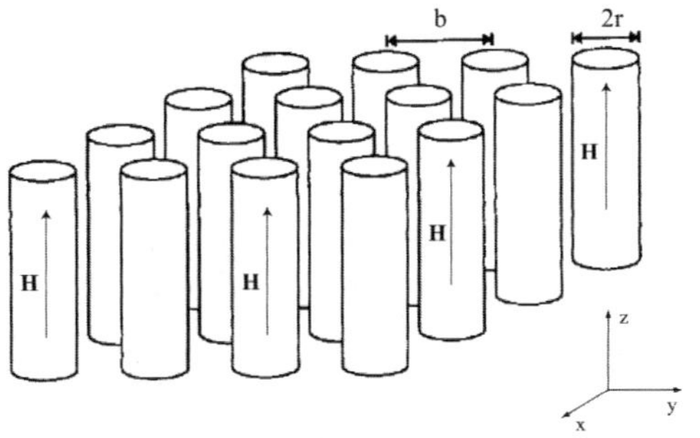

Figure 1. Sketch of the array of cylinders. The distance from center to center (the period) is b and each cylinder has radius r. For the sake of clarity, we don't show the structure of each cylinder in this figure.

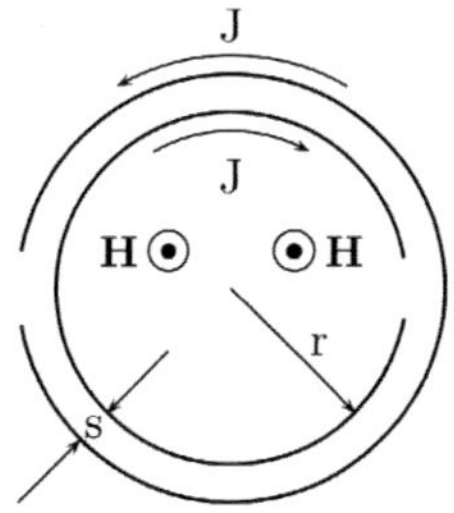

Figure 2. Structure of each cylinder, seen from the top.

to the density of field lines in that location, so the first contribution gives

$$H_I^{(1)} = \kappa\rho_{\rm in} = \frac{\kappa N}{\pi r^2}\,, \tag{2}$$

where N is the number of lines, $\rho_{\rm in}$ is the density of lines and κ is a proportionality constant. By following the same reasoning, in order to get the second contribution we must find the density of field lines "leaked" by all the cylinders, which is simply the density $\rho_{\rm out}$ "leaked" by one cylinder times the number of cylinders $N_{\rm cyl}$

$$|H_I^{(2)}| = \kappa\rho_{\rm out}N_{\rm cyl} = \kappa\frac{N}{A}\frac{A}{b^2} = \kappa\frac{N}{b^2}\,, \tag{3}$$

where A is the (infinite) area of a plane perpendicular to the cylinders and we used the fact that a given (finite) cylinder leaks all of its lines to its surroundings. From (3) we get

$$\frac{|H_I^{(2)}|}{H_I^{(1)}} = \frac{\pi r^2}{b^2}\,, \tag{4}$$

and, knowing that $H_I^{(2)}$ and $H_I^{(1)}$ have opposite signs, we may use our previous results to rewrite the total field inside a cylinder as

$$H_T^{\rm in} = H + \frac{1}{c}J - \frac{1}{c}\frac{\pi r^2}{b^2}J\,. \tag{5}$$

Since the field outside the cylinders may be worked out in a similar way, we shall just quote the result

$$H_T^{\rm out} = H - \frac{1}{c}\frac{\pi r^2}{b^2}J\,, \tag{6}$$

and point out that the only difference is that $H_I^{(1)}$ vanishes in this case. The key advantage of expressions (5) and (6) is that they are written in terms of the current density J, to what now we turn our attention. Each cylinder may be viewed as an RC circuit, and application of Faraday's law in one of them gives

$$\mathcal{E} = 2\pi r\alpha J - \frac{J}{i\omega C} = -\frac{\pi r^2}{c}\frac{\partial}{\partial t}\left[H_0 + \frac{1}{c}J - \frac{1}{c}\frac{\pi r^2}{b^2}J\right] , \tag{7}$$

where where α is the resistivity of the cylinder sheets and $\mathcal{C}$ is the effective circuit capacitance also per unit of length. By assuming that it evolves harmonically in time, i.e., that $J = J_0 e^{-i\omega t}$, we get

$$J = \frac{i\omega H}{\frac{2\pi\alpha c}{\pi r} - \frac{c}{i\omega \mathcal{C}\pi r^2} - \frac{i\omega}{c}\left(1 - \frac{\pi r^2}{b^2}\right)} , \tag{8}$$

which might be used in (5) and (6) to give explicit expressions to the fields. We are not, however, so interested in these absolute expressions as we are in their spatial average, since the latter presumably gives a better account of effective properties, specially at large scales. Starting from the integral form of Faraday's and Ampère's laws

$$\oint_C \mathbf{E}\cdot d\mathbf{l} = -\frac{1}{c}\frac{d}{dt}\int_S \mathbf{B}\cdot d\mathbf{s} \quad , \quad \oint_C \mathbf{H}\cdot d\mathbf{l} = \frac{1}{c}\int_S \mathbf{J}\cdot d\mathbf{s} , \tag{9}$$

we see that the important quantities in these equations are the *flux* of $\mathbf{B}$ and the *circulation* of $\mathbf{H}$. This clearly suggests that $\mathbf{B}$ is to be averaged over a surface while $\mathbf{H}$ should be averaged over a line. For a periodic system we need to average over an unit cell only, so, using the unit cell defined in Fig. 3 and the fact that H_T points in the z-direction, we conclude that a sensible definition of $\mathbf{B}_{ave}$ and $\mathbf{H}_{ave}$ is [16, 17]

$$\mathbf{B}_{\rm ave} = \frac{1}{b^2}\int_{S_z} \mathbf{B}\cdot d\mathbf{s} = \frac{1}{b^2}\int_{S_z} \mathbf{H}_T\cdot d\mathbf{s} = H \tag{10}$$

$$\mathbf{H}_{\rm ave} = \frac{1}{b}\int_{L_z} \mathbf{H}_T\cdot d\mathbf{l} = H_T^{\rm out} , \tag{11}$$

where S_z is either of the two faces perpendicular to the z-axis, and L_z is one of the edges parallel to $\hat{z}$. Now, following [16] and defining the effective magnetic permeability as

$$\mu_{\rm eff}(\omega) = \frac{|\mathbf{B}_{\rm ave}|}{|\mathbf{H}_{\rm ave}|} \tag{12}$$

we get finally

$$\mu_{\rm eff}(\omega) = \frac{H}{H_T^{\rm out}} = 1 - \frac{\pi r^2}{b^2}\frac{\omega^2}{\omega^2 - c^2/\pi r^2\mathcal{C} + 2i\alpha\omega c^2/r} = 1 - f\frac{\omega^2}{\omega^2 - \omega_m^2 + 2i\gamma_m\omega} , \tag{13}$$

where we defined of the filling factor f, the resonant frequency ω_m and the dissipation coefficient γ_m.

There are some peculiarities about the previous formula that we should point out. The first thing is the functional resemblance between expression (13) and the Drude-Lorentz model for the electric permittivity, the only difference between them being the presence of a ω^2 in the numerator of (13). As we shall see in the next section, this leads to interesting consequences in the Casimir force. In addition, it is easy to see that $\mu_{\rm eff}$ does not go to unity in the high frequency limit, as we should expect on physical grounds. This tells us that (13) cannot be valid up to arbitrarily high frequencies, what is makes perfect sense given that all our calculations are based on an effective medium picture that breaks down for very small wavelengths. Our

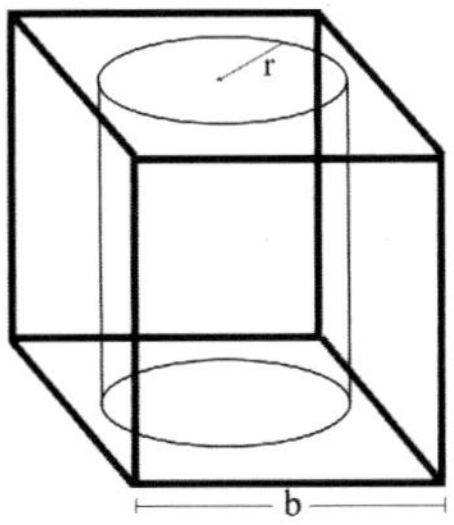

Figure 3. Illustration of the unit cell used in our array. Again, for the sake of simplicity, the cylinder is shown as a single sheet.

final remark is that, despite the problems in the high frequency range, expression (13) is totally acceptable from causality requirements, since it is analytic in the upper half plane of complex frequencies. Indeed, it may be shown that its real and imaginary part are linked together by a slightly modified Kramers-Kronig formulae

$$\epsilon'(\omega) = (1-f) + \frac{2}{\pi}\mathrm{P}\int_0^\infty dy y \frac{\epsilon''(y) - 1}{y^2 - \omega^2},$$
$$\epsilon''(\omega) = -\frac{2\omega}{\pi}\mathrm{P}\int_0^\infty dy \frac{\epsilon'(y) - (1-f)}{y^2-\omega^2} \tag{14}$$

where $\epsilon'(\omega) = \mathrm{Re}\,\epsilon(\omega)$ and $\epsilon''(\omega) = \mathrm{Im}\,\epsilon(\omega)$.

3. The Casimir Force

Now that we have an explicit formula for the magnetic activity of our MM, we may proceed with the evaluation of the Casimir force. There is, however, an important issue we have to deal with before going directly to calculations. The metamaterial described in the previous section is severely anisotropic, and (13) gives the right magnetic response only when $\mathbf{H}$ is parallel to the z-axis. As the Casimir force depends on virtual fluctuations coming from all directions, the inevitable conclusion is that μ_{eff} has to be generalized in order to take arbitrary propagation directions into account. Unfortunately, the simplicity present in the previous section disappears when we depart from the particular case where $\mathbf{H}//\hat{z}$, with effects like polarization mixing and spatial dispersion [18] coming into play. In addition, this high degree of complication obscures all the physics we are trying to analyze, namely, magnetic effects affecting the Casimir force. So, in order to avoid unnecessary distractions, from now on we consider a toy model that is based in the one we have been discussing, but with an isotropic magnetic permeability described by eq. (13) and an isotropic electric permittivity given by

$$\epsilon_{\mathrm{eff}}(\omega) = 1 - \frac{\Omega_e^2}{\omega^2 - \omega_e^2 + i\gamma_e\omega}, \tag{15}$$

where Ω_e, ω_e and γ_e are respectively the electric oscillating strength, resonance frequency and dissipation coefficient.

The Casimir pressure between two real (i.e., non-ideal) isotropic half-spaces, as shown in Fig. 4, is given by the Lifshitz formula [12]

$$\frac{F(d)}{A} = 2\hbar\int_0^\infty \frac{d\xi}{2\pi}\int\frac{d^2\mathbf{k}_\parallel}{(2\pi)^2} K_3 \sum_{j=\mathrm{TE,TM}} \frac{R_1^j R_2^j e^{-2K_3 d}}{1 - R_1^j R_2^j e^{-2K_3 d}}, \tag{16}$$

where $K_3 = \sqrt{k_\parallel^2 + \xi^2/c^2}$ and the reflection coefficients $R_n^{\rm TE}$, $R_n^{\rm TM}$, are given by

$$R_n^{\rm TE}(i\xi, \mathbf{k}_\parallel) = \frac{\mu_n(i\xi)K_3 - \sqrt{k_\parallel^2 + \mu_n(i\xi)\epsilon_n(i\xi)\xi^2/c^2}}{\mu_n(i\xi)K_3 + \sqrt{k_\parallel^2 + \mu_n(i\xi)\epsilon_n(i\xi)\xi^2/c^2}}$$
$$R_n^{\rm TM}(i\xi, \mathbf{k}_\parallel) = \frac{\epsilon_n(i\xi)K_3 - \sqrt{k_\parallel^2 + \mu_n(i\xi)\epsilon_n(i\xi)\xi^2/c^2}}{\epsilon_n(i\xi)K_3 + \sqrt{k_\parallel^2 + \mu_n(i\xi)\epsilon_n(i\xi)\xi^2/c^2}}\,, \tag{17}$$

where ϵ_n and μ_n are respectively the permittivity and permeability of the n-th half-space. In our conventions, a positive (negative) value of the force means attraction (repulsion). We are interested in the case where one of the half-spaces (see fig. 4) is purely metallic, implying in

$$\epsilon_1(\omega) = 1 - \frac{\Omega_1^2}{(\omega^2 + i\gamma_1\omega)} \quad , \quad \mu_1(\omega) = 1\,, \tag{18}$$

where Ω_1 is the metallic plasma frequency and γ_1 is the dissipation coefficient, and the other half-space is a MM characterized by $\epsilon_2(\omega) = \epsilon_{\rm eff}(\omega)$ and $\mu_2(\omega) = \mu_{\rm eff}(\omega)$, given respectively by (15) and (12).

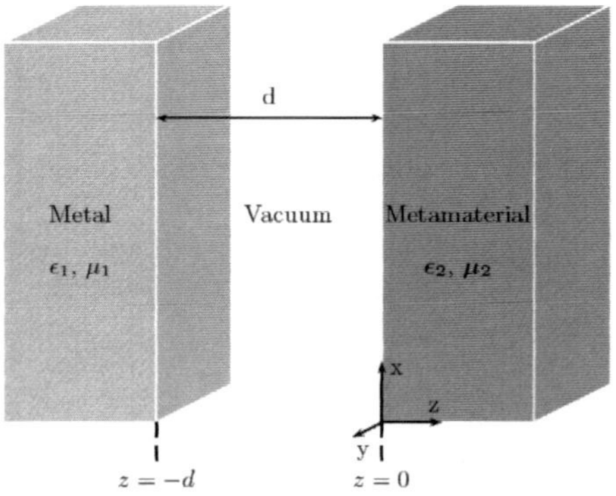

Figure 4. Illustration of the setup used in this paper to calculate the Casimir force.

Before going straight to the results we would like to point out a small subtlety brought by the use of $\mu_{\rm eff}$ in the Lifshitz formula. Buried in expression (16) is the assumption that the reflection coefficients are analytic in the whole upper plane of complex frequencies, which is based itself on the assumption that $\epsilon_{1,2}(i\xi), \mu_{1,2}(i\xi) \geq 0$. It is easy to see that functions $\epsilon_{1,2}(i\xi)$ meet this condition for any choice of parameters, but for $\mu_2(i\xi)$ it only holds when $r/b < 1/\sqrt{\pi}$. Fortunately, in our case we have necessarily $r/b < 1/2 < 1/\sqrt{\pi}$ so we don't have to worry, but this little analysis shows that one must be careful with the validity conditions of the Lifshitz formula when considering alternative permittivity and permeability models.

In Figs. 5 and 6 we compare the Casimir pressure as a function of the separation d for different MMs. We see that when the magnetic part is given by a Drude-Lorentz contribution, namely

$$\mu_{\rm DL}(\omega) = 1 - \frac{\Omega_m^2}{\omega^2 - \omega_m^2 + i\gamma_m\omega}\,, \tag{19}$$

the force decreases at least as fast as $1/d^3$, while for MMs modeled by (13) we have a transition region where F/d^3 actually increases before it goes to zero for large distances. We see also that

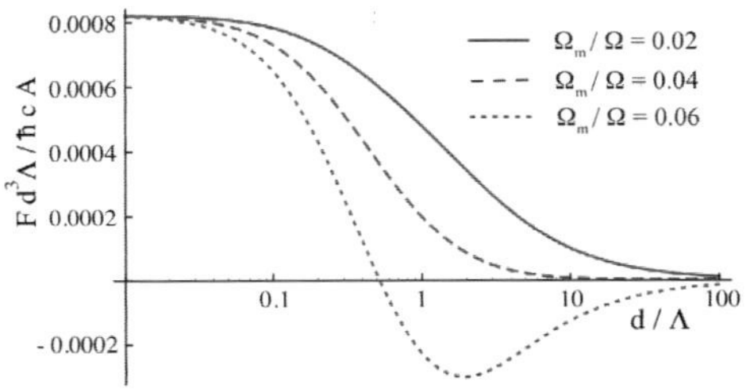

Figure 5. The Casimir pressure between a metallic half-space and a MM modeled by permittivity (15) and permeability (19), for several values of the magnetic oscillating strength Ω_m. The parameters used are $\Omega_1/\Omega = 0.96$, $\gamma_1/\Omega = 0.004$, $\Omega_e/\Omega = 0.04$, $\omega_e/\Omega = \omega_m/\Omega = 0.1$, $\gamma_e/\Omega = \gamma_m/\Omega = 0.005$. The frequency scale Ω is chosen as 1.43×10^{16} rad/s and Λ is defined as $\Lambda = 2\pi c/\Omega$.

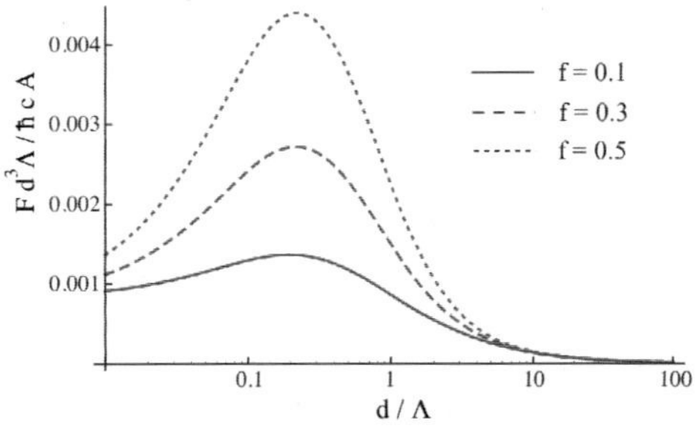

Figure 6. The Casimir pressure between a metallic half-space and a MM modeled by permittivity (15) and permeability (13), for several values of the filling factor f. The parameters used are the same as in figure 5.

in the first scenario the force might even assume negative values for strong enough magnetic activity, while on the second case the enhancement of the magnetic part just makes attraction stronger and stronger in the transitional region. This is a clear indication that it is impossible to achieve repulsion with a MM described by (13), and it is actually not difficult to check that this is indeed the case. As we can see from (16), a necessary (but not sufficient) condition for a negative force to occur is that, for at least one polarization, $R_1^j(i\xi)$ and $R_2^j(i\xi)$ must have different signs. This information is entirely contained in the numerators of both reflection coefficients in (17), and a simple (but tedious) analysis shows that when the permittivities and permeabilities of the materials involved are given by (15), (18) and (13) we have

$$\begin{aligned} \mu_n(i\xi)K_3 - \sqrt{k_\parallel^2 + \mu_n(i\xi)\epsilon_n(i\xi)\xi^2/c^2} \; &< \; 0 \\ \epsilon_n(i\xi)K_3 - \sqrt{k_\parallel^2 + \mu_n(i\xi)\epsilon_n(i\xi)\xi^2/c^2} \; &> \; 0 \quad , \quad \forall \, k_\parallel \, , \, \xi \, , \, n \, , \end{aligned} \tag{20}$$

what prevents any difference in sign between R_1^j and R_2^j. This is to be compared with the case where the MM permeability is described by (19), since in this instance it is possible to have $R_1^{TE}(i\xi) \cdot R_2^{TE}(i\xi) < 0$ and therefore Casimir repulsion is not ruled out.

4. Conclusions

In this paper we discussed how to obtain the magnetic permeability of a specific MM from a theoretical/phenomenological point of view, and how the Casimir force is affected by it. Unfortunately, the kind of magnetic activity we obtained does not lead to Casimir repulsion, nor even to reduced attraction. Incidentally, these results also show that for Casimir purposes, extrapolations of models for electromagnetic properties must be done in a very careful fashion, as these extrapolations may be extremely important to the determination of the Casimir force. We therefore conclude that, as the description of practical MMs is typically known only on a narrowband of frequencies and incidence angles, much has yet to be learned before a definitive statement about the possibility of repulsive Casimir forces can be made.

5. Acknowledgements
I am greatly indebted to D A R Dalvit and P W Milonni for their encouragement and constructive criticism. I would also like to thank J F O'Hara for useful discussions, and acknowledge the support of the U.S. Department of Energy through the LANL/LDRD program for this work.

6. References
[1] Casimir H B G 1948 *Proc. K. Ned. Akad. Wet.* **51**, 793.
[2] Bordag M, Mohideen U and Mostepanenko V M 2001 *Phys. Rep.* **353** 1
Milton K A 2004 *J. Phys. A* **24**, R209
Lamoreaux S K 2005 *Rep. Prog. Phys.* **68** 201
[3] Lamoreaux S K 1997 *Phys. Rev. Lett.* **78**, 5
Mohideen U and Roy A 1998 *ibid.* **81** 4549
Chan H B et al. 2001 *Science* **291** 1941
Bressi G et al. 2002 *Phys. Rev. Lett.* **88** 041804
Decca R S et al. 2003 *ibid.* **91** 050402 (2003)
[4] Ramakrishna S A 2005 *Rep. Prog. Phys.* **68** 449
Shalaev V M and Boardman A 2006 *J. Opt. Soc. Am. B* **23** 386.
[5] Veselago V G 1967 *Sov. Phys. Solid State* **8**, 2854
Smith D R et al. 2000 *Phys. Rev. Lett.* **84** 4184
Shelby R A et al. 2001 *Science* **292** 77.
[6] Pendry J B 2000 *Phys. Rev. Lett.* **85**, 3966 (2000)
[7] Schurig D et al. 2006 *Science* **314** 977
[8] Boyer T H 1974 *Phys. Rev. A* **9**, 2078
[9] Kenneth O et al. 2002 *Phys. Rev. Lett.* **89** 033001
[10] Grigorenko A N et al. 2005 *Nature* **438** 335
Dolling G et al. 2006 *Opt. Lett.* **31**, 1800
Shalaev V M 2007 *Nature Photonics* **1**, 41 (2007)
[11] Henkel C and Joulain K 2005 *Europhys. Lett.* **72**, 929
Leonhardt U and Philbin T G 2007 *New J. Phys.* **9**, 254
Pirozhenko I G and Lambrecht A 2008 *J. Phys. A: Math. Theor.* **41**, 164015
Yang Y et al. 2008 Phys. Rev. A **77** 015803
[12] Lifshitz E M 1955 *Zh. Eksp. Teor. Fiz.* **29** 94 [1956 *Sov. Phys. JETP* **2**, 73]
[13] Rosa F S S, Dalvit D A R and Milonni P W 2008 *Phys. Rev. Lett.* **100** 183602
[14] Rosa F S S, Dalvit D A R and Milonni P W 2008 *Phys. Rev A* **78** 032117
[15] Intravaia F and Henkel C 2007 arXiv:0711.0655
[16] Pendry J B et al. 1999 *IEEE Trans. Micro. Theory Tech.* **47** 2075
[17] Pendry J B and Smith D R 2006 *J. Opt. Soc. Am. B* **23** 391
[18] Belov P A et al. 2003 *Phys. Rev. B* **67** 113103 and references therein

On the manifestation of Casimir effects in intermolecular interactions via the method of induced moments

A Salam

Department of Chemistry, Wake Forest University, Winston-Salem, NC 27109, USA

E-mail: salama@wfu.edu

Abstract. A versatile, alternative method for calculating long-range interactions between atomic and molecular systems within the framework of non-relativistic quantum electrodynamics is the induced multipole moment approach. In the present contribution it is employed to compute the dispersion energy shift between a pair of neutral polarizable molecules in either ground or excited electronic states, and the change in mutual interaction energy caused by the action of an intense external radiation field. In this method the moments induced at each centre are coupled to each other via the resonant electric dipole-electric dipole interaction tensor, giving rise to an energy shift. On evaluating the expectation values for states corresponding to (i) both molecules in the ground state and the field in the vacuum state, (ii) both species excited and the field containing no photons, and (iii) both molecules in the ground state with the field containing *N* photons, respectively result in energy shifts for each of the three processes described above. For the modification of the dispersion interaction by laser light, two distinct mechanisms are identified. One, the dynamic contribution, is proportional to the polarizability of each molecule, and a second, the static term, involves the product of the static moment at one centre and the molecular first hyperpolarizability of the other entity. In both cases the energy shift has a linear dependence on radiation field intensity. After molecular averaging, the static contribution is found to vanish.

1. Background

Sixty years ago Casimir and Polder [1] performed their seminal calculation of the van der Waals dispersion interaction between two neutral polarizable molecules in the ground electronic state. A novel aspect of their approach was the application of the techniques of quantum field theory to the problem at hand. A characteristic feature of this fundamental methodology when applied to the interactions occurring between charged particles, which are known to be electromagnetic in origin, is that the radiation field, in addition to matter, is subject to the rules of quantum mechanics – the formulation being more commonly known as quantum electrodynamics (QED) [2]. A particular consequence of a second quantized description is that the electron emerges from quantization of the matter field, and the photon from quantization of the electromagnetic field.

More specifically, Casimir and Polder employed the minimal-coupling Hamiltonian [3] for the non-relativistic interaction of charged particles with electromagnetic radiation together with time-dependent perturbation theory to evaluate the dispersion energy shift. In this coupling scheme the total Hamiltonian operator for the system of interest comprises a sum of molecular Hamiltonians for the two particles A and B, the radiation field Hamiltonian, and the interaction Hamiltonian coupling the molecules to the electromagnetic field. This last component is given explicitly in minimal-coupling by

$$H_{int}^{min} = -\sum_{\xi}\sum_{\alpha} \frac{e_\alpha}{m_\alpha} \vec{p}_\alpha(\xi) \cdot \vec{a}(\vec{q}_\alpha(\xi)) + \sum_{\xi}\sum_{\alpha} \frac{e_\alpha^2}{2m_\alpha} \vec{a}^2(\vec{q}_\alpha(\xi)) + V_{inter}, \tag{1.1}$$

where e_α is the charge of particle α, of mass m_α, whose position vector is $\vec{q}_\alpha(\xi)$ and linear momentum vector is $\vec{p}_\alpha(\xi)$, belonging to molecule $\xi = A, B$, with $\vec{a}(\vec{r})$ the vector potential at field point $\vec{r}$ in the Coulomb gauge. The final term appearing in Eq. (1.1) is the intermolecular potential energy between molecules ξ and ξ', $V_{inter} = \sum_{\xi<\xi'} V(\xi,\xi')$. The intra-molecular part, $\sum_{\xi} V(\xi)$, of the total electrostatic potential energy, V, is included in the molecular Hamiltonian [3]. This partitioning of the potential energy is facilitated by the choice of gauge, namely the one in which the divergence of the vector potential vanishes, allowing the Coulombic and solenoidal fields to be completely separated. In the leading electric dipole approximation, which is employed throughout this work, the third term of Eq. (1.1) is given by the static dipolar Coulomb interaction

$$V_{inter} = \frac{1}{4\pi\varepsilon_0 R^3} \mu_i(A)\mu_j(B)(\delta_{ij} - 3\hat{R}_i\hat{R}_j), \tag{1.2}$$

where $\mu_i(\xi)$ is the i-th Cartesian component of the electric dipole moment operator of species ξ, with Einstein summation convention being adopted for repeated suffixes, and $\hat{R}$ is the unit separation distance vector between bodies B and A, $\vec{R} = \hat{R}\,|\,\vec{R}_B - \vec{R}_A\,|$. A consequence of the long-wavelength approximation is that there are no spatial variations of the vector potential over the extent of the molecules, so that it is convenient to replace $\vec{a}(\vec{q}(\xi))$ by $\vec{a}(\vec{R}_\xi)$. In a technically involved computation in which all terms correct to the fourth power of electronic charge were retained, and which necessitated contributions of second, third and fourth order in perturbation theory, the energy shift was calculated by Casimir and Polder for the state in which both molecules were in the ground electronic level, $|0^\xi>$, $\xi = A$, B, with no radiation field present [1,4]. The resulting dispersion potential, which now bears their name, has the functional form

$$\Delta E = -\frac{\hbar c}{16\pi^3\varepsilon_0^2} \int_0^\infty du\, u^6 e^{-2uR} \alpha(A;icu)\alpha(B;icu) \left[\frac{1}{u^2R^2} + \frac{2}{u^3R^3} + \frac{5}{u^4R^4} + \frac{6}{u^5R^5} + \frac{3}{u^6R^6}\right] \tag{1.3}$$

and which holds for all R beyond overlap of molecular charge distributions for isotropic A and B as indicated by the presence of the rotationally averaged electric dipole polarizability of species ξ at imaginary frequency $\omega = icu$,

$$\alpha(\xi;icu) = \frac{2}{3}\sum_t \frac{|\vec{\mu}^{t0}(\xi)|^2 E_{t0}}{E_{t0}^2 + (\hbar cu)^2}. \tag{1.4}$$

It is interesting that when R is much less than reduced characteristic molecular transition wavelengths, the energy shift Eq. (1.3) may be approximated in the near-zone to

$$\Delta E^{NZ} = -\frac{1}{24\pi^2\varepsilon_0^2 R^6}\sum_{r,s}\frac{|\vec{\mu}^{0r}(A)|^2|\vec{\mu}^{0s}(B)|^2}{E_{r0}+E_{s0}}, \tag{1.5}$$

which is instantly recognizable as the London dispersion energy formula [5], having R^{-6} dependence. This is the result predicted by semi-classical theory in which the dipolar coupling potential Eq. (1.2) is evaluated in second order of perturbation theory. In this description, light signals are taken to propagate instantaneously between the pair – which is clearly unphysical. Of much greater significance is the asymptotic behaviour of Eq. (1.3) at large intermolecular separation distances. In the far-zone the potential has the limiting form

$$\Delta E^{FZ} = -\frac{23\hbar c}{64\pi^3\varepsilon_0^2 R^7}\alpha(A;0)\alpha(B;0), \tag{1.6}$$

exhibiting an inverse seventh power distance dependence, and which is proportional to the product of static ($\omega \to 0$) polarizabilities of each molecule. The weakening of the potential in the wave-zone relative to the near-zone limit was attributed to the effect of retardation. Electromagnetic signals due to dipolar fluctuations in the charge distribution at one centre propagate at the correct speed of c in the QED treatment, and arrive at the second site after a time delay t-R/c, by which further time the dipole moment in the first species will have reoriented itself compared with its configuration at some initial time. This is one key aspect of phenomena said to exhibit so called "Casimir effects". Another, more widely known feature beautifully illustrated by the calculation of the dispersion force, and also a consequence of quantization of the radiation field, is the interpretation of the dispersion interaction as due to fluctuations of the vacuum electromagnetic field. This is the state of the radiation field in which there is an absence of photons. Yet the energy of the field in such a situation is non-zero, there being a zero-point energy of $\frac{1}{2}\hbar\omega$ associated with each mode of the radiation field of circular frequency ω. While the sum over zero-point energies of all modes is infinite in any finite volume, differences in the zero-point energy of the electromagnetic vacuum with and without the presence of a polarizable body lead to finite results and observable phenomena. Hence the concept of quantum vacuum fluctuations [6] has been a powerful one in atomic, molecular and optical physics, having been used to explain spontaneous emission from systems in excited electronic states and the Lamb shift, in addition to dispersion forces.

In an effort to bypass many of the complications associated with the use of the interaction Hamiltonian in minimal-coupling framework Eq. (1.1) in the calculation of Eq. (1.3), alternative physical viewpoints and simpler computational techniques have been put forward instead. Even though many of the new methods led to the Casimir-Polder potential being derived more easily, the most successful among them retained the key ingredients that give rise to Casimir effects, namely zero-point energy of the field and electromagnetic influences travelling at the speed of light. Chief among these alternatives is the multipolar Hamiltonian [7,8] of molecular QED, whose interaction Hamiltonian is expressed in the electric dipole approximation as

$$H_{int}^{mult} = -\varepsilon_0^{-1}\vec{\mu}(A)\cdot\vec{d}^{\perp}(\vec{R}_A) - \varepsilon_0^{-1}\vec{\mu}(B)\cdot\vec{d}^{\perp}(\vec{R}_B). \tag{1.7}$$

In this framework molecules couple directly to the radiation field as exemplified by the interaction of the electric dipole moment with the transverse electric displacement vector field operator $\vec{d}^{\perp}(\vec{r})$ in Eq. (1.7). The latter is typically expanded as a sum over radiation field modes $(\vec{k},\lambda)$ in terms of annihilation and creation operators $a^{(\lambda)}(\vec{k})$ and $a^{\dagger(\lambda)}(\vec{k})$ that decrease and increase by one the number of photons with direction of propagation vector $\vec{k}$ and index of polarization λ, respectively, according to

$$\vec{d}^{\perp}(\vec{r}) = i \sum_{\vec{k},\lambda} \left(\frac{\hbar c k \varepsilon_0}{2V}\right)^{1/2} [\vec{e}^{(\lambda)}(\vec{k}) a^{(\lambda)}(\vec{k}) e^{i\vec{k}\cdot\vec{r}} - \bar{\vec{e}}^{(\lambda)}(\vec{k}) a^{\dagger(\lambda)}(\vec{k}) e^{-i\vec{k}\cdot\vec{r}}], \tag{1.8}$$

where $\vec{e}^{(\lambda)}(\vec{k})$ is the unit electric polarization vector, and V is the quantization volume. Employing Eq. (1.7) greatly simplifies the calculation of the Casimir-Polder potential. Not only is the intermolecular Coulomb term absent, but H_{int}^{mult} is retarded at the outset, since $\vec{d}^{\perp}(\vec{r})$ in explicitly time dependent form is causal, and no static terms contribute. Further, the dispersion potential is understood as being mediated by the exchange of two virtual photons between A and B, and to leading order requires only the fourth order perturbation theory term for its evaluation [3,9].

Another prominent method within the framework of non-relativistic QED for computing interactions between molecules, in particular dispersion forces, is response theory [10-14]. In this approach, the electronic charge distribution associated with an atomic or molecular particle is expanded as a sum of multipole moments and viewed as a source of electromagnetic radiation. The quantum electrodynamical Maxwell field operators in the proximity of the source are calculated in the Heisenberg representation as a series expansion in powers of the multipole moments. A second, test body then responds via its polarizability to the fields of the first species at the position of the second particle, and vice versa. For isotropic source and test molecules in the electric dipole approximation, the interaction energy is calculated from

$$\Delta E = -\frac{1}{2\varepsilon_0^2} \sum_{\omega} \alpha(A;\omega) \vec{d}^{\perp 2}(B;\vec{R}_A) - \frac{1}{2\varepsilon_0^2} \sum_{\omega} \alpha(B;\omega) \vec{d}^{\perp 2}(A;\vec{R}_B), \tag{1.9}$$

where the expansion of the electric dipole dependent source field is of the form [11]

$$\vec{d}^{\perp}(\xi;\vec{r}) = \vec{d}^{(0)}(\xi;\vec{r}) + \vec{d}^{(1)}(\xi;\vec{r};\vec{\mu}) + \vec{d}^{(2)}(\xi;\vec{r};\vec{\mu}\vec{\mu}) + \ldots, \tag{1.10}$$

where the first term of the series is the vacuum field, independent of the source moment and is given by the interaction picture analogue of Eq. (1.8) with the boson operators specified at the initial time, $\vec{d}^{(1)}(\xi;\vec{r};\vec{\mu})$ is the electric displacement field linear in the electric dipole moment, while the third term of Eq. (1.10) is the $\vec{d}^{\perp}$-field dependent quadratically on $\vec{\mu}$, etc. Explicit functional forms for these three operators, as well as their magnetic field counterparts may be found in Refs. [11,14]. As expected, the Casimir-Polder potential Eq. (1.3) results on taking the expectation value of formula (1.9) over molecular ground electronic states and the vacuum state of the radiation field. Interestingly, terms due to the interference of the zeroth order field with the quadratic field contribute to the dispersion potential. Effects due to retardation are once again automatically included since the displacement field in the neighbourhood of a source is strictly causal [14].

In addition to the alternative approaches briefly reviewed above, several other methods have been developed, mostly within the context of multipolar coupling, in order to shed further insight into the manifestation and computation of the retarded dispersion potential. These include examination of differences in zero-point energy [15], radiation reaction [16], source theory [17,18], covariant scattering theory [19], and changes in the electromagnetic energy density in the vicinity of polarizable molecules [20].

One last method, which explicitly retains the salient features associated with Casimir effects, and which offers a number of calculational advantages, as well as having intuitive physical appeal, is the induced multipole moment approach [21-24]. This computational method is the focus of the present contribution. A brief introduction and outline of the approach is given in the next Section. This is followed by briefly recapitulating its application to the calculation of the Casimir-Polder potential in Section 3. It is then shown in Section 4 how the method can be used in a facile way to calculate the dispersion energy shift when one or both of the molecular pair are in excited states. This is followed in Section 5 by a presentation of the calculation of the change in mutual interaction energy when an external radiation field is applied to a pair of coupled molecules [25] in a straightforward extension of the induced moment method. This application is of current high interest in the context of light-induced optical forces and their subsequent control and manipulation [26].

2. Method of Induced Multipole Moments

The induced multipole moment method has recently proved to be extremely efficient and successful in describing a number of intermolecular interactions [21-24, 27-29]. It is closely related to the idea originally used to calculate London dispersion forces, namely the induction of dipoles by quantum fluctuations, but differs from the picture adopted by Spruch and Kelsey in which the effect of one dipole due to the electromagnetic field of another was evaluated [30]. Underlying the present approach are a few simple concepts and principles. Firstly, that application of an electric displacement field to an electrically polarizable molecule induces an electric dipole moment in leading order. Secondly, the dipoles induced at each molecular centre couple via an interaction potential. Thirdly, taking the expectation value of this product of induced moments and coupling factor over specific molecule and field states gives the energy shift for the particular intermolecular process under study. These steps are first applied to recalculate the Casimir-Polder interaction energy in order to illustrate the applicability of the approach. It should be remarked that the induced multipole moment method to be outlined and applied in the present contribution may be considered complementary to the approach adopted by Compagno, Passante and Persico and co-workers, in which the virtual photon cloud perpetually surrounding an atom is viewed as dressing the source. This field-dressed atom induces a change in the radiation field from which interatomic energy shifts may then be calculated by evaluating correlations of the vacuum electromagnetic field. Details of this technique may be found in Ref. [31].

3. Ground State Dispersion Interaction

Consider an isotropic polarizable molecule ξ positioned at $\vec{R}_\xi$ subject to the action of a transverse displacement field $\vec{d}^{\perp}(\vec{r})$ of mode $(\vec{k},\lambda)$ as given by Eq. (1.8). The i-th component of the induced moment is

$$\mu_i^{ind}(\xi;\vec{k}) = \varepsilon_0^{-1}\alpha(\xi;k)d_i^{\perp}(\vec{R}_\xi;\vec{k},\lambda), \tag{3.1}$$

where $\alpha(\xi;k)$ is the isotropic dynamic polarizability of ξ. The induced dipole moments associated with the two molecules A and B interact through the dipolar coupling potential, resulting in an energy shift

$$\Delta E = \sum_{\vec{k},\lambda} \mu_i^{ind}(A;\vec{k})\mu_j^{ind}(B;\vec{k})\mathrm{Re}V_{ij}(k,\vec{R}), \tag{3.2}$$

where $V_{ij}(k,\vec{R})$ is the single frequency, retarded resonant interaction tensor [32]

$$V_{ij}(k,\vec{R}) = \frac{1}{4\pi\varepsilon_0 R^3}[(\delta_{ij} - 3\hat{R}_i\hat{R}_j)(1-ikR) - (\delta_{ij} - \hat{R}_i\hat{R}_j)k^2R^2]e^{ikR}, \tag{3.3}$$

and $|\vec{R}_B - \vec{R}_A| = R$ as before.

An expression for the energy shift written explicitly in terms of the dynamic electric dipole polarizability of each species is easily obtained on substituting Eq. (3.1) into (3.2). The result is

$$\Delta E = \sum_{\vec{k},\lambda} \varepsilon_0^{-2}\alpha(A;k)\alpha(B;k)d_i^{\perp}(\vec{R}_A;\vec{k},\lambda)d_j^{\perp}(\vec{R}_B;\vec{k},\lambda)\,\mathrm{Re}\,V_{ij}(k,\vec{R}). \tag{3.4}$$

It is worth pointing out that appearing in formula (3.4) is the product of displacement fields at two different positions, which is immediately recognizable as the field-field spatial correlation function. In order to obtain the dispersion potential, the expectation value is taken of relation (3.4) over the ground state of each molecule and the vacuum state of the radiation field. The first of these is rudimentary and yields the molecular ground state polarizability of species $\xi = A, B$, whose form equivalent to the polarizability Eq. (1.4) at real frequency $\omega = ck$ is

$$\alpha(\xi;k) = \frac{2}{3}\sum_t \frac{|\vec{\mu}^{t0}(\xi)|^2 E_{t0}}{E_{t0}^2 - (\hbar ck)^2}. \tag{3.5}$$

For the factor involving the radiation field, the required quantity is

$$< 0(\vec{k},\lambda)\,|\,d_i^{\perp}(\vec{R}_A;\vec{k},\lambda)d_j^{\perp}(\vec{R}_B;\vec{k},\lambda)\,|\,0(\vec{k},\lambda) >= \left(\frac{\hbar ck\varepsilon_0}{2V}\right)e_i^{(\lambda)}(\vec{k})\bar{e}_j^{(\lambda)}(\vec{k})e^{-i\vec{k}\cdot\vec{R}}, \tag{3.6}$$

on employing the operator for the mode specific transverse displacement field Eq. (1.8). The energy shift Eq. (3.4) now becomes

$$\Delta E = \sum_{\vec{k},\lambda}\left(\frac{\hbar ck}{2\varepsilon_0 V}\right)\alpha(A;k)\alpha(B;k)e_i^{(\lambda)}(\vec{k})\bar{e}_j^{(\lambda)}(\vec{k})e^{-i\vec{k}\cdot\vec{R}}\,\mathrm{Re}\,V_{ij}(k,\vec{R}). \tag{3.7}$$

To proceed further, the mode sum has to be carried out. The standard way is to evaluate the polarization index sum using the identity

$$\sum_{\lambda} e_i^{(\lambda)}(\vec{k})\bar{e}_j^{(\lambda)}(\vec{k}) = \delta_{ij} - \hat{k}_i\hat{k}_j, \tag{3.8}$$

convert the wavevector sum to an integral via the prescription $\frac{1}{V}\sum_{\vec{k}} \Rightarrow \frac{1}{(2\pi)^3}\int d^3\vec{k}$, followed by integration over angular variables and integration over wavevector since in spherical polar coordinates the volume element is $d^3\vec{k} = k^2dkd\Omega$, with $d\Omega$ an element of solid angle. For the angular integral, relation (3.9) is used,

$$\frac{1}{4\pi}\int(\delta_{ij} - \hat{k}_i\hat{k}_j)e^{-i\vec{k}\cdot\vec{R}}d\Omega = \left[(\delta_{ij} - \hat{R}_i\hat{R}_j)\frac{\sin kR}{kR} + (\delta_{ij} - 3\hat{R}_i\hat{R}_j)\left(\frac{\cos kR}{k^2R^2} - \frac{\sin kR}{k^3R^3}\right)\right] \tag{3.9}$$

giving for ΔE, the expression

$$\Delta E = -\frac{\hbar c}{16\pi^3\varepsilon_0^2 R^2}\int_0^\infty dk\alpha(A;k)\alpha(B;k)k^4$$
$$\times\begin{bmatrix} 2\sin kR\cos kR + \frac{2}{kR}(\cos^2 kR - \sin^2 kR) - \frac{10}{k^2R^2}\sin kR\cos kR \\ -\frac{6}{k^3R^3}(\cos^2 kR - \sin^2 kR) + \frac{6}{k^4R^4}\sin kR\cos kR \end{bmatrix}, \quad (3.10)$$

after inserting the coupling tensor Eq. (3.3). Simplifying the trigonometric terms and transforming the integral to one over imaginary wavevector $k = iu$ results in the Casimir-Polder potential Eq. (1.3).

4. Coupling Between Excited Molecules

When both molecules are in ground electronic states, only upward transitions to intermediate states are possible. If either one or both of A and B are in excited electronic states, however, downward transitions as well as excitation to higher lying levels can now occur, giving contributions additional to the u-integral term of the Casimir-Polder potential. Let A and B now both be in excited electronic states $|m>$ and $|n>$, respectively making electric dipole allowed transitions to higher or lower lying intermediate states $|r>$ and $|s>$. The u-integral contribution to the total intermolecular interaction energy is similar to Eq. (1.3), having the functional form

$$\Delta E = -\frac{1}{36\pi^3\varepsilon_0^2\hbar c}\sum_{r,s}|\vec{\mu}^{mr}(A)|^2|\vec{\mu}^{ns}(B)|^2$$
$$\times\int_0^\infty duu^6 e^{-2uR}\frac{k_{rm}k_{sn}}{(k_{rm}^2+u^2)(k_{sn}^2+u^2)}\left[\frac{1}{u^2R^2}+\frac{2}{u^3R^3}+\frac{5}{u^4R^4}+\frac{6}{u^5R^5}+\frac{3}{u^6R^6}\right] \quad (4.1)$$

which can be expressed in terms of the excited state polarizability at imaginary frequency of each molecule using Eq. (1.4) with replacement of appropriate state labels. Note that Eq. (4.1) contains contributions from both upward and downward transitions from the initial state of each molecule as exemplified by the unrestricted nature of the summation over intermediate states.

For the contribution to the energy shift due solely to downward transitions, in which a real photon is emitted from the excited state, the expression analogous to Eq. (3.4) is

$$\Delta E^{RES} = \sum_{\vec{k},\lambda}\varepsilon_0^{-2}\alpha(A;k)\alpha(B;k)d_i^{\perp}(\vec{R}_A;\vec{k},\lambda)d_j^{\perp}(\vec{R}_B;\vec{k},\lambda)[V_{ij}^{RES}(k_{ns},\vec{R})+V_{ij}^{RES}(k_{mr},\vec{R})], \quad (4.2)$$

where $V_{ij}^{RES}(k_{pq},\vec{R})$ is the retarded interaction tensor Eq. (3.3) evaluated at the resonant frequency $\omega_{pq} = ck_{pq}$ corresponding to a downward transition $|q> \leftarrow |p>$. The evaluation of Eq. (4.2) is carried out in a manner similar to that used in the calculation of the ground state potential. Now the expectation value is taken over the molecular states $|m^A, n^B>$, with the radiation field in the vacuum state as before, with the expectation value of the field-field spatial correlation function given by Eq. (3.6). Converting the mode sum to an integral produces

$$\Delta E^{RES} = \frac{\hbar c}{16\pi^3\varepsilon_0}\int d^3\vec{k}k(\delta_{ij}-\hat{k}_i\hat{k}_j)e^{-i\vec{k}\cdot\vec{R}}\alpha(A;k)\alpha(B;k)[V_{ij}^{RES}(k_{ns},\vec{R})+V_{ij}^{RES}(k_{mr},\vec{R})]. \quad (4.3)$$

In carrying out the angular average it is convenient to apply a formula alternate to Eq. (3.9)

$$\frac{1}{4\pi}\int(\delta_{ij}-\hat{k}_i\hat{k}_j)e^{-i\vec{k}\cdot\vec{R}}d\Omega=\frac{1}{2ik^3}(-\nabla^2\delta_{ij}+\nabla_i\nabla_j)\frac{1}{R}(e^{ikR}-e^{-ikR}), \tag{4.4}$$

giving,

$$\Delta E^{RES}=\frac{1}{12\pi^2\varepsilon_0 i}(-\nabla^2\delta_{ij}+\nabla_i\nabla_j)\frac{1}{R}$$
$$\times\left\{\begin{array}{l}\int dk\alpha(A;k)\sum\limits_s\frac{k_{sn}}{k_{sn}^2-k^2}|\vec{\mu}^{sn}(B)|^2\,V_{ij}^{RES}(k_{ns},\vec{R})\\+\int dk\alpha(B;k)\sum\limits_r\frac{k_{rm}}{k_{rm}^2-k^2}|\vec{\mu}^{rm}(A)|^2\,V_{ij}^{RES}(k_{mr},\vec{R})\end{array}\right\}(e^{ikR}-e^{-ikR}). \tag{4.5}$$

On performing the k-integral, Eq. (4.5) becomes

$$\Delta E^{RES}=\frac{1}{12\pi\varepsilon_0}(-\nabla^2\delta_{ij}+\nabla_i\nabla_j)\frac{1}{R}\left\{\begin{array}{l}\sum\limits_{\substack{s\\E_n>E_s}}\alpha(A;k_{sn})|\vec{\mu}^{sn}(B)|^2\,e^{-ik_{ns}R}V_{ij}^{RES}(k_{ns},\vec{R})\\+\sum\limits_{\substack{r\\E_m>E_r}}\alpha(B;k_{rm})|\vec{\mu}^{rm}(A)|^2\,e^{-ik_{mr}R}V_{ij}^{RES}(k_{mr},\vec{R})\end{array}\right\}. \tag{4.6}$$

Inserting the alternate form for the resonant coupling tensor,

$$V_{ij}(k,\vec{R})=-\frac{1}{4\pi\varepsilon_0}(-\nabla^2\delta_{ij}+\nabla_i\nabla_j)\frac{e^{ikR}}{R} \tag{4.7}$$

at the appropriate resonant frequency into Eq. (4.6), yields for the additional contribution to the interaction energy the result

$$\Delta E^{RES}=-\frac{1}{48\pi^2\varepsilon_0^2}$$
$$\times\left\{\begin{array}{l}\sum\limits_{\substack{s\\E_n>E_s}}\alpha(A;k_{sn})|\vec{\mu}^{sn}(B)|^2\,[(-\nabla^2\delta_{ij}+\nabla_i\nabla_j)\frac{e^{-ik_{ns}R}}{R}][(-\nabla^2\delta_{ij}+\nabla_i\nabla_j)\frac{e^{ik_{ns}R}}{R}]\\+\sum\limits_{\substack{r\\E_m>E_r}}\alpha(B;k_{rm})|\vec{\mu}^{rm}(A)|^2\,[(-\nabla^2\delta_{ij}+\nabla_i\nabla_j)\frac{e^{-ik_{mr}R}}{R}][(-\nabla^2\delta_{ij}+\nabla_i\nabla_j)\frac{e^{ik_{mr}R}}{R}]\end{array}\right\}, \tag{4.8}$$

which after evaluating the gradients and simplifying produces

$$\Delta E^{RES}=-\frac{1}{24\pi^2\varepsilon_0^2}\sum_{\substack{s\\E_n>E_s}}\alpha(A;k_{sn})|\vec{\mu}^{sn}(B)|^2\,k_{ns}^6\left[\frac{1}{k_{ns}^2R^2}+\frac{1}{k_{ns}^4R^4}+\frac{3}{k_{ns}^6R^6}\right]$$
$$-\frac{1}{24\pi^2\varepsilon_0^2}\sum_{\substack{r\\E_m>E_r}}\alpha(B;k_{rm})|\vec{\mu}^{rm}(A)|^2\,k_{mr}^6\left[\frac{1}{k_{mr}^2R^2}+\frac{1}{k_{mr}^4R^4}+\frac{3}{k_{mr}^6R^6}\right], \tag{4.9}$$

with the total energy shift given by the sum of Eqs. (4.1) and (4.9), and is in agreement with previous perturbation [33] and response [34] theory calculations.

Only the first term of Eq. (4.9) and the u-integral term Eq. (4.1) survive if only B is excited, while if only A is excited the first term of Eq. (4.9) vanishes, with the appropriate ground state polarizability appearing in both situations.

At separation distances R large relative to characteristic molecular transition wavelengths, the dominant contribution to the energy shift when both A and B are excited arises from expression Eq. (4.9), exhibiting inverse square dependence, as to be expected for emission of a real photon from the excited electronic state,

$$\Delta E^{FZ} = -\frac{1}{36\pi^2\varepsilon_0^2(\hbar c)^4 R^2}\sum_{\substack{r,s \\ E_m > E_r \\ E_n > E_s}} \frac{|\vec{\mu}^{rm}(A)|^2|\vec{\mu}^{sn}(B)|^2}{(E_{mr}+E_{ns})} E_{mr}E_{ns}(E_{mr}^2 + E_{mr}E_{ns} + E_{ns}^2), \tag{4.10}$$

while the far-zone limit of the u-integral term Eq. (4.1) has an R^{-7} dependence as given by Eq. (1.6), with excited state polarizabilities replacing their ground state counterparts. In the near-zone, on the other hand, the asymptotic behaviour is R^{-6} and arises from both Eqs. (4.1) and (4.9), their sum yielding

$$\Delta E^{NZ} = -\frac{1}{24\pi^2\varepsilon_0^2 R^6}\sum_{\substack{r,s \\ E_m > E_r \\ E_n > E_s}} \frac{|\vec{\mu}^{rm}(A)|^2|\vec{\mu}^{sn}(B)|^2}{(E_{mr}+E_{ns})}. \tag{4.11}$$

Dispersion potentials involving one or two excited molecules are harder to obtain using diagrammatic perturbation theory due to the presence of resonant energy denominators which arise when a real photon is emitted or absorbed. In the response theory calculation, this aspect is dealt with more easily as the total Maxwell field operators in the proximity of a molecule already include contributions from photons on the energy shell. The induced moment approach, however, proves to be the most facile.

5. Modification of Dispersion Force by External Radiation

It is remarkable that the method of induced moments may be used to calculate the change in dispersion potential between a pair of polarizable molecules when subject to an applied electromagnetic field. In the leading electric dipole approximation, the change in mutual interaction energy is again given by formula (3.2) on omitting the mode sum since the external electric displacement field is mode specific.

If one or both molecules is polar, the induced moment given by expression Eq. (3.1) contains terms additional to that proportional to the electric dipole polarizability when the inducing field is of the electric displacement variety. For the present application these extra contributions include the permanent electric dipole moment of the molecule, $\mu^s(\xi)$, and a contribution dependent on the first hyperpolarizability, $\beta_{ijk}(\xi;k)$. The extension of Eq. (3.1) is now given by

$$\mu_i^{ind}(\xi;k) = \mu_i^s(\xi) + \varepsilon_0^{-1}\alpha_{ij}(\xi;k)d_j^{\perp}(\vec{R}_\xi;\vec{k},\lambda) + \varepsilon_0^{-2}\beta_{ijk}(\xi;k)d_j^{\perp}(\vec{R}_\xi;\vec{k},\lambda)d_k^{\perp}(\vec{R}_\xi;\vec{k},\lambda). \tag{5.1}$$

Inserting Eq. (5.1) into Eq. (3.2) and retaining terms correct up to the fourth power in the electronic charge yields for the energy shift the formula

$$\begin{aligned}\Delta E = \varepsilon_0^{-2}[&\alpha_{ik}(A;k)\alpha_{jl}(B;k)d_k^{\perp}(\vec{R}_A;\vec{k},\lambda)d_l^{\perp}(\vec{R}_B;\vec{k},\lambda)\,\mathrm{Re}V_{ij}(k,\vec{R}) \\ &+\tfrac{1}{2}(\mu_i^s(A)\beta_{jkl}(B;k)d_k^{\perp}(\vec{R}_B;\vec{k},\lambda)d_l^{\perp}(\vec{R}_B;\vec{k},\lambda) \\ &+\beta_{jkl}(A;k)\mu_i^s(B)d_k^{\perp}(\vec{R}_A;\vec{k},\lambda)d_l^{\perp}(\vec{R}_A;\vec{k},\lambda))V_{ij}(0,\vec{R})],\end{aligned} \tag{5.2}$$

where $V_{ij}(k,\vec{R})$ is the retarded coupling tensor given by Eq. (3.3), $V_{ij}(0,\vec{R})$ is its zero frequency form,

$$V_{ij}(0,\vec{R}) = \frac{1}{4\pi\varepsilon_0 R^3}(\delta_{ij} - 3\hat{R}_i\hat{R}_j), \tag{5.3}$$

and represents static (k independent) coupling. As in the computation of the dispersion potential between systems in the ground electronic state, the expectation value of the operator for the energy shift involving matter is taken over the state $|0^A,0^B>$. Instead of taking the expectation value over the vacuum state of the radiation field, as was done for the calculation of the Casimir-Polder potential in Section 3, now the expectation value of the factor involving the product of the electric displacement fields is taken over a state of the field which contains N photons of mode $(\vec{k},\lambda)$, $|N(\vec{k},\lambda)>$. The extension of Eq. (3.6) is easily seen to be

$$\begin{aligned} &< N(\vec{k},\lambda)| d_i^{\perp}(\vec{R}_A;\vec{k},\lambda) d_j^{\perp}(\vec{R}_B;\vec{k},\lambda)| N(\vec{k},\lambda)> \\ &= \left(\frac{\hbar ck\varepsilon_0}{2V}\right)[(N+1)e_i^{(\lambda)}(\vec{k})\bar{e}_j^{(\lambda)}(\vec{k})e^{-i\vec{k}\cdot\vec{R}} + N\bar{e}_i^{(\lambda)}(\vec{k})e_j^{(\lambda)}(\vec{k})e^{i\vec{k}\cdot\vec{R}}], \end{aligned} \tag{5.4}$$

with $\vec{R}$ as defined earlier to be the separation distance vector of B relative to A. The quantity Eq. (5.4) is used in the first term of Eq. (5.2). Interestingly, Eq. (5.4) reduces to Eq. (3.6) on letting $N = 0$. For the radiation field factor occurring in the second part of Eq. (5.2), where the product of the fields is evaluated at the same point in space, the relevant expectation value is

$$< N(\vec{k},\lambda)| d_i^{\perp}(\vec{r};\vec{k},\lambda) d_j^{\perp}(\vec{r};\vec{k},\lambda)| N(\vec{k},\lambda)> = \left(\frac{\hbar ck\varepsilon_0}{2V}\right)[(N+1)e_i^{(\lambda)}(\vec{k})\bar{e}_j^{(\lambda)}(\vec{k}) + N\bar{e}_i^{(\lambda)}(\vec{k})e_j^{(\lambda)}(\vec{k})], \tag{5.5}$$

which is a special case of Eq. (5.4) in which $\vec{R}_B = \vec{R}_A$, giving $\vec{R} = 0$. Thus evaluating Eq. (5.2) over the state $|0^A,0^B;N(\vec{k},\lambda)>$ and using the last two relations produces for the radiation-induced intermolecular energy shift the expression

$$\Delta E = \frac{I}{2\varepsilon_0 c} e_k^{(\lambda)}(\vec{k})\bar{e}_l^{(\lambda)}(\vec{k}) \left\{ \begin{array}{l} \alpha_{ik}(A;k)\alpha_{jl}(B;k)(e^{i\vec{k}\cdot\vec{R}} + e^{-i\vec{k}\cdot\vec{R}})\mathrm{Re}V_{ij}(k,\vec{R}) \\ +[\mu_i^{00}(A)\beta_{jkl}(B;k) + \beta_{jkl}(A;k)\mu_i^{00}(B)]V_{ij}(0,\vec{R}) \end{array} \right\}, \tag{5.6}$$

which can be decomposed as

$$\Delta E = \Delta E_{dyn} + \Delta E_{stat}, \tag{5.7}$$

where the so-called "dynamic" contribution to the interaction energy – for reasons which will become clear later on, is given by

$$\Delta E_{dyn} = \frac{I}{\varepsilon_0 c} e_k^{(\lambda)}(\vec{k})\bar{e}_l^{(\lambda)}(\vec{k})\alpha_{ik}(A;k)\alpha_{jl}(B;k)\cos(\vec{k}\cdot\vec{R})\mathrm{Re}V_{ij}(k,\vec{R}), \tag{5.8}$$

and similarly the "static" term is

$$\Delta E_{stat} = \frac{I}{2\varepsilon_0 c} e_k^{(\lambda)}(\vec{k})\bar{e}_l^{(\lambda)}(\vec{k})[\mu_i^{00}(A)\beta_{jkl}(B;k) + \beta_{jkl}(A;k)\mu_i^{00}(B)]V_{ij}(0,\vec{R}). \tag{5.9}$$

In these results, the irradiance of the incident laser beam is defined by $I = N\hbar c^2 k / V$. The frequency dependent polarizability tensor appearing in the dynamic contribution is given by

$$\alpha_{ij}(\xi;k) = \sum_r \left\{ \frac{\mu_i^{0r}(\xi)\mu_j^{r0}(\xi)}{E_{r0} - \hbar ck} + \frac{\mu_j^{0r}(\xi)\mu_i^{r0}(\xi)}{E_{r0} + \hbar ck} \right\}, \tag{5.10}$$

and the molecular first hyperpolarizability tensor that contributes to the static term is

$$\begin{aligned}\beta_{ijk}(\xi;k) = \sum_{r,s} \{ & \frac{\mu_i^{0s}(\xi)\mu_j^{sr}(\xi)\mu_k^{r0}(\xi)}{(E_{s0} + \hbar ck)E_{r0}} + \frac{\mu_i^{0s}(\xi)\mu_k^{sr}(\xi)\mu_j^{r0}(\xi)}{(E_{r0} + \hbar ck)(E_{s0} + \hbar ck)} + \frac{\mu_j^{0s}(\xi)\mu_i^{sr}(\xi)\mu_k^{r0}(\xi)}{(E_{s0} - \hbar ck)E_{r0}} \\ & + \frac{\mu_j^{0s}(\xi)\mu_k^{sr}(\xi)\mu_i^{r0}(\xi)}{(E_{r0} - \hbar ck)(E_{s0} - \hbar ck)} + \frac{\mu_k^{0s}(\xi)\mu_i^{sr}(\xi)\mu_j^{r0}(\xi)}{(E_{r0} + \hbar ck)E_{s0}} + \frac{\mu_k^{0s}(\xi)\mu_j^{sr}(\xi)\mu_i^{r0}(\xi)}{(E_{r0} - \hbar ck)E_{s0}} \},\end{aligned} \tag{5.11}$$

with $\mu_i^{00}(\xi)$ the *i*-th component of the ground state static dipole moment. Note that the form of Eq. (5.6) allows the cases in which *A* and *B* are the same or different species to be treated. Moreover, ΔE_{stat} is zero if either *A* or *B* are centrosymmetric since μ_i^{00} and β_{ijk} are then both vanishing quantities.

The result (5.6) has been obtained straightforwardly using the induced moment approach, and with considerably less arduousness than if carried out using diagrammatic perturbation theory techniques. In the latter method, each of the contributions to Eq. (5.6) are obtained by summing over forty-eight time-ordered diagrams [25,27,29] or state sequence pathways [35], with the initial and final states being equal to each other, and which are also the same for both mechanisms, namely $|i\rangle = |f\rangle = |0^A, 0^B; N(\vec{k},\lambda)\rangle$. In one mechanism, termed the dynamic one, scattering of a real photon takes place at different centres, with a virtual photon propagating between the pair, carrying electronic excitation energy with it. Hence two (real and virtual photon) emission events, or two absorption events, or one of each kind occur at both *A* and *B*, and the overall contribution requires fourth order perturbation theory for its evaluation, giving rise to term (5.8) which is proportional to the electric dipole polarizability of each molecule. For the second term contributing to the energy shift in Eq. (5.7), the fourth order formula of perturbation theory may again be employed. This time absorption and emission of a real photon from the incident laser field occurs at the same site, with a virtual photon again migrating between *A* and *B*. Because no energy is relayed in this transfer process, the mechanism is called the static one, and $V_{ij}(0,\vec{R})$ appears. In the language of time-dependent perturbation theory, since there are three electric dipole interaction vertices at one centre, and one at the other, and vice versa, ΔE_{stat} is proportional to the product of the first hyperpolarizability tensor and the permanent moment, as given explicitly by Eq. (5.9).

The energy shift result Eq. (5.6) holds for all separation distances *R* beyond overlap of molecular charge distributions for anisotropic *A* and *B*, and with the molecular pair having a fixed orientation relative to the direction of motion of the applied electric displacement field. For *A* and *B* randomly oriented with respect to each other, a molecular average can be carried out using standard tensor averaging techniques [36]. ΔE_{stat} is found to vanish after such a procedure. Further, for freely tumbling pairs, as occurs in the fluid phase, an average over all directions of $\vec{R}$ for a fixed $\vec{k}$, or vice versa must be performed. Carrying out both of the averages in the order stated on the remaining dynamic contribution to the energy shift Eq. (5.8) results in

$$\Delta E_{dyn} = -\frac{I}{8\pi\varepsilon_0^2 cR^3}\alpha(A;k)\alpha(B;k)[2kR\sin kR\cos kR + 2(\cos^2 kR - \sin^2 kR) - \frac{10}{kR}\sin kR\cos kR$$
$$-\frac{6}{k^2R^2}(\cos^2 kR - \sin^2 kR) + \frac{6}{k^3R^3}\sin kR\cos kR], \tag{5.12}$$

which is independent of polarization, is linearly proportional to the intensity of the incident beam, and agrees with the result obtained using perturbation theory [25,35]. From Eq. (5.12) it is seen that the force induced by the laser is an attractive one, and constitutes an optical binding force. The asymptotic limits are readily derived from Eq. (5.12). In the near-zone the energy shift has an inverse dependence on R, which can be appreciable for a molecular assembly. At large separations, the interaction energy varies by a modulated inverse square behaviour.

6. Summary

In conclusion it has been shown how the method of induced multipole moments provides an alternative physical viewpoint and an efficient calculational scheme for the evaluation of a number of intermolecular interactions which exhibit all of the key features associated with Casimir effects. In this method, an electric dipole moment induced in a polarizable molecule by the application of an external electric displacement field interacts with a similarly induced electric dipole in a second molecule. The coupling takes place via the retarded resonant interaction tensor. An expression for the interaction energy shift can then be written in terms of the electric dipole polarizability of each species, the spatial field-field correlation function and the coupling tensor. On evaluating the expectation value of this quantity over the ground electronic state of each molecule and the state of the field containing no photons, remarkably the Casimir-Polder potential results. A particular advantage of the approach is that the energy shift when one or both molecules are electronically excited is obtained in a straightforward manner. Now the expectation value is taken over the appropriate molecular excited state, with the radiation field in the vacuum state as before. In addition to a Casimir-Polder like u-integral term which contains excited state polarizabilities, a contribution due to real photon emission from the excited state is present, with the coupling tensor now evaluated at the resonant frequency of the downward transition from the higher lying energy state to a lower one. In a further facile extension of the method, the change in the dispersion energy shift caused by the application of an intense external radiation field is also readily calculable. Once again the underlying equation for the energy shift is the same as for the dispersion force. But now the expectation value is taken over a state of the field containing N photons instead of zero, with both molecules in the ground state. For this process, two terms are found to contribute to the interaction energy for molecules that are polar. One contribution is proportional to the product of electric dipole polarizability of each species, while the second depends on the product of the permanent moment of one body and the first hyperpolarizability of the other. All of the results presented agree with previous diagrammatic perturbation theory calculations, but have been obtained with far greater ease. Perturbation theory evaluation of the ground state dispersion potential requires summation over twelve time ordered graphs, its modification by a radiation field involves addition of contributions from ninety-six diagrams, while when one or both entities is excited, intermediate state resonances have to be identified and tacked properly to yield the correct dispersion energy, and in all three computations, virtual photon summation has to be performed. While the induced moment approach has been shown to reproduce in a facile way leading order contributions to the dispersion potential and its modification by external radiation, calculational advantages offered by the method show up most clearly when higher order processes are tackled such as those involving terms arising from the inclusion of multipole moments in addition to the leading electric dipole interaction term, such as magnetic dipole and electric quadrupole couplings, which are crucial when evaluating interactions between chiral molecules, as well as when three-body energy shifts are computed [37]. By including fields that are explicitly time-dependent, which thereby induce moments that are a function of t, it is anticipated that time-dependent energy shifts may be computed in future using this approach. Hence the induced multipole method offers a viable and practical alternative to conventional techniques for the computation of various intermolecular processes.

Acknowledgement

Financial support in the form of travel funds provided by the International Centre for Condensed Matter Physics at the University of Brasilia and the Office of the Provost, Wake Forest University, are gratefully acknowledged. This paper is dedicated to the memory of T. Thirunamachandran who sadly passed away on 13 October, 2008 and from whom the author learnt molecular quantum electrodynamics.

References

[1] Casimir H B G and Polder D 1948 *Phys. Rev.* **73** 360
[2] Schwinger J S 1958 *Selected Papers on Quantum Electrodynamics* (New York: Dover)
[3] Craig D P and Thirunamachandran T 1998 *Molecular Quantum Electrodynamics* (New York: Dover)
[4] Power E A 1964 *Introductory Quantum Electrodynamics* (London: Longmans) Chapter 7 Section 5
[5] London F 1930 *Z. Phys.* **63** 245
[6] Milonni P W 1994 *The Quantum Vacuum* (San Diego: Academic)
[7] Power E A and Zienau S 1959 *Philos. Trans. Roy. Soc. Lond.* A **251** 427
[8] Woolley R G 1971 *Proc. Roy. Soc. Lond.* A **321** 557
[9] Alligood B W and Salam A 2007 *Mol. Phys.* **105** 395
[10] Power E A and Thirunamachandran T 1983 *Phys. Rev.* A **28** 2671
[11] Power E A and Thirunamachandran T 1983 *Phys. Rev.* A **28** 2663
[12] Jenkins J K, Salam A and Thirunamachandran T 1994 *Phys. Rev.* A **50** 4767
[13] Salam A 1996 *Mol. Phys.* **87** 919
[14] Salam A 2008 *Int. Rev. Phys. Chem.* **27** 405
[15] Boyer T H 1969 *Phys. Rev.* **180** 19
[16] Milonni P W 1982 *Phys. Rev.* A **25** 1315
[17] Schwinger J S, DeRaad L L and Milton K A 1978 *Ann. Phys. (NY)* **115** 1
[18] Milonni P W and Shih M L 1992 *Phys. Rev.* A **45** 4241
[19] Feinberg G and Sucher J 1980 *Phys. Rep.* **180** 1
[20] Compagno G, Passante R and Persico F 1983 *Phys. Lett.* A **98** 253
[21] Power E A and Thirunamachandran T 1993 *Phys. Rev.* A **48** 4761
[22] Craig D P and Thirunamachandran T 1999 *Theor. Chem. Acc.* **102** 112
[23] Salam A 2006 *J. Phys. B: At. Mol. Opt. Phys.* **39** S651
[24] Salam A 2006 *J. Phys. B: At. Mol. Opt. Phys.* **39** S661
[25] Thirunamachandran T 1980 *Mol. Phys.* **40** 393
[26] Chen F, Klimchitskaya G L, Mostapanenko V M and Mohideen U 2007 *Opt. Express* **15** 4823
[27] Salam A 2006 *Phys. Rev.* A **73** 013406
[28] Salam A 2006 *J. Chem. Phys.* **124** 014302
[29] Salam A 2007 *Phys. Rev.* A **76** 063402
[30] Spruch L and Kelsey E J 1978 *Phys. Rev.* A **18** 845

[31] Compagno G, Passante R and Persico F 1995 *Atom-Field Interactions and Dressed Atoms* (Cambridge University Press: Cambridge)
[32] Craig D P and Thirunamachandran T 1989 *Chem. Phys.* **135** 37
[33] Power E A and Thirunamachandran T 1995 *Phys. Rev.* A **51** 3660
[34] Power E A and Thirunamachandran T 1993 *Phys. Rev.* A **47** 2539
[35] Bradshaw D S and Andrews D L 2005 *Phys. Rev.* A **72** 033816; 2006 *Phys. Rev.* A **73** 033903(E)
[36] Andrews D L and Thirunamachandran T 1977 *J. Chem. Phys.* **67** 5026
[37] Cirone M and Passante R 1996 *J. Phys. B: At. Mol. Opt. Phys.* **29** 1871

Journal of Physics: Conference Series **161** (2009) 012041 doi:10.1088/1742-6596/161/1/012041

Casimir-Polder interaction between two atoms in electromagnetic fields

Yury Sherkunov

Department of Physics, University of Warwick, CV4 7AL, Coventry, UK

E-mail: y.b.sherkunov@warwick.ac.uk

Abstract. We reexamine the dispersion interaction between two atoms in external electromagnetic fields, provided the atoms are *not* in the equilibrium with the field. We show that the processes of absorption and spontaneous/stimulated emission should be taken into account in the case of non-equilibrium dynamics.

1. Introduction

Recent progress in experimental methods stimulated a new wave of research in Casimir, Casimir-Polder (CP), and van der Waals dispersion interaction under real experimental conditions, e.g. at finite temperature and at the presence of electromagnetic (EM) fields [1, 2, 3, 4, 5]. Although the theoretical base is quite robust, there are some problems to be resolved. One of these problems is the Casimir-Lifshitz interaction between real metals. Although, the interaction is described by the famous Lifshitz theory, the theoretical predictions dramatically depend on the model describing the real metal (see for example [6, 7, 8] and references therein). Another problem to be clarified is the non-equilibrium CP interaction between a single atom and a metal (or dielectric) plate if the temperature of the plate differs from the temperature of the atomic gas. This force was recently measured with the help of reflection spectroscopy of the atom interacting with the dielectric slub [9], but the agreement with the theoretical predictions has not been achieved [9].

The theoretical treatment of the CP interaction under non-equilibrium conditions is based either on the linear-responce (or Lifshitz-type) theory [3, 2, 10], or macroscopic quantum electrodynamics [11]. As it has been demonstrated by Buhmann and Scheel [12] one should distinguish two different problems. If the atom is thermalized and coupled to its thermal bath, the both approaches lead to the same result. If the atom is not thermalized, the results of the linear-responce theory could underestimate the the CP force [12]. The macroscopic quantum-electrodynamical (QED) approach takes into account the possible absorption and spontaneous/stimulated emission of a thermal photon by the interacting atom. These contributions result in the enhancement of the CP force even if the atom is in its ground state. For the case of thermalized atom, this terms cancel out and one arrives at the predictions of Lifshitz theory [13].

The dispersion interaction between two atoms embedded in an EM field was considered by Milonni and Smith [14]. They showed that the interaction potential between them can be obtained from the vacuum Casimir-Polder one by a simple substitution $\hbar\omega_{\mathbf{k}}/2 \rightarrow$

$\hbar\omega_{\mathbf{k}}(N_{\mathbf{k}\lambda}+1/2)$ [14], where $\omega_{\mathbf{k}}$ is the energy of the photon of the state $\{\mathbf{k},\lambda\}$, with $\mathbf{k}$ and λ the wave vector and the polarisation index of EM field, $N_{\mathbf{k}\lambda}$ is the number of photons in the state $\{\mathbf{k}\lambda\}$. But what happens at the initial stage of interaction, when the atoms are not in the equilibrium with the EM field? Will the Milonni-Smith result still hold, or should it be modified as in the case of atom-slab interaction considered in Ref.[12]? The aim of this paper is to answer these questions.

2. Interaction between two atoms

We consider dispersion interaction between two dissimilar atoms A and B embedded in an EM field. Let atom A be situated at a point with the radius-vector $\mathbf{R}_A$, and B in its arbitrary state (either excited or ground-state) and situated at a point $\mathbf{R}_B$. The distance between the atoms is large enough to neglect the exchange interaction.

The Hamiltonian of the system atom A-EM field -atom B is ($\hbar=1, c=1$)

$$H = H_A + H_B + H_{ph} + H_{int}, \tag{1}$$

where $H_A=\sum_i \epsilon_{Ai} b_i^\dagger b_i$, $H_B=\sum_i \epsilon_{Bi}\beta_i^\dagger \beta_i$ are the Hamiltonians of non-interacting atoms A and B, $\epsilon_{A(B)i}$ is the energy of i-th state of atom A (B). The annihilation operators of the i-th state of atom A (B) is $b_i(\beta_i)$. $H_{ph}=\sum_{\mathbf{k}\lambda} k(\alpha_{\mathbf{k}\lambda}^\dagger\alpha_{\mathbf{k}\lambda}+1/2)$ is the Hamiltonian of free photons. The interaction Hamiltonian is

$$H_{int} = -\int \psi_A^\dagger(X)\mathbf{dE}(X)\psi_A(X)d\mathbf{r} - \int \psi_B^\dagger(X)\mathbf{dE}(X)\psi_B(X)d\mathbf{r}, \tag{2}$$

where

$$\psi_A(X)=\sum_j \phi_j(\mathbf{r}-\mathbf{R}_A)e^{-i\epsilon_{Aj}t}b_j, \psi_B(X)=\sum_j \phi_j(\mathbf{r}-\mathbf{R}_B)e^{-i\epsilon_{Bj}t}\beta_j, \tag{3}$$

with $\phi_j(\mathbf{r}-\mathbf{R}_{A(B)})$ the wave function of state j of the corresponding atom. $\mathbf{d}$ is the operator of dipole moment,

$$\mathbf{E}(X)=i\sum_{\mathbf{k}\lambda}(2\pi k/V)^{1/2}\mathbf{e}_{\mathbf{k}\lambda}\left(\alpha_{\mathbf{k}\lambda}e^{i\mathbf{kr}}e^{-ikt}-\alpha_{\mathbf{k}\lambda}^\dagger e^{-i\mathbf{kr}}e^{ikt}\right) \tag{4}$$

is the operator of free EM field, V is the volume, $\mathbf{e}_{\mathbf{k}\lambda}$ is the polarisation unit vector of EM field, $X=\{\mathbf{r},t\}$.

To calculate the interaction potential we will first assume that atom A is in its ground state and the probability of excitation is negligible. In other words the number of photons at the transition frequency ω_A is negligible, although the number of photons at the transition frequency ω_B of atom B is sufficiently large $N(\omega_A) << N(\omega_B)$. This assumption enables us to calculate the interaction potential as the energy shift of atom A. Then, we will obtain a general result for arbitrary initial state of atom A and arbitrary $N(\omega_A)$ (but it still should not be too large to neglect the ionisation of the atoms)using symmetry properties of the potential.

If the atom does not change its initial state, the interaction potential, as one could easily show with the help of Hartree-Fock method, can be represented as the energy shift of the atomic level.

$$U=\Delta\epsilon_A \tag{5}$$

To calculate the energy shift, we will use the Keldysh-Green function approach adopted to the microscopic systems [15, 16]. The Keldysh Green function of atom A is [17]

$$G_{ll'}^A(X,X')=-i\langle T_c\psi_{Al}(X)\psi_{Al'}^\dagger(X')S_c\rangle \tag{6}$$

where T_c is the time-ordering operator given on the Keldysh contour c, which goes from $-\infty$ to $+\infty$ for $l = 1$ and from $+\infty$ to $-\infty$ for $l = 2$. So, for $l = 1$ T_c is a conventional ordering operator, while for $l = 2$ it is an antiordering operator.Further we omit index A. The operators are given in the interaction representation. $\langle ... \rangle$ means the averaging over the initial state of the system. The scattering matrix S_c is

$$S_c = T_c \exp\left(\sum_{l=1,2} (-1)^l i \int_{-\infty}^{\infty} H_{int}(t)dt\right) \tag{7}$$

Using the Green function (6), one can readily find the density matrix of atom A

$$\rho(X, X') = iG_{12}(X, X') \tag{8}$$

To calculate the density matrix ρ we expand Ex. (6) as a perturbation series

$$\rho(X, X') = \rho_0(X, X') - \frac{1}{2}\langle T_c \int dX_1 dX_2 \psi_1(X)\psi_2^\dagger(X')(-1)^{l_1+l_2}\psi_{l_1}^\dagger(X_1)\mathbf{E}_{l_1}(X_1)\mathbf{d}\psi_{l_1}(X_1)$$
$$\times\psi_{l_2}^\dagger(X_2)\mathbf{E}_{l_2}(X_2)\mathbf{d}\psi_{l_2}(X_2)\rangle + ... \tag{9}$$

Index "0" means non-interacting atoms. To tackle the EM field we will use the standard Keldysh formalism [17]. We use the exact Wick's theorem to average the atomic operators [18]:

$$T_c\psi_l(X)\psi_{l'}^\dagger(X') =: \psi_l(X)\psi_{l'}^\dagger(X') : +ig_{ll'}^0(X, X') \tag{10}$$

where : ... : means normal ordering,

$$g_{ll'}^0(X, X') = -i\langle T_c\psi_l(X)\psi_{l'}^\dagger(X')\rangle_{vacuum} \tag{11}$$

is the electron propagator, $\langle ... \rangle_{vacuum}$ means the averaging over vacuum state. The average of normal products over the initial state occupied by a single atom $\langle: \psi_l(X)\psi_{l'}^\dagger(X')\psi_{l_1}(X_1)\psi_{l_2}^\dagger(X_2)... :\rangle = 0$ for all the orders but the second one. For the second order it is just the density matrix of non-interacting atom A $\rho_0(X, X') = \langle: \psi_l(X)\psi_{l'}^\dagger(X') :\rangle$. Taking into account the processes, which do not change the initial state of atom A, we find

$$\rho(X, X') = \rho_0(X, X') + \int dX_1 dX_2 \left[g_{11}(X, X_1)M_{11}(X_1, X_2)\rho_0(X_2, X')\right.$$
$$\left.+\rho_0(X, X_1)M_{22}(X_1, X_2)g_{22}(X_2, X')\right]$$
$$+\int dX_1 dX_2 dX_3 dX_4 g_{11}(X, X_1)M_{11}(X_1, X_2)\rho_0(X_2, X_3)M_{22}(X_3, X_4)g_{22}(X_4, X') \tag{12}$$

where the electron propagator $g_{ll'}(X, X') = -i\langle T_c\psi_l(X)\psi_{l'}(X')S_c\rangle_{vacuum}$ obeys the Dyson equation

$$g_{ll'}(X, X') = g_{ll'}^0(X, X') + \int dX_1 dX_2 \sum_{l_1 l_2} g_{ll_1}^0(X, X_1)M_{l_1 l_2}(X_1, X_2)g_{l_2 l'}(X_2, X') \tag{13}$$

The mass operator M is

$$M_{ll'}(X, X') = i(-1)^{l+l'} g_{ll'}(X, X')d^\nu d^{\nu'} D_{l'l}^{\nu'\nu}(X', X) \tag{14}$$

D is the photonic Green tensor, ν is the tensor index.

$$D_{ll'}^{\nu\nu'}(X, X') = -i\langle T_c E_l^\nu(X)E_{l'}^{\nu'}(X')S_c\rangle \tag{15}$$

The Feynman diagrams describing the processes taken into account in Ex.(12) are given in Fig.1. Using the explicite expressions for the free electron propagators in energy space [18]

$$g^0_{12} = 0, \tag{16}$$

$$g^0_{11}(E, \mathbf{r}, \mathbf{r}') = \sum_i \frac{\phi_i(\mathbf{r} - \mathbf{R_A})\phi_i^*(\mathbf{r}' - \mathbf{R_A})}{E - \epsilon_i + i0}, \tag{17}$$

$$g^0_{22}(E, \mathbf{r}, \mathbf{r}') = -\sum_i \frac{\phi_i(\mathbf{r} - \mathbf{R_A})\phi_i^*(\mathbf{r}' - \mathbf{R_A})}{E - \epsilon_i - i0}, \tag{18}$$

we can solve the Dyson equation (13) in the pole approximation.

$$g_{11}(E, \mathbf{r}, \mathbf{r}') = \sum_i \frac{\phi_i(\mathbf{r} - \mathbf{R_A})\phi_i^*(\mathbf{r}' - \mathbf{R_A})}{E - \epsilon_i - \langle\phi_i|M_{11}(\epsilon_i)|\phi_i\rangle}, \tag{19}$$

$$g_{22}(E, \mathbf{r}, \mathbf{r}') = -\sum_i \frac{\phi_i(\mathbf{r} - \mathbf{R_A})\phi_i^*(\mathbf{r}' - \mathbf{R_A})}{E - \epsilon_i - \langle\phi_i|M_{22}(\epsilon_i)|\phi_i\rangle} \tag{20}$$

We suppose that the interaction between the atom A and the EM field was switched on at a moment t_0. Thus, the density matrix of the free atom A in its initial state p is

$$\rho_0(X, X') = \phi_p(\mathbf{r} - \mathbf{R_A})\phi_p^*(\mathbf{r}' - \mathbf{R_A})e^{-i\epsilon_p(t-t')}\theta(t - t_0)\theta(t' - t_0) \tag{21}$$

where θ is the Heavyside function. Substituting the Exs.(18),(20), and (21) into Ex.(12), we obtain

$$\langle\phi_p|\rho(X, X')|\phi_p\rangle = \exp\left[-i(\epsilon_p - \langle\phi_p|M_{11}(\epsilon_p)|\phi_p\rangle)(t - t_0)\right]\exp\left[i(\epsilon_p - \langle\phi_p|M_{22}(\epsilon_p)|\phi_p\rangle)(t' - t_0)\right] \tag{22}$$

Thus, the energy shift of the atomic level along with the interaction potential is

$$U = \Delta\epsilon_p = Re\langle\phi_p|M_{11}(\epsilon_p)|\phi_p\rangle \tag{23}$$

The imaginary part ImM_{11} describes the decay of the initial state as a result of excitation. Substituting Ex. (14) into (23), we get

$$U = Re\left[\frac{i}{2\pi}\int_{-\infty}^{\infty} d\omega \int d\mathbf{r}d\mathbf{r}'\phi_p^*(\mathbf{r}')g_{11}(\omega + \epsilon_p, \mathbf{r}, \mathbf{r}')d^{\nu'}D_{11}^{\nu'\nu}(\omega, \mathbf{r}', \mathbf{r})d^{\nu}\phi_p(\mathbf{r})\right] \tag{24}$$

After integration over spatial coordinates and making use of the Ex.(18), we obtain:

$$U = Re\left[\frac{i}{2\pi}\sum_j \int_{-\infty}^{\infty} d\omega \frac{d_{pj}^{\nu'}D_{11}^{\nu'\nu}(\omega, \mathbf{R_A}, \mathbf{R_A})d_{jp}^{\nu}}{\omega - \omega_{jp} + i0}\right] \tag{25}$$

where $\omega_{jp} = \epsilon_j - \epsilon_p$. If atom A is in its ground state

$$U = -Re\frac{i}{2\pi}\sum_j \int_0^{\infty} d\omega\alpha_A^{\nu\nu'}(\omega)D_{11}^{\nu'\nu}(\omega, \mathbf{R_A}, \mathbf{R_A}) \tag{26}$$

where

$$\alpha_A^{\nu\nu'}(\omega) = \sum_j\left(\frac{d_{pj}^{\nu}d_{jp}^{\nu'}}{\omega_{jp} - \omega - i0} + \frac{d_{jp}^{\nu}d_{pj}^{\nu'}}{\omega_{jp} + \omega + i0}\right) \tag{27}$$

is the polarisability of atom A.

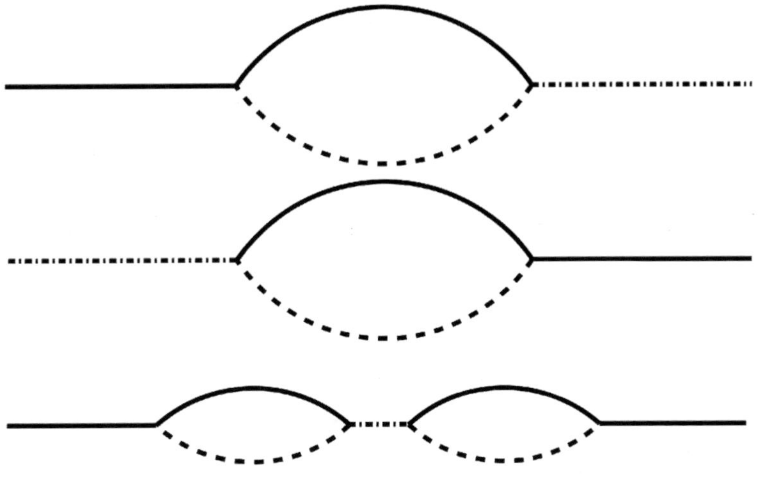

Figure 1. Feynman's diagrams for the interaction of atom A with EM field. The solid line corresponds to g, the dashed line corresponds to D, and the dashed-dotted line to ρ_0

Ex. (26) generalises the vacuum interaction potential obtained in Ref.[16] to the case of external EM field.

To calculate the Green tensor D_{11} we use the standard Keldysh technique [17, 19]. First we notice that the casual Green tensor D_{11} can be written as a sum of the retarded Green tensor D_r and the Green tensor D_{12} [17, 19]

$$D = D_r + D_{12} \tag{28}$$

where

$$D_{12}^{\nu\nu'}(X, X') = -i\langle \hat{E}^{\nu'}(X')\hat{E}^{\nu}(X)\rangle \tag{29}$$

Note, that here $\hat{\mathbf{E}}$ is in Heisenberg representation. Substituting (29) and (28) into (26) we find

$$U = -Re\frac{i}{2\pi}\int_0^\infty \alpha_A^{\nu'\nu}(\omega)D_r^{\nu\nu'}(\omega, \mathbf{R}_A, \mathbf{R}_A)d\omega$$
$$-Re\frac{i}{2\pi}\int_0^\infty \alpha_A^{\nu'\nu}(\omega)D_{12}^{\nu\nu'}(\omega, \mathbf{R}_A, \mathbf{R}_A)d\omega \tag{30}$$

The first term of the Eq.(30) describes the interaction with the vacuum EM field, modified by the presence of another atom or macroscopic body [20]. The second term describes the interaction with the photons of the external EM field.

For zero temperature the tensor D_{12} describes only the spontaneous emission by the medium (if it is excited) and Ex. (30) results in Ex. (41) of [16].

Let atom B at a position $\mathbf{R}_B$. It can be either in the excited state or ground one. For simplicity we suppose that the atoms are exposed to isotropic unpolarised EM field, *i.e.* $N_{\mathbf{k}\lambda}$ depends only on $|\mathbf{k}| = \omega$, $N_{\mathbf{k}\lambda} = N(\omega)$. We will attack the problem perturbatively. The density matrix D_{12} can be calculated with the help of Keldysh technique [17, 19]. According to (26), we should integrate the Green tensor with respect to positive frequencies $\omega > 0$. Thus

$$D_{12}(X, X') = D_{12}^0(X, X') + \int D_r^0(X, X_1)\Pi_r(X_1, X_2)D_{12}^0(X_2, X')dX_1dX_2$$
$$+\int D_{12}^0(X, X_1)\Pi_a(X_1, X_2)D_a^0(X_2, X')dX_1dX_2$$
$$-\int D_r^0(X, X_1)\Pi_{12}(X_1, X_2)D_a^0(X_2, X')dX_1dX_2, (\omega > 0) \tag{31}$$

Here we suppressed the tensor indices ν. D^0 means free photon Green tensor. The polarisation operators Π describe the interaction between the EM field and atom B which is in its initial state n [16]

$$\Pi_r^{\nu\nu'}(\omega,\mathbf{r},\mathbf{r}') = -\alpha_B^{\nu\nu'}(\omega)\delta(\mathbf{r}-\mathbf{r}') = \Pi_a^{\nu\nu'*}(\omega,\mathbf{r}',\mathbf{r})$$
$$\Pi_{12}^{\nu\nu'}(\omega,\mathbf{r},\mathbf{r}') = \sum_m 2\pi i d_{nm}^{\nu B} d_{mn}^{\nu' B}\delta(\omega-\omega_{nm})\delta(\mathbf{r}-\mathbf{r}') \tag{32}$$

The polarisation operators Π_r and Π_a describe the elastic scattering of light on the atom B, while Π_{12} describes the downward transition of the atom as a result of spontaneous or stimulated emission. For positive frequencies,the free field Green tensor D_{12}^0 is

$$D_{12}^0(\omega,\mathbf{r},\mathbf{r}') = N(\omega)(D_r^0(\omega,\mathbf{r},\mathbf{r}') - D_a^0(\omega,\mathbf{r},\mathbf{r}')) \tag{33}$$

The formula (31) along with (32) and (33) yield

$$\begin{aligned} D_{12}(\omega,\mathbf{R}_A,\mathbf{R}_A) = {} & D_{12}^0(\omega,\mathbf{R}_A,\mathbf{R}_A) - 2N(\omega)D_r^0(\omega,\mathbf{R}_A,\mathbf{R}_B)\alpha_B^{\nu\nu'}(\omega)D_r^0(\omega,\mathbf{R}_B,\mathbf{R}_A) \\ & + 2N(\omega)ReD_r^0(\omega,\mathbf{R}_A,\mathbf{R}_B)\alpha_B^{\nu\nu'}(\omega)D_r^0(\omega,\mathbf{R}_B,\mathbf{R}_A) \\ & + \sum_m 2\pi i d_{nm}^{\nu B} d_{mn}^{\nu' B}\delta(\omega-\omega_{mn})N(\omega)D_r^0(\omega,\mathbf{R}_A,\mathbf{R}_B)D_a^0(\omega,\mathbf{R}_B,\mathbf{R}_A) \\ & - \sum_m 2\pi i d_{nm}^{\nu B} d_{mn}^{\nu' B}\delta(\omega-\omega_{nm})(N(\omega)+1)D_r^0(\omega,\mathbf{R}_A,\mathbf{R}_B)D_a^0(\omega,\mathbf{R}_B,\mathbf{R}_A) \end{aligned} \tag{34}$$

The first term of the r.h.s of Eq.(34) describes the free EM field. This term results in the optical Stark shift of the energy levels of atom A and should be omitted. The second and the third terms describe the elastic scattering of light on atom B. The fourth term is responsible for the absorption of a photon by atom B and the last term is for the spontaneous and stimulated emission of a photon by atom B. For isotopic atom B ($\alpha^{\nu\nu'} \propto \delta_{\nu,\nu'}$) prepared in a mixed state with the probability to find the atom in n-th state p_n we obtain

$$\begin{aligned} U = {} & Re\frac{i}{\pi}\int_0^\infty d\omega\,(N(\omega)+1/2)\,\alpha_A(\omega)\alpha_B(\omega)(D_r^0(\omega,\mathbf{R}_A\mathbf{R}_B))^2 \\ & + \frac{1}{3}Re\sum_{mn}|d_{mn}^B|^2\alpha_A(\omega_{mn})|D_r^0(\omega_{mn},\mathbf{R}_A,\mathbf{R}_B)|^2\left[p_n N(\omega_{mn})\theta(\omega_{mn}) - p_n\left(N(\omega_{nm})+1\right)\theta(\omega_{nm})\right] \end{aligned} \tag{35}$$

The first term of the Eq. (35) coincides with the one obtained by Milonni and and Smith with the help of the source theory [14]. The second term describes the absorption of a photon by atom B (first term in the square brackets) as well as spontaneous and stimulated emission by atom B (second term in the square brackets). At thermal equilibrium the probability to find an atom in its n-th state p_n is given by the Boltzmann distribution, while the number of photons obeys the Bose statistics. One can easily check that in the equilibrium, the second term is zero due to the detailed balance, and we find that the equation (35) coincides with the Lifshitz formula [13].

Thus we arrive at a conclusion that the Milonni-Smith formula describes only the interaction between two atoms at equilibrium. But for the non-equilibrium case, if one or both atoms are not equilibrated, the Milonni-Smith formula, which does not take into account the absorption and emission of photons, underestimates the interaction potential between two atoms in the EM field even if the atoms are in their ground states.

Now to obtain quantitative results, we substitute the explicite expression for the retarded Green tensor [21]

$$\begin{aligned} D_r^{\nu\nu'0}(\omega,\mathbf{r},\mathbf{r}') = {} & -\omega^2\left[\delta_{\nu\nu'}\left(1+\frac{i}{\omega R}-\frac{1}{(\omega R)^2}\right)\right. \\ & \left. + s^\nu s^{\nu'}\left(\frac{3}{(\omega R)^2}-\frac{3i}{\omega R}-1\right)\right]\frac{\exp(i\omega R)}{R}, \end{aligned} \tag{36}$$

where $R = |\mathbf{r}-\mathbf{r}'|$ and $s^\nu = (r-r')^\nu/|\mathbf{r}-\mathbf{r}'|$, into (35). The interaction potential is the sum of the equilibrium potential U_{eq}, which coincides with the Milonni-Smith result, and non-equilibrium one U_{neq}.

$$\begin{aligned}
U &= U_{eq} + U_{neq} \\
U_{eq} &= -\frac{1}{\pi R^2}\int_0^\infty \alpha_A(iu)\alpha_B(iu)u^4\left(1+\frac{2}{uR}+\frac{5}{(uR)^2}+\frac{6}{(uR)^3}+\frac{3}{(uR)^4}\right) \\
&\quad \times (2N(iu)+1)\exp[-2iuR]du \qquad (37)\\
U_{neq} &= \frac{4}{9R^2}\sum_{mnj}\frac{|d^A_{pj}|^2|d^B_{mn}|^2\omega_{jp(A)}\omega^4_{mn(B)}}{\omega^2_{jp(A)}-\omega^2_{mn(B)}}\left(1+\frac{1}{(\omega_{mn(B)}R)^2}+\frac{3}{(\omega_{mn(B)}R)^4}\right) \\
&\quad \times \left[p_n N(\omega_{mn(B)})\theta(\omega_{mn(B)}) - p_n(N(\omega_{nm(B)})+1)\theta(\omega_{nm(B)})\right]. \qquad (38)
\end{aligned}$$

The interaction potential between two atoms in EM vacuum has been calculated by Power and Thirunamachandran [22, 23].

$$\begin{aligned}
U_{neq} &= -\frac{4}{9R^2}\sum_{mj}\frac{|d^A_{pj}|^2|d^B_{mn}|^2\omega_{jp(A)}\omega^4_{mn(B)}}{\omega^2_{jp(A)}-\omega^2_{mn(B)}} \\
&\times\left(1+\frac{1}{(\omega_{mn(B)}R)^2}+\frac{3}{(\omega_{mn(B)}R)^4}\right)\theta(\omega_{nm(B)}). \qquad (39)
\end{aligned}$$

Eqs (38) and (39) obviously coincide if $N(\omega) = 0$ in Eq.(38). Thus, Eq. (38) generalizes the Power-Thirunamachandran theory to the case of atoms embedded in a EM field. For the limit $R >> \lambda$ (λ is the characteristic wavelength of the atomic transition), the potential drops with the distance between the atoms as $U_{neq} \propto R^{-2}$ (compare to the Casimir-Polder formula $U \propto R^{-7}$), this peculiar behaviour of the potential was discussed by Power and Thirunamachandran [22, 23]. (see also [16], where the interaction potential between an excited atom and a ground-state one embedded in a dielectric medium was calculated for vacuum state of EM field). In the presence of EM field, the non-equilibrium contribution to the interaction between a ground-state atom and an excited one is enhanced by factor $N(\omega_B)+1$ compared to the interaction potential in EM vacuum.

If both atoms are in their ground states, we should take into account only the first term in the square brackets of (40) For the retarded regime ($R >> \lambda$)

$$U_{neq} = \frac{4}{9R^2}\sum_{mj}\frac{|d^A_{pj}|^2|d^B_{mn}|^2\omega_{jp(A)}\omega^4_{mn(B)}}{\omega^2_{jp(A)}-\omega^2_{mn(B)}}N(\omega_{mn(B)})\theta(\omega_{mn(B)}). \qquad (40)$$

This potential drops as R^{-2} with the distance. It means that for non-equilibrium case, the contribution of the non-equilibrium term due to photon absorption could be much greater then the contribution of the equilibrium one described by the Milonni-Smith formula. We should mention, that the approach we developed in this presentation is valid for the initial stage of interaction provided the atoms and the EM field are not in the equilibrium. Obviously, after the equilibration the non-equilibrium term is zero and we obtain the Milonni-Smith result.

The results obtained above are valid only if the probe atom A is in its ground state during the interaction process. It means that the number of photons at the transition energies of atom A is negligible ($N(\omega_A) \approx 0$), while the number of photons at the frequency of atom B is large $N(\omega_B) >> N(\omega_A)$. Our formalism does not allow us to calculate the interaction potential for the case when atom A can be excited by the electromagnetic field. But to extand our results to the case free from the above mentioned assumption, we can use the symmetry properties of the potential: (i) the poptential is obviously symmetric with respect to atomic indices; (ii) if both atoms are thermalized the non-equilibrium contribution to the potential should be equal to zero. Using these properties, we get the non-equilibrium potential in general case

$$U_{neq} = \sum_{mnjp} S(|\omega_A|, |\omega_B|) \left[p_n N(\omega_B)\theta(\omega_B) - p_n(N(-\omega_B)+1)\theta(-\omega_B)\right] \left[p_p\theta(\omega_A) - p_p\theta(-\omega_A)\right]$$
$$+ \sum_{mnjp} S(|\omega_B|, |\omega_A|) \left[p_p N(\omega_A)\theta(\omega_A) - p_p(N(-\omega_A)+1)\theta(-\omega_A)\right] \left[p_n\theta(\omega_B) - p_n\theta(-\omega_B)\right] \quad (41)$$

with $\omega_A = \omega_{jp(A)}, \omega_B = \omega_{mn(B)}$ (note that the frequencies ω_A and ω_B are positive for upward transitions and negative otherwise. S is given by

$$S(\omega_A, \omega_B) = \frac{4}{9R^2} \frac{|d^A_{pj}|^2 |d^B_{mn}|^2 \omega_A \omega_B^4}{\omega_A^2 - \omega_B^2} \left(1 + \frac{1}{(\omega_B R)^2} + \frac{3}{(\omega_B R)^4}\right) \quad (42)$$

For vacuum state of the EM field we recover all the corresponding results by Power and Thirunamachandran [22, 23].

As an example we will consider two two-level atoms embedded in a thermal EM field at temperature T.

1. Let both atoms be in their ground-states.

If the temperature is low compared to the transition frequencies ($T << \omega_{A,B}$), but high compared to the frequency difference ($T >> |\omega_A - \omega_B|$), we find

$$U_{neq} = \frac{4}{9R^2} \frac{\exp(-\omega_A/T)|d_A|^2 |d_B|^2 \omega_A \omega_B}{\omega_A + \omega_B} \left(-(\omega_B^2 + \omega_A\omega_B + \omega_A^2) - \frac{1}{R^2} + \frac{3}{\omega_A\omega_B R^4}\right) \quad (43)$$

This result is not resonant and exponentially small compared to the equilibrium contribution, but the assimptotic behaviour of Ex.(43) for large distances ($R >> \lambda$) is $1/R^2$, while the assimptotics of the equilibrium contribution ($R >> 1/T$) is $1/R^6$ [21].

For $T << |\omega_A - \omega_B|$,

$$U_{neq} = \frac{4}{9R^2} \frac{\exp(-\omega_B/T)|d_A|^2 |d_B|^2 \omega_A \omega_B^4}{\omega_A^2 - \omega_B^2} \left(1 + \frac{1}{(\omega_B R)^2} + \frac{3}{(\omega_B R)^4}\right), (\omega_A > \omega_B); \quad (44)$$

$$U_{neq} = -\frac{4}{9R^2} \frac{\exp(-\omega_A/T)|d_A|^2 |d_B|^2 \omega_B \omega_A^4}{\omega_A^2 - \omega_B^2} \left(1 + \frac{1}{(\omega_A R)^2} + \frac{3}{(\omega_A R)^4}\right), (\omega_A < \omega_B). \quad (45)$$

The contributions of the non-equilibrium potential are resonant now.

For high temperature case ($T >> \omega_{A,B}$) we obtain a universal expression which is valid independent of the initial states of the atoms.

$$U_{neq} = \frac{4}{9R^2} T |d_A|^2 |d_B|^2 \omega_A \omega_B \left(-1 + \frac{3}{(\omega_B\omega_A)^2 R^4}\right) \quad (46)$$

The equilibrium contribution at high temperatures ($T >> 1/R$) is [21]

$$U_{eq} = -\frac{4}{3R^6} T |d_A|^2 |d_B|^2 \frac{3}{\omega_B \omega_A R^6} \quad (47)$$

Journal of Physics: Conference Series **161** (2009) 012041 doi:10.1088/1742-6596/161/1/012041

The total potential is

$$U = U_{eq} + U_{neq} = \frac{4}{9R^2} T |d_A|^2 |d_B|^2 \omega_A \omega_B \tag{48}$$

So for $T >> \omega_{A,B}, T >> 1/R$ the total interaction potential drops like $1/R^2$ with the distance R, even though $R << \lambda$.

2. Two excited atoms.

If $T << \omega_{A,B}$,

$$U_{neq} = \frac{4}{9R^2} \frac{|d_A|^2 |d_B|^2 \omega_A \omega_B}{\omega_A + \omega_B} \left(-(\omega_B^2 + \omega_A \omega_B + \omega_A^2) - \frac{1}{R^2} + \frac{3}{\omega_A \omega_B R^4} \right) \tag{49}$$

This result is no more exponentially small (although is not resonant) compared to the equilibrium potential and could be greater than the latter. For large distances we once again obtained the R^{-2} dependence of the potential. If $T >> \omega_{A,B}$, we obtain Ex.(48).

3. Atom A is ground-state, atom B is excited.

$T << \omega_{A,B}$

$$U_{neq} = \frac{4}{9R^2} \frac{|d_A|^2 |d_B|^2 \omega_A \omega_B^4}{\omega_A^2 - \omega_B^2} \left(1 + \frac{1}{(\omega_B R)^2} + \frac{3}{(\omega_B R)^4} \right) \tag{50}$$

This non-equilibrium contribution is resonant and exceeds the equilibrium contribution.

For $T >> \omega_{A,B}$ we again obtain the result (48).

3. Conclusions

We reexamined the dispersion interaction between two atom in external EM field. We showed that the Milonni-Smith approach is valid only if the atoms are in equilibrium with the EM field. To study the initial stage of the non-equilibrium dynamics, one should take into account absorption and spontaneous/stimulated emission. The latter significantly modifies the interaction potential, resulting in some cases in resonance contributions. We generalised the Power- Thirunamachandran theory for two atoms in EM vacuum and included the effects of external EM field. As en example we considered the interaction between two atoms in thermal EM field. For low temperatures, the result dramatically depends on the initial state of the atoms, while for large temperatures the interaction potential is universal, at least for the pure initial states of atoms.

References

[1] Cohen A E and Mukamel S 2003 *Phys. Rev. Lett.* **91** 233202
[2] Henkel C, Joulain K, Mulet J P and Greffet J J 2002 *J. Opt. A: Pure Appl. Opt.* **4** S109–S114
[3] Antezza M, Pitaevskii L P and Stringari S 2005 *Phys. Rev. Lett.* **95** 113202
[4] Antezza M, Pitaevskii L P, Stringari S and Svetovoy V B 2008 *Phys. Rev. A* **77** 022901
[5] Obrecht J M, Wild R J, Antezza M, Pitaevskii L P, Stringari S and Cornell E A 2007 *Phys. Rev. Lett.* **98** 063201
[6] Brevik I, Aarseth J B, Hoye J S and Milton K A 2005 *Phys. Rev. E* **71** 056101
[7] Bezerra V B, Decca R S, Fischbach E, Geyer B, Klimchitskaya G L, Krause D E, Lopez D, Mostepanenko V M and Romero C 2006 *Phys. Rev. E* **73** 028101
[8] Mostepanenko V M, Bezerra V B, Decca R, Geyer B, Fischbach E, Klimchitskaya G L, Krause D E, Lopez D and Romero C 2006 *J. Phys. A: Math. Gen.* **39** 6589
[9] Chaves de Souza Segundo P, Hamdi I, Fichet M, Bloch D and Ducloy M 2007 *Laser Phys.* **17** 983
[10] McLachlan A D 1963 *Proc. R. Soc. Lond. Ser. A* **274** 80
[11] Gorza M P and Ducloy M 2006 *Eur. Phys. J. D* **40** 343
[12] Buhmann S Y and Scheel S 2008 *Phys. Rev. Lett.* **100** 253201
[13] Lifshitz E M 1956 *Zh. Eksp. Teor. Fiz.* **2** 73
[14] Milonni P W and Smith A 1996 *Phys. Rev. A* **53** 3484–3489

[15] Sherkunov Y 2005 *Phys. Rev. A* **72** 052703
[16] Sherkunov Y 2007 *Phys. Rev. A* **75** 012705
[17] Keldysh L 1964 *Zh. Eksp. Teor. Fiz.* **47** 1515
[18] Berestetskii V, Lifshitz E M and Pitaevskii L P 1982 *Quantum Electrodynamics. Course on theoretical physics v.4* (Pergamon, Oxford)
[19] Lifshitz E M and Pitaevskii L P 1981 *Physical kinetics. Course on theoretical physics v.10* (Pergamon, Oxford)
[20] Wylie J M and Sipe J E 1985 *Phys. Rev. A* **32** 2030–2043
[21] Lifshitz E M and Pitaevskii L P 1980 *Statistical physics, part 2, Course on theoretical physics v.9* (Pergamon, Oxford)
[22] Power E A and Thirunamachandran T 1993 *Phys. Rev. A* **47** 2539–2551
[23] Power E A and Thirunamachandran T 1995 *Phys. Rev. A* **51** 3660–3666

AUTHOR INDEX